T0190325

Lecture Notes in Computer Science

Lecture Notes in Artificial Intelligence   **13714**

Founding Editor

Jörg Siekmann

Series Editors

Randy Goebel, *University of Alberta, Edmonton, Canada*
Wolfgang Wahlster, *DFKI, Berlin, Germany*
Zhi-Hua Zhou, *Nanjing University, Nanjing, China*

The series Lecture Notes in Artificial Intelligence (LNAI) was established in 1988 as a topical subseries of LNCS devoted to artificial intelligence.

The series publishes state-of-the-art research results at a high level. As with the LNCS mother series, the mission of the series is to serve the international R & D community by providing an invaluable service, mainly focused on the publication of conference and workshop proceedings and postproceedings.

Massih-Reza Amini · Stéphane Canu ·
Asja Fischer · Tias Guns · Petra Kralj Novak ·
Grigorios Tsoumakas
Editors

Machine Learning and Knowledge Discovery in Databases

European Conference, ECML PKDD 2022
Grenoble, France, September 19–23, 2022
Proceedings, Part II

 Springer

Editors
Massih-Reza Amini
Grenoble Alpes University
Saint Martin d'Hères, France

Stéphane Canu
INSA Rouen Normandy
Saint Etienne du Rouvray, France

Asja Fischer
Ruhr-Universität Bochum
Bochum, Germany

Tias Guns
KU Leuven
Leuven, Belgium

Petra Kralj Novak
Central European University
Vienna, Austria

Grigorios Tsoumakas
Aristotle University of Thessaloniki
Thessaloniki, Greece

ISSN 0302-9743   ISSN 1611-3349 (electronic)
Lecture Notes in Artificial Intelligence
ISBN 978-3-031-26389-7   ISBN 978-3-031-26390-3 (eBook)
https://doi.org/10.1007/978-3-031-26390-3

LNCS Sublibrary: SL7 – Artificial Intelligence

Preface

The European Conference on Machine Learning and Principles and Practice of Knowledge Discovery in Databases (ECML–PKDD 2022) in Grenoble, France, was once again a place for in-person gathering and the exchange of ideas after two years of completely virtual conferences due to the SARS-CoV-2 pandemic. This year the conference was hosted for the first time in hybrid format, and we are honored and delighted to offer you these proceedings as a result.

The annual ECML–PKDD conference serves as a global venue for the most recent research in all fields of machine learning and knowledge discovery in databases, including cutting-edge applications. It builds on a highly successful run of ECML–PKDD conferences which has made it the premier European machine learning and data mining conference.

This year, the conference drew over 1080 participants (762 in-person and 318 online) from 37 countries, including 23 European nations. This wealth of interest considerably exceeded our expectations, and we were both excited and under pressure to plan a special event. Overall, the conference attracted a lot of interest from industry thanks to sponsorship, participation, and the conference's industrial day.

The main conference program consisted of presentations of 242 accepted papers and four keynote talks (in order of appearance):

- Francis Bach (Inria), Information Theory with Kernel Methods
- Danai Koutra (University of Michigan), Mining & Learning [Compact] Representations for Structured Data
- Fosca Gianotti (Scuola Normale Superiore di Pisa), Explainable Machine Learning for Trustworthy AI
- Yann Le Cun (Facebook AI Research), From Machine Learning to Autonomous Intelligence

In addition, there were respectively twenty three in-person and three online workshops; five in-person and three online tutorials; two combined in-person and one combined online workshop-tutorials, together with a PhD Forum, a discovery challenge and demonstrations.

Papers presented during the three main conference days were organized in 4 tracks, within 54 sessions:

- Research Track: articles on research or methodology from all branches of machine learning, data mining, and knowledge discovery;
- Applied Data Science Track: articles on cutting-edge uses of machine learning, data mining, and knowledge discovery to resolve practical use cases and close the gap between current theory and practice;
- Journal Track: articles that were published in special issues of the journals *Machine Learning* and *Data Mining and Knowledge Discovery*;

– Demo Track: short articles that propose a novel system that advances the state of the art and include a demonstration video.

We received a record number of 1238 abstract submissions, and for the Research and Applied Data Science Tracks, 932 papers made it through the review process (the remaining papers were withdrawn, with the bulk being desk rejected). We accepted 189 (27.3%) Research papers and 53 (22.2%) Applied Data science articles. 47 papers from the Journal Track and 17 demo papers were also included in the program. We were able to put together an extraordinarily rich and engaging program because of the high quality submissions.

Research articles that were judged to be of exceptional quality and deserving of special distinction were chosen by the awards committee:

– Machine Learning Best Paper Award: "*Bounding the Family-Wise Error Rate in Local Causal Discovery Using Rademacher Averages*", by Dario Simionato (University of Padova) and Fabio Vandin (University of Padova)
– Data-Mining Best Paper Award: "*Transforming PageRank into an Infinite-Depth Graph Neural Network*", by Andreas Roth (TU Dortmund), and Thomas Liebig (TU Dortmund)
– Test of Time Award for highest impact paper from ECML–PKDD 2012: "*Fairness-Aware Classifier with Prejudice Remover Regularizer*", by Toshihiro Kamishima (National Institute of Advanced Industrial Science and Technology AIST), Shotaro Akashi (National Institute of Advanced Industrial Science and Technology AIST), Hideki Asoh (National Institute of Advanced Industrial Science and Technology AIST), and Jun Sakuma (University of Tsukuba)

We sincerely thank the contributions of all participants, authors, PC members, area chairs, session chairs, volunteers, and co-organizers who made ECML–PKDD 2022 a huge success. We would especially like to thank Julie from the Grenoble World Trade Center for all her help and Titouan from Insight-outside, who worked so hard to make the online event possible. We also like to express our gratitude to Thierry for the design of the conference logo representing the three mountain chains surrounding the Grenoble city, as well as the sponsors and the ECML–PKDD Steering Committee.

October 2022

Massih-Reza Amini
Stéphane Canu
Asja Fischer
Petra Kralj Novak
Tias Guns
Grigorios Tsoumakas
Georgios Balikas
Fragkiskos Malliaros

Organization

General Chairs

Massih-Reza Amini   University Grenoble Alpes, France
Stéphane Canu   INSA Rouen, France

Program Chairs

Asja Fischer   Ruhr University Bochum, Germany
Tias Guns   KU Leuven, Belgium
Petra Kralj Novak   Central European University, Austria
Grigorios Tsoumakas   Aristotle University of Thessaloniki, Greece

Journal Track Chairs

Peggy Cellier   INSA Rennes, IRISA, France
Krzysztof Dembczyński   Yahoo Research, USA
Emilie Devijver   CNRS, France
Albrecht Zimmermann   University of Caen Normandie, France

Workshop and Tutorial Chairs

Bruno Crémilleux   University of Caen Normandie, France
Charlotte Laclau   Telecom Paris, France

Local Chairs

Latifa Boudiba   University Grenoble Alpes, France
Franck Iutzeler   University Grenoble Alpes, France

Proceedings Chairs

Wouter Duivesteijn   Technische Universiteit Eindhoven,
                                     the Netherlands
Sibylle Hess   Technische Universiteit Eindhoven,
                                     the Netherlands

Industry Track Chairs

Rohit Babbar   Aalto University, Finland
Françoise Fogelmann   Hub France IA, France

Discovery Challenge Chairs

Ioannis Katakis   University of Nicosia, Cyprus
Ioannis Partalas   Expedia, Switzerland

Demonstration Chairs

Georgios Balikas   Salesforce, France
Fragkiskos Malliaros   CentraleSupélec, France

PhD Forum Chairs

Esther Galbrun   University of Eastern Finland, Finland
Justine Reynaud   University of Caen Normandie, France

Awards Chairs

Francesca Lisi   Università degli Studi di Bari, Italy
Michalis Vlachos   University of Lausanne, Switzerland

Sponsorship Chairs

Patrice Aknin   IRT SystemX, France
Gilles Gasso   INSA Rouen, France

Web Chairs

Martine Harshé	Laboratoire d'Informatique de Grenoble, France
Marta Soare	University Grenoble Alpes, France

Publicity Chair

Emilie Morvant	Université Jean Monnet, France

ECML PKDD Steering Committee

Annalisa Appice	University of Bari Aldo Moro, Italy
Ira Assent	Aarhus University, Denmark
Albert Bifet	Télécom ParisTech, France
Francesco Bonchi	ISI Foundation, Italy
Tania Cerquitelli	Politecnico di Torino, Italy
Sašo Džeroski	Jožef Stefan Institute, Slovenia
Elisa Fromont	Université de Rennes, France
Andreas Hotho	Julius-Maximilians-Universität Würzburg, Germany
Alípio Jorge	University of Porto, Portugal
Kristian Kersting	TU Darmstadt, Germany
Jefrey Lijffijt	Ghent University, Belgium
Luís Moreira-Matias	University of Porto, Portugal
Katharina Morik	TU Dortmund, Germany
Siegfried Nijssen	Université catholique de Louvain, Belgium
Andrea Passerini	University of Trento, Italy
Fernando Perez-Cruz	ETH Zurich, Switzerland
Alessandra Sala	Shutterstock Ireland Limited, Ireland
Arno Siebes	Utrecht University, the Netherlands
Isabel Valera	Universität des Saarlandes, Germany

Program Committees

Guest Editorial Board, Journal Track

Richard Allmendinger	University of Manchester, UK
Marie Anastacio	Universiteit Leiden, the Netherlands
Ira Assent	Aarhus University, Denmark
Martin Atzmueller	Universität Osnabrück, Germany
Rohit Babbar	Aalto University, Finland

Jaume Bacardit   Newcastle University, UK
Anthony Bagnall   University of East Anglia, UK
Mitra Baratchi   Universiteit Leiden, the Netherlands
Francesco Bariatti   IRISA, France
German Barquero   Universität de Barcelona, Spain
Alessio Benavoli   Trinity College Dublin, Ireland
Viktor Bengs   Ludwig-Maximilians-Universität München,
                                  Germany
Massimo Bilancia   Università degli Studi di Bari Aldo Moro, Italy
Ilaria Bordino   Unicredit R&D, Italy
Jakob Bossek   University of Münster, Germany
Ulf Brefeld   Leuphana University of Lüneburg, Germany
Ricardo Campello   University of Newcastle, UK
Michelangelo Ceci   University of Bari, Italy
Loic Cerf   Universidade Federal de Minas Gerais, Brazil
Vitor Cerqueira   Universidade do Porto, Portugal
Laetitia Chapel   IRISA, France
Jinghui Chen   Pennsylvania State University, USA
Silvia Chiusano   Politecnico di Torino, Italy
Roberto Corizzo   Università degli Studi di Bari Aldo Moro, Italy
Bruno Cremilleux   Université de Caen Normandie, France
Marco de Gemmis   University of Bari Aldo Moro, Italy
Sebastien Destercke   Centre National de la Recherche Scientifique,
                                  France
Shridhar Devamane   Global Academy of Technology, India
Benjamin Doerr   Ecole Polytechnique, France
Wouter Duivesteijn   Technische Universiteit Eindhoven,
                                  the Netherlands
Thomas Dyhre Nielsen   Aalborg University, Denmark
Tapio Elomaa   Tampere University, Finland
Remi Emonet   Université Jean Monnet Saint-Etienne, France
Nicola Fanizzi   Università degli Studi di Bari Aldo Moro, Italy
Pedro Ferreira   University of Lisbon, Portugal
Cesar Ferri   Universität Politecnica de Valencia, Spain
Julia Flores   University of Castilla-La Mancha, Spain
Ionut Florescu   Stevens Institute of Technology, USA
Germain Forestier   Université de Haute-Alsace, France
Joel Frank   Ruhr-Universität Bochum, Germany
Marco Frasca   Università degli Studi di Milano, Italy
Jose A. Gomez   Universidad de Castilla-La Mancha, Spain
Stephan Günnemann   Institute for Advanced Study, Germany
Luis Galarraga   Inria, France

Corrado Loglisci   Università degli Studi di Bari Aldo Moro, Italy
Nuno Lourenço   University of Coimbra, Portugal
Claudio Lucchese   Ca'Foscari University of Venice, Italy
Brian MacNamee   University College Dublin, Ireland
Davide Maiorca   University of Cagliari, Italy
Giuseppe Manco   National Research Council, Italy
Elio Masciari   University of Naples Federico II, Italy
Andres Masegosa   University of Aalborg, Denmark
Ernestina Menasalvas   Universidad Politecnica de Madrid, Spain
Lien Michiels   Universiteit Antwerpen, Belgium
Jan Mielniczuk   Polish Academy of Sciences, Poland
Paolo Mignone   Università degli Studi di Bari Aldo Moro, Italy
Anna Monreale   University of Pisa, Italy
Giovanni Montana   University of Warwick, UK
Gregoire Montavon   Technische Universität Berlin, Germany
Amedeo Napoli   LORIA, France
Frank Neumann   University of Adelaide, Australia
Thomas Nielsen   Aalborg Universitet, Denmark
Bruno Ordozgoiti   Aalto-yliopisto, Finland
Panagiotis Papapetrou   Stockholms Universitet, Sweden
Andrea Passerini   University of Trento, Italy
Mykola Pechenizkiy   Technische Universiteit Eindhoven,
                                     the Netherlands
Charlotte Pelletier   IRISA, France
Ruggero Pensa   University of Turin, Italy
Nico Piatkowski   Technische Universität Dortmund, Germany
Gianvito Pio   Università degli Studi di Bari Aldo Moro, Italy
Marc Plantevit   Université Claude Bernard Lyon 1, France
Jose M. Puerta   Universidad de Castilla-La Mancha, Spain
Kai Puolamaki   Helsingin Yliopisto, Finland
Michael Rabbat   Meta Platforms Inc, USA
Jan Ramon   Inria Lille Nord Europe, France
Rita Ribeiro   Universidade do Porto, Portugal
Kaspar Riesen   University of Bern, Switzerland
Matteo Riondato   Amherst College, USA
Celine Robardet   INSA Lyon, France
Pieter Robberechts   KU Leuven, Belgium
Antonio Salmeron   University of Almería, Spain
Jorg Sander   University of Alberta, Canada
Roberto Santana   University of the Basque Country, Spain
Michael Schaub   Rheinisch-Westfälische Technische Hochschule,
                                     Germany

Erik Schultheis   Aalto-yliopisto, Finland
Thomas Seidl   Ludwig-Maximilians-Universität München,
                                   Germany
Moritz Seiler   University of Münster, Germany
Kijung Shin   KAIST, South Korea
Shinichi Shirakawa   Yokohama National University, Japan
Marek Smieja   Jagiellonian University, Poland
James Edward Smith   University of the West of England, UK
Carlos Soares   Universidade do Porto, Portugal
Arnaud Soulet   Université de Tours, France
Gerasimos Spanakis   Maastricht University, the Netherlands
Giancarlo Sperli   University of Campania Luigi Vanvitelli, Italy
Myra Spiliopoulou   Otto von Guericke Universität Magdeburg,
                                   Germany
Jerzy Stefanowski   Poznan University of Technology, Poland
Giovanni Stilo   Università degli Studi dell'Aquila, Italy
Catalin Stoean   University of Craiova, Romania
Mahito Sugiyama   National Institute of Informatics, Japan
Nikolaj Tatti   Helsingin Yliopisto, Finland
Alexandre Termier   Université de Rennes 1, France
Luis Torgo   Dalhousie University, Canada
Leonardo Trujillo   Tecnologico Nacional de Mexico, Mexico
Wei-Wei Tu   4Paradigm Inc., China
Steffen Udluft   Siemens AG Corporate Technology, Germany
Arnaud Vandaele   Université de Mons, Belgium
Celine Vens   KU Leuven, Belgium
Herna Viktor   University of Ottawa, Canada
Marco Virgolin   Centrum Wiskunde en Informatica,
                                   the Netherlands
Jordi Vitria   Universität de Barcelona, Spain
Jilles Vreeken   CISPA Helmholtz Center for Information
                                   Security, Germany
Willem Waegeman   Universiteit Gent, Belgium
Markus Wagner   University of Adelaide, Australia
Elizabeth Wanner   Centro Federal de Educacao Tecnologica de
                                   Minas, Brazil
Marcel Wever   Universität Paderborn, Germany
Ngai Wong   University of Hong Kong, Hong Kong, China
Man Leung Wong   Lingnan University, Hong Kong, China
Marek Wydmuch   Poznan University of Technology, Poland
Guoxian Yu   Shandong University, China
Xiang Zhang   University of Hong Kong, Hong Kong, China

Ye Zhu   Deakin University, USA
Arthur Zimek   Syddansk Universitet, Denmark
Albrecht Zimmermann   Université de Caen Normandie, France

Area Chairs

Fabrizio Angiulli   DIMES, University of Calabria, Italy
Annalisa Appice   University of Bari, Italy
Ira Assent   Aarhus University, Denmark
Martin Atzmueller   Osnabrück University, Germany
Michael Berthold   Universität Konstanz, Germany
Albert Bifet   Université Paris-Saclay, France
Hendrik Blockeel   KU Leuven, Belgium
Christian Böhm   LMU Munich, Germany
Francesco Bonchi   ISI Foundation, Turin, Italy
Ulf Brefeld   Leuphana, Germany
Francesco Calabrese   Richemont, USA
Toon Calders   Universiteit Antwerpen, Belgium
Michelangelo Ceci   University of Bari, Italy
Peggy Cellier   IRISA, France
Duen Horng Chau   Georgia Institute of Technology, USA
Nicolas Courty   IRISA, Université Bretagne-Sud, France
Bruno Cremilleux   Université de Caen Normandie, France
Jesse Davis   KU Leuven, Belgium
Gianmarco De Francisci Morales   CentAI, Italy
Tom Diethe   Amazon, UK
Carlotta Domeniconi   George Mason University, USA
Yuxiao Dong   Tsinghua University, China
Kurt Driessens   Maastricht University, the Netherlands
Tapio Elomaa   Tampere University, Finland
Sergio Escalera   CVC and University of Barcelona, Spain
Faisal Farooq   Qatar Computing Research Institute, Qatar
Asja Fischer   Ruhr University Bochum, Germany
Peter Flach   University of Bristol, UK
Eibe Frank   University of Waikato, New Zealand
Paolo Frasconi   Università degli Studi di Firenze, Italy
Elisa Fromont   Université Rennes 1, IRISA/Inria, France
Johannes Fürnkranz   JKU Linz, Austria
Patrick Gallinari   Sorbonne Université, Criteo AI Lab, France
Joao Gama   INESC TEC - LIAAD, Portugal
Jose Gamez   Universidad de Castilla-La Mancha, Spain
Roman Garnett   Washington University in St. Louis, USA
Thomas Gärtner   TU Wien, Austria

Aristides Gionis	KTH Royal Institute of Technology, Sweden
Francesco Gullo	UniCredit, Italy
Stephan Günnemann	Technical University of Munich, Germany
Xiangnan He	University of Science and Technology of China, China
Daniel Hernandez-Lobato	Universidad Autonoma de Madrid, Spain
José Hernández-Orallo	Universität Politècnica de València, Spain
Jaakko Hollmén	Aalto University, Finland
Andreas Hotho	Universität Würzburg, Germany
Eyke Hüllermeier	University of Munich, Germany
Neil Hurley	University College Dublin, Ireland
Georgiana Ifrim	University College Dublin, Ireland
Alipio Jorge	INESC TEC/University of Porto, Portugal
Ross King	Chalmers University of Technology, Sweden
Arno Knobbe	Leiden University, the Netherlands
Yun Sing Koh	University of Auckland, New Zealand
Parisa Kordjamshidi	Michigan State University, USA
Lars Kotthoff	University of Wyoming, USA
Nicolas Kourtellis	Telefonica Research, Spain
Danai Koutra	University of Michigan, USA
Danica Kragic	KTH Royal Institute of Technology, Sweden
Stefan Kramer	Johannes Gutenberg University Mainz, Germany
Niklas Lavesson	Blekinge Institute of Technology, Sweden
Sébastien Lefèvre	Université de Bretagne Sud/IRISA, France
Jefrey Lijffijt	Ghent University, Belgium
Marius Lindauer	Leibniz University Hannover, Germany
Patrick Loiseau	Inria, France
Jose Lozano	UPV/EHU, Spain
Jörg Lücke	Universität Oldenburg, Germany
Donato Malerba	Università degli Studi di Bari Aldo Moro, Italy
Fragkiskos Malliaros	CentraleSupelec, France
Giuseppe Manco	ICAR-CNR, Italy
Wannes Meert	KU Leuven, Belgium
Pauli Miettinen	University of Eastern Finland, Finland
Dunja Mladenic	Jožef Stefan Institute, Slovenia
Anna Monreale	Università di Pisa, Italy
Luis Moreira-Matias	Finiata, Germany
Emilie Morvant	University Jean Monnet, St-Etienne, France
Sriraam Natarajan	UT Dallas, USA
Nuria Oliver	Vodafone Research, USA
Panagiotis Papapetrou	Stockholm University, Sweden
Laurence Park	WSU, Australia

Andrea Passerini	University of Trento, Italy
Mykola Pechenizkiy	TU Eindhoven, the Netherlands
Dino Pedreschi	University of Pisa, Italy
Robert Peharz	Graz University of Technology, Austria
Julien Perez	Naver Labs Europe, France
Franz Pernkopf	Graz University of Technology, Austria
Bernhard Pfahringer	University of Waikato, New Zealand
Fabio Pinelli	IMT Lucca, Italy
Visvanathan Ramesh	Goethe University Frankfurt, Germany
Jesse Read	Ecole Polytechnique, France
Zhaochun Ren	Shandong University, China
Marian-Andrei Rizoiu	University of Technology Sydney, Australia
Celine Robardet	INSA Lyon, France
Sriparna Saha	IIT Patna, India
Ute Schmid	University of Bamberg, Germany
Lars Schmidt-Thieme	University of Hildesheim, Germany
Michele Sebag	LISN CNRS, France
Thomas Seidl	LMU Munich, Germany
Arno Siebes	Universiteit Utrecht, the Netherlands
Fabrizio Silvestri	Sapienza, University of Rome, Italy
Myra Spiliopoulou	Otto-von-Guericke-University Magdeburg, Germany
Yizhou Sun	UCLA, USA
Jie Tang	Tsinghua University, China
Nikolaj Tatti	Helsinki University, Finland
Evimaria Terzi	Boston University, USA
Marc Tommasi	Lille University, France
Antti Ukkonen	University of Helsinki, Finland
Herke van Hoof	University of Amsterdam, the Netherlands
Matthijs van Leeuwen	Leiden University, the Netherlands
Celine Vens	KU Leuven, Belgium
Christel Vrain	University of Orleans, France
Jilles Vreeken	CISPA Helmholtz Center for Information Security, Germany
Willem Waegeman	Universiteit Gent, Belgium
Stefan Wrobel	Fraunhofer IAIS, Germany
Xing Xie	Microsoft Research Asia, China
Min-Ling Zhang	Southeast University, China
Albrecht Zimmermann	Université de Caen Normandie, France
Indre Zliobaite	University of Helsinki, Finland

Program Committee Members

Amos Abbott	Virginia Tech, USA
Pedro Abreu	CISUC, Portugal
Maribel Acosta	Ruhr University Bochum, Germany
Timilehin Aderinola	Insight Centre, University College Dublin, Ireland
Linara Adilova	Ruhr University Bochum, Fraunhofer IAIS, Germany
Florian Adriaens	KTH, Sweden
Azim Ahmadzadeh	Georgia State University, USA
Nourhan Ahmed	University of Hildesheim, Germany
Deepak Ajwani	University College Dublin, Ireland
Amir Hossein Akhavan Rahnama	KTH Royal Institute of Technology, Sweden
Aymen Al Marjani	ENS Lyon, France
Mehwish Alam	Leibniz Institute for Information Infrastructure, Germany
Francesco Alesiani	NEC Laboratories Europe, Germany
Omar Alfarisi	ADNOC, Canada
Pegah Alizadeh	Ericsson Research, France
Reem Alotaibi	King Abdulaziz University, Saudi Arabia
Jumanah Alshehri	Temple University, USA
Bakhtiar Amen	University of Huddersfield, UK
Evelin Amorim	Inesc tec, Portugal
Shin Ando	Tokyo University of Science, Japan
Thiago Andrade	INESC TEC - LIAAD, Portugal
Jean-Marc Andreoli	Naverlabs Europe, France
Giuseppina Andresini	University of Bari Aldo Moro, Italy
Alessandro Antonucci	IDSIA, Switzerland
Xiang Ao	Institute of Computing Technology, CAS, China
Siddharth Aravindan	National University of Singapore, Singapore
Héber H. Arcolezi	Inria and École Polytechnique, France
Adrián Arnaiz-Rodríguez	ELLIS Unit Alicante, Spain
Yusuf Arslan	University of Luxembourg, Luxembourg
André Artelt	Bielefeld University, Germany
Sunil Aryal	Deakin University, Australia
Charles Assaad	Easyvista, France
Matthias Aßenmacher	Ludwig-Maximilians-Universität München, Germany
Zeyar Aung	Masdar Institute, UAE
Serge Autexier	DFKI Bremen, Germany
Rohit Babbar	Aalto University, Finland
Housam Babiker	University of Alberta, Canada

Antonio Bahamonde	University of Oviedo, Spain
Maroua Bahri	Inria Paris, France
Georgios Balikas	Salesforce, France
Maria Bampa	Stockholm University, Sweden
Hubert Baniecki	Warsaw University of Technology, Poland
Elena Baralis	Politecnico di Torino, Italy
Mitra Baratchi	LIACS - University of Leiden, the Netherlands
Kalliopi Basioti	Rutgers University, USA
Martin Becker	Stanford University, USA
Diana Benavides Prado	University of Auckland, New Zealand
Anes Bendimerad	LIRIS, France
Idir Benouaret	Université Grenoble Alpes, France
Isacco Beretta	Università di Pisa, Italy
Victor Berger	CEA, France
Christoph Bergmeir	Monash University, Australia
Cuissart Bertrand	University of Caen, France
Antonio Bevilacqua	University College Dublin, Ireland
Yaxin Bi	Ulster University, UK
Ranran Bian	University of Auckland, New Zealand
Adrien Bibal	University of Louvain, Belgium
Subhodip Biswas	Virginia Tech, USA
Patrick Blöbaum	Amazon AWS, USA
Carlos Bobed	University of Zaragoza, Spain
Paul Bogdan	USC, USA
Chiara Boldrini	CNR, Italy
Clément Bonet	Université Bretagne Sud, France
Andrea Bontempelli	University of Trento, Italy
Ludovico Boratto	University of Cagliari, Italy
Stefano Bortoli	Huawei Research Center, Germany
Diana-Laura Borza	Babes Bolyai University, Romania
Ahcene Boubekki	UiT, Norway
Sabri Boughorbel	QCRI, Qatar
Paula Branco	University of Ottawa, Canada
Jure Brence	Jožef Stefan Institute, Slovenia
Martin Breskvar	Jožef Stefan Institute, Slovenia
Marco Bressan	University of Milan, Italy
Dariusz Brzezinski	Poznan University of Technology, Poland
Florian Buettner	German Cancer Research Center, Germany
Julian Busch	Siemens Technology, Germany
Sebastian Buschjäger	TU Dortmund Artificial Intelligence Unit, Germany
Ali Butt	Virginia Tech, USA

Narayanan C. Krishnan	IIT Palakkad, India
Xiangrui Cai	Nankai University, China
Xiongcai Cai	UNSW Sydney, Australia
Zekun Cai	University of Tokyo, Japan
Andrea Campagner	Università degli Studi di Milano-Bicocca, Italy
Seyit Camtepe	CSIRO Data61, Australia
Jiangxia Cao	Chinese Academy of Sciences, China
Pengfei Cao	Chinese Academy of Sciences, China
Yongcan Cao	University of Texas at San Antonio, USA
Cécile Capponi	Aix-Marseille University, France
Axel Carlier	Institut National Polytechnique de Toulouse, France
Paula Carroll	University College Dublin, Ireland
John Cartlidge	University of Bristol, UK
Simon Caton	University College Dublin, Ireland
Bogdan Cautis	University of Paris-Saclay, France
Mustafa Cavus	Warsaw University of Technology, Poland
Remy Cazabet	Université Lyon 1, France
Josu Ceberio	University of the Basque Country, Spain
David Cechák	CEITEC Masaryk University, Czechia
Abdulkadir Celikkanat	Technical University of Denmark, Denmark
Dumitru-Clementin Cercel	University Politehnica of Bucharest, Romania
Christophe Cerisara	CNRS, France
Vítor Cerqueira	Dalhousie University, Canada
Mattia Cerrato	JGU Mainz, Germany
Ricardo Cerri	Federal University of São Carlos, Brazil
Hubert Chan	University of Hong Kong, Hong Kong, China
Vaggos Chatziafratis	Stanford University, USA
Siu Lun Chau	University of Oxford, UK
Chaochao Chen	Zhejiang University, China
Chuan Chen	Sun Yat-sen University, China
Hechang Chen	Jilin University, China
Jia Chen	Beihang University, China
Jiaoyan Chen	University of Oxford, UK
Jiawei Chen	Zhejiang University, China
Jin Chen	University of Electronic Science and Technology, China
Kuan-Hsun Chen	University of Twente, the Netherlands
Lingwei Chen	Wright State University, USA
Tianyi Chen	Boston University, USA
Wang Chen	Google, USA
Xinyuan Chen	Universiti Kuala Lumpur, Malaysia

Yuqiao Chen	UT Dallas, USA
Yuzhou Chen	Princeton University, USA
Zhennan Chen	Xiamen University, China
Zhiyu Chen	UCSB, USA
Zhqian Chen	Mississippi State University, USA
Ziheng Chen	Stony Brook University, USA
Zhiyong Cheng	Shandong Academy of Sciences, China
Noëlie Cherrier	CITiO, France
Anshuman Chhabra	UC Davis, USA
Zhixuan Chu	Ant Group, China
Guillaume Cleuziou	LIFO, France
Ciaran Cooney	AflacNI, UK
Robson Cordeiro	University of São Paulo, Brazil
Roberto Corizzo	American University, USA
Antoine Cornuéjols	AgroParisTech, France
Fabrizio Costa	Exeter University, UK
Gustavo Costa	Instituto Federal de Goiás - Campus Jataí, Brazil
Luís Cruz	Delft University of Technology, the Netherlands
Tianyu Cui	Institute of Information Engineering, China
Wang-Zhou Dai	Imperial College London, UK
Tanmoy Dam	University of New South Wales Canberra, Australia
Thi-Bich-Hanh Dao	University of Orleans, France
Adrian Sergiu Darabant	Babes Bolyai University, Romania
Mrinal Das	IIT Palakaad, India
Sina Däubener	Ruhr University, Bochum, Germany
Padraig Davidson	University of Würzburg, Germany
Paul Davidsson	Malmö University, Sweden
Andre de Carvalho	USP, Brazil
Antoine de Mathelin	ENS Paris-Saclay, France
Tom De Schepper	University of Antwerp, Belgium
Marcilio de Souto	LIFO/Univ. Orleans, France
Gaetan De Waele	Ghent University, Belgium
Pieter Delobelle	KU Leuven, Belgium
Alper Demir	Izmir University of Economics, Turkey
Ambra Demontis	University of Cagliari, Italy
Difan Deng	Leibniz Universität Hannover, Germany
Guillaume Derval	UCLouvain - ICTEAM, Belgium
Maunendra Sankar Desarkar	IIT Hyderabad, India
Chris Develder	University of Ghent - iMec, Belgium
Arnout Devos	Swiss Federal Institute of Technology Lausanne, Switzerland

Laurens Devos	KU Leuven, Belgium
Bhaskar Dhariyal	University College Dublin, Ireland
Nicola Di Mauro	University of Bari, Italy
Aissatou Diallo	University College London, UK
Christos Dimitrakakis	University of Neuchatel, Switzerland
Jiahao Ding	University of Houston, USA
Kaize Ding	Arizona State University, USA
Yao-Xiang Ding	Nanjing University, China
Guilherme Dinis Junior	Stockholm University, Sweden
Nikolaos Dionelis	University of Edinburgh, UK
Christos Diou	Harokopio University of Athens, Greece
Sonia Djebali	Léonard de Vinci Pôle Universitaire, France
Nicolas Dobigeon	University of Toulouse, France
Carola Doerr	Sorbonne University, France
Ruihai Dong	University College Dublin, Ireland
Shuyu Dong	Inria, Université Paris-Saclay, France
Yixiang Dong	Xi'an Jiaotong University, China
Xin Du	University of Edinburgh, UK
Yuntao Du	Nanjing University, China
Stefan Duffner	University of Lyon, France
Rahul Duggal	Georgia Tech, USA
Wouter Duivesteijn	TU Eindhoven, the Netherlands
Sebastijan Dumancic	TU Delft, the Netherlands
Inês Dutra	University of Porto, Portugal
Thomas Dyhre Nielsen	AAU, Denmark
Saso Dzeroski	Jožef Stefan Institute, Ljubljana, Slovenia
Tome Eftimov	Jožef Stefan Institute, Ljubljana, Slovenia
Hamid Eghbal-zadeh	LIT AI Lab, Johannes Kepler University, Austria
Theresa Eimer	Leibniz University Hannover, Germany
Radwa El Shawi	Tartu University, Estonia
Dominik Endres	Philipps-Universität Marburg, Germany
Roberto Esposito	Università di Torino, Italy
Georgios Evangelidis	University of Macedonia, Greece
Samuel Fadel	Leuphana University, Germany
Stephan Fahrenkrog-Petersen	Humboldt-Universität zu Berlin, Germany
Xiaomao Fan	Shenzhen Technology University, China
Zipei Fan	University of Tokyo, Japan
Hadi Fanaee	Halmstad University, Sweden
Meng Fang	TU/e, the Netherlands
Elaine Faria	UFU, Brazil
Ad Feelders	Universiteit Utrecht, the Netherlands
Sophie Fellenz	TU Kaiserslautern, Germany

Stefano Ferilli	University of Bari, Italy
Daniel Fernández-Sánchez	Universidad Autónoma de Madrid, Spain
Pedro Ferreira	Faculty of Sciences University of Porto, Portugal
Cèsar Ferri	Universität Politècnica València, Spain
Flavio Figueiredo	UFMG, Brazil
Soukaina Filali Boubrahimi	Utah State University, USA
Raphael Fischer	TU Dortmund, Germany
Germain Forestier	University of Haute Alsace, France
Edouard Fouché	Karlsruhe Institute of Technology, Germany
Philippe Fournier-Viger	Shenzhen University, China
Kary Framling	Umeå University, Sweden
Jérôme François	Inria Nancy Grand-Est, France
Fabio Fumarola	Prometeia, Italy
Pratik Gajane	Eindhoven University of Technology, the Netherlands
Esther Galbrun	University of Eastern Finland, Finland
Laura Galindez Olascoaga	KU Leuven, Belgium
Sunanda Gamage	University of Western Ontario, Canada
Chen Gao	Tsinghua University, China
Wei Gao	Nanjing University, China
Xiaofeng Gao	Shanghai Jiaotong University, China
Yuan Gao	University of Science and Technology of China, China
Jochen Garcke	University of Bonn, Germany
Clement Gautrais	Brightclue, France
Benoit Gauzere	INSA Rouen, France
Dominique Gay	Université de La Réunion, France
Xiou Ge	University of Southern California, USA
Bernhard Geiger	Know-Center GmbH, Germany
Jiahui Geng	University of Stavanger, Norway
Yangliao Geng	Tsinghua University, China
Konstantin Genin	University of Tübingen, Germany
Firas Gerges	New Jersey Institute of Technology, USA
Pierre Geurts	University of Liège, Belgium
Gizem Gezici	Sabanci University, Turkey
Amirata Ghorbani	Stanford, USA
Biraja Ghoshal	TCS, UK
Anna Giabelli	Università degli studi di Milano Bicocca, Italy
George Giannakopoulos	IIT Demokritos, Greece
Tobias Glasmachers	Ruhr-University Bochum, Germany
Heitor Murilo Gomes	University of Waikato, New Zealand
Anastasios Gounaris	Aristotle University of Thessaloniki, Greece

Antoine Gourru   University of Lyon, France
Michael Granitzer   University of Passau, Germany
Magda Gregorova   Hochschule Würzburg-Schweinfurt, Germany
Moritz Grosse-Wentrup   University of Vienna, Austria
Divya Grover   Chalmers University, Sweden
Bochen Guan   OPPO US Research Center, USA
Xinyu Guan   Xian Jiaotong University, China
Guillaume Guerard   ESILV, France
Daniel Guerreiro e Silva   University of Brasilia, Brazil
Riccardo Guidotti   University of Pisa, Italy
Ekta Gujral   University of California, Riverside, USA
Aditya Gulati   ELLIS Unit Alicante, Spain
Guibing Guo   Northeastern University, China
Jianxiong Guo   Beijing Normal University, China
Yuhui Guo   Renmin University of China, China
Karthik Gurumoorthy   Amazon, India
Thomas Guyet   Inria, Centre de Lyon, France
Guillaume Habault   KDDI Research, Inc., Japan
Amaury Habrard   University of St-Etienne, France
Shahrzad Haddadan   Brown University, USA
Shah Muhammad Hamdi   New Mexico State University, USA
Massinissa Hamidi   PRES Sorbonne Paris Cité, France
Peng Han   KAUST, Saudi Arabia
Tom Hanika   University of Kassel, Germany
Sébastien Harispe   IMT Mines Alès, France
Marwan Hassani   TU Eindhoven, the Netherlands
Kohei Hayashi   Preferred Networks, Inc., Japan
Conor Hayes   National University of Ireland Galway, Ireland
Lingna He   Zhejiang University of Technology, China
Ramya Hebbalaguppe   Indian Institute of Technology, Delhi, India
Jukka Heikkonen   University of Turku, Finland
Fredrik Heintz   Linköping University, Sweden
Patrick Hemmer   Karlsruhe Institute of Technology, Germany
Romain Hérault   INSA de Rouen, France
Jeronimo Hernandez-Gonzalez   University of Barcelona, Spain
Sibylle Hess   TU Eindhoven, the Netherlands
Fabian Hinder   Bielefeld University, Germany
Lars Holdijk   University of Amsterdam, the Netherlands
Martin Holena   Institute of Computer Science, Czechia
Mike Holenderski   Eindhoven University of Technology,
                                        the Netherlands
Shenda Hong   Peking University, China

Yupeng Hou	Renmin University of China, China
Binbin Hu	Ant Financial Services Group, China
Jian Hu	Queen Mary University of London, UK
Liang Hu	Tongji University, China
Wen Hu	Ant Group, China
Wenbin Hu	Wuhan University, China
Wenbo Hu	Tsinghua University, China
Yaowei Hu	University of Arkansas, USA
Chao Huang	University of Hong Kong, China
Gang Huang	Zhejiang Lab, China
Guanjie Huang	Penn State University, USA
Hong Huang	HUST, China
Jin Huang	University of Amsterdam, the Netherlands
Junjie Huang	Chinese Academy of Sciences, China
Qiang Huang	Jilin University, China
Shangrong Huang	Hunan University, China
Weitian Huang	South China University of Technology, China
Yan Huang	Huazhong University of Science and Technology, China
Yiran Huang	Karlsruhe Institute of Technology, Germany
Angelo Impedovo	University of Bari, Italy
Roberto Interdonato	CIRAD, France
Iñaki Inza	University of the Basque Country, Spain
Stratis Ioannidis	Northeastern University, USA
Rakib Islam	Facebook, USA
Tobias Jacobs	NEC Laboratories Europe GmbH, Germany
Priyank Jaini	Google, Canada
Johannes Jakubik	Karlsruhe Institute of Technology, Germany
Nathalie Japkowicz	American University, USA
Szymon Jaroszewicz	Polish Academy of Sciences, Poland
Shayan Jawed	University of Hildesheim, Germany
Rathinaraja Jeyaraj	Kyungpook National University, South Korea
Shaoxiong Ji	Aalto University, Finland
Taoran Ji	Virginia Tech, USA
Bin-Bin Jia	Southeast University, China
Yuheng Jia	Southeast University, China
Ziyu Jia	Beijing Jiaotong University, China
Nan Jiang	Purdue University, USA
Renhe Jiang	University of Tokyo, Japan
Siyang Jiang	National Taiwan University, Taiwan
Song Jiang	University of California, Los Angeles, USA
Wenyu Jiang	Nanjing University, China

Zhen Jiang	Jiangsu University, China
Yuncheng Jiang	South China Normal University, China
François-Xavier Jollois	Université de Paris Cité, France
Adan Jose-Garcia	Université de Lille, France
Ferdian Jovan	University of Bristol, UK
Steffen Jung	MPII, Germany
Thorsten Jungeblut	Bielefeld University of Applied Sciences, Germany
Hachem Kadri	Aix-Marseille University, France
Vana Kalogeraki	Athens University of Economics and Business, Greece
Vinayaka Kamath	Microsoft Research India, India
Toshihiro Kamishima	National Institute of Advanced Industrial Science, Japan
Bo Kang	Ghent University, Belgium
Alexandros Karakasidis	University of Macedonia, Greece
Mansooreh Karami	Arizona State University, USA
Panagiotis Karras	Aarhus University, Denmark
Ioannis Katakis	University of Nicosia, Cyprus
Koki Kawabata	Osaka University, Tokyo
Klemen Kenda	Jožef Stefan Institute, Slovenia
Patrik Joslin Kenfack	Innopolis University, Russia
Mahsa Keramati	Simon Fraser University, Canada
Hamidreza Keshavarz	Tarbiat Modares University, Iran
Adil Khan	Innopolis University, Russia
Jihed Khiari	Johannes Kepler University, Austria
Mi-Young Kim	University of Alberta, Canada
Arto Klami	University of Helsinki, Finland
Jiri Klema	Czech Technical University, Czechia
Tomas Kliegr	University of Economics Prague, Czechia
Christian Knoll	Graz, University of Technology, Austria
Dmitry Kobak	University of Tübingen, Germany
Vladimer Kobayashi	University of the Philippines Mindanao, Philippines
Dragi Kocev	Jožef Stefan Institute, Slovenia
Adrian Kochsiek	University of Mannheim, Germany
Masahiro Kohjima	NTT Corporation, Japan
Georgia Koloniari	University of Macedonia, Greece
Nikos Konofaos	Aristotle University of Thessaloniki, Greece
Irena Koprinska	University of Sydney, Australia
Lars Kotthoff	University of Wyoming, USA
Daniel Kottke	University of Kassel, Germany

Anna Krause	University of Würzburg, Germany
Alexander Kravberg	KTH Royal Institute of Technology, Sweden
Anastasia Krithara	NCSR Demokritos, Greece
Meelis Kull	University of Tartu, Estonia
Pawan Kumar	IIIT, Hyderabad, India
Suresh Kirthi Kumaraswamy	InterDigital, France
Gautam Kunapuli	Verisk Inc, USA
Marcin Kurdziel	AGH University of Science and Technology, Poland
Vladimir Kuzmanovski	Aalto University, Finland
Ariel Kwiatkowski	École Polytechnique, France
Firas Laakom	Tampere University, Finland
Harri Lähdesmäki	Aalto University, Finland
Stefanos Laskaridis	Samsung AI, UK
Alberto Lavelli	FBK-ict, Italy
Aonghus Lawlor	University College Dublin, Ireland
Thai Le	University of Mississippi, USA
Hoàng-Ân Lê	IRISA, University of South Brittany, France
Hoel Le Capitaine	University of Nantes, France
Thach Le Nguyen	Insight Centre, Ireland
Tai Le Quy	L3S Research Center - Leibniz University Hannover, Germany
Mustapha Lebbah	Sorbonne Paris Nord University, France
Dongman Lee	KAIST, South Korea
John Lee	Université catholique de Louvain, Belgium
Minwoo Lee	University of North Carolina at Charlotte, USA
Zed Lee	Stockholm University, Sweden
Yunwen Lei	University of Birmingham, UK
Douglas Leith	Trinity College Dublin, Ireland
Florian Lemmerich	RWTH Aachen, Germany
Carson Leung	University of Manitoba, Canada
Chaozhuo Li	Microsoft Research Asia, China
Jian Li	Institute of Information Engineering, China
Lei Li	Peking University, China
Li Li	Southwest University, China
Rui Li	Inspur Group, China
Shiyang Li	UCSB, USA
Shuokai Li	Chinese Academy of Sciences, China
Tianyu Li	Alibaba Group, China
Wenye Li	The Chinese University of Hong Kong, Shenzhen, China
Wenzhong Li	Nanjing University, China

Xiaoting Li	Pennsylvania State University, USA
Yang Li	University of North Carolina at Chapel Hill, USA
Zejian Li	Zhejiang University, China
Zhidong Li	UTS, Australia
Zhixin Li	Guangxi Normal University, China
Defu Lian	University of Science and Technology of China, China
Bin Liang	UTS, Australia
Yuchen Liang	RPI, USA
Yiwen Liao	University of Stuttgart, Germany
Pieter Libin	VUB, Belgium
Thomas Liebig	TU Dortmund, Germany
Seng Pei Liew	LINE Corporation, Japan
Beiyu Lin	University of Nevada - Las Vegas, USA
Chen Lin	Xiamen University, China
Tony Lindgren	Stockholm University, Sweden
Chen Ling	Emory University, USA
Jiajing Ling	Singapore Management University, Singapore
Marco Lippi	University of Modena and Reggio Emilia, Italy
Bin Liu	Chongqing University, China
Bowen Liu	Stanford University, USA
Chang Liu	Institute of Information Engineering, CAS, China
Chien-Liang Liu	National Chiao Tung University, Taiwan
Feng Liu	East China Normal University, China
Jiacheng Liu	Chinese University of Hong Kong, China
Li Liu	Chongqing University, China
Shengcai Liu	Southern University of Science and Technology, China
Shenghua Liu	Institute of Computing Technology, CAS, China
Tingwen Liu	Institute of Information Engineering, CAS, China
Xiangyu Liu	Tencent, China
Yong Liu	Renmin University of China, China
Yuansan Liu	University of Melbourne, Australia
Zhiwei Liu	Salesforce, USA
Tuwe Löfström	Jönköping University, Sweden
Corrado Loglisci	Università degli Studi di Bari Aldo Moro, Italy
Ting Long	Shanghai Jiao Tong University, China
Beatriz López	University of Girona, Spain
Yin Lou	Ant Group, USA
Samir Loudni	TASC (LS2N-CNRS), IMT Atlantique, France
Yang Lu	Xiamen University, China
Yuxun Lu	National Institute of Informatics, Japan

Massimiliano Luca	Bruno Kessler Foundation, Italy
Stefan Lüdtke	University of Mannheim, Germany
Jovita Lukasik	University of Mannheim, Germany
Denis Lukovnikov	University of Bonn, Germany
Pedro Henrique Luz de Araujo	University of Brasília, Brazil
Fenglong Ma	Pennsylvania State University, USA
Jing Ma	University of Virginia, USA
Meng Ma	Peking University, China
Muyang Ma	Shandong University, China
Ruizhe Ma	University of Massachusetts Lowell, USA
Xingkong Ma	National University of Defense Technology, China
Xueqi Ma	Tsinghua University, China
Zichen Ma	The Chinese University of Hong Kong, Shenzhen, China
Luis Macedo	University of Coimbra, Portugal
Harshitha Machiraju	EPFL, Switzerland
Manchit Madan	Delivery Hero, Germany
Seiji Maekawa	Osaka University, Japan
Sindri Magnusson	Stockholm University, Sweden
Pathum Chamikara Mahawaga	CSIRO Data61, Australia
Saket Maheshwary	Amazon, India
Ajay Mahimkar	AT&T, USA
Pierre Maillot	Inria, France
Lorenzo Malandri	Unimib, Italy
Rammohan Mallipeddi	Kyungpook National University, South Korea
Sahil Manchanda	IIT Delhi, India
Domenico Mandaglio	DIMES-UNICAL, Italy
Panagiotis Mandros	Harvard University, USA
Robin Manhaeve	KU Leuven, Belgium
Silviu Maniu	Université Paris-Saclay, France
Cinmayii Manliguez	National Sun Yat-Sen University, Taiwan
Naresh Manwani	International Institute of Information Technology, India
Jiali Mao	East China Normal University, China
Alexandru Mara	Ghent University, Belgium
Radu Marculescu	University of Texas at Austin, USA
Roger Mark	Massachusetts Institute of Technology, USA
Fernando Martínez-Plume	Joint Research Centre - European Commission, Belgium
Koji Maruhashi	Fujitsu Research, Fujitsu Limited, Japan
Simone Marullo	University of Siena, Italy

Elio Masciari	University of Naples, Italy
Florent Masseglia	Inria, France
Michael Mathioudakis	University of Helsinki, Finland
Takashi Matsubara	Osaka University, Japan
Tetsu Matsukawa	Kyushu University, Japan
Santiago Mazuelas	BCAM-Basque Center for Applied Mathematics, Spain
Ryan McConville	University of Bristol, UK
Hardik Meisheri	TCS Research, India
Panagiotis Meletis	Eindhoven University of Technology, the Netherlands
Gabor Melli	Medable, USA
Joao Mendes-Moreira	INESC TEC, Portugal
Chuan Meng	University of Amsterdam, the Netherlands
Cristina Menghini	Brown University, USA
Engelbert Mephu Nguifo	Université Clermont Auvergne, CNRS, LIMOS, France
Fabio Mercorio	University of Milan-Bicocca, Italy
Guillaume Metzler	Laboratoire ERIC, France
Hao Miao	Aalborg University, Denmark
Alessio Micheli	Università di Pisa, Italy
Paolo Mignone	University of Bari Aldo Moro, Italy
Matej Mihelcic	University of Zagreb, Croatia
Ioanna Miliou	Stockholm University, Sweden
Bamdev Mishra	Microsoft, India
Rishabh Misra	Twitter, Inc, USA
Dixant Mittal	National University of Singapore, Singapore
Zhaobin Mo	Columbia University, USA
Daichi Mochihashi	Institute of Statistical Mathematics, Japan
Armin Moharrer	Northeastern University, USA
Ioannis Mollas	Aristotle University of Thessaloniki, Greece
Carlos Monserrat-Aranda	Universität Politècnica de València, Spain
Konda Reddy Mopuri	Indian Institute of Technology Guwahati, India
Raha Moraffah	Arizona State University, USA
Pawel Morawiecki	Polish Academy of Sciences, Poland
Ahmadreza Mosallanezhad	Arizona State University, USA
Davide Mottin	Aarhus University, Denmark
Koyel Mukherjee	Adobe Research, India
Maximilian Münch	University of Applied Sciences Würzburg, Germany
Fabricio Murai	Universidade Federal de Minas Gerais, Brazil
Taichi Murayama	NAIST, Japan

Stéphane Mussard	CHROME, France
Mohamed Nadif	Centre Borelli - Université Paris Cité, France
Cian Naik	University of Oxford, UK
Felipe Kenji Nakano	KU Leuven, Belgium
Mirco Nanni	ISTI-CNR Pisa, Italy
Apurva Narayan	University of Waterloo, Canada
Usman Naseem	University of Sydney, Australia
Gergely Nemeth	ELLIS Unit Alicante, Spain
Stefan Neumann	KTH Royal Institute of Technology, Sweden
Anna Nguyen	Karlsruhe Institute of Technology, Germany
Quan Nguyen	Washington University in St. Louis, USA
Thi Phuong Quyen Nguyen	University of Da Nang, Vietnam
Thu Nguyen	SimulaMet, Norway
Thu Trang Nguyen	University College Dublin, Ireland
Prajakta Nimbhorkar	Chennai Mathematical Institute, Chennai, India
Xuefei Ning	Tsinghua University, China
Ikuko Nishikawa	Ritsumeikan University, Japan
Hao Niu	KDDI Research, Inc., Japan
Paraskevi Nousi	Aristotle University of Thessaloniki, Greece
Erik Novak	Jožef Stefan Institute, Slovenia
Slawomir Nowaczyk	Halmstad University, Sweden
Aleksandra Nowak	Jagiellonian University, Poland
Eirini Ntoutsi	Freie Universität Berlin, Germany
Andreas Nürnberger	Magdeburg University, Germany
James O'Neill	University of Liverpool, UK
Lutz Oettershagen	University of Bonn, Germany
Tsuyoshi Okita	Kyushu Institute of Technology, Japan
Makoto Onizuka	Osaka University, Japan
Subba Reddy Oota	IIIT Hyderabad, India
María Óskarsdóttir	University of Reykjavík, Iceland
Aomar Osmani	PRES Sorbonne Paris Cité, France
Aljaz Osojnik	JSI, Slovenia
Shuichi Otake	National Institute of Informatics, Japan
Greger Ottosson	IBM, France
Zijing Ou	Sun Yat-sen University, China
Abdelkader Ouali	University of Caen Normandy, France
Latifa Oukhellou	IFSTTAR, France
Kai Ouyang	Tsinghua University, France
Andrei Paleyes	University of Cambridge, UK
Pankaj Pandey	Indian Institute of Technology Gandhinagar, India
Guansong Pang	Singapore Management University, Singapore
Pance Panov	Jožef Stefan Institute, Slovenia

Apostolos Papadopoulos	Aristotle University of Thessaloniki, Greece
Evangelos Papalexakis	UC Riverside, USA
Anna Pappa	Université Paris 8, France
Chanyoung Park	UIUC, USA
Haekyu Park	Georgia Institute of Technology, USA
Sanghyun Park	Yonsei University, South Korea
Luca Pasa	University of Padova, Italy
Kevin Pasini	IRT SystemX, France
Vincenzo Pasquadibisceglie	University of Bari Aldo Moro, Italy
Nikolaos Passalis	Aristotle University of Thessaloniki, Greece
Javier Pastorino	University of Colorado, Denver, USA
Kitsuchart Pasupa	King Mongkut's Institute of Technology, Thailand
Andrea Paudice	University of Milan, Italy
Anand Paul	Kyungpook National University, South Korea
Yulong Pei	TU Eindhoven, the Netherlands
Charlotte Pelletier	Université de Bretagne du Sud, France
Jaakko Peltonen	Tampere University, Finland
Ruggero Pensa	University of Torino, Italy
Fabiola Pereira	Federal University of Uberlandia, Brazil
Lucas Pereira	ITI, LARSyS, Técnico Lisboa, Portugal
Aritz Pérez	Basque Center for Applied Mathematics, Spain
Lorenzo Perini	KU Leuven, Belgium
Alan Perotti	CENTAI Institute, Italy
Michaël Perrot	Inria Lille, France
Matej Petkovic	Institute Jožef Stefan, Slovenia
Lukas Pfahler	TU Dortmund University, Germany
Nico Piatkowski	Fraunhofer IAIS, Germany
Francesco Piccialli	University of Naples Federico II, Italy
Gianvito Pio	University of Bari, Italy
Giuseppe Pirrò	Sapienza University of Rome, Italy
Marc Plantevit	EPITA, France
Konstantinos Pliakos	KU Leuven, Belgium
Matthias Pohl	Otto von Guericke University, Germany
Nicolas Posocco	EURA NOVA, Belgium
Cedric Pradalier	GeorgiaTech Lorraine, France
Paul Prasse	University of Potsdam, Germany
Mahardhika Pratama	University of South Australia, Australia
Francesca Pratesi	ISTI - CNR, Italy
Steven Prestwich	University College Cork, Ireland
Giulia Preti	CentAI, Italy
Philippe Preux	Inria, France
Shalini Priya	Oak Ridge National Laboratory, USA

Ricardo Prudencio	Universidade Federal de Pernambuco, Brazil
Luca Putelli	Università degli Studi di Brescia, Italy
Peter van der Putten	Leiden University, the Netherlands
Chuan Qin	Baidu, China
Jixiang Qing	Ghent University, Belgium
Jolin Qu	Western Sydney University, Australia
Nicolas Quesada	Polytechnique Montreal, Canada
Teeradaj Racharak	Japan Advanced Institute of Science and Technology, Japan
Krystian Radlak	Warsaw University of Technology, Poland
Sandro Radovanovic	University of Belgrade, Serbia
Md Masudur Rahman	Purdue University, USA
Ankita Raj	Indian Institute of Technology Delhi, India
Herilalaina Rakotoarison	Inria, France
Alexander Rakowski	Hasso Plattner Institute, Germany
Jan Ramon	Inria, France
Sascha Ranftl	Graz University of Technology, Austria
Aleksandra Rashkovska Koceva	Jožef Stefan Institute, Slovenia
S. Ravi	Biocomplexity Institute, USA
Jesse Read	Ecole Polytechnique, France
David Reich	Universität Potsdam, Germany
Marina Reyboz	CEA, LIST, France
Pedro Ribeiro	University of Porto, Portugal
Rita P. Ribeiro	University of Porto, Portugal
Piera Riccio	ELLIS Unit Alicante Foundation, Spain
Christophe Rigotti	INSA Lyon, France
Matteo Riondato	Amherst College, USA
Mateus Riva	Telecom ParisTech, France
Kit Rodolfa	CMU, USA
Christophe Rodrigues	DVRC Pôle Universitaire Léonard de Vinci, France
Simon Rodríguez-Santana	ICMAT, Spain
Gaetano Rossiello	IBM Research, USA
Mohammad Rostami	University of Southern California, USA
Franz Rothlauf	Mainz Universität, Germany
Celine Rouveirol	Université Paris-Nord, France
Arjun Roy	Freie Universität Berlin, Germany
Joze Rozanec	Josef Stefan International Postgraduate School, Slovenia
Salvatore Ruggieri	University of Pisa, Italy
Marko Ruman	UTIA, AV CR, Czechia
Ellen Rushe	University College Dublin, Ireland

Dawid Rymarczyk	Jagiellonian University, Poland
Amal Saadallah	TU Dortmund, Germany
Khaled Mohammed Saifuddin	Georgia State University, USA
Hajer Salem	AUDENSIEL, France
Francesco Salvetti	Politecnico di Torino, Italy
Roberto Santana	University of the Basque Country (UPV/EHU), Spain
KC Santosh	University of South Dakota, USA
Somdeb Sarkhel	Adobe, USA
Yuya Sasaki	Osaka University, Japan
Yücel Saygın	Sabancı Universitesi, Turkey
Patrick Schäfer	Humboldt-Universität zu Berlin, Germany
Alexander Schiendorfer	Technische Hochschule Ingolstadt, Germany
Peter Schlicht	Volkswagen Group Research, Germany
Daniel Schmidt	Monash University, Australia
Johannes Schneider	University of Liechtenstein, Liechtenstein
Steven Schockaert	Cardiff University, UK
Jens Schreiber	University of Kassel, Germany
Matthias Schubert	Ludwig-Maximilians-Universität München, Germany
Alexander Schulz	CITEC, Bielefeld University, Germany
Jan-Philipp Schulze	Fraunhofer AISEC, Germany
Andreas Schwung	Fachhochschule Südwestfalen, Germany
Vasile-Marian Scuturici	LIRIS, France
Raquel Sebastião	IEETA/DETI-UA, Portugal
Stanislav Selitskiy	University of Bedfordshire, UK
Edoardo Serra	Boise State University, USA
Lorenzo Severini	UniCredit, R&D Dept., Italy
Tapan Shah	GE, USA
Ammar Shaker	NEC Laboratories Europe, Germany
Shiv Shankar	University of Massachusetts, USA
Junming Shao	University of Electronic Science and Technology, China
Kartik Sharma	Georgia Institute of Technology, USA
Manali Sharma	Samsung, USA
Ariona Shashaj	Network Contacts, Italy
Betty Shea	University of British Columbia, Canada
Chengchao Shen	Central South University, China
Hailan Shen	Central South University, China
Jiawei Sheng	Chinese Academy of Sciences, China
Yongpan Sheng	Southwest University, China
Chongyang Shi	Beijing Institute of Technology, China

Youming Tao	Shandong University, China
Martin Tappler	Graz University of Technology, Austria
Garth Tarr	University of Sydney, Australia
Mohammad Tayebi	Simon Fraser University, Canada
Anastasios Tefas	Aristotle University of Thessaloniki, Greece
Maguelonne Teisseire	INRAE - UMR Tetis, France
Stefano Teso	University of Trento, Italy
Olivier Teste	IRIT, University of Toulouse, France
Maximilian Thiessen	TU Wien, Austria
Eleftherios Tiakas	Aristotle University of Thessaloniki, Greece
Hongda Tian	University of Technology Sydney, Australia
Alessandro Tibo	Aalborg University, Denmark
Aditya Srinivas Timmaraju	Facebook, USA
Christos Tjortjis	International Hellenic University, Greece
Ljupco Todorovski	University of Ljubljana, Slovenia
Laszlo Toka	BME, Hungary
Ancy Tom	University of Minnesota, Twin Cities, USA
Panagiotis Traganitis	Michigan State University, USA
Cuong Tran	Syracuse University, USA
Minh-Tuan Tran	KAIST, South Korea
Giovanni Trappolini	Sapienza University of Rome, Italy
Volker Tresp	LMU, Germany
Yu-Chee Tseng	National Yang Ming Chiao Tung University, Taiwan
Maria Tzelepi	Aristotle University of Thessaloniki, Greece
Willy Ugarte	University of Applied Sciences (UPC), Peru
Antti Ukkonen	University of Helsinki, Finland
Abhishek Kumar Umrawal	Purdue University, USA
Athena Vakal	Aristotle University, Greece
Matias Valdenegro Toro	University of Groningen, the Netherlands
Maaike Van Roy	KU Leuven, Belgium
Dinh Van Tran	University of Freiburg, Germany
Fabio Vandin	University of Padova, Italy
Valerie Vaquet	CITEC, Bielefeld University, Germany
Iraklis Varlamis	Harokopio University of Athens, Greece
Santiago Velasco-Forero	MINES ParisTech, France
Bruno Veloso	Porto, Portugal
Dmytro Velychko	Carl von Ossietzky Universität Oldenburg, Germany
Sreekanth Vempati	Myntra, India
Sebastián Ventura Soto	University of Cordoba, Portugal
Rosana Veroneze	LBiC, Brazil

Jan Verwaeren	Ghent University, Belgium
Vassilios Verykios	Hellenic Open University, Greece
Herna Viktor	University of Ottawa, Canada
João Vinagre	LIAAD - INESC TEC, Portugal
Fabio Vitale	Centai Institute, Italy
Vasiliki Voukelatou	ISTI - CNR, Italy
Dong Quan Vu	Safran Tech, France
Maxime Wabartha	McGill University, Canada
Tomasz Walkowiak	Wroclaw University of Science and Technology, Poland
Vijay Walunj	University of Missouri-Kansas City, USA
Michael Wand	University of Mainz, Germany
Beilun Wang	Southeast University, China
Chang-Dong Wang	Sun Yat-sen University, China
Daheng Wang	Amazon, USA
Deng-Bao Wang	Southeast University, China
Di Wang	Nanyang Technological University, Singapore
Di Wang	KAUST, Saudi Arabia
Fu Wang	University of Exeter, UK
Hao Wang	Nanyang Technological University, Singapore
Hao Wang	Louisiana State University, USA
Hao Wang	University of Science and Technology of China, China
Hongwei Wang	University of Illinois Urbana-Champaign, USA
Hui Wang	SKLSDE, China
Hui (Wendy) Wang	Stevens Institute of Technology, USA
Jia Wang	Xi'an Jiaotong-Liverpool University, China
Jing Wang	Beijing Jiaotong University, China
Junxiang Wang	Emory University, USA
Qing Wang	IBM Research, USA
Rongguang Wang	University of Pennsylvania, USA
Ruoyu Wang	Shanghai Jiao Tong University, China
Ruxin Wang	Shenzhen Institutes of Advanced Technology, China
Senzhang Wang	Central South University, China
Shoujin Wang	Macquarie University, Australia
Xi Wang	Chinese Academy of Sciences, China
Yanchen Wang	Georgetown University, USA
Ye Wang	Chongqing University, China
Ye Wang	National University of Singapore, Singapore
Yifei Wang	Peking University, China
Yongqing Wang	Chinese Academy of Sciences, China

Yuandong Wang	Tsinghua University, China
Yue Wang	Microsoft Research, USA
Yun Cheng Wang	University of Southern California, USA
Zhaonan Wang	University of Tokyo, Japan
Zhaoxia Wang	SMU, Singapore
Zhiwei Wang	University of Chinese Academy of Sciences, China
Zihan Wang	Shandong University, China
Zijie J. Wang	Georgia Tech, USA
Dilusha Weeraddana	CSIRO, Australia
Pascal Welke	University of Bonn, Germany
Tobias Weller	University of Mannheim, Germany
Jörg Wicker	University of Auckland, New Zealand
Lena Wiese	Goethe University Frankfurt, Germany
Michael Wilbur	Vanderbilt University, USA
Moritz Wolter	Bonn University, Germany
Bin Wu	Beijing University of Posts and Telecommunications, China
Bo Wu	Renmin University of China, China
Jiancan Wu	University of Science and Technology of China, China
Jiantao Wu	University of Jinan, China
Ou Wu	Tianjin University, China
Yang Wu	Chinese Academy of Sciences, China
Yiqing Wu	University of Chinese Academic of Science, China
Yuejia Wu	Inner Mongolia University, China
Bin Xiao	University of Ottawa, Canada
Zhiwen Xiao	Southwest Jiaotong University, China
Ruobing Xie	WeChat, Tencent, China
Zikang Xiong	Purdue University, USA
Depeng Xu	University of North Carolina at Charlotte, USA
Jian Xu	Citadel, USA
Jiarong Xu	Fudan University, China
Kunpeng Xu	University of Sherbrooke, Canada
Ning Xu	Southeast University, China
Xianghong Xu	Tsinghua University, China
Sangeeta Yadav	Indian Institute of Science, India
Mehrdad Yaghoobi	University of Edinburgh, UK
Makoto Yamada	RIKEN AIP/Kyoto University, Japan
Akihiro Yamaguchi	Toshiba Corporation, Japan
Anil Yaman	Vrije Universiteit Amsterdam, the Netherlands

Hao Yan	Washington University in St Louis, USA
Qiao Yan	Shenzhen University, China
Chuang Yang	University of Tokyo, Japan
Deqing Yang	Fudan University, China
Haitian Yang	Chinese Academy of Sciences, China
Renchi Yang	National University of Singapore, Singapore
Shaofu Yang	Southeast University, China
Yang Yang	Nanjing University of Science and Technology, China
Yang Yang	Northwestern University, USA
Yiyang Yang	Guangdong University of Technology, China
Yu Yang	The Hong Kong Polytechnic University, China
Peng Yao	University of Science and Technology of China, China
Vithya Yogarajan	University of Auckland, New Zealand
Tetsuya Yoshida	Nara Women's University, Japan
Hong Yu	Chongqing Laboratory of Comput. Intelligence, China
Wenjian Yu	Tsinghua University, China
Yanwei Yu	Ocean University of China, China
Ziqiang Yu	Yantai University, China
Sha Yuan	Beijing Academy of Artificial Intelligence, China
Shuhan Yuan	Utah State University, USA
Mingxuan Yue	Google, USA
Aras Yurtman	KU Leuven, Belgium
Nayyar Zaidi	Deakin University, Australia
Zelin Zang	Zhejiang University & Westlake University, China
Masoumeh Zareapoor	Shanghai Jiao Tong University, China
Hanqing Zeng	USC, USA
Tieyong Zeng	The Chinese University of Hong Kong, China
Bin Zhang	South China University of Technology, China
Bob Zhang	University of Macau, Macao, China
Hang Zhang	National University of Defense Technology, China
Huaizheng Zhang	Nanyang Technological University, Singapore
Jiangwei Zhang	Tencent, China
Jinwei Zhang	Cornell University, USA
Jun Zhang	Tsinghua University, China
Lei Zhang	Virginia Tech, USA
Luxin Zhang	Worldline/Inria, France
Mimi Zhang	Trinity College Dublin, Ireland
Qi Zhang	University of Technology Sydney, Australia

Qiyiwen Zhang	University of Pennsylvania, USA
Teng Zhang	Huazhong University of Science and Technology, China
Tianle Zhang	University of Exeter, UK
Xuan Zhang	Renmin University of China, China
Yang Zhang	University of Science and Technology of China, China
Yaqian Zhang	University of Waikato, New Zealand
Yu Zhang	University of Illinois at Urbana-Champaign, USA
Zhengbo Zhang	Beihang University, China
Zhiyuan Zhang	Peking University, China
Heng Zhao	Shenzhen Technology University, China
Mia Zhao	Airbnb, USA
Tong Zhao	Snap Inc., USA
Qinkai Zheng	Tsinghua University, China
Xiangping Zheng	Renmin University of China, China
Bingxin Zhou	University of Sydney, Australia
Bo Zhou	Baidu, Inc., China
Min Zhou	Huawei Technologies, China
Zhipeng Zhou	University of Science and Technology of China, China
Hui Zhu	Chinese Academy of Sciences, China
Kenny Zhu	SJTU, China
Lingwei Zhu	Nara Institute of Science and Technology, Japan
Mengying Zhu	Zhejiang University, China
Renbo Zhu	Peking University, China
Yanmin Zhu	Shanghai Jiao Tong University, China
Yifan Zhu	Tsinghua University, China
Bartosz Zieliński	Jagiellonian University, Poland
Sebastian Ziesche	Bosch Center for Artificial Intelligence, Germany
Indre Zliobaite	University of Helsinki, Finland
Gianlucca Zuin	UFM, Brazil

Program Committee Members, Demo Track

Hesam Amoualian	WholeSoft Market, France
Georgios Balikas	Salesforce, France
Giannis Bekoulis	Vrije Universiteit Brussel, Belgium
Ludovico Boratto	University of Cagliari, Italy
Michelangelo Ceci	University of Bari, Italy
Abdulkadir Celikkanat	Technical University of Denmark, Denmark

Zhirong Yang   Norwegian University of Science and Technology,
                                        Norway
Xiangyu Zhao   City University of Hong Kong, Hong Kong, China

Sponsors

Contents – Part II

Social Network Analysis

Graph Neural Networks

Natural Language Processing and Text Mining

Conversational Systems

Networks and Graphs

Algorithmic Tools for Understanding the Motif Structure of Networks

Tianyi Chen[1], Brian Matejek[2,3], Michael Mitzenmacher[2],
and Charalampos E. Tsourakakis[1,2,4(✉)]

[1] Boston University, Boston, MA, USA
tsourolampis@gmail.com
[2] Harvard University, Cambridge, MA, USA
[3] Computer Science Laboratory, SRI International, Washington D.C., USA
[4] ISI Foundation, Turin, Italy

Abstract. Motifs are small subgraph patterns that play a key role towards understanding the structure and the function of biological and social networks. The current *de facto* approach towards assessing the statistical significance of a motif $\mathcal{M}$ relies on counting its occurrences across the network, and comparing that count to its expected count under some null generative model. This approach can be misleading due to *combinatorial artifacts*. That is, there may be a large count for a motif due to multiple copies sharing many vertices and edges connected to a subgraph, such as a clique, that completes the multiple copies of the motif.

In this work we introduce the novel concept of an (f, q)-spanning motif. A motif $\mathcal{M}$ is (f, q)-spanning if there exists a q-fraction of the nodes that induces an f-fraction of the occurrences of $\mathcal{M}$ in G. Intuitively, when f is close to 1, and q close to 0, most of the occurrences of $\mathcal{M}$ are localized in a small set of nodes, and thus its statistical significance is likely to be due to a combinatorial artifact. We propose efficient heuristics for finding the maximum f for a given q and minimum q for a given f for which a motif is (f, q)-spanning and evaluate them on real-world datasets. Our methods successfully identify combinatorial artifacts that otherwise go undetected using the standard approach for assessing statistical significance.

Finally, we leverage the motif structure of a network to design MOTIF-SCOPE, an algorithm that takes as input a graph and two motifs $\mathcal{M}_1, \mathcal{M}_2$, and finds subgraphs of the graph where $\mathcal{M}_1, \mathcal{M}_2$ occur infrequently and frequently respectively. We show that a good selection of $\mathcal{M}_1, \mathcal{M}_2$ allows us to find anomalies in large networks, including bipartite cliques in social graphs, and subgraphs rated with distrust in Bitcoin markets.

Keywords: Motifs · Graph mining · Statistical significance · Anomaly detection

Supplementary Information The online version contains supplementary material available at https://doi.org/10.1007/978-3-031-26390-3_1.

M.-R. Amini et al. (Eds.): ECML PKDD 2022, LNAI 13714, pp. 3–19, 2023.
https://doi.org/10.1007/978-3-031-26390-3_1

1   Introduction

Network motifs, or small induced subgraph patterns, are known to play a key role in understanding the structure and function of various real-world networks, especially biological [28,40], and social networks [47]. For example the feed-forward loop (FFL) is one of the most significant subgraphs in the transcription network of the bacteria Escherichia coli. The FFL has three nodes corresponding to transcription factors. The transcription factor X regulates a second transcription factor Y, and together they bind the regulatory region of a target gene Z, jointly modulating its transcription rate [27]. In social networks, triangles (K_3s) are known to appear frequently despite the edge sparsity of the network [49]. Ugander, Backstrom, and Kleinberg [47] showed that on the other hand social networks have very few cycles of length 4 (C_4s). This sheer contrast in the counts of K_3s and C_4s relates to human nature. Specifically, friends of friends are typically friends themselves, thus introducing edges that create K_3s but remove C_4s [49]. An FFL and a C_4 are shown in Fig. 1(a).

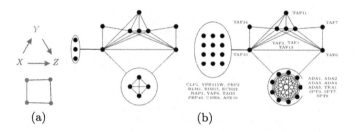

(a)   (b)

Fig. 1. (a) A feed-forward loop (FFL, top) and a C_4 (bottom). (b) Figure source [17]: the subgraph $\mathcal{M}$ on the left appears to be statistically significant in the network G on the right due to the presence of a large independent set, and a large clique in G. The independent set creates $\binom{12}{3}$ stars with three leaves, while the large clique creates $\binom{9}{4}$ smaller cliques of order 4, resulting in a total count of $\binom{12}{3} \times \binom{9}{4}$ occurrences, leading to the misleading conclusion that $\mathcal{M}$ is a statistically significant motif. We refer to this phenomenon as a combinatorial artifact, see also [17,32].

The *de facto* current approach towards assessing the statistical significance of a motif $\mathcal{M}$ involves two steps: (i) counting the occurrences of $\mathcal{M}$ in the input graph, and (ii) comparing that count to the expected number of occurrences of $\mathcal{M}$ under a null generative model. This approach has been widely used in the literature since the early 2000s [28,40], but nonetheless has significant drawbacks. The proper choice of the null model is a concern that was raised soon after the publication of the seminal work of Milo et al. [28], see the comment by Artzy et al. [1]. A suitable null model should generate networks similar to the input graph, as otherwise there is a danger of incorrectly assessing a motif as statistically significant (or not) due to an ill-posed null hypothesis. Also importantly, the current approach suffers from *combinatorial artifacts*. As observed originally by Lior Pachter in his blog [32], as well as by Grochow and Kellis [17], the existence of large independent sets and large cliques can obfuscate the relevance of the count of a motif. Consider the motif $\mathcal{M}$ with fifteen nodes corresponding to proteins shown in Fig. 1(b) on the left as originally shown in [17]. A node connected with a line to a set of nodes enclosed by a circle/oval denotes that the node is connected to all the nodes within that set. The closed circle/oval shows the topology of the set of nodes within it. For example, we observe that the node in the middle left is connected to three isolated nodes, whereas the two nodes in the middle (both left and right) are connected to four nodes that form a K_4. Figure 1(b) on the right shows the input network. Due to the existence of a large independent set, and a large clique, the number of occurrences of $\mathcal{M}$ is equal to $\binom{12}{3} \times \binom{9}{4}$. Such a high count may lead to the misleading assessment that $\mathcal{M}$ is statistically significant. Indeed, combinatorial artifacts occur frequently in real-world networks, which often contain large cliques and independent sets, similar to Fig. 1(b).

In this work we contribute towards understanding the motif structure of a network (directed or undirected) in the following ways:

- We propose the novel concept of an (f, q)-spanning motif. Specifically, a motif is (f, q)-spanning if there exists a subset of nodes S that induces an f-fraction of the motifs, while being a q-fraction of the node set V. Intuitively, if f is close to 1, and q is close to 0, the motif is likely to be a combinatorial artifact. Based on dense subgraph discovery tools [15], we propose a heuristic algorithm that allows us to test in near-linear time whether a motif is (f, q)-spanning.
- We propose MOTIFSCOPE, a novel framework that leverages frequently and infrequently appearing motifs to find anomalies in real-world networks. Our framework uses heuristics to find a subgraph that induces many copies of a motif $\mathcal{M}_2$ and few copies of a motif $\mathcal{M}_1$. We show that our framework allows us to find anomalies in social and trust networks.
- We perform an extensive experimental evaluation of various classical and state-of-the-art generative models as null models for assessing statistical significance, which highlights their similarities and differences, as well as the importance of choosing the models.

2   Related Work

Motifs. A motif is typically a subgraph of constant size. The goal of understanding the motif structure of a network spans numerous disciplines, ranging from systems biology [51] to social network analysis [47] and socio-economics [55], as it sheds light into the building blocks of networks [28]. Motifs have found various algorithmic and machine learning applications, under the umbrella of higher order methods [2, 23, 46, 52].

Assessing the Statistical Significance of a Motif. The *de facto* approach for deciding if a motif $\mathcal{M}$ is statistically significant or not relies on comparing its frequency

$f_{\mathcal{M}}$ to its expected frequency in a null random graph model [28]. While other approaches to assessing the statistical significance of motifs have been proposed, e.g., [4]; in this work we focus on the prevalent approach as introduced by Milo et al. [28]. Given the null model, one samples a large number of networks with the same number of nodes, and counts the frequency of $\mathcal{M}$; let $\bar{f}_{\mathcal{M}}$, $\sigma_{\mathcal{M}}$ be the average number of occurrences of $\mathcal{M}$ and the sample standard deviation, respectively. The z-score is defined as

$$z\text{-score}(\mathcal{M}) = z_{\mathcal{M}} = \frac{f_{\mathcal{M}} - \bar{f}_{\mathcal{M}}}{\sigma_{\mathcal{M}}}.$$

Observe that the z-score of a motif can be negative; motifs that have a large negative score, and thus appear less often than expected, are sometimes referred in the literature as *anti-motifs* [28,29].

An important issue is the choice of the null model. A common choice is the configuration model, or one of its variants [5,10,14]. This family of models generates a random (di)graph with a given (in-, out-)degree sequence(s). The configuration model was used in the influential works of Milo et al. [28,29]. However, their approach has received valid critique for a variety of reasons, such as the lack of spatial characteristics [1,20].

The densest subgraph problem aims to find the subgraph with the maximum average degree over all possible subgraphs [8,16]. Higher-order extensions have been recently proposed that maximize the average density of a small motif such as a triangle [30,44]. For this problem, as long as the number of nodes in the small subgraph is constant, there exist both efficient polynomial time exact algorithms [44], and faster greedy approximation algorithms [6,8].

Graph-based Anomaly Detection is an intensively active area of graph mining [31], with diverse industrial and scientific applications. We discuss related works in greater detail in the Appendix.

3   How to Address Combinatorial Artifacts?

Problem Definition. As discussed in Fig. 1(b), the significance of the motif on the left hand side does not truly represent statistically significant recurring independent motifs, but rather this motif arises because of a combinatorial artifact [32]. It appears around 30 000 times in a PPI network of S. *cerevisiae*, while its occurrences are concentrated into less than 30 nodes. To help clarify such situations, we provide the following definition.

Definition 1. A motif $\mathcal{M}$ is (f, q)-spanning in graph $G(V, E)$ if there exists a set of nodes $S \subseteq V$ such that $|S| \leq q|V|$ and the induced subgraph $G[S]$ contains an (at least) f-fraction of the occurrences of $\mathcal{M}$ in G.

We will (loosely) say the statistical significance of a motif $\mathcal{M}$ according to some null generative model is a *combinatorial artifact* if it is an (f, q)-spanning motif in $G(V, E)$ with $q \ll 1$, and f close to 1.[1]

Our definition of an (f, q)-spanning motif naturally introduces the following optimization problem.

> **Problem 1.** Given a motif $\mathcal{M}$ and a graph $G(V, E)$, what is the largest possible fraction f of occurrences of $\mathcal{M}$ among all subgraphs with (at most) $q|V|$ nodes for a given value of q?

We implicitly assume that the motif $\mathcal{M}$ appears frequently in the graph, and has been assessed statistically significant according to some null generative model; our goal is to understand whether its (apparent) significance is due to a combinatorial artifact or not.

Hardness. Problem 1 is NP-hard, and this holds both when we require $S \subseteq V$ to have exactly $k = q|V|$ nodes, and at most k nodes. The reduction is straightforward, and we omit all details. The idea of the proof is that if we could solve Problem 1, then by setting the motif $\mathcal{M}$ to be a simple undirected edge, we would be able to solve densest-k-subgraph (DkS) problem, and the densest-at-most-k-subgraph (DamkS) problems respectively. Furthermore, we know that these two problems are close in terms of approximation guarantees: if there exists an α-approximation algorithm for the DamkS problem, then there exists an $O(\alpha^2)$ approximation algorithm for the DkS problem. The best known approximation factor for the DkS is $O(n^{-1/4})$ due to Bhaskara et al. [3].

Theorem 1. *Problem 1 is NP-hard.*

We also provide a formulation which aims to optimize q for a given f, stated as the next problem.

> **Problem 2.** Given a motif $\mathcal{M}$ with $m(V)$ total occurrences in a graph $G(V, E)$, what is the smallest possible size $q|V|$ of the union of a set of $f \cdot m(V)$ occurrences for a given value of f?

The results of Chlamtáč et al. [9] yield the following corollary.

---

[1] It is worth outlining that forcing $f = 1$, and thus simplifying the definition above to a $(1, q)$- or just q-spanning motif is not a robust in the following sense. Consider a graph that is the union of a linear number of node disjoint triangles, and a clique of order $\sqrt{n}$. Each node in the graph participates in a triangle, and thus when $f = 1$, then $q = 1$. However, notice that most of the triangle occurrences appear in the small clique, i.e., $O(\sqrt{n})^3) = O(n^{3/2}) \gg O(n)$. Thus for $f = O(\frac{n^{3/2}}{n+n^{3/2}}) = 1 - o(1)$, q suddenly becomes $O(\frac{\sqrt{n}}{n}) = o(1)$. Similarly, a graph could have multiple distinct smaller combinatorial artifacts, in which case f might be a constant further from 1 (e.g., 3 small subgraphs with each around $1/3$ of the motif copies).

Corollary 1 (Theorem 1.1 [9]). *Problem 2 is NP-hard. Furthermore, there exists an $O(\sqrt{m(V)})$-approximation algorithm that runs in polynomial time.*

This corollary relates to their results for the minimum p-union problem (MpU). Consider a hypergraph where each hyperedge corresponds to an occurrence of a motif. Problem 2 can be restated as a minimum p-union problem (MpU), with $p = f \cdot m(V)$. However, their approximation algorithm is not practical for our purposes as it relies on computing maximum flows or solving linear programs, and we are interested in motifs with a large number of occurrences. We therefore propose a more efficient heuristic that works for both problem variants.

Algorithm 1: COMBART$(G(V,E), \mathcal{M}, f)$

---

1 Initialize $S_f^* = \emptyset$;
2 Count the total number m of occurrences of $\mathcal{M}$ in G;
3 **while** $m(S_f^*)/m < f \wedge m(V) > 0$ **do**
4 $\quad$ $S \leftarrow GreedyPeeling(G, \mathcal{M})$;
5 $\quad$ $S_f^* \leftarrow S_f^* \cup S$;
6 $\quad$ $E \leftarrow E \backslash E[S_f^*]$;
7 $\quad$ Update the motif count $m(V)$;
8 $\quad$ Compute $m(S_f^*)$;
9 $/ * E[S_f^*]$ is the set of edges in the induced subgraph $G[S_f^*] * /$;
10 $q \leftarrow |S_f^*|/|V|$;
11 **return** q ;

Proposed Heuristic. Our heuristic is based on the polynomially time solvable higher-order extension of the densest subgraph problem (DSP) due to Tsourakakis et al. [30,44]. Our algorithm is shown in pseudocode as Algorithm 1. The algorithm[2] runs as a black-box a greedy peeling algorithm until an f-fraction of the motif occurrences in the graph have been covered by the subgraph S_f^*. In each round, the greedy algorithm provides a $\frac{1}{|V(\mathcal{M})|}$-approximation to the optimization problem $\rho^* = \max_{S \subseteq V} \frac{m(S)}{|S|}$, see Appendix for its pseudocode. Here, $m(S)$ is the number of induced occurrences of motif $\mathcal{M}$ in S. Once the algorithm has covered an f-fraction of $\mathcal{M}$-occurrences in G, we compute q as $|S_f^*|/n$ where n is the number of nodes in G.

4   MOTIFSCOPE: Anomaly Detection via Motif Contrasting

A reason statistical significance of motifs is considered a worthwhile issue for study is because it gives us important information about graph structure. Indeed,

---

[2] While it aims to solve Problem 2, with minor changes it becomes a heuristic for Problem 1.

the existence of subgraphs that occur either frequently or infrequently can have interesting algorithmic implications and applications. Here we consider the problem of using motif counts to determine anomalies in a graph structure, such as a social network. Our results utilize the following natural problem.

Problem 3 Given a frequent motif $\mathcal{M}_1$, and an occurring but infrequent motif $\mathcal{M}_2$ in a graph G, find the subset of nodes $S \subseteq V$ that maximizes the average density difference

$$\max_{S \subseteq V} \frac{m_2(S)}{|S|} - \frac{m_1(S)}{|S|}.$$

Intuitively, an induced subgraph $G[S]$ that contains many induced copies of $\mathcal{M}_2$, but few induced copies of $\mathcal{M}_1$ differs significantly from the global network G with respect to those two motifs, and therefore possibly in other interesting ways. To solve Problem 3, we use the dense subgraph discovery framework of Tsourakakis et al. [45] with negative weights. We provide an extension of this approach for contrast of motif structures as follows: each node v is associated with a score $score(v)$ that is equal to $m_2(v) - m_1(v)$. Intuitively, we want to remove nodes that have a large negative score, and keep nodes with a high positive score. The pseudocode is shown in Algorithm 2. Assuming a method MOTIFCOUNT with time complexity $f(\mathcal{M})$ for motif $\mathcal{M}$, our algorithm runs in $O(n \log n + m + f(\mathcal{M}))$ time in the standard RAM model.

Algorithm 2: MOTIFSCOPE $(G, \mathcal{M}_1, \mathcal{M}_2)$

1 $m_i(v) = \#$ motifs of type $\mathcal{M}_i$ node v is contained in ($i = 1, 2, v \in V(G)$);
2 $n \leftarrow |V|$;
3 $H_n \leftarrow G$;
4 **for** $i \leftarrow n$ *to* 2 **do**
5   Let v be the vertex of G_i of minimum score, i.e., $score(v) = m_2(v) - m_1(v)$ (break ties arbitrarily);
6      $H_{i-1} \leftarrow H_i \backslash v$;
7   Update counts $m_1(v), m_2(v)$ for all $v \in V$;
8 **return** H_j that achieves maximum average density $\frac{m_2(S)-m_1(S)}{|S|}$ among H_is, $i = 1, \dots, n$.;

Implications and Applications. As a specific and important example of the MOTIFSCOPE algorithm, we explain how it can be used to find dense (near-)bipartite subgraphs. In general, the problem of detecting a dense bipartite subgraph in a graph is NP-hard [25]. Finding such subgraphs is important in practice since large bipartite subgraphs in social and trust networks are known to be rare, and frequently correspond to anomalies, such as a collection of manufactured accounts for illicit uses such as money laundering [33, 43]. To attack this problem using MOTIFSCOPE we leverage the fact that a bipartite subgraph does not contain any triangles (K_3s), which are otherwise common in social networks, but will probably contain several induced cycles of length 4 (C_4s), which are otherwise rare in social networks [47]. Therefore we set $\mathcal{M}_1 = K_3$ and $\mathcal{M}_2 = C_4$. While our approach is not guaranteed to output a bipartite graph (or even a near-bipartite graph), we show that on real data optimizing for minimizing K_3s while maximizing C_4s often yields a bipartite subgraph in practice. As a rule-of-thumb for using MOTIFSCOPE for anomaly detection applications, we propose either using prior knowledge of important subgraphs (such as with the K_3 and C_4 example above), or by choosing $\mathcal{M}_1$ to be one of the motifs with high z-score and $\mathcal{M}_2$ to be one of the motifs with low z-score.

5   Experiments

Datasets and Code. Table 1 summarizes the datasets that we use. We use publicly available datasets from a variety of domains, including biological, social, power, and trust networks. The code was written in Python3. We provide both the code and the datasets anonymously at https://github.com/tsourakakis-lab/motifscope.

Table 1. Summary of datasets.

Dataset	$|V|$	$|E|$	Description	Directed
S. cerevisiae [54]	759	1 593	PPI	×
C. elegans-PPI [54]	2 018	2 930	PPI	×
C. elegans-brain [51]	219	2 416	Connectome	✓
hamsterster [36]	2 426	1 593	Social	×
Eris1176 [36]	1 176	18 552	Power	×
Bitcoin-OTC [22]	5 881	35 592	Trust	✓
Bitcoin-Alpha [21]	3 783	24 186	Trust	✓
LastFM [38]	7 624	27 806	Social	×
Twitch-EN [37]	7 126	35 324	Social	×

Experimental Setup. The experiments are performed on a single machine, with an Intel i7-10850H CPU @ 2.70 GHz and 32 GB of main memory. The motif listing algorithm we use is due to Wernicke [50]. We focus on small-sized subgraphs. Figure 2 presents the 13 possible directed motifs of order 3; we shall refer to each motif with their id, for example $motif_{13}$ is the triangle with all six possible directed edges.

5.1   Combinatorial Artifacts

Table 2 summarizes the performance of COMBART algorithm on five different networks. The second column of the table visualizes a motif of interest $\mathcal{M}$. We

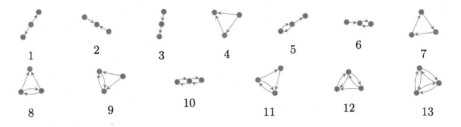

Fig. 2. There exist 13 possible directed motifs of order 3.

use a similar notation as [17], where a large node annotated as $S - c$ $(K - c)$ represents an independent set (clique) with c nodes. We observe that real-world networks typically contain large cliques and independent sets, and thus there exist various motifs whose significance will be a combinatorial artifact. The third column summarizes the subgraph which causes the combinatorial artifact, while the fourth and fifth columns show the motif count which happens to be also the global count $(f = 1)$, and the (f, q) values. As we observe, our novel definition sheds light into assessing the significance of those motifs, by noting that $f = 1$ and q is a small fraction of the node set. In contrast, the FFL motif, which is known to play a biological role, is $(0.8, 0.61)$-spanning, indicating statistical significance is not due to a combinatorial artifact. We believe these examples show our proposed method can be a significant enhancement to the current approach of assessing the statistical significance of motifs.

5.2   MOTIFSCOPE **Case Studies**

We show two case studies of MOTIFSCOPE. The first is an algorithmic application that attacks an NP-hard problem using prior knowledge about the appearance of motifs $\mathcal{M}_1, \mathcal{M}_2$, while the second application first analyzes the network to choose $\mathcal{M}_1, \mathcal{M}_2$.

Bipartite Subgraphs in Social Networks. As we mentioned in Sect. 4, we run MOTIFSCOPE using $\mathcal{M}_1 = K_3, \mathcal{M}_2 = C_4$, aiming to find a subgraph that induces many cycles of length 4, and few triangles. Our results are summarized in Table 3 for four datasets. We report the total number of induced edges, and the number of nodes in the bi-partition (L, R) of the output node set. Even though our method is not guaranteed to output bipartite subgraphs, the output subgraphs here were in fact all bipartite, i.e., all reported edges having one endpoint in L and one in R.

Anomaly Detection in Trust Networks. We use the Bitcoin-OTC network to illustrate the use of MOTIFSCOPE for anomaly detection on real-world networks. In the Appendix we provide additional results for the Bitcoin-alpha network and camouflage behaviors discovered by MOTIFSCOPE. Since we have no prior knowledge about the motifs in Bitcoin-OTC, we consider all motifs of order 3, and we compute their z-scores. Figure 3a shows the z-scores of all 13

Table 2. Motifs that are statistically significant from different networks due to combinatorial artifacts. Subgraphs the motifs are clustered in are also listed together with other statistics.

Dataset	Motif	Artifact source	Count	(f, q)
S. cerevisiae			$\binom{37}{19} \times \binom{71}{35}$	$(1, 0.06)$
C. Elegans-PPI			$\binom{23}{11} \times \binom{6}{3}$	$(1, 0.063)$
hamsterster			$\binom{21}{10} \times \binom{20}{10}$	$(1, 0.027)$
Eris1176			$\binom{55}{27} \times \binom{79}{40}$	$(1, 0.117)$
C. Elegans-Brain		-	1554	$(0.8, 0.61)$

Table 3. Bipartite subgraph found by contrasting C_4 and K_3.

Dataset	# edges	# nodes in L	# nodes in R
LastFM	124	21	37
Bitcoin-Alpha	24	5	9
Bitcoin-OTC	31	6	10
Twitch-EN	61	7	23

motifs. We observe that motif 3 has the most negative z-score indicating that it appears significantly less often than what we would expect in the directed configuration model. On the contrary, motifs 11, and 13 appear significantly more often. Thus, we use each of motifs 11 and 13 for $\mathcal{M}_1$, and motif 3 for $\mathcal{M}_2$.

The whole Bitcoin-OTC network contains 11% negative edges, which denote distrust. Figure 3b shows the precision and recall for MOTIFSCOPE, and popular graph anomaly detection methods that use dense subgraph discovery methods, including Core-A and Truss-A from Corescope [41], EigenSpokes [34], Holoscope [26], and Fraudar [18]. Here, we measure the quality of a subgraph S, using: (i) the precision, namely the fraction of negative edges induced by S over the total number of edges in S, and (ii) the recall, namely the fraction of negative

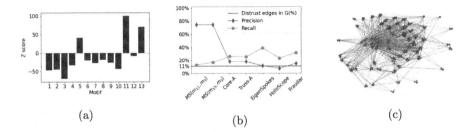

(a)   (b)   (c)

Fig. 3. Results on the Bitcoin-OTC network. (a) When no prior knowledge is available, we use the z-scores. Here, we show the z-scores of the 13 motifs of order 3. (b) Precision and recall for various anomaly detection methods and MOTIFSCOPE (MS) using as $(\mathcal{M}_1, \mathcal{M}_2)$ motifs $(motif_{11}, motif_3)$, and $(motif_{13}, motif_3)$, see Fig. 2 for the actual motifs. (c) Subgraph found by MOTIFSCOPE for $(motif_{11}, motif_3)$. Distrust relations are colored red, and trust relations are colored green. (Color figure online)

edges in S over the number of negative edges in the whole graph. We observe that our method outperforms competitors, finding subgraphs that induce a lot of distrust. Figure 3c visualizes one such subgraph. It is worth noting that motifs 11 and 13 are strongly connected, indicating that in this dataset reciprocal edges correlate with trust, whereas motif 3 is a directed chain that lacks reciprocity and correlates with distrust.

Running Times. Since our graphs are small to medium size, the main computational bottleneck comes from computing motifs on a large ensemble of sampled graphs from the null models. For instance, for Bitcoin-OTC, listing all motifs of order 3 takes around 20 s per sampled graph, and the dense subgraph discovery process (greedy peeling [8]) takes around 17 s.

6   Motif Significance and Null Models

As we have seen, the calculation of statistical significance depends on an underlying null model. In this section we study the following questions, to better understand similarities and differences among frequently used null models.

Q1. How robust is the significance (or lack thereof) of a given motif $\mathcal{M}$ across different null models? Is there a consensus between different null models on whether a motif is significant or not?

Q2. What are the sets of motifs that are statistically significant for different null models, and how do these sets compare to each other? How similar are they with respect to ranking motifs according to their z-scores?

Q3. How many samples do we need to generate from a null model, in order to obtain a concentrated estimate of the expected motif count? Is this sample size motif-dependent?

In looking at these questions, We consider seven null models summarized in Table 4 and all 13 motifs of order three in Fig. 2. The answer for Q3 is provided in the Appendix due to space constraints. We compare the null models to the well studied *C. elegans* connectome. The network consists of 219 neurons and 2 416 synapses that are represented as nodes and edges respectively, see also Table 1. The network we use corresponds to the adulthood of the *C. elegans*, and was obtained via high-resolution electron microscopy by [51]. All seven generative models we use are well-established in the literature, and they span a period of time from the origins of random graph theory to the most recent advances that involve deep-learning inspired models. Furthermore, we use graph models with independent edge probabilities and dependent edge probabilities. Considering both types of models is important as it was recently shown that random graph models where each edge is added to the graph independently with some probability are inherently limited in their ability to generate graphs with high triangle and other subgraph densities [7]. Furthermore, for any sparse graph, the configuration model is unlikely to generate a large clique. In contrast, it is known that biological networks tend to contain cliques and independent sets [32]. For this reason, we also use state-of-the-art non-independent models including the prescribed k-core model (KC) [48], and GraphRNN [53]. For a detailed description of the models, see the Appendix (supplementary material).

Table 4. Null models used in our experiments, along with their abbreviation. The first five models are *edge independent*, i.e., each edge $\{i, j\}$ exists independently from the rest with some probability p_{ij}, while KC and GRNN are not.

Null Models
Directed Erdős-Rényi model (ER) [13]
Edge swap configuration model (ES) [19]
Chung-Lu model (CL) [11]
Partially directed configuration model (PD) [42]
Stochastic Kronecker graphs (KG) [24]
Prescribed k-core model (KC) [48]
GraphRNN (GRNN) [53]

Is there Consensus Among Null Models? Mostly no. We use the *de facto* approach as described in Sect. 2 to test whether a motif $\mathcal{M}$ appears more often than expected (i.e., $\mathcal{M}$ is a statistically significant motif), or less often than expected (i.e., $\mathcal{M}$ is a statistically significant anti-motif) with respect to each of the seven null models. For each null model, we ensure that we have obtained enough samples for a concentrated estimate of the expectation of each motif $\mathcal{M}$ in Fig. 2, by requiring that the coefficient of variation $CV^2 = \frac{\sigma_{\mathcal{M}}^2}{f_{\mathcal{M}}}$ is at most 10^{-2}; the weak law of large numbers guarantees concentration, and is a direct application of Chebyshev's inequality.

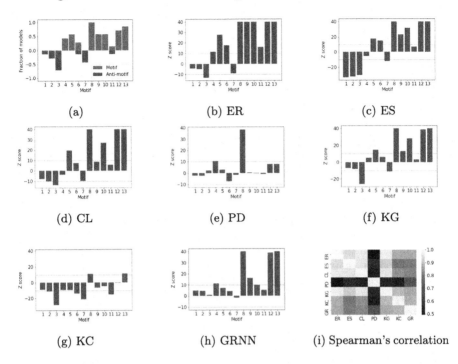

Fig. 4. (a) Histogram of models report each subgraph of size 3 as motif or anti-motif. (b)-(h) Motif significance with respect to z-score by different random graph models. Plots are clipped at a max value of 40. (i) Pairwise Spearman's correlation coefficient of motif z-scores of seven models.

For each motif $motif_i, i = 1, \ldots, 13$ we compute the percentage of the null models that assess it as a statistically significant motif (type A), and anti-motif (type B) respectively. Figure 4(a) summarizes our results. For example, motif 11 is assessed as a type A motif by one model, and similarly as type B by one model. According to the five other models, it is not statistically significant in either sense. Figures 4(b)–(g) provide a detailed overview of the assessment of each model. Perhaps surprisingly, motif 8 is the single motif that is assessed as statistically significant by all seven models. Previous research on other *C. elegans* datasets have identified motif 8 as statistically significant in both the male and hermaphrodite sexes [12]. One can construct motif 8 from motif 4, the feedforward loop (FFL), by introducing one reciprocal connection. Analysis of several species has shown that reciprocal connections are over-represented in connectomes [39]. Interestingly, we do not find feedforward loops [28] being statistically significant by several null models, and this can serve as a criterion for the quality of null models but with caution. The absence of several motor neurons in the analyzed connectomes could in part explain the reduced significance of FFLs. There is a general hierarchy of neurons in *C. elegans* with sensory neurons often connecting to interneurons and interneurons often connecting to motor neurons. Although prior research finds the significance of FFLs within each layer, many of the FFLs did contain one neuron of each type [35].

Do Null Models' Rankings Agree? Figure 4(i) shows Spearman's correlation coefficient of the z-scores respectively for all pairs of null models. The results are illustrated as a heatmap with the similarity scale on the right. We see that the partially directed configuration model is distinctively different from the rest of the 6 models. We explain this difference due to the fact that *C. elegans* has lots of reciprocal directed arcs, i.e., undirected edges, and thus it can model this aspect better than other models in sparse graphs. We observe that variants of the configuration model are not necessarily similar, a point raised by [14]. GraphRNN produces qualitatively similar results to the partially directed configuration model, but the z-scores are larger due to the fact that the directed version does not capture the frequency of reciprocal edges, despite the wide search of hyperparameters we performed (all details are included in the code).

In a nutshell, caution is required when choosing a null model. Non-independent models, such as the KC and GRNN models, can possibly model complex dependencies that create independent sets and cliques, as described in [7]. GraphRNN seems to be a promising null model for modeling connectomes, although it may not scale well to larger graphs.

7   Conclusion

Understanding the importance of motifs in networks is a key problem in connectomics, with a wide range of applications ranging from social network analysis to machine learning. In this work we introduce the novel concept of an (f, q)-spanning motif that addresses the major issue of *combinatorial artifacts*. We show that determining the smallest value of q for which there exists a node set of cardinality (at most) $q|V|$ that induces an f fraction of the motifs is NP-hard, and we design an efficient heuristic based on dense subgraph discovery methods. Furthermore, we provide new insights into the importance of the null model choice by an extensive empirical analysis of classic and state-of-the-art generative models. Finally, we design the MOTIFSCOPE framework that uses the motif structure of a graph to detect anomalies.

Our work opens several interesting directions. What are the best non-independent edge models as a null model choice? There is an ongoing line of research, with graph RNNs being a recent example [7,53]. Can we develop new generative models that leverage motifs for *C. Elegans* and model its temporal evolution, see also [47]?

References

1. Artzy-Randrup, Y., Fleishman, S.J., Ben-Tal, N., Stone, L.: Comment on "network motifs: simple building blocks of complex networks" and "superfamilies of evolved and designed networks". Science **305**(5687), 1107–1107 (2004)
2. Benson, A.R., Gleich, D.F., Leskovec, J.: Higher-order organization of complex networks. Science **353**(6295), 163–166 (2016)
3. Bhaskara, A., Charikar, M., Chlamtac, E., Feige, U., Vijayaraghavan, A.: Detecting high log-densities: an $o(n^{-1/4})$ approximation for densest k-subgraph. In: Proceedings of STOC 2010, pp. 201–210 (2010)
4. Bloem, P., de Rooij, S.: Large-scale network motif analysis using compression. Data Min. Knowl. Disc. **34**(5), 1421–1453 (2020). https://doi.org/10.1007/s10618-020-00691-y
5. Bollobás, B.: A probabilistic proof of an asymptotic formula for the number of labelled regular graphs. Eur. J. Comb. **1**(4), 311–316 (1980)
6. Boob, D., et al.: Flowless: extracting densest subgraphs without flow computations. In: Proceedings of TheWebConf 2020, pp. 573–583 (2020)
7. Chanpuriya, S., Musco, C., Sotiropoulos, K., Tsourakakis, C.: On the power of edge independent graph models. Adv. Neural Inf. Process. Syst. **34**, 24418–24429 (2021)
8. Charikar, M.: Greedy approximation algorithms for finding dense components in a graph. In: Jansen, K., Khuller, S. (eds.) APPROX 2000. LNCS, vol. 1913, pp. 84–95. Springer, Heidelberg (2000). https://doi.org/10.1007/3-540-44436-X_10
9. Chlamtáč, E., Dinitz, M., Konrad, C., Kortsarz, G., Rabanca, G.: The densest k-subhypergraph problem. arXiv preprint arXiv:1605.04284 (2016)
10. Chung, F., Chung, F.R., Graham, F.C., Lu, L., Chung, K.F., et al.: Complex graphs and networks, no. 107, American Mathematical Society (2006)
11. Chung, F., Lu, L.: The average distances in random graphs with given expected degrees. PNAS **99**(25), 15879–15882 (2002)
12. Cook, S.J., et al.: Whole-animal connectomes of both caenorhabditis elegans sexes. Nature **571**(7763), 63–71 (2019)
13. Erdős, P., Rényi, A.: On the evolution of random graphs. Publ. Math. Inst. Hung. Acad. Sci **5**(1), 17–60 (1960)
14. Fosdick, B.K., Larremore, D.B., Nishimura, J., Ugander, J.: Configuring random graph models with fixed degree sequences. Siam Rev. **60**(2), 315–355 (2018)
15. Gionis, A., Tsourakakis, C.E.: Dense subgraph discovery: KDD 2015 tutorial. In: Proceedings of KDD 2015, pp. 2313–2314 (2015)
16. Goldberg, A.V.: Finding a maximum density subgraph. University of California Berkeley, CA (1984)
17. Grochow, J.A., Kellis, M.: Network motif discovery using subgraph enumeration and symmetry-breaking. In: Speed, T., Huang, H. (eds.) RECOMB 2007. LNCS, vol. 4453, pp. 92–106. Springer, Heidelberg (2007). https://doi.org/10.1007/978-3-540-71681-5_7
18. Hooi, B., Song, H.A., Beutel, A., Shah, N., Shin, K., Faloutsos, C.: Fraudar: bounding graph fraud in the face of camouflage. In: Proceedings of KDD 2016, pp. 895–904 (2016)
19. Kannan, R., Tetali, P., Vempala, S.: Simple markov-chain algorithms for generating bipartite graphs and tournaments. Random Struct. Algor. **14**(4), 293–308 (1999)
20. King, O.D.: Comment on "subgraphs in random networks". Phys. Rev. E **70**(5), 058101 (2004)

21. Kumar, S., Hooi, B., Makhija, D., Kumar, M., Faloutsos, C., Subrahmanian, V.: Rev2: fraudulent user prediction in rating platforms. In: Proceedings of WSDM 2018, pp. 333–341. ACM (2018)
22. Kumar, S., Spezzano, F., Subrahmanian, V., Faloutsos, C.: Edge weight prediction in weighted signed networks. In: ICDM, pp. 221–230. IEEE (2016)
23. Lee, J.B., Rossi, R.A., Kong, X., Kim, S., Koh, E., Rao, A.: Graph convolutional networks with motif-based attention. In: Proceedings of CIKM 2019, pp. 499–508 (2019)
24. Leskovec, J., Chakrabarti, D., Kleinberg, J., Faloutsos, C., Ghahramani, Z.: Kronecker graphs: an approach to modeling networks. J. Mach. Learn. Res (JMLR) **11**, 985–1042 (2010)
25. Lin, B.: The parameterized complexity of the k-biclique problem. J. ACM (JACM) **65**(5), 1–23 (2018)
26. Liu, S., Hooi, B., Faloutsos, C.: Holoscope: topology-and-spike aware fraud detection. In: Proceedings of CIKM 2017, pp. 1539–1548 (2017)
27. Mangan, S., Alon, U.: Structure and function of the feed-forward loop network motif. PNAS **100**(21), 11980–11985 (2003)
28. Milo, R., Shen-Orr, S., Itzkovitz, S., Kashtan, N., Chklovskii, D., Alon, U.: Network motifs: simple building blocks of complex networks. Science **298**(5594), 824–827 (2002). https://doi.org/10.1126/science.298.5594.824
29. Milo, R., et al.: Superfamilies of evolved and designed networks. Science **303**(5663), 1538–1542 (2004). https://doi.org/10.1126/science.1089167
30. Mitzenmacher, M., Pachocki, J., Peng, R., Tsourakakis, C., Xu, S.C.: Scalable large near-clique detection in large-scale networks via sampling. In: Proceedings of KDD 2015, pp. 815–824. ACM (2015)
31. Noble, C.C., Cook, D.J.: Graph-based anomaly detection. In: Proceedings of KDD 2003, pp. 631–636 (2003)
32. Pachter, L.: Why i read the network nonsense papers. https://liorpachter.wordpress.com/2014/02/12/why-i-read-the-network-nonsense-papers/
33. Pandit, S., Chau, D.H., Wang, S., Faloutsos, C.: Netprobe: a fast and scalable system for fraud detection in online auction networks. In: WWW (2007)
34. Prakash, B.A., Sridharan, A., Seshadri, M., Machiraju, S., Faloutsos, C.: Eigen-Spokes: surprising patterns and scalable community chipping in large graphs. In: Zaki, M.J., Yu, J.X., Ravindran, B., Pudi, V. (eds.) PAKDD 2010. LNCS (LNAI), vol. 6119, pp. 435–448. Springer, Heidelberg (2010). https://doi.org/10.1007/978-3-642-13672-6_42
35. Reigl, M., Alon, U., Chklovskii, D.B.: Search for computational modules in the c. elegans brain. BMC Biol. **2**(1), 1–12 (2004)
36. Rossi, R.A., Ahmed, N.K.: The network data repository with interactive graph analytics and visualization. In: AAAI (2015). https://networkrepository.com
37. Rozemberczki, B., Allen, C., Sarkar, R.: Multi-scale attributed node embedding (2019)
38. Rozemberczki, B., Sarkar, R.: Characteristic functions on graphs: birds of a feather, from statistical descriptors to parametric models. In: Proceedings of CIKM 2020, pp. 1325–1334 (2020)
39. Scheffer, L.K., et al.: A connectome analysis of the adult drosophila central brain. Elife **9**, e57443 (2020)
40. Shen-Orr, S., Milo, R., Mangan, S., Alon, U.: Network motifs in the transcriptional regulation network of escherichia coli. Nat. Genet. **31**, 64–8 (2002)
41. Shin, K., Eliassi-Rad, T., Faloutsos, C.: Corescope: graph mining using k-core analysis: patterns, anomalies and algorithms. In: ICDM 2016, pp. 469–478 (2016)

42. Spricer, K., Britton, T.: The configuration model for partially directed graphs. J. Stat. Phys. **161**, 965–985 (2015)
43. Starnini, M., et al.: Smurf-based anti-money laundering in time-evolving transaction networks. In: Dong, Y., Kourtellis, N., Hammer, B., Lozano, J.A. (eds.) ECML PKDD 2021. LNCS (LNAI), vol. 12978, pp. 171–186. Springer, Cham (2021). https://doi.org/10.1007/978-3-030-86514-6_11
44. Tsourakakis, C.: The k-clique densest subgraph problem. In: Proceedings of WWW 2015, pp. 1122–1132 (2015)
45. Tsourakakis, C.E., Chen, T., Kakimura, N., Pachocki, J.: Novel dense subgraph discovery primitives: risk aversion and exclusion queries. In: Brefeld, U., Fromont, E., Hotho, A., Knobbe, A., Maathuis, M., Robardet, C. (eds.) ECML PKDD 2019. LNCS (LNAI), vol. 11906, pp. 378–394. Springer, Cham (2020). https://doi.org/10.1007/978-3-030-46150-8_23
46. Tsourakakis, C.E., Pachocki, J., Mitzenmacher, M.: Scalable motif-aware graph clustering. In: Proceedings of WWW 2017, pp. 1451–1460 (2017)
47. Ugander, J., Backstrom, L., Kleinberg, J.: Subgraph frequencies: mapping the empirical and extremal geography of large graph collections. In: Proceedings of WWW 2013, pp. 1307–1318 (2013)
48. Van Koevering, K., Benson, A., Kleinberg, J.: Random graphs with prescribed k-core sequences: a new null model for network analysis. In: Proceedings of TheWebConf 2021, pp. 367–378 (2021)
49. Wasserman, S., Faust, K., et al.: Social network analysis: methods and applications (1994)
50. Wernicke, S., Rasche, F.: Fanmod: a tool for fast network motif detection. Bioinformatics **22**(9), 1152–1153 (2006)
51. Witvliet, D.E.A.: Connectomes across development reveal principles of brain maturation. Nature **596**(7871), 257–261 (2021)
52. Yin, H., Benson, A.R., Leskovec, J., Gleich, D.F.: Local higher-order graph clustering. In: Proceedings of KDD 2017, pp. 555–564 (2017)
53. You, J., Ying, R., Ren, X., Hamilton, W.L., Leskovec, J.: Graphrnn: generating realistic graphs with deep auto-regressive models. In: ICML (2018)
54. Yu, H., et al.: High-quality binary protein interaction map of the yeast interactome network. Science (New York, N.Y.) **322**, 104–110 (2008)
55. Zhang, X., Shao, S., Stanley, H., Havlin, S.: Dynamic motifs in socio-economic networks. EPL (Europhys. Lett.) **108**, 58001 (2014)

Anonymity can Help Minority: A Novel Synthetic Data Over-Sampling Strategy on Multi-label Graphs

Yijun Duan[1]([⊠]), Xin Liu[1], Adam Jatowt[2], Hai-tao Yu[3], Steven Lynden[1], Kyoung-Sook Kim[1], and Akiyoshi Matono[1]

[1] AIRC, AIST, Tosu, Japan
{yijun.duan,xin.liu,steven.lynden,ks.kim,a.matono}@aist.go.jp
[2] Department of Computer Science, University of Innsbruck, Innsbruck, Austria
jatowt@acm.org
[3] Faculty of Library, Information and Media Science, University of Tsukuba, University of Tsukuba, Japan
yuhaitao@slis.tsukuba.ac.jp

Abstract. In many real-world networks (e.g., social networks), nodes are associated with multiple labels and node classes are imbalanced, that is, some classes have significantly fewer samples than others. However, the research problem of imbalanced multi-label graph node classification remains unexplored. This non-trivial task challenges existing graph neural networks (GNNs) because the majority class could dominate the loss functions of GNNs and result in overfitting to those majority class features and label correlations. On non-graph data, minority over-sampling methods (such as SMOTE and its variants) have been demonstrated to be effective for the imbalanced data classification problem. This study proposes and validates a new hypothesis with unlabeled data oversampling, which is meaningless for imbalanced non-graph data; however, feature propagation and topological interplay mechanisms between graph nodes can facilitate representation learning of imbalanced graphs. Furthermore, we determine empirically that ensemble data synthesis through the creation of virtual minority samples in the central region of a minority, and the generation of virtual unlabeled samples in the boundary region between a minority and majority is the best practice for the imbalanced multi-label graph node classification task. Our proposed novel data over-sampling framework is evaluated using multiple real-word network datasets, and it outperforms diverse, strong benchmark models by a large margin.

Keywords: Imbalanced learning · Graph representation learning · Data over-sampling · Generative adversarial network

1 Introduction

Graphs are becoming ubiquitous across a large spectrum of real-world applications in the forms of social networks, citation networks, telecommunication

M.-R. Amini et al. (Eds.): ECML PKDD 2022, LNAI 13714, pp. 20–36, 2023.
https://doi.org/10.1007/978-3-031-26390-3_2

networks, biological networks, etc. [32]. For a considerable number of real-world graph node classification tasks, the training data follows a *long-tail* distribution, and the node classes are *imbalanced*. In other words, a few majority classes have a significant fraction of samples, while most classes only contain a handful of instances. Taking the NCI chemical compound graph as an example, only about 5% of molecules are labeled as active in the anticancer bioassay test [25]. On the other hand, graph nodes are associated with multiple labels in many real-world networked data instead of a single one. Many social media sites, such as Flickr and YouTube, allow users to join diverse groups representing their various interests. A person can join several interest groups on Flickr, such as *Landscape* and *Travel*, and different video genres on YouTube, such as *Cooking* and *Wrestling*.

To date, a large body of work has been focused on the representation learning of graphs with balanced node classes and simplex labels [8,11,23,29]. However, these models do not perform well on the widely-existing imbalanced and multi-label graphs because of the following reasons. (1) *Problem caused by the imbalanced setting*: The imbalanced data makes the classifier overfit the majority class, and the features of the minority class cannot be sufficiently learned [9]. Furthermore, the above problem is aggravated by the presence of the *topological interplay* effect [25] between graph nodes, making the feature propagation dominated by the majority classes. (2) *Problem caused by the multi-label setting*: Multi-label graph architectures typically encode significantly more complex interactions between nodes with shared labels [25], which is challenging to capture. Therefore, it is essential to develop a specific graph learning method for class imbalanced multi-label graph data. However, research in this direction is still in its infancy. Thus, in this study, we propose *imbalanced multi-label graph representation learning* to address this challenge while also contributing to graph learning theory.

Many past studies [2,34,35] have demonstrated that for imbalanced data, *minority over-sampling* is an effective measure to improve classification accuracy. This strategy has recently been confirmed to be still effective for graph data [33]. Traditional over-sampling techniques mainly consist of two steps: (1) selecting some minority instances as "seed examples"; (2) generating synthetic data with features and label similar to the seed examples and adding them into the training set. For example, the most popular over-sampling technique SMOTE [2] addresses the problem of minority generation by performing interpolation between randomly-selected minority instances and their nearest neighbors. However, mainstream over-sampling techniques have the following shortcomings when applied to graph data: (1) the selection of seed examples prioritizes global minority nodes while ignoring local minority nodes; (2) each synthetic instance is always assigned a label based on some specific strategy, which may be incorrect. Different from i.i.d. non-graph data, because the relationship between graph nodes are explicitly expressed by the edge connecting them, the representation learning of a node can be heavily dependent on its neighboring *unlabeled* nodes through the feature propagation mechanism on graphs.

Motivated by the observations above, we propose and validate the following assumption. In addition to synthetic minority samples, synthetic *unlabeled* samples can also facilitate the debiasing of GNNs on an imbalanced training set. In particular, for nearby global minority samples which are a local majority, we can "safely" produce virtual samples of the same class and add them into the training sets to balance class distribution. Global minority samples, which are also a local minority, are more likely to be local outliers and thus risky for selection as seed examples for further over-sampling; for nearby global minority samples whose neighbors are class-balanced, it is difficult to determine the labels of virtual samples. Thus, the production of *unlabeled* virtual nodes should be encouraged, which can help minorities by "blocking" the over-aggregation of majority features delivered through edges. This idea is illustrated in Fig. 1. **We argue that the key to over-sampling on an imbalanced multi-label graph is to flexibly combine the synthesis of both labeled and unlabeled instances enriched by label correlations.**

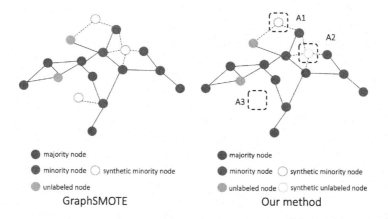

Fig. 1. A comparison between our method and the current state-of-the-art graph over-sampling method GraphSMOTE [33]. The latter's idea is to generate new minority instances near randomly selected minority nodes and create virtual edges (dotted lines in the figure) between those synthetic nodes and real nodes. Instead, we synthesize minority instances in safe areas (i.e., A1), generate *unlabeled* instances in locally balanced areas (i.e., A2), and do not conduct data over-sampling near minority nodes which are outliers (i.e., A3). For the simplicity of illustration, only a single-label scenario is shown.

We extend the existing over-sampling algorithms to a novel framework for the imbalanced multi-label graph node classification task based on the above considerations. We extend the classic global minority-based seed examples selection to the local minority perspective (see Sect. 4.1). Distinct from interpolation that is commonly-used in mainstream over-sampling techniques [18], we use a generative adversarial network (GAN) [7] to generate new instances. As a representative deep generative model, GAN can capture label correlation information by estimating the probability distribution of seed examples [31]. We propose an ensemble architecture of GAN and cGAN [16] for the flexible generation of both unlabeled and labeled synthetics (see Sect. 4.2). To make use of the graph topology information, we propose to obtain new edges between generated samples and existing data with an edge predictor (see Sect. 4.3). The augmented graph is finally sent to a graph convolutional network (GCN) [11] for representation learning, together with the learned label correlations (see Sect. 4.4). We name our proposed framework as **SORAG**, which is abbreviated from **S**ynthetic data **O**versampling St**RA**tegy on **G**raph.

In summary, our contribution is three-fold:

- We advance the traditional simplex-label graph learning to an imbalanced multi-label graph learning setting, which is more general and common in real-world applications. To the best of our knowledge, this study is the first to focus on this task.
- We propose a novel and general framework which extends a previous oversampling algorithm to adapt to graph data. It flexibly ensembles the synthesis of labeled and unlabeled nodes to support the minority classes and leverage label correlations to generate more natural nodes.
- Extensive experiments on multiple real-world datasets demonstrate the high effectiveness of our approach. Compared with the current state-of-the-art model GraphSMOTE [33], our method has an improvement of 1.5% in terms of Micro-F1 and 3.3% in terms of Macro-F1 on average.

2   Related Works

2.1   Graph Neural Networks

Graph representation learning (GRL) has evolved considerably in recent years. GNN can be broadly regarded as the third (and latest) generation of GRL after traditional graph embedding and modern graph embedding [15]. GNNs can be classified into spatial and spectral types based on their graph filter. Spatial-based graph filters explicitly leverage the graph structure. Representative works in this field include the GraphSAGE filter [8], GAT-filter [29], the ECC-filter [26], GGNN-filter [12], Mo-filter [17], and so on. Spectral-based graph filters use graph spectral theory to design filtering operations in the spectral domain. An early work [1] deals with the eigendecomposition of the Laplacian matrix and the matrix multiplication between dense matrices, thus being computationally expensive. To overcome this problem, the Poly-Filter [3], Cheby-Filter [3], and GCN-Filter [11] have been successively proposed. In particular, our task is semi-supervised, which means we need to learn the representation of all nodes from a small portion of labeled nodes. Some recent works on semi-supervised graph node classification can be found in [15].

2.2 Imbalanced Learning

Learning from imbalanced data has been a long-standing challenge in machine learning. With an imbalanced class distribution, existing methods addressing this issue can be grouped into three categories [9]: (1) pre-processing the training data, (2) post-processing the output, and (3) direct learning methods. Data pre-processing aims to make the classification results on the new training set equivalent to imbalance-aware classification decisions on the original training set, typically like sampling [5] and weighting [35]. Post-processing the output makes the classifier biased toward minority classes by adjusting the classifier decision threshold [4,24]. Direct learning methods embed class distribution information into the component (e.g., objective function) of the learning algorithm, with typical methods being cost-sensitive decision tree [14], cost-sensitive SVM [19], and so on. Studies on multi-class single-label imbalanced GRL have emerged only recently [25,30,33]. However, different from these works, our proposal is the first to utilize synthetic unlabeled nodes to weaken the tendency of GNNs to overfit to majority without introducing contradictory labels. Additionally, our proposed model is also applicable to multi-label datasets.

3 Problem Formulation

Input. The input is a graph $G = \{V, A, X, L, B\}$. $V = \{v_1, v_2, ..., v_n\}$ denotes the set of nodes. $A \in \mathbb{R}^{n \times n}$ is the adjacency matrix. $A_{ij} = 1$ when there is an undirected edge between nodes v_i and v_j; otherwise $A_{ij} = 0$. The self-loops in G have been removed, so $A_{ii} = 0$, $i \in \{1, 2, ..., n\}$. $X \in \mathbb{R}^{n \times k}$ is the feature matrix, where $x_i \in \mathbb{R}^{1 \times k}$ is the feature vector of node v_i. $L = \{c_1, c_2, ..., c_m\}$ is a set of unique labels. B is a $n \times m$ affiliation matrix of labels with $B_{ij} = 1$ if v_i has label c_j; otherwise $B_{ij} = 0$. Our task is in a semi-supervised transductive manner. Only a tiny portion of the nodes is used for training, which we denote as V^{train}.

Output. Our goal is to learn a graph neural network f that maps the input graph G into a dense vector representation $Z \in \mathbb{R}^{n \times d}$, where $z_i \in \mathbb{R}^{1 \times d}$ is the vector of node v_i, and predicts the class labels for the test nodes set V^{test}.

Imbalanced Learning. Let $|c_i|$ represent the number of samples associated with the label c_i. The distribution of $\{|c_1|, |c_2|, ..., |c_m|\}$ is imbalanced. That is, a few labels contain most samples, and most labels contain only a few samples. When presented with imbalanced data, existing GNNs tend to bias toward majority groups, leaving minority instances under-trained. We aim to learn a neural network classifier f that can work well for both majority and minority classes.

4 Methodology

An illustration of the proposed framework is shown in Fig. 2. We elaborate on each component as follows.

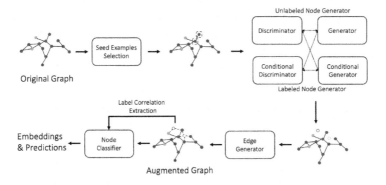

Fig. 2. Overview of the proposed method.

4.1   Imbalance Measurement

In multi-label learning, a commonly used measure that evaluates the global imbalance of a particular label is *IRLbl*. Let $|C_i|$ be the number of instance whose i-th label value is 1; *IRLbl* is then defined as follows.

$$IRLbl_i = \frac{max\left\{|c_1|, |c_2|, ..., |c_m|\right\}}{c_i} .\tag{1}$$

Therefore, the larger the value of *IRLbl* for a label, the more minority class it is. For a node v_i, its GMD is defined as follows.

$$GMD_i = \frac{IRLbl_j \cdot [B_{ij} = 1]}{\sum_{j=1}^{m}[B_{ij} = 1]},\tag{2}$$

where $[B_{ij} = 1]$ means v_i has the j-th label, and $\sum_{j=1}^{m}[B_{ij} = 1]$ counts the number of labels v_i has.

The local minority degree (LMD) of a node can be measured by the proportion of opposite class values in its local neighborhood. For v_i, let N_i^k denote its K-hop neighbor nodes. Then, for label c_j, the proportion of neighbors having an opposite class to the class of v_i is computed as

$$S_{ij} = \frac{\sum_{v_m \in N_i^k}[B_{ij} \neq B_{mj}]}{|N_i^k|},\tag{3}$$

where $S \in \mathbb{R}^{n \times m}$ is a matrix defined to store the local imbalance of all nodes for each label. Given S, a straightforward way to compute LMD for v_i is to average its S_{ij} for all labels as follows.

$$LMD_i = \frac{\sum_{j=1}^{m} S_{ij}[B_{ij} = g_j]}{m},\tag{4}$$

where $g_j \in \{0, 1\}$ denotes the minority class of j-th label. Namely, if $|c_j| \geq 0.5 \cdot n$, $g_j = 1$; else, $g_j = 0$. Here, n is the total number of vertices. Further, we group global minority nodes into different types based on LMD, and each type is identified correctly by the classifier with different difficulties. Following [13, 20], we discretize the range [0,1] of LMD_i to define four types of nodes, namely safe (SF), borderline (BD), rare (RR) and outlier (OT), according to their local imbalance.

- SF: $0 \leq LMD_i < 0.3$. Safe nodes are basically surrounded by nodes containing similar labels.
- BD: $0.3 \leq LMD_i < 0.7$. Borderline nodes are located in the decision boundary between different classes.
- RR: $0.7 \leq LMD_i < 1.0$. Rare nodes are located in the region overwhelmed by different nodes and distant from the decision boundary.
- OT: $LMD_i = 1.0$. Outliers are totally connected to different nodes.

Furthermore, for v_i, we define two metrics: labeled seed probability (LSP) and unlabeled seed probability (USP) to describe the probability of being selected as a seed example to generate labeled synthetic nodes and unlabeled synthetic nodes, respectively. The LSP and USP are calculated as follows.

$$LSP_i = GMD_i \cdot LMD_i, v_i \in SF \tag{5}$$

$$USP_i = GMD_i \cdot LMD_i, v_i \in BD \tag{6}$$

We compute the LSP and USP scores for all nodes and sort them in descending order. The top-ranked nodes (controlled by the hyper-parameter seed example rate ρ) will be selected as seed examples. A min-max normalization processes all the GMD and LMD scores to improve the computation stability.

4.2   Node Generator

We denote the joint distribution of node feature x and label y in SF region as $P_{SF}(x, y)$, the marginal distribution of y as $P_{SF}(y)$, and the marginal distribution of x in BD region as $P_{BD}(x)$. Generator G_l is expected to generate labeled instances in the SF region, while generator G_u should output unlabeled synthetics in the BD region. Let the data distribution produced by G_l and G_u be denoted as $P_l(x, y)$ and $P_u(x)$, respectively; then, we expect $P_{BD}(x) \approx P_u(x)$ and $P_{SF}(x, y) \approx P_l(x, y)$. Furthermore, a more flexible goal is to have $P_{BD}(x) \approx \alpha \cdot P_u(x) + (1-\alpha) \cdot P_l(x)$, $P_{SF}(x, y) \approx \beta \cdot P_l(x, y) + (1-\beta) \cdot P_u(x, y)$, $\alpha \approx 1, \beta \approx 1$. By adjusting the values of α and β, we can control G_l and G_u to produce various data distributions to fit the original data. Here, $P_u(x, y)$ is the joint distribution of $P_u(x)$ and $P_{SF}(y)$, and $P_l(x)$ is the marginal distribution of $P_l(x, y)$.

To achieve the above goal, we propose a node generator, which is essentially an ensemble of a GAN [7] and a conditional GAN (cGAN) [16]. The GAN is responsible for generating unlabeled synthetic nodes, whose generator and discriminator are respectively denoted as G_u and D_u. The cGAN is used for generating labeled synthetic instances, where its generator and discriminator are denoted as G_l and D_l, respectively. Our loss function for training the GAN is

$$\min_{G_u} \max_{D_u} \mathcal{L}_{GAN} = \mathbb{E}_{x \sim P_{BD}(x)} log D_u(x) + \alpha \cdot \mathbb{E}_{x \sim P_u(x)} log(1 - D_u(x)) \tag{7}$$

For cGAN, our objective is given as

$$\min_{G_l} \max_{D_l} \mathcal{L}_{cGAN} = \mathbb{E}_{(x,y) \sim P_{SF}(x,y)} log D_l(x,y) + \beta \cdot \mathbb{E}_{(x,y) \sim P_l(x,y)} log(1 - D_l(x,y)) \tag{8}$$

To achieve flexible control over G_l and G_u, we design the following loss function based on the interaction of GAN and cGAN

$$\min_{G_u,G_l} \max_{D_u,D_l} \mathcal{L}_{GAN-cGAN} = (1 - \alpha) \cdot \mathbb{E}_{x \sim P_l(x)} log(1 - D_u(x))$$
$$+ (1 - \beta) \cdot \mathbb{E}_{(x,y) \sim P_u(x,y)} log(1 - D_l(x,y)) \tag{9}$$

Putting all these together, our final loss for node generation $\mathcal{L}_{node}$ is

$$\mathcal{L}_{node} = \min_{G_u,G_l} \max_{D_u,D_l} \mathcal{L}_{GAN} + \mathcal{L}_{cGAN} + \mathcal{L}_{GAN-cGAN} \tag{10}$$

For our proposed generator, the following theoretical analysis is performed.

Proposition 1. *For any fixed G_u and G_l, the optimal discriminator D_u and D_l of the game defined by $\mathcal{L}_{node}$ is*

$$D_u^*(x) = \frac{P_{BD}(x)}{P_{BD}(x) + P_\alpha(x)}, D_l^*(x,y) = \frac{P_{SF}(x,y)}{P_{SF}(x,y) + P_\beta(x,y)} \tag{11}$$

where $P_\alpha(x) = \alpha \cdot P_u(x) + (1 - \alpha) \cdot P_l(x)$, and $P_\beta(x,y) = \beta \cdot P_l(x,y) + (1 - \beta) \cdot P_u(x,y)$.

Proof. We have

$$\mathcal{L}_{node} = \int_x P_{BD}(x) log D_u(x) dx + \int_{x,y} P_{SF}(x,y) log D_l(x,y) dx dy$$
$$+ \alpha \cdot \int_x P_u(x) log(1 - D_u(x)) dx + \beta \cdot \int_{x,y} P_l(x,y) log(1 - D_l(x,y)) dx dy$$
$$+ (1 - \alpha) \cdot \int_x P_l(x) log(1 - D_u(x)) dx + (1 - \beta) \cdot \int_{x,y} P_u(x,y) log(1 - D_l(x,y)) dx dy \tag{12}$$
$$= \int_x P_{BD}(x) log D_u(x) + P_\alpha(x) \cdot log(1 - D_u(x)) dx$$
$$+ \int_{x,y} P_{SF}(x,y) log D_l(x,y) + P_\beta(x,y) \cdot log(1 - D_l(x,y)) dx dy$$

For any $(a, b) \in \mathbb{R}^2 \backslash \{0, 0\}$, the function $f(y) = a \log y + b \log(1 - y)$ achieves its maximum in $[0, 1]$ at $\frac{a}{a+b}$. This concludes the proof.

Proposition 2. *The equilibrium of $\mathcal{L}_{node}$ is achieved if and only if $P_{BD}(x) = P_\alpha(x)$ and $P_{SF}(x, y) = P_\beta(x, y)$ with $D_u^*(x) = D_l^*(x, y) = \frac{1}{2}$, and the optimal value of $\mathcal{L}_{node}$ is $-4 \log 2$.*

Proof. When $D_u(x) = D_u^*(x), D_l(x, y) = D_l^*(x, y)$, we have

$$
\begin{aligned}
\mathcal{L}_{node} &= \int_x P_{BD}(x) \log \frac{P_{BD}(x)}{P_{BD}(x) + P_\alpha(x)} dx + \int_{x,y} P_{SF}(x, y) \log \frac{P_{SF}(x, y)}{P_{SF}(x, y) + P_\beta(x, y)} dx dy \\
&+ \int_x P_\alpha(x) \log \frac{P_\alpha(x)}{P_{BD}(x) + P_\alpha(x)} dx + \int_{x,y} P_\beta(x, y) \log \frac{P_\beta(x, y)}{P_{SF}(x, y) + P_\beta(x, y)} dx dy \\
&= -4 \log 2 + 2 \cdot JSD(P_{BD}(x) \| P_\alpha(x)) + 2 \cdot JSD(P_{SF}(x, y) \| P_\beta(x, y)) \\
&\geq -4 \log 2
\end{aligned}
\tag{13}
$$

where the optimal value is achieved when the two Jensen-Shannon divergences are equal to 0, namely, $P_{BD}(x) = P_\alpha(x)$, and $P_{SF}(x, y) = P_\beta(x, y)$. When $\alpha = \beta = 1$, we have $P_{BD}(x) = P_u(x), P_{SF}(x, y) = P_l(x, y)$.

In the implementation, both G_u and G_l are designed as a 3-layer feed-forward neural network. In contrast, D_u and D_l are designed with a relatively weaker structure: a 1-layer feed-forward neural network for facilitating the training.

4.3   Edge Generator

The edge generator described in this section is responsible for estimating the relation between virtual nodes and real nodes, which facilitates feature propagation, feature extraction, and node classification. Such edge generators will be trained on real nodes and existing edges. Following a previous work [33], the inter-node relation is embodied in the weighted inner product of node features. Specifically, for two nodes v_i and v_j, let E_{ij} denote the probability of the existence of an edge between them, which is computed as

$$
E_{ij} = \sigma(x_i \cdot W^{edge} \cdot x_j^T)
\tag{14}
$$

where x_i and x_j are the feature vectors of v_i and v_j, respectively. $W^{edge} \in \mathbb{R}^{k \times k}$ is the weight parameter matrix to be learned, and $\sigma = Sigmoid()$. Then, the extended adjacency matrix A' is defined as follows

$$
A'_{ij} = \begin{cases} A_{ij}, & \text{if } v_i \text{ and } v_j \text{ are real nodes} \\ E_{ij}, & \text{if } v_i \text{ or } v_j \text{ is synthetic node} \end{cases}
\tag{15}
$$

Compared to A, A' contains new information about virtual nodes and edges, which will be sent to the node classifier in Sect. 4.4. As the edge generator is expected to be partially trained based on the final node classifier (see Sect. 4.5), predicted edges should be set as continuous so that the gradient can be calculated and propagated from the node classifier. Thus, E_{ij} is not discretized to some value in $\{0,1\}$. The edge generator should be capable of predicting real edges accurately to generate realistic virtual nodes. Then, the pre-trained loss function for training the edge generator is

$$\mathcal{L}_{edge} = \|E - A\|^2 \tag{16}$$

where E refers to predicted edges between real nodes.

4.4   Node Classifier

We now obtain an augmented balanced graph $G' = \{V', A', X', B'\}$, where V' consists of both real nodes and synthetic labeled and unlabeled nodes; further, A', X', and B' denote the edge, feature, and label information of the enlarged vertex set, respectively. A classic two-layer GCN structure [11] is adopted for node classification, given its high accuracy and efficiency. Its first and second layers are denoted as L^1 and L^2, respectively, and their corresponding outputs $\{O^1, O^2\}$ are

$$O^1 = ReLU(\tilde{D}^{-\frac{1}{2}}\tilde{A}'\tilde{D}^{-\frac{1}{2}}X'W^1) \tag{17}$$

$$O^2 = \sigma(F\tilde{D}^{-\frac{1}{2}}\tilde{A}'\tilde{D}^{-\frac{1}{2}}O^1W^2) \tag{18}$$

where $\tilde{A}' = A' + I$, I is an identity matrix of the same size as A'. $\tilde{D}$ is a diagonal matrix and $\tilde{D}_{ii} = \sum_j \tilde{A}'_{ij}$. $\tilde{D}^{-\frac{1}{2}}\tilde{A}'\tilde{D}^{-\frac{1}{2}}$ is the normalized adjacency matrix. Further, W^1 and W^2 are the learnable parameters in the first and second layers, respectively. $ReLU$ and σ are the respective activation functions of the first and the second layer, where $ReLU(Z)_i = max(0, Z_i)$, $\sigma(Z)_i = Sigmoid(Z)_i = \frac{1}{1+exp(-Z_i)}$. O^2 is the posterior probability of the class to which the node belongs. F is the label correlation matrix that is computed in the same way as in [25], which provides helpful extra-label correlation and interaction information. Eventually, given the training labels B^{train}, we minimize the following cross-entropy error to learn the classifier, where p is the number of training samples, m is the size of the label set, and nc stands for node classifier.

$$\mathcal{L}_{nc} = -\sum_{i=1}^{p}\sum_{j=1}^{m} B_{ij}^{train} ln O_{ij}^2 \tag{19}$$

4.5   Optimization Objective

Based on the above content, the final objective function of our framework is given as

$$\min_{\Theta,\Phi,\Psi} \mathcal{L}_{nc} + \lambda \cdot \mathcal{L}_{node} + \mu \cdot \mathcal{L}_{edge} \tag{20}$$

where Θ, Φ, and Ψ are the sets of parameters for the synthetic node generator (Sect. 4.2), edge generator (Sect. 4.3) and node classifier (Sect. 4.4), respectively. λ and μ are weight parameters. The best training strategy in our experiments is to pre-train the node generator and the edge generator first, and then minimize Eq. (20) to train the node classifier and fine-tune the node generator and edge generator at the same time. Our entire framework is easy to implement, general, and flexible. Different structural choices can be adopted for each component, and different regularization terms can be enforced to provide prior knowledge.

4.6 Training Algorithm

The 1 algorithm illustrates the proposed framework. **SORAG** is trained through the following components: (1) the selection of seed examples based on node LSP and USP scores; (2) the pre-training of the node generator (i.e., the ensemble of GAN and cGAN) for synthetic data generation; (3) the pre-training of the edge generator to produce new relation information; and finally, (4) the training of the node classifier on top of the over-sampled graph and the fine-tuning of node generator and edge generator.

5   Experimental Settings

5.1   Datasets

We use three multi-label networks: BLOGCATALOG3, FLICKR, and YOUTUBE as benchmark datasets. In Table 1, we list the statistical information of all datasets used, including the number of nodes, the number of edges, the number of node classes, and the tuned optimal value of key parameters of $\mathbf{SORAG}_F$: {learning rate, weight decay, dropout rate, k (Sect. 4.1), ρ (Sect. 4.1), α (Sect. 4.2), β (Sect. 4.2), λ (Sect. 4.5), μ (Sect. 4.5)}. For each dataset, we assume that a majority class is one with more samples than the average class size, while a minority class is one with less samples. Below is a brief description of each dataset used.

- BLOGCATALOG3 [27] is a network of social relationships provided by blogger authors. The labels represent the topic categories provided by the authors, such as *Education*, *Food*, and *Health*. This network contains 10,312 nodes, 333,983 edges, and 39 labels.
- FLICKR [27] is a network of contacts between users of the photo-sharing website. The labels represent the interest groups of the users, such as *black and white photos*. This network contains 80,513 nodes, 5,899,882 edges, and 195 labels.
- YOUTUBE [28] is a social network between users of the popular video sharing website. The labels represent groups of viewers that enjoy common video genres such as *anime* and *wrestling*. This network contains 1,138,499 nodes, 2,990,443 edges and 47 labels.

Algorithm 1 Full Training Algorithm

Inputs: Graph data: $G = \{V, A, X, L, B\}$
Outputs: Network parameters, node representations, and predicted node class
1: Initialize the node generator, edge generator, and node classifier
2: Compute the node LSP and USP scores based on Eq. (5) and Eq. (6), respectively
3: Select the fraction of nodes with the highest LSP and USP scores as seed examples for D_l and D_u, respectively
4: **while** Not Converged **do** ▷ Pre-train the node generator
5:    Update D_l by ascending along its gradient based on $\mathcal{L}_{node}$ (Eq. (10))
6:    Update G_l by descending along its gradient based on $\mathcal{L}_{node}$
7:    Update D_u by ascending along its gradient based on $\mathcal{L}_{node}$
8:    Update G_u by descending along its gradient based on $\mathcal{L}_{node}$
9: **end while**
10: **while** Not Converged **do** ▷ Pre-train the edge generator
11:    Update the edge generator by descending along its gradient based on $\mathcal{L}_{edge}$ (Eq. (16))
12: **end while**
13: Construct label-occurrence network and extract label correlations [25]
14: **while** Not Converged **do**▷ Train the node classifier and pre-train the other components
15:    Generate new unlabeled nodes using G_u
16:    Generate new labeled nodes using G_l
17:    Generate the new adjacency matrix A' using the edge generator
18:    Update the full model based on $\mathcal{L}_{nc} + \lambda \cdot \mathcal{L}_{node} + \mu \cdot \mathcal{L}_{edge}$ (Eq. (20))
19: **end while**
20: Predict the test set labels with the trained model

For all datasets, we attribute each node with a 64-dim embedding vector obtained by performing dimensionality reduction on the adjacency matrix using PCA [6], similar to [25,33]. All of the above datasets are available at http://zhang18f.myweb.cs.uwindsor.ca/datasets/.

5.2 Analyzed Methods and Metrics

To validate the performance of our approach, we compare it against a number of state-of-the-art and representative methods for multi-label graph learning and imbalanced graph learning, which include **GCN** [11], **ML-GCN** [25], **SMOTE** [2], **GraphSMOTE** [33], and **RECT** [30]. Additionally, three variants of our proposed method are implemented, which are $\mathbf{SORAG}_F$ (the full model), $\mathbf{SORAG}_L$ (only labeled nodes are generated), and $\mathbf{SORAG}_U$ (only unlabeled nodes are generated).

It is necessary to mention that all the baselines above except **ML-GCN** (which is intrinsically designed as a multi-label classifier) are manually set to

Table 1. Dataset statistics.

Dataset	BLOGCATALOG3	FLICKR	YOUTUBE
#Nodes	10,312	333,983	39
#Edges	80,513	5,899,882	195
#Classes	1,138,499	2,990,443	47
Learning rate	0.05	0.01	0.1
Weight decay	5e–4	1e–4	1e–3
Dropout rate	0.5	0.5	0.9
k	2	2	2
ρ	0.5	0.5	0.5
α	0.9	0.5	0.8
β	0.8	0.9	0.8
λ	0.1	1	1
μ	1	1	1

conduct the multi-label node classification by modifying the last layer of their network structure. The implementation of the baseline approaches relies on publicly released code from relevant sources[1234]. We adopt Micro-F1 and Macro-F1 to evaluate the model performance, which are commonly used in imbalanced data classification.

5.3 Training Configurations

Following the semi-supervised learning setting, we randomly sample a portion of the labeled nodes (i.e., sampling ratio) of each dataset and use them for evaluation. Then, we randomly split the sampled nodes into 60%/20%/20% for training, validation, and testing, respectively. Similar to [22], the sampling ratios for the BLOGCATALOG3 network, the FLICKR network, and the YOUTUBE network are set as 10%, 1%, and 1%, respectively. To make the class size balanced, we experiment with different over-sampling rates, and finally they are set as those in Table 2. All the analyzed models are trained using Adam optimizer [10] in PyTorch (2020.2.1, community edition) [21]. Each result is presented as a mean based on 10 replicated experiments. All models are trained until they converge, with a typical number of training epochs as 200.

---

[1] **SMOTE**: https://github.com/analyticalmindsltd/smote_variants.
[2] **GraphSmote**: https://github.com/TianxiangZhao/GraphSmote.
[3] **RECT**: https://github.com/zhengwang100/RECT.
[4] **GCN**: https://github.com/tkipf/pygcn.

Table 2. The optimal over-sampling rates for the synthetic unlabeled nodes (denoted as $Rate_U$) and the synthetic labeled nodes (denoted as $Rate_L$) on each dataset. N/A abbreviates for "not applicable".

	BLOGCATALOG3		FLICKR		YOUTUBE	
	$Rate_U$	$Rate_L$	$Rate_U$	$Rate_L$	$Rate_U$	$Rate_L$
SORAG$_U$	0.9	N/A	0.6	N/A	0.7	N/A
SORAG$_L$	N/A	0.1	N/A	0.3	N/A	0.6
SORAG$_F$	0.1	0.9	0.2	0.9	0.2	0.4

6   Experimental Results

6.1   Imbalanced Multi-label Classification Performance

Table 3 shows the performance of all methods in terms of Micro-F1 and Macro-F1. The results are presented as a mean based on 10 repeated experiments. Based on the results, we reach the following conclusions.

Table 3. Imbalanced multi-label classification comparison. The 1st and 2nd best results are boldfaced and underscored, respectively.

Metrics	Micro-F1 (%)			Macro-F1 (%)		
Methods\Datasets	BLOGCATALOG3	FLICKR	YOUTUBE	BLOGCATALOG3	FLICKR	YOUTUBE
GCN	37.36	34.03	36.19	30.27	21.17	26.53
ML-GCN	37.51	38.91	37.64	30.39	21.56	27.52
SMOTE	40.24	39.30	39.01	30.65	23.08	28.53
GraphSMOTE	42.82	40.01	**43.70**	35.58	24.25	33.81
RECT	41.72	41.23	42.66	<u>38.66</u>	24.47	33.94
SORAG$_L$	<u>44.58</u>	<u>41.61</u>	41.98	38.45	<u>26.48</u>	<u>35.01</u>
SORAG$_U$	43.21	37.92	40.83	37.28	26.15	32.53
SORAG$_F$	**44.89**	**43.13**	<u>42.86</u>	**40.01**	**26.85**	**36.57**

- When compared with the GCN and ML-GCN methods, which do not consider class distribution, the three variants of **SORAG** show significant improvements. For example, compared with ML-GCN, the improvement brought by **SORAG$_F$** is 7.4%, 4.2%, and 5.2% in terms of Micro-F1 and 9.6%, 5.3%, and 9.1% in terms of Macro-F1, respectively. This demonstrates that our proposed data over-sampling strategy effectively enhances the classification performance of GNNs on imbalanced multi-label graph data.
- **SORAG** provides much more benefits than when applying the previous imbalanced graph node classifier (SMOTE, GraphSMOTE, RECT). On average, it outperforms earlier methods by 3.3%, 3.0%, and 1.1% in terms of

Micro-F1 and 2.5%, 2.9%, and 4.5% in terms of Macro-F1, respectively. This result validates the advantage of **SORAG** over previous over-sampling techniques in combining the generation of minority and unlabeled samples.

- Both minority over-sampling and unlabeled data over-sampling can improve classification performance. In particular, the former is more effective. A combination of the two strategies works the best. As a supporting evidence, **SORAG**$_F$ is the best performer in 5/6 tasks and the second-best performer in the remaining task.

7   Conclusions

This study investigated a new research problem: imbalanced multilabel graph node classification. In contrast to existing oversampling algorithms, which only generate new minority instances to balance the class distribution, we proposed a novel data generation strategy named **SORAG** which ensembles the synthesis of labeled instances in minority class centers and unlabeled instances in minority class borders. The new supervision information brought about by labeled synthetics and the blocking of over-propagated majority features by unlabeled synthetics facilitates balanced learning between different classes, taking advantage of the strong topological interdependence between nodes on a graph.

We conducted extensive comparative studies to evaluate the proposed framework on diverse naturally imbalanced multilabel networks. The experimental results demonstrated the high effectiveness and robustness of **SORAG** in handling imbalanced data. In the future, we will work on developing graph neural network models that are more adapted to the nature of real-world networks (e.g., scale-free and small-world features, etc.).

Acknowledgements. This paper is based on results obtained from a project, JPNP20006, commissioned by the New Energy and Industrial Technology Development Organization (NEDO). We would also like to acknowledge partial support from JSPS Grant-in-Aid for Scientific Research (Grant Number 21K12042).

References

1. Bruna, J., Zaremba, W., Szlam, A., LeCun, Y.: Spectral networks and locally connected networks on graphs. arXiv preprint arXiv:1312.6203 (2013)
2. Chawla, N.V., Bowyer, K.W., Hall, L.O., Kegelmeyer, W.P.: Smote: synthetic minority over-sampling technique. JAIR **16**, 321–357 (2002)
3. Defferrard, M., Bresson, X., Vandergheynst, P.: Convolutional neural networks on graphs with fast localized spectral filtering. Adv. Neural Inf. Process. Syst. **29**, 3844–3852 (2016)
4. Domingos, P.: Metacost: a general method for making classifiers cost-sensitive. In: Proceedings of the Fifth ACM SIGKDD International Conference on Knowledge Discovery and Data Mining, pp. 155–164 (1999)
5. Elkan, C.: The foundations of cost-sensitive learning. In: International Joint Conference on Artificial Intelligence, vol. 17, pp. 973–978. Lawrence Erlbaum Associates Ltd. (2001)

6. F.R.S., K.P.: Liii. on lines and planes of closest fit to systems of points in space. Lond. Edinburgh, Dublin Phil. Maga. J. Sci. **2**(11), 559–572 (1901). https://doi.org/10.1080/14786440109462720
7. Goodfellow, I., et al.: Generative adversarial nets. Adv. Neural Inf. Process. Syst. **27** (2014)
8. Hamilton, W.L., Ying, R., Leskovec, J.: Inductive representation learning on large graphs. In: Proceedings of the 31st International Conference on Neural Information Processing Systems, pp. 1025–1035 (2017)
9. He, H., Garcia, E.A.: Learning from imbalanced data. IEEE Trans. Knowl. Data Eng. **21**(9), 1263–1284 (2009)
10. Kingma, D.P., Ba, J.: Adam: a method for stochastic optimization. arXiv preprint arXiv:1412.6980 (2014)
11. Kipf, T.N., Welling, M.: Semi-supervised classification with graph convolutional networks. arXiv preprint arXiv:1609.02907 (2016)
12. Li, Y., Tarlow, D., Brockschmidt, M., Zemel, R.: Gated graph sequence neural networks. arXiv preprint arXiv:1511.05493 (2015)
13. Liu, B., Blekas, K., Tsoumakas, G.: Multi-label sampling based on local label imbalance. Pattern Recogn. **122**, 108294 (2022)
14. Lomax, S., Vadera, S.: A survey of cost-sensitive decision tree induction algorithms. ACM Comput. Surv. (CSUR) **45**(2), 1–35 (2013)
15. Ma, Y.T.: Deep Learning on Graphs. Cambridge University Press, Cambridge (2021)
16. Mirza, M., Osindero, S.: Conditional generative adversarial nets. arXiv preprint arXiv:1411.1784 (2014)
17. Monti, F., Bronstein, M.M., Bresson, X.: Geometric matrix completion with recurrent multi-graph neural networks. arXiv preprint arXiv:1704.06803 (2017)
18. More, A.: Survey of resampling techniques for improving classification performance in unbalanced datasets. arXiv preprint arXiv:1608.06048 (2016)
19. Morik, K., Brockhausen, P., Joachims, T.: Combining statistical learning with a knowledge-based approach: a case study in intensive care monitoring. Technical Report (1999)
20. Napierala, K., Stefanowski, J.: Types of minority class examples and their influence on learning classifiers from imbalanced data. J. Intell. Inf. Syst. **46**(3), 563–597 (2016)
21. Paszke, A., et al.: Pytorch: an imperative style, high-performance deep learning library. Adv. Neural Inf. Process. Syst. **32**, 8026–8037 (2019)
22. Perozzi, B., Al-Rfou, R., Skiena, S.: Deepwalk: online learning of social representations. In: Proceedings of the 20th ACM SIGKDD International Conference on Knowledge Discovery and Data Mining, pp. 701–710 (2014)
23. Scarselli, F., Gori, M., Tsoi, A.C., Hagenbuchner, M., Monfardini, G.: The graph neural network model. IEEE Trans. Neural Netw. **20**(1), 61–80 (2008)
24. Sheng, V.S., Ling, C.X.: Thresholding for making classifiers cost-sensitive. In: AAAI, vol. 6, pp. 476–481 (2006)
25. Shi, M., Tang, Y., Zhu, X., Liu, J.: Multi-label graph convolutional network representation learning. IEEE Trans. Big Data **8**, 1169–1181 (2020)
26. Simonovsky, M., Komodakis, N.: Dynamic edge-conditioned filters in convolutional neural networks on graphs. In: Proceedings of the IEEE Conference on Computer Vision and Pattern Recognition, pp. 3693–3702 (2017)
27. Tang, L., Liu, H.: Relational learning via latent social dimensions. In: Proceedings of the 15th ACM SIGKDD International Conference on Knowledge Discovery and Data Mining, pp. 817–826 (2009)

28. Tang, L., Liu, H.: Scalable learning of collective behavior based on sparse social dimensions. In: Proceedings of the 18th ACM Conference on Information and Knowledge Management, pp. 1107–1116 (2009)
29. Veličković, P., Cucurull, G., Casanova, A., Romero, A., Lio, P., Bengio, Y.: Graph attention networks. arXiv preprint arXiv:1710.10903 (2017)
30. Wang, Z., Ye, X., Wang, C., Cui, J., Yu, P.: Network embedding with completely-imbalanced labels. IEEE Trans. Knowl. Data Eng. **33**, 3634–3647 (2020)
31. Xu, L., Skoularidou, M., Cuesta-Infante, A., Veeramachaneni, K.: Modeling tabular data using conditional gan. In: Advances in NIPS (2019)
32. Zhang, D., Yin, J., Zhu, X., Zhang, C.: Network representation learning: a survey. IEEE Trans. Big Data **6**(1), 3–28 (2018)
33. Zhao, T., Zhang, X., Wang, S.: Graphsmote: imbalanced node classification on graphs with graph neural networks. In: Proceedings of the 14th ACM International Conference on Web Search and Data Mining, pp. 833–841 (2021)
34. Zhou, Z.H., Liu, X.Y.: Training cost-sensitive neural networks with methods addressing the class imbalance problem. IEEE Trans. Knowl. Data Eng. **18**(1), 63–77 (2005)
35. Zhou, Z.H., Liu, X.Y.: On multi-class cost-sensitive learning. Comput. Intell. **26**(3), 232–257 (2010)

Understanding the Benefits of Forgetting When Learning on Dynamic Graphs

Julien Tissier and Charlotte Laclau$^{(\boxtimes)}$

Univ. Lyon, UJM Saint-Etienne CNRS, Lab Hubert Curien UMR 5516,
42023 Saint-Etienne, France
{julien.tissier,charlotte.laclau}@univ-st-etienne.fr

Abstract. In order to solve graph-related tasks such as node classification, recommendation or community detection, most machine learning algorithms are based on node representations, also called embeddings, that allow to capture in the best way possible the properties of these graphs. More recently, learning node embeddings for dynamic graphs attracted significant interest due to the rich temporal information that they provide about the appearance of edges and nodes in the graph over time. In this paper, we aim to understand the effect of taking into account the static and dynamic nature of graph when learning node representations and the extent to which the latter influences the success of such learning process. Our motivation to do this stems from empirical results presented in several recent papers showing that static methods are sometimes on par or better than methods designed specifically for learning on dynamic graphs. To assess the importance of temporal information, we first propose a similarity measure between nodes based on the time distance of their edges with an explicit control over the decay of forgetting over time. We then devise a novel approach that combines the proposed time distance with static properties of the graph when learning temporal node embeddings. Our results on 3 different tasks (link prediction, node and edge classification) and 6 real-world datasets show that finding the right trade-off between static and dynamic information is crucial for learning good node representations and allows to significantly improve the results compared to state-of-the-art methods.

Keywords: Node vectors · Embedding · Dynamic graph

1 Introduction

Data in the form of graphs have become ubiquitous for describing complex information or structures from a large variety of domains of application such as a social network where users can *follow* and communicate with each other; webpages *linking* to other webpages; a group of cities connected by roads or rails; protein-protein interaction network to study genetic interactions in biology. In all these examples, a graph is defined by a set of entities (named *nodes* or *vertices*) and a set of pairs of related nodes (named *edges* or *links*). For instance,

M.-R. Amini et al. (Eds.): ECML PKDD 2022, LNAI 13714, pp. 37–52, 2023.
https://doi.org/10.1007/978-3-031-26390-3_3

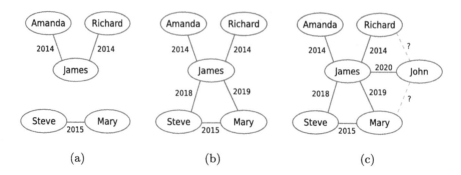

Fig. 1. A dynamic graph representing working relationships.

in the case of a social network, the nodes represent users and the edges can be any relation between two nodes such as sending messages to one another or being friends. Many different problems can be solved with graph modelization: searching for the most relevant webpages given a query, being able to predict whether two people will start a relationship [33] or finding who should collaborate together [3,8]. A common approach to solve these kinds of problems is to associate each node of the graph with an *embedding*, a numeric vector reflecting the properties of this node such as its neighborhood, and more generally the structure of the overall graph. Node embeddings are then fed into a downstream machine learning model which is trained and optimized for a given task, e.g., link prediction [5,10] or node classification [4]. Several methods have been proposed to learn node embeddings directly from a graph [9,12,14,27]. Most of these methods are designed for *static* graphs, where there is no temporal information about the relations between nodes. However, most of real-world problems are represented by *dynamic* or *time-evolving* graphs where edges are ordered in time and nodes can be added or removed. Therefore, not using the temporal information during training prevents from capturing the evolution of the interactions between nodes inside their embeddings and can lead to poor predictions.

To illustrate this, let us consider a group of people and their work relationships where our task is to suggest a new collaboration to John based on Fig. 1. One may see that a good suggestion here for John's new collaboration is Mary as she is the most recent collaboration (2019) of John's only connection James. However, a static method that doesn't take into account the information about the temporal evolution of this graph (i.e., years over edges) will suggest Amanda, Richard, Steve and Mary to John as they are connected to his sole connection James. On the other hand, a dynamic method will take into account the temporal information *over all timestamps* even though only timestamp Fig. 1b provides helpful information in this case, while Fig. 1a is uninformative to this task. While for one uninformative timestamp this may not lead to a failure of the model, real-world graphs may have thousands of timestamps and attributing the same importance to all of them may have a dramatic effect on the overall performance. Ideally, one would like to have a method that allows to control the forgetting and its speed when learning node representations in dynamic graphs to take into account only relevant information contained in them.

This paper addresses the problem of learning node representations in dynamic graphs where the temporal information is encoded inside their embeddings. We use the intuition presented above to provide a model with a way to forget the past timestamps with an explicit control over the speed of this forgetting. The main contributions are: **(1)** a novel approach to compute similarities between nodes based on static or dynamic information, suitable for both continuous and discrete dynamic graphs; **(2)** a model that learns nodes embeddings using the computed similarities and generates vectors that reflect the temporal characteristics of the graph; **(3)** an evaluation on 6 real-world datasets and 3 different tasks showing that a good trade-off between static and dynamic parts of the graph lead to the best performance in most cases.

2   Related Work

2.1   Node Embeddings in Static Graphs

In a static graph, all nodes and edges exist at the same time and no new edges appear over time. In this context, the goal of a node embedding method is to learn a function that takes a network as an input and maps each node to a low-dimensional vector. The learned vectors should reflect the structure of the network and the relations between nodes, *i.e.*, similar nodes in the graph have similar vectors. In [27], authors simulate random walks from one node to another using the edges and optimize node embeddings such that nodes that co-occur often in random walks of fixed length should be close in the embedding space. Node2Vec [9] uses a similar approach to build walks in the graph but it selects the next node based on a biased sampling. The generated paths are then fed into a Word2Vec model [21]. In the same vein, [1] propose to extend another word embedding model, namely GloVe [26], to learn node representations. Other methods factorize the adjacency matrix to learn a vector representation for each node with either SGD or SVD [2,23] , or train a model that learns how to combine the features of a node and its neighborhood [11]. For more details on this topic, we refer the interested reader to [12].

2.2   Node Embeddings in Dynamic Graphs

In dynamic graphs, edges and nodes can appear (or disappear) over time. They can be separated into two categories: *time-continuous graphs*, where each change in the graph happens at a specific time t, and *discrete graphs* where a batch of changes happens during a time interval. The former can be transformed into the latter by grouping together all the changes that happen during $[t, t + T]$ where T is the duration of the interval, but not the other way around. A naive way to learn node embeddings from a dynamic graph would be to use static methods on the final state of the graph, but the temporal information would not be captured in this case.

Several temporal node embedding techniques directly follows Node2Vec [22,35]. CTDNE generates paths in a time-continuous graph where the order of visited nodes respects the order of appearance of edges [22]. tNodeEmbed uses Node2Vec on each interval of a dynamic graph, aligns the node embeddings and passes them into a LSTM to obtain a unique vector for each node [30]. DynGEM trains autoencoders to reconstruct the adjacency matrix of each interval but initializes their weights with the weights of the autoencoder trained on the previous interval. The embedding of a node is the latent layer of the autoencoder after the final interval [7]. This method was extended in dyngraph2vec [6] where the adjacency matrices of multiple previous intervals are passed into the autoencoder. Finally, recent approaches such as EvolveGCN [25] or GAEN [29] learn embeddings at each timestep with Graph Convolution Networks or Attention models, and combine it with RNN or GRU to capture the graph evolution.

Although these methods operate on dynamic graphs, they do not take into account the activity history of an edge, *i.e.*, how often an edge appeared in the past and whether or not an edge has recently been used between two nodes. With the example in Fig. 1, learning embeddings with those methods would make the vector of James close to the vectors of Richard, Mary and John, but the vectors of Mary and John should be closer because their relation with James is more recent. That is, since new edges appear over time, they should have more weight during training. The method we propose uses time difference between edges to select and weight more importantly the most recent edges during training.

Another drawback of these approaches is that they only focus on the temporal aspect of the graph and do not allow one to balance between static and dynamic structural aspects. As a result, they can be outperformed by purely static approaches on graphs where the dynamics do not carry as much information as the original structure of the graph. To tackle this problem, JODIE was proposed by [19]. This model focuses on bipartite graphs and learns two embeddings (a static and a dynamic one) for each node separately using RNNs. In this article, we are interested in non-bipartite graphs and we devise an objective function that allows one to learn node embeddings using both the static and the temporal information simultaneously, with an explicit control on the weight given to each part during the training.

3   Learning Node Embeddings Using Time Distance

Below, we present the learning setup considered in this paper and our general framework that learns node embeddings by taking into account both static and dynamic information of a graph. This latter is then equipped with a forgetting mechanism that attributes more relevance to recent events.

3.1   Problem Setup

Given a graph $G = (\mathcal{V}, \mathcal{E})$ where $\mathcal{V} = v_1, \cdots, v_n$ is the set of vertices ($|\mathcal{V}| = n$) and $\mathcal{E} \subseteq \mathcal{V} \times \mathcal{V}$ is the set of edges, we aim to learn a function that maps vertices $v \in \mathcal{V}$ into a d-dimensional vector space, with $d \ll n$. This mapping function outputs a node embedding vector denoted by $z \in \mathbb{R}^d, \forall v \in \mathcal{V}$. $Z \in \mathbb{R}^{(n \times d)}$ is the matrix storing all node embeddings. In the context of dynamic graphs, we further assume that each edge $e_{ij} \in \mathcal{E}$ is characterized by both static and temporal attributes: the static attribute, denoted by a_{ij}, corresponds to the number of edges which occurred between a pair of nodes (v_i, v_j); the temporal attribute, denoted by $t_{ij} = t_{ij}^{(1)}, \cdots, t_{ij}^{(a_{ij})}$, is a list containing the timestamps associated with each link.

3.2   General Framework

Our first goal is to design a general framework for learning node embeddings from both static and temporal attributes simultaneously. Given two nodes (v_i, v_j), we propose to minimize the error between their similarity given by the dot product of their embeddings and their static $sim_{\mathcal{S}} : \mathcal{V} \times \mathcal{V} \to \mathbb{R}$ and temporal $sim_{\mathcal{T}} : \mathcal{V} \times \mathcal{V} \to \mathbb{R}$ similarities in the original graph. As for each edge associated to a pair of nodes, these nodes can either be seen as the source or as the target node (i.e., either e_{ij} or e_{ji}), we learn two vectors for each node and store them in two matrices, Z and $\tilde{Z}$. When the graph is undirected, both matrices are equivalent; when the graph is directed, the two embedding vectors allow one to differentiate edges for which a given node is the source node from edges for which it is a target node. In the end, we take the average of these embeddings to obtain one single vector for each node. Putting it all together, we consider the following optimization objective:

$$J = \sum_{i,j=1}^{n} \lambda \left[z_i^T \cdot \tilde{z}_j - \log(sim_{\mathcal{S}}(v_i, v_j) \right]^2 + (1 - \lambda) \left[z_i^T \cdot \tilde{z}_j - \log(sim_{\mathcal{T}}(v_i, v_j)) \right]^2,$$

(1)

where λ is a hyperparameter allowing us to control the balance between the static and the temporal component in the final node embeddings. Since the *log* function is not defined for 0, we discard the zero-values of $sim_{\mathcal{S}}(v_i, v_j)$ and $sim_{\mathcal{T}}(v_i, v_j)$ to be used in the objective function. If two nodes have a static or a temporal similarity of 0, it means there is no edge between them and therefore their vectors should not be trained to be moved closer. We now proceed to defining the last two missing ingredients $sim_{\mathcal{S}}(v_i, v_j)$ and $sim_{\mathcal{T}}(v_i, v_j)$.

3.3   Static and Temporal Similarities Between Nodes

To illustrate our proposal for static and dynamic similarity functions, we use a running example given in Fig. 2 throughout this section. The latter is given by a small dynamic graph composed of 7 nodes ($n = 7$). In this dynamic graph, an

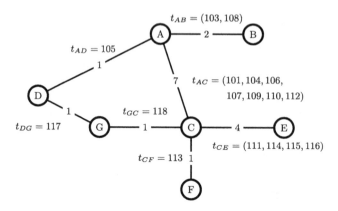

Fig. 2. Example of a temporal graph. In red, the number of edges between nodes. In blue, the times at which they appear. (Color figure online)

edge between two nodes can appear multiple times (*e.g.*, two people can send several emails to the other one). For instance, between A and B, there are 2 edges, that appear at time $t = 103$ and $t = 108$ (blue numbers). Red numbers in Fig. 2 correspond to the number of edges between two related nodes (which does not depend on the time of appearance of edges). We are now ready to define $sim_S(v_i, v_j)$ and $sim_T(v_i, v_j)$.

Static Similarities. Given the abundance of different node embedding techniques for static graphs, one can define $sim_S(v_i, v_j)$ in many different ways. In this work, we propose to define a similarity based on the normalized adjacency matrix similar to the LINE embedding model [31] where $sim_S(v_i, v_j)$ denotes the probability of going from a node v_i to a node v_j in a random walk. These probabilities rely on both first and second-order proximity statistics and are computed using the static edge attributes. We have, $\forall i, j = 1, \cdots, n$:

$$sim_S(v_i, v_j) = \begin{cases} \frac{a_{ij}}{a_i} & \text{if } e_{ij} \in \mathcal{E} \\ \sum_k \frac{a_{ik}}{a_i} \cdot \frac{a_{kj}}{(a_k - a_{ik})} & \text{if } e_{ij} \notin \mathcal{E} \cap \exists v_k : e_{ik}, e_{kj} \in \mathcal{E} \\ 0 & \text{otherwise,} \end{cases} \quad (2)$$

where a_{ij} is the number of edges that occur between a pair of nodes v_i and v_j; $a_i = \sum_j a_{ij}$. This similarity can be easily computed for either directed or undirected graphs. One should note that the role of the first term is to capture first-order proximity between two nodes in the graph (*i.e.*, the existence of an edge between two nodes) while the second term captures second-order proximity for pair of nodes separated by a distance of 2 in the graph.

In Fig. 2, we have $s_{CA} = 7/13$ and $s_{AC} = 7/10$. One should note that static similarities are not symmetric. For nodes with a distance of 2, we have $s_{CB} = s_{CA} \times s_{AB} = 14/39$. When several paths are available to join v_i and v_j, we sum the probabilities of all paths (so $s_{CD} = 10/39$).

We want to stress out that the first part of the objective function is versatile (see [32]), as one can define the static similarity using other popular methods such as, for instance, Personalized PageRank [24] or SimRank [13].

Temporal Similarities. As explained above, the temporal similarity should take into account the information from different timestamps of a dynamic graph but also allow the model to forget the past that became irrelevant. To this end, we propose to define sim_T as a function of the time delta between the most recent and the other edges of a node and its neighbors. Formally, it is defined as:

$$sim_T(v_i, v_j) = \begin{cases} \sum_k f(\Delta_{t_{ij}^{(k)}}) & \text{if } e_{ij} \in \mathcal{E} \\ 0 & \text{otherwise,} \end{cases} \tag{3}$$

where f is a decreasing function, allowing us to give more weight to recent edges (*i.e.*, when Δ is small) and $\Delta_{t_{ij}} = \max_j(t_{ij}) - t_{ij}$ (*i.e.*, the time difference between the timestamp t_{ij} and the most recent timestamp among all edges starting from node v_i). Choosing f to be a decreasing function indicates that we assume that as times passes, the strength of the relation between two nodes becomes weaker. We believe that for social networks, or co-citation networks this is a reasonable assumption. However for applications such as Protein-Protein interaction where this assumption might not be ideal, one can easily relax this condition and choose a function f that suits best their need.

Going back to our example, blue numbers in Fig. 2 are the times at which edges appear between the nodes. A model learning node embeddings using only static information would make the vector of C more similar to A than to E because it has more edges with A ($7 > 4$). However, edges between C and E are more recent than between C and A ($t = 114, 115, 116$ vs. $t = 110, 112$). The intuition behind temporal similarities is to bring closer the vectors of nodes having the most recent interactions. In Fig. 2, we have $d_{CE} = f(\Delta_{111}) + f(\Delta_{114}) + f(\Delta_{115}) + f(\Delta_{116})$. The most recent edge of C appears at $t = 118$, so $\Delta_{111} = 118 - 111 = 7$ and therefore $d_{CE} = f(7) + f(4) + f(3) + f(2)$. One should note that temporal similarities are also not symmetric. Indeed, for d_{EC}, the most recent edge of E appears at $t = 116$, so we would have $\Delta_{111} = 116 - 111 = 5$.

In the following, we choose f to be a survival function of the form:

$$f : x \rightarrow e^{[-\alpha*(x/x_{max})^2]},$$

where α is a hyperparameter that controls the decay rate of this weight function and x_{max} is the maximum value that can be passed to f (*i.e.*, the largest time distance). This function models the probability for a given relation (i.e. an edge) to survive past a certain time, here represented by time-steps (see Fig. 3). For high values of α this probability decreases faster to 0, than for small values of α. In that case, our model strongly favors the short-term rather than the long-term dynamic of the graph, a reasonable assumption when dealing with a graph presenting important structural changes between two time-steps. On the other hand, as α decreases, our model will take into account all the edges from the past, hence capturing long-term dynamics. This particular setting will suit graphs presenting a smooth evolution over time, with all edges being relevant at any time. In the asymptotic limit, this function can be "flat" enough to behave as a dynamic model that takes into account all timestamps of a dynamic graph.

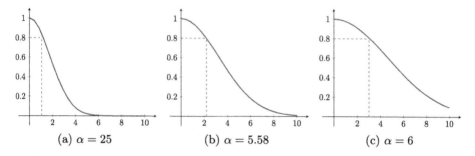

(a) $\alpha = 25$ (b) $\alpha = 5.58$ (c) $\alpha = 6$

Fig. 3. Survival function $f : x \rightarrow exp^{[-\alpha*(x/x_{max})^2]}$ for different values of α and with $x_{max} = 16$. (b) satisfies the Pareto's law meaning that 20% most recent edges have a weight among the highest 80%, i.e., $f(0.2 * \Delta_{t_{i_{max}}}) = 0.8$.

3.4   Complexity Analysis

For $sim_{\mathcal{T}}$, we have to compute at most $|\mathcal{E}|$ values (corresponding to the non-zero entries of the adjacency matrix) because only direct edges. For $sim_{\mathcal{S}}$ we have to compute on average $|\mathcal{V}| \times \kappa^2$ values where κ is the average node degree in the graph (so each node can reach on average κ^2 other nodes with a distance of 2). Therefore, the complexity of our model is $\mathcal{O}(|\mathcal{V}| \times \kappa^2 + |\mathcal{E}|)$ because it iterates only over the non-zero similarities.

4   Experiments

4.1   Datasets

We evaluate our hypothesis about the importance of combining both static and dynamic information for learning good node embeddings on 6 real-world datasets representing dynamic graphs: messages sent between people (Radoslaw[1] [28], ENRON[2] [16]), links between webpages (Subreddit[3] [17]), network of routers

---

[1] https://networkrepository.com/ia-radoslaw-email.php.
[2] https://networkrepository.com/ia-enron-employees.php.
[3] https://snap.stanford.edu/data/soc-RedditHyperlinks.html.

(Autonomous Systems[4] [20]), rating of Bitcoin users (BTC-Alpha[5] and BTC-OTC[6] [18]). As explained in Subsect. 2.2, dynamic graphs can be continuous or discrete. We use both types of graphs in our experiments to demonstrate that our model works regardless of the nature of the dynamic graph. Table 1 reports statistics about each dataset. For some datasets, we use a smaller version because some baselines were not able to run on the full version (for Subreddit, we consider only nodes with at least 10 edges; for Autonomous Systems (AS), we use only the 100 first steps).

Table 1. Statistics about the datasets used for experiments. (∗) indicates datasets that have been shrunk. $\overline{deg}(v_i)$ (resp. $\overline{C_i}$ coef.) is the average degree (resp. clustering coefficient) of all nodes.

Dataset	Nodes	Edges	Type	$\overline{deg}(v_i)$	$\overline{C_i}$ coef
Radoslaw	167	82,876	Continuous	992.5	0.592
ENRON	150	47,088	Continuous	627.8	0.521
Subreddit (∗)	6,340	223,457	Continuous	70.5	0.364
Auto. Sys. (∗)	3,569	561,139	Discrete	314.5	0.257
BTC-Alpha	3,783	24,186	Continuous	12.8	0.177
BTC-OTC	5,881	35,592	Continuous	12.1	0.178

4.2 Evaluation Tasks

Link predictions. This task consists in predicting if a link between two nodes exists or not in the graph. We follow the same protocol as in [15]. For each edge (u, v) in $\mathcal{T}$, the set of all unique test edges, we generate the list $\mathcal{N}_v$ (resp. $\mathcal{N}_u$) of negative edges obtained by replacing v (resp. u) by another node from the graph such that the negative edge does not exist. Then, the cosine similarity between the embeddings of u and v is computed, as well as its rank r_v (resp. r_u) against the cosine similarities of all edges in $\mathcal{N}_v$ (resp. $\mathcal{N}_u$).

The Mean Reciprocal Rank (MRR) is the mean of the inverse of r_v and r_u for all edges (u, v) in $\mathcal{T}$:

$$MRR = \frac{1}{2 \times |\mathcal{T}|} \sum_{(u,v) \in \mathcal{T}} \left(\frac{1}{r_u} + \frac{1}{r_v} \right).$$

In addition to the MRR, we also compute Hits@K metrics (the percentages of ranks r_v or r_u which are less than K).

---

[4] https://snap.stanford.edu/data/as-733.html.
[5] https://snap.stanford.edu/data/soc-sign-bitcoin-alpha.html.
[6] https://snap.stanford.edu/data/soc-sign-bitcoin-otc.html.

Node classification. This task only applies to the Subreddit dataset. It consists in predicting the correct label of nodes in the graph. Plenty of graphs with labeled nodes exist in the static configuration (*i.e.* when there is no evolution in the network over time) but dynamic graphs with node class information are almost non-existent. To overcome this problem, we generate labels for each node in the Subreddit dataset. It contains 6,340 nodes. We use an automatic method to generate the labels for similar nodes with a clustering algorithm and manually verify that related subreddits are within the same cluster. Each node in this dataset represents a subreddit, the name of a topic-specific discussion forum (*e.g.* chess, Olympics). Using the property of word embeddings to encapsulate semantic information, we generate the word embedding of each subreddit name using Fasttext library[7]. Since subreddits such as "skiing"or"skateboarding" are words related to the same semantic field, their respective word embeddings should be similar thanks to the Fasttext learning scheme. We then use a K-Means clustering algorithm to group together similar word embeddings, thus grouping related subreddit. We try different values for K, from 20 to 80. When K is too low, there are not enough clusters and unrelated subreddits fall into the same category when they should not. When K is too high, related subreddits are often separated into different groups. We find that using 50 clusters is a good trade-off. Each node is then associated with the ID of the group it belongs to. Table 2 shows some examples of subreddits with the same label.

Table 2. Examples of subreddits with the Autona for the node classification

Label = 3	Label = 17	Label = 32	Label = 41	Label = 43
Altcoin	Judo	Blizzard	Albania	Amazon
Bitcoin	Olympics	Blood borne	Finland	Ebay
Bitcoin mining	Skate boarding	Counter strike	Italy	Netflix
Crypto currency	Skiing	Halo	Poland	Silkroad
Dogecoin	Swimming	Overwatch	Spain	Spotify
Ethereum	Tennis	Streetfighter	Usa	Tumblr

For the node classification task, we train a logistic regression classifier on the learned node embeddings (the embeddings learned from our dynamic graph method, not those generated with Fasttext) to predict their corresponding class. We report the accuracy.

Edge classification. This task only applies to the Bitcoin datasets and consists in predicting the class of edges. In BTC-Alpha and BTC-OTC, an edge indicates that one user of the Bitcoin trading platform rated another user on a scale from -10 to +10. We generate a binary label for each edge: 0 if the rating is negative, 1 otherwise. We compute an embedding for each edge by averaging the embeddings of the involved nodes. We train a linear regression classifier on

---

[7] https://fasttext.cc/.

the edge embeddings to predict their corresponding binary label. The dataset is unbalanced, about 90% of edges have a label of 1. We report the F1 score.

4.3 Training Settings

Each graph is split into a train (75%) and a test set (25%) according to timestamps. The same train and test sets are used for all models. We train vectors of 20 dimensions for small datasets (ENRON, Radoslaw) and 100 dimensions otherwise (same as in [7]). Our hyperparameters are tuned with a grid search to maximize the MRR on the train set, with α ranging from 2 to 60 with steps of 3, λ ranging from 0 to 1 with steps of 0.05 and the number of epochs ranging from 20 to 300 by steps of 20. All experiments are done with an Intel Xeon E3-1246 CPU, a NVIDIA Titan X GPU and 32 GB of RAM.

4.4 Baselines

Our model uses both static and temporal information during training. Therefore, we compare it against methods that learn node embeddings from static graphs (Node2Vec, GraphSage [11]) and other methods from dynamic graphs (CTDNE, tNodeEmbed, DynGEM, dyngraph2vec (AERNN version), EvolveGCN). We set a walk length of 40 for Node2Vec and CTDNE, and a length of 5 for GraphSage. We generate 10 walks per node for these methods. Other hyperparameters are set as indicated in their respective papers. For GraphSage, we use one-hot vectors as a replacement for node feature vectors when they are not present, as advised by the authors. We do not use DynamicTriad [34] as a baseline because it has been reported to have lower scores than tNodeEmbed and dyngraph2vec. Baselines are trained on the same machine as our model.

5 Results and Analysis of the Model

In this section[8], our goal is to answer two following questions:

Q1. *Does learning from both static and dynamic information lead to better node embeddings?*
To this end, we compare our method with several static and dynamic node embedding methods on 6 real-world datasets and on 3 different tasks.

Q2. *What insights such framework can provide about the graphs on which it is applied?*
To answer this question, we analyze the optimal values of λ and α revealing the importance of static component for learning node embeddings and the effect of forgetting when taking into account temporal information.

The proposed method has both a static and a dynamic component. We first compare its results against dynamic methods as we are working with dynamic graphs, and then against static methods. Thereafter, we analyze the role of each component depending on the dataset and we evaluate its complexity.

---

[8] Code to reproduce our results and access datasets can be found here: https://github.com/laclauc/DynSimilarity.

Table 3. MRR and Hits@K metrics on 5 datasets for our method and other baselines on a link prediction task. Bold and underline results indicate respectively the best and the second best value for each metric.

	Radoslaw			ENRON			Subreddit			AS		
	MRR	Hits@5	Hits@50	MRR	Hits@5	Hits@50	MRR	Hits@5	Hits@50	MRR	Hits@5	Hits@50
Static												
Node2Vec	.088	15.82%	62.18%	.247	52.65%	91.27%	.123	21.07%	37.04%	**.221**	**44.94%**	**70.12%**
GraphSage	.100	17.42%	71.12%	.230	48.85%	86.24%	.100	15.60%	31.21%	.146	27.79%	56.04%
Dynamic												
CTDNE	.111	19.28%	82.46%	.235	48.32%	86.80%	.112	18.11%	31.34%	.215	42.53%	69.88%
tNodeEmbed	.169	33.93%	71.22%	.228	47.24%	86.35%	.105	16.32%	31.29%	.020	2.24%	6.19%
DynGEM	.123	16.93%	55.62%	.177	34.18%	74.44%	.080	9.29%	11.56%	.077	8.37%	11.10%
dyngraph2vec	.180	36.51%	81.02%	.145	27.46%	72.05%	.088	9.69%	17.79%	.031	3.84%	8.19%
EvolveGCN	.097	16.96%	64.43%	.124	22.54%	67.01%	.053	6.02%	10.40%	.017	1.64%	4.48%
Our	**0.329**	**69.33%**	**96.71%**	**0.298**	**62.24%**	**92.16%**	**0.132**	**22.54%**	**37.58%**	0.146	28.47%	50.80%

5.1 Against Dynamic Graph Methods

Table 3 reports the scores of all modelsfor the task of link prediction. The MRR is the average of the inverse rank of a true edge against false random edges. Higher scores indicate that a model is able to better differentiate true edges against negative ones. On 3 out of 5 datasets, our method strongly outperforms all the other dynamic methods in terms of MRR. The Hits@K metrics in Table 3 indicate the percentage of true edges whose rank is among the first K when compared to several hundreds negative edges. We observe similar results as for the MRR. For instance, on Radoslaw, our method is able to retrieve almost 70% of true edges in the top 5 while other methods can only retrieve 36% at best. This means that our method ranks the majority of true edges with a much better rank than the other methods (top 5 vs. top 50), which is useful in a recommendation task. Note that Radoslaw and ENRON are the two datasets with the highest clustering coefficient and average degree (Table 1). The temporal information is crucial in these datasets because a change in the network topology spreads faster than for other datasets as they both represent email communications between people of a company, and are well suited to evaluate dynamic methods [5]. Our method outperforms other dynamic methods on those graphs, demonstrating that it is well appropriate for graphs where temporal information is important.

Results on the two other tasks are reported in Table 4. For node classification, the proposed model improves the classification accuracy over the other dynamic baselines. The results notably show an important improvement over the auto-encoder and GCN-based approaches, with results almost 4 times better than EvolveGCN for instance. Finally, for the edge classification task, we outperform all dynamic baselines on both Bitcoin datasets by an important margin.

Table 4. Results on a node classification task (accuracy of correct predictions among 50 classes) and an edge binary classification task (F1 score).

	Node classif	Edge classif.	
	Subreddits	BTC-Alpha	BTC-OTC
Static methods			
Node2Vec	18.24%	0.254	0.290
GraphSage	19.67%	0.293	0.312
Dynamic methods			
CTDNE	19.21%	0.192	0.251
tNodeEmbed	22.20%	0.173	0.280
DynGEM	7.03%	0.080	0.321
Dyngraph2vec	8.24%	0.398	0.273
EvolveGCN	6.04%	0.361	0.268
Our	**22.86%**	**0.429**	**0.360**

5.2   Against Static Graph Methods

Table 3 reports the scores of static baselines (Node2Vec and GraphSage). Our method outperforms GraphSage and Node2Vec on 4 out of 5 graphs in terms of MRR (*e.g.* 0.329 vs. 0.100 on Radoslaw), and Hits@5 (*e.g.* 62.24% vs. 52.65% on Subreddit). It is on par on AS against GraphSage but looses against Node2Vec.

One should note that AS and Subreddits are the biggest graphs among the datasets, and most of the edges do not vary over time. Therefore, their behavior is similar to a static graph, which explains why Node2Vec is only slightly under or better than our dynamic method for them. Our best results for these datasets are obtained when the dynamic component of our model driven by $1-\lambda$ is almost zero , validating the assumption that they have a static behavior. However, our method doing better on the other datasets demonstrates that the dynamic component in our model is important to learn embeddings when the dataset evolves greatly over time, which Node2Vec cannot achieve. Our conclusions on node and edge classification are similar to the one made against dynamic approaches.

Overall, the obtained results demonstrate that our model is able to produce node embeddings capturing both the dynamic and the static aspects of an evolving graph. They are versatile, as we obtain consistently good results for various tasks, including link prediction, node classification and edge classification. Our approach is also robust to various graphs topology as we either outperform all other baselines or rank second at worst on all datasets. This shows the benefit brought by the two components of our objective function.

5.3   Influence of the Hyperparameters of the Model

We are now ready to address the second question. Our model naturally provide a way to gain insights regarding the dynamic of the graphs, through its two

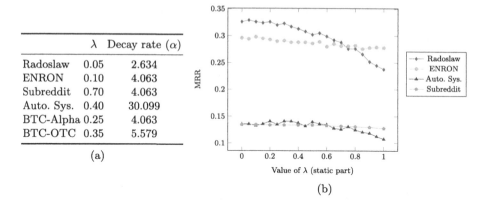

	λ	Decay rate (α)
Radoslaw	0.05	2.634
ENRON	0.10	4.063
Subreddit	0.70	4.063
Auto. Sys.	0.40	30.099
BTC-Alpha	0.25	4.063
BTC-OTC	0.35	5.579

(a)

(b)

Fig. 4. Evolution of MRR versus static coefficient used in our model.

hyper-parameters λ and α. Our model has both a static and a dynamic component, each one with its own coefficient in the objective function (resp. λ and $(1-\lambda)$). The two coefficients are inter-dependent: increasing one decreases the other. Therefore, they behave as a cursor that one can set to favour the static or the dynamic part of the method. We notice in our experiments that the value of λ must be selected depending on the nature of the dataset. Figure 4(b) shows the evolution of the MRR according to the value of λ on some datasets. We can see that when λ increases, the MRR drops for most datasets. Indeed, a value of λ close to 1 means that the dynamic part in our objective function is almost non-existent, which is detrimental for dynamic graphs. Unsurprisingly, the highest decrease is on Radoslaw, two datasets with a strong evolution over time that therefore require a large dynamic component in the model to capture temporal information. For Subreddit, the drop is smaller because this dataset does not vary a lot over time, so the static part is more important in this case. This is confirmed by Fig. 4(a), which reports the λ and α hyperparameters that gives the best scores for each dataset. Radoslaw needs a dynamic oriented model (($1-\lambda$) close to 1) while Subreddit needs a dominant static part ($\lambda = 0.7$) to achieve good results. We also notice that a large value of α makes the temporal similarities more focused on very short-term interactions while a smaller α allows to consider a longer history of previous events.

5.4   Training Times

Table 5 reports the time required by each method to train on all 6 datasets. Due to its simplicity (no complex architectures like RNN/LSTM nor convolution networks), our method only needs 40 min, which makes it the fastest among all the dynamic methods (*e.g.* 6h36 for DynGEM, 24h01 for dyngraph2vec).

Table 5. Times required to train all models on the 6 graphs. Minimum time is highlighted by boldface.

	Node2vec	GraphSage	tNodeEmbed	CTDNE	DynGEM	dyngraph2vec	EvolveGCN	Our
Training time	48:53	14:43:27	1:34:08	4:00:56	6:36:10	24:01:18	5:33:11	**40:36**

6 Conclusion

This paper studies the importance of combining both static and dynamic information when learn node embeddings in dynamic graphs. It introduces a novel temporal similarity measure between nodes based on time distance of edges and a model that uses it in addition to static similarities to learn embeddings that reflect the structure of the dynamic graph. This method allows one to emphasize either the static or the dynamic component of the model to adapt to different kinds of graphs. It obtains better scores than other dynamic methods on 6 real-world datasets for various tasks thus suggesting that fully dynamic approaches may be too rigid for efficient learning in dynamic graphs. Further research directions of this work are many. First, we would like to explore new types of node similarities to train on different specific graphs such as bipartite graphs or knowledge-based graphs. tasks. We also plan to investigate how to integrate temporal node attributes into this framework by leveraging adapted Graph Neural Networks (GNN) architectures [36].

References

1. Brochier, R., Guille, A., Velcin, J.: Global vectors for node representations. In: WWW, pp. 2587–2593. ACM (2019)
2. Cao, S., Lu, W., Xu, Q.: GraRep: learning graph representations with global structural information. In: CIKM, pp. 891–900 (2015)
3. Chuan, P.M., Ali, M., Khang, T.D., Dey, N., et al.: Link prediction in co-authorship networks based on hybrid content similarity metric. Appl. Intell. **48**(8), 2470–2486 (2018)
4. Dalmia, A., Gupta, M.: Towards interpretation of node embeddings. In: Companion Proceedings of the The Web Conference 2018, pp. 945–952 (2018)
5. De Winter, S., Decuypere, T., Mitrović, S., Baesens, B., De Weerdt, J.: Combining temporal aspects of dynamic networks with node2vec for a more efficient dynamic link prediction. In: ASONAM, pp. 1234–1241. IEEE (2018)
6. Goyal, P., Chhetri, S.R., Canedo, A.: dyngraph2vec: capturing network dynamics using dynamic graph representation learning. KBS **187**, 104816 (2020)
7. Goyal, P., Kamra, N., He, X., Liu, Y.: DynGEM: deep embedding method for dynamic graphs. arXiv preprint arXiv:1805.11273 (2018)
8. Goyal, P., Sapienza, A., Ferrara, E.: Recommending teammates with deep neural networks. In: Hypertext and Social Media, pp. 57–61 (2018)
9. Grover, A., Leskovec, J.: node2vec: scalable feature learning for networks. In: KDD, pp. 855–864 (2016)
10. Haghani, S., Keyvanpour, M.R.: A systemic analysis of link prediction in social network. Artif. Intell. Rev. **52**(3), 1961–1995 (2019)

11. Hamilton, W., Ying, Z., Leskovec, J.: Inductive representation learning on large graphs. In: NeurIPS (2017)
12. Hamilton, W.L., Ying, R., Leskovec, J.: Representation learning on graphs: methods and applications. IEEE Data Eng. Bull. **40**(3), 52–74 (2017)
13. Jeh, G., Widom, J.: Simrank: a measure of structural-context similarity. In: Proceedings KDD, pp. 538–543 (2002)
14. Kazemi, S.M., et al.: Representation learning for dynamic graphs: a survey. JMLR **21**(70), 1–73 (2020)
15. Kazemi, S.M., Poole, D.: Simple embedding for link prediction in knowledge graphs. In: NeurIPS (2018)
16. Klimt, B., Yang, Y.: Introducing the enron corpus. In: CEAS (2004)
17. Kumar, S., Hamilton, W.L., Leskovec, J., Jurafsky, D.: Community interaction and conflict on the web. In: WWW, pp. 933–943 (2018)
18. Kumar, S., Spezzano, F., Subrahmanian, V., Faloutsos, C.: Edge weight prediction in weighted signed networks. In: ICDM, pp. 221–230. IEEE (2016)
19. Kumar, S., Zhang, X., Leskovec, J.: Predicting dynamic embedding trajectory in temporal interaction networks. In: KDD, pp. 1269–1278 (2019)
20. Leskovec, J., Kleinberg, J., Faloutsos, C.: Graphs over time: densification laws, shrinking diameters and possible explanations. In: KDD, pp. 177–187 (2005)
21. Mikolov, T., Chen, K., Corrado, G., Dean, J.: Efficient estimation of word representations in vector space. arXiv preprint arXiv:1301.3781 (2013)
22. Nguyen, G.H., Lee, J.B., Rossi, R.A., Ahmed, N.K., Koh, E., Kim, S.: Continuous-time dynamic network embeddings. In: WWW (2018)
23. Ou, M., Cui, P., Pei, J., Zhang, Z., Zhu, W.: Asymmetric transitivity preserving graph embedding. In: KDD, pp. 1105–1114 (2016)
24. Page, L., Brin, S., Motwani, R., Winograd, T.: The pagerank citation ranking: bringing order to the web. Technical Report 1999–66, Stanford InfoLab (1999)
25. Pareja, A., et al.: EvolveGCN: evolving graph convolutional networks for dynamic graphs. In: AAAI, pp. 5363–5370 (2020)
26. Pennington, J., Socher, R., Manning, C.D.: Glove: global vectors for word representation. In: EMNLP, pp. 1532–1543. ACL (2014)
27. Perozzi, B., Al-Rfou, R., Skiena, S.: Deepwalk: online learning of social representations. In: KDD, pp. 701–710 (2014)
28. Rossi, R.A., Ahmed, N.K.: The network data repository with interactive graph analytics and visualization. In: AAAI (2015). http://networkrepository.com
29. Shi, M., Huang, Y., Zhu, X., Tang, Y., Zhuang, Y., Liu, J.: GAEN: graph attention evolving networks. In: IJCAI, pp. 1541–1547 (2021)
30. Singer, U., Guy, I., Radinsky, K.: Node embedding over temporal graphs. In: IJCAI, pp. 4605–4612 (2019)
31. Tang, J., Qu, M., Wang, M., Zhang, M., Yan, J., Mei, Q.: Line: large-scale information network embedding. In: Proceedings of WWW, pp. 1067–1077 (2015)
32. Tsitsulin, A., Mottin, D., Karras, P., Müller, E.: Verse: versatile graph embeddings from similarity measures. In: Proceedings of WWW, pp. 539–548 (2018)
33. Wang, P., Xu, B., Wu, Y., Zhou, X.: Link prediction in social networks: the state-of-the-art. Sci. China Inf. Sci. **58**(1), 1–38 (2015)
34. Zhou, L., Yang, Y., Ren, X., Wu, F., Zhuang, Y.: Dynamic network embedding by modeling triadic closure process. In: AAAI (2018)
35. Haddad, M., Bothorel, C., Lenca, P., Bedart, D.: TemporalNode2vec: temporal node embedding in temporal networks. In: Complex Networks (2019)
36. Rossi, E., Chamberlain, B., Frasca, F., Eynard, D., Monti, F., Bronstein, M.: Temporal graph networks for deep learning on dynamic graphs. In: arxiv

Summarizing Labeled Multi-graphs

Dimitris Berberidis[1], Pierre J. Liang[2], and Leman Akoglu[1(✉)]

[1] Carnegie Mellon University,
Heinz College of Information Systems and Public Policy, Pittsburgh, USA
{dbermper,lakoglu}@andrew.cmu.edu
[2] Carnegie Mellon University, Tepper School of Business, Pittsburgh, USA
liangj@tepper.cmu.edu

Abstract. Real-world graphs can be difficult to interpret and visualize beyond a certain size. To address this issue, graph summarization aims to simplify and shrink a graph, while maintaining its high-level structure and characteristics. Most summarization methods are designed for homogeneous, undirected, simple graphs; however, many real-world graphs are *ornate*; with characteristics including node labels, directed edges, edge multiplicities, and self-loops. In this paper we propose TG-SUM, a *versatile* yet rigorous graph summarization model that (to the best of our knowledge, for the first time) can handle graphs with *all* the aforementioned characteristics (and *any* combination thereof). Moreover, our proposed model captures basic sub-structures that are prevalent in real-world graphs, such as cliques, stars, etc. Experiments demonstrate that TG-SUM facilitates the visualization of real-world complex graphs, revealing interpretable structures and high-level relationships. Furthermore, TG-SUM achieves better trade-off between compression rate and running time, relative to existing methods (only) on comparable settings.

Keywords: Graph summarization · Super graph · Labeled multi-graph

1 Introduction

Given a directed labeled multi-graph G, how can we construct a small summary graph g that reflects the high-level structures and relationships in G? How can we find a succinct g that is yet an accurate representation, which requires a small amount of corrections to recover the original G? With the advent of technology, not only the size but also the complexity of real-world graphs have grown immensely. Today graph data often contains node labels, multi-edges, etc. Graph summarization aims to find high-level structural patterns and most salient information in large complex graphs to enable efficient storage, processing, visualization and interpretation.

A large body of existing graph summarization techniques is for *plain* graphs with homogeneous unlabeled nodes [14–16, 20, 23, 24, 26]. However there exist

---

Supplementary Information The online version contains supplementary material available at https://doi.org/10.1007/978-3-031-26390-3_4.

M.-R. Amini et al. (Eds.): ECML PKDD 2022, LNAI 13714, pp. 53–68, 2023.
https://doi.org/10.1007/978-3-031-26390-3_4

numerous real-world graphs with multiple node labels; including transaction networks containing nodes (i.e. accounts) of various types (cash, revenue, expense, etc.) or heterogeneous graphs such as publication records among entities of various types (paper, author, venue, etc.). We refer to both kinds as node-labeled, or simply *labeled graphs*. Moreover, a vast majority of prior work are for summarizing *simple* [9, 14–16, 19, 20, 24, 25, 28] *undirected* [14–16, 20, 23–26, 28] graphs, whereas the edges in real-world graphs may repeat (e.g., multiple transactions between two accounts, multiple exchanges between two email addresses, etc.) which are called *multi-graphs*. As is the case for transaction and email graphs, among others, the edges can also be *directed*.

In this work we propose (to the best of our knowledge; see, Table ??) the *first* method called TG-SUM, for *multi-Type* (i.e. node-labeled) *multi-Graph SUMmarization*, with directed edges and possible self-loops. (See Sect. 2, and a recent comprehensive survey [18].) Besides, TG-SUM is *versatile* in that it can also handle graphs with any combination of those properties (i.e., (un)directed, plain/labeled, simple/multi- or weighted graphs).

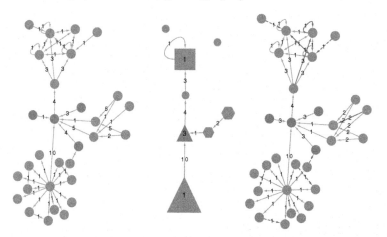

Fig. 1. (best in color) Ex. input graph (left), its summary/super-graph (middle), and the decompressed graph (right) w/ edge corrections in red , where dashed - - (solid —) are edges that need to be added (removed) for lossless reconstruction. See text for description of the scalars, node color, size, and shape. (Color figure online)

Our goal is to output a small yet representative summary that facilitates the visualization, by which, improves the understanding of the overall structure of an input graph. To this end, we model a *summary graph (or super-graph)* as a collection of labeled super-nodes and weighted super-edges. As illustrated in Fig. 1, we merge structurally similar nodes of the same type/label (depicted by color) into super-nodes (size reflecting the number of constituent nodes). Supernodes capture prevalent structural constructs found in real-world graphs, such as stars and cliques [15] (depicted by shape). A super-node is also marked with a scalar (i.e., weight), representative of the edge multiplicities among its nodes. A super-edge is placed between two super-nodes whose constituent nodes are sufficiently well-connected, and is also marked with a scalar (i.e., weight) that best represents the edge multiplicities inbetween.

We aim to construct a small summary graph, which accurately reflects the input graph. Here, succinctness and accuracy are in trade-off; the coarser the summary graph, the more information about the original graph is lost. We design a novel two-part *lossless* encoding scheme, describing (i) the summary graph and (ii) the corrections required to reconstruct the input graph losslessly. Treating the total number of encoding bits as a cost function, we design algorithms to find a summary with a small total cost. In summary, our main contributions are:

- The first method for Summarizing LMDS-Graphs.
- A Novel Super-graph Model, in Sect. 3.1
- A Novel Two-part Lossless Encoding Scheme, in Sect. 3.2.
- Efficient Search Algorithms, in Sect. 4.
- Extensive experiments on real-world graphs, in Sect. 5

Reproducibility: Source code for TG-SUM and all public-domain datasets are shared at https://bit.ly/3d4vogt.

2   Related Work

Graph *summarization* and graph *compression* techniques, while related, exhibit a key distinction. The former typically aims to simplify an input graph into a coarser one, while reflecting its prominent structure. On the other hand, the latter aims at reducing the storage requirements of a graph, often enabling speedy querying, while maintaining a certain level of query accuracy [2,6,7,13,17]. (See [5] for a recent survey.) In this work we focus on graph summarization, with a goal to extract a simplified overview of key structural patterns within an input graph. Most graph summarization techniques are designed for unlabeled, undirected, and simple graphs without edge multiplicities, weights, or self-loops [14–16,24]. Closely-related are graph-pooling methods used within graph neural networks to gradually reduce the dimension of the layers; see. e.g. [27]. Riondato *et al.* [23] and Toivonen *et al.* [26] are some of the few summarization methods that can accommodate weighted edges, but not labeled nodes or directed edges. Among the methods that can handle graphs with multiple node labels, CoSum [28], Liu *et al.* [19], and SNAP [25] build a coarser graph by only merging the nodes of the same label into super-nodes. Differently, Subdue [9] replaces frequent sub-graphs that potentially contain different labels with a super-node, which makes the interpretation of the summary graph harder. Closest to our work is the approach by Navlakha *et al.* [20], which iteratively merges nodes into super-nodes as long as the description length of the input graph decreases. Thanks to its simple model and algorithm, it can be modified to handle labeled graphs, specifically by restricting the node merges to same-label nodes. However, its model is unable, nor is it trivial to modify, to accommodate edge weights/multiplicities. All in all, there is *no existing work* that can summarize labeled multi-graphs – with labeled nodes, directed and multi-edges and self-loops. (See [18] for an extended survey and Table 1 therein.) While our TG-SUM is the first of its kind, it is *versatile* in that it can also accommodate graphs with any combination of those properties. Besides input graph properties, prior work can also be classified w.r.t. the properties of the summary. Here, we focus on summaries where the output is itself a (coarser) graph, called the summary or super graph. VoG [15] identifies key sub-structures (stars, (near)cliques, etc.) however does not provide any super-edges, i.e., its summary graph is disconnected. Second, the summary may be lossy; including only the coarse summary graph [9,19,23,25,26,28], or lossless; consisting of both the summary and the corrections necessary to fully reconstruct the input graph [14–16,20,24]. Finally, a desired characteristic of a summary is multi-granularity; where the coarseness or resolution of the summary graph can be adjusted on demand [9,16,19,20,23,25,26], via appropriately altering some of the model parameters. Notably, TG-SUM exhibits all of these three properties: super graph output, lossless and multi-resolution summary.

3   Graph Summary Design and Encoding

3.1   Summary and Decompression

Given a directed graph $\mathcal{G} = \{\mathcal{V}, \mathcal{E}, \mathcal{T}\}$ with edge multiplicities $m(e) \in \mathbb{N}, \forall e \in \mathcal{E}$, node labels/types $\ell(u) \in \mathcal{T}, \forall u \in \mathcal{V}$, and self-loops, we define the *summary* and *decompressed* graphs as follows.

3.1.1 Summary Graph (or Super-Graph) Model
Let $\mathcal{G}_s = \{\mathcal{V}_s, \mathcal{E}_s\}$ be the sets of super-nodes and directed super-edges that define the summary graph topology. Each super-node $v \in \mathcal{V}_s$ is annotated by four components: (i) its label $\ell(v)$ (depicted by color), (ii) the number $|\mathcal{S}_v|$, $\mathcal{S}_v \subset \mathcal{V}$, of nodes it contains (depicted by size), (iii) the *glyph* $\mu(v) \in \mathcal{M}$ it represents (depicted by shape), and (iv) the *representative* multiplicity $m(v)$ of the edges it summarizes (depicted as a scalar inside the glyph). For each super-edge $e \in \mathcal{E}_s$, we let $m(e)$ be the *representative* multiplicity of the edges it captures, depicted as a scalar on the super-edge. (We describe how to find the "representative" multiplicity of a set of edges in Sect. 4.2.)

Figure 1 (left) and (middle) respectively depict an example input graph and its corresponding summary graph. Apart from unmerged simple nodes that are depicted as plain circles, the set $\mathcal{M}$ of possible glyphs that TG-SUM supports contains: 1) `Clique` (square), 2) `In-star` (triangle), 3) `Out-star` (triangle), and 4) `Disconnected` set (hexagon). Such structures are commonly found in real-word graphs [15]. For instance, a clique can represent a tightly-knit group of friends in a social network, while an out-star can capture spam-like activity in an email or call network. Moreover, using glyphs has been shown to yield easily interpretable visualizations [10].

3.1.2 Decompression

The summary graph $\mathcal{G}_s$ decompresses *uniquely* and *unambiguously* into $\mathcal{G}' = \text{dec}(\mathcal{G}_s) = \{\mathcal{V}, \mathcal{E}'\}$ according to simple and intuitive rules (e.g., Fig. 1 (right)). First, every super-node expands to the set of nodes it contains, all of which also inherit the super-node's label. The nodes are then connected according to the super-node's glyph: for out(in)-stars a node defined as the hub points to (is pointed by) all other nodes, for cliques all possible directed edges are added between the nodes, and for disconnected sets no edges are added. Moreover, a super-node self-loop expands to self-loops on every node it contains. On the other hand, super-edges expand to sets of edges that have the same direction.

Apart from enabling a clear interpretation of a given summary, the decompression rules help quantify how well the summary represents the original graph. For example, the pink triangle with representative multiplicity 3 in Fig. 1 (middle) expands to an in-star with all edges having multiplicity 3 as shown in Fig. 1 (right). While the topology is perfectly captured (pink nodes form a perfect in-star), the expanded multiplicities are not always equal to the original ones. On the other hand, expanding the green triangle perfectly captures the edge multiplicities (all are 1), but only approximates the topology, as the original green subgraph also contains some edges between the spokes of the hub node.

3.2   Model Encoding

Following the two-part Minimum Description Length paradigm [12], we aim to identify a summary graph $\mathcal{G}_s$ that minimizes the total description cost of the full graph, that is,

$$\mathcal{G}_s^* := \arg\min_{\mathcal{G}_s}\ L(\mathcal{G}_s) + L(\mathcal{G}|\mathcal{G}'), \quad \text{s.t.} \quad \mathcal{G}' = \text{dec}(\mathcal{G}_s) \tag{1}$$

where $L(\mathcal{G}_s)$ measures the number of bits required to encode the summary graph, and $L(\mathcal{G}|\mathcal{G}')$ the bits needed to encode the corrections (or extra-information) for reconstructing the original graph $\mathcal{G}$ from the (uniquely and unambiguously) decompressed $\mathcal{G}'$. These costs can be quantified as follows.

3.2.1 Encoding the Summary Graph

We first encode the size of the summary graph $L_{\mathbb{N}}(|\mathcal{V}_s|)$, and the number of labels $L_{\mathbb{N}}(|\mathcal{T}|)$.[1] For each super-node, $\log_2 |\mathcal{T}|$ bits are used to record its label, $\log_2 |\mathcal{M}|$ for its glyph, $L_{\mathbb{N}}(|\mathcal{S}_v|)$ for its size, $\mathcal{L}_{\mathbb{N}}(m(v))$ for the within-glyph representative multiplicity, $\log_2(|\mathcal{V}_s|)$ for the number of super-nodes in $\mathcal{G}_s$ that it points to, and $\log_2 \binom{|\mathcal{V}_s|}{|\mathcal{N}(v)|}$ to identify the specific set of super-nodes it points to, where $\mathcal{N}(v)$ denotes the set of direct (out)neighbors. For each super-edge, $L_{\mathbb{N}}(m(e))$ bits are used for the representative multiplicity. In total, the number of bits required to encode a summary graph is given as

---

[1] $L_{\mathbb{N}}(k) = 2\log k + 1$ bits are required to encode an arbitrarily large natural number k, using the variable-length prefix-free encoding; see, Ex. 2.4 in [12].

$$L(\mathcal{G}_s) = L_\mathbb{N}(|\mathcal{V}_s|) + L_\mathbb{N}(|\mathcal{T}|) + \sum_{v \in \mathcal{V}_s} L_{\text{SNODE}}(v), \tag{2}$$

where

$$L_{\text{SNODE}}(v) = \log_2 |\mathcal{T}| + \log_2 |\mathcal{M}| + L_\mathbb{N}(|\mathcal{S}_v|) + \mathcal{L}_\mathbb{N}(m(v)) + \log_2(|\mathcal{V}_s|)$$
$$+ \log_2 \binom{|\mathcal{V}_s|}{|\mathcal{N}(v)|} + \sum_{z \in \mathcal{N}(v)} L_\mathbb{N}(m(v,z)) \tag{3}$$

3.2.2 Encoding the Corrections

For the overall cost for corrections, we first compute the number of bits used to correct the *topology* of the expanded (i.e., decompressed) graph, followed by the number of bits needed to represent the true *multiplicities*. Regarding the topology, we first map the expanded nodes back to the original node-set $\mathcal{V}$. This costs $L_{\text{MAP}}(v) = \log_2 \binom{|\mathcal{V}|}{|\mathcal{S}_v|} + \mathbb{1}_{\{\mu(v)=\text{STAR}\}} \log_2 |\mathcal{S}_v|$ bits per super-node v (the latter term identifying the hub of a star). Subsequently, we have two types of edge corrections: Either adding edges that exist in the original graph but not in the expanded graph (i.e., positive corrections) or removing edges from the expanded graph because they do not exist in the original graph (i.e., negative corrections).

These costs are compactly encoded for every expanded super-edge and every expanded super-node (glyph), using the binomial encoding $L(\mathcal{E}_{\text{COR}}) = L_\mathbb{N}(|\mathcal{E}_{\text{COR}}|) + \log_2 \binom{|\mathcal{E}_{\text{COR}}^{\max}|}{|\mathcal{E}_{\text{COR}}|}$, where $\mathcal{E}_{\text{COR}}$ denotes the possible set of corrections (positive or negative), and $\mathcal{E}_{\text{COR}}^{\max}$ the largest set that $\mathcal{E}_{\text{COR}}$ can possibly be. For example, for positive edge corrections in a disconnected set, we have $\mathcal{E}_{\text{COR}}^{\max} = \mathcal{S}_v \times \mathcal{S}_v$, and similarly for negative edge corrections in a clique. For super-edges, corrections are computed according to the decompression rules (see Sect. 3.1). For the (few) edges in the original graph between super-nodes v and z that are not represented by a super-edge, the corrections are always positive, and $\mathcal{E}_{\text{COR}}^{\max} = \mathcal{S}_v \times \mathcal{S}_z$.

The binomial encoding arises from using the uniform code over all the lexicographically ordered *subsets* of possible corrections. An alternative to this, as suggested in [16,23], would be to encode each correction *individually* using an optimal prefix code. Then, interpreting $p = |\mathcal{E}_{\text{COR}}|/|\mathcal{E}_{\text{COR}}^{\max}|$ as the "probability" of each correction, we would need $L_{\text{ENTR}} = H(\text{Ber}[p]) \cdot |\mathcal{E}_{\text{COR}}^{\max}|$ bits, where $H(\cdot)$ is the Shannon entropy, and $\text{Ber}[p]$ is a Bernoulli with parameter p. Denoting $|\mathcal{E}_{\text{COR}}| = n'$ and $|\mathcal{E}_{\text{COR}}^{\max}| = n$, we can show that our binomial encoding is more efficient.

Theorem 1. *It holds that $L_{\text{ENTR}} \geq \log_2 \binom{n}{n'}$, for all $n > n' > 0$.*

Proof. See Appendix.

Theorem 1 establishes that the binomial encoding always gives a more compact measure of information required for corrections. Having corrected the edge topology, we compute the cost of correcting the edge multiplicities. Since any edge e not included in a glyph or super-edge does not have a representative multiplicity, its multiplicity correction is encoded by $L_\mathbb{N}(m(e))$, encoding its true value. The reason for using $L_{\mathbb{N}}(\cdot)$ to encode multiplicities is the fact that, for most real graphs, multiplicities follow a power-law distribution. Since the vast majority has small values, $L_{\mathbb{N}}(\cdot)$ will generally be a more "compact" encoding compared to a uniform code based on the maximum multiplicities. For expanded super-nodes and super-edges with representative multiplicity m, we obtain the cost of correcting the multiplicities as

$$L_{\mathrm{DIFF}}(\mathcal{E}_{\mathrm{sup}}, m) = \sum_{e \in \mathcal{E}_{\mathrm{sup}}} \ell_{\mathrm{diff}}(m(e), m), \tag{4}$$

where $\mathcal{E}_{\mathrm{sup}}$ in this context is the set of all edges contained in said super-node or super-edge, and

$$\ell_{\mathrm{diff}}(m', m) = \begin{cases} 1 \,, & m = m' \\ 2 \log_2(|m - m'|) + 3 \,, & m \neq m' \end{cases}, \tag{5}$$

bits are needed to encode the *difference* between a true multiplicity m' and its representative m. Note that, since $L_{\mathbb{N}}(\cdot)$ only holds for natural numbers (see footnote 3), one extra bit is required to indicate whether the difference is 0, and one more for the sign of the difference.

4   Graph Summary Search

The discrete optimization problem in (1) has a very large set of feasible solutions, and needs to be approximated efficiently. Towards this goal, we follow a two-step process, where we first generate a list of (possibly overlapping) groups of nodes, which we term *candidate* node-sets (see Sect. 4.1), and then decide which ones to merge into super-nodes. These candidates have varying size and quality (i.e., structural-similarity). Larger candidates with low quality compress the graph more (reduced $L(\mathcal{G}_{\mathrm{s}})$), but also typically require more corrections (increased $L(\mathcal{G}|\mathcal{G}')$). Clearly, the best candidates have both high quality and large size. For this reason, we first sort the candidate sets in descending order with respect to the product of their size and quality. We then process the sorted list from top to bottom, and merge the candidate sets into super-nodes, updating the summary graph accordingly (see Sect. 4.2). To ensure the quality of summarization, we only monitor the overall total cost, and only commit to a given candidate if $\Delta_{\mathrm{cost}} = \mathrm{Cost\_After} - \mathrm{Cost\_Before} < 0$. This offers two benefits: (1) We avoid the cumbersome process of merging nodes in pairs (i.e. two at a time) and instead merge *in groups*, and (2) We achieve ability to summarize at *multiple resolutions*. The overview is given in Algorithm 1.

4.1   Candidate Set Generation

4.1.1 Measuring Candidate Quality
To quantify a candidate set's quality, we first need to define a proper metric of structural node similarity. For undirected graphs, the Jaccard similarity between

---

Algorithm 1. TG-SUM: Summarizing Labeled Multi-Graphs

Input: directed labeled multi-graph $\mathcal{G}$

1 Construct candidate node-sets (Sect. 4.1);
2 Sort candidates w.r.t. (size × quality);
3 **for** *every candidate set in list* **do**
4   Merge unmarked nodes in set and decide glyph (Sect. 4.2.1);
5   Decide super-edges (Sect. 4.2.2) ;
6   Compute representative multiplicities (Sect. 4.2.3);
7   Mark candidate node-set as merged;
8   **if** $\Delta_{\mathrm{cost}} < 0$ **then**
9   Commit to merged super-node and its super-edges;

10 **Return** summary graph $\mathcal{G}_{\mathrm{s}}$;

---

two nodes v and v' is given as $J^U(v, v') = \frac{|\mathcal{N}^U(v) \cap \mathcal{N}^U(v')|}{|\mathcal{N}^U(v) \cup \mathcal{N}^U(v')|}$, and simply measures the proportion of common neighbors that they share. Naïvely using $J^U(\cdot, \cdot)$ on directed graphs is straightforward by ignoring the directions of the edges, however, it may yield misleading results by often over-estimating the true node similarity. To mitigate such inconsistencies, we introduce the following extension of Jaccard that may also accommodate directed graphs, by taking into account the similarity of both *I*ncoming and *O*utgoing edges.

Definition 1. *The* directed Jaccard *similarity between any two pair of nodes* v, v' *of a directed graph is given as*

$$J^D(v, v') = \frac{|\mathcal{N}^I(v) \cap \mathcal{N}^I(v')| + |\mathcal{N}^O(v) \cap \mathcal{N}^O(v')|}{|\mathcal{N}^I(v) \cup \mathcal{N}^I(v')| + |\mathcal{N}^O(v) \cup \mathcal{N}^O(v')|} \tag{6}$$

First, it can easily be observed that for undirected graphs, $J^D(v, v') = J^U(v, v')$, since $\mathcal{N}^I(v) = \mathcal{N}^O(v) = \mathcal{N}^U(v)$. Note however, that in our example, $J^D(v, v')$ becomes 0 for all cross-pairs between $\{B, C, D\}$ and $\{E, F\}$, effectively creating two separate groups. In general for directed graphs, $J^D(v, v')$ will be more "informed" than $J^U(v, v')$, typically yielding lower similarity scores. We then define

Definition 2. *Any set* $\mathcal{C} \subseteq \mathcal{V}$, *is* $t-$*bounded if* $J^D(v, v') \geq t \quad \forall (v, v') \in \mathcal{C} \times \mathcal{C}$.

We use the $t-$bounded-ness of a candidate to serve as a pessimistic valuation of its quality. In addition, given that we are interested in a *collection* of candidate sets, we would like the sets to be *non-redundant* defined as follows.

Definition 3. *Let* $\mathcal{C}_{\mathrm{S}}$ *be a collection of candidate sets, each one accompanied by a bound* t. *We call* $\mathcal{C}_{\mathrm{S}}$ non-redundant, *if for any* $\mathcal{C} \in \mathcal{C}_{\mathrm{S}}$ *that is* $t-$*bounded, there exists no* $t'-$*bounded* $\mathcal{C}' \in \mathcal{C}_{\mathrm{S}}$, *such that* $t' \geq t$ *and* $\mathcal{C} \subset \mathcal{C}'$.

Simply put, non-redundancy ensures that none of the candidate sets is a strict subset of another set of higher or equal quality.

4.1.2 Incremental LSH

To group nodes according to their similarity, we first utilize Locality Sensitive Hashing (LSH) [4]. Specifically for every node v, we generate a set of r *minhash* signatures

$$h_j(v) := \min_{z \in \mathcal{N}^D(v)} f_j(z) \quad \forall j = 1, \ldots, r \tag{7}$$

where f_j's are independent and uniform hash functions (see, e.g., [4] for implementation details), and $\mathcal{N}^D(v) := \mathcal{N}^I(v) \| \mathcal{N}^O(v)$ is the concatenated adjacency list of node v that includes all incoming and outgoing neighbors separately. It can then be shown that $\Pr\{h_j(v) = h_j(v')\} = J^D(v, v')$; that is, two nodes share a minhash signature with probability proportional to their directed Jaccard similarity. Since the r hash functions are independent, it follows that $\Pr\{\mathbf{h}(v) = \mathbf{h}(v')\} = (J^D(v, v'))^r$, where $\mathbf{h}(v) := [h_1(v), \ldots, h_r(v)]^T$ is the r−length minhash signature vector of node v. If the nodes are hashed into buckets according to their r minhash signatures, the equality gives the probability that two nodes hash into the same bucket. By collecting b hash-tables corresponding to b bands of r minhash signatures, the probability that v and v' hash to the same bucket at least once is

$$\Pr\{\mathbf{h}_i(v) = \mathbf{h}_i(v') \; \exists i = 1, \ldots, b\} = 1 - \left(1 - (J^D(v, v'))^r\right)^b \tag{8}$$

Interestingly, for sufficiently large r and b, the RHS expression in Eq. (8) when viewed as a function of $(J^D(v, v'))$ approximates a step function around the threshold $t = \left(\frac{1}{b}\right)^{\frac{1}{r}} \in (0, 1]$, meaning that with high probability v and v' will belong in a t−bounded set. To avoid repeating the entire process for different values of b, we incrementally generate and add more bands of minhash node signatures, that in turn hash nodes into new buckets. The new buckets are then merged with any overlapping existing buckets, gradually coalescing into larger clusters that are *approximately* t−bounded, with $t = \left(\frac{1}{b}\right)^{\frac{1}{r}}$ decreasing as b increases. This is exactly how we obtain larger candidate sets, albeit of lower quality, incrementally by the addition of new bands.

4.1.3 Filtered LSH

While the incremental LSH described in the previous section efficiently guides the process of forming candidate sets, merged buckets are not guaranteed to be t−bounded due to the false alarm probability. For this purpose, we maintain an undirected similarity graph $\mathcal{G}_{\text{sim}}$, where an edge (v, v') is *guaranteed* to appear if and only if $J^D(v, v') \geq t$. Intuitively, $\mathcal{G}_{\text{sim}}$ serves as a data structure where large t−bounded candidates appear as maximal cliques. As new LSH buckets appear and clusters are updated, we compute $J^D(v, v')$ for newly coalesced pairs of nodes (v, v'), and add the latter as an edge to $\mathcal{G}_{\text{sim}}$ if $J^D(v, v') \geq t$. If the threshold is not satisfied, the computed value $J^D(v, v')$ is not discarded, but cached into a max-heap since it may satisfy a lower t in one of the subsequent iterations as b is increased.

As mentioned earlier, candidate sets are collected as maximal cliques in $\mathcal{G}_{\mathrm{sim}}$. To ensure that the set of candidates is *non-redundant* (cf. Def.n 3), we maintain for every node the size of the maximum clique that it has been found to belong in. Every time a new clique is discovered, we update the maximum-sizes for all the nodes it contains using the clique's size. As new edges are added to $\mathcal{G}_{\mathrm{sim}}$, we examine every node for newly emerged cliques, and we rely on the heuristic in [22] to prune the search by avoiding the evaluation of cliques that cannot exceed the size of the previously-found maximum clique.

4.2   Merging Candidates: Glyphs, Super-Edges, Multiplicities

Every time a candidate set $\mathcal{C}$ is tested, we deploy subroutines that efficiently update the summary graph, by making decisions regarding (1) the glyph that will be assigned to the merged set of nodes, (2) super-edges that emerge (or disappear) due to the merging, as well as (3) representative multiplicities for the set of edges summarized by the glyph and its super-edges.

4.2.1 Glyph Decision Rules

To preserve super-node label homogeneity, a candidate set that contains nodes of different labels is first split into same-label subsets. Each subset is merged into a separate super-node using the procedure described below. Hereafter, the term candidate set refers to such a label-homogeneous subset. For the glyph decision, we first identify the number of directed edges $E_{\mathcal{C}}$ that are included in the subgraph induced on nodes that corresponds to $\mathcal{C}$ in the candidate set. Consequently, if $E_{\mathcal{C}} \geq |\mathcal{C}|(|\mathcal{C}| - 1)/2$, i.e., at least half of all possible directed edges are present, then we decide Clique since it most likely is the best glyph in terms of number of edge corrections. For sparsely-edged candidate sets that do not pass the clique threshold, we proceed to test for the presence of stars. If there is a suitable out-/in-star present in $\mathcal{C}$, then its hub will be the highest out/in-degree node in $\mathcal{C}$. We use the followin proxy correction cost for encoding an in-star

$$\mathrm{Cost}_{\mathrm{IN}} = (|\mathcal{C}| - 1 - d_{\mathrm{max}}^{I}) + (E_{\mathcal{C}} - d_{\mathrm{max}}^{I}) , \qquad (9)$$

and similarly $\mathrm{Cost}_{\mathrm{OUT}}$ for out-star using d_{max}^{O}. Intuitively, the first term of Eq. (9) is the number of edges that will have to be removed from the full decompressed star, while the second part is the number of edges that cannot be "explained" by the star and will have to be added. We then compare $\mathrm{Cost}_{\mathrm{IN}}$ and $\mathrm{Cost}_{\mathrm{OUT}}$ with $E_{\mathcal{C}}$, i.e., the number of edges that will have to be added if we decide that $\mathcal{C}$ is a disconnected set. If only $\mathrm{Cost}_{\mathrm{IN}}$ (or only $\mathrm{Cost}_{\mathrm{OUT}}$) is smaller than $E_{\mathcal{C}}$, then we decide In-star (or Out-star). If both $\mathrm{Cost}_{\mathrm{IN}}$ and $\mathrm{Cost}_{\mathrm{OUT}}$ are smaller than $E_{\mathcal{C}}$, then we choose the smallest of the two. Finally, if neither $\mathrm{Cost}_{\mathrm{IN}}$ nor $\mathrm{Cost}_{\mathrm{OUT}}$ are smaller than $E_{\mathcal{C}}$, we decide $\mathcal{C}$ is a Disconnected set.

4.2.2 Super-edge Decision Rule

Having decided the glyph of $\mathcal{C}$, we merge any outgoing and incoming edges and/or super-edges into "bundles" of edges and their corresponding multiplicities.

We then obtain the topology-based correction costs of merging or not merging each bundle into a super-edge (recall Sect. 3.2). If the total cost (topology and multiplicities) of representing each bundle of edges with a super-edge is lower than the cost of *not* representing it, then the corresponding super-edge (and its representative multiplicity) is added to the summary.

4.2.3 Finding Representative Multiplicities

For every newly-formed super-node as well as each potential super-edge, we find the representative multiplicity m^* as $m^* := \arg\min_m \; L_{\mathrm{DIFF}}(\mathcal{E}_{\mathrm{sup}}, m)$ where $L_{\mathrm{DIFF}}(\mathcal{E}_{\mathrm{sup}}, m)$ is defined in Eq. (4), and $\mathcal{E}_{\mathrm{sup}}$ is the set of all edges contained in a given super-node or bundled by a super-edge. Although this 1D-optimization problem is not convex, we find that the dichotomous search algorithm [8] finds the optimal solution in most cases, and runs in $\mathcal{O}(|\mathcal{E}_{\mathrm{sup}}| \log_2 R)$ time, where $R = \max_{e \in \mathcal{E}_{\mathrm{sup}}} m(e) - \min_{e \in \mathcal{E}_{\mathrm{sup}}} m(e)$, i.e., the dynamic range of multiplicities.

5   Experiments

5.1   Setup

We experimented with real-world graphs of a wide variety of sizes and characteristics, including a senator-to-senator network extracted from the 2009–2010 US Congress dataset [3] and the Political Blogs network [1], both with political affiliation labels; the Cora and Citeseer citation networks [11] that are labeled by publication venue; and finally, transaction networks from 3 (anonymous) corporations that we collaborated with. See Table 1 for a summary of network characteristics. There is *no existing method* for LMDS-graph summarization, thus we compare only under simplified settings, w.r.t. running time and compression rate. Moreover, TG-SUM is only comparable to lossless methods. We modify the *Randomized* algorithm of Navlakha *et al.* [20] to accommodate node labels and edge directions, and compare on all graphs, ignoring the edge multiplicities.

Table 1. Real-world graphs used in experiments. ∗ depicts naturally directed graphs that are typically treated as undirected. For SH, HW, KD #labels is given for EB/FS labeling.

Name	#nodes	#m-edges	#labels	Lbl.	Dir.	Mult.	S-loop
senate	0.1K	2.4K	2	✓			
polblogs	1.5K	19K	2	✓	∗		
cora	2.7K	10.6K	6	✓	∗		
citeseer	3.3K	9.2K	7	✓	∗		
SH trans	0.25K	301K	11/27	✓	✓	✓	✓
HW trans	0.32K	268K	11/60	✓	✓	✓	✓
KD trans	2.3K	648K	10/29	✓	✓	✓	✓

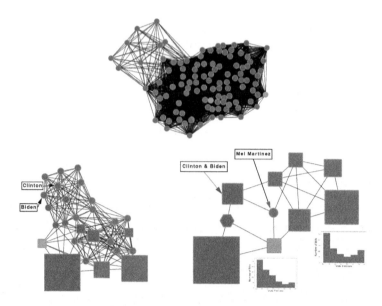

Fig. 2. (left) original US Senate graph, (middle) high resolution ($b = 2$) summary, (right) low resolution ($b = 5$) summary. (Color figure online)

5.2   Qualitative Evaluation: TG-SUM at Work

The US Senate dataset contains the (positive or negative) votes of 108 senators for 696 congressional bills. The senators are labeled as Republicans (red), Democrats (blue), or independent (green). We construct an undirected graph where two senators are connected by an edge if the cosine similarity of their votes is larger than 0.3. The graph is plotted in Fig. 2, along with two summaries at different resolutions, leading to the following observations. Interestingly, while most democratic senators eventually form a clique, there is a smaller group of East coast senators, including prominent Democratic figures such as Joe Biden, Hillary Clinton, and Ted Kennedy that do not merge with the main body and form their own separate clique. Furthermore, this clique of Democrats is directly linked to certain Republicans, such as the Florida-based Mel Martinez, who has most recently opposed Trump openly and explicitly expressed his preference for Joe Biden[2]. The second observation is that Republican senators overall exhibit a more fragmented voting behavior, splitting into multiple cliques of comparable size. This is corroborated by computing the entropies of the votes for all the bills, for Democrats and Republicans separately. Intuitively, bills with high entropy indicate a low degree of agreement on the subject. By plotting the histograms of the voting entropies (see Fig. 2 (right)) for the two groups, it becomes apparent that Republican votes exhibit higher entropy (median = 0.21) than Democrats (median = 0.16).

---

[2] https://bit.ly/3qwc9zu.

5.3 Quantitative Evaluation: Evaluating Financial Accounts Labeling

In this section, we show how we employed TG-SUM to quantitatively address a domain-specific problem, specifically, evaluating a pre-existing *labeling*, i.e., the set of types pre-assigned to the nodes in an accounting network that connects business accounts via credit/debit transaction relations. A business entity's Chart of Accounts (COA) lists, and also pre-assigns a label to, each distinct account used in its ledgers. Such labeling helps companies prepare their aggregate financial statements (FS). For example, the FS caption "Cash and Cash Equivalents" is used to describe the total sum of all liquid assets tracked in a number of accounts; e.g., currencies, checking accounts, etc. In the US, FS captions are not uniform across corporations. In fact, the data we have from 3 different companies (anonymized as SH, HW, and KD in Table 1) each contains different FS captions.

How suitable is a given FS labeling? Can a different labeling be shown to be quantitatively better than another?

To this end, our collaborator (an accounting expert) designed a new labeling (referred as EB for economic bookkeeping), relabeling the accounts based on their primary economic nature. Specifically, EB organizes them into operating versus financing and long- versus short-term accounts. Expert knowledge suggests that EB improves over FS captions by categorizing the accounts such that accounts of the same label should "behave" similarly in the system. This behavior can be discerned from the real-world usage data, in particular the transactions graph, where accounts are connected through credit/debit relations. Under a more suitable labeling, the accounts with the same label should have more structural similarity and yield better compression.

Table 2. Evaluating account labelings in financial networks

Dataset	Labeling	Shuffled	Actual	Norm. gain (%)
SH	EB	0.28	0.32	**5.6 %**
	FS	0.25	0.27	2.7 %
HW	EB	0.36	0.47	**17.0 %**
	FS	0.16	0.27	13.0 %
KD	EB	0.33	0.42	**13.7 %**
	FS	0.31	0.39	12.0 %

To compare EB vs. FS, we employ TG-SUM on each graph using one or the other labeling separately, and record the compression rate. Next, we shuffle the labels (within each setting) randomly, and employ TG-SUM again. "Shuffled" and "Actual" compression rates are reported in Table 2 (the former averaged over 20 random shuffles). EB rates are higher—this is not surprising as EB has fewer labels as compared to FS (See $|\mathcal{T}|$ in Table 1), and hence TG-SUM has higher degree of freedom to merge nodes on EB-labeled graphs. As such, Actual values are not directly comparable. What is comparable is the difference from Shuffled, that is, how much the labeling can improve on top of the random assignment of the *same* set of labels. Here, the absolute difference is always equal or larger for EB. However, even the absolute difference of compression rates is not fair to compare—it is harder to compress a graph that has been compressed quite a bit even further. For EB, Shuffled rates are already high. Improving over Shuffled even by the same amount proves EB superior to FS. Therefore, we report the normalized gain; defined as (Actual−Shuffled)/(1−Shuffled), which shows that our expert-designed EB labeling is better, for the aforementioned reasons.

5.4 Quantitative Evaluation: Compression Rate, Running Time, Scalability

Quantitatively, we measure summarization performance in terms of both (1) running time, as well as (2) the size reduction achieved in terms of bits (including bits required for correction). Specifically, upon obtaining the total number of bits (as given by the encoding scheme of each method), we measure the compression ratio as Compress Ratio $= \frac{\text{Bits\_Before} - \text{Bits\_After}}{\text{Bits\_Before}} \in [0, 1)$, that is the fraction of the encoding cost that has been reduced by summarization. We compare with the Navlakha algorithm [20], which we modified to handle edge directions and node labels. We run TG-SUM by gradually increasing b, to increase the number of candidate sets and obtain multi-resolution summaries. A larger number of candidates is expected to yield higher compression ratio, albeit at the cost of increased running time—hence enabling the user to choose a suitable trade-off in practice.

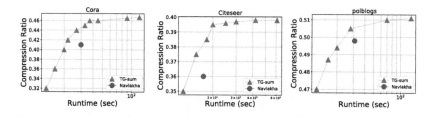

Fig. 3. Compression ratio vs. runtime on undirected graphs

Results are given in Fig. 3, where TG-SUM remarkably outperforms the alternative in terms of compression ratio in almost all cases. In absolute terms, it achieves roughly 30–60% compression across these various real-world graphs with up to hundreds of thousands of edges and tens of distinct labels. Finally, we measure the scalability of TG-SUM by first generating an increasing size synthetic directed k-out graph, where nodes are incrementally added and connected to $k = 10$ of the existing nodes, simulating a preferential attachment process. Results in Fig. 4 (left) show that, unlike the modified Navlakha, the runtime of TG-SUM grows in a near-linear fashion.

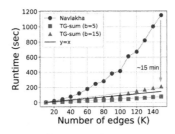

Fig. 4. TG-SUM complexityscales linearly with the number of edges.

6 Conclusion

We introduced TG-SUM, a versatile graph summarization algorithm that (for the first time) can handle *directed*, *node-labeled*, *multi*-graphs with possible self-loops (or *any* combination). Built on a novel encoding scheme, TG-SUM seeks to minimize the total encoding cost of (*i*) a summary graph, and (*ii*) the corrections to reconstruct the input graph losslessly. It efficiently finds structurally-similar nodes to create super-nodes of larger sizes incrementally, producing multi-resolution summaries. Extensive experiments show that TG-SUM (1) provides insights into the high-level structure of real-world graphs, (2) achieves better trade-off between compression and runtime relative to baselines (only) on comparable settings, and (3) scales linearly in the number of edges.

Acknowledgements. This work has been sponsored by the U.S. National Science Foundation CAREER 1452425 and the PwC Risk and Regulatory Services Innovation Center at Carnegie Mellon University. Any conclusions expressed in this material are those of the author and do not necessarily reflect the views, expressed or implied, of the funding parties.

References

1. Adamic, L.A., Glance, N.: The political blogosphere and the 2004 us election: divided they blog. In: Proceedings of the 3rd International Workshop on Link Discovery, pp. 36–43 (2005)
2. Adler, M., Mitzenmacher, M.: Towards compressing web graphs. In: Data Compression Conference IEEE Computer Society (2001)
3. Akoglu, L.: Quantifying political polarity based on bipartite opinion networks. In: ICWSM (2014)
4. Andoni, A., Indyk, P.: Near-optimal hashing algorithms for approximate nearest neighbor in high dimensions. Commun. ACM, pp. 117–122 (2008)
5. Besta, M., Hoefler, T.: Survey and taxonomy of lossless graph compression and space-efficient graph representations (2018). arxiv:1806.01799
6. Boldi, P., Vigna, S.: The webgraph framework i: compression techniques. In: WWW, pp. 595–602 (2004)
7. Buehrer, G., Chellapilla, K.: A scalable pattern mining approach to web graph compression with communities. In: WSDM, pp. 95–106. ACM (2008)

8. Chong, E.K., Zak, S.H.: An introduction to optimization. John Wiley & Sons (2004)
9. Cook, D.J., Holder, L.B.: Substructure discovery using minimum description length and background knowledge. J. Artif. Intell. Res. **1**, 231–255 (1993)
10. Dunne, C., Shneiderman, B.: Motif simplification: improving network visualization readability with fan, connector, and clique glyphs. In: SIGCHI, pp. 3247–3256 (2013)
11. Giles, C.L., Bollacker, K.D., Lawrence, S.: CiteSeer: an automatic citation indexing system. In: Proceedings of the Conference on Digital Libraries, pp. 89–98 (1998)
12. Grünwald, P.D.: The Minimum Description Length Principle. The MIT Press, Cambridge, MA (2007)
13. Kang, U., Faloutsos, C.: Beyond 'caveman communities': hubs and spokes for graph compression and mining. In: ICDM, pp. 300–309 (2011)
14. Khan, K.U., Nawaz, W., Lee, Y.K.: Set-based approximate approach for lossless graph summarization. Computing **97**(12) (2015)
15. Koutra, D., Kang, U., Vreeken, J., Faloutsos, C.: Summarizing and understanding large graphs. Statist. Anal. Data Min. **8**(3), 183–202 (2015)
16. LeFevre, K., Terzi, E.: Grass: graph structure summarization. In: SDM, pp. 454–465. SIAM (2010)
17. Liakos, P., Papakonstantinopoulou, K., Sioutis, M.: Pushing the envelope in graph compression. In: CIKM, pp. 1549–1558 (2014)
18. Liu, Y., Safavi, T., Dighe, A., Koutra, D.: Graph summarization methods and applications: a survey. ACM Comput. Surv. **51**(3), 1–34 (2018)
19. Liu, Z., Yu, J.X., Cheng, H.: Approximate homogeneous graph summarization. Info. Media Tech. **7**(1), 32–43 (2012)
20. Navlakha, S., Rastogi, R., Shrivastava, N.: Graph summarization with bounded error. In: SIGMOD, pp. 419–432. ACM (2008)
21. Oughtred, R., et al.: The biogrid interaction database **47**(D1), D529–D541 (2019)
22. Pattabiraman, B., Patwary, M.M.A., Gebremedhin, A.H., Liao, W.K., Choudhary, A.: Fast algorithms for the maximum clique problem on massive sparse graphs. In: International Workshop. AMWG (2013)
23. Riondato, M., García-Soriano, D., Bonchi, F.: Graph summarization with quality guarantees. Data Min. Knowl. Disc. **31**(2), 314–349 (2017)
24. Shin, K., Ghoting, A., Kim, M., Raghavan, H.: SWeG: lossless and lossy summarization of web-scale graphs. In: WWW, pp. 1679–1690. ACM (2019)
25. Tian, Y., Hankins, R., Patel, J.: Efficient aggregation for graph summarization. In: SIGMOD (2008)
26. Toivonen, H., Zhou, F., Hartikainen, A., Hinkka, A.: Compression of weighted graphs. In: KDD (2011)
27. Ying, Z., You, J., Morris, C., Ren, X., Hamilton, W., Leskovec, J.: Hierarchical graph representation learning with differentiable pooling. In: NeurIPS, pp. 4800–4810 (2018)
28. Zhu, L., Ghasemi-Gol, M., Szekely, P., Galstyan, A., Knoblock, C.A.: Unsupervised entity resolution on multi-type graphs. In: International Semantic Web Conference, pp. 649–667 (2016)

Inferring Tie Strength in Temporal Networks

Lutz Oettershagen[1]([✉]), Athanasios L. Konstantinidis[2],
and Giuseppe F. Italiano[2]

[1] Institute of Computer Science, University of Bonn, Bonn, Germany
`lutz.oettershagen@cs.uni-bonn.de`
[2] Luiss University, Rome, Italy
{`akonstantinidis,gitaliano`}`@luiss.it`

Abstract. Inferring tie strengths in social networks is an essential task in social network analysis. Common approaches classify the ties as *weak* and *strong* ties based on the *strong triadic closure (STC)*. The STC states that if for three nodes, A, B, and C, there are strong ties between A and B, as well as A and C, there has to be a (weak or strong) tie between B and C. So far, most works discuss the STC in static networks. However, modern large-scale social networks are usually highly dynamic, providing user contacts and communications as streams of edge updates. *Temporal networks* capture these dynamics. To apply the STC to temporal networks, we first generalize the STC and introduce a weighted version such that empirical a priori knowledge given in the form of edge weights is respected by the STC. The weighted STC is hard to compute, and our main contribution is an efficient 2-approximative streaming algorithm for the weighted STC in temporal networks. As a technical contribution, we introduce a fully dynamic 2-approximation for the minimum weight vertex cover problem, which is a crucial component of our streaming algorithm. Our evaluation shows that the weighted STC leads to solutions that capture the a priori knowledge given by the edge weights better than the non-weighted STC. Moreover, we show that our streaming algorithm efficiently approximates the weighted STC in large-scale social networks.

Keywords: Triadic closure · Temporal network · Tie strength inference

1 Introduction

Due to the explosive growth of online social networks and electronic communication, the automated inference of tie strengths is critical for many applications, e.g., advertisement, information dissemination, or understanding of complex human behavior [11, 20]. Users of large-scale social networks are commonly

Giuseppe F. Italiano is partially supported by MUR, the Italian Ministry for University and Research, under PRIN Project AHeAD (Efficient Algorithms for HArnessing Networked Data).

M.-R. Amini et al. (Eds.): ECML PKDD 2022, LNAI 13714, pp. 69–85, 2023.
https://doi.org/10.1007/978-3-031-26390-3_5

connected to hundreds or even thousands of other participants [26,30]. It is the typical case that these ties are not equally important. For example, in a social network, we can be connected with close friends as well as casual contacts. Since a pioneering work of Granovetter [12], the topic of tie strength inference has gained increasing attention fueled by the advent of online social networks and ubiquitous contact data. Nowadays, ties strength inference in social networks is an extensively studied topic in the graph-mining community [10,20,38]. A recent work by Sintos and Tsaparas [39] introduced the *strong triadic closure (STC)* property, where edges are classified as either *strong* or *weak*—for three persons with two *strong* ties, there has to be a *weak* or *strong* third tie. Hence, if person A is strongly connected to B, and B is strongly connected to C, A and C are at least weakly connected. The intuition is that if A and B are good friends, and B and C are good friends, A and C should at least know each other.

(a) Example for optimal weighted STC. The edge weights correspond the amount of communications.

(b) Example for optimal non-weighted STC. Ignoring the weights leads to three strong edges.

Fig. 1. Example for the difference between weighted and non-weighted STC. Strong edges are highlighted in red, and weak edges are dashed. (Color figure online)

We first generalize the ideas of [39] such that edge weights representing empirical tie strength are included in the computation of the STC. The idea is to consider edge weights that correspond to the empirical strength of the tie, e.g., the frequency or duration of communication between two persons. If this weight is high, we expect the tie to be strong and weak otherwise. However, we still want to fulfill the STC, and simple thresholding would not lead to correct results. Figure 1 shows an example where we have a small social network consisting of four persons A, B, C, and D. In Fig. 1a, the edge weights correspond to some empirical a priori information of the tie strength like contact frequency or duration, e.g., A and B chatted for ten hours and B and D for only one hour. The optimal weighted solution classifies the edges between A and B as well as between C and D as strong (highlighted in red). However, if we ignore the weights, as shown in Fig. 1b, the optimal (non-weighted) solution has three strong edges. Even though the non-weighted solution has more strong edges, the weighted version agrees more with our intuition and the empirical a priori knowledge.

We employ this generalization of the STC for inferring the strength of ties between nodes in temporal networks. A temporal network consists of a fixed set of vertices and a chronologically ordered stream of appearing and disappearing temporal edges, i.e., each temporal edge is only available at a specific discrete point in time. Temporal networks can naturally be used as models for real-life scenarios, e.g., communication [4,7], contact [33], and social networks [14, 17,31]. In contrast to static graphs, temporal networks are not simple in the sense that between each pair of nodes, there can be several temporal edges, each corresponding to, e.g., a contact or communication at a specific time [18]. Hence, there is no one-to-one mapping between edges and ties. Given a temporal network, we map it to a weighted static graph such that the edge weights are a function of the empirical tie strength. We then classify the edges using the weighted STC, respecting the a priori information given by the edge weights.

A major challenge is that the weighted STC is hard to compute and real-world temporal networks are often provided as large or possibly infinite streams of graph updates. To tackle this computational challenge, we employ a sliding time window approach and introduce a streaming algorithm that can efficiently update a 2-approximate of the minimum weighted STC, i.e., the problem that asks for the minimum total weight of weak edges. Our contributions are:

1. We generalize the STC for weighted graphs and apply the weighted STC for determining tie strength in temporal networks. To this end, we use temporal information to infer the edge strengths of the underlying static graph.
2. We provide a streaming algorithm to efficiently approximate the weighted STC over time with an approximation factor of two. As a technical contribution, we propose an efficient dynamic 2-approximation for the minimum weighted vertex cover problem, a key ingredient of our streaming algorithm.
3. Our evaluation shows that the weighted STC leads to strong edges with higher weights consistent with the given empirical edge weights. Furthermore, we show the efficiency of our streaming algorithm, which is orders of magnitude faster than the baseline.

Omitted proofs and the appendix can be found in the extended version [32].

2   Related Work

There are various studies on predicting the strength of ties given different features of a network, e.g., [10,44]. However, these works do not classify edges with respect to the STC. In contrast, our work is based on the STC property, which was introduced by Granovetter [12]. An extensive analysis of the STC can be found in the book of Easley and Kleinberg [6]. Recently, Sintos and Tsaparas [39] proposed an optimization problem by characterizing the edges of the network as strong or weak using only the structure of the network. They proved that the problem of maximizing the strong edges is NP-hard, although they provided two approximation algorithms to solve the dual problem of minimizing the weak edges. In the following works, the authors of [13,24,25] focused on restricted networks to further explore the complexity of STC maximization.

Rozenshtein et al. [38] discuss the STC with additional community connectivity constraints. Adriaens et al. [1] proposed integer linear programming formulations and corresponding relaxations. Very recently, Matakos et al. [29] proposed a new problem that uses the strong ties of the network to add new edges and increase its connectivity. The mentioned works only consider static networks and do not include edges weights in the computation of the STC. We propose a weighted variant and use it to infer ties strength in temporal networks.

Even though temporal networks are a quite recent research field, there are some comprehensive surveys introduce the notation, terminology, and applications [18,27]. Additionally, there are systematic studies into the complexity of well-known graph problems on temporal networks (e.g. [16,21]). The problem of finding communities and clusters, which can be considered as a related problem, has been studied on temporal networks [5,41]. Furthermore, Zhou et al. [45] studied dynamic network embedding based on a model of the triadic closure process, i.e., how open triads evolve into closed triads. Huang et al. [19] studied the formation of closed triads in dynamic networks. The authors of [2] introduce a probabilistic model for dynamic graphs based on the triadic closure.

Wei et al. [42] introduced a dynamic $(2 + \varepsilon)$-approximation for the minimum weight vertex cover problem with $\mathcal{O}(\log n/\varepsilon^2)$ amortized update time based on a vertex partitioning scheme [3]. However, the algorithm does not support updates of the vertex weights, which is an essential operation in our streaming algorithm.

3   Preliminaries

An *undirected* (static) *graph* $G = (V, E)$ consists of a finite set of nodes V and a finite set $E \subseteq \{\{u, v\} \subseteq V \mid u \neq v\}$ of undirected edges. We use $V(G)$ and $E(G)$ to denote the sets of nodes and edges, respectively, of G. The set $\delta(v) = \{e = \{v, w\} \mid e \in E(G)\}$ contains all edges incident to $v \in V(G)$, and we use $d(v) = |\delta(v)|$ to denote the degree of $v \in V$. An *edge-weighted* undirected graph $G = (V, E, w_E)$ is an undirected graph with additional weight function $w : E \to \mathbb{R}$. Analogously, we define a *vertex-weighted* undirected graph $G = (V, E, w_V)$ with a weight function for the vertices $w : V \to \mathbb{R}$. If the context is clear, we omit the subscript of the weight function.

A *wedge* is defined as a triplet of nodes $u, v, w \in V$ such that $\{\{u, v\}, \{v, w\}\} \subseteq E$ and $\{u, w\} \notin E$. We denote such a wedge by $(v, \{u, w\})$, and with $\mathcal{W}(G)$ the set of wedges in a graph G. Next, we define the *weighted wedge graph*. The non-weighted version is also known as the *Gallai* graph [8].

Definition 1. *Let $G = (V, E, w_E)$ be a edge-weighted graph. The* weighted wedge graph $W(G) = (U, H, w_V)$ *consists of the vertex set $U = \{n_{uv} \mid \{u, v\} \in E\}$, the edges set $H = \{\{n_{uv}, n_{vw}\} \mid (v, \{u, w\}) \in \mathcal{W}(G)\}$, and the vertex weight function $w_V(n_{uv}) = w_E(\{u, v\})$.*

Temporal Networks. A *temporal network* $\mathcal{G} = (V, \mathcal{E})$ consists of a finite set of nodes V, a possibly infinite set $\mathcal{E}$ of undirected *temporal edges* $e = (\{u, v\}, t)$ with u and v in V, $u \neq v$, and *availability time* (or *timestamp*) $t \in \mathbb{N}$. For ease of notation, we may denote temporal edges $(\{u, v\}, t)$ with (u, v, t). We use $t(e)$ to denote the availability time of e. We do not include a duration in the definition of temporal edges, but our approaches can easily be adapted for temporal edges with duration parameters. We define the underlying static, weighted, *aggregated* graph $A_\phi(\mathcal{G}) = (V, E, w)$ of a temporal network $\mathcal{G} = (V, \mathcal{E})$ with the edges set $E = \{\{u, v\} \mid (\{u, v\}, t) \in \mathcal{E}\}$ and edge weight function $w : E \rightarrow \mathbb{R}$. The edge weights are given by the function $\phi : 2^{\mathcal{E}} \rightarrow \mathbb{R}$ such that $w(e) = \phi(\mathcal{F}_e)$ with $\mathcal{F}_e = \{e \mid (e, t) \in \mathcal{E}\}$. We discuss various weighting functions in Sect. 5.1. Finally, we denote the lifetime of a temporal network $\mathcal{G} = (V, \mathcal{E})$ with $T(\mathcal{G}) = [t_{min}, t_{max}]$ with $t_{min} = \min\{t \mid e = (u, v, t) \in \mathcal{E}\}$ and $t_{max} = \max\{t \mid e = (u, v, t) \in \mathcal{E}\}$.

Strong Triadic Closure. Given a (static) graph $G = (V, E)$, we can assign one of the labels *weak* or *strong* to each edge in $e \in E$. We call such a labeling a *strong-weak labeling*, and we specify the labeling by a subset $S \subseteq E$. Each edge $e \in S$ is called *strong*, and $e \in E \setminus S$ *weak*. The *strong triadic closure (STC)* of a graph G is a strong-weak labeling $S \subseteq E$ such that for any two strong edges $\{u, v\} \in S$ and $\{v, w\} \in S$, there is a (weak or strong) edge $\{u, w\} \in E$. We say that such a labeling *fulfills* the strong triadic closure. In other words, in a strong triadic closure there is no pair of strong edges $\{u, v\}$ and $\{v, w\}$ such that $\{u, w\} \notin E$. Consequently, a labeling $S \subseteq E$ fulfills the STC if and only if at most one edge of any wedge in $\mathcal{W}(G)$ is in S, i.e., there is no wedge with two strong edges [39]. The decision problem for the STC is denoted by MaxSTC and is stated as follows: Given a graph $G = (V, E)$ and a non-negative integer k. Does there exist $S \subseteq E$ that fulfills the strong triadic closure and $|S| \geq k$?

See Fig. 1 in Appendix C.1 for examples of a temporal, aggregated, and wedge graph, and the STC.

4   Weighted Strong Triadic Closure

Let $G = (V, E, w)$ be a graph with edge weights reflecting the importance of the edges. We determine a *weighted* strong triadic closure that takes the weights of the edges by the importance given by w into account. To this end, let $S \subseteq E$ be a strong-weak labeling. The labeling S fulfills the weighted STC if (1) for any two strong edges $\{u, v\}, \{v, w\} \in S$ there is a (weak or strong) edge $\{u, w\} \in E$, i.e., fulfills the unweighted STC, and (2) maximizes $\sum_{e \in S} w(e)$.

The corresponding decision problem WEIGHTEDMAXSTC has as input a graph $G = (V, E)$ and $U \in \mathbb{R}$, and asks for the existence of a strong-weak labeling that fulfills the strong triadic closure and for which $\sum_{e \in S} w(e) \geq U$. Sintos and Tsaparas [39] showed that MaxSTC is NP-complete using a reduction from Maximum Clique. The reduction implies that we cannot approximate the MaxSTC with a factor better than $\mathcal{O}(n^{1-\epsilon})$. Because MaxSTC is a special case of WEIGHTEDMAXSTC, these negative results also hold for the latter.

Instead of maximizing the weight of strong edges, we can equivalently minimize the weight of weak edges resulting in the corresponding problem WEIGHTEDMINSTC[1]. Here, we search a strong-weak labeling that fulfills the STC and

---

[1] We use WEIGHTEDMINSTC for the decision and the optimization problem in the following if the context is clear.

minimizes the weight of the edges not in S. Both the maximation and the minimization problems can be solved exactly using integer linear programming (ILP). We provide the corresponding ILP formulations in Appendix A. The advantage of WEIGHTEDMINSTC is that we can obtain a 2-approximation.

To approximate WEIGHTEDMINSTC in an edge-weighted graph $G = (V, E, w)$, we first construct the weighted wedge graph $W(G) = (V_W, E_W, w_{V_W})$. Solving the minimum weighted vertex cover (MWVC) problem on $W(G)$ leads then to a solution for the minimum weighted STC of G, where MWVC is defined as follows. Given a vertex-weighted graph $G = (V, E, w)$, the minimum weighted vertex cover asks if there exists a subset of the vertices $C \subseteq V$ such that each edge $e \in E$ is incident to a vertex $v \in C$ and the sum $\sum_{v \in C} w(v)$ is minimal.

Lemma 1. *Solving the MWVC on $W(G)$ leads to a solution of the minimum weight STC on G.*

Proof. It is known that a (non-weighted) vertex cover $C \subseteq V(W)$ in $W(G)$ is in one-to-one correspondence to a (non-weighted) STC in G, see [39]. The idea is the following. Recall that the wedge graph $W(G)$ contains for each edge $\{i, j\} \in E(G)$ one vertex $n_{ij} \in V(W)$. Two vertices n_{uv}, n_{uw} in $W(G)$ are only adjacent if there exists a wedge $(u, \{v, w\}) \in \mathcal{W}(G)$. If we choose the weak edges $E \setminus S$ to be the edges $\{i, j\} \in E$ such that $n_{ij} \in C$ each wedge has at least one weak edge. Now for the weighted case, by definition, the weight of the STC $\sum_{e \in E \setminus S} w(e)$ equals the weight of a minimum vertex cover $\sum_{v \in C} w_{V_W}(v)$.   □

Lemma 1 implies that an approximation for MWVC yields an approximation for WEIGHTEDMINSTC. A well-known 2-approximation for the MWVC is the pricing method which we briefly describe in the following. The idea of the pricing algorithm is to assign to each edge $e \in E$ a price $p(e)$ initialized with zero. We say a vertex is *tight* if the sum of the prices of its incident edges equals the weight of the vertex. We iterate over the edges, and if for $e = \{u, v\}$ both u and v are not tight, we increase the price of $p(e)$ until at least one of u or v is tight. Finally, the vertex cover is the set of tight vertices. See, e.g., [22] for a detailed introduction. In Sect. 5.3, we generalize the pricing algorithm for fully dynamic updates of edge insertions and deletions, and vertex weight updates.

5 Strong Triadic Closure in Temporal Networks

We first present meaningful weighting functions to obtain an edge-weighted aggregated graph from the temporal network. Next, we discuss the approximation of the WEIGHTEDMINSTC in the non-streaming case. Finally, we introduce the 2-approximation streaming algorithm for temporal networks.

5.1 Weighting Functions for the Aggregated Graph

A key step in the computation of the STC for temporal networks is the aggregation and weighting of the temporal network to obtain a weighted static network.

Recall that the weighting of the aggregated graph $A_\phi(\mathcal{G})$ is determined by the weighting function $\phi : 2^{\mathcal{E}} \to \mathbb{R}$ such that $w(e) = \phi(\mathcal{F}_e)$ with $\mathcal{F}_e = \{e \mid (e,t) \in \mathcal{E}\}$. Naturally, the weighting function ϕ needs to be meaningful in terms of tie strength; hence, we propose the following variants of ϕ.

– *Contact frequency:* We set $\phi(\mathcal{F}_e) = |\mathcal{F}_e|$, i.e., the weight $w(e)$ of edge e in the aggregated graph equals the number of temporal edges between the endpoints of e. Contact frequency is a popular and common substitute for tie strength [10,12,28].
– *Exponential decay:* The authors of [28] proposed to measure tie strength in terms of the recency of contacts. We propose the following weighting variant to capture this property where $\phi(\mathcal{F}_e) = \sum_{i=1}^{|\mathcal{F}_e|-1} e^{-(t(e_{i+1})-t(e_i))}$ if $|\mathcal{F}_e| \geq 2$ and else $\phi(\mathcal{F}_e) = 0$. Here, we interpret $\mathcal{F}_e$ as a chronologically ordered sequence of the edges.
– *Duration:* Temporal networks can include durations as an additional parameter of the temporal edges, i.e., each temporal edge e has an assigned value $\lambda(e) \in \mathbb{N}$ that describes, e.g., the duration of a contact [18]. The duration is also commonly used as an indicator for tie strength [10]. We can define ϕ in terms of the duration, e.g., $\phi(\mathcal{F}_e) = \sum_{f \in \mathcal{F}_e} \lambda(f)$.

Other weighting functions are possible, e.g., combinations of the ones above or weighting functions that include node feature similarities.

5.2 Approximation of WeightedMinSTC

Before introducing our streaming algorithm, we discuss how to compute and approximate the WEIGHTEDMINSTC in a temporal network $\mathcal{G} = (V, \mathcal{E})$ in the non-streaming case. Consider the following algorithm:

1. Compute $A_\phi(\mathcal{G}) = (V, E, w)$ using an appropriate weighting function ϕ.
2. Compute the vertex-weighted wedge graph $W(A_\phi(\mathcal{G})) = (V_W, E_W, w_{V_W})$.
3. Compute an MWVC C on $W(A_\phi(\mathcal{G}))$.

The nodes n_{uv} in C then correspond to the weak ties $\{u, v\}$ in $\mathcal{G}$. Depending on how we solve step three, we can either compute an optimal or approximate solution, e.g., using the pricing approximation for the MWVC, we obtain a 2-approximation for WEIGHTEDMINSTC. Using the pricing approximation, we have linear running time in the number of edges in the wedge graph. The problem with this direct approach is its limited scalability. The reason is that the number of vertices in the wedge graph $|V_W| = |E(A_\phi(\mathcal{G}))|$ and the number of edges equals the number of wedges in A, which is bounded by $\mathcal{O}(|V|^3)$, see [36], leading to a total running time and space complexity of $\mathcal{O}(|V|^3)$.

5.3 Streaming Algorithm for WeightedMinSTC

In the previous section, we saw that the size of the wedge graph could render the direct approximation approach infeasible for large temporal networks. We use a

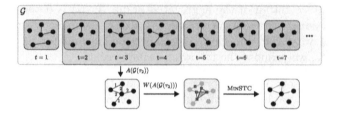

Fig. 2. Example for computing the weighted STC of a sliding time window.

sliding time window of size $\Delta \in \mathbb{N}$ to compute the changing STC for each time window to overcome this obstacle. The advantage is two-fold: (1) By considering limited time windows, the size of the wedge graphs for which we have to compute the MWVC is reduced because usually not all participants in a network have contact in the same time window. (2) If we consider temporal networks spanning a long (possibly infinite) time range, the relationships, and thus, tie strengths, between participants change over time. Using the sliding time window approach, we are able to capture such changes.

The following discussion assumes the weighting function ϕ to be linear in the contact frequency, and we omit the subscript. But, our results are general and can be applied to other weighting functions. Let τ be a time interval and let $A(\mathcal{G}(\tau))$ be the aggregated graph of $\mathcal{G}(\tau)$, i.e., the temporal network that only contains edges starting and arriving during the interval τ. For a time window size of $\Delta \in \mathbb{N}$, we define the sliding time window τ_t at timestamp t with $t \in [1, T(\mathcal{G}) - \Delta + 1]$ as $\tau_t = [t, t + \Delta - 1]$.

Figure 2 shows an example of our streaming approach for $\Delta = 3$. The first seven timestamps of temporal network $\mathcal{G}$ are shown as static slices. The time window τ_2 of size three starts at $t = 2$. First, the static graph $A(\mathcal{G}(\tau_2))$ is aggregated, and the wedge graph $W(A(\mathcal{G}(\tau_2)))$ is constructed. The wedge graph is used to compute the weighted STC. After this, the time window is moved one time stamp further, i.e., it starts at $t = 3$ and ends at $t = 5$, and the aggregation and STC computation are repeated (not shown in Fig. 2). In the following, we describe how the aggregated and wedge graphs are updated when the time window is moved forward, how the MWVC is updated for the changes of the wedge graph, and how the final streaming algorithm proceeds.

Updating the Aggregated and Wedge Graphs. Let τ_{t_1} and τ_{t_2} be to consecutive time windows, i.e., $t_2 = t_1 + 1$. Furthermore, let $A_i = A(\mathcal{G}(\tau_{t_i}))$ and $W_i = W(A(\mathcal{G}(\tau_{t_i})))$ with $i \in \{1, 2\}$ be the corresponding aggregated and wedge graphs. The sets of edges appearing in the time windows $\mathcal{G}(\tau_{t_1})$ and $\mathcal{G}(\tau_{t_2})$ might differ. For each temporal edge that is in $\mathcal{G}(\tau_{t_1})$ but not in $\mathcal{G}(\tau_{t_2})$, we reduce the weight of the corresponding edge in the aggregated graph A_1. If the weight reaches zero, we delete the edge from A_1. Analogously, for each temporal edge that is in $\mathcal{G}(\tau_{t_2})$ but not in $\mathcal{G}(\tau_{t_1})$, we increase the weight of the corresponding edge in A_1. If the edge is missing, we insert it. This way, we obtain A_2 from A_1 by a sequence of update operations. Now, we map these edge removals, additions, and edge weight changes between A_1 and A_2 to updates on W_1 to obtain W_2. For each edge removal (addition) $e = \{u, v\}$ between A_1 and A_2, we remove (add) the corresponding vertex (and incident edges) in W_1. We also have to add or remove edges in W_1 depending on newly created or removed wedges. More precisely, for every new wedge in A_1, we add an edge between the corresponding vertices in W_1, and for each removed wedge, i.e., by deleting an edge or creating a new triangle, in A_1, we remove the edges between the corresponding vertices in W_1. Furthermore, for each edge weight change between A_1 and A_2, we decrease (increase) the weight of the corresponding vertex in W_1. Hence, the wedge graph W_1 is edited by a sequence σ of vertex and edge insertions, vertex and edge removals, and weight changes to obtain W_2. Because we only need to insert or remove a vertex in the wedge graph W_1 if the degree changes between zero and a positive value, we do not consider vertex insertion and removal in W_1 as separate operations in the following. The number of vertices and edges in W_1 is bounded by the current numbers of edges and wedges in A_1. Furthermore, we bound the number of changes in W_1 after inserting or deleting edges from A_1.

Lemma 2. *The number of new edges in W_1 after inserting $\{v, w\}$ into A_1 is at most $d(v) + d(w)$, and the number of edges removed from W is at most $\min(d(v), d(w))$. The number of new edges in W_1 after removing $\{v, w\}$ from A_1 is at most $\min(d(v), d(w))$, and the number of edges removed from W_1 is at most $d(v) + d(w)$.*

Updating the MWVC. If the sliding time window moves forward, the current wedge graph W is updated by the sequence σ. We consider the updates occurring one at a time and maintain a 2-approximation of an MWVC in W. Algorithm 1 shows our dynamic pricing approximation based on the non-dynamic 2-approximation for the MWVC. The algorithm supports the needed operations of inserting and deleting edges, as well as increasing and decreasing vertex weights. When called for the first time, an empty vertex cover C and wedge graph W are initialized (line 1f.), which will be maintained and updated in subsequent calls of the algorithm. In the following, we show that our algorithm gives a 2-approximation of the MWVC after each of the update operations.

Definition 2. *We assign to each edge $e \in E(W)$ a price $p(e) \in \mathbb{R}$. We call prices fair, if $s(v) = \sum_{e \in \delta(v)} p(e) \leq w(v)$ for all $v \in V(W)$. And, we say a vertex $v \in V(W)$ is tight if $s(v) = w(v)$.*

Let W be the current wedge graph and σ a sequence of dynamic update requests, i.e., inserting or deleting edges and increasing or decreasing vertex weights in W. Algorithm 1 calls for each request $r \in \sigma$ a corresponding procedure to update W and the current vertex cover C (line 35f.). We show that after each processed request, the following invariant holds.

Invariant 1. *The prices are fair, i.e., $s(v) \leq w(v)$ for all vertices $v \in V(W)$, and $C \subseteq V(W)$ is a vertex cover.*

Lemma 3. *If Invariant 1 holds, after calling one of the procedures* INSEDGE, DELEDGE, DECWEIGHT, *or* INCWEIGHT *Invariant 1 still holds.*

Theorem 1. *Algorithm 1 maintains a vertex cover with $w(C) \leq 2w(OPT)$.*

Proof. Lemma 3 ensures that C is a vertex cover and after each dynamic update the prices are fair, i.e., $\sum_{e \in \delta(v)} p(e) \leq w(v)$. Furthermore, (1) for an optimal MWVC OPT and fair prices, it holds that $\sum_{e \in E(W)} p(e) \leq w(OPT)$, and (2) for the vertex cover C and the computed prices, it holds that $\frac{1}{2}w(C) \leq \sum_{e \in E(W)} p(e)$. Hence, the result follows. □

We now discuss the running times of the dynamic update procedures. For each of the four operations, the running time is in $\mathcal{O}(F)$, i.e., the size of the set for which we call the UPDATE procedure.

Theorem 2. *Let d_{max} be the largest degree of any vertex in $V(W)$. The running time of INSEDGE is in $\mathcal{O}(1)$, and DELEDGE is in $\mathcal{O}(d_{max})$. DECWEIGHT is in $\mathcal{O}(d_{max}^2)$, and INCWEIGHT is in $\mathcal{O}(d_{max})$.*

Algorithm 1: Dynamic Pricing Approximation

Input: Sequence σ of dynamic update requests
Output: Algorithm maintains a 2-approximation of MWVC C

1 Initialize and maintain vertex cover C
2 Initialize and maintain wedge graph W

3 **Procedure** UPDATE(*set of edges F*):
4   **foreach** $e = \{u, v\} \in F$ **do**
5   **if** u *or* v *is tight* **then**
6   continue
7   increase $p(e)$ until u or v tight
8   add newly tight vertices to C

9 **Procedure** DECWEIGHT(v, w_n):
10     $w(v) \leftarrow w_n$
11     $C \leftarrow C \setminus \{v\}$
12     $F' \leftarrow \delta(v)$
13   initialize $F = \emptyset$
14   **foreach** $e = \{v, x\} \in F'$ **do**
15         $p(e) \leftarrow 0$
16   **if** x *not tight* **then**
17             $C \leftarrow C \setminus \{x\}$
18             $F \leftarrow F \cup \{\{x, y\} \in E(G) \mid y$ is not tight $\} \cup \{e\}$
19   UPDATE(F)

20 **Procedure** DELEDGE($e_d = \{u, v\} \in E(W)$):
21     $E(W) \leftarrow E(W) \setminus \{e_d\}$
22   update $s(u)$ and $s(v)$
23     $C \leftarrow C \setminus e_d$
24     $F \leftarrow \{\{x, y\} \in E(W) \mid y \in e_d$ and x is not tight$\}$
25   UPDATE(F)

26 **Procedure** INSEDGE($e_n = \{u, v\}$):
27     $E(W) \leftarrow E(W) \cup \{e_n\}$
28   UPDATE($\{e_n\}$)

29 **Procedure** INCWEIGHT(v, w_n):
30     $w(v) \leftarrow w_n$
31   **if** $v \in C$ **then**
32         $C \leftarrow C \setminus \{v\}$
33         $F \leftarrow \{\{x, v\} \in E(W) \mid x$ is not tight$\}$
34   UPDATE(F)

35 **foreach** *update request* $r \in \sigma$ **do**
36   Call the to r corresponding funct. $f \in \{$INSEDGE, DELEDGE, INCWEIGHT, DECWEIGHT$\}$.

The Streaming Algorithm. Algorithm 2 shows the final streaming algorithm that expects as input a stream of chronologically ordered temporal edges and the time window size Δ. As long as edges are arriving, it iteratively updates the time windows and uses Algorithm 1 to compute the MINWEIGHTSTC approximation for the current time window τ_t with $t \in [1, T(\mathcal{G}) - \Delta]$. Algorithm 2 outputs the strong edges based on the computed vertex cover C_t in line 9. It skips lines 7–9 if there are no changes in E_τ.

Theorem 3. Let d_t^W (d_t^A) be the maximal degree in W $(A$, resp.$)$ after iteration t of the while loop in Algorithm 2. The running time of iteration t is in $\mathcal{O}(\xi \cdot d_t^A \cdot (d_t^W)^2)$, with $\xi = \max\{|E_t^-|, |E_t^+|\}$.

Algorithm 2: Streaming algorithm for the STC in temporal networks

---

Input: Stream of edges arriving in chronological order, $\Delta \in \mathbb{N}$
Output: 2-Approx. of MINWEIGHTSTC for each time window of size Δ

1 Initialize $t_s = 1$, $t_e = t_s + \Delta - 1$
2 Initialize empty list of edges E_τ and empty aggregated graph A
3 **while** *temporal edges are incoming* **do**
4   Update E_τ for time window $\tau_t = [t_s, t_e]$ such that $\forall e \in E_\tau$ it holds $t(e) \in \tau_t$
5   Let E_τ^- (E_τ^+) be the edges removed from (inserted to) E_τ
6   **if** $E_\tau^- \neq \emptyset$ *or* $E_\tau^+ \neq \emptyset$ **then**
7   Use E_τ^- and E_τ^+ to update A and to obtain the update sequence σ_t
8   Call Algorithm 1 with σ_t to obtain the the MWVC approximation C_t
9   Output $S_t = \{\{u, v\} \in E(A) \mid n_{u,v} \notin C_t\}$
10   Move time window forward by increasing t_s and t_e

---

6 Experiments

We compare the weighted and unweighted STC on real-world temporal networks and evaluate the efficiency of our streaming algorithm. We use the following algorithms for computing the weighted STC.

- ExactW is the weighted exact computation using the ILP (see Appendix A).
- Pricing uses the non-dynamic pricing approximation in the wedge graph.
- DynAppr is our dynamic streaming Algorithm 2.
- STCtime is a baseline streaming algorithm that recomputes the MWVC with the pricing method for each time window.

Moreover, we use the following algorithms for computing the non-weighted STC.

- ExactNw is the exact computation using an ILP (see [1]).
- Matching is the matching-based approximation of the unweighted vertex cover in the (non-weighted) wedge graph, see [39].
- HighDeg is a $\mathcal{O}(\log n)$ approximation by iteratively adding the highest degree vertex to the vertex cover, and removing all incident edges, see [39].

We implemented all algorithms in C++, using GNU CC Compiler 9.3.0 with the flag --O2 and Gurobi 9.5.0 with Python 3 for solving ILPs. All experiments ran on a workstation with an AMD EPYC 7402P 24-Core Processor with 3.35 GHz and 256 GB of RAM running Ubuntu 18.04.3 LTS, and with a time limit of twelve hours. Our source code is available at https://gitlab.com/tgpublic/tgstc.

Data Sets. Table 1 shows the statistics of the real-world data set used for our experiments. Note that for a wedge graph W of an aggregated graph A, $|V(W)| = |E(A)|$, and the number of edges $|E(W)|$ equals the number of wedges in A. For *Reddit* and *StackOverflow* the size of $|E(W)|$ and the number of triangles are estimated using vertex sampling from [43]. Further details of the data sets are available in Appendix B.

Table 1. Statistics of the data sets (*estimated).

Data set	Properties							
	$|V|$	$|\mathcal{E}|$	$|\mathcal{T}(\mathcal{G})|$	$|V(W)|$	$|E(W)|$	#Triangles	Domain	Ref.
Malawi	86	102 293	43 438	347	2 254	441	Human contact	[34]
Copresence	219	1 283 194	21 536	16 725	549 449	713 002	Human contact	[9]
Primary	242	125 773	3 100	8 317	337 504	103 760	Human contact	[40]
Enron	87 101	1 147 126	220 312	298 607	45 595 540	1 234 257	Communication	[23]
Yahoo	100 001	3 179 718	1 498 868	594 989	18 136 435	590 396	Communication	[37]
StackOverflow	2 601 977	63 497 050	41 484 769	28 183 518	*33 898 217 240	*110 670 755	Social network	[35]
Reddit	5 279 069	116 029 037	43 067 563	96 659 109	*86 758 743 921	*901 446 625	Social network	[15]

6.1   Comparing Weighted and Non-weighted STC

First, we count the number of strong edges and the mean edge weight of strong edges of the first five data sets. *StackOverflow* and *Reddit* are too large for the direct computation. We use the contact frequency as the weighting function for the aggregated networks. Table 2a shows the percentage of strong edges computed using the different algorithms. The exact computation for *Enron* and *Yahoo* could not be finished within the given time limit. For the remaining data sets, we observe for the exact solutions that the number of strong edges in the non-weighted case is higher than for the weighted case. This is expected, as for edge weights of at least one, the number of strong edges in the non-weighted STC is an upper bound for the number of strong edges in the weighted STC. However, when we look at the quality of the STC by considering how the strong edge weights compare to the empirical strength of the connections, we see the benefits of our new approach.

Table 2. Comparison of the weighted and non-weighted STC (OOT—out of time)

(a) Percentage of strong edges in aggregated graph.

Data set	Weighted		Non-weighted		
	ExactW	Pricing	ExactNw	Matching	HighDeg
Malawi	30.83	29.97	37.75	27.38	36.31
Copresence	31.12	21.37	37.95	29.20	35.31
Primary	27.17	21.94	27.83	18.99	27.35
Enron	OOT	2.75	OOT	3.28	4.61
Yahoo	OOT	9.86	OOT	9.98	14.29

(b) Mean edge weights.

Data set	Weighted				Non-weighted					
	ExactW		Pricing		ExactNw		Matching		HighDeg	
	Weak	Strong	Weak	Strong	Weak	Strong	Weak	Strong	Weak	Strong
Malawi	23.87	902.46	24.40	926.58	218.08	421.27	255.33	399.48	242.84	385.92
Copresence	20.30	78.32	46.13	189.31	27.22	56.56	58.85	120.07	57.13	112.63
Primary	2.73	20.48	6.58	45.50	3.34	18.49	9.32	39.88	6.19	38.84
Enron	OOT	OOT	3.69	9.33	OOT	OOT	3.77	6.01	3.76	5.50
Yahoo	OOT	OOT	4.37	14.23	OOT	OOT	4.78	10.42	4.60	9.84

An STC labeling with strong edges with high average weights and weak edges with low average weights is favorable. The mean weights of the strong and weak edges are shown in Table 2b. `Pricing` leads to the highest mean edge weight for strong edges in almost all data sets. The mean weight of the strong edges for the exact methods is always significantly higher for `ExactW` than `ExactNw`. The reason is that `ExactNw` does not consider the edge weights. Furthermore, it shows the effectiveness of our approach and indicates that the empirical a priori knowledge given by the edge weights is successfully captured by the weighted STC. To further verify this claim, we evaluated how many of the highest-weight edges are classified as strong. To this end, we computed the *precision* and *recall* for the top-100 weighted edges in the aggregated graph and the set of strong edges. Let H be the set of edges with the top-100 highest degrees. The precision is defined as $p = |H \cap S|/|S|$ and the recall as $r = |H \cap S|/|H|$. Figure 3 shows the results. Note that the y-axis of precision uses a logarithmic scale. The results show that the algorithms for the weighted STC lead to higher precision and recall values for all data sets.

6.2 Efficiency of the Streaming Algorithm

In order to evaluate our streaming algorithm, we measured the running times on the *Enron, Yahoo, StackOverflow,* and *Reddit* data sets with time window sizes Δ

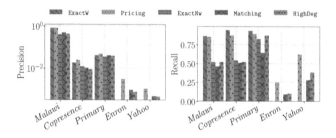

Fig. 3. Precision and recall for classifying the top-100 highest weighted edges in the aggregated graph as strong edges. The y-axis of precision is logarithmic.

Table 3. Running times in seconds of the streaming alg. (OOT—out of time).

Data set	$\Delta = 1$ hour		$\Delta = 1$ day		$\Delta = 1$ week	
	DynAppr	STCtime	DynAppr	STCtime	DynAppr	STCtime
Enron	264.74	**89.18**	**306.13**	1 606.09	**352.01**	20 870.77
Yahoo	**15.99**	767.40	**91.46**	OOT	**144.52**	OOT
StackOverflow	**170.38**	2 298.58	**971.22**	OOT	**16 461.53**	OOT
Reddit	**1 254.66**	13 244.84	**37 627.79**	OOT	OOT	OOT

of one hour, one day, and one week, respectively. Table 3 shows the results. In almost all cases, our streaming algorithm `DynAppr` beats the baseline `STCtime` with running times that are often orders of magnitudes faster. The reason is that `STCtime` uses the non-dynamic pricing approximation, which needs to consider all edges of the current wedge graph in each time window. Hence, the baseline is often not able to finish the computations within the given time limit, i.e., for seven of the twelve experiments, it runs out of time. The only case in which the baseline is faster than `DynAppr` is for the *Enron* data set and a time window size of one hour. Here, the computed wedge graphs of the time windows are, on average, very small (see Fig. 2 (a) in Appendix C.2), and the dynamic algorithm can not make up for its additional complexity due to calling Algorithm 1. However, we also see for *Enron* that for larger time windows, the running times of the baseline strongly increase, and for `DynAppr`, the increase is slight. In general, the number of vertices and edges in the wedge graphs increases with larger time window sizes Δ (see Fig. 2 in Appendix C.2). Hence, the running times increase for both algorithms with increasing Δ. In the case of *Reddit* and a time window size of one week, the sizes of the wedge graphs are too large to compute all solutions within the time limit, even for DYNAPPR.

7   Conclusion and Future Work

We generalized the STC to a weighted version to include a priori knowledge in the form of edge weights representing empirical tie strength. We applied our new STC variant to temporal networks and showed that we obtained meaningful results. Our main contribution is our 2-approximative streaming algorithm for the weighted STC in temporal networks. We empirically validated its efficiency in our evaluation. Furthermore, we introduced a fully dynamic 2-approximation of the MWVC problem with efficient update routines as part of our streaming algorithm. It might be of interest in itself.

As an extension of this work, discussion of further variants of the STC can be interesting. For example, [39] introduced variants with edge additions and multiple relationship types. Efficient streaming algorithms for weighted versions of these variants are planned as future work.

References

1. Adriaens, F., De Bie, T., Gionis, A., Lijffijt, J., Matakos, A., Rozenshtein, P.: Relaxing the strong triadic closure problem for edge strength inference. Data Min. Knowl. Disc. **34**, 1–41 (2020)
2. Ahmadian, S., Haddadan, S.: A theoretical analysis of graph evolution caused by triadic closure and algorithmic implications. In: International Conference on Big Data (Big Data), pp. 5–14. IEEE (2020)
3. Bhattacharya, S., Henzinger, M., Italiano, G.F.: Deterministic fully dynamic data structures for vertex cover and matching. J. Comput. **47**(3), 859–887 (2018)
4. Candia, J., González, M.C., Wang, P., Schoenharl, T., Madey, G., Barabási, A.L.: Uncovering individual and collective human dynamics from mobile phone records. J. Phys. A Math. Theor. **41**(22), 224015 (2008)
5. Chen, J., Molter, H., Sorge, M., Suchý, O.: Cluster editing in multi-layer and temporal graphs. In: International Symposium on Algorithms and Computation, ISAAC, LIPIcs, vol. 123, pp. 24:1–24:13. Schloss Dagstuhl-LZI (2018)
6. Easley, D.A., Kleinberg, J.M.: Networks, Crowds, and Markets - Reasoning About a Highly Connected World. Cambridge University Press, Cambridge (2010)
7. Eckmann, J.P., Moses, E., Sergi, D.: Entropy of dialogues creates coherent structures in e-mail traffic. Proc. Natl. Acad. Sci. **101**(40), 14333–14337 (2004)
8. Gallai, T.: Transitiv orientierbare graphen. Acta M. Hung. **18**(1–2), 25–66 (1967)
9. Génois, M., Barrat, A.: Can co-location be used as a proxy for face-to-face contacts? EPJ Data Sci. **7**(1), 11 (2018)
10. Gilbert, E., Karahalios, K.: Predicting tie strength with social media. In: Proceedings of the 27th International Conference on Human Factors in Computing Systems, CHI, pp. 211–220. ACM (2009)
11. Gilbert, E., Karahalios, K., Sandvig, C.: The network in the garden: an empirical analysis of social media in rural life. In: Proceedings of the SIGCHI Conference on Human Factors in Computing Systems, pp. 1603–1612 (2008)
12. Granovetter, M.S.: The strength of weak ties. A. J. Soc. **78**(6), 1360–1380 (1973)
13. Grüttemeier, N., Komusiewicz, C.: On the relation of strong triadic closure and cluster deletion. Algorithmica **82**, 853–880 (2020)
14. Hanneke, S., Xing, E.P.: Discrete temporal models of social networks. In: Airoldi, E., Blei, D.M., Fienberg, S.E., Goldenberg, A., Xing, E.P., Zheng, A.X. (eds.) ICML 2006. LNCS, vol. 4503, pp. 115–125. Springer, Heidelberg (2007). https://doi.org/10.1007/978-3-540-73133-7_9

15. Hessel, J., Tan, C., Lee, L.: Science, askscience, and badscience: on the coexistence of highly related communities. In: Proceedings of the International AAAI Conference on Web and Social Media, vol. 10, pp. 171–180 (2016)

16. Himmel, A., Molter, H., Niedermeier, R., Sorge, M.: Adapting the bron-kerbosch algorithm for enumerating maximal cliques in temporal graphs. Soc. Netw. Anal. Min. **7**(1), 35:1–35:16 (2017)

17. Holme, P., Edling, C.R., Liljeros, F.: Structure and time evolution of an internet dating community. Social Netw. **26**(2), 155–174 (2004)

18. Holme, P., Saramäki, J.: Temporal networks. Phys. Rep. **519**(3), 97–125 (2012)

19. Huang, H., Tang, J., Wu, S., Liu, L., Fu, X.: Mining triadic closure patterns in social networks. In: International Conference on World Wide Web, WWW, pp. 499–504. ACM (2014)

20. Kahanda, I., Neville, J.: Using transactional information to predict link strength in online social networks. In: Proceedings of the International AAAI Conference on Web and Social Media, vol. 3, pp. 74–81 (2009)

21. Kempe, D., Kleinberg, J.M., Kumar, A.: Connectivity and inference problems for temporal networks. J. Comput. Syst. Sci. **64**(4), 820–842 (2002)

22. Kleinberg, J., Tardos, E.: Algorithm Design. Pearson Education, Noida (2006)

23. Klimt, B., Yang, Y.: The enron corpus: a new dataset for email classification research. In: Boulicaut, J.-F., Esposito, F., Giannotti, F., Pedreschi, D. (eds.) ECML 2004. LNCS (LNAI), vol. 3201, pp. 217–226. Springer, Heidelberg (2004). https://doi.org/10.1007/978-3-540-30115-8_22

24. Konstantinidis, A.L., Nikolopoulos, S.D., Papadopoulos, C.: Strong triadic closure in cographs and graphs of low maximum degree. Theor. Comp. Sci. **740**, 76–84 (2018)

25. Konstantinidis, A.L., Papadopoulos, C.: Maximizing the strong triadic closure in split graphs and proper interval graphs. Disc. Appl. Math. **285**, 79–95 (2020)

26. Kossinets, G., Watts, D.J.: Empirical analysis of an evolving social network. Science **311**(5757), 88–90 (2006)

27. Latapy, M., Viard, T., Magnien, C.: Stream graphs and link streams for the modeling of interactions over time. Soc. Netw. Anal. Min. **8**(1), 61:1–61:29 (2018)

28. Lin, N., Dayton, P.W., Greenwald, P.: Analyzing the instrumental use of relations in the context of social structure. Sociol. Meth. Res. **7**(2), 149–166 (1978)

29. Matakos, A., Gionis, A.: Strengthening ties towards a highly-connected world. Data Min. Knowl. Disc. **36**(1), 448–476 (2022)

30. Mislove, A., Marcon, M., Gummadi, K.P., Druschel, P., Bhattacharjee, B.: Measurement and analysis of online social networks. In: Proceedings of the 7th ACM SIGCOMM Conference on Internet Measurement, pp. 29–42 (2007)

31. Moinet, A., Starnini, M., Pastor-Satorras, R.: Burstiness and aging in social temporal networks. Phys. Rev. Lett. **114**(10), 108701 (2015)

32. Oettershagen, L., Konstantinidis, A.L., Italiano, G.F.: Inferring tie strength in temporal networks (2022). https://arxiv.org/abs/2206.11705

33. Oettershagen, L., Kriege, N.M., Morris, C., Mutzel, P.: Classifying dissemination processes in temporal graphs. Big Data **8**(5), 363–378 (2020)

34. Ozella, L., et al.: Using wearable proximity sensors to characterize social contact patterns in a village of rural malawi. EPJ Data Sci. **10**(1), 46 (2021)

35. Paranjape, A., Benson, A.R., Leskovec, J.: Motifs in temporal networks. In: Proceedings of the tenth ACM International Conference on Web Search and Data Mining, pp. 601–610 (2017)

36. Pyatkin, A., Lykhovyd, E., Butenko, S.: The maximum number of induced open triangles in graphs of a given order. Optim. Lett. **13**(8), 1927–1935 (2019)

37. Rossi, R.A., Ahmed, N.K.: The network data repository with interactive graph analytics and visualization. In: AAAI (2015). https://networkrepository.com/
38. Rozenshtein, P., Tatti, N., Gionis, A.: Inferring the strength of social ties: a community-driven approach. In: Proceedings of the 23rd ACM SIGKDD International Conference on Knowledge Discovery and Data Mining, pp. 1017–1025. ACM (2017)
39. Sintos, S., Tsaparas, P.: Using strong triadic closure to characterize ties in social networks. In: Proceedings of the 20th ACM SIGKDD International Conference on Knowledge Discovery and Data Mining, pp. 1466–1475 (2014)
40. Stehlé, J., et al.: High-resolution measurements of face-to-face contact patterns in a primary school. PloS One **6**(8), e23176 (2011)
41. Tantipathananandh, C., Berger-Wolf, T.Y.: Finding communities in dynamic social networks. In: International Conference on Data Mining, ICDM, pp. 1236–1241. IEEE (2011)
42. Wei, H.T., Hon, W.K., Horn, P., Liao, C.S., Sadakane, K.: An O(1)-approximation algorithm for dynamic weighted vertex cover with soft capacity. In: Approximation, Random, and Combinatorial Optics Algorithm and Techniques (APPROX/RANDOM 2018). LIPIcs, vol. 116, pp. 27:1–27:14. Schloss Dagstuhl-LZI (2018)
43. Wu, B., Yi, K., Li, Z.: Counting triangles in large graphs by random sampling. IEEE Trans. Knowl. Data Eng. **28**(8), 2013–2026 (2016)
44. Xiang, R., Neville, J., Rogati, M.: Modeling relationship strength in online social networks. In: International Conference on World Wide Web, WWW, pp. 981–990. ACM (2010)
45. Zhou, L., Yang, Y., Ren, X., Wu, F., Zhuang, Y.: Dynamic network embedding by modeling triadic closure process. In: Proceedings of the Thirty-Second AAAI Conference on Artificial Intelligence, pp. 571–578. AAAI Press (2018)

Joint Learning of Hierarchical Community Structure and Node Representations: An Unsupervised Approach

Ancy Sarah Tom[1]([✉]), Nesreen K. Ahmed[2], and George Karypis[1]

[1] University of Minnesota, Twin Cities, MN, USA
{tomxx030,karypis}@umn.edu
[2] Intel Labs, Santa Clara, CA, USA
nesreen.k.ahmed@intel.com

Abstract. Graph representation learning has demonstrated improved performance in tasks such as link prediction and node classification across a range of domains. Research has shown that many natural graphs can be organized in hierarchical communities, leading to approaches that use these communities to improve the quality of node representations. However, these approaches do not take advantage of the learned representations to also improve the quality of the discovered communities and establish an iterative and joint optimization of representation learning and community discovery. In this work, we present *Mazi*, an algorithm that jointly learns the hierarchical community structure and the node representations of the graph in an unsupervised fashion. To account for the structure in the node representations, *Mazi* generates node representations at each level of the hierarchy, and utilizes them to influence the node representations of the original graph. Further, the communities at each level are discovered by simultaneously maximizing the modularity metric and minimizing the distance between the representations of a node and its community. Using multi-label node classification and link prediction tasks, we evaluate our method on a variety of synthetic and real-world graphs and demonstrate that *Mazi* outperforms other hierarchical and non-hierarchical methods.

Keywords: Networks · Network embedding · Unsupervised learning · Graph representation learning · Hierarchical clustering · Community detection

1 Introduction

Representation learning in graphs is an important field, demonstrating good performance in many tasks in diverse domains, such as social network analysis,

Supplementary Information The online version contains supplementary material available at https://doi.org/10.1007/978-3-031-26390-3_6.

M.-R. Amini et al. (Eds.): ECML PKDD 2022, LNAI 13714, pp. 86–103, 2023.
https://doi.org/10.1007/978-3-031-26390-3_6

user modeling and profiling, brain modeling, and anomaly detection [7]. Graphs arising in many domains are often characterized by a hierarchical community structure [13], where the communities (i.e., clusters) at the lower (finer) levels of the hierarchy are better connected than the communities at the higher (coarser) levels of the hierarchy. For instance, in a large company, the graph that captures the relations (edges) between the different employees (nodes) will tend to form communities at different levels of granularity. The communities at the lowest levels will be tightly connected corresponding to people that are part of the same team or project, whereas the communities at higher levels will be less connected corresponding to people that are part of the same product line or division.

In recent years, researchers have conjectured that when present, the hierarchical community structure of a graph can be used as an inductive bias in unsupervised node representation learning. This has led to various methods that learn node representations by taking into account a graph's hierarchical community structure. *HARP* [3] advances from the coarsest level to the finest level to learn the node representations of the graph at the coarser level, and then uses it as an initialization to learn the representations of the finer level graph. *LouvainNE* [1] uses a modularity-based [13] recursive decomposition approach to generate a hierarchy of communities. For each node, it then proceeds to generate representations for the different sub-communities that it belongs to. These representations are subsequently aggregated in a weighted fashion to form the final node representation, wherein the weights progressively decrease with coarser levels in the hierarchy. *SpaceNE* [11] constructs sub-spaces within the feature space to represent different levels of the hierarchical community structure, and learns node representations that preserves proximity between vertices as well as similarities within communities and across communities. Further, in recent times, certain GNN-based approaches [10,18] have also been proposed which exploit the hierarchical community structure while learning node representations. However, these methods use supervised learning and require more information to achieve good results.

Though all of the above methods are able to produce better representations by taking into account the hierarchical community structure, the information flow is unidirectional—from the hierarchical communities to the node representations. We postulate that the quality of the node representations can be improved if we allow information to also flow in the other direction—from the node representations to hierarchical communities—which can be used to improve the discovered hierarchical communities. Moreover, this allows for an iterative and joint optimization of both the hierarchical community structure and the representation of the nodes.

We present *Mazi*[1], an algorithm that performs a joint unsupervised learning of the hierarchical community structure of a graph and the representations of its nodes. The key difference between *Mazi* and prior methods is that the community structure and the node representations help improve each other. *Mazi* estimates node representations that are designed to encode both local information and information about the graph's hierarchical community structure. By taking into account local information, the estimated representations of nodes that

---

[1] Mazi is Greek for together.

are topologically close will be similar. By taking into account the hierarchical community structure, the estimated representations of nodes that belong to the same community will be similar and that similarity will progressively decrease for nodes that are together only in progressively coarser-level communities.

Mazi forms successively smaller graphs by coarsening the original graph using the hierarchical community structure such that the communities at different levels represent nodes in the coarsened graphs. Then, iterating over all levels, *Mazi* learns node representations at each level by maximizing the proximity of the representation of a node to that of its adjacent nodes while also drawing it closer to the representation of its community. Furthermore, at each level, *Mazi* learns the communities by taking advantage both of the graph topology and the node representations. This is done by simultaneously maximizing the *modularity* of the communities, maximizing the affinity among the representations of near-by nodes by using a Skip-gram [12] objective, and minimizing the distance between the representations that correspond to a node and its parent in the next-level coarser graph.

We evaluate *Mazi* on the node classification and the link prediction tasks on synthetic and real-world graphs. Our experiments demonstrate that *Mazi* achieves an average gain of 215.5% and 9.3% over competing approaches on the link prediction and node classification tasks, respectively. The contributions of our paper are the following:

1. We develop an unsupervised approach to simultaneously organize a graph into hierarchical communities and to learn node representations that account for that hierarchical community structure. We achieve this by introducing and jointly optimizing an objective function that contains (i) modularity- and skip-gram-based terms for each level of the hierarchy and (ii) inter-level node-representation consistency terms.
2. We present a flexible synthetic generator for graphs that contain hierarchically structured communities and community-derived node properties. We use this generator to study the effectiveness of different node representation learning algorithms.
3. We show that our method learns node representations that outperform competing approaches on synthetic and real-world datasets for the node classification and link prediction tasks.

2   Definitions and Notation

Let $G = (V, E)$ be an undirected graph where V is its set of n nodes and E is its set of m edges. Let $\mathbf{X} \in \mathbb{R}^{n \times d}$ store the representation vector x_i at the ith row for $v_i \in V$. A *community* refers to a group of nodes that are better connected with each other than with the rest of the nodes in the graph. A graph is said to have a *community structure*, if it can be decomposed into communities. In many natural graphs, communities often exist at different levels of granularity. At the upper (coarser) levels, there is a small number of large communities,

Table 1. Summary of notation.

Notation	Description
l	A level in the hierarchical structure
L	The number of levels in the hierarchical communities
G	The graph $G = (V, E, W)$, where V is the set of n nodes, E is the set of m edges, and W stores the edge weights
v_i	A vertex in G
$\deg(v_i)$	The degree of node v_i
X	The node representations of G
$\mathbb{C}$	A community decomposition of G
H	The community membership indicator vector of G
C_i	A community in $\mathbb{C}$
$\deg_{int}(C_i)$	The internal degree of community C_i, i.e., the number of edges that connect nodes in C_i to other nodes in C_i
$\deg_{ext}(C_i)$	The external degree of community C_i, i.e., the number of edges that connect C_i to nodes in other communities
$\deg(C_i)$	The overall degree of community C_i, i.e., the sum of $\deg_{int}(C_i)$ and $\deg_{ext}(C_i)$
ID	An array containing the vertex internal degrees
ED	An array containing the vertex external degrees
Q	The modularity of G for a given $\mathbb{C}$ (cf. Eq. 1)
G^l	The graph $G^l = (V^l, E^l, W^l)$ at level l
X^l	The node representations at level l
H^l	The community structure at level l
d	The dimension of X^l, where $l \in 1, \ldots, L$
ne^l	The number of epochs at level l
lr^l	The learning rate at level l
k	The context size extracted from walks
wl	The length of random-walk
r	The number of walks per node
α	The weight of the contribution of node neighborhood to the overall loss
β	The weight of the contribution of proximity to a node's community to the overall loss
γ	The weight of the contribution of Q to the overall loss

whereas at the lower (finer) levels, there is a large number of small communities. In general, the communities at the coarser levels are less well-connected than the finer level communities. When the communities at different levels of granularity form a hierarchy, that is, a community at a particular level is fully contained within a community at the next level up, then we will say that the graph has a *hierarchical community structure*.

Let $\mathbb{C} = \{C_0, \ldots, C_{k-1}\}$, with $V = \cup_i C_i$ and $C_i \cap C_j = \emptyset$ for $0 \leq i, j < k$ be a k-way *community decomposition* of G with C_i indicating its ith community. Let H be the *community membership* indicator vector where $0 \leq H[v_i] < k$ indicates v_i's community. Given a k-way community decomposition $\mathbb{C}$ of $G^l = (V^l, E^l)$, its *coarsened* graph $G^{l+1} = (V^{l+1}, E^{l+1})$ is obtained by creating k vertices—one for each community in $\mathbb{C}$—and adding an edge $(v_i, v_j) \in E^{l+1}$ if there are edges $(u_p, u_q) \in E^l$ such that $u_p \in C_i$ and $u_q \in C_j$. The weight of the (v_i, v_j) edge is set equal to the sum of the weights of all such (u_p, u_q) edges in E^l. In addition, each $v_i \in V^{l+1}$ is referred to as the parent node to all $u \in C_i$. Given $\mathbb{C}$, the *modularity* of G is defined as

$$Q = \frac{1}{2m} \left(\sum_{C_i \in \mathbb{C}} \left(\deg_{int}(C_i) - \frac{\deg(C_i)^2}{2m} \right) \right). \tag{1}$$

Q measures the difference between the actual number of edges within C_i and the expected number of edges within C_i, aggregated over all $C_i \in \mathbb{C}$. Q ranges from -0.5, when all the edges in G are between C_i and C_j, where $i \neq j$, and approaches 1.0 if all the edges are within any C_i and k is large.

Let the hierarchical community structure of G, with L levels, be represented by a sequence of successively coarsened graphs, denoted by $G, G^2, \cdots, G^L$, such that $|V| > |V^2| > \cdots > |V^L|$, wherein at each $l \in L$, the communities in G^l are collapsed to form the nodes in G^{l+1}. Every $v_i^l \in V^l$ is collapsed to a single parent node, v_j^{l+1}, in the next level coarser graph, G^{l+1}. Let us denote a model that takes the hierarchical community structure into account as hierarchical models and those that do not as flat models. Finally, we summarize all the notations in Table 1.

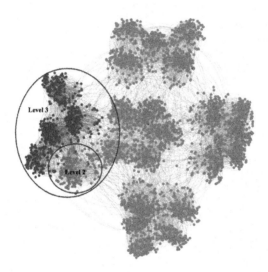

Fig. 1. A visualization of a synthetic 3K-node graph with a hierarchical community structure created by the proposed generator in Sect. 4. A common-ratio of 3.0 and a max. degree of 7.5 are used. A branching factor of 5 is used except at the finest level, which uses 30. Nodes in the hierarchical community structure are depicted using communities. A community in Level 3 and a sub-community in Level 2 are marked. A node's Level 1 community is itself.

3 *Mazi*

Given a graph G, *Mazi* seeks to jointly learn its node representations and its hierarchical community structure organized in L levels. *Mazi* coarsens the graphs at all levels of the hierarchy and learns representations for all nodes. At any given level, the node representation is learned such that it is similar to those of the nodes in its neighborhood, to its community and to the nodes it serves as a community to. This ensures the node representations at all levels align with the hierarchical community structure. Further, the communities at all levels are learned by utilizing node representations along with the graph topology. *Mazi* utilizes Skip-gram to model the similarity in the representations of a node and its neighbors. To model the similarity in the representations of node and its associated community, *Mazi* minimizes the distance between the respective two representations. Finally, to learn the communities, *Mazi* maximizes the *modularity* metric along with the above objectives.

Figure 1 illustrates a graph with a hierarchical community structure. From the figure, we see that the original graph (nodes of the graph is level 1 in the hierarchical structure) contains 5 large communities (level 3) in its coarsest level, each of which can be further split into 5 sub-communities (level 2). The nodes in the graph are marked such that the figure illustrates the level it belongs to in the hierarchical community structure. *Mazi* learns the representation of a node belonging to the level 2 community such that it will be similar to other nodes in that community over others. Furthermore, it will also be similar in representation to the nodes in its upper-level community at level 3, although this similarity value will be progressively lower as compared to that of the nodes in the level 2 community.

3.1 Objective Function

Mazi defines the objective function used for learning node representations using three major components. First, at each level, for each node, *Mazi* maximizes the proximity of its representation to the representation of the nodes belonging to its neighborhood using the Skip-gram objective. Second, iterating over all levels, the proximity of the representation of a node to that of its direct lineage in the embedding space is maximized. Third, the communities at each level are learned and refined by maximizing the *modularity* metric.

Modeling Node Proximity to its Neighborhood. As previously studied, see [6], to capture the neighbourhood of a node in the representations, we seek to maximize the log-likelihood of observing the neighbors of a node conditioned on its representation using the Skip-gram model with negative sampling. Utilizing the concept of sequence-based representations, neighboring nodes of a node v_i, represented by $N(v_i)$, are sampled to form its context. Let the negative sampling distribution of v_i be denoted by P_n and the number of negative samples considered for training the loss be denoted by R. We use L_{nbr_pos} and L_{nbr_neg} to denote the loss of v_i to its neighbors and to its negative samples, respectively. Using the above, we define

$$L_{nbr\_pos} = \frac{1}{|N(v_i)|} \sum_{v_j \in N(v_i)} \log \sigma(x_i^\top x_j), \tag{2a}$$

$$L_{nbr\_neg} = R \cdot E_{v_n \sim P_n(v_i)} \log(1 - \sigma(x_i^\top x_j)). \tag{2b}$$

Taken together, we model the neighbourhood proximity of v_i as:

$$L_{nbr} = L_{nbr\_pos} + L_{nbr\_neg}. \tag{3}$$

Modeling Node Proximity to its Community. In many domains, nodes belonging to a community tend to be functionally similar to each other in comparison to nodes lying outside the community [4]. As a consequence, we expect the representation of a node to be similar to the representation of its lineage in the hierarchy. Consider a level, l, in the hierarchical community structure of G. At l, for v_i^l, with representation x_i^l, we let the representation of its associated community (parent-node), $H^l(v_i^l)$, in the next level coarser graph, G^{l+1}, be denoted by $x_{H^l(v_i^l)}^{l+1}$. To model the relationship between v_i^l and $H^l(v_i^l)$, we use:

$$L_{comm}^l = \log \sigma\left(x_i^{l\top} x_{H^l(v_i^l)}^{l+1}\right). \tag{4}$$

As we iterate over the levels in the hierarchy of the graph, we bring together nodes in each level closer to its parent node in the next-level coarser graph in the embedding space. Consequently, the representation of a node is influenced by the communities the node belongs to at different levels.

Jointly Learning the Hierarchical Community Structure and Node Representations. Typically, community detection algorithms utilize the topological structure of a graph to discover communities. However, we may also take advantage of the information contained within the node representations while forming the communities at each level in the hierarchy. *Mazi* discovers the communities in the graph by jointly maximizing the *modularity* metric, described in Eq. 1, at each level and minimizing the distance between the representations of a node and its community in the next level coarser graph. The communities that we learn at each level, thus, better align with the structural and the functional components of the graph at that level. At each level in the hierarchical community structure, we use Eq. 3 and Eq. 4 to model and learn the node representations.

Consequently, putting all the components together, we get the following coupled objective function:

$$\max_\theta \sum_{l=1}^{L} \left(\frac{1}{|V^l|} \left(L_{nbr\_pos}^l + \alpha^l L_{nbr\_neg}^l + \beta^l L_{comm}^l \right) + \gamma^l Q^l \right), \tag{5}$$

$$\theta = x_i^l, H^l, \ i \in 1 \ldots |V|^l \ \forall l \in 1 \ldots L.$$

Since the order of the three terms that contribute to the overall objective is different, the terms are normalized with its respective order of contribution. Further, α^l, β^l and γ^l serve as regularization parameters and are added to Sub-Eqs. (2b), (4) and (1) in the overall objective for each level l, respectively.

3.2 Algorithm

An initial hierarchical community structure of the graph at level 1, denoted by $G^1 = (V^1, E^1, W^1)$, is constructed and node representations are computed for all the levels in the hierarchy. Then, using an alternating optimization approach in a level-by-level fashion, the objective, defined previously, is optimized. The optimization updates step through the levels from the finest level graph to the coarsest level graph and then from the coarsest level graph to the finest level graph in multiple iterations. This enables the node representations at each level to align itself to its direct lineage in the embedding space, additionally refining the community structure by the information contained within this space. An outline of the overall algorithm and its complexity can be found in Algorithm 1, Appendix A.1 and Appendix A.3, respectively, in the supplementary materials.

Initializing the Hierarchical Community Structure and Node Representations. A hierarchical community structure with L levels and their associated community membership vectors for G is initialized by successively employing existing community detection algorithms, such as *Metis* [9] at each level $l \in L$. The node representations at the finest level of the graph, denoted by X^1, are initialized by using existing representation learning methods such as *node2vec*, *DeepWalk* [6,14]. Node representations of coarser level graphs are then initialized by computing the average of the representations of nodes that belong to a community in the previous level finer graph, G^{l-1}.

Optimization Strategy. At each level, *Mazi* utilizes an alternating optimization (AO) approach to optimize its objective function. *Mazi* performs AO in a level-by-level fashion, by fixing variables belonging to all the levels except one, say denoted by l, and optimizing the variables associated with that level. At l, the community membership vector, H^l, is held fixed and the node representations, X^l, is updated. Then, X^l is fixed, and H^l is updated. Let us denote the node representation update as the X^l sub-problem, and the community membership update as the H^l sub-problem for further reference.

Node Representation Learning and Community Structure Refinement. At each level l, *Mazi* computes the gradient updates for the X^l sub-problem. By holding H^l fixed, *Mazi* updates x_i^l to be closer to the representation of $v_j \in N(v_i^l)$, x_j^l, and its parent node, $x_{H^l(v_i^l)}^{l+1}$ (see Eq. 3 and 4). The H^l sub-problem is then optimized using the updated X^l at l. To maximize the *modularity* objective (Q^l) in the H^l sub-problem, *Mazi* utilizes an efficient move-based approach. Consider reassigning v_i^l from its existing community C_a^l to a candidate destination community C_b^l. We note that Q^l, in Eq. 1, depends on $\deg_{int}(C_i^l)$ and $\deg_{ext}(C_i^l)$, where $C_i^l \in \mathbb{C}$. Instead of computing the contribution of each community to determine Q^l, we only modify the internal and the external degrees of C_a^l and C_b^l by computing how the contribution of v_i^l to C_a^l and C_b^l changes. Utilizing this, the new community of v_i^l is determined such that it maximizes Q^l and minimizes the distance between x_i^l and $x_{H^l(v_i)}^{l+1}$.

After alternatively solving for the sub-problems X^l and H^l at level l, *Mazi* optimizes level $l + 1$. These steps proceed up the hierarchy in this fashion until it reaches level $L - 1$. Starting at $L - 1$, the sub-problems X^{L-1} and H^{L-1} is optimized in the backward direction level-by-level using the updated representations, that is, $l = L - 1, L - 2, \ldots, 1$. By performing the optimization in the backward direction such as above, the node representations at the finer levels of the hierarchy are influenced by the updated representations at the coarser levels. After W such iterations, the refined node representations and community membership vectors for all levels are returned as the result of the algorithm.

4   Experiments

In order to evaluate the proposed algorithm, *Mazi*, we design experiments on real-world as well as synthetic graphs. We test *Mazi* on two major tasks: (i) link prediction and (ii) node classification. We compare *Mazi* against the following state-of-the-art baseline methods: (i) *node2vec* [6], a flat embedding model, (ii) *ComE* [2], a model respecting only a single-level in the hierarchy, (iii) *HARP* [3], (iv) *LouvainNE* [1], which are both hierarchical models, and (v) variations of the above mentioned models.

4.1   Experimental Setup

Link Prediction Task Setup. We divide the original graph into validation (sample 5% of the edges) and test ((sample 10% of the edges)) sets, and train graph. For each positive sample (existent edge in the graph), we sample 99 negative samples (non-existent edges). We use the train graph to generate node representations. Then, for every edge in the validation and test sets, we compute its prediction score using the representations of the edge's node pairs along with that of its corresponding negative samples and determine the mean average precision. Further, to test our algorithm on link prediction using learnable decoders, we implement the *DistMult* model [16] and a *2-layer multi-layer perceptron* (MLP). We provide the element-wise product of the representations of the nodes that comprise an edge as input to train the above models. We use 2% of the edges as the train set and 1% each for the validation and test set, with 20 negative samples for each positive edge, and report the average precision (AP) score of the test set for the best performing score on the validation set.

We run an elaborate search on the random-walk hyper-parameters. In *node2vec*, `context_size`, `walk_length`, `walks_per_node`, `p`, `q`, and #epochs select values between $\{2 - 5\}$, $\{4 - 10\}$, $\{5 - 60\}$, $\{0.1 - 10\}$, $\{0.1 - 10\}$, and, $\{1 - 4\}$, respectively. For *ComE* and *HARP*, the `context_size`, `walk_length`

Table 2. Real-world graph dataset statistics. Experiments are conducted on the induced subgraph formed by the nodes in the largest connected component in the graph. Label rate is the fraction of nodes in the training set. The #communities in each coarsened level is equal to $\sqrt{n}$, where, n is the current level graph's number of nodes. We stop coarsening when #communities ≤ 10. The last level is the all-encompassing node.

Dataset	#nodes	#edges	#labels	#communities in coarsened levels	Label rate
BlogCatalog	10312	667966	39	$\{100, 10, 1\}$	0.17
CS-CoAuthor	18333	163788	15	$\{135, 12, 1\}$	0.08
DBLP	20111	115016	4	$\{142, 12, 1\}$	0.49

and `walks_per_node` is chosen from $\{2-6\}$, $\{5-50\}$, and $\{5-30\}$, respectively. We choose parameters specific to *Mazi*, β and γ, from $\{0.25-2.5\}$ and $\{1.0-3.0\}$, respectively. We use the stochastic variation of *LouvainNE*, which is reported to obtain the best performance. We search all partitioning schemes of *LouvainNE*, used for generating the hierarchy, and use values $0.0001, 0.001, 0.01, 0.1, 1.0$ for the damping parameter. The #dimensions for all methods is 128.

Multi-label Classification Task Setup. We use a One-vs-Rest Logistic Regression model (implemented using LibLinear [5]) with L2 regularization. We split the nodes in a graph (real-world and synthetic datasets) into train, validation and test sets. We sample a fixed number of instances, s, from each class to form a representative train set. The validation and the test set is, thereafter, formed by almost equally splitting the remaining samples. In Table 2, we detail the exact fraction of nodes (label rate) in the real-world graphs that were used to the train the model. In case of synthetic datasets, we choose 45 samples per class, leading to a label rate of 0.6. We choose the regularizer weight from the range $\{0.1, 1.0, 10.0\}$, such that it gives the best average macro F1 score on the validation set for the different methods. To generate the best performing model of the approaches for evaluation, we conduct a search over the different hyper-parameters for the synthetic and the real-world graphs.

For the synthetic graphs, in *node2vec*, `context_size`, `walk_length`, `walks_per_node`, and #epochs are chosen from $\{5, 10, 15\}$, $\{10, 20, 30\}$, $\{10, 20, 30\}$, and $1, 2$, respectively. `p` and `q` are chosen from $\{0.25 - 4\}$. *ComE, HARP* and *Mazi* also use the above for its parameters. #*clusters* in *ComE* has been chosen from $\sqrt{\#nodes}$ and #*labels*. Additionally, specific to *Mazi*, we choose both β and γ from $\{0.0, 1.0, 2.0\}$. For the real world graphs, the `context_size`, `walk_length`, and `walks_per_node` parameters have been varied between $\{5, 10, 15\}$, $\{10, 20, 30, 40\}$, and $\{10, 20, 30, 40\}$. `p` and `q` are chosen from $\{0.25, 0.50, 1, 2, 4\}$. The #dimensions for all methods is 128.

Table 3. Link prediction on real-world graphs. Link prediction task performance of the methods is listed in the table. All *HARP* variants use *node2vec* as the base model. The mean average precision score is reported. The results are the average of 3 runs. The observed standard deviation was less than 0.01.

Method	Mean average precision		
	BlogCatalog	CS-CoAuth	DBLP
node2vec	0.534	0.797	0.914
HARP w. 2 lvls	0.532	0.755	0.881
HARP w. 3 lvls	0.460	0.732	0.874
HARP w. all lvls	0.126	0.647	0.769
LouvainNE	0.035	0.270	0.397
ComE	0.389	0.745	0.896
Mazi	**0.587**	**0.824**	**0.930**

4.2   Evaluation

Real World Datasets. We evaluate the proposed algorithm on three real world networks: *BlogCatalog*, *CS-CoAuthor*, and *DBLP*. *BlogCatalog* is a social network illustrating connections between bloggers while *CS-CoAuthor* and *DBLP* are co-authorship networks. Both *DBLP* and *CS-CoAuthor* exhibit high values of *modularity*, that is, 0.83 and 0.75, respectively, while *BlogCatalog* has a relatively lower *modularity* value of 0.23. More information about each dataset is detailed in Table 2.

Evaluation on the Link Prediction Task. *Mazi* demonstrates good performance over competing approaches as shown in Table 3 on the real-world datasets. We observe, in general, that *Mazi* demonstrates higher gains on datasets with higher *modularity* values. Over *node2vec*, the gains observed by *Mazi* in mean average precision (MAP) varies between 1.6% in *DBLP* to 10% in *BlogCatalog*. In comparison to *HARP*, referred to as *HARP w. all lvls* in Table 3, *Mazi* shows gains as high as 366% in *BlogCatalog*. To further study the behaviour of *HARP*, we evaluate its performance by restricting the total number of levels to 2 (*HARP w. 2 lvls*), and 3 (*HARP w. 3 lvls*). We note that both these approaches demonstrate higher MAP scores in comparison to *HARP*. We reason that since *HARP* collapses random edges and star-like structures to coarsen the graph in multiple levels, the coarsened graph in the last level may not be indicative of the global structure of the network and could serve as poor initializations. *Mazi* demonstrates gains between 3.79% and 50.89% against *ComE*. *ComE*, using gaussian mixtures to model its single-level communities, may not well capture the defining structural characteristics of the graph while generating representations. *LouvainNE*'s best performing version, as per the authors, uses random vectors for node representations at all levels in the graph's extracted hierarchy. Although LouvainNE may capture the hierarchical structure in a node's representation

Table 4. Link prediction using learnable decoders on *BlogCatalog*. We report average precision score on link prediction task of the methods using **learnable decoders** - *DistMult* and 2-*layer multi-layer perceptron.* σ is short for the sigmoid function.

Method σ	Using	DistMult	2-layer MLP
node2vec	0.62	0.62	0.59
HARP w. 2 lvls	0.62	0.62	0.62
HARP w. 3 lvls	0.05	0.56	0.57
HARP w. all lvls	0.05	0.32	0.43
LouvainNE	0.07	0.08	0.12
ComE	0.47	0.47	0.46
Mazi	**0.70**	**0.70**	**0.69**

by performing a weighted aggregation of vectors belonging to its hierarchy, we note that it may not well capture its local neighborhood. Thus, nodes that are in close proximity may not be represented similarly, and may indicate its low performance on the task. In Table 4, we report the average precision (AP) scores using learnable models, *DistMult* and a 2-*layer MLP*, on *BlogCatalog*. We note very similar trends as in Table 3 and observe that despite using learnable decoders, *Mazi* outperforms all other approaches in this task.

Evaluation on the Multi-label Node Classification Task. Table 5 reports micro and macro F1 score obtained by the methods on real-world datasets. We note that *Mazi* obtains a gain of up to 4.19% and 7.55% in macro F1 on *BlogCatalog* against *node2vec* and *HARP*, respectively. Against *LouvainNE*, *Mazi* achieves great gains on *BlogCatalog* (137%) and *CS-CoAuthor* (13%), respectively. While the gain obtained in *CS-CoAuthor* against *node2vec* and *HARP* is 0.43% and 0.85%, respectively, in macro F1, we observe that in *DBLP*, with a higher *modularity* value, the performance of *Mazi* is comparable with other approaches. *ComE* obtains a slightly better micro F1 score in *BlogCatalog*. Its choice of using gaussian mixtures to model community distributions appears to capture the weak community structure in *BlogCatalog* (*modularity* value of 0.23) well.

Synthetic Datasets. We design a novel synthetic graph generator that is capable of generating graphs with a hierarchical community structure and real-world structural properties. This is achieved by modeling the hierarchical community structure using a hierarchical tree (see Fig. 2). Each level in the hierarchical tree is a level in the hierarchical community structure, wherein the nodes of the tree forms the communities in the graph at that level. The nodes in the last level form the nodes of the generated graph. A node in the graph is generated such that, in expectation, it is able to form edges with other nodes in communities in the upper levels of the hierarchical community structure. For this, we accept a parameter, referred to as **common-ratio**, to generate L terms in geometric progression, for

Table 5. Multi-label node classification performance. Multi-label classification performance of the different methods are listed. The micro and macro F1 scores are reported. We report the scores achieved on the test set such that it achieves the best macro F1 score in the validation set chosen from the relevant hyper-parameters associated with each method. The results report the average of 3 runs. The standard deviation up to 2 decimal points is reported within the parentheses.

Method	BlogCatalog		CS-CoAuth		DBLP	
	Micro F1	Macro F1	Micro F1	Macro F1	Micro F1	Macro F1
node2vec	0.3718 (0.00)	0.2430 (0.00)	0.8670 (0.00)	0.8213 (0.00)	0.2499 (0.00)	0.2314 (0.00)
HARP (n2v)	0.3602 (0.00)	0.2418 (0.00)	0.8634 (0.00)	0.8153 (0.00)	0.2515 (0.00)	0.2326 (0.00)
LouvainNE	0.2275 (0.00)	0.1051 (0.00)	0.7790 (0.00)	0.7317 (0.00)	**0.2578 (0.01)**	**0.2367 (0.01)**
Mazi	0.3874 (0.00)	**0.2499 (0.00)**	**0.8708 (0.00)**	**0.8266 (0.00)**	0.2510 (0.00)	0.2317 (0.00)
ComE	**0.4016 (0.00)**	0.2464 (0.00)	0.8696 (0.00)	0.8238 (0.00)	0.2517 (0.00)	0.2323 (0.00)

each level in the hierarchical community structure (refer to Appendix A.2 for detailed descriptions). With these terms, we compute a probability distribution for a node to form an edge with another. Higher values of the `common-ratio` result in fewer edges between nodes belonging to different communities and thus, increases *modularity* of the graph as computed by the communities in the second last level. Further, we use a power distribution to model the graph's node degrees to capture the behavior of real-world networks. Other properties that we tune are the maximum degree, number of levels, branching factor of nodes, number of leaves, among others. To aid us in the node classification task, we generate labels for nodes such that they correlate with the hierarchical structure of the graph. We discuss further details of the proposed generator in the Appendix A.2.

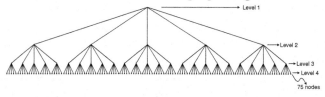

Fig. 2. The hierarchical tree structure used to generate our synthetic datasets. It has 4 levels. While the finest level uses a branching factor of 75, all other levels use 5.

In our experiments, we create a 5-level hierarchical tree with a branching factor of 5 in most levels. The branching factor in the level before the leaves is 75, thus, resulting in a total of 9375 nodes. We range `common-ratio` between $\{1.05, 1.2, 1.4, 1.6, 1.8, 2.0\}$. On average, the *modularity* of the graph for the corresponding `common-ratio` is $0.23, 0.28, 0.33, 0.37, 0.41, 0.44$. The power-law distribution parameter is 4.5 for the node degree. The maximum and the average degree of a node in the (directed) graphs we study are 187 and 33, respectively.

Evaluation on the Multi-label Node Classification Task. Figure 3 plots the micro and macro F1 scores obtained by the methods on the synthetic datasets on node

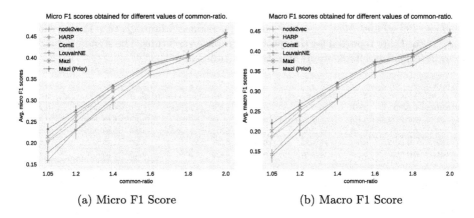

(a) Micro F1 Score   (b) Macro F1 Score

Fig. 3. Average micro and macro F1 scores on the synthetic graphs. Results are obtained over 3 runs with standard deviation. *HARP* method is built on the *node2vec* model, *Mazi (Prior)* uses the community structure generated by the hierarchical clustering tree, and *Mazi* uses the community structure generated by Metis. *Mazi (Metis)*, uses 4 levels in the hierarchy. The #communities in next coarser level is generated using $\sqrt{n}$, where, n is the number of nodes in the graph in the current level.

classification. The average gains observed in the macro F1 scores by *Mazi (Prior)* against *node2vec* range from over 50% to 5% for the `common-ratio` value of 1.05 to 2.0. As the *modularity* of the graph, as defined by the finest level community structure, decreases, the random-walks in *node2vec* will tend to stray outside the community. The labels are, however, distributed in accordance with the community structure, and thus, could indicate its lowered performance. *Mazi (Metis)* achieves similar performance as *Mazi (Prior)* against *node2vec*, ranging from 42% to 5% for `common-ratio` 1.05 to 2.0. Further, *Mazi (Prior)* and *Mazi (Metis)* both are able to demonstrate significant benefits in comparison to *HARP* for graphs with `common-ratio` ranging from 1.05 to 1.6. The average gain obtained by *Mazi (Prior)* and *Mazi (Metis)* are as high as 19% and 9.5%, respectively, for `common-ratio` 1.05. We reason that for the graphs whose *modularity*, as defined by the prior hierarchical community structure is low, the coarsening scheme of *HARP* may not be able to capture a fitting hierarchical community structure. Thus, the representations learnt on the coarsest level may not serve as good initializations for finer levels. Against *ComE*, we observe similar gains at about 20% with `commmon-ratio` 1.05 in the F1 scores. We also observe similar trends with *Mazi* over *LouvainNE*. For the lower values of common ratio, and thus, the *modularity*, we believe that *LouvainNE* may not be able to capture the local neighborhood of a node well.

4.3   Ablation Study

We study the effect of two important parameters, β and γ, in the performance of *Mazi*. We set $\beta = 0.0$ to fully ignore the contribution of the proximity between

Table 6. Ablation study on node classification. Macro F1 scores and %gain achieved by *Mazi (Prior)* and *Mazi* against respective versions without Q^l (Eq. 1) and without L^l_{comm} (Eq. 4) are reported for the synthetic graphs over 3 runs. The standard deviation up to 2 decimal points is reported in the parentheses.

CR	Mazi (Prior)					Mazi				
	$\gamma = 0.0$	%gain w/o Q^l	$\beta = 0.0$	%gain w/o L^l_{comm}	$\beta, \gamma \neq 0.0$	$\gamma = 0.0$	%gain w/o Q^l	$\beta = 0.0$	%gain w/o L^l_{comm}	$\beta, \gamma \neq 0.0$
1.2	**0.267**	-0.42 (0.61)	0.256	4.26 (0.56)	0.266	0.258	0.28 (0.89)	0.256	0.97 (0.91)	**0.259**
1.4	0.321	0.07 (0.10)	0.316	1.83 (1.12)	**0.321**	0.314	0.63 (0.04)	0.314	0.69 (0.08)	**0.316**
1.6	**0.374**	-0.10 (0.13)	0.369	1.14 (0.10)	0.373	0.371	0.03 (0.55)	0.369	0.49 (1.01)	**0.371**
1.8	0.394	0.09 (0.12)	0.387	1.77 (0.98)	**0.394**	0.388	0.27 (0.41)	0.387	0.39 (0.58)	**0.389**
2.0	0.444	0.00 (0.00)	0.437	1.48 (0.63)	**0.444**	0.439	0.20 (0.33)	0.438	0.35 (0.02)	**0.440**

Table 7. Ablation study on link prediction. Mean average precision and %gain, averaged over 3 runs, achieved by *Mazi* on link prediction over *Mazi* without Q^l (Eq. 1) and *Mazi* without L^l_{comm} (Eq. 4) is reported for the real-world graphs. The standard deviation up to 2 decimal points is reported in the parentheses.

	$\gamma = 0.0$	%gain w/o Q^l	$\beta = 0.0$	%gain w/o L^l_{comm}	$\beta, \gamma \neq 0.0$
BlogCatalog	0.586	0.15 (0.03)	0.564	4.09 (0.09)	**0.587**
CS_CoAuthor	0.823	0.03 (0.12)	0.821	0.29 (0.09)	**0.824**
DBLP	**0.930**	−0.02 (0.08)	0.929	0.08 (0.17)	0.930

the node and its community representations while optimizing the objective. We set $\gamma = 0.0$ to fix the hierarchical community structure to its initial value and optimize only the node representations. In node classification, a non-zero value of β plays a crucial role in ensuring *Mazi*'s good performance (see Table 6). Since the representations learned are benefited by the knowledge of a hierarchical community structure, performance achieved by $\beta = 0.0$ is consistently lower than when $\beta \neq 0.0$. The effect of γ is more apparent in *Mazi* using the *Metis* community structure. Since the hierarchical community structure generated using *Metis* does not fully conform to the prior community structure and the label distribution on the synthetic graphs correlate with the finest level community structure, we note that refining the hierarchical community structure and thereby, using it to improve the representations lead to better performance of the model.

We also report the effectiveness of β and γ in link prediction in Table 7. All the datasets achieve better performance when accounting for non-zero values of the β. This is especially evident in the *BlogCatalog* dataset, wherein *Mazi* shows a gain as high as 4.09%. Further, the community structure refinement in *BlogCatalog* and *CS_CoAuthor* leads to better performance when $\gamma \neq 0.0$, whereas in *DBLP*, the results obtained are comparable in both cases.

5   Related Work

Several methods model node representations using deep learning losses in supervised, semi-supervised and unsupervised settings. Amongst the unsupervised methods, the Skip-gram model is a popular approach used in the literature [6,14] to model the local neighborhood of a node using random walks while learning its representation. However, unlike our method, these representations are inherently flat and do not account for the hierarchical community structure that is present in the network.

Existing methods have also explored jointly learning communities at a single level and the representations of the nodes in the graph [2,15]. *ComE* [2] models the community and the node representations using a gaussian mixture formulation. *vGraph* [15] assumes each node to belong to multiple communities and a community to contain multiple nodes, and parametrizes the node-community distributions using the representations of the nodes and communities. Unlike these approaches, our approach utilizes the inductive bias introduced by the hierarchical community structure in the representations.

Recently, many unsupervised hierarchical representation learning methods, such as *HARP* [3] and *LouvainNE* [1], have been explored that leverage the multiple levels formed by hierarchical community structure in the graph. *HARP* uses an existing methods, such as *node2vec*, to generate node representations for graphs at coarser levels and initializes node representations at finer levels using these. *LouvainNE* recursively partitions each community in a graph to form sub-communities. The representations for a node in all the different sub-communities are generated and subsequently aggregated in a weighted fashion to form the final node representation. *SpaceNE* [11] represents the hierarchical community structure using sub-spaces in the feature space and learns node representations that preserves proximity between nodes as well as similarities within communities and across communities. However, all these approaches consider a static hierarchical community structure to influence the representations. In comparison, we jointly learn the node representations and the hierarchical community structure that is influenced by the node representations. In a parallel line, some GNN-based methods have been suggested to model the hierarchical structure present in the graphs while learning network representations [8,10,17,18]. While *DiffPool* [17] and *AttPool* [8] learn graph representations, *HC-GNN* [18] and *GXN* [10] target node representation learning. However, these are supervised methods and use task specific losses while considering static hierarchical community structures.

6 Conclusion

This paper develops a novel algorithm, *Mazi*, for joint unsupervised learning of a given graph's node representations and hierarchical community structure. At each level in the hierarchy, *Mazi* coarsens the graph and learns its node representations and leverages them to discover communities in the hierarchy. In turn, *Mazi* uses the hierarchy to learn the representations. Experiments conducted on synthetic and real-world graph datasets in the node classification and link prediction demonstrate the competitive performance of *Mazi* compared to competing approaches.

Acknowledgements. This work was supported in part by NSF (1447788, 1704074, 1757916, 1834251), Army Research Office (W911NF1810344), Intel Corp, and the Digital Technology Center (DTC) at the University of Minnesota. Many thanks to the reviewers for their helpful inputs. Also, many thanks to Saurav Manchanda for his helpful discussions. Access to research facilities was provided by the DTC & the Minnesota Supercomputing Institute.

References

1. Bhowmick, A.K., Meneni, K., Danisch, M., Guillaume, J.L., Mitra, B.: Louvainne: hierarchical louvain method for high quality and scalable network embedding. In: Proceedings of 13th International Conference on Web Search and Data Mining (2020)
2. Cavallari, S., Zheng, V.W., Cai, H., Chang, K.C.C., Cambria, E.: Learning community embedding with community detection & node embedding on graphs. In: Proceedings of 2017 ACM on CIKM (2017)
3. Chen, H., Perozzi, B., Hu, Y., Skiena, S.: Harp: hierarchical representation learning for networks. In: Thirty-Second AAAI Conference on Artificial Intelligence (2018)
4. Clauset, A., Moore, C., Newman, M.E.: Hierarchical structure and the prediction of missing links in networks. Nature **453**(7191), 98–101 (2008)
5. Fan, R.E., Chang, K.W., Hsieh, C.J., Wang, X.R., Lin, C.J.: Liblinear: a library for large linear classification. J. Mach. Learn. Res. **9**, 1871–1874 (2008)
6. Grover, A., Leskovec, J.: node2vec: scalable feature learning for networks. In: Proceedings of the 22nd ACM SIGKDD Conference on KDD, pp. 855–864 (2016)
7. Hamilton, W.L., Ying, R., Leskovec, J.: Representation learning on graphs: methods and applications. arXiv:1709.05584 (2017)
8. Huang, J., Li, Z., Li, N., Liu, S., Li, G.: Attpool: towards hierarchical feature representation in graph convolutional networks via attention mechanism. In: Proceedings of the IEEE ICCV, pp. 6480–6489 (2019)
9. Karypis, G., Kumar, V.: Multilevel graph partitioning schemes. In: ICPP, no. 3 (1995)
10. Li, M., Chen, S., Zhang, Y., Tsang, I.W.: Graph cross networks with vertex infomax pooling. arXiv preprint arXiv:2010.01804 (2020)
11. Long, Q., Wang, Y., Du, L., Song, G., Jin, Y., Lin, W.: Hierarchical community structure preserving network embedding: subspace approach. In: 28th ACM CIKM (2019)
12. Mikolov, T., Chen, K., Corrado, G., Dean, J.: Efficient estimation of word representations in vector space. arXiv:1301.3781 (2013)
13. Newman, M.E.: Modularity and community structure in networks. Proc. Natl. Acad. Sci. **103**(23), 8577–8582 (2006)
14. Perozzi, B., Al-Rfou, R., Skiena, S.: Deepwalk: online learning of social representations. In: Proceedings of the 20th ACM SIGKDD International Conference on Knowledge Discovery and Data Mining (2014)
15. Sun, F.Y., Qu, M., Hoffmann, J., Huang, C.W., Tang, J.: vgraph: a generative model for joint community detection and node representation learning. arXiv preprint arXiv:1906.07159 (2019)
16. Yang, B., Yih, W.T., He, X., Gao, J., Deng, L.: Embedding entities and relations for learning and inference in knowledge bases. arXiv preprint arXiv:1412.6575 (2014)

17. Ying, Z., You, J., Morris, C., Ren, X., Hamilton, W., Leskovec, J.: Hierarchical graph representation learning with differentiable pooling. In: Advances in Neural Information Processing Systems, pp. 4800–4810 (2018)
18. Zhong, Z., Li, C.T., Pang, J.: Hierarchical message-passing graph neural networks. arXiv preprint arXiv:2009.03717 (2020)

Knowledge Graphs

ProcK: Machine Learning
for Knowledge-Intensive Processes

Tobias Jacobs[1]([⊠])[ID], Jingyi Yu[1,2][ID], Julia Gastinger[1][ID], and Timo Sztyler[1][ID]

[1] NEC Laboratories Europe GmbH, Heidelberg, Germany
{tobias.jacobs,julia.gastinger,timo.sztyler}@neclab.eu
[2] Faculty of Electrical Engineering and Information Technology, RWTH Aachen,
Aachen, Germany
jingyi.yu@rwth-aachen.de

Abstract. We present a novel methodology to build powerful predictive process models. Our method, denoted *ProcK (**Proc**ess & **K**nowledge)*, relies not only on sequential input data in the form of event logs, but can learn to use a knowledge graph to incorporate information about the attribute values of the events and their mutual relationships. The idea is realized by mapping event attributes to nodes of a knowledge graph and training a sequence model alongside a graph neural network in an end-to-end fashion. This hybrid approach enhances the flexibility and applicability of predictive process monitoring, as both the static and dynamic information residing in the databases of organizations can be taken as input data. We demonstrate the potential of ProcK by applying it to a number of predictive process monitoring tasks, including tasks with knowledge graphs available as well as an existing process monitoring benchmark where no such graph is given. The experiments provide evidence that our methodology achieves state-of-the-art performance and improves predictive power when a knowledge graph is available.

Keywords: Predictive process management · Neural networks

1 Introduction

We introduce *ProcK (**Proc**ess & **K**nowledge)*, a pipeline for predictive process monitoring. ProcK combines the usage of two complementary data representations in a novel way.

Predictive process monitoring deals with the task of forecasting properties of business processes that are currently under execution. This includes the type and occurrence time of future events as well as the process outcome. The primary input to predictive process models are logs recorded during process execution, and it is best practice in the process mining community to model them as sets of discrete *events*. Each event is characterized by its case identifier, activity type, timestamp, and potentially further data; see Van Der Aalst *et al.* (2012).

Machine learning methods for predictive process monitoring described in literature can be separated into two main approaches. The more traditional approach is to hand-engineer a set of feature extraction functions that operate on

© The Author(s), under exclusive license to Springer Nature Switzerland AG 2023
M.-R. Amini et al. (Eds.): ECML PKDD 2022, LNAI 13714, pp. 107–121, 2023.
https://doi.org/10.1007/978-3-031-26390-3_7

top of event sequences. The set of features, after some further pre-processing, is then used as input to train a machine learning model like e.g. an SVM (Leontjeva *et al.* 2016). The second approach relies on deep learning, feeding the raw event log directly into a deep neural network which builds meaningful features automatically during training. Because of the sequential nature of the input data, it is a natural choice to apply a recurrent neural network, like done by Tax *et al.* (2017). More recently, feedforward networks have been demonstrated by Mauro *et al.* (2019) and other authors to achieve superior performance in many cases.

Our work follows the deep learning paradigm, and, to the best of our knowledge, we are the first to complement the event sequence data model with an additional representation of the available input data as a knowledge graph. This idea is rooted in the first fundamental step of typical practical process mining projects: to extract the event log from the data lake of an organization (see Reinkemeyer 2020), which is often structured in the form of one or more relational databases. When selecting data with the goal of building a high-quality prediction model, the limitations of the event sequence view become apparent: only a subset of the relevant data can be naturally expressed in the form of events with case identifier and timestamp.

Example 1. For predicting the success of a loan repayment process at the time when the first rate has been paid, it might be relevant to take into account the bank account from which the rate was transferred. Specifically, the economic stability of the bank account's country might be an indicator.

In Example 1, the relevant piece of information (economic stability) is not an event, and it is neither a primary attribute of the event. It is rather an indirect attribute of the transfer event, which has to be derived via a specific semantic path (transfer → bank → country → economic stability). Domain experts could re-define such derived attributes by hand as primary event information, but this counteracts the benefit of deep learning to be applicable on top of the raw data.

To address that issue, ProcK takes a knowledge graph as additional input. It stacks a sequence model for events on top of a graph neural network in order to compute meaningful event representations. In Example 1, information about the *economic stability* can be propagated backwards across the path *economic stability ← country ← bank ← transfer*, where *economic stability, bank, country* are knowledge graph nodes and *transfer* is a time-stamped event of a particular process. Having been integrated into the representation of *transfer*, the information is then further processed by the event sequence model in order to predict the success probability of the repayment process.

The contributions of this work are summarized as follows: (1) We present the conceptual architecture of ProcK, which combines the usage of two complementary data representations in a novel way. (2) We describe an implementation of ProcK based on deep learning models for graph-structured and sequential data. (3) We document an experimental study based on four datasets, three of them including a knowledge graph, from different application domains. Our experiments demonstrate that ProcK achieves state-of-the-art predictive performance, which is further improved by utilizing the additional knowledge graph input.

2   Related Work

2.1   Predictive Process Monitoring

Our work presents a new approach for *predictive process monitoring*, the task to predict future properties of processes from their execution logs. A considerable range of machine learning techniques have been studied in the context of process predictions. A review of seven methods that fall into the category of traditional machine learning (decision trees, random forests, support-vector machines, boosted regression, all with heavy feature engineering) has been published by Teinemaa *et al.* (2019).

Considering deep learning methods, due to the sequential nature of process logs, it is a straightforward approach to apply models designed for sequences. Tax *et al.* (2017) study the usage of LSTM neural networks for various prediction tasks, including the next activities and the remaining process time. Further approaches based on RNN and LSTM networks have been presented by Evermann *et al.* (2017), Tello-Leal *et al.* (2018), Camargo *et al.* (2019), Lin *et al.* (2019).

More recently, it has been demonstrated that feedforward networks often outperform recurrent neural networks for predictive process monitoring tasks. Al-Jebrni *et al.* (2018) employ 1D-convolutional networks, Mauro *et al.* (2019) study stacked inception CNNs, and Pasquadibisceglie *et al.* (2019) present a method where traces and their prefixes are first mapped onto a 2D image-like structure and then 2-dimensional CNNs are applied. Finally, Taymouri *et al.* (2020) present a prediction approach leveraging generative adversarial networks.

Although the main idea presented and evaluated in this work is independent from the particular choice of the sequence processing model, we share the experience of Al-Jebrni *et al.* (2018) and other authors that feedforward networks are more reliable to achieve good results in the process monitoring domain.

A direction that has been followed by several researchers is to make use of explicit process models. More than two decades of research on process mining has yielded sophisticated algorithms to create graph-shaped models of processes from event log data, often in the form of Petri nets (Van Der Aalst *et al.* 2012). From the viewpoint of such a process model, events trigger state changes of process instances, and predictive process monitoring models can take the current process state as input. Van der Aalst *et al.* (2011) propose a solution to the problem of predicting the completion time, simply by calculating the mean remaining time for each state. Prediction from a combination of partial process models and event annotations has been performed by Ceci *et al.* (2014). Folino *et al.* (2014) combine some of the aforementioned ideas, first clustering events and processes to achieve more abstract process models and then applying cluster-specific prediction models. Recently, Theis and Darabi (2019) have presented a methodology to annotate states with time and other information, and use this information as input to a deep learning model.

What makes the idea of using a graphical process model powerful is that it builds upon a mature topic in the process mining domain, where the graph can be constructed on top of the existing event log. Our work, in contrast, is built on the hypothesis that, in concrete applications, *additional* non-sequential data is available, and that data is naturally modeled as a graph. In other words, process modeling is a specific type of feature pre-processing, while our approach taps a previously unused resource of data.

While there is a large body of work on predictive process mining, the task of *prescriptive* process mining is less well-studied. In a very recent work, Fahrenkrog- Petersen *et al.* (2022) propose a framework to prevent the undesired process outcomes based on formalized notions of alarms and interventions.

2.2   Combined Sequence and Graph Models

The machine learning model we employ in our work uses a combination of graph-structured and sequential data as input. To our knowledge, we are the first to apply this approach to the domain of predictive process monitoring. Nevertheless, the benefits of combining such complementary views have been demonstrated in other domains. Fang *et al.* (2018) construct a graph based on spatial distances between mobile cells. Their model for cellular demand prediction uses an LSTM to compute a feature vector for each cell based on its demand history, then a graph convolutional network is used to model influences between nearby cells. Hu *et al.* (2019) address the problem of predicting the freezing-of-gaits symptom of Parkinson discase patients from video segments, where the first layer of their model is used to compute a representation of anatomic joints and their interactions as a graph. The authors introduce specialized LSTM cells to model both time-based interactions between subsequent video segments and interactions between joints. Wang *et al.* (2022) combine the graph-based representation of molecules with the SMILES string representation. Two individual models are trained for those two input representations, and the models are combined using ensemble techniques. An application in the retail domain has been proposed by Chang *et al.* (2021). Here the first step is to convert the sequence of past user interactions with items into a graph with nodes representing items, and graph-based models are then applied to predict the interest of users.

While a variety of approaches to combine graphical and sequential input have been proposed in the body of work mentioned above, our approach has some unique and novel features. Firstly, each of the above methods performs some sort of node or graph classification which is only assisted by the sequential input. In our work, event sequences constitute the primary input, and the knowledge graph is used to assist the model to interpret the event data. Secondly and most importantly, the knowledge graph is used in our work to capture overall knowledge about the domain and application context, while information about each instance of the prediction problem is represented by its event sequence.

Another direction of related work is machine learning for dynamic knowledge graphs, called temporal knowledge graphs, where nodes, attributes, or relations change over time. The area of temporal knowledge graph reasoning can be divided into the *interpolation* and the *extrapolation* setting. As described by Jin *et al.* (2020), in the interpolation setting, new facts are predicted for time steps up until the current time step, taking into account time information from past and current time steps; see García-Durán *et al.* (2018). The methods in the extrapolation setting predict facts for future time steps. Recent work in the extrapolation setting includes RE-NET by Jin *et al.* (2020) and CluSTeR by Li *et al.* (2021).

The predictive process management application in our work requires to treat events as independent input. Modeling each event as a knowledge graph element would be technically possible, but only at the price of scalability, as typical applications include tens or hundreds of millions of events. Thus, in our work events do not have a direct interpretation as nodes or edges, but they instead contain attributes in the form of references to graph nodes. Conversely, triples in our knowledge graphs are in general not necessarily interpretable as events.

To summarize the discussion of our work in light of state-of-the-art, we are the first to utilize a knowledge graph as additional input to predictive process monitoring models to help interpreting the event data. The idea is realized by a new type of neural network architecture which takes events as primary input and learns to utilize an additional knowledge graph to interpret the event data.

3   Preliminaries

Following the process mining terminology, an *event log* $\mathcal{L} = (L, C, T, A)$ consists of an event set L, a set C of cases, a set T of possible event types, and a set A of additional event attributes. Each event $\ell = (c_\ell, t_\ell, \tau_\ell, \alpha_\ell) \in L$ is a 4-tuple characterized by its case identifier $c_\ell \in C$, its event type $t_\ell \in T$, a timestamp $\tau_\ell \in \mathbb{N}$, and a partial assignment function $\alpha_\ell : A \to V$ which specifies the values of a subset of attributes. For each case $c \in C$ we define $L_c := \{\ell \in L \mid c_\ell = c\}$ as the subset of events belonging to case c.

The domain V of possible attribute values is arbitrary in general; in this work it will be assumed that V is the node set of a knowledge graph. This assumption represents only a mild limitation, because categorical attribute values that do not appear in the given knowledge graph can simply be interpreted as isolated nodes. In fact, our experimental study includes one dataset where no knowledge graph is given at all. Extending ProcK with the ability to incorporate numerical attributes *directly* (i.e. without discretizing them to categorical attributes) remains for future work.

A knowledge graph (also called *knowledge base*) $\mathcal{G} = (V, R, E)$ is a directed graph defined by the node set V, relation types R, and edges E, where each edge $(v, r, v') \in E$ is a triple containing the head node $v \in V$, relation type $r \in R$, and tail node $v' \in V$.

As mentioned above, in typical practical applications, the input data will originate from the databases of the organization that performs predictive process monitoring. Relational databases are likely to contain time-stamped records that can become events. At the same time, mutual references between different tables are a fundamental element of relational data models, which makes it straightforward to interpret a part of the database as a knowledge graph. For the purpose of our experimental study we have developed tools to extract both event data and a knowledge graph from a database dump.

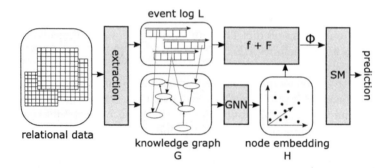

Fig. 1. ProcK conceptual architecture. From relational or tabular data, an event log and a knowledge graph is extracted. Then, the four components GNN, F, f, and SM of the neural network model subsequently compute node embedding H, event embeddings Φ, and the final prediction.

4   ProcK Architecture

We first specify, in Sect. 4.1, the conceptual architecture which is composed of four functional components GNN, f, F, and SM. Then, in Sect. 4.2, our implementation of each of the components is described.

4.1   Conceptual Architecture

The architecture of ProcK, depicted in Fig. 1, combines a graph neural network with a model for sequential data. From the bottom to the top of the network, input elements like nodes, edges, timestamps, and events will be encoded by embedding vectors. We employ a fixed embedding dimensionality $d \in \mathbb{N}$ across all layers.

The first component of the ProcK model is a graph neural network GNN which computes embedding vectors $H = (h_v)_{v \in V}$ containing an embedding $h_v \in \mathbb{R}^d$ for every node v of the knowledge graph. Formally,

$$h_v := \text{GNN}(\mathcal{G}, v), v \in V, \tag{1}$$

where $\mathcal{G} = (V, R, E)$ is the given knowledge graph.

Next, consider a single event $\ell = (c_\ell, t_\ell, \tau_\ell, \alpha_\ell)$ from the event log $\mathcal{L}$. Let F be an aggregation function for sets of d-dimensional vectors. The first step of constructing the event embedding is to compute

$$\beta_\ell := F(\{h_v \mid v \in \alpha_\ell(A)\}) \in \mathbb{R}^d. \tag{2}$$

Recall from the previous section that α_ℓ is a partial assignment function which specifies values for a subset of the attributes from A. In the above equation, $a_\ell(A)$ is the set of those attribute values.

Further, we employ a timestamp embedding function $f : \mathbb{N} \to \mathbb{R}^d$ and compute the final event embedding as

$$\phi_\ell := \beta_\ell + f(\tau_\ell). \tag{3}$$

Having computed the sequence representation as a series of d-dimensional vectors $\Phi = (\phi_\ell)_{\ell \in L_c}$ for the given case $c \in C$, we feed the event representations into a sequence model SM to compute the prediction for the given case:

$$P_c := \mathrm{SM}(\{\phi_\ell \mid \ell \in L_c\}). \tag{4}$$

The structure of P_c depends on the prediction target; e.g. it can be a single real number for regression tasks or a vector of probabilities for a classification task. All four functions GNN, $F, f,$ SM are potentially parameterized by trainable vectors $\Theta_{\mathrm{GNN}}, \Theta_F, \Theta_f, \Theta_{\mathrm{SM}}$, respectively.

4.2   Implementation

The bottom layer of our GNN implementation is a trainable embedding vector $h_v^0 \in \mathbb{R}^d$ for each node $v \in V$, as well as an embedding vector $h_r^0 \in \mathbb{R}^d$ for every relation $r \in R$. We compute higher-level node embeddings using graph convolution layers, where we adopt a simplified version of the compGCN architecture proposed by Vashishth $et\ al.$ (2020) as described below. Given the layer i embeddings $(h_v^i)_{v \in V}, (h_r^i)_{r \in R}$, two transformations are applied for each $v \in V$:

$$h_{v,\mathrm{self}}^{i+1} := W_{\mathrm{self}}^{i+1} \cdot h_v^i \tag{5}$$

$$h_{v,\mathrm{adj}}^{i+1} := W_{\mathrm{adj}}^{i+1} \cdot \sum_{(v,r,v') \in E} \mathrm{cmp}(h_r^i, h_{v'}^i), \tag{6}$$

where cmp $: \mathbb{R}^d \times \mathbb{R}^d \to \mathbb{R}^d$ is the $composition\ operator$. Our implementation supports the composition operators of addition and element-wise multiplication; we employ the latter throughout our experiments. The node and relation embeddings on layer $i + 1$ are then computed via

$$h_v^{i+1} = \mathrm{relu}(h_{v,\mathrm{self}}^{i+1} + h_{v,\mathrm{adj}}^{i+1}), v \in V \tag{7}$$

$$h_r^{i+1} = W_{\mathrm{rel}}^{i+1} \cdot h_r^i, r \in R. \tag{8}$$

The computations on each layer are parameterized by $W_{\mathrm{self}}^{i+1}, W_{\mathrm{adj}}^{i+1}, W_{\mathrm{rel}}^{i+1} \in \mathbb{R}^{d \times d}$. Equation 6 specifies backward flow across the edges of the directed graph. In Vashishth $et\ al.$ (2020), forward flow with independent parameterization is additionally specified. This is also supported by our implementation, but we only consider backward flow in our experiments. The reason is that the knowledge graphs in our datasets contain nodes with a huge number of incoming links, and we experienced that summing up over them during the calculation of forward flow de-stabilizes the model and does not scale well.

Having computed k layers of graph convolution, the final node embeddings are the GNN output:

$$h_v = \text{GNN}(\mathcal{G}, v) := h_v^k, v \in V. \tag{9}$$

For the aggregation function F we employ mean pooling across the nodes referenced in each event:

$$\beta_\ell = F(\{h_v \mid v \in \alpha_\ell(A)\}) := \frac{1}{|\alpha_\ell(A)|} \sum_{v \in \alpha_\ell(A)} h_v. \tag{10}$$

For the timestamp embedding function f our implementation supports parameterized embedding and non-parameterized embedding based on sinusoids as described by Vaswani et $al.$ (2017). We have however found that, for the prediction tasks included by our experimental study, the timestamp input is not essential; thus we applied the constant zero function there for most datasets.

We now describe our implementation of the sequence model SM. First, a linear transformation is applied to the embeddings $(\phi_\ell)_{\ell \in L_c}$:

$$\phi'_\ell = W_1 \cdot \phi_\ell, \ell \in L_c. \tag{11}$$

After this initial transformation, we aggregate over the events of the sequence using mean pooling:

$$\phi'' - \frac{1}{|L_c|} \sum_{\ell \in L_c} \phi'_\ell. \tag{12}$$

A final fully connected hidden layer connects the aggregated events with the output:

$$\phi''' = \text{relu}\left(W_2 \cdot \phi''\right), \tag{13}$$

$$P_c = \text{SM}(\{\phi_\ell \mid \ell \in L_c\}) := g(W_3 \cdot \phi'''). \tag{14}$$

The dimensionality of the matrices is $W_1, W_2 \in \mathbb{R}^{d \times d}$ and $W_3 \in \mathbb{R}^{o \times d}$. For binary classification problems, $o = 1$, and g is the sigmoid activation function. For multi-class classification, o corresponds to the number of classes and g is the softmax function. Finally, for regression problems, $o = 1$ and g is the identity function.

We remark that the design choice of the sequence model is a result of exploratory experiments with various architectures. During those experiments we observed that, throughout the datasets, more sophisticated architectures (recurrent networks, transformer) did not lead to better results and introduced stability problems. This is in line with the finding, reported by Al-Jebrni et $al.$ (2018) and Mauro et $al.$ (2019), that feedforward networks outperform the LSTM architecture for predictive process monitoring tasks.

5   Experiments

5.1   Data

Our experiments encompass six prediction tasks using four datasets; see Table 1. Three tasks are based on the Open University Learning Analytics

Table 1. Summary of the event logs and knowledge graphs extracted from the datasets.

Dataset	Knowledge graph	Event log	Reference
OULAD	240K nodes, 1.1M edges	33K cases, 11M events	Kuzilek *et al.* (2017)
PKDD99	200K nodes, 1,1M edges	4.5K cases, 1M events	Berka (1999)
BPI12	–	13K cases, 180K events	van Dongen (2012)
DBLP	370K nodes, 1.2M edges	29K cases, 1.7M events	Tang *et al.* (2008)

(OULAD) dataset provided by Kuzilek *et al.* (2017). The dataset has the structure of a relational database dump, consisting of seven tables that represent information about students registering for courses, interacting with the study material, taking assessments and exams. From this data we extracted a knowledge graph as described in Fig. 2. We further extracted an event log where each case $c \in C$ represents one student taking part in one course. There are five event types: *case info*, containing links to student, module, and semester, *assessment*, containing the submission date as the timestamp, and a link to the corresponding student assessment node in the knowledge graph, *student registration* and *deregistration*, containing the date of (de)registration as the timestamp, and a link to the student registration node, and finally *VLE interaction*, containing as timestamp the time of interaction with material of the Virtual Learning Environment (VLE), and a link to the knowledge graph node representing the material.

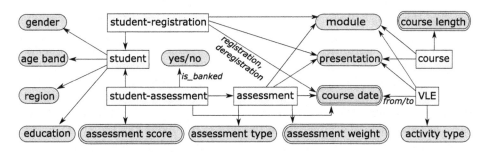

Fig. 2. Schema of the knowledge graph generated from the OULAD dataset. Box-shaped meta-nodes represent nodes generated from table rows, single-lined oval meta-nodes originate from categorical values, and double-lined meta-nodes represent discretized numerical values. Each arrow represents a distinct relation type (sometimes two); annotations have been added only where the type is not self-explanatory.

For the OULAD dataset we distinguish between three prediction targets: *dropout* (predict whether the student will drop out from the course), *success* (predict whether the student will finish the course successfully), and *exam score* (regression task to predict the final exam score, a number between 0 and 100). We further consider three different time horizons: *late prediction* is a variant of the prediction task where all events that happened during the course (except for the final exam and the deregistration event) are available as input. *Early prediction* refers to predictions taking into account only the events that happened before the 60th day of the course (all course modules take between 234 and 269 days), while in *very early prediction* no events have been recorded and only the *case info* is available. Across all tasks and variants, we uniformly selected 20% of the data for validation and 5% for testing, like implemented by Jha *et al.* (2019).

A dataset from the financial domain was provided by Berka (1999) for the *PKDD99* challenge. This dataset also comes in the form of multiple inter-connected tables, representing bank accounts, financial transactions, clients, geo-graphical districts, and loans. The task is to predict the status of a loan (non-critical or critical) given the history of transactions and information about the loan and the client. The triples of the knowledge graph relate loans to accounts, accounts to districts, and bank orders as well as transactions to banks. Categorical and numerical attributes (after discretization) of those entities are additionally represented as neighbor nodes of them. The two event types extracted from the dataset are *case info*, including links to the loan and the account node, and *transaction*, containing a link to the node representing the transaction in the knowledge graph. We chose 20% of the data for validation and another 20% for testing. Because the number of samples in this dataset is comparably small and the dataset is rather imbalanced with only 10% of the loans having a critical status, we used stratified sampling to enforce the same balance of positive and negative examples in the training, validation, and test set.

Our second dataset from the financial domain represents an established process mining benchmark, but it comes only with an event log. The events contained in the *BPI12* dataset (van Dongen 2012) were recorded during the application procedure at a financial institution. Due to the lack of a knowledge graph, we treat all event attributes as isolated nodes of a graph without edges. To make our experiments comparable to Tax *et al.* (2017), we only consider events that are marking the completion of manually executed subprocesses, and we used the latest 30% among the sequences as the test set. Another 20% of the sequences were chosen as the validation set uniformly at random. We treat every prefix of every sequence as one sample where the task is to predict the type of the next event. This step was done separately for the training, validation, and test set.

Our final prediction problem is the number of future citations of papers published in the year 2000 as reported in the DBLP dataset introduced by Tang *et al.* (2008). We extracted a knowledge graph consisting of relations between papers, authors, and venues. To prevent information leakage, events related to a publication after 2000 are not considered for the knowledge graph construction. For each paper, the history of previous publications of all authors is used as the event sequence. We selected 20% of the data for validation and 5% for testing.

5.2   Setup

The machine learning models were implemented in Tensorflow 2.5.0, and the computational experiments were performed on GPUs (Nvidia GeForce GTX

Table 2. Hyperparameters used in the experiments. Note that the LSTM model implicitly takes into account the event position by design.

Task	Dropout rate	l2-weight	GC layers	Time embedding
OULAD (dropout)	0.1	0.01	1	None
OULAD (success)	0.1	0.01	1	None
OULAD (score)	0.7	0	1	None
PKDD99	0.1	0.03	3	None
BPI12	0.1	0.01	1	Parameterized
DBLP	0.5	0	0	None
All tasks, LSTM model	0.25	0.01	–	Implicit

1080 Ti). For all classification tasks we used the cross-entropy loss function for training and selected the model having the highest accuracy on the validation set among all 200 training epochs. The learning rate was set to 0.01, and we chose an embedding width of $d = 100$. We also applied dropout and l2-regularization, which required different strategies for different tasks. Our implementation supports dropout with uniform rate after each aggregation layer in the graph convolutional network and after the final fully-connected dense layer. Table 2 lists the chosen strategy used for ProcK for every task, as well as for the LSTM baseline, where we found a different configuration to work best.

5.3   Results

The results of our experiments are displayed in Table 3. The top part contains results for classification problems, where the accuracy and the Area Under the Curve (AUC, only for binary classification) metric are reported. On most datasets a knowledge graph is available, enabling us to compare models trained with such a graph to models trained without one. As a baseline we also evaluated an LSTM model which was trained using the event sequence as input. Whenever available, the table also contains the best results reported in literature.

For the AUC metric, it turns out that the availability of additional data in form of a graph improves the ability of the model to correctly separate positive and negative test samples, both in comparison with ProcK without knowledge graph and the LSTM model. The improvement is observable across most of the problems. For dropout prediction on the OULAD dataset, all three deep learning models outperform the results of Gradient Boosting Machine (GMB), reported by Jha *et al.* (2019), while GBM performs better for the success prediction task.

When looking at the accuracy metric, the advantage of the knowledge graph input is not that clearly visible; only on two out of seven problems the full ProcK model with knowledge graph exhibits the best performance, on one problem it is outperformed by ProcK without knowledge graph input, and two tasks the LSTM model performs best. The only classification problem with more than two classes is next event type prediction on the BPI12 dataset. Here no knowledge

Table 3. Experimental results

Prediction task	Model	Accuracy	AUC
OULAD (dropout, late)	ProcK	0.86	**0.93**
	ProcK (no KG)	0.86	**0.93**
	LSTM	0.86	0.92
	GBM (Jha *et al.* 2019)	–	0.91
OULAD (dropout, early)	ProcK	**0.83**	**0.84**
	ProcK (no KG)	0.81	0.82
	LSTM	0.81	0.82
OULAD (dropout, very early)	ProcK	0.68	0.58
	ProcK (no KG)	0.69	**0.60**
	LSTM	**0.70**	0.57
OULAD (success, late)	ProcK	0.87	0.91
	ProcK (no KG)	**0.88**	0.88
	LSTM	0.86	0.86
	GBM (Jha *et al.* 2019)	–	**0.93**
OULAD (success, early)	ProcK	0.73	**0.74**
	ProcK (no KG)	0.73	0.73
	LSTM	**0.75**	0.73
OULAD (success, very early)	ProcK	**0.69**	**0.58**
	ProcK (no KG)	**0.69**	0.56
	LSTM	0.68	0.57
PKDD99	ProcK	**0.89**	**0.71**
	ProcK (no KG)	**0.89**	**0.71**
	LSTM	**0.89**	0.50
BPI12 (KG not available)	ProcK	**0.83**	–
	LSTM	0.71	–
	LSTM (Tax *et al.* 2017)	0.76	–
Prediction target	Model	RMSE	
OULAD (score, late)	ProcK	**18.93**	
	ProcK (no KG)	18.95	
	LSTM	20.35	
OULAD (score, early)	ProcK	**19.88**	
	ProcK (no KG)	**19.88**	
	LSTM	21.08	
OULAD (score, very early)	ProcK	**20.10**	
	ProcK (no KG)	20.44	
	LSTM	20.13	
DBLP	ProcK	**3.98**	
	LSTM	4.01	

graph is given, and ProcK outperforms the LSTM model by a large margin. This finding is consistent to results found in other works e.g. by Mauro *et al.* (2019).

We are the first to study variants of OULAD with different points of prediction time (early and very early prediction). It turns out that the length of the event log makes a significant difference, with AUC values decreasing to less than 0.60 when only the initial case information is available. However, the benefit of using knowledge graph input does not seem to depend on the length of the event log, which can be explained with the fact that less event input also means less information from the knowledge graph.

The bottom part of Table 3 contains the results for regression problems, including exam score (OULAD dataset) and number of citations (DBLP dataset). We do not include ProcK without knowledge graph input for the latter, because here the number of graph convolution layers is set to zero (see Table 2), making models with and without knowledge graph input equivalent. The table reports the root mean square error, and here again the benefits of the knowledge graph input can be demonstrated across the variants of the score prediction task.

6   Summary and Conclusion

In this work, we introduced ProcK, a novel machine learning pipeline for data from knowledge-intensive processes. Within the pipeline, two complementary views of the available information are first extracted from raw tabular data and then re-combined as input to the downstream prediction model. We implemented prototypes of each pipeline component, and we tested their interplay on six prediction tasks on four datasets. We could demonstrate on the majority of classification tasks that ProcK achieves improved AUC values when having a knowledge graph available as input, but further investigation of the accuracy metric remains a task for future work. Also for regression tasks ProcK exhibits a small but consistent advantage in terms of the RMSE metric.

One interesting future question from a practical viewpoint is how machine learning can be employed to extract the knowledge base and event log from the source databases in a configuration-free manner. Furthermore, while ProcK has been applied for prediction tasks so far, the ability to compute recommendations to positively influence processes will be an important next step.

Ethics Discussion. All experiments reported in this work are based on anonymized datasets (OULAD, PKDD, BPI12) or data actively published by the data subjects (DBLP). Nevertheless, the presented technology is applicable to ethically sensitive tasks, including assessment of loan applications and performance prediction of humans. A careful assessment of potential ethical issues has to be carried out prior to bringing this work to application.

References

Al-Jebrni, A., Cai, H., Jiang, L.: Predicting the next process event using convolutional neural networks. In: 2018 IEEE International Conference on Progress in Informatics and Computing (PIC), pp. 332–338. IEEE (2018)

Berka, P.: Workshop Notes on Discovery Challenge, PKDD 1999 (1999)

Camargo, M., Dumas, M., González-Rojas, O.: Learning accurate LSTM models of business processes. In: Hildebrandt, T., van Dongen, B.F., Röglinger, M., Mendling, J. (eds.) BPM 2019. LNCS, vol. 11675, pp. 286–302. Springer, Cham (2019). https://doi.org/10.1007/978-3-030-26619-6_19

Ceci, M., Lanotte, P.F., Fumarola, F., Cavallo, D.P., Malerba, D.: Completion time and next activity prediction of processes using sequential pattern mining. In: Džeroski, S., Panov, P., Kocev, D., Todorovski, L. (eds.) DS 2014. LNCS (LNAI), vol. 8777, pp. 49–61. Springer, Cham (2014). https://doi.org/10.1007/978-3-319-11812-3_5

Chang, J., et al.: Sequential recommendation with graph neural networks. In: Proceedings of the 44th International ACM SIGIR Conference on Research and Development in Information Retrieval, pp. 378–387 (2021)

Evermann, J., Rehse, J.-R., Fettke, P.: Predicting process behaviour using deep learning. Decis. Support Syst. **100**, 129–140 (2017)

Fahrenkrog-Petersen, S.A., et al.: Fire now, fire later: alarm-based systems for prescriptive process monitoring. Knowl. Inf. Syst. **64**(2), 559–587 (2022). https://doi.org/10.1007/s10115-021-01633-w

Fang, L., Cheng, X., Wang, H., Yang, L.: Mobile demand forecasting via deep graph-sequence spatiotemporal modeling in cellular networks. IEEE Internet Things J. **5**(4), 3091–3101 (2018)

Folino, F., Guarascio, M., Pontieri, L.: Mining predictive process models out of low-level multidimensional logs. In: Jarke, M., et al. (eds.) CAiSE 2014. LNCS, vol. 8484, pp. 533–547. Springer, Cham (2014). https://doi.org/10.1007/978-3-319-07881-6_36

García-Durán, A., Dumancic, S., Niepert, M.: Learning sequence encoders for temporal knowledge graph completion. In: Proceedings of the 2018 Conference on Empirical Methods in Natural Language Processing (EMNLP), pp. 4816–4821. Association for Computational Linguistics (2018)

Hu, K., et al.: Graph sequence recurrent neural network for vision-based freezing of gait detection. IEEE Trans. Image Process. **29**, 1890–1901 (2019)

Jha, N.I., Ghergulescu, I., Moldovan, A.-N.: OULAD MOOC dropout and result prediction using ensemble, deep learning and regression techniques. In: CSEDU (2), pp. 154–164 (2019)

Jin, W., Qu, M., Jin, X., Ren, X.: Recurrent event network: autoregressive structure inference over temporal knowledge graphs. In: Proceedings of the 2020 Conference on Empirical Methods in Natural Language Processing (EMNLP), pp. 6669–6683 (2020)

Kuzilek, J., Hlosta, M., Zdrahal, Z.: Open university learning analytics dataset. Sci. Data **4**(1), 1–8 (2017)

Leontjeva, A., Conforti, R., Di Francescomarino, C., Dumas, M., Maggi, F.M.: Complex symbolic sequence encodings for predictive monitoring of business processes. In: Motahari-Nezhad, H.R., Recker, J., Weidlich, M. (eds.) BPM 2015. LNCS, vol. 9253, pp. 297–313. Springer, Cham (2015). https://doi.org/10.1007/978-3-319-23063-4_21

Li, Z., et al.: Search from history and reason for future: two-stage reasoning on temporal knowledge graphs. In: Proceedings of the 59th Annual Meeting of the Association for Computational Linguistics and the 11th International Joint Conference on Natural Language Processing, ACL/IJCNLP 2021 (Volume 1: Long Papers), pp. 4732–4743. Association for Computational Linguistics (2021)

Lin, L., Wen, L., Wang, J.: MM-Pred: a deep predictive model for multi-attribute event sequence. In: Proceedings of the 2019 SIAM International Conference on Data Mining, pp. 118–126. SIAM (2019)

Di Mauro, N., Appice, A., Basile, T.M.A.: Activity prediction of business process instances with inception CNN models. In: Alviano, M., Greco, G., Scarcello, F. (eds.) AI*IA 2019. LNCS (LNAI), vol. 11946, pp. 348–361. Springer, Cham (2019). https://doi.org/10.1007/978-3-030-35166-3_25

Pasquadibisceglie, V., Appice, A., Castellano, G., Malerba, D.: Using convolutional neural networks for predictive process analytics. In: 2019 International Conference on Process Mining (ICPM), pp. 129–136. IEEE (2019)

Reinkemeyer, L.: Process Mining in Action. Springer, Cham (2020). https://doi.org/10.1007/978-3-030-40172-6

Tang, J., Zhang, J., Yao, L., Li, J., Zhang, L., Su, Z.: ArnetMiner: extraction and mining of academic social networks. In: Proceedings of the 14th ACM SIGKDD International Conference on Knowledge Discovery and Data Mining, pp. 990–998 (2008)

Tax, N., Verenich, I., La Rosa, M., Dumas, M.: Predictive business process monitoring with LSTM neural networks. In: Dubois, E., Pohl, K. (eds.) CAiSE 2017. LNCS, vol. 10253, pp. 477–492. Springer, Cham (2017). https://doi.org/10.1007/978-3-319-59536-8_30

Taymouri, F., Rosa, M.L., Erfani, S., Bozorgi, Z.D., Verenich, I.: Predictive business process monitoring via generative adversarial nets: the case of next event prediction. In: Fahland, D., Ghidini, C., Becker, J., Dumas, M. (eds.) BPM 2020. LNCS, vol. 12168, pp. 237–256. Springer, Cham (2020). https://doi.org/10.1007/978-3-030-58666-9_14

Teinemaa, I., Dumas, M., Rosa, M.L., Maggi, F.M.: Outcome-oriented predictive process monitoring: review and benchmark. ACM Trans. Knowl. Discov. Data (TKDD) **13**(2), 1–57 (2019)

Tello-Leal, E., Roa, J., Rubiolo, M., Ramirez-Alcocer, U.M.: Predicting activities in business processes with LSTM recurrent neural networks. In: 2018 ITU Kaleidoscope: Machine Learning for a 5G Future (ITU K), pp. 1–7. IEEE (2018)

Theis, J., Darabi, H.: Decay replay mining to predict next process events. IEEE Access **7**, 119787–119803 (2019)

van der Aalst, W., et al.: Process mining manifesto. In: Daniel, F., Barkaoui, K., Dustdar, S. (eds.) BPM 2011. LNBIP, vol. 99, pp. 169–194. Springer, Heidelberg (2012). https://doi.org/10.1007/978-3-642-28108-2_19

Van der Aalst, W.M.P., Schonenberg, M.H., Song, M.: Time prediction based on process mining. Inf. Syst. **36**(2), 450–475 (2011)

van Dongen, B.: BPI challenge 2012, April 2012

Vashishth, S., Sanyal, S., Nitin, V., Talukdar, P.P.: Composition-based multi-relational graph convolutional networks. In: 8th International Conference on Learning Representations, ICLR 2020, Addis Ababa, Ethiopia, 26–30 April 2020. OpenReview.net (2020)

Vaswani, A., et al.: Attention is all you need. In: Advances in Neural Information Processing Systems, pp. 5998–6008 (2017)

Wang, Z., et al.: Advanced graph and sequence neural networks for molecular property prediction and drug discovery. Bioinformatics **38**(9), 2579–2586 (2022)

Enhance Temporal Knowledge Graph Completion via Time-Aware Attention Graph Convolutional Network

Haohui Wei, Hong Huang[(✉)], Teng Zhang, Xuanhua Shi, and Hai Jin

National Engineering Research Center for Big Data Technology and System,
Services Computing Technology and System Lab, Cluster and Grid Computing Lab,
School of Computer Science and Technology,
Huazhong University of Science and Technology, Wuhan 430074, China
{weihh77,honghuang,tengzhang,xhshi,hjin}@hust.edu.cn

Abstract. Previous works on knowledge graph representation learning focus on static knowledge graph and get fully developed. However, task on temporal knowledge graph is far from consummation because of its late start. Recent researches have shifted to the temporal knowledge graph relying on the extension of static ones. Most of these methods seek approaches to incorporate temporal information but neglect the potential adjacent impact merged in temporal knowledge graphs. Meanwhile, different temporal information of involved facts evoke impact with different extent on the concerned entity, which is always overlooked in the previous works. In our paper, we propose a Time-aware Attention Graph Convolutional Network, named *TAGCN*, for temporal knowledge graph completion. Entity completion can be turned into interactions between entity and associated neighborhood. We utilize a graph convolutional network with a novel temporal attention layer to obtain neighboring information at all timestamps to avoid diachronic sparsity. We conduct extensive experiments on various datasets to evaluate our model performance. The results illustrate that our model outperforms the state-of-the-art baselines on entity prediction.

Keywords: Temporal knowledge graph · Representation learning · Graph neural networks

1 Introduction

Temporal Knowledge Graph (TKG) stores structured data in quadruples to extract interactions between entities on specific timestamps to represent fact, which is the information of events that have occurred in the reality. Temporal information is a crucial element in the real world because some facts are only valid at some timestamps. Due to its capability to contain rich information, TKG benefits numerous downstream applications, like transaction recommendation, social relation inferring, and event process prediction, all of which are dependent

© The Author(s), under exclusive license to Springer Nature Switzerland AG 2023
M.-R. Amini et al. (Eds.): ECML PKDD 2022, LNAI 13714, pp. 122–137, 2023.
https://doi.org/10.1007/978-3-031-26390-3_8

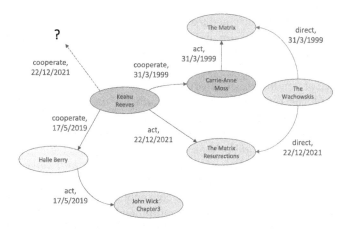

Fig. 1. A toy example for entity prediction in TKG, and the deep green candidate is the most probable entity for the query *(Keanu_Reeves, cooperate, ?, 22/12/2021)*. (Color figure online)

on the quality of the knowledge graph. Nonetheless, TKG is inevitably incomplete and sparse due to corruption and some other irresistible factors, which will undermines the performance of downstream tasks. Therefore, there is a growing need of valid approaches for completion in TKG.

Recently, more and more works pay attention to TKG completion. Majority of researches, like TTransE [1], TNTComplEx [2], neglect rich information that is inherent within the vicinity of entities and limit their semantic capturing because they expand to TKG domain through a straightforward extension from static methods, such as TransE [3], focusing on independent validation of each quadruple. Taking Fig. 1 as an example, our purpose is to predict who *Keanu Reeves* would *cooperate* with on *22/12/2021*. It is obvious that *Carrie-Anne Moss* and *Halle Berry* share the same entity interaction *cooperate* with *Keanu Reeves*. However, the neighborhood of *Carrie-Anne Moss* is quite different from that of *Halle Berry*. Characterizing this would provide abundant information when predicting the missing tail entity in *(Keanu_Reeves, cooperate, ?, 22/12/2021)*. To conclude, neighboring information has large impact on representation learning because learning quadruples independently would lose the huge amount of interactions among entities from the neighboring structure of TKG.

Furthermore, fact in TKG relies on long-term dependency which means fact with long interval may still determine the future prediction. Using the same query in Fig. 1, although *Halle Berry* has closer interactive behavior *cooperate* with *Keanu Reeves* than *Carrie-Anne Moss*, it is not definite that *Halle Berry* has a bigger chance to be the correct prediction in the query *(Keanu_Reeves, cooperate, ?, 22/12/2021)*. But due to the difficulty of simulating complex temporal influence in TKG, some of relevant works, like RE-NET [4] and RTFE [5], only pay more attention to recent fact.

Considering the aforementioned features, it is prominent to comprehend TKG both from the neighbor structure and the temporal dependency of TKG. Therefore, it evokes several challenges when tackling these features. First, *how to extract temporal neighboring information from the neighbor structure of TKG?* The significance of neighborhood encoding has been revealed in several static KG methods [6], but the extension to TKG is difficult due to the additional time dimension. Utilizing the Message Passing Process approaches, it is critical to integrate temporal information with interactions between relations and entities when encapsulating the neighborhood surrounding entities.

The second challenge is *how to encode time to maintain long-term dependency?* Encoding time by separating time snapshots and training through the temporal order may give rise to the same consequence of sequence-based models mentioned above. It is also difficult to comprehend the influence of all time on prediction by simply projecting time into the same space of entity and relation. Therefore, to distinguish different temporal influence on a certain timestamp, it is needed to implement a specific approach to learn the dependency between temporal information and the certain timestamp of the concerned query.

To this end, we propose our *Time-aware Attention Graph Convolutional Network* (TAGCN). We decouple temporal knowledge graph completion into two phase, neighboring temporal message aggregation and entity temporal focus attachment. To alleviate the above two challenges, TAGCN is used to encode neighboring information to contextualize the representation of entities. Inspired by self-attention mechanism [7], we devise a novel *temporal self-attention* (TeA) layer to locally extract the temporal influence between timestamps and involved fact on concerned query. The *temporal message aggregation* (TMA) module is used to extract neighborhood structure in TKG. Moreover, a time-aware decoder is applied and uses a simple way to activate different attention to neighboring impact regarding to the time of query. Our contributions can be summarized as follows:

1. We propose TAGCN, which introduces a knowledge graph convolutional network to learn the representation of entity capturing temporal dependency and complex interactions between entities and temporal facts. We decouple temporal knowledge graph completion into two phases, neighboring temporal message aggregation and entity temporal focus attachment.
2. Our work initiates a well-designed temporal self-attention layer, which is leveraged to encode locally temporal impact between target entity and involved facts. Thus, enhancing the estimation of the temporal influence of neighborhood.
3. We conduct extensive experiments on several real-world datasets. Experimental results show that the proposed method has achieved the state-of-the-art results in the task of entity prediction for TKG.

2   Related Works

2.1   Static KG Completion

There have been a large number of researches giving insight into static knowledge graph completion. Traditional KG completion methods map entities and relations into a low-dimensional vector with a score function measuring the possibility of the candidates. They can be classified into three categories in general: translational models, factorization based models, and *convolutional neural network* (CNN) models. TransE [3] is the most well-known method using translation to embed entities and relations. Following TransE, several methods [8–10] using different mapping methodology come out in succession and achieve better results. Rescal [11] and DistMult [12] are two factorization based models. Simultaneously, ComplEx [13] also projects entities and relations into different spaces using the above-mentioned evaluation function. In this way, these models can further develop the expressiveness beyond the limitation of Euclidean space and learn more complex interaction between entities and relations. Besides these works, ConvE [14] applies convolutional filters to process the vector of entity and relation. Its success ignites further application of other neural networks [15]. However, completion task on TKG reaps poor effect with static methods because of the lack of temporal information processing.

2.2   Temporal KG Completion

Lots of previous works on TKG mainly pay attention to the independent validation of quadruples and lay more focus on the incorporation of static methods and temporal information. The main distinction lies in the representation of timestamps. As mentioned in the above section, TTransE [1] views time as a new element and adds timestamp embedding into the relation embedding to fulfill the conventional score function TransE [3]. HyTE [16] adopts the idea of TransH [8], viewing different hyperplanes as different temporal spaces. Then it projects entities and relations to these hyperplanes and uses static score function to measure the possibility. TA-DistMult [17] learns time embedding by encoding the timestamp string sequence while DE-SimplE [18] leverages diachronic embedding by concatenating it with enduring one. TNTComplEx [2] makes a decomposition of 4 tensors in complex domain, adding the timestamp representation compared with ComplEx [13]. Although these methods extend to TKG successfully, none of them considers the neighborhood information within knowledge graph. RTFE [5] proposes a new training framework to enhance the further boost of conventional methods. T-GAP [19] adopts query-relevant temporal displacement to process the whole TKG, but the time overhead is too high to follow.

3   Proposed Model

3.1   Problem Definition

Temporal knowledge graph $\mathcal{G}$ is composed of quadruples $\mathcal{G} = \{(h, r, t, \tau) \mid h, t \in \mathcal{E}, r \in \mathcal{R}, \tau \in \mathcal{T}\}$, where h refers to head entity, r denotes the relation, t is the

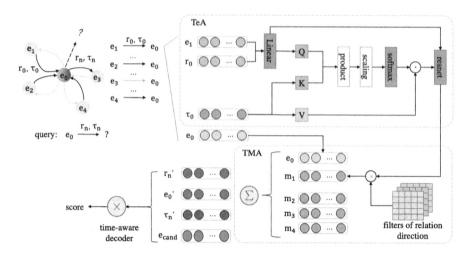

Fig. 2. Overview of our work. Arrows in gray refer to inverse directions, and self loop is omitted for clarity. All entities in $\mathcal{E}$ are considered as candidates.

tail entity, and τ is the timestamp. $\mathcal{E}$, $\mathcal{R}$ represent the entity set and relation set contained in $\mathcal{G}$, and $\mathcal{T}$ stands for the known timestamps. To distinguish different concept, we use lower-case letters to represent object in dataset, e to represent specific entity in TKG, such as $e_1, e_2 \in \mathcal{E}$ in Fig. 2, z to represent embedding of each object, like $z_r, z_h \in \mathbb{R}^d$ where d is the embedding dimension, and z_x^τ to represent entity embedding z_x under timestamp τ.

The aim of TKGC is to find the missing entity of an incomplete query. Given a query lacking in tail entity, such as $(h, r, ?, \tau)$ where τ is within the observed set $\mathcal{T}$, we hope to learn a mapping function $f_{map}: e \to \mathbb{R}^d$, where $e \in \mathcal{E}$ and d represents the dimension of embedding $d \ll |\mathcal{E}|$ and a score function f_{score} to infer the most probable $t \in \mathcal{E}$ to fulfill the missing component in the query based on the known information of $\mathcal{G}$. What's more, the mapping function f_{map} not only needs to consider the temporal information of each fact but also consolidate neighbor structure.

3.2 Model Overview

In this section, we describe our proposed model TAGCN, which can simultaneously extracts structural information and temporal dependency. Our model is in the architecture of encoder-decoder, and the framework is shown in Fig. 2. Quadruple sets in TKG are mapped into low-dimensional spaces at the beginning. Then, two stages are entailed for entity prediction. With the novel *temporal attention layer* (TeA), we capture the temporal dependency of the neighboring fact locally and dispose with *temporal message aggregation* (TMA) to increase the expressiveness of entities. Last, the time-aware decoder integrates temporal information to model the temporal impact on entity and gives the probability of candidates.

3.3   TAGCN

Preprocess. To best improve the connectivity and information transmission efficiency, we allow information in TKG flows along three directions: original, inverse, and self-loop. Among which, self-loop $loop_e$ is constructed for each entity $e \in \mathcal{E}$, quadruple with self-loop is extended with a self-loop relevant timestamp τ_{loop}. Therefore, TKG $\mathcal{G}$ grows into $\mathcal{G}'$,

$$\mathcal{G}' = \mathcal{G} \cup \{(t, r^{-1}, h, \tau) \mid (h, r, t, \tau) \in \mathcal{G}\} \cup \{(e, loop_e, e, \tau_{loop}) \mid e \in \mathcal{E}\}. \tag{1}$$

$\mathcal{R}$ and $\mathcal{T}$ are also extended to $\mathcal{R}' = \mathcal{R} \cup \mathcal{R}_{inv} \cup \{loop_e\}$, $\mathcal{T}' = \mathcal{T} \cup \{\tau_{loop}\}$. Meanwhile, since $d(rel)$ short for the direction of relations is divided into three types, we adopt three different filters, and the relational direction filter is defined as follows:

$$\mathbf{W}_{d(rel)} = \begin{cases} \mathbf{W}_{ori} & rel \in \mathcal{R}, \\ \mathbf{W}_{inv} & rel \in \mathcal{R}_{inv}, \\ \mathbf{W}_{loop} & rel \in \{loop_e\}. \end{cases} \tag{2}$$

Temporal Attention Layer. Attention is of great importance in nowadays researches. In order to encode temporal dependency between entity and fact, we utilize self-attention layer for better usage of adequate information among available edge attributes. We compute the implicit attention score when an entity concerns its surrounding neighborhood. We treat neighboring messages as two part for decoupling the structural and temporal information. For a corresponding fact (h, r, t, τ), we use a linear layer to combine the representations of head entity and relation for semantic information $m_{h,r}^s$ of the fact:

$$m_{h,r}^s = \mathbf{W}_s([z_h | z_r]), \tag{3}$$

where $[\cdot | \cdot]$ denotes concatenate operation, and $\mathbf{W}_s$ is to project the embedding size to the standard dimension space.

Then we use two weight matrices to get the query and key of m^s and z_τ. The intermediate representation $\hat{m}$ refers to the combination of temporal information τ and semantic information of the fact, which could be explored in the following step. Following previous work [7], the scaling operation is employed to alleviate the over inflation of dot products, thus avoiding the extremely small gradient:

$$\hat{m}_{h,r}^\tau = \frac{\mathbf{W}_K(z_\tau) \otimes \mathbf{W}_Q(m_{h,r}^s)}{\sqrt{d}}, \tag{4}$$

where $\mathbf{W}_K$, $\mathbf{W}_Q$ are weight matrices. $\hat{m}_{h,r}^\tau$ is an implicit temporal representation of neighborhood. Thus we get the temporal fact attention generated from temporal and semantic representation after softmax. To strengthen the potential impact of fact on specific timestamp, we model the temporal attention by using temporal information and the fact temporal representation:

$$m_{h,r}^\tau = \text{softmax}(\hat{m}_{h,r}^\tau)\mathbf{W}_V(z_\tau). \tag{5}$$

Up to now, we obtain the temporal $m_{h,r}^\tau$ information of neighboring fact. To enhance the semantic influence, we utilize the residual network for numerical stability. The neighboring message of the certain fact is formulated as follows:

$$m_f = \text{FCN}(m_{h,r}^\tau + m_{h,r}^s), \tag{6}$$

where FCN denotes the fully connected net with norm and dropout layer to enhance the generalization. We also use a residual layer to maintain layer wise fact information to deal with some long maintaining facts.

Temporal Message Aggregation Module. With the process of TeA layer, we obtain the input message by modeling the temporal dependency with semantic and temporal information when concerning the central entity e. To improve the representation, we adopt the aggregation and take the relation direction into account. The new entity feature is computed by combining all incoming messages to e:

$$z_e = \sum_{f \in \mathcal{N}_e} \mathbf{W}_{d(r)} m_f, \tag{7}$$

where $\mathcal{N}_e$ is the neighboring facts with e as the tail entity, and the facts are all in the original temporal knowledge graph $\mathcal{G}$.

After conducting TAGCN, we use the output as the ultimate representations of entities. These representations well integrate the temporal dependency on certain facts with interactions between entities and the involved facts from the neighboring structure of the knowledge graph.

3.4 Time-Aware Decoder

Using the encoder, we obtain representations with perception of the whole TKG. When considering one query, the temporal effect on entity should be activated in decoder for better comprehension. To implement conventional decoders, we combine the entity and timestamp embedding with weight matrix $\mathbf{W} \in \mathbb{R}^{2d \times d}$, which enables the entity embedding to comprehend diversity of fact influence on different timestamps:

$$z_h^\tau = \mathbf{W}[z_h | z_\tau], \tag{8}$$

where z_h and z_t are the embeddings of head entity and timestamp selected from embedding matrix according to the query.

In our work, we choose ConvE [14] as decoder to estimate probability for quadruples. ConvE employs convolutional and fully-connected layers to model the interactions between entities and relations input. It stacks the embeddings of head entity and relation, uses convolution operation to acquire the score of quadruple as follows:

$$\boldsymbol{p}_{(h,r,t,\tau)} = \text{ReLU}(vec(\text{ReLU}([\boldsymbol{z}_h^\tau | \boldsymbol{z}_r] * \omega))\mathbf{W})\mathbf{E}^{\text{T}}, \tag{9}$$

where vec$(\cdot)$ represents flattening tensor into vector, $*$ is the convolution operation, ω is the convolutional filter, and $\mathbf{W}$ denotes a parameter matrix to project the flattened result to the embedding dimension. $\mathbf{E}$ is the entity embedding matrix without temporal embedding, so that all entities are treated as candidates.

The model is trained using standard cross entropy loss:

$$\mathcal{L} = -\frac{1}{N}\sum_{i\in\mathcal{G}}(t_i \cdot \log(p_i) + (1 - t_i)\log(1 - p_i)), \tag{10}$$

where t_i is the gold label of fact i while p_i is the inferred probability. At last, we train our model using the optimizer Adam.

4   Experimental Setup

4.1   Datasets

In our work, we evaluate the proposed model on several public datasets for TKG completion, namely, ICEWS14, ICEWS05-15, YAGO11k, and Wikidata12k. iCEWS14 and ICEWS05-15 both come from the *Integrated Crisis Early Warning System* (ICEWS)[1]. ICEWS14 covers facts occurred in 2014, while ICEWS05-15 collects facts occurred between 2005 and 2015. ICEWS datasets are stored in the form of (h, r, t, τ). Wikidata12k and YAGO11k are subsets of Wikidata[2] and YAGO[3], formatted as $(h, r, t, \tau_{start}, \tau_{end})$. Following the same data splitting strategy in HyTE [16], we discretize them into each snapshot. The details of above-mentioned datasets are listed in Table 1.

4.2   Evaluation Metrics

Generated from reality, facts may evolve over time. Two quadruples may appear sharing same factors, like (h, r, t, τ) and (h, r, t', τ). However, results can be flawed once one quadruple end up with testing ones, while the other is from the training set. To avoid the misleading of these corrupted facts, we remove from the dataset the corrupted facts, corresponding to the former work [18]. Thus all metrics in our work are filtered ones. *Mean reciprocal rank* (MRR), Hits@1, Hits@3, and Hits@10 are formally used in knowledge graph entity prediction to compare the performance against other baselines. In our work, we evaluate TAGCN in

---

[1] https://dataverse.harvard.edu/dataverse/icews.

[2] https://www.wikidata.org.

[3] https://yago-knowledge.org/.

Table 1. Statistics of datasets. $|\mathcal{E}|$, $|\mathcal{R}|$, and $|\mathcal{T}|$ are the total number of entities, relations, and timestamps. Meanwhile, #train, #test, and #valid refer to the quadruple numbers of train, test, and valid set respectively.

Dataset	$|\mathcal{E}|$	$|\mathcal{R}|$	$|\mathcal{T}|$	#train	#test	#valid	Granularity
ICEWS14	7,128	230	365	72,826	8,941	8,963	1 day
ICEWS05-15	10,488	251	4,017	368,962	46,275	46,092	1 day
YAGO11k	10,623	10	60	203,858	21,763	21,159	1 year
Wikidata12k	12,254	24	77	239,928	18,633	17,616	1 year

tail and head entity prediction respectively. The test set is represented by s_{test}, MRR is defined as $\text{MRR}_x = \frac{1}{|S_{test}|} \sum_{(h,r,t,\tau) \in S_{test}} \frac{1}{\text{rank}_x}$, where $x \in \{h, t\}$, and we calculate both rankings of head entity $rank_h$ and tail entity $rank_t$, consisting with the evaluation method of previous work. Meanwhile, Hits@n, $n = 1, 3, 10$, is defined as $\text{Hits@n}_x = \frac{1}{|S_{test}|} \sum_{(h,r,t,\tau) \in S_{test}} \mathbb{I}(\text{rank}_x \leq n)$, where $\mathbb{I}(\cdot)$ is an indicator function equaling to one if the condition holds, and zero otherwise.

4.3   Baselines

To show the competitiveness of our model, we make a comparison with numbers of temporal and static KG completion models. In order to meet the requirements of static models, we ignore the time information when training these models. T-GAP [19] is not set as our baseline because of the high time overhead, and it only uses tail prediction as its results in the paper, while the tail prediction performance is usually better than the head one. Sequence-based models like RE-NET [4] and RE-GCN [20] are excluded because they are assigned for extrapolation task.

Static Baselines are Listed as Follows

- *TransE* [3]. In TransE, entities and relations are mapped into the same embedding space, and use translation method to infer the tuples.
- *DistMult* [12]. With the same projection as TransE, this work uses factorization method to calculate the score of each tuple.
- *ComplEx* [13]. ComplEx proposes a method based on complex representation. It divides the embeddings of entities and relations into real and imaginary parts. Lastly it adopts a complex multiplication, and maintains the real part as the final score.
- *ConvE* [14]. This method construct the score function with the same projection as TransE, and uses convolution operation to obtain the interactions between entities and relations.

Temporal Baselines are Listed as Follows

- *TTransE* [1]. TTransE is an extension of TransE with an additional dimension of timestamp, it uses a linear add operation to encode the temporal information into relation.

- *HyTE* [16]. Using the same projection as TTransE, but HyTE projects all entities and relations to the space of timestamps by linear transition associated with timestamps.
- *TA-DistMult* [17]. TA-DistMult treats timestamp as a string, and maps each character to a vector. When dealing with a tuple, the temporal relation embedding $z_{r,\tau}$ is generated by feeding r and characters in τ as a sequence into a LSTM to encode temporal information.
- *DE-SimplE* [18]. Focusing on temporal encoding, DE-SimplE puts forward a novel entity embedding function by considering that entities are combined with temporal and static features, using diachronic embeddings to estimate the probability of the incomplete tuples.
- *TNTComplEx* [2]. This work focuses on the integration of timestamp with ComplEx, and projects timestamps the same way as other objects in ComplEx. To be more clear, TNTComplEx embeds temporal features through linear representation in complex space.
- *RTFE* [5]. RTFE is a framework to enhance the performance of existing methods. It chooses a method $\hat{f}$, and trains along time after pretraining with setting the embedding of former snapshot $\mathcal{G}_{n-1}$ as the initial embedding of the latter one $\mathcal{G}_n$, where snapshots are separated on account of timestamps.

4.4   Implementation Details

We conduct all experiments of our model and baselines using Pytorch on a Intel(R) Xeon(R) Gold 5117 CPU, Tesla V100 GPUs, and 250 GB Memory server. The software of experiment environment is Ubuntu 18.04 with CUDA 11.4. We evaluate our model with setting the learning rate of Adam as 0.001, batch size as 512 in ICEWS05-15 and 128 in other datasets. Moreover, the initiate embedding dimension is set as 100 for both entities and timestamps while output dimension is set as 200, label smooth in two datasets of ICEWS is set as 0 and 0.01 in the other datasets. We use the released code of baselines and the parameters in baselines are set as their default settings.

4.5   Results and Comparison

In this part, we show our performance against other baselines and make an analysis on the results.

Table 2 demonstrates entity prediction results comparing with other models on the popular datasets for TKG completion task. Different from previous works, we use respective head and tail entity prediction to evaluate for rigorous comparison. For saving space, H@1, H@3, and H@10 are used to replace Hits@1, Hits@3, Hits@10 in this section. Our work shows outstanding performance against other baselines on ICEWS14 and ICEWS05-15. TAGCN delivers an increment of 2% on MRR on these two datasets. Although we still observe that TAGCN does not always achieve the best results from Table 2 on YAGO11k, we achieve the best results in mean metrics. Further analysis shows that this attributes to the

Table 2. Entity prediction results on several popular datasets. The best results of each metric are in bold, and the second ones are underlined. The percent sign is omitted for all data.

Metrics	MRR		H@1		H@3		H@10		MRR		H@1		H@3		H@10	
	t	h	t	h	t	h	t	h	t	h	t	h	t	h	Tail	Head
	ICEWS14								ICEWS05-15							
TransE	33.4	29.4	17.2	12.0	45.3	38.1	67.2	61.0	34.2	30.8	17.5	13.3	45.7	40.3	68.4	62.6
DistMult	50.7	37.1	36.4	28.6	52.1	46.5	73.1	59.9	47.1	44.1	36.9	30.7	55.1	47.9	71.9	66.3
ComplEx	50.7	37.7	42.4	37.6	45.7	40.1	68.9	64.1	49.5	42.9	36.8	32.2	54.9	49.7	72.7	66.1
ConvE	51.2	40.8	37.2	31.0	54.2	49.4	74.1	65.7	48.9	44.5	37.3	31.1	56.2	49.6	73.1	68.1
TTransE	27.2	23.8	11.1	3.7	43.7	37.2	62.1	57.9	32.0	22.2	10.0	6.8	42.8	38.6	64.1	59.1
HyTE	30.5	25.7	12.8	7.3	45.1	36.7	65.4	63.5	33.9	28.1	15.2	8.1	46.9	40.5	70.5	65.8
TA-DistMult	49.1	46.3	38.4	34.2	46.7	40.9	71.1	66.1	50.2	44.6	38.3	30.9	50.1	44.9	75.8	69.8
DE-SimplE	52.1	47.5	40.1	37.3	59.7	52.9	75.5	66.9	54.3	48.7	43.2	37.8	60.3	56.5	74.2	69.8
TNTComplEx	59.1	53.9	54.1	40.1	66.1	57.7	78.1	69.9	62.1	58.3	54.7	45.9	66.8	59.2	77.3	70.7
RTFE	<u>62.1</u>	<u>56.5</u>	<u>56.8</u>	<u>43.8</u>	<u>68.4</u>	<u>58.8</u>	<u>78.9</u>	<u>70.0</u>	<u>64.2</u>	<u>61.2</u>	<u>56.7</u>	<u>53.1</u>	<u>70.4</u>	<u>61.0</u>	<u>81.7</u>	<u>76.9</u>
TAGCN	**63.7**	**59.1**	**58.3**	**49.9**	**70.4**	**61.8**	**80.1**	**71.9**	**66.3**	**62.0**	**57.2**	**53.4**	**71.2**	**65.7**	**85.1**	**77.3**
	Wikidata12k								YAGO11k							
TransE	21.1	17.3	12.4	8.2	21.2	17.0	42.4	25.2	14.2	6.4	1.9	0.7	18.2	9.2	36.2	13.4
DistMult	24.1	20.1	13.8	10.1	27.1	20.3	48.1	44.3	17.2	13.4	12.4	8.8	19.7	12.7	33.3	20.1
ComplEx	25.6	21.2	14.9	10.0	28.3	22.1	47.7	39.3	18.1	14.5	12.8	8.2	17.5	13.5	34.4	21.8
ConvE	24.3	20.7	14.8	10.6	25.9	18.7	42.1	39.7	15.8	11.8	10.1	7.3	15.4	11.6	33.2	17.4
TTransE	21.1	13.3	11.4	7.8	22.8	14.0	40.9	24.9	14.0	7.1	3.2	0.8	22.3	7.7	34.6	15.6
HyTE	21.2	15.1	11.5	7.7	24.1	13.5	40.9	27.1	14.0	6.5	2.8	0.7	21.5	8.0	35.2	11.7
TA-DistMult	24.5	18.9	14.1	10.3	25.7	20.7	47.1	42.7	18.4	14.2	13.1	9.5	22.9	10.5	39.7	19.3
DE-SimplE	31.3	19.7	17.9	11.7	27.4	22.8	55.6	42.6	17.3	13.5	10.1	6.5	31.9	14.0	33.8	19.6
TNTComplEx	45.1	26.7	28.7	24.3	57.2	30.0	66.9	49.6	31.7	18.3	21.9	11.8	34.5	16.3	50.9	26.8
RTFE	<u>52.5</u>	<u>37.3</u>	<u>41.9</u>	<u>29.3</u>	<u>58.7</u>	<u>41.8</u>	<u>68.3</u>	<u>54.8</u>	<u>32.5</u>	<u>18.3</u>	<u>23.4</u>	<u>14.2</u>	<u>35.3</u>	<u>16.9</u>	<u>51.9</u>	**27.6**
TAGCN	**53.4**	**37.3**	**43.2**	**29.5**	**59.5**	**41.9**	**70.3**	**55.4**	**33.4**	**18.5**	**23.7**	**14.3**	**35.7**	**17.1**	**53.1**	<u>27.4</u>

continuity of these two datasets. YAGO11k contains facts with their starting and ending time. This characteristic accommodates for the snapshot separation of RTFE. To be more specific, the continuity makes the timestamps between the time span have a better understanding of the maintaining facts. While our model only aggregate the temporal information after the preprocessing of temporal data, our training strategy brings out the fluctuations. To conclude, the properties of dataset have a large impact on the processing procedure.

Meanwhile, it is also interesting to find that methods for static knowledge graph are better than the previous temporal knowledge graph [1,16]. From our analysis, it is explicit that encoding temporal information into the same space of entities with linear function cannot help the integration. It is crucial to improve the expressiveness, and the temporal effect should be emphasized.

Table 3. Ablation experiments on ICEWS14. TAGCN is implemented with TeA layer, 1-layer GCN, and temporal modified decoder.

Method	MRR	H@1	H@3	H@10
TMA + time-aware decoder	56.0	43.1	61.3	72.6
Only time-aware decoder	54.8	42.8	60.7	71.3
TAGCN with conventional decoder	48.1	37.8	55.7	70.7
TAGCN	61.8	54.1	66.1	76.0

4.6 Ablation Study

To examine the effect of our model, we further conduct ablation study on each module. The results are displayed in Table 3. For saving place, we use mean metrics of head and tail entity predictions as the final results in the further experiments. Consisted of encoder-decoder framework shown in the Fig. 2, we assign the ablation study as follows:

Firstly, we hide the TeA layer and only use a linear function to integrate timestamp, relation, and head entity embeddings as the neighbor message to update all representations of entities, it is obvious that performance on ICEWS14 degrades about 10% in H@1. The result attests that our proposed temporal information integration method has a better comprehension on facts impact and encodes temporal dependency more effectively.

Secondly, we construct a model only utilizing the time-aware decoder to train for representation learning. From Table 3, we can find that the performance is behind the best performance of TAGCN, which denotes that the temporal encoder could definitely capture neighboring structure information and lift the expressive capability. Meanwhile, the accuracy reduces slightly comparing with the former study, we can interpret that temporal dependency in TKG has greater impact than adjacent information.

Lastly, to analyse the importance of our decoder, we resort to a new structure of TAGCN with a conventional decoder. In other words, we use the original ConvE [14] as the decoder, thus overlooking the temporal impact on entity. The result in Table 3 implies that the modification on decoder helps the entity concentrates on the within-time facts impact and chooses the most appropriate candidate.

4.7 Parameter Analysis

Besides, to evaluate our model, we analyze important parameters in TAGCN: number of aggregation layers l, the embedding dimension d, and temporal integration variant operation ope. We conduct these experiments on ICEWS14 and ICEWS05-15.

Number of Layers l. Considering multi-hop neighbors could enhance full graph perception and increase the global understanding. The results are shown in Fig. 3.

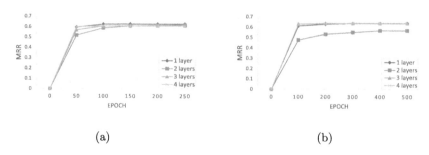

(a)   (b)

Fig. 3. Experimental results of influence with layer numbers, we conduct this experiment on ICEWS14 (a) and ICEWS05-15 (b). The X-axis represents epochs, while the Y-axis is the MRR of entity prediction on each dataset.

We show how the layer of aggregation can influence TAGCN. Experimental results in Fig. 3 indicate that training with two layers undermines the accuracy in both datasets. More training layers only bring little gain at the cost of time overhead and mainly accelerate the convergence in the early stage of the model.

The Embedding Dimension d. It is obvious that the expressiveness is related to the embedding dimension d. As shown in Fig. 4(a), TAGCN does not need too little or too superfluous embedding dimension to capture the features in TKG. Therefore, the proper size of embedding dimension is around 200.

Temporal Integration Operation ope. Further, we evaluate the effectiveness of TAGCN with different operations, such as linear calculations and convolution operation, to integrate timestamp and entity embeddings in the decoder. The experimental results are listed in Fig. 4(b), Sum and $Mult$ have better performance than $Conv$, which denotes that too complex approach will sabotage the representation and integration. To balance the performance and the complexity, we use MLP as temporal integration operation.

4.8 Further Analysis

As mentioned in the former section, we hypothesize that the temporal change in the relation is too small. Considering the example mentioned in Sect. 1, the meaning and usage of relation *cooperate* has been maintained for a long time. However, the characteristics of figures change with time. It is obvious that *Keanu Reeves* was the actor in different movie in *31/3/1999* and *17/5/2019*. So the temporal change in entity is explicitly more significant than that in relation. To verify this point, we employ the temporal embedding in three approaches respectively to compare the accuracy. This experiment is conducted on ICEWS14. The result is shown in Table 4.

From Table 4, we can confirm that decoder with only temporal embedding on entities outperforms the other approaches. Temporal embedding only on entities is 2% ahead of that on relations. Temporal embeddings on both entities and relations is 1.8% behind because the temporal changes in relations mislead

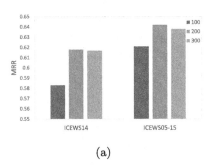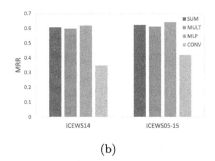

(a)   (b)

Fig. 4. Comparison of different parameters: the embedding dimension (a), temporal integration operation (b). The X-axis represents different datasets, while the Y-axis is the MRR of entity prediction.

Table 4. Further analysis of temporal embedding

Method	MRR	H@1	H@3	H@10
Temporal embedding on both relations and entities	60.0	52.1	64.3	74.6
Temporal embedding on relations only	59.8	51.8	64.7	73.3
Temporal embedding on entities only	61.8	54.1	66.1	76.0

the representation learning for entity prediction. This experiment corroborates the hypothesis we suggest: entity is more sensitive to temporal impact while relation may evolve at a very low rate, and forcing relations to give reflect to temporal information will only degrade the performance. Thus only using static representations to model relations is more sufficient.

4.9 Time Prediction

The visualization of time prediction on three facts is shown in Table 5. For all timestamps in the time set $\mathcal{T}$, we calculate the probability of the object entity among all candidates. For a more intuitive comparison, we reserve the non-zero position in the same order of magnitude. We only pick the top three scores for display to save place.

From Table 5, it is explicit that the true timestamp has the highest score in the second fact and among the top three scores in the other two facts. We collate the dataset and find that in ICEWS14, some facts appear in different timestamps, which is consistent with our results. For example, the fact (*Angela_Merkel, Express_intent_to_cooperate, Barack_Obama*) occurred three times corresponding to listed timestamps in ICEWS14 and same for the other facts. So our model can obviously focus on the appropriate timestamps and improve the decoder performance.

Table 5. Case study of three time prediction on ICEWS14

Gold facts	Timestamps	Scores
(Merkel, Intent_to_cooperate, Obama, 19/03/2014)	03/07/2014	5.1949
	19/03/2014	4.6959
	09/08/2014	3.6903
(Portugal, Consult, European_Central_Bank, 24/10/2014)	24/10/2014	3.9880
	21/10/2014	1.7212
	21/02/2014	1.0325
(Juan_Carlos_I, Make_statement, Felipe_de_Borbon, 03/06/2014)	03/06/2014	3.6341
	03/06/2014	2.5742
	11/02/2014	1.3227

5   Conclusion

In this paper, we propose a time-aware attention graph convolutional network TAGCN for link prediction in TKG. Inspired by self-attention layer in Transformer, we bring out a novel message generator for neighboring temporal message. To accomplish entity prediction in temporal knowledge graph, we decouple this task into two phases, using the encoder to aggregate neighboring semantic and temporal information, and acquire different temporal impact in decoding phase on account of the query. The proposed TeA layer is used to capture the neighboring information of all involved facts when considering the central entity. We conduct abundant experiments on real-world datasets, and the results show that TAGCN achieves best performance. In the future, we will investigate the better approach to encode potential temporal information from a novel angle.

Acknowledgements. The work is supported by the National Key Research and Development Program of China (Grant No. 2020AAA0108501) and the National Natural Science Foundation of China (Grant No. 62172174, 62127808).

References

1. Jiang, T., et al.: Towards time-aware knowledge graph completion. In: Proceedings of COLING, pp. 1715–1724 (2016)
2. Lacroix, T., Obozinski, G., Usunier, N.: Tensor decompositions for temporal knowledge base completion. In: Proceedings of ICLR (2020)
3. Bordes, A., Usunier, N., García-Durán, A., Weston, J., Yakhnenko, O.: Translating embeddings for modeling multi-relational data. In: Proceedings of NeurIPS (2013)
4. Jin, W., Qu, M., Jin, X., Ren, X.: Recurrent event network: autoregressive structure inference over temporal knowledge graphs. In: Proceedings of EMNLP, pp. 6669–6683 (2020)
5. Xu, Y., et al.: RTFE: a recursive temporal fact embedding framework for temporal knowledge graph completion. In: Proceedings of NAACL-HLT, pp. 5671–5681 (2021)

6. Nathani, D., Chauhan, J., Sharma, C., Kaul, M.: Learning attention-based embeddings for relation prediction in knowledge graphs. In: Proceedings of ACL, pp. 4710–4723 (2019)

7. Vaswani, A., et al.: Attention is all you need. In: Proceedings of NeurIPS (2017)

8. Wang, Z., Zhang, J., Feng, J., Chen, Z.: Knowledge graph embedding by translating on hyperplanes. In: Proceedings of AAAI, pp. 1112–1119 (2014)

9. Ji, G., He, S., Xu, L., Liu, K., Zhao, J.: Knowledge graph embedding via dynamic mapping matrix. In: Proceedings of ACL, pp. 687–696 (2015)

10. Lin, Y., Liu, Z., Sun, M., Liu, Y., Zhu, X.: Learning entity and relation embeddings for knowledge graph completion. In: Proceedings of AAAI, pp. 2181–2187 (2015)

11. Nickel, M., Tresp, V., Kriegel, H.P.: A three-way model for collective learning on multi-relational data. In: Proceedings of ICLR, pp. 809–816 (2011)

12. Yang, B., Yih, W.T., He, X., Gao, J., Deng, L.: Embedding entities and relations for learning and inference in knowledge bases. In: Proceedings of ICLR (2015)

13. Trouillon, T., Welbl, J., Riedel, S., Gaussier, É., Bouchard, G.: Complex embeddings for simple link prediction. In: Proceedings of ICML, pp. 2071–2080 (2016)

14. Dettmers, T., Minervini, P., Stenetorp, P., Riedel, S.: Convolutional 2D knowledge graph embeddings. In: Proceedings of AAAI, pp. 1811–1818 (2018)

15. Vashishth, S., Sanyal, S., Nitin, V., Agrawal, N., Talukdar, P.: InteractE: improving convolution-based knowledge graph embeddings by increasing feature interactions. In: Proceedings of AAAI, pp. 3009–3016 (2020)

16. Dasgupta, S., Ray, S., Talukdar, P.: HyTE: hyperplane-based temporally aware knowledge graph embedding. In: Proceedings of EMNLP, pp. 2001–2011 (2018)

17. García-Durán, A., Dumancic, S., Niepert, M.: Learning sequence encoders for temporal knowledge graph completion. In: Proceedings of EMNLP, pp. 4816–4821 (2018)

18. Goel, R., Kazemi, S., Brubaker, M., Poupart, P.: Diachronic embedding for temporal knowledge graph completion. In: Proceedings of AAAI, pp. 3988–3995 (2020)

19. Jung, J., Jung, J., Kang, U.: Learning to walk across time for interpretable temporal knowledge graph completion. In: Proceedings of KDD (2021)

20. Li, Z., et al.: Temporal knowledge graph reasoning based on evolutional representation learning. In: Proceedings of SIGIR (2021)

Start Small, Think Big: On Hyperparameter Optimization for Large-Scale Knowledge Graph Embeddings

Adrian Kochsiek$^{(\boxtimes)}$ ⓘ, Fritz Niesel ⓘ, and Rainer Gemulla ⓘ

University of Mannheim, Mannheim, Germany
{akochsiek,fniesel,rgemulla}@uni-mannheim.de

Abstract. Knowledge graph embedding (KGE) models are an effective and popular approach to represent and reason with multi-relational data. Prior studies have shown that KGE models are sensitive to hyperparameter settings, however, and that suitable choices are dataset-dependent. In this paper, we explore hyperparameter optimization (HPO) for very large knowledge graphs, where the cost of evaluating individual hyperparameter configurations is excessive. Prior studies often avoided this cost by using various heuristics; e.g., by training on a subgraph or by using fewer epochs. We systematically discuss and evaluate the quality and cost savings of such heuristics and other low-cost approximation techniques. Based on our findings, we introduce GRASH, an efficient multi-fidelity HPO algorithm for large-scale KGEs that combines both graph and epoch reduction techniques and runs in multiple rounds of increasing fidelities. We conducted an experimental study and found that GRASH obtains state-of-the-art results on large graphs at a low cost (three complete training runs in total). Source code and auxiliary material at https://github.com/uma-pi1/GraSH.

Keywords: Knowledge graph embedding · Multi-fidelity hyperparameter optimization · Low-fidelity approximation

1 Introduction

A knowledge graph (KG) is a collection of facts describing relationships between a set of entities. Each fact can be represented as a (subject, relation, object)-triple such as (*Rami Malek, starsIn, Mr. Robot*). Knowledge graph embedding (KGE) models [4,8,16,21,23,28] represent each entity and each relation of the KG with an *embedding*, i.e., a low-dimensional continuous representation. The embeddings are used to reason about or with the KG; e.g., to predict missing facts in an incomplete KG [15], for drug discovery in a biomedical KG [14], for question answering [18,19], or visual relationship detection [2].

Prior studies have shown that embedding quality is highly sensitive to the hyperparameter choices used when training the KGE model [1,17]. Moreover,

M.-R. Amini et al. (Eds.): ECML PKDD 2022, LNAI 13714, pp. 138–154, 2023.
https://doi.org/10.1007/978-3-031-26390-3_9

the search space is large and hyperparameter choices are dataset- and model-dependent. For example, the best configuration found for one model may perform badly for a different one. As a consequence, we generally cannot transfer suitable hyperparameter configurations from one dataset to another or from one KGE model to another. Instead, a separate hyperparameter search is often necessary to achieve high-quality embeddings.

While using an extensive hyperparameter search may be feasible for smaller datasets—e.g., the study of Ruffinelli et al. [17] uses 200 configurations per dataset and model—, such an approach is generally not cost-efficient or even infeasible on large-scale KGs, where KGE training is expensive in terms of runtime, memory consumption, and storage cost. For example, the Freebase KG consists of $\approx 86\,M$ entities and more than $300\,M$ triples. A single training run of a 512-dimensional ComplEx embedding model on Freebase takes up to $50\,min$ per epoch utilizing 4 GPUs and requires $\approx 164\,GB$ of memory to store the model.

To reduce these excessive costs, prior studies on large-scale KGE models either avoid hyperparameter optimization (HPO) altogether or reduce runtime and memory consumption by employing various heuristics. The former approach leads to suboptimal quality, whereas the impact in terms of quality and cost of the heuristics used in the latter approach has not been studied in a principled way. The perhaps simplest of such heuristics is to evaluate a given hyperparameter configuration using only a small number of training epochs (e.g., [11] uses only 20 epochs for HPO on the Wikidata5M dataset). Another approach is to use a small subset of the large KG (e.g., the small FB15k benchmark dataset instead of full Freebase) to obtain a suitable hyperparameter configuration [11,13,30] or a set of candidate configurations [29]. The general idea behind these heuristics is to employ *low-fidelity approximations* (fewer epochs, smaller graph) to compare the performance of different hyperparameter configuration during HPO, before training the final model on *full fidelity* (many epochs, entire graph).

In this paper, we explore how to effectively use a given HPO budget to obtain a high-quality KGE model. To do so, we first summarize and analyze both cost and quality of various low-fidelity approximation techniques. We found that there are substantial differences between techniques and that a combination of reducing the number of training epochs and the graph size is generally preferable. To reduce KG size, we propose to use its *k-core* subgraphs [20]; this simple approach worked best throughout our study.

Building upon these results, we present GRASH, an efficient HPO algorithm for large-scale KGE models. At its heart, GRASH is based on successive halving [10]. It uses multiple fidelities and employs several KGE-specific techniques, most notably, a simple cost model, negative sample scaling, subgraph validation, and a careful choice of fidelities. We conducted an extensive experimental study and found that GRASH achieved state-of-the-art results on large-scale KGs with a low overall search budget corresponding to only three complete training runs. Moreover, both the use of multiple reduction techniques simultaneously and of multiple fidelity levels was key for reaching high quality and low resource consumption.

2   Preliminaries and Related Work

A general discussion of KGE models and training is given in [15,25]. Here we summarize key points and briefly discuss prior approaches to HPO.

Knowledge Graph Embeddings. A *knowledge graph* $\mathcal{G} = (\mathcal{E}, \mathcal{R}, \mathcal{K})$ consists of a set $\mathcal{E}$ of entities, a set $\mathcal{R}$ of relations, and a set $\mathcal{K} \subseteq \mathcal{E} \times \mathcal{R} \times \mathcal{E}$ of triples. *Knowledge graph embedding* models [4,8,16,21,23,28] represent each entity $i \in \mathcal{E}$ and each relation $p \in \mathcal{R}$ with an *embedding* $\boldsymbol{e}_i \in \mathbb{R}^d$ and $\boldsymbol{e}_p \in \mathbb{R}^d$, respectively. They model the plausibility of each subject-predicate-object triple (s, p, o) via a model-specific scoring function $f(\boldsymbol{e}_s, \boldsymbol{e}_p, \boldsymbol{e}_o)$, where high scores correspond to more, low scores to less plausible triples.

Training and Training Cost. KGE models are trained [25] to provide high scores for the positive triples in $\mathcal{K}$ and low scores for negative triples by minimizing a loss such as cross-entropy loss. Since negatives are typically unavailable, KGE training methods employ *negative sampling* to generate *pseudo-negative triples*, i.e., triples that are likely but not guaranteed to be actual negatives. The number N^- of generated pseudo-negatives per positive is an important hyperparameter influencing both model quality and training cost. In particular, during each epoch of training a KGE model, all positives and their associated negatives are scored, i.e., the overall number of per-epoch score computations is $(|\mathcal{K}| + 1)N^-$. We use this number as a proxy for computational cost throughout. The size of the KGE model itself scales linearly with the number of entities and relations, i.e., $O(|\mathcal{E}|d + |\mathcal{R}|d)$ if all embeddings are d-dimensional.

Evaluation and Evaluation Cost. The standard approach to evaluate KGE model quality for link prediction task is to use the *entity ranking* protocol and a filtered metric such as mean reciprocal rank (MRR). For each (s, p, o)-triple in a held-out test set $\mathcal{K}^{\text{test}}$, this protocol requires to score all triples of form $(s, p, ?)$ and $(?, p, o)$ using all entities in $\mathcal{E}$. Overall, $|\mathcal{K}^{\text{test}}||\mathcal{E}|$ scores are computed so that evaluation cost scales linearly with the number of entities. Since this cost can be substantial, sampling-based approximations have been used in some prior studies [13,30]. We do not use such approximations here since they can be misleading in that they do not reflect model quality faithfully [11].

Hyperparameters. The hyperparameter space for KGE models is discussed in detail in [1,17]. Important hyperparameters include embedding dimensionalities, training type, number N^- of negatives, sampling type, loss function, optimizer, learning rate, type and weight of regularization, and amount of dropout.

Full-Fidelity HPO. Recent studies analyzed the impact of hyperparameters and training techniques for KGE models using full-fidelity HPO [1,17]. In these studies, the vast hyperparameter search space was explored using a random search and Bayesian optimization with more than 200 full training runs per model and dataset. The studies focus on smaller benchmark KGs, however; such an approach is excessive for large-scale knowledge graphs.

Low-Fidelity HPO. Current work on large-scale KGE models circumvented the high cost of full-fidelity HPO by relying on low-fidelity approximations such as epoch reduction [5,11] and using smaller benchmark graphs [11,13,30] in a heuristic fashion. The best performing hyperparameters in low-cost approximations were directly applied to train a single full-fidelity model. Our experimental study suggests that such an approach may neither be cost-efficient nor produce high-quality results.

Two-Stage HPO. AutoNE [24] is an HPO approach for training large-scale network embeddings that optimizes hyperparameters in two stages. It first approximates hyperparameter performance on subgraphs created by random walks, a technique that we will explore in Sect. 4. Subsequently, AutoNE transfers these results to the full graph using a meta learner. In the context of KGs, this approach was outperformed by KGTuner [29],[1] which uses a multi-start random walk (fixed to 20% of the entities) in the first stage and evaluates the top-performing configurations (fixed to 10) at full fidelity in the second stage. Such fixed heuristics often limit flexibility in terms of budget allocation and lead to an expensive second stage on large KGs. In contrast, GRASH makes use of multiple fidelity levels, carefully constructs and evaluates low-fidelity approximations, and adheres to a prespecified overall search budget. These properties are key for large KGs; see Sect. 5.3 for an experimental comparison with KGTuner.

3   Successive Halving for Knowledge Graphs (GRASH)

GRASH is a multi-fidelity HPO algorithm for KGE models based on successive halving [10]. As successive halving, GRASH proceeds in multiple *rounds* of increasing fidelity; only the best configurations from each round are transferred to the next round. In contrast to the HPO techniques discussed before, this approach allows to discard unpromising configurations at very low cost. GRASH differs from successive halving mainly in its parameterization and its use of KG-specific reduction and validation techniques.

Parameterization. GRASH is summarized as Alg. 1. Given knowledge graph $\mathcal{G}$, GRASH outputs a single optimized hyperparameter configuration. GRASH is parameterized as described in Alg. 1; default parameter values are given in parentheses if applicable. The most important parameters are the maximal number E of epochs and the overall search budget B. The search budget B is relative to the cost of a full training run, which in turn is determined by E. The default choice $B = 3$, for example, corresponds to an overall search cost of three full training runs. We chose this parameterization because it is independent of utilized hardware and both intuitive and well-controllable. The reduction factor η controls the number of configurations (starts at n, decreases by factor of η per round) and fidelity (increases by factor of η) of each round. Note that GRASH does not train at full fidelity, i.e., its final configuration still needs to be trained

---

[1] KGTuner was proposed in parallel to this work.

Algorithm 1. GRASH: Successive Halving for Knowledge Graph Embeddings

Require:

KG $\mathcal{G} = (\mathcal{E}, \mathcal{R}, \mathcal{K})$, max. epochs E, search budget B (=3), num. configurations n (=64), reduction factor η (=4), variant $v \in \{epoch, graph, combined\}$ (=combined)

Ensure: Hyperparameter configuration

1: $s \leftarrow \lceil \log_\eta(n) \rceil$ ▷ Number of rounds
2: $R \leftarrow B/s$ ▷ Per-round budget
3: $\Lambda_1 \leftarrow \{\lambda_1, ..., \lambda_n\}$ ▷ Generate n hyperparameter configurations
4: **for** $i \in \{1, ..., s\}$ **do**   ▷ i-th round
5:     $f_i \leftarrow R/|\Lambda_i|$ **if** $v \neq combined$ **else** $R/\sqrt{|\Lambda_i|}$ ▷ Target fidelity
6:     $E_i \leftarrow f_i E$ **if** $v \neq graph$ **else** E ▷ Epochs in round i
7:     $\mathcal{G}_i \leftarrow$ reduced KG with $f_i|\mathcal{K}|$ triples **if** $v \neq epoch$ **else** $\mathcal{G}$ ▷ Graph in round i
8:     $\mathcal{G}_i^{\text{train}}, \mathcal{G}_i^{\text{valid}} \leftarrow$ random train-valid split of $\mathcal{G}_i$
9:     $V_i \leftarrow$ train each $\lambda \in \Lambda_i$ on $\mathcal{G}_i^{\text{train}}$ for E_i epochs and validate using $\mathcal{G}_i^{\text{valid}}$
10:     $\Lambda_{i+1} \leftarrow$ best $\lceil |\Lambda_i|/\eta \rceil$ configurations from Λ_i according to V_i
11: **end for**
12: **return** Λ_{s+1} ▷ Only single configuration left

---

on the full KG (not part of budget B). Finally, GRASH is parameterized by a *variant* v. This parameter controls which reduction technique to use (only epoch, only graph, or combined).

Algorithm Overview. Like successive halving, GRASH proceeds in rounds. Each round has approximately the same overall budget, but differs in the number of configurations and fidelity. For example, using the default settings of $B = 3$, $n = 64$ and $\eta = 4$, GRASH uses three rounds with 64, 16, and 4 configurations and a fidelity of $1/64, 1/16, 1/4$, respectively. The hyperparameter configurations in the first round are sampled randomly from the hyperparameter space. Depending on the variant being used, GRASH reduces the number of epochs, the graph size, or both to reach the desired fidelity. If no reduced graph corresponds to the fidelity, the next smaller one is used. After validating each configuration (see below), the best performing $1/\eta$-th of the configurations is passed on to the next round. This process is repeated until only one configuration remains.

Validation on Subgraphs. Care must be taken when validating a KGE model trained on a subgraph, e.g., $\mathcal{G}_i = (\mathcal{E}_i, \mathcal{R}_i, \mathcal{K}_i)$ in round i. Since $\mathcal{G}_i$ typically contains a reduced set of entities $\mathcal{E}_i \subseteq \mathcal{E}$, a full validation set for $\mathcal{G}$ cannot be used. This is because no embedding is learned for the "unseen" entities in $\mathcal{E} \setminus \mathcal{E}_i$, so that we cannot score any triples containing these entities (as required by the entity ranking protocol). To avoid this problem, we explicitly create new train and valid splits $\mathcal{G}_i^{\text{train}}$ and $\mathcal{G}_i^{\text{valid}}$ in round i. Here, $\mathcal{K}_i^{\text{valid}}$ is sampled randomly from $\mathcal{K}_i$ and $\mathcal{K}_i^{\text{train}} = \mathcal{K}_i \setminus \mathcal{K}_i^{\text{valid}}$. Although this approach is very simple, it worked well in our study. An alternative is the construction of "hard" validation sets as in [22]. We leave the exploration of such techniques to future work.

Negative Sample Scaling. Recall that the number N^- of negative samples is an important hyperparameter for KGE model training. Generally (and assuming without-replacement sampling), each entity is sampled as a negative with probability $N^-/|\mathcal{E}|$. When we use a subgraph $\mathcal{G}_i$ as in GRASH, this probability increases to $N^-/|\mathcal{E}_i|$, i.e., each entity is more likely to act as a negative sample due to the reduction of the number of entities. To correctly assess hyperparameter configurations in such cases, GRASH scales the number of negative examples and uses $N_i^- = \frac{|\mathcal{E}_i|}{|\mathcal{E}|} N^-$ in round i. This choice preserves the probability of sampling each entity as a negative and provides additional cost savings since the total number of scored triples is further reduced in low-fidelity experiments.

Cost Model and Budget allocation. To distribute the search budget B over the rounds, we make use of a simple cost model to estimate the relative runtime of low-fidelity approximations. This cost model drives the choice of f_i in Algorithm 1. In particular, we assume that training cost is linear in both the number of epochs (E_i) and the number of triples ($|\mathcal{K}_i|$). For example, this implies that training five configurations for one epoch has the same cost as training one configuration for five epochs. Likewise, training five configurations with 20% of the triples has the same cost as training one configuration on the whole KG. Using this assumption, the relative cost of evaluating a single hyperparameter configuration in round i is given by $\frac{E_i}{E} \frac{|\mathcal{K}_i|}{|\mathcal{K}|}$. More elaborate cost models are conceivable, but this simple approach already worked well in our experimental study. Note, for example, that our simple cost model neglects negative sample scaling and thus tends to overestimate (but avoids underestimation) of training cost.

4   Low-fidelity Approximation Techniques

In this section, we summarize and discuss various low-fidelity approximation techniques. As discussed previously, the two most common types are *graph reduction* (i.e. training on a reduced graph) and *epoch reduction* (i.e., training for fewer epochs). Note that although graph reduction is related to dataset reduction techniques used in other machine learning domains, it represents a major challenge since the relationships between entities need to be taken into account.

Generally, good low-fidelity approximations satisfy the following criteria:

1. **Low cost.** Computational and memory costs for model training (including model initialization) and evaluation should be low. Recall that computational costs are mainly determined by the number of triples, whereas memory and evaluation cost are determined by the number of entities. Ideally, both quantities are reduced.
2. **High transferability.** Low-fidelity approximations should transfer to the full KG in that they provide useful information. E.g., rankings of low-fidelity approximations should match or correlate with the rankings at full-fidelity.

Table 1. Comparison of low-fidelity approximation techniques.

Technique	Low Cost	High Transferability	Flexibility
Triple sampling	o	−	+
Random walk	o	o	+
k-core decomposition	+	+	o
Epoch reduction	−	o	+

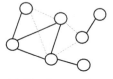  

(a) Triple sampling (60%).

(b) Random walk ($s = 2$, $l = 3$).

(c) k-core decomposition ($k = 3$).

Fig. 1. Schematic illustration of selected graph reduction techniques. All reduced graphs contain 6 of the 10 original triples but a varying number of entities.

3. **Flexibility.** It should be possible to flexibly trade-off computational cost and transferability.

All three points are essential for cost-effective and practical multi-fidelity HPO.

In the following, we present the graph reduction approaches *triple sampling*, *multi-start random walk*, and *k-core decomposition*, as well as epoch reduction. A high-level comparison of these approaches w.r.t. the above desiderata is provided in Table 1. The assessment given in the table is based on our experimental results (Sect. 5.2).

4.1   Graph Reduction

Graph reduction techniques produce a reduced KG $\mathcal{G}_i = (\mathcal{E}_i, \mathcal{R}_i, \mathcal{K}_i)$ from the full KG $\mathcal{G} = (\mathcal{E}, \mathcal{R}, \mathcal{K})$. This is commonly done by first determining the reduced set $\mathcal{K}_i$ of triples and subsequently retaining only those entities (in $\mathcal{E}_i$) and relations (in $\mathcal{R}_i$) that occur in $\mathcal{K}_i$.[2] A reduction in triples thus may lead to a reduction in the number of entities and relations as well. This consequently results in further savings in computational cost, evaluation cost, and memory consumption. The graph reduction techniques discussed here are illustrated in Fig. 1.

Triple Sampling (Fig. 1a). The perhaps simplest approach to reduce graph size is to sample triples randomly from the graph. As shown in Fig. 1a, many entities

---

[2] All other entities/relations do not occur in the reduced training data so that we cannot learn useful embeddings for them.

with sparse interconnections can remain in the resulting subgraphs (e.g., the two entities at the top right) so that $\mathcal{E}_i$ tends to be large. The cost in terms of model size and evaluation time is consequently only slightly reduced. We also observed (see Sect. 5.2) that triple sampling leads to low transferability, most likely due to this sparsity. Triple sampling does offer very good flexibility, however, since triple sets of any size can be constructed easily.

Random Walk (Fig. 1b). In multi-start random walk, which is used in AutoNE [24], a set of s random entities is samples from $\mathcal{E}$. A random walk of length l is started from each of these entities and the resulting triples form $\mathcal{K}_i$. Empirically, many entities may ultimately remain so that the reduction of memory consumption and evaluation cost is limited. Although the resulting subgraph tends to be better connected than the ones obtained by triple sampling, transferability is still low and close to triple sampling (again, see Sect. 5.2). As triple sampling, the approach is very flexible though. KGTuner [29] improves on the basic random walk considered here by using biased starts and adding all connections between the retained entities (even if they do not occur in a walk). The k-core decomposition, which we discuss next, offers a more direct approach to obtain such a highly-connected graph.

k-core Decomposition (Fig. 1c). The k-core decomposition [20] allows for the construction of subgraphs with increasing cohesion. The k-core subgraph of $\mathcal{K}$, where $k \in \mathbb{N}$ is a parameter, is defined as the largest induced subgraph in which every retained entity (i.e., $\mathcal{E}_i$) occurs in at least k retained triples (i.e., in $\mathcal{K}_i$). The computation of k-cores is cheap and supported by common graph libraries. Generally, k-cores contain only a small number of entities because long-tail entities with infrequent connections are removed. Moreover, they are highly interconnected by construction. As a consequence, we found that computational cost and memory consumption is low and transferability high. The approach is less flexible than the other graph reduction techniques, as the choice of k and the graph structure determines the resulting fidelity. One may interpolate between k-cores for improved flexibility but we do not explore this approach in this work.

4.2 Epoch Reduction

Epoch reduction is the most common form of fidelity control used in HPO [3, 26]. As the set $\mathcal{E}$ of entities does not change with varying fidelity, memory and evaluation cost are very large even when using low-fidelity approximations. We observed good transferability as long as the number of epochs is not too small (Sect. 5.2); otherwise, transferability is often considerably worse than graph reduction techniques. This limits flexibility: Especially on large-scale graphs, the overall training budget often consists of only a small number of epochs in the first place (e.g., 10 as in [11, 30]). Note that the available budget in low-fidelity approximations can be smaller than the cost of one complete epoch (when $f_i < 1/E$ in Algorithm 1). Although partial epochs can be used easily, epoch reduction then corresponds to a form of triple sampling (with the additional disadvantage of not reducing the set of entities).

Table 2. Dataset statistics.

Scale	Dataset	Entities	Relations	|Train|	|Valid|	|Test|
Small	Yago3-10	123 182	37	1 079 040	5 000	5 000
Medium	Wikidata5M	4 594 485	822	21 343 681	5 357	5 321
Large	Freebase	86 054 151	14 824	304 727 650	1 000	10 000

4.3   Summary

In summary, as long the desired fidelity is sufficiently high, epoch reduction offers high-quality approximations and high flexibility. It does not improve memory consumption and evaluation cost, however, and it leads to high cost and low quality on large-scale graphs with limited budget. Graph reduction approaches, on the other hand, reduce the number of entities and hence memory consumption and evaluation cost. Compared to triple sampling and random walks, the k-core decomposition has the highest transferability and lowest cost. In GRASH, we use a combination of epoch reduction and k-core decomposition by default to avoid training for partial epochs and the use of very small subgraphs with low-fidelity.

5   Experimental Study

We conducted an experimental study to investigate (i) to what extent hyperparameter rankings obtained with low-fidelity approximations correlate with the ones obtained at full fidelity (Sect. 5.2); (ii) the performance of GRASH in terms of quality (Sect. 5.3), resource consumption (Sect. 5.4) and robustness (Sect. 5.5). In summary, we found that:

1. GRASH was cost-effective and produced high-quality hyperparameter configurations. It reached state-of-the-art results on a large-scale graph with a small overall search budget of three complete training runs (Sect. 5.3).
2. Using multiple reduction techniques was beneficial. In particular, a combination of graph- and epoch-reduction performed best (Sect. 5.2 and 5.3).
3. Low-fidelity approximations correlated best to full fidelity for graph reduction using the k-core decomposition and, as long as the budget was sufficiently large, second-best for epoch reduction (Sect. 5.2).
4. Graph reduction was more effective than epoch reduction in terms of reducing computational and memory cost. Evaluation using small subgraphs had low memory consumption and short runtimes (Sect. 5.4).
5. Using multiple rounds with increasing fidelity levels was beneficial (Sect. 5.5).
6. GraSH was robust to changes in budget allocation across rounds (Sect. 5.5).

5.1   Experimental Setup

Source code, search configurations, resulting hyperparameters, and an online appendix can be found at https://github.com/uma-pi1/GraSH.

Datasets. We used common KG benchmark datasets of varying sizes with a focus on larger datasets; see Table 2. *Yago3-10* [8] is a subset of Yago 3 containing only entities that occur at least ten times in the complete graph. *Wikidata5M* [27] is a large-scale benchmark and the induced graph of the five million most-frequent entities of Wikidata. The largest dataset is *Freebase* as used in [11,30]. For all datasets except Freebase, we use the validation and test sets that accompany the datasets to evaluate the final model. For Freebase, we used the sub-sampled validation (1 000 triples) and test sets (10 000 triples) from [11].[3]

Hardware. All runtime, GPU memory, and model size measurements were taken on the same machine (40 Intel Xeon E5-2640 v4 CPUs @ 2.4GHz; 4 NVIDIA GeForce RTX 2080 Ti GPUs).

Implementation and Models. GRASH uses DISTKGE [11] for parallel training of large-scale graphs and HpBandSter [9] for the implementation of SH. We considered the models ComplEx [23], RotatE [21] and TransE [4]. ComplEx and RotatE are among the currently best-performing KGE models [1,12,17,21] and represent semantic matching and translational distance models, respectively. All three models are commonly used for large-scale KGEs [11,13,30].

Hyperparameters. We used the hyperparameter search space of [11]. The search space consists of nine continuous and two categorical hyperparameters. The upper bound on the number of negative samples for ComplEx is 10 000 and for RotatE and TransE 1 000 (since these models are more memory-hungry). We set the maximum training epochs on Yago3-10 to 400, on Wikidata5M to 64, and on Freebase to 10.

Methodology. For the GRASH search, we used the default settings ($B = 3$, $\eta = 4$, $n = 64$). Apart from the upper bound of negatives, we used the same 64 initial hyperparameter settings for all models and datasets to allow for a fair comparison. For graph reduction, we used k-core decomposition unless mentioned otherwise. Subgraph validation sets generated by GRASH consisted of 5 000 triples. The resulting best configurations are published along with our online appendix.

Metrics. We used the common filtered MRR metric to evaluate KGE model quality on the link prediction task as described in Sect. 2. Results for Hits@k are given in our online appendix.

5.2   Comparison of Low-Fidelity Approximation Techniques (Fig. 2)

In our first experiment, we studied and compared the transferability of low-fidelity approximations to full-fidelity results. To do so, we first ran a full-fidelity hyperparameter search consisting of 30 pseudo-randomly generated trials.

---

[3] The original test set contains $\approx$17 M triples, which leads to excessive evaluation costs. For the purpose of MRR computation, a much smaller test set is sufficient.

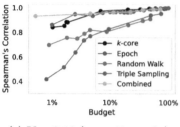

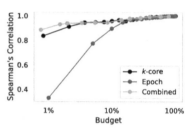

(a) Yago3-10 (max. 40 epochs).   (b) Wikidata5M (max. 20 epochs).

Fig. 2. Comparison of low-fidelity approximations techniques. Shows Spearman's rank correlation between low-fidelity approximations and a full-fidelity baseline. Budget (log-scale) corresponds to the relative amount of epochs and/or triples.

We then trained and evaluated the same 30 trials using the approximation techniques described in Sect. 4 at various budgets. To keep computational cost feasible, this experiment was only performed on the two smaller datasets and with a smaller number of epochs.

Since the validation sets used with graph reductions differ from the one used at full fidelity (see Sect. 3), we compared the ranking of hyperparameter configurations instead of their MRR metrics. In particular, we used Spearman's rank correlation coefficient [31] between the low-fidelity and the high-fidelity results. A higher value corresponds to a better correlation.

Our results on Yago3-10 are visualized in Fig. 2a. We found high transferability for the k-core decomposition and epoch reduction. Graph reduction based on triple sampling and random walks led to clearly inferior results and was not further considered. A combination of k-core subgraphs and reduced epochs (each contributing 50% to the savings) further improved low-budget results.

To investigate the behavior on a larger graph, we evaluated the three best techniques on Wikidata5M, see Fig. 2b. Recall that due to the high cost, a small number of epochs is often used for training on large KGs. This has a detrimental effect on the transferability of epoch reduction, as partial epochs need to be used for low-fidelity approximations (see Sect. 4.2). In particular, there is a considerable drop in transferability for epoch reduction below the 10% budget. This drop in performance is neither visible for the k-core approximations nor for the combined approach.

Note that even for the best low-fidelity approximation, the rank correlation increased with budget. This suggests that using multiple fidelities (as in GRASH) instead of a single fidelity is beneficial. In our study, this was indeed the case (see Sect. 5.5).

5.3 Final Model Quality (Table 3)

In our next experiment, we analyzed the performance of GRASH in terms of the quality of its selected hyperparameter configuration. Table 3a shows the test-data

Table 3. Model quality in terms of MRR. State-of-the-art results underlined. Best reduction variant in bold. Note that best prior results often use a considerably larger budget and/or model dimensionality.

(a) GRASH with default settings ($B = 3$, $n = 64$, $\eta = 4$).   (b) Prior results

	Dataset	Variant → Model ↓	Epoch Dim 128	Graph Dim 128	Comb. Dim 128	Comb. Dim 512	MRR	Dim	Epochs	
Small	Yago 3-10 ($E = 400$)	ComplEx	**0.536**	0.463	0.528	0.552	0,551	128	400	[5][b]
		RotatE	0.432	0.432	**0.434**	0.453[a]	0,495[a]	1 000	?	[21]
		TransE	**0.499**	0.422	**0.499**	0.496	0,510[c]	350	4 000	[6]
Medium	Wiki-data5M ($E = 64$)	ComplEx	**0.300**	**0.300**	**0.300**	0.294	0,308	128	300	[11]
		RotatE	**0.241**	0.232	**0.241**	0.261	0,290	512	1 000	[27]
		TransE	0.263	0.263	0.268	0.249	0,253	512	1 000	[27]
Large	Free-base ($E = 10$)	ComplEx	0.572	**0.594**	**0.594**	0.678	0,612	400	10	[11]
		RotatE	0.561	**0.613**	**0.613**	0.615	0,567	128	10	[11]
		TransE	0.261	**0.553**	**0.553**	0.559	-	-	-	

[a] RotatE benefits from self-adversarial sampling as used in [21]. We did not use this technique to keep the search space consistent across all models. An adapted GRASH search space led to an MRR of 0.494 (combined, $d = 512$), matching the prior result.
[b] Published in the online appendix of [5].
[c] Published with the AmpliGraph library [6], which ignores unseen entities during evaluation. This inflates the MRR so that results are not directly comparable.

performance of this resulting configuration trained at full fidelity. We report results for different datasets, different reduction techniques, different KGE models, and different model dimensionalities.

Results (Table 3a). The combined variant of GRASH offered best or close to best results across all datasets and models. In comparison to the other variants, it avoided the drawbacks of training partial epochs (e.g., epoch reduction on Freebase) as well as using subgraphs that are too small (e.g., graph reduction on Yago3-10).

Comparison to prior results (Table 3b). We compared the results obtained by GRASH to the best published prior results known to us, see Table 3b. Note that prior models were often trained at substantially higher cost. For example, on Wikidata5M, GRASH used an overall budget of $4 \cdot 64 = 256$ epochs for HPO and training, whereas some prior methods used 1 000 epochs for a single training run. Likewise, dimensionalities of up to 1 000 were sometimes used. For a slightly more informative comparison, we performed a GRASH search with an increased dimensionality of 512, but kept the low search and training budgets. Even with this low budget, we found that on small to midsize graphs, GRASH performed either similarly (ComplEx, Yago3-10 & Wikidata5M) and sometimes slightly worse (RotatE, Wikidata5M) than the best prior results. On the large-scale Freebase KG, where low-fidelity hyperparameter search is a necessity, GRASH outperformed state-of-the-art results by a large margin.

Table 4. Resource consumption per round (ComplEx). The time needed to compute the k-core decompositions is excluded. It is negligible compared to the overall search time (e.g., ≈ 28 min for Freebase with igraph [7]).

		Round Time (min)			Model Size (MB)		
		Epoch	Graph	Comb.	Epoch	Graph	Comb.
Yago3-10	Round 1	43.9	24.7	15.9	60.2	0.3	2.0
	Round 2	34.8	13.3	27.1	60.2	0.4	6.3
	Round 3	38.7	28.1	33.5	60.2	6.3	16.7
	Total	117.4	**66.1**	76.5			
Wikidata5M	Round 1	182.3	60.1	82.3	2 353.3	1.0	71.3
	Round 2	134.2	87.4	88.6	2 353.3	36.0	182.0
	Round 3	126.9	92.5	95.3	2 353.3	182.0	454.7
	Total	443.4	**240.0**	266.2			
Freebase	Round 1	915.9	250.7	179.7	42 025.9	87.3	1 322.2
	Round 2	507.9	172.0	151.2	42 025.9	520.1	2 667.7
	Round 3	423.4	197.5	207.0	42 025.9	2 667.7	6 571.3
	Total	1 847.2	620.2	**537.9**			

Comparison to KGTuner. KGTuner [29] was developed in parallel to this work and follows similar goals as GRASH. We compared the two approaches on the smaller Yago3-10 KGE with ComplEx; a comparison on the larger datasets was not feasible since KGTuner has large computational costs. We ran both GraSH and KGTuner with the default settings of KGTuner ($n = 50$ trials, $E = 50$ epochs, dim. 1 000) to obtain a fair comparison. KGTuner reached an MRR of 0.505 in about 5 d (its search budget corresponds to $B \approx 20$). GRASH reached an MRR of 0.530 in about 1.5 hours ($B = 3$, sequential search on 1 GPU), i.e., a higher quality result at lower cost. The high computational cost of KGTuner mainly stems from its inflexible and inefficient budget allocation (e.g., always 10 full-fidelity evaluations). The higher quality of GRASH stems from its use of multiple fidelities (vs. two in KGTuner) and by using a combination of k-cores and epoch reduction (vs. random walks in KGTuner).

5.4   Resource Consumption (Table 4)

Next, we investigated the computational cost and memory consumption of each round of GRASH. We used 4 GPUs in parallel, evaluating one trial per GPU with the same settings as used in Sect. 5.3. Our results are summarized in Table 4.

Table 5. Influence of the number of rounds on model quality in terms of MRR (ComplEx, graph reduction, $n = 64$ trials, $B = 3$). The number of rounds is directly controlled by the choice of n and η.

Dataset	$\eta = 2$ 6 rounds	$\eta = 4$ 3 rounds (default)	$\eta = 8$ 2 rounds	$\eta = 64$ 1 round $B = 3$	$\eta = 64$ 1 round $B = 1$
Yago3-10	0.463	0.463	0.485	0.427	0.427
Wikidata5M	0.300	0.300	0.300	0.300	0.285
Freebase	0.594	0.594	0.594	0.572	0.572

Memory Consumption. Epoch reduction was less effective than graph reduction and a combined approach in terms of memory usage. With epoch reduction, training is performed on the full graph in every round and therefore performed with full model size. Due to the large model sizes on the largest graph Freebase, the model could not be kept in GPU memory introducing further overheads for parameter management. Graph reduction with k-core decomposition reduced the number of entities contained in a subgraph considerably. As the model size is mainly driven by the number of entities, the resulting model sizes were small.

Runtime. Similarly to memory consumption, a GRASH search based on epoch reduction was less effective in terms of runtime compared to graph reduction and a combined approach. With epoch reduction, runtime was mainly driven by the cost of model evaluation and model initialization (see Sect. 2 and 4.2). This is especially visible in the first round of the search on large graphs. Here, the number of trials and therefore the number of model initializations and evaluations is high. Additionally, on the largest graph, the overhead for parameter management for training on the full KG increased runtime further. In contrast, small model sizes and low GPU utilization with graph reduction would even allow further performance gains. For example, improving on the presented results, the runtime of the first round on Wikidata5M can be reduced from 60.1 to 22.9 minutes by training three models per GPU instead of one.

5.5   Influence of Number of Rounds (Table 5)

In our final experiment, we investigated the sensitivity of GRASH with respect to the number of rounds being used as well as whether using multi-fidelity optimization is beneficial. Our results are summarized in Table 5. All experiments were conducted at the same budget ($B = 3$) and number of trials ($n = 64$). Note that the number of rounds used by GRASH is given by $\log_{\eta}(n)$, where n denotes the number of trials and η the reduction factor. The smaller η, the more rounds are used and the lower the (initial) fidelity.

We found that on the two larger graphs, the search was robust to changes in budget allocation and η did not influence the final trial selection (as long as at least 2 rounds were used). Only on the smaller Yago3-10 KG, the final model quality differed with varying values of η. Here, low-fidelity approximation (small η) was riskier since the subgraphs used in the first rounds were very small.

To investigate whether multi-fidelity HPO—i.e., multiple rounds—are beneficial, we (i) used the best configuration of the first round directly ($\eta = 64$, $B = 1$) and (ii) performed an additional single-round search with a comparable budget to all other settings ($\eta = 64$, $B = 3$). As shown in Table 5, both settings did not reach the performance achieved via multiple rounds. We conclude that the use of multiple fidelity levels is essential for cost-effective HPO.

6   Conclusion

We first presented and experimentally explored various low-fidelity approximation techniques for evaluating hyperparameters of KGE models. Based on our findings, we proposed GRaSH, an open-source, multi-fidelity hyperparameter optimizer for KGE models based on successive halving. We found that GRaSH often reproduced or outperformed state-of-the-art results on large knowledge graphs at very low overall cost, i.e., the cost of three complete training runs. We argued that the choice of low-fidelity approximation is crucial (k-core reduction combined with epoch reduction worked best), as is the use of multiple fidelities.

References

1. Ali, M., et al.: Bringing light into the dark: a large-scale evaluation of knowledge graph embedding models under a unified framework. IEEE Trans. Pattern Anal. Mach. Intell. (2021)
2. Baier, S., Ma, Y., Tresp, V.: Improving visual relationship detection using semantic modeling of scene descriptions. In: d'Amato, C., et al. (eds.) ISWC 2017. LNCS, vol. 10587, pp. 53–68. Springer, Cham (2017). https://doi.org/10.1007/978-3-319-68288-4_4
3. Baker, B., Gupta, O., Raskar, R., Naik, N.: Accelerating neural architecture search using performance prediction. In: International Conference on Learning Representations (Workshop) (2018)
4. Bordes, A., Usunier, N., Garcia-Duran, A., Weston, J., Yakhnenko, O.: Translating embeddings for modeling multi-relational data. Adv. Neural. Inf. Process. Syst. **26**, 2787–2795 (2013)
5. Broscheit, S., Ruffinelli, D., Kochsiek, A., Betz, P., Gemulla, R.: Libkge a knowledge graph embedding library for reproducible research. In: Proceedings of the 2020 Conference on Empirical Methods in Natural Language Processing (2020)
6. Costabello, L., Pai, S., Van, C.L., McGrath, R., McCarthy, N., Tabacof, P.: AmpliGraph: a Library for Representation Learning on Knowledge Graphs (Mar 2019)
7. Csardi, G., Nepusz, T., et al.: The igraph software package for complex network research. Int. J. Complex Syst. **1695**(5), 1–9 (2006)

8. Dettmers, T., Minervini, P., Stenetorp, P., Riedel, S.: Convolutional 2D knowledge graph embeddings. In: Proceedings of the 32nd AAAI Conference on Artificial Intelligence, pp. 1811–1818 (2018)

9. Falkner, S., Klein, A., Hutter, F.: BOHB: robust and efficient hyperparameter optimization at scale. In: International Conference on Machine Learning, pp. 1437–1446. PMLR (2018)

10. Jamieson, K., Talwalkar, A.: Non-stochastic best arm identification and hyperparameter optimization. In: Artificial Intelligence and Statistics, pp. 240–248. PMLR (2016)

11. Kochsiek, A., Gemulla, R.: Parallel training of knowledge graph embedding models: a comparison of techniques. Proc. VLDB Endowment **15**(3), 633–645 (2021)

12. Lacroix, T., Usunier, N., Obozinski, G.: Canonical tensor decomposition for knowledge base completion. In: Proceedings of 35th International Conference on Machine Learning, pp. 2863–2872. PMLR (2018)

13. Lerer, A., et al.: PyTorch-BigGraph: a large-scale graph embedding system. In: Proceedings of the 2nd SysML Conference (2019)

14. Mohamed, S.K., Nounu, A., Nováček, V.: Drug target discovery using knowledge graph embeddings. In: Proceedings of the 34th ACM/SIGAPP Symposium on Applied Computing, pp. 11–18 (2019)

15. Nickel, M., Murphy, K., Tresp, V., Gabrilovich, E.: A review of relational machine learning for knowledge graphs. In: Proceedings of the IEEE (2015)

16. Nickel, M., Tresp, V., Kriegel, H.P.: A three-way model for collective learning on multi-relational data. In: Proceedings of the 28th International Conference on Machine Learning, vol. 11, pp. 809–816 (2011)

17. Ruffinelli, D., Broscheit, S., Gemulla, R.: You CAN teach an old dog new tricks! on training knowledge graph embeddings. In: International Conference on Learning Representations (2020)

18. Saxena, A., Kochsiek, A., Gemulla, R.: Sequence-to-sequence knowledge graph completion and question answering. In: Proceedings of the 60th Annual Meeting of the Association for Computational Linguistics (Volume 1: Long Papers), pp. 2814–2828 (2022)

19. Saxena, A., Tripathi, A., Talukdar, P.: Improving multi-hop question answering over knowledge graphs using knowledge base embeddings. In: Proceedings of the 58th Annual Meeting of the Association for Computational Linguistics, pp. 4498–4507 (2020)

20. Seidman, S.B.: Network structure and minimum degree. Social Netw. **5**(3), 269–287 (1983)

21. Sun, Z., Deng, Z.H., Nie, J.Y., Tang, J.: Rotate: Knowledge graph embedding by relational rotation in complex space. In: International Conference on Learning Representations (2019)

22. Toutanova, K., Chen, D.: Observed versus latent features for knowledge base and text inference. In: Proceedings of the 3rd Workshop on Continuous Vector Space Models and Their Compositionality, pp. 57–66 (2015)

23. Trouillon, T., Welbl, J., Riedel, S., Gaussier, É., Bouchard, G.: Complex embeddings for simple link prediction. In: International Conference on Machine Learning, pp. 2071–2080. PMLR (2016)

24. Tu, K., Ma, J., Cui, P., Pei, J., Zhu, W.: AutoNE: hyperparameter optimization for massive network embedding. In: Proceedings of the 25th ACM SIGKDD International Conference on Knowledge Discovery & Data Mining, pp. 216–225 (2019)

25. Wang, Q., Mao, Z., Wang, B., Guo, L.: Knowledge graph embedding: a survey of approaches and applications. IEEE Trans. Knowl. Data Eng. **29**(12), 2724–2743 (2017)
26. Wang, R., Chen, X., Cheng, M., Tang, X., Hsieh, C.J.: RANK-NOSH: efficient predictor-based architecture search via non-uniform successive halving. In: Proceedings of the IEEE/CVF International Conference on Computer Vision, pp. 10377–10386 (2021)
27. Wang, X., Gao, T., Zhu, Z., Liu, Z., Li, J., Tang, J.: KEPLER: a unified model for knowledge embedding and pre-trained language representation. Trans. Assoc. Comput. Linguist. (2021)
28. Yang, B., Yih, S.W.T., He, X., Gao, J., Deng, L.: Embedding entities and relations for learning and inference in knowledge bases. In: Proceedings of the International Conference on Learning Representations (2015)
29. Zhang, Y., Zhou, Z., Yao, Q., Li, Y.: Efficient hyper-parameter search for knowledge graph embedding. In: Proceedings of the 60th Annual Meeting of the Association for Computational Linguistics (Volume 1: Long Papers), pp. 2715–2735 (2022)
30. Zheng, D., et al.: DGL-KE: training knowledge graph embeddings at scale. In: Proceedings of the 43rd International ACM SIGIR Conference on Research and Development in Information Retrieval (2020)
31. Zwillinger, D., Kokoska, S.: CRC standard probability and statistics tables and formulae. CRC Press (1999)

Multi-source Inductive Knowledge Graph Transfer

Junheng Hao[1(✉)], Lu-An Tang[2], Yizhou Sun[1], Zhengzhang Chen[2],
Haifeng Chen[2], Junghwan Rhee[3], Zhichuan Li[4], and Wei Wang[1]

[1] University of California Los Angeles (UCLA), Los Angeles, CA 90095, USA
{jhao,yzsun,weiwang}@cs.ucla.edu
[2] NEC Laboratories America, Inc. (NEC Labs), Princeton, NJ 08540, USA
{ltang,zchen,haifeng}@nec-labs.com
[3] University of Central Oklahoma, Edmond, OK 73034, USA
jrhee2@uco.edu
[4] Stellar Cyber, Santa Clara, CA 95054, USA

Abstract. Large-scale information systems, such as knowledge graphs (KGs), enterprise system networks, often exhibit dynamic and complex activities. Recent research has shown that formalizing these information systems as graphs can effectively characterize the entities (nodes) and their relationships (edges). Transferring knowledge from existing well-curated source graphs can help construct the target graph of newly-deployed systems faster and better which no doubt will benefit downstream tasks such as link prediction and anomaly detection for new systems. However, current graph transferring methods are either based on a single source, which does not sufficiently consider multiple available sources, or not selectively learns from these sources. In this paper, we propose MSGT-GNN, a graph knowledge transfer model for efficient graph link prediction from multiple source graphs. MSGT-GNN consists of two components: the *Intra-Graph Encoder*, which embeds latent graph features of system entities into vectors; and the graph transferor, which utilizes graph attention mechanism to learn and optimize the embeddings of corresponding entities from multiple source graphs, in both node level and graph level. Experimental results on multiple real-world datasets from various domains show that MSGT-GNN outperforms other baseline approaches in the link prediction and demonstrate the merit of attentive graph knowledge transfer and the effectiveness of MSGT-GNN.

Keywords: Knowledge graphs · Graph neural network · Transfer learning

1 Introduction

Various large-scale information systems, such as knowledge bases (KBs), enterprise security systems, IoT computing systems and social networks [4], exhibit

---

Supplementary Information The online version contains supplementary material available at https://doi.org/10.1007/978-3-031-26390-3_10.

comprehensive interactions and complex relationships among entities from multiple different and interrelated domains. For example, knowledge bases, such as DBpedia [1], contain rich information of real-world entities (people, geographic locations, etc.), normally from multiple domains and languages; and IoT systems contain thousands of mobile interrelated computing devices, mechanical and digital machines with various functions that constantly record surrounding physical environments and interact with each other. These systems can be formulated as heterogeneous graphs with nodes as system entities and edges as activities. Considering an enterprise security system as one example shown in Fig. 1 (right), processes, internet sockets, and files can be treated as different types of nodes. Activities between entities, such as a process accessing a destination port or importing system libraries, are treated as edges in the graph. They can be utilized for many downstream tasks including identifying active entities or groups in social networks, inferring new knowledge in KBs and detecting abnormal behaviors [3].

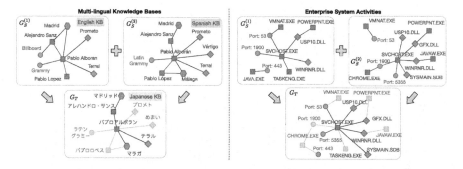

Fig. 1. Two examples of multi-source graph transfer in knowledge bases (left) and enterprise systems (right). By leveraging the entities and relations from sources $G_{S(1)}$ and $G_{S(2)}$, we can estimate the target graph $\hat{G}_T$ based on the current observation G_T. Grey nodes/links in $\hat{G}_T$ denote new predictions from graph knowledge transfer. (Best viewed in color)

Due to the complex nature of real-world systems, it normally takes a long time, sometimes even months for newly-deployed information systems to construct a reliable graph "profile" to identify featured entities and activities. Therefore, there is a need to transfer and migrate knowledge (potential entities with corresponding high-confidence interactions) from other available sources provided by existing multiple well-developed systems. However, directly transferring existing nodes and links by copying is not reasonable and reliable enough since the source and target systems are not necessary for the exact same domains (e.g., transferring knowledge from existing departments to a new department in a corporation). It may transfer irrelevant or even incorrect entities and activities to the target graph. Existing research work [15] mostly focuses on design learning frameworks for effective graph knowledge transfer between one source system and one target system and shows promising results on graph knowledge transfer. But in reality, it is quite common that multiple system sources are available. Simply using single-source graph knowledge transfer has its own limits: (1) the information from a single source is not sufficient in most cases; and (2) using only one source may lack generalization ability especially when the source and target are largely different, which leads to potential transfer failure. Learning graphs for newly-deployed systems through multi-source graphs will no doubt provide more comprehensive coverage of system entities and activities in multiple domains, and it will be more robust for downstream applications relying on learned target graph after selectively adapting knowledge from source graphs[1]. Two application scenarios are shown in Fig. 1. In the case of multi-lingual KBs, low-resource KB (such as Japanese) can be enriched and improved with other KBs, and especially in the case of `Pablo Alboran` (Spanish pop singer), Spanish KBs can provide better and more accurate knowledge facts than others. Similarly in the example of enterprise systems, after the observation that the system has similar patterns of `.dll` connections of `SVCHOST.EXE`, a reasonable interpretation is that the target graph G_T will more likely grow more closely related patterns shown in source graphs.

However, the aforementioned selective multi-source transfer faces several challenges: (i) *How to represent multiple source graphs and target graphs effectively i.e. set up connections to leverage the graph knowledge in source graphs to the target graphs.* Not all sources are equally related to the target and it is required to differentiate multiple input source graphs in the transferring process, which is a difficult but important task to handle and will significantly affect the transfer performance. (ii) *How to handle potential conflicts on entities and interactions observed in multiple graphs.* The same interactions may be observed in some sources, but are not in others. In other words, there are potentially conflicting observations that cannot be easily tackled by simple transfer. In other words, if all sources are credited equally (for example, using one combined graph to include all the nodes and edges) and other methods that concatenate multiple graphs, one inductive bias is incorrectly assumed that nodes and/or edges are transferred and learned without selectivity and the approaches are subject to noise and misinformation on part of the sources.

To address the aforementioned tasks and corresponding challenges, we proposed a novel type of graph neural network designed for Multi-Source Graph Knowledge Transfer named `MSGT-GNN` which contains two model components: *Intra-Graph Encoder* and *Attention-based Cross-graph Transfer*. The high-level idea is that the knowledge transfer between the source and target graphs is done in a controllable manner where they are selectively learned. We employ self graph encoder model to a variety of state-of-the-art graph neural networks (GNNs) to obtain the node representations, that is, node embeddings learned from the node features itself and neighborhood in the context of the same source/target graph.

---

[1] In this paper, we use the *source graph* as the graph profiles for existing well-observed systems and *target graph* as the graph profile for new systems, which is relatively smaller than source graphs in graph size (e.g. number of nodes/edges). We assume that the number of source graphs is at least 2 and that of the target graph is 1.

On top of the encoder model, the Cross-graph Transfer module adopts a novel attention mechanism based on both node level and graph level. This module can better learn the representations by attentively aggregating nodes in the broader context, which later applies in the graph decoder for link prediction. As a result, not only can we accelerate the process of graph enlargement to fast characterize the target graph, but we can also selectively and effectively leverage multiple sources in the information systems to estimate more reliable and accurate target graphs. Experimental results on target graph link prediction confirm that the effectiveness of MSGT-GNN and the performance of knowledge transfer significantly outperforms other state-of-the-art models including TINET.

2   Problem Statement

Given n multiple source domains $\mathcal{D}_S^{(i)}$ ($i = 1, 2, \ldots, m$) and one target domain $\mathcal{D}_T$ as input graphs have been on source domain for and these source graphs $G_S^{(i)}$ are stable already. Meanwhile, the system in $\mathcal{D}_T$ is possibly newly deployed and therefore the target graph G_T incomplete and of relatively small size. Our goal is to transfer the graph knowledge (entity and edges) from $G_S^{(i)}$ ($i = 1, 2, \ldots, n$) to G_T, and then help quickly enlarge and estimate an estimated complete graph $\hat{G}_T$ to fit the domain of $\mathcal{D}_T$, which should be as close to the ground truth $\bar{G}_T$ as possible. Note that in this paper, we assume that alignments of the same entity among source and target graphs are well established, though such alignments are not fully feasible especially in knowledge bases. Under such formulation, we also point out that our proposed problem focuses on the graph enhancement from its incomplete status, different from temporal graph modeling where graphs are dynamically changed with multiple timestamps. Notations of all symbols used in this paper are summarized in Table 1. Scalars, vectors and matrices are denoted with lowercase unbolded letters, lowercase bolded letters and uppercase bolded letters, if not explicitly specified.

We acknowledge that entity alignment may not be flawlessly given in many real-world applications and there are many existing research works lying on the direction of entity disambiguation, etc. As mentioned in Sect. 2, we point out that in this paper we do not cover the scope of the entity alignments [22,24] (or entity resolution, entity conflation), which essentially predicts the correspondences of the same entity among different graphs. For example, in enterprise graphs, entities are generally identifiable with their IDs; in encyclopedic KGs, some labeled-property graphs are equipped with UID (universal identifier), which significantly reduces the alignment challenge. However, we believe such assumption can be relaxed, that is, MSGT-GNN can be further adapted to partially-given alignment or cross-graph alignment can be jointly learned, corrected, and/or enhanced, which is left as one direction of our future work.

3   Methodology

In this section, we formally propose MSGT-GNN to tackle multi-source graph knowledge transfer problem inspired by multi-task learning. As the model archi-

Table 1. Summary of important notations.

Notation	Description
$\mathcal{D}_S^{(i)}$	i-th source domain
$\mathcal{D}_T$	Target domain
$G_S^{(i)}$	The graph of the i-th source from $\mathcal{D}_S^{(i)}$
$\bar{G}_T, G_T, \hat{G}_T$	The ground-truth complete/incomplete/estimated complete graph of the target system from $\mathcal{D}_T$
$A_S^{(i)}, A_T$	The adjacency matrix of the i-th source graph $G_S^{(i)}$/the target graph G_T
$\mathbf{Z}, \mathbf{Z}_S^{(i)}$	Embedding table for all N entities, or for $N_S^{(i)}$ entities from the i-th source graph (as output of graph encoders)
$\mathbf{h}_{S(m)_i}^l, \mathbf{h}_{T_i}^l$	Embedding of the i-th node in the m-th source graph (or target graph) at the l-th layer of GNN (node embeddings, with node index)
$\mathbf{h}_{S(m)}^l, \mathbf{h}_T^l$	Embedding of the m-th source graph (or target graph), at the l-th layer of GNN (graph embeddings, without node index)

tecture of `MSGT-GNN` shown in Fig. 2, it breaks down into two components: *Intra-graph encoder* and *Cross-Graph transfer*, which are explained in Sect. 3.1 and Sect. 3.2 respectively.

3.1   Intra-graph Encoder

Generally, a graph encoder serves a function to represent nodes by their embeddings, from the original node features (categorical attributes, textual descriptions, etc.), based on the graph features. Our proposed Intra-Graph Encoder, as the first component in `MSGT-GNN`, aims to learn the node features in the context of its own graph (source or target), i.e. the graph to which it originally belongs. As discussed in Sect. 5, graph neural networks (GNNs), deep learning based approaches that operate on graph-structured data, have recently shown effective for various applications such as node classification, link prediction and community detection. A generalized framework of GNNs consists of such a graph encoder, taking as input an adjacency matrix A, as well as original (optional) node features $X = \{X_N\}$. A typical graph encoder parameterized by $\mathbf{\Theta}_{\text{enc}}$ combines the graph structure with node features to produce node embeddings as, $Z = \text{ENC}\,(A, X, \mathbf{\Theta}_{\text{enc}})$, where Z is the learned comprehensive representation from GNNs and is used for downstream tasks with designated graph decoders.

More specifically, in `MSGT-GNN`, for homogeneous graphs, we choose the Intra-Graph Encoder as standard GCN [12], which can be described as,

$$\mathbf{H}_i^{(l+1)} = \sigma\left(\hat{D}^{-\frac{1}{2}}\hat{A}_G\hat{D}^{-\frac{1}{2}}\mathbf{H}_i^{(l)}W^{(l)}\right), \tag{1}$$

where $\mathbf{H}_i^{(l)} \in \mathbb{R}^{n \times d}$ are embedding of after l-th GCN layers and $\hat{A}_G = A_G + I$ where I is the identity matrix, A_G is adjacency matrix of given graph G, $\hat{D}$ is the

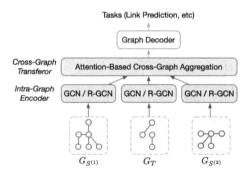

Fig. 2. Model architecture overview for MSGT-GNN (two source graphs are shown). Node embeddings across multiple graphs are learned through two-module framework, i.e. *Intra-graph encoder*, which learns node embeddings of its own graph context from initial node features; and *Cross-Graph transfer*, which enables learning through mulitple graphs and node embeddings are updated by its corresponding nodes in other source as well as the graph-level information.

diagonal node degree matrix of $\hat{A}$, as defined in [12]. Note that G can be either any source graph $G_{S(i)}$ or target graph G_T. For multi-relational heterogeneous graphs such as knowledge graphs and enterprise systems, we adopt R-GCN [18], which utilizes relation-wise weight matrix,

$$\mathbf{h}_i^{(l+1)} = \sigma \left(\mathbf{W}_0^l \mathbf{h}_i^{(l)} + \sum_{r \in \mathcal{R}} \sum_{j \in \mathcal{N}_i^r} \frac{1}{c_{i,r}} \mathbf{W}_r^l \mathbf{h}_j^{(l)} \right), \qquad (2)$$

where $\mathbf{W}_0^l$ is the weight matrix for the node itself and $\mathbf{W}_r^l$ is used specifically for the neighbors having relation r, i.e., $\mathcal{N}_i^r$, $\mathcal{R}$ is the relation set and $c_{i,r}$ is for normalization. Similarly, R-GCN applies both in the source graphs and the target graph. In both cases, the number of GNN layers L is one hyperparameter[2].

3.2 Attention-Based Cross-Graph Transfer

The goal of our proposed *Cross-graph Transfer* is to provide a valid transfer mechanism in the entity embedding space for multi-source graphs. It is built on top of the Intra-Graph Encoder to enable the node embeddings selectively updated by the cross-graph "neighborhood" in both node level and graph level attention mechanism. Details of *Cross-graph Transfer* are shown in Fig. 3.

To prepare for cross-graph transfer, one necessary module is Graph-level Aggregator, which takes the set of node representations and compute graph level representation, as $\mathbf{h}_G = f_G(\{\mathbf{h}_i^G\})$ where $\mathbf{h}_G \in \mathbb{R}^d$, for both source and target

---

[2] In this work, the performance is relatively insensitive to L where we fix $L = 2$ for GNN modules including baselines.

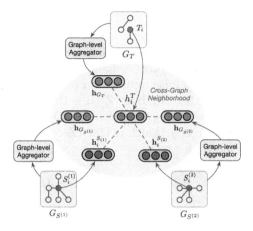

Fig. 3. Details about Cross-Graph Transfer Layer operating on the Node T_i, updated by itself and its corresponding cross-graph neighbors (node-level embeddings), attentively learned from graph-level embeddings (Best viewed in color)

graphs[3]. We use the MLP aggregator following the implementation in [13]. The aggregation function operating on a node i of the target graph is defined as,

$$\mathbf{h}_{T_i}^{l+1} = \sigma\left(\mathbf{W}_0^l\mathbf{h}_{T_i}^l + \sum_m \alpha_m \mathbf{W}_n^l \mathbf{h}_{S^{(m)}_i}^l\right), \tag{3}$$

where $\mathbf{W}_0^l$ is the weight matrix for the node itself and W_n^l is used specifically for the cross-graph neighbors (from the given alignments), of the l-th layer. $h_{S^{(m)}_i}^l$ denotes the l-th layer's hidden representation of node i in $G_{S^{(m)}}$. α_m is attention weight computed over all m cross-graph neighbors as,

$$\alpha_m = \mathrm{softmax}\left(\left[\mathbf{h}_i^{S^{(m)}}; \mathbf{h}_{G_{S_{(m)}}}\right]^T \cdot \mathbf{W}_{\mathrm{att}} \cdot \mathbf{h}_{T_i}^l\right), \tag{4}$$

where $\mathbf{W}_{\mathrm{att}} \in \mathbb{R}^{2d \times d}$ and $\left[\mathbf{h}_i^{S^{(m)}}; \mathbf{h}_{G_{S_{(m)}}}\right]$ is the concatenation of node-level cross-graph neighbor embedding and the graph-level embedding. By such cross-graph transfer, the node in one graph will be consequently updated and optimized attentively by nodes from other associated graphs. It is noteworthy to point out that our proposed MSGT-GNN does not explicitly differentiate source graphs and target graphs, which means the learned embeddings are not limited to make predictions over the target graph.

---

[3] Theoretically the embedding dimension of graph-level representation can be different from that of the node-level. For simplicity, we choose both dimensions are the same, that is, $\dim(\mathbf{h}_G) = \dim(\mathbf{h}_{G_i}^l)$, where G refers to either source or target graph.

3.3 Graph Decoder

Graph Decoder and Training Objective. The graph decoder use the learned representation from `MSGT-GNN` for link prediction during the inference stage. For homogeneous graph, we apply inner product to represent the edge plausibility, which is $\text{DEC}(\mathbf{Z}) = \mathbf{h}_i^T \mathbf{h}_j$ where $\mathbf{h}_i, \mathbf{h}_j \in \mathbf{Z}$ ($\mathbf{h}$ is the learned embedding table for all nodes). For multi-relational graph, we apply DistMult score function [32] to represent the edge plausibility, which is $\text{DEC}(\mathbf{Z}) = \mathbf{h}_i^T D \mathbf{h}_j$ where $\mathbf{h}_i, \mathbf{h}_j \in \mathbf{Z}$ and $\mathbf{D}_r$ is a diagonal matrix for relation r. Therefore, the training objective is,

$$\mathcal{L}_G(\mathbf{Z_G}) = \left(\mathbf{Z_G} \mathbf{D}_r \mathbf{Z_G}^T - A_G \right)^\theta + \Omega(\mathbf{Z_G}), \tag{5}$$

where $\theta = 2$ in practice and $\Omega(\mathbf{Z}_G, \mathbf{w}) = \lambda \|\mathbf{Z_G}\|_F$ is regularization term. $\mathbf{D_r} = I$ for homogeneous graph.

3.4 Training, Inference and Complexity

Joint Training on Source and Target Graphs. Considering all the source and target graphs, `MSGT-GNN` minimizes the joint loss with meta-path similarity matrices for multiple graphs, $\mathcal{L} = \mu \sum_i \mathcal{L}_{S^{(i)}} + (1 - \mu)\mathcal{L}_T$, where $\mu \in (0, 1)$ is a hyperparameter that explicitly balances the importance of source and target graphs. We use the Adam [11] to optimize the joint loss.

Inference. During the inference stage, similar to other graph neural networks with downstream link prediction task, two steps of graph encoders (intra-graph and cross-graph) encodes pairs of nodes (from the target graph only for valid testing) into their representations through the trained GNN with the neighbor nodes (both inside its own graph and other sources) weighted by the graph-level representations. Later such embeddings are forwarded to graph decoder for link prediction which outputs plausibility scores of the given potential edges, as link prediction results.

Complexity Analysis. For `MSGT-GNN` with the direct encoder, the overall runtime complexity is $\mathcal{O}(tnd|E|)$, which is linear to the size of total edges in multiple source graphs ($|E|$ is the total number of links in source/target graph). As for model parameter complexity, including all embeddings and transformation functions, the result is $\mathcal{O}(|V|d + nd^2)$ ($|V|$ is the total number of nodes in source/target graphs).

4 Experiments

4.1 Datasets

Three datasets on the knowledge bases, enterprise security and academic scholar community are used in the experiments. Data from a real-world enterprise system are collected from 145 machines from 4 departments (3 used as sources and 1

Table 2. Dataset statistics.

Dataset	Scholar	Enterprise		DBpedia
		Windows	Linux	
# Graphs	3	5	5	5
# Rel. Types	1	3	3	96
# Nodes	2.1k	10.7k	8.9k	12.5k
# Edges	9.0k	87.9k	62.5k	278.1k

used as a target) in a period of 30 days, with a size of 3.45 GB after integration and filtering. The entire enterprise security system contains both Windows and Linux machines and we consider they are disjoint graphs as datasets (named as **Windows and Linux Dataset**). Similar to the example in Fig. 1, the entities (nodes) in all graphs are processes, internet sockets and libraries (mostly .dll files) and interactions (edges) between the process to file, process to process and process to internet sockets are observed as links in the dataset.

We also consider alternative datasets that are publicly available and from diverse domains are, (i) encyclopedia knowledge bases i.e. **DBpedia** [1][4], extracted from five languages (en, es, de, fr, ja) of variant graph sizes and completeness; and (ii) **Aminer**, as one academic scholar community dataset [23][5] from Aminer on five data mining/machine learning related research communities in the past years. The nodes are authors and links are simply co-author relationships, which is essentially a homogeneous graph. More specifically dataset, we consider different languages as different domains in the context of MSGT-GNN, and given the graph size of these languages, we adopt two disjoint settings: {en, fr, de}→ja[6] and {en, fr, de}→es. This results in a total of 5 datasets from 3 domains in our experiments. More details are listed in Table 2.

4.2   Baseline Methods

We compare our proposed model MSGT-GNN with the following baseline methods:

No Transfer (NT) directly uses the original observed incomplete target graph without any knowledge transfer, that is, $\hat{G}_T = G_T$.

Direct Union Transfer (DUT) directly combines all source graphs and the incomplete target graph, as prediction ("union" graph). That is, DUT outputs a union set on entities and links from all observed graphs without any selection, which means, $\hat{G}_T = G_T + \left(\bigcup_i G_S^{(i)} \right)$.

---

[4] Processed DBpedia dataset are downloadable at: Link.

[5] We use a subset of the co-author networks, which is available at https://aminer.org/data#Topic-coauthor.

[6] {en, fr, de}→es means the source graphs are from DBpedia English, French and German KBs and the target is Spanish KB.

Table 3. Results of target graph completion task on 5 different transfer settings from 3 different domains (scholar, enterprise and encyclopedia). The best scores are **bolded**.

Dataset	Scholar	Enterprise		Encyclopedia	
		Windows	Linux	{en, fr, de}→ja	{en, fr, de}→es
NT	0.526 ± 0.000	0.664 ± 0.000	0.656 ± 0.000	0.475 ± 0.000	0.545 ± 0.000
DT	0.398 ± 0.000	0.480 ± 0.000	0.578 ± 0.000	0.299 ± 0.000	0.408 ± 0.000
C-TINET	0.635 ± 0.009	0.727 ± 0.008	0.759 ± 0.009	0.596 ± 0.010	0.764 ± 0.013
U-TINET	0.618 ± 0.015	0.718 ± 0.012	0.733 ± 0.008	0.617 ± 0.014	0.750 ± 0.012
W-TINET	0.644 ± 0.017	0.739 ± 0.011	**0.772 ± 0.017**	0.645 ± 0.022	0.779 ± 0.018
O-TINET	0.622 ± 0.014	0.715 ± 0.012	0.740 ± 0.014	0.620 ± 0.009	0.766 ± 0.011
UT-GCN/RGCN	0.606 ± 0.025	0.700 ± 0.030	0.722 ± 0.019	0.576 ± 0.022	0.756 ± 0.026
UT-GAT/KGAT	0.635 ± 0.018	0.744 ± 0.023	0.750 ± 0.015	0.559 ± 0.012	0.710 ± 0.014
Insta-Only GCN/RGCN	0.597 ± 0.014	0.745 ± 0.012	0.734 ± 0.014	0.661 ± 0.015	0.739 ± 0.021
Insta-Only GAT/KGAT	0.624 ± 0.020	0.742 ± 0.018	0.738 ± 0.021	0.656 ± 0.016	0.724 ± 0.016
UDA-GCN	0.652 ± 0.017	0.735 ± 0.013	0.727 ± 0.016	0.610 ± 0.024	0.688 ± 0.022
MSGT-GNN	**0.668 ± 0.016**	**0.776 ± 0.021**	0.768 ± 0.018	**0.685 ± 0.018**	**0.801 ± 0.028**

TINET applies the single graph knowledge transfer framework [15]. To fit the multi-source setting, we choose three variations about TINET models: (i) to use the closest[7] source graph as the transfer source, named **C-TINET**; (ii) to use the union graph as defined in DUT, as the single transfer source, named **U-TINET**; iii to use TINET iteratively on multiple sources, i.e. transferring one source once in an order, named **O-TINET**. Best performance is reported among all transfer orders.

W-TINET. This method uses the weighted version of TINET for source and target graphs. Extending the single-source graph knowledge transfer model to multi-source, we adopt the same sub-model components (EEM, DCM) but adjust the objective function to be the sum of all source graphs.

Intra-Only GNN only uses *Intra-Graph Encoder* component in MSGT-GNN and discards the *Cross-Graph Transfer*. That is, standard GCN [12] is applied for homogeneous graphs and R-GCN [18] is applied for multi-relational graphs which preceded the graph decoder. Alternatively, we also consider existing attention-based graph neural networks (applied on a single graph) i.e. GAT [26]/KGAT [28] as replacement of GCN/R-GCN (Denoted as "Intra-Only GCN/RGCN" and "Intra-Only GAT/KGAT" respectively).

UT-GNN. Similar to Intra-Only GNN, this method applies *Intra-Graph Encoder* component only on the "union graph" from the DUT method which forms one combined graph instead of multiple sources and target graphs. Two options (GCN/RGCN, GAT/KGAT) are still considered except the different graph inputs (Denoted as "UT-GCN/RGCN" and "UT-GAT/KGAT" respectively).

UDA-GCN. It develops a dual attention-based graph convolutional network component and domain adaptive learning module, which jointly exploits local

---

[7] Default similarity between the source and target graph is based on the Jaccard index.

and global consistency for feature aggregation to produce unified representation for nodes. We replace the decoder module[8] for link prediction instead of node classification in the original paper [30].

4.3 Experiment Setup

Evaluation Protocol. Similar to [15], we adopt F1 score to evaluate the accuracy of the graph completion task on the target system instead of Hit@K or MRR score in knowledge graph completion[9]. In our experiment for multi-graph knowledge transfer, the main result is reported as the average and standard deviation of link prediction (edge) F1 score. As F1 score generally is the harmonic mean of precision and recall, we hereby define the *precision* and *recall* by comparing the estimated links between entities with the ground truth. The precision and recall are defined as: $Precision = N_C/N_E$ and $Recall = N_C/N_T$, where N_C is the number of correctly estimated links, N_E is the number of estimated links in total, and N_T is the number of the ground-truth links. For training, as mentioned in Sect. 2, we choose one incomplete target graph as the "new" system and complete source graphs from the rest as "old" systems and for training. In addition, e use $m = l/l_{full}$ as an index of *"graph maturity"*, which is defined as the observed number of edges (in training set) l of the target graph and the total number of edges l_{full} recorded in the ground truth target graph.

Hyperparameters. In the experiment, we set $m = 0.4$ and $d = 128$ if not specified. The number of GCN/R-GCN layers in Intra-Graph Encoder is set as 2 and The number of Cross-Graph Transfer layers is set as 1. Default node embeddings are initialized by either node categorical features (scholar and enterprise dataset) or BERT sentence embeddings from entity descriptions (KB datasets). Hyperparameters are discussed in Sect. 4.5 and the supplementary material.

4.4 Results

In this section, we investigate the sensitivity of target graph input maturity m, embedding dimension d and balance weight μ between the source and target graphs, as three key hyperparameters of MSGT-GNN, compared with some of the strongest baseline methods. Results on the target graph completion task are shown in Table 3. We observe that MSGT-GNN outperforms other baselines in terms of average F-1 score. Especially compared with non-transfer, MSGT-GNN achieves an average increase of 0.05 on F1 score among all datasets, which proves that

---

[8] Original code implementation: https://github.com/GRAND-Lab/UDAGCN.

[9] We point out the thread of KG embedding in Sect. 5, including TransE and recent variants [27]. The limitation of such methods is that they are transductive methods. This is generally not applicable to our inductive learning and its downstream link prediction. However, as for evaluation metrics, we follow the metrics adopted in previous work [15] for target-adapted edge prediction instead of MRR or Hit score for a different triple completion task.

MSGT-GNN transfers useful graph knowledge to the target. Also, MSGT-GNN out-performs all the TINET variants in the average F1 score especially on U-TINET and W-TINET which indicates that MSGT-GNN adopts a more effective strategy to use multiple sources and learn better latent feature representations of entities with the process graph encoding and domain transferring. Since TINET follows a two-stage (entity selection and edge prediction), the performances significantly decrease when wrong or incomplete entity set is selected for subsequent link prediction. Unlike TINET and its variants, MSGT-GNN adopt end-to-end model architecture without explicit steps of entity/node selection. Comparing MSGT-GNN and standard GCN/R-GCN or GAT/KGAT, we also observe that MSGT-GNN achieves better link prediction performance with a relative gain of 4.9%, which shows the benefit of Cross-Graph Attention Transfer, which can better char-acterize node latent representations from actively and selectively aggregating useful information from the cross-graph neighborhood. It is noteworthy that NT directly uses the currently observed target graph (incomplete) as output; DT means the union set of all G_S and G_T without any selection. Typically DT includes much more noise and unwanted information into the target graph com-pared "beneficial section of transfer", i.e., lots of links/edges are falsely predicted as positive. A similar observation is also reported in one of our baselines, TINET. Furthermore, we observe that GAT/KGAT variants almost have similar perfor-mance on the task (sometimes even worse). We hypothesize that the attention mechanism adopted by the original GAT/KGAT cannot best selectively learn the knowledge transfer in the cross-graph setting, although recent research shows that they outperform GCN/RGCN on the intra-graph node classification task. It is also noticed that UT-GNN generally performs worse than the Insta-Only setting which indicates that the union graph which equally combines the source graphs without selection has inductive biases which compromise the knowledge transfer in link prediction on the target graph.

4.5   Hyperparameters

In this section, we primarily investigate the sensitivity of target graph input maturity m. Other hyperparameters such as embedding dimension d and balance weight μ between the source and target graphs are discussed in the supplemen-tary material.

Graph Maturity m. We vary the target graph input by controlling the graph maturity m (let $m = \{0.2, 0.4, 0.6, 0.8, 1.0\}$). From Fig. 4, we observe that, for both Windows and DBpedia: {en, fr, en}→es graph, the performance of all models increases when the graph maturity m increases. As other approaches achieve F1 score of 1 when m gets close to 1, direct transfer only achieves around 0.60 as F-1 score, which seems not effective because all the irrelevant entities and links are adopted in the output target graph prediction. On the other hand, given the same level of graph maturity, MSGT-GNN achieves the best performance among all other methods on all datasets.

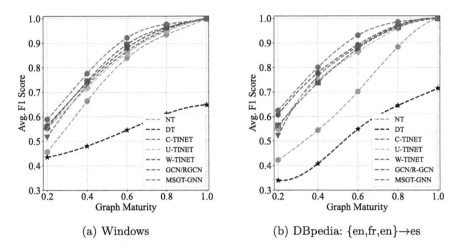

(a) Windows   (b) DBpedia: {en,fr,en}→es

Fig. 4. Performances with graph maturity. Most models achieve average F1 score close to 1 as the maturity of input observed target graph grows, while MSGT-GNN outperforms other baselines.

5 Related Work

Transfer Learning, Graph Transfer and Multi-source Adaption. Transfer learning, domain adaption, and translation [29] have been widely studied in the past decade and played an important role in real-life applications [19] especially on deep transfer learning [14]. Existing transfer learning research is mostly done on the numeric, grid and sequential data, especially image (specific domain classification, style transfer) and text (translation), but research on graphs, networks, or structured data, whose format are relatively less ordered. Some representative work includes *TrGraph* [5], which leverages information via common signature subgraphs. [15] is state-of-the-art and most related research aligned with this direction with two-step learning on entity estimation and dependency reconstruction. The aforementioned methods are mostly based on single-graph knowledge transfer. Note that there is some related work on multi-source adaption that has the same goal of reliable knowledge transfer from multiple sources [16]. However, they are still limited within the domain of images and text rather than graphs. Thus their frameworks cannot be directly applied on graph knowledge transfer. Despite the usage of an attention-based model in transfer, one related work [30] focuses on the node classification task and substantial changes are necessary to make for link prediction in target graphs. We clarify the term of "graph transfer" in Sect. 2 and distinguish it from other research on the concept of "knowledge transfer" to avoid confusion.

Representation Learning on Knowledge Graphs. Graph link prediction is a basic research topic on network analysis. For transfer purposes, [33] presented a transfer learning algorithm to address the edge sign prediction problem using latent topological features from the target and sources. Collective matrix factorization [20] is another major technique. However, these methods are not suitable for dynamics among multiple different domains and the target domain. Another important branch of research related to graph link prediction is network embedding (network representation learning) and similarity search. By representing high-dimensional structured data with embedding vectors, link prediction can be easily performed by node similarity search. These methods can be categorized as meta-path based [21], random walk based [6], matrix factorization based [17] and graph neural networks based methods [7,9]. Similar techniques are applied in multi-relational heterogeneous graphs, i.e. knowledge graphs [25] and their applications [8–10]. These embedding based methods (for example [25]) provide insights for representing node features by gathering neighborhood (multi-relational) connections and/or meta-paths and designing graph encoders and decoders. It is worth noted that the most common task over knowledge graphs is triple completion, different from link prediction where focuses on the existence of relations over pairs of nodes in the graph. Another recent research thread along this direction increasingly focuses more on temporal/dynamic graph representation learning [31], which specifically models the graph evolving patterns over time. However, we emphasize that in this work, though it is assumed that the target graphs are relatively incomplete and sparse, we temporarily do not incorporate the time information, as one of the future directions.

Multitask Learning. Multitask learning [34] is one emerging active research topic with the rise of artificial intelligence. With the goal of "one model for all tasks", it is widely applied in the area of computer vision and natural language processing. One of the most common approaches in multitask learning is parameter sharing [2]. MSGT-GNN is inspired by the similar multi-task learning mechanism considering each graph as one "task", however these frameworks themselves in multitask learning is not applicable for our settings.

6   Conclusion and Future Work

In this paper, we formulate a challenging problem on the necessity and benefits of transferring from multi-source graphs into the target graph and then propose MSGT-GNN, with the intra-graph Encoder and attention-based cross-graph transfer as major model components. MSGT-GNN addresses the challenges and accelerates high-quality knowledge transfer and graph enhancement in the target newly-observed system. Experiments show that MSGT-GNN can successfully transfer useful graph knowledge from multiple sources and enable fast target graph construction. For future improvements, one important extension is to temporal graph modeling where we can dive deep into how target graphs grow on newly-deployed systems can grow with the development from multiple sources, which significantly improves explainability on the graph knowledge transfer.

Acknowledgement. This work was primarily done and supported during the internship at NEC Laboratories America, Inc (NEC Labs). We thank Dr. Zong Bo for research discussions. We also would like to thank the anonymous reviewers for their insightful and constructive comments.

References

1. Auer, S., Bizer, C., Kobilarov, G., Lehmann, J., Cyganiak, R., Ives, Z.: DBpedia: a nucleus for a web of open data. In: Aberer, K., et al. (eds.) ASWC/ISWC -2007. LNCS, vol. 4825, pp. 722–735. Springer, Heidelberg (2007). https://doi.org/10.1007/978-3-540-76298-0_52
2. Caruana, R.: Multitask learning. Mach. Learn. **28**(1), 41–75 (1997)
3. Cheng, W., Zhang, K., Chen, H., Jiang, G., Chen, Z., Wang, W.: Ranking causal anomalies via temporal and dynamical analysis on vanishing correlations. In: Proceedings of the 22nd ACM SIGKDD International Conference on Knowledge Discovery and Data Mining, pp. 805–814 (2016)
4. Dong, B., et al.: Efficient discovery of abnormal event sequences in enterprise security systems. In: Proceedings of the 2017 ACM on Conference on Information and Knowledge Management, pp. 707–715 (2017)
5. Fang, M., Yin, J., Zhu, X., Zhang, C.: TrGraph: cross-network transfer learning via common signature subgraphs. IEEE Trans. Knowl. Data Eng. **27**(9), 2536–2549 (2015)
6. Grover, A., Leskovec, J.: node2vec: scalable feature learning for networks. In: Proceedings of the 22nd ACM SIGKDD International Conference on Knowledge Discovery and Data Mining, pp. 855–864 (2016)
7. Hamilton, W., Ying, Z., Leskovec, J.: Inductive representation learning on large graphs. In: Advances in Neural Information Processing Systems, vol. 30 (2017)
8. Hao, J., Ju, C.J.T., Chen, M., Sun, Y., Zaniolo, C., Wang, W.: Bio-JOIE: joint representation learning of biological knowledge bases. In: Proceedings of the 11th ACM International Conference on Bioinformatics, Computational Biology and Health Informatics, pp. 1–10 (2020)
9. Hao, J., et al.: MEDTO: medical data to ontology matching using hybrid graph neural networks. In: Proceedings of the 27th ACM SIGKDD Conference on Knowledge Discovery & Data Mining, pp. 2946–2954 (2021)
10. Hao, J., et al.: P-companion: a principled framework for diversified complementary product recommendation. In: Proceedings of the 29th ACM International Conference on Information & Knowledge Management, pp. 2517–2524 (2020)
11. Kingma, D.P., Ba, J.: Adam: a method for stochastic optimization. arXiv preprint arXiv:1412.6980 (2014)
12. Kipf, T.N., Welling, M.: Semi-supervised classification with graph convolutional networks. In: International Conference on Learning Representations (2017)
13. Li, Y., Gu, C., Dullien, T., Vinyals, O., Kohli, P.: Graph matching networks for learning the similarity of graph structured objects. In: International Conference on Machine Learning, pp. 3835–3845. PMLR (2019)
14. Long, M., Zhu, H., Wang, J., Jordan, M.I.: Deep transfer learning with joint adaptation networks. In: International Conference on Machine Learning, pp. 2208–2217. PMLR (2017)
15. Luo, C., et al.: TINET: learning invariant networks via knowledge transfer. In: Proceedings of the 24th ACM SIGKDD International Conference on Knowledge Discovery & Data Mining, pp. 1890–1899 (2018)

16. Mansour, Y., Mohri, M., Rostamizadeh, A.: Domain adaptation with multiple sources. In: Advances in Neural Information Processing Systems, vol. 21 (2008)

17. Qiu, J., Dong, Y., Ma, H., Li, J., Wang, K., Tang, J.: Network embedding as matrix factorization: unifying DeepWalk, LINE, PTE, and node2vec. In: Proceedings of the Eleventh ACM International Conference on Web Search and Data Mining, pp. 459–467 (2018)

18. Schlichtkrull, M., Kipf, T.N., Bloem, P., van den Berg, R., Titov, I., Welling, M.: Modeling relational data with graph convolutional networks. In: Gangemi, A., et al. (eds.) ESWC 2018. LNCS, vol. 10843, pp. 593–607. Springer, Cham (2018). https://doi.org/10.1007/978-3-319-93417-4_38

19. Shin, H.C., et al.: Deep convolutional neural networks for computer-aided detection: CNN architectures, dataset characteristics and transfer learning. IEEE Trans. Med. Imaging **35**(5), 1285–1298 (2016)

20. Singh, A.P., Gordon, G.J.: Relational learning via collective matrix factorization. In: Proceedings of the 14th ACM SIGKDD International Conference on Knowledge Discovery and Data Mining, pp. 650–658 (2008)

21. Sun, Y., Han, J., Yan, X., Yu, P.S., Wu, T.: PathSim: meta path-based top-k similarity search in heterogeneous information networks. Proc. VLDB Endow. **4**(11), 992–1003 (2011)

22. Sun, Z., et al.: A benchmarking study of embedding-based entity alignment for knowledge graphs. Proc. VLDB Endow. **13**(11), 2326–2340 (2020)

23. Tang, J., Sun, J., Wang, C., Yang, Z.: Social influence analysis in large-scale networks. In: Proceedings of the 15th ACM SIGKDD International Conference on Knowledge Discovery and Data Mining, pp. 807–816 (2009)

24. Trivedi, R., Sisman, B., Dong, X.L., Faloutsos, C., Ma, J., Zha, H.: LinkNBed: multi-graph representation learning with entity linkage. In: Proceedings of the 56th Annual Meeting of the Association for Computational Linguistics (Volume 1: Long Papers), pp. 252–262 (2018)

25. Vashishth, S., Sanyal, S., Nitin, V., Talukdar, P.: Composition-based multi-relational graph convolutional networks. In: ICLR (2020)

26. Veličković, P., Cucurull, G., Casanova, A., Romero, A., Liò, P., Bengio, Y.: Graph attention networks. In: International Conference on Learning Representations (2018)

27. Wang, Q., Mao, Z., Wang, B., Guo, L.: Knowledge graph embedding: a survey of approaches and applications. IEEE Trans. Knowl. Data Eng. **29**(12), 2724–2743 (2017)

28. Wang, X., He, X., Cao, Y., Liu, M., Chua, T.S.: KGAT: knowledge graph attention network for recommendation. In: Proceedings of the 25th ACM SIGKDD International Conference on Knowledge Discovery & Data Mining, pp. 950–958 (2019)

29. Weiss, K., Khoshgoftaar, T.M., Wang, D.D.: A survey of transfer learning. J. Big Data **3**(1), 1–40 (2016). https://doi.org/10.1186/s40537-016-0043-6

30. Wu, M., Pan, S., Zhou, C., Chang, X., Zhu, X.: Unsupervised domain adaptive graph convolutional networks. In: Proceedings of the Web Conference 2020, pp. 1457–1467 (2020)

31. Wu, Z., Pan, S., Chen, F., Long, G., Zhang, C., Philip, S.Y.: A comprehensive survey on graph neural networks. IEEE Trans. Neural Netw. Learn. Syst. **32**(1), 4–24 (2020)

32. Yang, B., Yih, W., He, X., Gao, J., Deng, L.: Embedding entities and relations for learning and inference in knowledge bases. In: International Conference on Learning Representations (2015)

33. Ye, J., Cheng, H., Zhu, Z., Chen, M.: Predicting positive and negative links in signed social networks by transfer learning. In: Proceedings of the 22nd International Conference on World Wide Web, pp. 1477–1488 (2013)
34. Zhang, Y., Yang, Q.: A survey on multi-task learning. IEEE Trans. Knowl. Data Eng. **34**(12), 5586–5609 (2021)

MULTIFORM: Few-Shot Knowledge Graph Completion via Multi-modal Contexts

Xuan Zhang, Xun Liang$^{(\boxtimes)}$, Xiangping Zheng, Bo Wu, and Yuhui Guo

Renmin University of China, Beijing, China
{zhangxuanalex,xliang,xpzheng,wubochn,yhguo}@ruc.edu.cn

Abstract. Knowledge Graphs (KGs) have been applied to many downstream applications such as semantic web, recommender systems, and natural language processing. Previous research on Knowledge Graph Completion (KGC) usually requires a large number of training instances for each relation. However, considering the accelerated growth of online information, there can be some relations that do not have enough training examples. In fact, in most real-world knowledge graph datasets, instance frequency obeys a long-tail distribution. Existing knowledge embedding approaches suffer from the lack of training instances. One approach to alleviating this issue is to incorporate few-shot learning. Despite the progress they bring, they sorely depend on entities' local graph structure and ignore the multi-modal contexts, which could make up for the lack of training information in the few-shot scenario. To this end, we propose a multi-modal few-shot relational learning framework, which utilizes the entities' multi-modal contexts to connect few instances to the knowledge graphs. For the first stage, we encode entities' images, text descriptions, and neighborhoods to acquire well-learned entity representations. In the second stage, our framework learns a matching metric to match the query triples with few-shot reference examples. The experimental results on two newly constructed datasets show the superiority of our framework against various baselines.

Keywords: Few-shot learning · Meta-learning · Knowledge graphs · Attention aggregation function · Multi-modal contexts

1 Introduction

Knowledge Graphs (KGs) encode structured information of entities and their relations in the form of triples (h, r, t), where h represents some head entity and r represents some relation that connects h to some tail entity t. For example, a statement like *"Isaac Newton worked at the University of Cambridge"* can be represented as (*Isaac Newton, Work location, University of Cambridge*). KGs are the key components of various practical applications such as visual transfer learning [19], recommender systems [32] and so on. Despite their usefulness

M.-R. Amini et al. (Eds.): ECML PKDD 2022, LNAI 13714, pp. 172–187, 2023.
https://doi.org/10.1007/978-3-031-26390-3_11

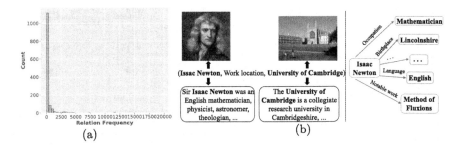

(a)

(b)

Fig. 1. (a) The distribution of relation frequencies in FB15K. (b) An example of multi-modal contexts of KGs: The left presents the images and textual descriptions of the entities in the triple (*Issac Newton, Work location, University of Cambridge*); The right presents the one-hop graph structure of the entity *Issac Newton*.

and popularity, KGs are often highly incomplete. Extensive research, termed as knowledge embedding [2,24,29], has made great progress in automatically completing missing links in KGs.

However, research on Knowledge Graph Completion (KGC) for KGs usually assume that sufficient training examples for each relation are available and cannot cope with few-shot relations. In the real world, the KGs evolve quickly with new entities and relations being added by the second and some new relations may not have enough training examples. Even in the classic knowledge graph FB15K, long-tail relations (few-shot relations), which have very few training triples, are actually very common as shown in Fig. 1(a). To be more specific, FB15K contains 1345 relations and about 0.6 million instances, but over 36% of these relations contain no more than 10 instances.

There are also some few-shot learning methods, such as GMatching [37] and FAAN [21], concentrating on alleviating the challenge of the lack of training examples for the long-tail relations. These models aim at predicting new links given only few training triples in a meta-learning scenario. Their main ideas are devising a neighbor encoder to acquire well-represented entities from the neighbors, and then represent few-shot relations with the learned entities. One of the key challenges is to learn the accurate entity representations with very few training information available.

While the few-shot learning models focus on developing various complicated algorithms, they depend on limited training information sorely from the entities' neighborhoods and ignore other crucial multi-modal contexts widely existing in KGs and Freebase [1], such as images and the text descriptions. As Fig. 1(b) shows, these additional multi-modal contexts contain abundant information, which could be helpful during training and make up for the lack of training information in the few-shot scenario.

With the aforementioned statements, we go back to the original KGs, and extract the entities' images, text descriptions and neighborhoods as additional visual, textual and topological information respectively. To predict new links with only few-shot given instances, we propose a <u>MULTI</u>-modal <u>Fe</u>w-sh<u>O</u>t

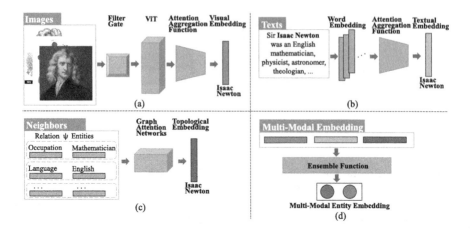

Fig. 2. Multi-modal context encoder for entities: (a) Image encoder; (b) Text encoder; (c) Neighbor encoder; (d) Multi-modal embedding fusion model.

Relational learning fraMework (MULTIFORM). In contrast to previous few-shot learning models sorely depending on entities' neighborhoods, MULTIFORM is able to benefit from all multi-modal contexts. MULTIFORM consists of a multi-modal context encoder and a metric learning module. The multi-modal context encoder produces well-learned representations of entities via multi-modal contexts. We separately encode image embedding, text descriptions and one-hop neighbors of entities and leverage an ensemble function to produce new accurate embeddings containing multi-modal information of entities. Our metric learning module aims at learning a matching function that can be used to discover more similar triples given few-shot reference triples. With two newly constructed datasets, i.e., MM-FB15K and MM-DBpedia, we show that our model can achieve consistent improvements over various state-of-the-art baselines on the few-shot KGC task. In summary, the present work makes the following contributions:

- As far as we know, this paper is the first to study few-shot KGC tasks with multi-modal contexts. We design three encoders to extract crucial information from different multi-modal data.
- We explore the impact of different multi-modal contexts, which is empirically important but ignored by the previous studies on multi-modal KGs.
- We construct two new datasets MM-FB15K and MM-DBpedia from FB15K and DBpedia for multi-modal few-shot KGC evaluation. We evaluate our model in the few-shot scenario and the experimental results show the superiority of our model against various state-of-the-art baselines.

2   Related Work

Here we survey three topics relevant to our research: unimodal knowledge embedding models, multi-modal knowledge embedding models, and few-shot learning.

2.1   Unimodal Knowledge Embedding Models

Unimodal knowledge embedding models aim at modeling multi-relational data and automatically inferring missing facts in KGs. Many of them encode both entities and relations into a continuous low dimensional vector space. RESCAL [17] utilizes tensor operations to model relations. TransE [2] is a classic work that encodes both entities and relations into a 1-D vector space. Following this line of research, more effective models such as DistMult [38], ComplEx [29], ConvE [5], Rotate [25], and Rot-Pro [24] have been proposed for further improvements. These embedding-based models heavily rely on extensive collections of training examples, and they are not qualified to deal with sparse triples, as presented in [2] and [37].

2.2   Multi-modal Knowledge Embedding Models

Multi-modal knowledge embedding models mainly focus on encoding visual and structural contexts. IKRL [36] separately trains visual information and structural information on TransE [2]. Mousselly et al. [15] uses three different ensemble function, i.e., simple concatenation, DeViSE [8], and Imagined [4] to fuse multi-modal context embeddings. TransAE [34] utilizes an auto-encoder to integrate them. RSME [33] evaluates different image encoders for multi-modal KGC and verify the effectiveness of Visual Transformer (ViT), so we adopt ViT as image encoder in this paper. There are several models [22,35] taking rich text descriptions into consideration to handle unseen entities.

2.3   Few-Shot Learning

Few-shot learning methods seek to learn novel concepts with only a small number of labeled examples. Recent deep learning based few-shot learning models can be classified into three groups. The first group is *model-based approaches*, which depend on a specially designed part like memory to quickly optimize the model parameters given few-shot training examples. MetaNet [16], a typical model-based approach, learns meta knowledge across tasks and generalizes rapidly via its fast parameterization. The second group is *metric-based approaches*, which try to learn a generalizable metric and the corresponding matching functions among a set of training examples. For example, prototypical networks [23] classify each instance by calculating the similarity to prototype representation of each class, whose idea is similar to some nearest neighbor algorithms. GMatching [37], FSRL [39], and FAAN [21] can also be considered as a metric-based approach. The third group is *optimization-based approaches* [7,13,20], which aim to learn faster by changing the optimization methods on few-shot reference instances. One example is model-agnostic meta-learning (MAML) [7], which first proposed the framework of updating parameters of a task-specific learner and performing meta optimization across tasks by using the above updated parameters. MetaR [3], which transfers relation-specific meta information from support set to query set, can also be regarded as an optimization-based approach for knowledge graph.

Previous few-shot learning research mainly focuses on vision [27], sentiment analysis [12] domains. As for few-shot learning on KGC, Bordes et al. [2] first realized the number of training examples for each relation in KGs have a great impact on the accuracy of the embedding model. However, he did not formulate it as a few-shot learning task. Existing few-shot learning models [21,37,39] on KGC tasks all sorely depend on local graph structures. In contrast to their approaches, we intend to leverage visual, textual and topological context to improve the quality of entity embeddings.

3   Preliminaries

3.1   Task Formulation

Here we give the definition of the few-shot KGC task via multi-modal contexts as follows:

Definition 1. *Given an incomplete KG $\mathcal{G} = (\mathbf{E}, \mathbf{R}, \mathbf{T})$, where $\mathbf{E}$, $\mathbf{R}$ and $\mathbf{T}$ are the entity set, relation set, and triple set, respectively, the few-shot KGC task completes $\mathcal{G}$ by finding a set of missing triples $\mathbf{T}' = \{(h, r, t) \mid (h, r, t) \notin \mathbf{T}, h, t \in \mathbf{E}, r \in \mathbf{R}\}$ when only few-shot entity pairs (h, t) and their multi-modal contexts are known for each relation r.*

In Definition 1, it is also called the K-shot KGC task when K training examples are given for each relation. In contrast to previous work, which usually assumes the availability of enough triples for training, this work studies the case where only few training triples are available. To be more specific, the goal is to rank the true tail entity higher than other candidate entities, given only K example triples $(h_i', r, t_i')_{i=1}^{K}$ for relation r. The candidate set is constructed using the entity type constraint [28].

3.2   Few-Shot Learning Settings

Following the standard meta-learning pipelines [7,20], we describe the settings for training and evaluation of our few-shot learning model. We have different sets for meta-training, meta-validation, and meta testing ($D_{\text{meta-train}}$, $D_{\text{meta-validation}}$, and $D_{\text{meta-test}}$) respectively. Note that none of the above share the same relation label space. On $D_{\text{meta-train}}$, we are interested in training a learning procedure (the meta-learner) that can take few examples as input and produce a matching metric (the learner) that could be used to predict new facts. Using $D_{\text{meta-validation}}$ we can perform hyper-parameter selection of the meta-learner and evaluate its generalization performance on $D_{\text{meta-test}}$.

More specifically, a $D_{\text{meta-train}}$ corresponding to a certain relation $r \in \mathcal{R}$, consists of support and query triples: $D_r = \{D_{\text{s\_r}}, D_{\text{q\_r}}\}$. There are K triples in $D_{\text{s\_r}}$ for K-shot KGC tasks. $D_{\text{q\_r}} = \{h_i, r, t_i, C_{h_i, r}\}$ consists of the query triples of r with ground-truth tail entities t_i for each query (h_i, r), and the corresponding tail entity candidates $C_{h_i, r} = \{t_{ij}\}$ where each t_{ij} is an entity in the KGs.

Then the metric model can be tested on this set by ranking the candidate set $C_{h_i,r}$, given the test query (h_i, r) and the labeled triple in $D_{\text{s\_r}}$. $D_{\text{meta-validation}}$ and $D_{\text{meta-test}}$ are composed of $D_{\text{s\_r}}$, $D_{\text{q\_r}}$. We denote the ranking loss of relation r as $\ell_\theta (h_i, r, t_i \mid C_{h_i,r}, D_{\text{s\_r}})$, where θ represents the parameters of our model. Thus, the objective of model training can be defined as:

$$\min_\theta \mathbb{E}_{D_r} \left[\sum_{(h_i, r, t_i, C_{h_i,r}) \in D_{\text{q\_r}}} \frac{\ell_\theta (h_i, r, t_i \mid C_{h_i,r}, D_{\text{s\_r}})}{|D_{\text{q\_r}}|} \right] \tag{1}$$

where D_r is sampled from the meta-training set $D_{\text{meta-train}}$ and $|D_{\text{q\_r}}|$ denotes the number of tuples in $D_{\text{q\_r}}$.

After sufficient training, we are able to predict facts of each new relation $r' \in \mathcal{R}'$. Due to the assumption of K-shot learning, the relation label space of the above meta-sets is disjoint with each other, i.e., $\mathcal{R} \cap \mathcal{R}' = \phi$. Otherwise, the metric model will actually see more than K-shot labeled data during meta-testing, thus the few-shot assumption is violated. Finally, we construct a subset $\mathcal{G}^*$ from $\mathcal{G}$ by removing all relations in $D_{\text{meta-train}}$, $D_{\text{meta-validation}}$ and $D_{\text{meta-test}}$ to construct entities' neighborhoods.

4   Model

Our model MULTIFORM consists of two modules: a multi-modal context encoder and a metric learning module. The core of our proposed model is a similarity function $f_\mathcal{S} ((h,t), (h',t') \mid \mathcal{V}^*, \mathcal{T}^*, \mathcal{G}^*)$, where $\mathcal{V}^*, \mathcal{T}^*, \mathcal{G}^*$ is the set of entities' visual context, textual context, and topological context, respectively. Given K known facts $(h'_i, r, t'_i)_{i=1}^K$ for any query relation r, the model could predict the likelihood of testing triples $\{(h_i, r, t_{ij}) \mid t_{ij} \in C_{h_i,r}\}$, based on the matching score between each (h_i, t_{ij}) and its semantic average of $(h'_i, t'_i)_{i=1}^K$. The implementation of the above matching function involves two sub-tasks: (1) the representations of entity pairs; and (2) the comparison function between two entity-pair representations.

4.1   Multi-modal Context Encoder

Multi-modal context encoder aims at utilizing the multi-modal contexts to learn well-represented entities. Specifically, it can be split into four parts: an image encoder, a text encoder, a neighbor encoder and a multi-modal embedding fusion model as illustrated in Fig. 2. The image encoder aims to extract the visual representations of entities' images and acquire visual embeddings for entities. The text encoder takes textual descriptions as input and output entities' textual embeddings. The neighbor encoder learns from entities' neighborhoods and produces topological embedding. The multi-modal embedding fusion model concatenates on integrating various multi-modal context embeddings and acquiring the accurate entity embeddings.

Image Encoder. Since most entities have more than one image collected in various scenarios, the image set is very possible to contain wrong images, which do not match the corresponding entities. It is essential to find out which images better represent their corresponding entities and filter out the noisy images. [33] shows that incorrect images account for only a small proportion of all images in KGs. Inspired by [33], we utilize a filter gate based on the empirical analysis that the incorrect images have low similarity with the right images. To be more specific, given an entity h, its multiple images can be presented as $V = \{v_1, v_2, \ldots, v_n\}$, where $V \in \mathcal{V}^*$. The filter gate selects the image with the highest similarity to the other images of the given entity to learn the visual embeddings:

$$v_h = \arg\max_{v_i \in V} \left\| \sum_j S\left(v_i, v_j\right) \right\|, \tag{2}$$

where S is the function to measure the visual similarity of two images. We adopt pHash [18] for simplicity. As ViT achieves the best performance over the Convolutional Neural Network (CNN) based models according to [33], we adopt ViT to encode the selected right images to obtain the corresponding embeddings of images in V as $\{z_{v_1}, z_{v_2}, \ldots, z_{v_n}\}$. Finally, we devise an attention aggregation function f_{aggre} to model representations of different images of the given entity and obtain the visual embedding z_V:

$$f_{aggre}(V) = \sigma\left(\sum_i \alpha_i z_{v_i}\right), \tag{3}$$

$$\alpha_i = \frac{\exp\left\{u_v^T\left(W_v z_{v_i} + b_v\right)\right\}}{\sum_j \exp\left\{u_v^T\left(W_v z_{v_j} + b_v\right)\right\}}, \tag{4}$$

where *sigma* denotes activation unit (we use Tanh); $z_{v_i} \in \mathbb{R}^{d \times 1}$ is the output representations of ViT and d is dimension of representation vectors; $u_v \in \mathbb{R}^{d \times 1}, W_v \in \mathbb{R}^{d \times d}, b_v \in \mathbb{R}^{d \times 1}$ are learnable parameters.

Text Encoder. Given a certain entity and its text description $X = \{x_1, x_2, \ldots, x_n\}$ where x is the word in the sentence, we first use BERT [6] to generate the word embedding $\{z_{x_1}, z_{x_2}, \ldots, z_{x_n}\}$. Similarly, we adopt the attention aggregation function f_{aggre} to obtain the textual embedding z_X:

$$f_{aggre}(X) = \sigma\left(\sum_i \beta_i z_{x_i}\right), \tag{5}$$

$$\beta_i = \frac{\exp\left\{u_x^T\left(W_x z_{x_i} + b_x\right)\right\}}{\sum_j \exp\left\{u_x^T\left(W_x z_{x_j} + b_x\right)\right\}}, \tag{6}$$

where $z_{x_i} \in \mathbb{R}^{d \times 1}$ is the output representations of BERT and d is dimension of word embedding vectors; $u_x \in \mathbb{R}^{d \times 1}, W_x \in \mathbb{R}^{d \times d}, b_x \in \mathbb{R}^{d \times 1}$ are learnable parameters.

Neighbor Encoder. Recently, Xiong et al. [37] and Zhang et al. [39] have demonstrated the effectiveness of encoding local graph structures as entity representations. Following their researches and inspired by the progress in Graph Convolutional Network (GCN), we consider CompGCN [30] to model the local heterogeneous feature of the neighborhoods. Specifically, for each given head entity h, its neighborhoods forms a set of $\{relation, tail\ entity\}$ tuples. As shown in Fig. 2(c), for the entity *Issac Newton*, one of such tuples is $\{Occupation, Mathematician\}$. Thus, the neighbor set can be denoted as $\mathcal{N}_h = \{r_i, t_i\}_{i=1}^{I}$, where r_i and t_i represent the i-th relation and corresponding tail entity of h, respectively. I is the number of such neighbors and $(h, r_i, t_i) \in \mathcal{G}^*$.

Our CompGCN-based neighbor encoder aims at encoding $\mathcal{N}_h$ and generating a well-learned vector as the feature representation of local connections of h. The details are as follows:

$$y_h^{(k)} = \sigma \left(\sum_{(r_i, t_i) \in \mathcal{N}_h} W_{\lambda(r)}^{(k)} \psi \left(y_{r_i}^{(k-1)}, y_{t_i}^{(k-1)} \right) \right), \tag{7}$$

where $W_{\lambda(r)}^{(k)}$ is a relation-specific shared parameter to learn; ψ a composition function of the relation r_i with its respective tail entity t_i. The composition $\psi : \mathbb{R}^d \times \mathbb{R}^d \to \mathbb{R}^d$ can be any entity-relation function akin to TransE [2] or RotatE [25] (We choose RotatE according to experimental results); y_h, y_r, y_t is the embeddings of h, r, t respectively and can random initialized or pretrained by existing embedding-based models; $y_h^{(k)}$ is the final topological embedding.

Multi-modal Entity Embedding Fusion Model. With multi-modal context information encoded, an embedding fusion model is developed to improve the representations of the given entity. Among various ensemble functions, [15] point out that simple concatenation works better than DeViSE [8] and Imagined [4] on multi-modal KGC tasks, and taking limited computational resources and scalability of MULTIFORM, we use simple concatenation to aggregate the visual embedding, textual embedding and topological embedding.

4.2 Metric Learning Module

This module is designed to do effective similarity matching given the output of feature fusion module. For K-shot learning scenario, we get two sets of entity pairs: the query entity pair set (h_i, t_{ij}) and the support pair set $(h_i', t_i')_{i=1}^{K}$. We obtain well represented entity embeddings for each set: $\left[o\left(\mathcal{N}_{h_i}\right) ; o\left(\mathcal{N}_{t_{ij}}\right) \right]$ and $\left[o\left(\mathcal{N}_{h'}\right) ; o\left(\mathcal{N}_{t'}\right) \right]$ via the multi-modal context encoder. When $K > 1$, we employ a simple sematic averaging function to get $\mathcal{N}_{h'}$ and $\mathcal{N}_{t'}$:

$$\mathcal{N}_{h'} = \frac{\sum_{i=1}^{K} \mathcal{N}_{h_i'}}{K} \tag{8}$$

$$\mathcal{N}_{t'} = \frac{\sum_{i=1}^{K} \mathcal{N}_{t_i'}}{K}. \tag{9}$$

Table 1. Statistics of the Datasets. # Entities denotes the number of unique entities and # Relations denotes the number of all relations. # Tasks denotes the number of relations we use as few-shot tasks.

Dataset	#Entities	# Relations	# Triples	# Tasks
MM-FB15K	14951	1345	592,213	356
MM-DBpedia	12842	279	297,084	69

We can simply concatenate $o(\mathcal{N}_{h'})$ and $o(\mathcal{N}_{t'})$ and calculate similarity between pairs in the two sets. For our model's scalability, we use the same multi-step matching processor as [37]. Every process step is defined as follows:

$$h'_{k+1}, c_{k+1} = \text{LSTM}\left(p, [h_k \oplus s, c_k]\right) \tag{10}$$

$$h_{k+1} = h'_{k+1} + p \tag{11}$$

$$score_{k+1} = \frac{h_{k+1} \odot s}{|h_{k+1}| |s|} \tag{12}$$

where $s = o(\mathcal{N}_{h'}) \oplus o(\mathcal{N}_{t'})$, $p = o(\mathcal{N}_{h_i}) \oplus o(\mathcal{N}_{t_{ij}})$ are concatenated well-learned embeddings of the support pair and query pair. After n processing steps, we use $score_k$ as the final similarity score between the query and support entity pair.

4.3 Loss Function

For a selected query relation r and its support triples $(h'_i, r, t'_i)_{i=1}^{K}$, we employ negative sampling methods to construct query triples, i.e., we collect a group of positive query triples $\left\{ (h_i, r, t_i^+) \mid (h_i, r, t_i^-) \notin \mathcal{G} \right\}$ and corrupt the tail entities to construct another group negative query triples $\left\{ (h_i, r, t_i^-) \mid (h_i, r, t_i^-) \notin |\mathcal{G}| \right\}$. Following previous few-shot learning models, we utilize a hinge loss function for our model:

$$l_\theta = \max\left(0, \gamma + score_\theta^- - score_\theta^+\right) \tag{13}$$

where $score_\theta^+$ and $score_\theta^-$ are scalars calculated by comparing the query triple $\left(h_i, r, t_i^+/t_i^-\right)$ with the support triples $(h'_i, r, t'_i)_{i=1}^{K}$ using our metric learning model, and the margin γ is a hyperparameter to tune. For each training episode, we first sample D_r from the meta-training set $D_{\text{meta-train}}$. Then we sample K triple as the support triple D_{s_r} and a batch of other triples as the positive query/test triples D_{q_r} from all known triples in D_r.

5 Experiments

With MULTIFORM, we investigate three issues: (1) Will the incorporation of multi-modal contexts help the few-shot KGC tasks? (2) How much visual context, textual context and topological context contribute to MULTIFORM's performance, respectively? (3) Does the number of multi-modal training triples affect the performance of MULTIFORM? To explore these questions, we conduct a series of experiments on two few-shot multi-modal knowledge graph datasets and systematically analyze the corresponding results.

5.1 Datasets

Our constructed multi-modal datasets MM-FB15K and MM-DBpedia are based on FB15K [1,2] and DBpedia [11,14,22]. The dataset statistics are shown in Table 1. Figure 1(b) shows an example of visual and textual contexts. Each entity in MM-FB15K and MM-DBpedia has at least one image and a description of no less than 15 words. Following [37], we construct few-shot multi-modal KGs by selecting those relations that do not have too many training triples. Specifically, to guarantee enough triples for evaluation, we select the relations with less than 500 but more than 50 triples as few-shot tasks, i.e., we obtain 356 and 69 few-shot relations in MM-FB15K and MM-DBpedia, respectively. The rest of the relations are referred to as background relations and their triples provide neighborhoods to learn topological information. In addition, For MM-FB15K, we use 267/18/71 and 51/6/12 task relations for training/validation/testing in MM-FB15K and MM-DBpedia, respectively. The division ratio is about 15:1:4, similar to the data split in [37,39].

5.2 Baseline Methods

For fair comparison, we select three kinds of baseline methods including unimodal knowledge embedding models, multi-modal knowledge embedding models, and few-shot learning models.

- **Unimodal Knowledge Embedding Models.** This line of research models multi-relational structures in KGs and encodes both entities and relations into a continuous low dimensional vector space. We consider the four widely used baseline methods as follows: TransE [2], DistMult [38], ComplEx [29] and Rot-Pro [24]. For implementation, we use an Open Toolkit [9] released by Xu Han et al. which provides the above knowledge embedding models. We also select RotatE [25], which has been reported very robust under different evaluation protocols in the extensive conducted experiments, comparing with a series of state-of-the-art knowledge embedding methods [26]. For fair comparison, all triples of background relations, training triples, and support triples of validation and test relations, are used during training.
- **Multi-modal Knowledge Embedding Models.** The models mainly focus on encoding visual and structural contexts. We select two state-of-the-art methods, i.e., TransAE [34] and RSME [33] as our baselines.
- **Few-Shot Learning Models.** This type of model concentrates on predicting new facts in KGs with only few-shot reference triples. For fair comparison, we select three typical neighbor encoder based models, i.e., GMatching [37], FSRL [39], FAAN [21].

Table 2. The 5-shot KGC results on the testing dataset. The best baseline results are indicated by underline and the best results of all methods are highlighted in bold.

Model	MM-FB15K				MM-DBpedia			
	MRR	Hits@10	Hits@5	Hits@1	MRR	Hits@10	Hits@5	Hits@1
TransE	0.116	0.164	0.139	0.089	0.103	0.155	0.120	0.077
DistMult	0.083	0.132	0.095	0.037	0.091	0.141	0.118	0.088
ComplEx	0.067	0.147	0.089	0.05	0.121	0.17	0.123	0.109
RotatE	0.131	0.189	0.160	0.101	0.150	0.242	0.179	0.120
Rot-Pro	0.099	0.145	0.112	0.061	0.139	0.200	0.154	0.107
TransAE	0.130	0.243	0.155	0.116	0.156	0.237	0.185	0.131
RSME	0.188	0.308	0.249	0.152	0.177	0.280	<u>0.219</u>	<u>0.145</u>
GMatching	0.261	0.377	0.340	0.189	0.176	0.293	0.231	0.116
FSRL	0.162	0.289	0.197	0.085	0.158	0.304	0.220	0.071
FAAN	<u>0.341</u>	<u>0.458</u>	<u>0.382</u>	<u>0.279</u>	<u>0.195</u>	<u>0.310</u>	0.217	0.136
MULTIFORM	**0.437**	**0.550**	**0.461**	**0.305**	**0.303**	**0.425**	**0.334**	**0.279**

Table 3. Results of model variants on MM-FB15K dataset. The best results are highlighted in bold.

Model variants	MRR	Hits@10	Hits@5	Hits@1
AS_1	0.401	0.499	0.450	0.293
AS_2	0.383	0.482	0.443	0.288
AS_3	0.351	0.472	0.397	0.282
MULTIFORM	**0.437**	**0.550**	**0.461**	**0.305**

5.3   Implementation Details

The embedding size d is set to 128 and 256 for MM-FB15K and MM-DBpedia datasets, respectively. The number of local neighbors used in the neighbor encoder is set to 45, which works the best for both datasets. As for image encoder and text encoder, we use the open resource from huggingface to implement ViT[1] and BERT[2] and keep their default settings about transformer layers. Besides, the LSTM cell is utilized in the matching function as a matching processor. The dimension of LSTM's hidden state is set to 128 and 256 for MM-FB15K and MM-DBpedia datasets, respectively. The optimal matching step is 2. For parameter updates, we use Adam [10] with the initial learning rate of 0.001 and we have the learning rate decay 0.2 for each 50k training step. The margin γ used in the base loss function is 5.0.

---

[1] https://huggingface.co/docs/transformers/model_doc/bert.
[2] https://huggingface.co/docs/transformers/model_doc/vit.

5.4   Results

We first evaluate our model on the few-shot KGC task, which predicts new facts on a query set given only few support triples and their multi-modal contexts. As shown in Table 2, MULTIFORM shows a significant margin over all three types of baselines in the 5-shot scenario. Taking the experimental results (testing MRR and Hits@10) on MM-FB15K as an example, the relative improvement (%) of MULTIFORM against RotatE (the best-performing knowledge embedding models) is up to 233.59% and 191.01%; MULTIFORM outperforms RSME (the best-performing multi-modal knowledge embedding models) by 132.45% and 78.37%; MULTIFORM shows a significant improvement margin over FAAN (the best-performing few-shot learning models) by 28.15% and 20.09%. These results, to some extent, confirm the effectiveness of the idea that incorporating multi-modal contexts can be helpful to few-shot KGC tasks since multi-modal contexts shape more accurate and well-represented entities' embeddings. Thus, we have so far answered the first question, i.e., MULTIFORM can be well adapted into the few-shot KGC task and produce consistent improvements over all types of baselines by incorporating multi-modal contexts. We also observe that most multi-modal knowledge embedding models have better performance than unimodal knowledge embedding models, which verifies the benefit of utilizing multi-modal contexts. We also noticed that unimodal/multi-modal knowledge embedding models have a big gap in performance compared with few-shot learning models. We guess unimodal/multi-modal knowledge embedding models are designed for transductive learning with sufficient training data and can not be adapted into the few-shot scenario where only few training data are available. By the way, this demonstrates that the few-shot KGC task is a very challenging problem.

5.5   Ablation Study

Here We seek the answer to our second question in this section, i.e., investigating the effectiveness of each context of the proposed model. We consider the following ablation studies:

- **(AS_1)** We evaluate the effectiveness of images. We use randomly initialized vectors as visual embeddings and keep the other two encoders.
- **(AS_2)** We use randomly initialized vectors as the output of the text encoder to verify the effectiveness of text descriptions.
- **(AS_3)** We use randomly initialized vectors as topological embeddings to evaluate the effectiveness of entities' graph structure.

As shown in Table 3, our model has better performance than all model variants. The comparison between MULTIFORM and AS_1, AS_2, and AS_3 indicates that all visual context, textual context and topological context contribute to improvements of our model. By comparison among AS_1, AS_2, and AS_3, we also notice that topological context contributes most to the model's performance,

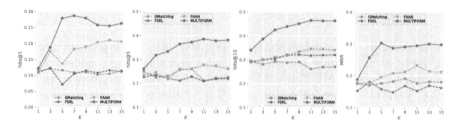

Fig. 3. Impact of few-shot size K.

since MULTIFORM shows the largest decrease when randomly initializing topological embeddings (refer to AS_3); We think it is because the knowledge of the same modality can be absorbed by neural networks more easily. The next largest contribution is made by textual context (refer to AS_2). We guess it is because KGs are originally extracted from the text so there exists semantic similarity. In summary, these results demonstrate that all contexts are important and contribute to MULTIFORM (Fig. 3).

5.6   Impact of Few-Shot Size

Since this work studies few-shot learning for KGC tasks, we conduct experiments to analyze the impact of few-shot size K. MULTIFORM consistently outperforms all few-shot baselines under different K, indicating the effectiveness of our model on few-shot link prediction on KGs. We also notice that as K increases, MULTIFORM gets relatively stable improvements compared to GMatching and FSRL, which demonstrates MULTIFORM's stability and robustness.

6   Conclusion and Future Work

In the present work, we introduce a multi-modal few-shot learning framework named MULTIFORM for KGC tasks. MULTIFORM aims at predicting new facts with only several training data and their multi-modal contexts, which is a challenging problem. MULTIFORM leverages visual, textual, and topological information of entities to produce well-learned representations and uses a metric learning method to match entity pairs. The experiment results demonstrate that MULTIFORM can outperform the state-of-the-art baselines. We also analyze the impact of few-shot size and conduct ablation studies on multi-modal contexts, which verify the effectiveness of each context. The goal of our future work is to incorporate external text content of relations and try more feature fusion methods to extend our model in the zero-shot scenario.

Acknowledgement. This work was supported by the National Natural Science Foundation of China (62072463, 71531012), and the National Social Science Foundation of China (18ZDA309).

References

1. Bollacker, K., Evans, C., Paritosh, P., Sturge, T., Taylor, J.: Freebase: a collaboratively created graph database for structuring human knowledge. In: Proceedings of the 2008 ACM SIGMOD International Conference on Management of Data (2008)
2. Bordes, A., Usunier, N., García-Durán, A., Weston, J., Yakhnenko, O.: Translating embeddings for modeling multi-relational data. In: Advances in Neural Information Processing Systems 26: 27th Annual Conference on Neural Information Processing Systems 2013. Proceedings of a Meeting Held 5–8 December 2013, Lake Tahoe, Nevada, United States (2013)
3. Chen, M., Zhang, W., Zhang, W., Chen, Q., Chen, H.: Meta relational learning for few-shot link prediction in knowledge graphs. In: Proceedings of the 2019 Conference on Empirical Methods in Natural Language Processing and the 9th International Joint Conference on Natural Language Processing (EMNLP-IJCNLP) (2019)
4. Collell, G., Zhang, T., Moens, M.: Imagined visual representations as multimodal embeddings. In: Proceedings of the Thirty-First AAAI Conference on Artificial Intelligence, San Francisco, California, USA, 4–9 February 2017 (2017)
5. Dettmers, T., Minervini, P., Stenetorp, P., Riedel, S.: Convolutional 2D knowledge graph embeddings. In: Proceedings of the Thirty-Second AAAI Conference on Artificial Intelligence (AAAI-2018), the 30th Innovative Applications of Artificial Intelligence (IAAI-2018), and the 8th AAAI Symposium on Educational Advances in Artificial Intelligence (EAAI-2018), New Orleans, Louisiana, USA, 2–7 February 2018 (2018)
6. Devlin, J., Chang, M.W., Lee, K., Toutanova, K.: BERT: pre-training of deep bidirectional transformers for language understanding. In: Proceedings of the 2019 Conference of the North American Chapter of the Association for Computational Linguistics: Human Language Technologies, Volume 1 (Long and Short Papers) (2019)
7. Finn, C., Abbeel, P., Levine, S.: Model-agnostic meta-learning for fast adaptation of deep networks. In: Proceedings of the 34th International Conference on Machine Learning, ICML 2017, Sydney, NSW, Australia, 6–11 August 2017. Proceedings of Machine Learning Research (2017)
8. Frome, A., et al.: DeViSE: a deep visual-semantic embedding model. In: Advances in Neural Information Processing Systems 26: 27th Annual Conference on Neural Information Processing Systems 2013. Proceedings of a Meeting Held 5–8 December 2013, Lake Tahoe, Nevada, United States (2013)
9. Han, X., et al.: OpenKE: an open toolkit for knowledge embedding. In: Proceedings of the 2018 Conference on Empirical Methods in Natural Language Processing: System Demonstrations (2018)
10. Kingma, D.P., Ba, J.: Adam: a method for stochastic optimization. In: 3rd International Conference on Learning Representations, ICLR 2015, San Diego, CA, USA, 7–9 May 2015, Conference Track Proceedings (2015)
11. Lehmann, J., et al.: DBpedia-a large-scale, multilingual knowledge base extracted from Wikipedia. Semant. Web $6(2)$, 167–195 (2015)
12. Li, Z., Li, X., Wei, Y., Bing, L., Zhang, Y., Yang, Q.: Transferable end-to-end aspect-based sentiment analysis with selective adversarial learning. In: Proceedings of the 2019 Conference on Empirical Methods in Natural Language Processing and the 9th International Joint Conference on Natural Language Processing (EMNLP-IJCNLP) (2019)

13. Li, Z., Zhou, F., Chen, F., Li, H.: Meta-SGD: learning to learn quickly for few-shot learning. arXiv preprint arXiv:1707.09835 (2017)

14. Liu, Y., Li, H., Garcia-Duran, A., Niepert, M., Onoro-Rubio, D., Rosenblum, D.S.: MMKG: multi-modal knowledge graphs. In: Hitzler, P., et al. (eds.) ESWC 2019. LNCS, vol. 11503, pp. 459–474. Springer, Cham (2019). https://doi.org/10.1007/978-3-030-21348-0_30

15. Mousselly-Sergieh, H., Botschen, T., Gurevych, I., Roth, S.: A multimodal translation-based approach for knowledge graph representation learning. In: Proceedings of the Seventh Joint Conference on Lexical and Computational Semantics (2018)

16. Munkhdalai, T., Yu, H.: Meta networks. In: Proceedings of the 34th International Conference on Machine Learning, ICML 2017, Sydney, NSW, Australia, 6–11 August 2017. Proceedings of Machine Learning Research (2017)

17. Nickel, M., Tresp, V., Kriegel, H.: A three-way model for collective learning on multi-relational data. In: Proceedings of the 28th International Conference on Machine Learning, ICML 2011, Bellevue, Washington, USA, 28 June–2 July 2011 (2011)

18. Niu, X., Jiao, Y.: An overview of perceptual hashing. Acta Electronica Sin. **36**(7), 1405 (2008)

19. Peng, Z., Li, Z., Zhang, J., Li, Y., Qi, G., Tang, J.: Few-shot image recognition with knowledge transfer. In: 2019 IEEE/CVF International Conference on Computer Vision, ICCV 2019, Seoul, Korea (South), 27 October–2 November 2019 (2019)

20. Ravi, S., Larochelle, H.: Optimization as a model for few-shot learning. In: 5th International Conference on Learning Representations, ICLR 2017, Toulon, France, 24–26 April 2017, Conference Track Proceedings (2017)

21. Sheng, J., et al.: Adaptive attentional network for few-shot knowledge graph completion. In: Proceedings of the 2020 Conference on Empirical Methods in Natural Language Processing (EMNLP) (2020)

22. Shi, B., Weninger, T.: Open-world knowledge graph completion. In: Proceedings of the Thirty-Second AAAI Conference on Artificial Intelligence (AAAI-2018), the 30th Innovative Applications of Artificial Intelligence (IAAI-2018), and the 8th AAAI Symposium on Educational Advances in Artificial Intelligence (EAAI-2018), New Orleans, Louisiana, USA, 2–7 February 2018 (2018)

23. Snell, J., Swersky, K., Zemel, R.S.: Prototypical networks for few-shot learning. In: Advances in Neural Information Processing Systems 30: Annual Conference on Neural Information Processing Systems 2017, 4–9 December 2017, Long Beach, CA, USA (2017)

24. Song, T., Luo, J., Huang, L.: Rot-Pro: modeling transitivity by projection in knowledge graph embedding. In: Advances in Neural Information Processing Systems (2021)

25. Sun, Z., Deng, Z., Nie, J., Tang, J.: RotatE: knowledge graph embedding by relational rotation in complex space. In: 7th International Conference on Learning Representations, ICLR 2019, New Orleans, LA, USA, 6–9 May 2019 (2019)

26. Sun, Z., Vashishth, S., Sanyal, S., Talukdar, P., Yang, Y.: A re-evaluation of knowledge graph completion methods. In: Proceedings of the 58th Annual Meeting of the Association for Computational Linguistics (2020)

27. Sung, F., Yang, Y., Zhang, L., Xiang, T., Torr, P.H.S., Hospedales, T.M.: Learning to compare: relation network for few-shot learning. In: 2018 IEEE Conference on Computer Vision and Pattern Recognition, CVPR 2018, Salt Lake City, UT, USA, 18–22 June 2018 (2018)

28. Toutanova, K., Chen, D., Pantel, P., Poon, H., Choudhury, P., Gamon, M.: Representing text for joint embedding of text and knowledge bases. In: Proceedings of the 2015 Conference on Empirical Methods in Natural Language Processing (2015)

29. Trouillon, T., Welbl, J., Riedel, S., Gaussier, É., Bouchard, G.: Complex embeddings for simple link prediction. In: Proceedings of the 33nd International Conference on Machine Learning, ICML 2016, New York City, NY, USA, 19–24 June 2016. JMLR Workshop and Conference Proceedings (2016)

30. Vashishth, S., Sanyal, S., Nitin, V., Talukdar, P.P.: Composition-based multi-relational graph convolutional networks. In: 8th International Conference on Learning Representations, ICLR 2020, Addis Ababa, Ethiopia, 26–30 April 2020 (2020)

31. Vinyals, O., Blundell, C., Lillicrap, T., Kavukcuoglu, K., Wierstra, D.: Matching networks for one shot learning. In: Advances in Neural Information Processing Systems 29: Annual Conference on Neural Information Processing Systems 2016, Barcelona, Spain, 5–10 December 2016 (2016)

32. Wang, H., et al.: RippleNet: propagating user preferences on the knowledge graph for recommender systems. In: Proceedings of the 27th ACM International Conference on Information and Knowledge Management, CIKM 2018, Torino, Italy, 22–26 October 2018 (2018)

33. Wang, M., Wang, S., Yang, H., Zhang, Z., Chen, X., Qi, G.: Is visual context really helpful for knowledge graph? A representation learning perspective. In: Proceedings of the 29th ACM International Conference on Multimedia (2021)

34. Wang, Z., Li, L., Li, Q., Zeng, D.: Multimodal data enhanced representation learning for knowledge graphs. In: 2019 International Joint Conference on Neural Networks (IJCNN). IEEE (2019)

35. Xie, R., Liu, Z., Jia, J., Luan, H., Sun, M.: Representation learning of knowledge graphs with entity descriptions. In: Proceedings of the Thirtieth AAAI Conference on Artificial Intelligence, Phoenix, Arizona, USA, 12–17 February 2016 (2016)

36. Xie, R., Liu, Z., Luan, H., Sun, M.: Image-embodied knowledge representation learning. In: Proceedings of the Twenty-Sixth International Joint Conference on Artificial Intelligence, IJCAI 2017, Melbourne, Australia, 19–25 August 2017 (2017)

37. Xiong, W., Yu, M., Chang, S., Guo, X., Wang, W.Y.: One-shot relational learning for knowledge graphs. In: Proceedings of the 2018 Conference on Empirical Methods in Natural Language Processing (2018)

38. Yang, B., Yih, W., He, X., Gao, J., Deng, L.: Embedding entities and relations for learning and inference in knowledge bases. In: 3rd International Conference on Learning Representations, ICLR 2015, San Diego, CA, USA, 7–9 May 2015, Conference Track Proceedings (2015)

39. Zhang, C., Yao, H., Huang, C., Jiang, M., Li, Z., Chawla, N.V.: Few-shot knowledge graph completion. In: The Thirty-Fourth AAAI Conference on Artificial Intelligence, AAAI 2020, The Thirty-Second Innovative Applications of Artificial Intelligence Conference, IAAI 2020, The Tenth AAAI Symposium on Educational Advances in Artificial Intelligence, EAAI 2020, New York, NY, USA, 7–12 February 2020 (2020)

RDF Knowledge Base Summarization by Inducing First-Order Horn Rules

Ruoyu Wang[1,2], Daniel Sun[1,3]([⊠]), and Raymond Wong[1]

[1] University of New South Wales, Sydney, Australia
ruoyu.wang2@unsw.edu.au, wong@cse.unsw.edu.au
[2] Shanghai Jiao Tong University, Shanghai, China
wang.ruoyu@sjtu.edu.cn
[3] Enhitech LLC., Shanghai, China
danielwsun@gmail.com

Abstract. RDF knowledge base summarization produces a compact and faithful abstraction for entities, relations, and ontologies. The summary is critical to a wide range of knowledge-based applications, such as query answering and KB indexing. The patterns of graph structure and/or association are commonly employed to summarize and reduce the number of triples. However, knowledge coverage is low in state-of-the-art techniques due to limited expressiveness of patterns, where variables are under-explored to capture matched arguments in relations. This paper proposes a novel summarization technique based on first-order logic rules where quantified variables are extensively taken into account. We formalize this new summarization problem to illustrate how the rules are used to replace triples. The top-down rule mining is also improved to maximize the reusability of cached results. Qualitative and quantitative analyses are comprehensively done by comparing our technique against state-of-the-art tools, with showing that our approach outperforms the rivals in conciseness, completeness, and performance.

Keywords: Data summarization · RDF KB summarization · Knowledge graphs · Logic rule mining · Rule-based approaches

1 Introduction

Data summarization [1] is to extract, from the source, a subset or a compact abstraction that includes the most representative features or contents. Summarization of RDF Knowledge Bases (KBs) are also being studied for over a decade [3], especially after the concepts of semantic web and linked data are widely accepted, and the online data amount grows unexpectedly large.

To serve the purpose of concise and faithful summarization, structural methods [7,16] are among the first attempts where techniques are borrowed from general graph mining approaches. Statistical and deep learning techniques [10,15] are also welcome in the research to alleviate the impact of noise and capture latent correlations. However, the above methodologies cannot provide the

M.-R. Amini et al. (Eds.): ECML PKDD 2022, LNAI 13714, pp. 188–204, 2023.
https://doi.org/10.1007/978-3-031-26390-3_12

overview in an interpretable way and, in the meantime, be dependable in reasoning and deduction. Thus, approaches based on association patterns and logic rules are studied in more recent works [2,14,20].

Current pattern-based and rule-based methods summarize KGs and produce schematic views of the data. A technique for Logical Linked Data Compression (LLC) [14] has been proposed to extract association rules that represent repeated entities or relation-entity pairs at a lower cost. Labeled frequent graph structures are encoded as bit strings in KGIST [2] and summarized from the perspective of bit compression. Nevertheless, the extracted patterns fail to conclude general patterns with arbitrary variables and thus cover only a tiny part of the factual knowledge. First-order logic rules, such as Horn rules, are a promising upgrade where universally and existentially quantified variables are extensively supported, but the rules have not yet been used for the summarization purpose. First-order logic rules have been proved useful to KGs in knowledge-based applications, such as KG completion [9], and show competitive capabilities. However, the performance turns out to be the cost of expressiveness. For example, first-/higher-order logic rule mining techniques [19,23] cannot scale to databases consisting of thousands of records without parallelization [8,26]. Current techniques usually limit the expressiveness for high performance [9], and this decreases the completeness of induced semantics. Moreover, the selection of best semantics is also challenging, for the number of applicable rules induced from a knowledge base is much larger than required for the summarization.

This paper bridges the gap between RDF KB summarization and first-order logic rule mining. We propose a novel summarization technique based on first-order Horn rules where quantified variables are extensively taken into account. The formal definitions illustrate a new summarization problem: inducing Horn rules from an RDF KB, such that the KB is separated into two parts, where one is inferable (thus removable) by the other with respect to the rules. The top-down rule mining mechanism is also improved to maximize the reusability of cached contents. Contributions of this paper include:

- We are the first to employ first-order Horn rules in RDF KB summarization. Variables are explored to extend the coverage and the completeness of semantic patterns. The new approach is also applicable to relational databases.
- We refine the extension operations in top-down rule mining to a smaller step size, such that the conciseness and performance are both improved.
- We qualitatively analyze the superiority of our approach and demonstrate the reasons with quantitative experimental results. The experiments show that our technique summarizes a database to less than 40% of the size, covering more than 70% contents with induced rules. The performance of our technique is up to two orders faster than the rivals.

The remains of the paper are organized as follows: Sect. 2 reviews major studies in RDF KB summarization. Definitions and details of our approach are proposed in Sect. 3. Section 4 evaluates the performance of our technique and shows evidence of the improvement from a quantitative perspective. Finally, Sect. 5 concludes the entire paper.

2   Related Work

RDF KB summarization aims to extract concise and precise abstraction from facts and ontologies, providing a preview and overall understanding of large-scale knowledge data. Structural, statistical, and pattern/rule-based approaches have been studied for over a decade.

Structural approaches represent the summary as a smaller graph, where vertices and edges are either fragments of the original graph or converted according to some mapping criteria. *Quotient Graphs* [25] are widely applied in many structural approaches. Vertices in a quotient graph represent collections of vertices in the original graph according to an equivalence relation over the vertices. An edge in the quotient graph represents shared edges between the adjacent vertex collections. Forward and backward (bi)simulation [7] properties guarantee that a query on the quotient summary of a knowledge graph returns non-empty results if the results are non-empty from the original database. Indexing [16] is the major benefit of the structural approaches.

Statistical approaches focus on quantitative summaries for visualization [6], query answering [22], selective data access [13], and description generation [10]. The approaches are motivated by the source selection problem, where quantitative statistics reports on how relevant a knowledge base is to a query [13]. Query sensitive information, such as the existence or quantity of relevant entities, triples, or schematic rules, is calculated and stored [22]. To evaluate the relevance to a topic, centrality and frequency analyses within a neighborhood are employed to entities and ontological schema [10]. Summarization techniques for other data types, such as text, are also used to rank objects in different circumstances [18].

Pattern/rule-based approaches employ data mining approaches to extract frequent patterns, in the form of graph structures or rules, from the RDF graph. [28] summarizes with a set of approximate graph patterns in accordance with SPARQL query evaluations. KGIST [2] encodes RDF graph structures into bit strings and takes advantage of information theory to minimize the description length of the entire bit string. The codebook for bit compression represents sub-structures in the original graph. Meier [17] studied an RDF minimization problem under user-defined constraints via Datalog programs. The constrained minimization problem has been proved intractable, and the author identifies a tractable fragment solvable in polynomial time.

LLC [14] summarizes and compresses Linked Open Data (LOD) via association rules, and Fig. 1 shows the overall workflow. LLC converts an RDF knowledge base into a transactional database, and the itemsets consist of objects or relation-object pairs for every subject in the graph. Then FP-growth [12] is used to extract a list of frequent itemsets, and association rules are ranked and selected according to a measure representing the capability of replacement. The original graph is separated into two parts: G_A and G_D, by matching the rules on each itemset. G_D contains triples that cannot be replaced from the knowledge base, and triples in G_A are the replacement of those in $R(G_A)$. Thus, G_A, G_D, and the set of rules R make up the summary. The recovery of the original KG is accomplished by applying R on G_A.

Rules (1) and (2) are two types of association rules extracted by LLC:

$$\wedge_{i=1}^{n} < X, p, o_i > \leftarrow < X, p, o > \quad (1)$$
$$\wedge_{i=1}^{n} < X, p_i, o_i > \leftarrow < X, p, o > \quad (2)$$

where $< s, p, o >$ refers to a triple, p, p_i are relations, o, o_i are entities, and X is a universally quantified variable. The above rules can be converted to the following Horn rules:

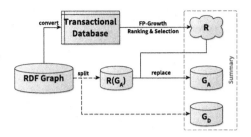

Fig. 1. LLC workflow

$$p(X, o_i) \leftarrow p(X, o), i = 1, \ldots, n \quad (3)$$
$$p_i(X, o_i) \leftarrow p(X, o), i = 1, \ldots, n \quad (4)$$

Inductive Logic Programming (ILP) provides top-down [9,23,27] and bottom-up [19] solutions to logic rule mining. Probability can also be used for noise tolerance [24]. Parallelization is often employed when inducing from large-scale databases [8,26].

3   Summarization via First-Order Horn Rules

This section presents the formal definition of the summarization with first-order Horn rules and shows how the rules are used in the solution framework. The logic rule mining process is also improved to extensively explore quantified variables and maximize the reusability of cached contents. The advantages of our technique are demonstrated by comparisons against LLC.

3.1   Preliminaries, Definitions and Notation Conventions

Let Σ be a finite set of constant symbols, e.g. $\{a, b, c, \ldots\}$. Let Γ be a finite set of variable symbols, e.g. $\{X, Y, Z, \ldots\}$. Let $\mathcal{P}^n (n \geq 0)$ be a finite set of n-ary predicate symbols (i.e. relations), and $\mathcal{P} = \bigcup_{i \geq 0} \mathcal{P}^i$. A **first-order predicate** (or simply, predicate) is composed of a predicate symbol and a list of arguments enclosed in parentheses, written as $p(t_1, \ldots, t_k)$, where $p \in \mathcal{P}^k, t_i \in \Sigma \cup \Gamma$. Let P be a predicate, $\phi(P)$ is the **arity** of P. P is a **ground predicate** if all arguments are constants. The above definitions do not break those in First-order Predicate Logic (FOL). In the context of RDF knowledge bases, all predicate symbols are binary, although the formalization and solution to the summarization problem fit in the broader domain of the relational data model.

Formally, an RDF knowledge base is a finite set of binary ground predicates. In FOL, the truth value of a ground predicate is determined by the interpretation and domain. In this paper, the interpretation of non-logic symbols is the definition of relations in databases, and the domain is the set of relation names and constant values. Therefore, a ground predicate P is $True$ according to some database $\mathcal{D}$ if and only if $P \in \mathcal{D}$.

A **first-order Horn rule** is of the form: $Q \leftarrow \wedge_i P_i$, where Q, P_i are **atoms** (predicates or the negations). In this paper, only non-negative atoms are considered in the rules. Q is called the **head** of the rule, and predicates P_i make up the **body**. Q is entailed by P_i if P_i are all $True$, that is, $(\wedge_i P_i) \wedge (Q \leftarrow \wedge_i P_i) \models Q$. Thus, by binding the variables in the entailment, the grounded predicate Q' is entailed by grounded predicates P_i' w.r.t. the rule r and a database $\mathcal{D}$ if every $P_i' \in \mathcal{D}$, written as $\{P_i'\} \models_r Q'$. Let $\mathcal{S}, \mathcal{T}$ be sets of ground predicates, $\mathcal{H}$ be a set of first-order Horn rules, $\mathcal{S} \models_{\mathcal{H}} \mathcal{T}$ if $\forall T \in \mathcal{T}, \exists \mathcal{S}' \subseteq \mathcal{S}, r \in \mathcal{H}$, such that $\mathcal{S}' \models_r T$. Suppose T is entailed by a set of predicates w.r.t. a rule r. If $T \in \mathcal{D}$, T is said to be **positively entailed** by $\mathcal{S}$ w.r.t. r; otherwise, T is **negatively entailed**. If a predicate is positively entailed by some grounding of r, the grounding is called an **evidence** of the predicate. The set of positive and negative entailments w.r.t. rule r is denoted by $\mathcal{E}_r^+$ and $\mathcal{E}_r^-$, and $\mathcal{E}_r = \mathcal{E}_r^+ \cup \mathcal{E}_r^-$.

Notation Conventions. Capital letters refer to variables, such as X, Y. **Unlimited Variables** *(UVs)* are variables assigned to only one argument in some rule; **Limited Variables** *(LVs)* are those assigned to at least two arguments. A question mark ('?') refers to a unique UV in a rule. Uncapitalized words as arguments refer to constants, e.g., *tom*. Uncapitalized words before the parenthesis or a period are predicate symbols, and the number after the period is the index of the argument in the predicate, starting from 0, such as $father.0$. For example, the following two rules are the same. Variables X and Y are LVs, while Z and W are UVs and can be simplified to two question marks.

$$p(X, Y, Z) \leftarrow q(X, Y), s(Y, W) \tag{5}$$
$$p(X, Y, ?) \leftarrow q(X, Y), s(Y, ?) \tag{6}$$

Definition 1 (Knowledge Graph Summarization). *Let $\mathcal{D}$ be an RDF KB. The summarization on $\mathcal{D}$ is a triple $(\mathcal{H}, \mathcal{N}, \mathcal{C})$ with minimal size, where $\mathcal{H}$ (for "Hypothesis") is a set of inference rules, both $\mathcal{N}$ (for "Necessaries") and $\mathcal{C}$ (for "Counterexamples") are sets of predicates. $\mathcal{D}, \mathcal{H}, \mathcal{N}, \mathcal{C}$ satisfies: 1) $\mathcal{N} \subseteq \mathcal{D}$; 2) $\mathcal{N} \models_{\mathcal{H}} (\mathcal{D} \setminus \mathcal{N}) \cup \mathcal{C}$; 3) $\forall e \notin \mathcal{D} \cup \mathcal{C}, \nexists r \in \mathcal{H}, \mathcal{N} \models_r e$.*

The size of $(\mathcal{H}, \mathcal{N}, \mathcal{C})$ is $\|\mathcal{H}\| + |\mathcal{N}| + |\mathcal{C}|$. $|\mathcal{N}|$ is the number of predicates in $\mathcal{N}$, and so be $|\mathcal{C}|$. $\|\mathcal{H}\|$ is defined as the sum of lengths of all rules in it.

In LLC, the total size of the patterns is the number of rules-that is-the length of each rule is one, no matter what form and pattern it describes. The coarse definition does not reflect the complexity of identifying semantic patterns. Other rule mining studies [11,23,27] intuitively count in the number of terms or different variables, which emphasizes to some extent the complexity of identifying a pattern, but it is still not convincing enough.

In our technique, the length of a Horn rule is measured by the total arity of the rule and the number of different variables at the same time:

$$|r| = \left(\sum_{P \in r} \phi(P) \right) - var(r) \tag{7}$$

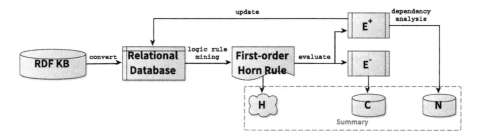

Fig. 2. The workflow of our technique

Table 1. An example knowledge base

s	p	o	s	p	o	s	p	o
tom	father	jerry	tom	gender	male	tom	type	man
bob	father	alice	bob	gender	male	bob	type	man
matt	father	adam	matt	gender	male	matt	type	man
daniel	father	felix	daniel	gender	male	felix	type	man

$var(r)$ is the number of different variables in rule r. Intuitively speaking, the above definition reflects the *minimum* number of equivalence conditions that identify the pattern in r. For example, the length of Rule (8) is 2, because the pattern is characterized by two conditions: $gender.0 = father.0$, $gender.1 = male$. The UV in Rule (8) is existentially quantified.

The size of every triple in Definition 1 is one no matter what relation and entity are represented by the triple because the logic-rule-based summarization cover and remove each triple as a whole. The comparison (Fig. 4a) between summarization ratios and compression ratios has justified that it is proper to define the size of a triple as one.

3.2 Summarization Workflow and the Recovery

Figure 2 shows the overall workflow of our technique. An RDF KB is converted to a relational database, where the subjects and objects are the two arguments in the relations. Each triple in the KB is converted to a single record in the relational database. Labels and types are converted to unary relations where relation names are from the label or type value. Then, logic rules are iteratively induced from each relation until no proper rule is returned. Each Horn rule is evaluated on the database to find the entailments and corresponding evidence. Negatively entailed records are simply collected in the counterexample set $\mathcal{C}$. Positively entailed records and the corresponding evidence are further analyzed to finally determine the set $\mathcal{N}$, in case that there are circular entailments in the summarization.

Table 2. Converted Relational Database of Table 1

tom	**jerry**	tom	male	tom
bob	**alice**	bob	male	bob
matt	**adam**	matt	male	matt
daniel	**felix**	daniel	male	**felix**

Table 1 shows an example RDF knowledge base, and Table 2 shows the converted version. The original size of the KB is 12, and the following rules are induced for the summarization:

$$gender(X, male) \leftarrow father(X, ?) \quad (8)$$

$$man(X) \leftarrow father(X, ?) \quad (9)$$

The total size of the rules is 3, and only 1 counterexample man(daniel) is generated by the rules. The 5 records in bold font remain in $\mathcal{N}$, and the others can be entailed from $\mathcal{N}$ w.r.t. the above rules thus are removable. Therefore, the total size of the summarization is 9.

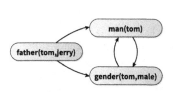

The evidence of positively entailed triples can be represented as a graph, where edges refer to the inference dependency from the body to the head. Therefore, circular dependencies occur as directed cycles in the graph. Figure 3 shows an example where the following rules are included in $\mathcal{H}$ and cause the cycles:

Fig. 3. An example of circular entailment

$$man(X) \leftarrow gender(X, male)$$
$$gender(X, male) \leftarrow man(X)$$

Minimum Feedback Vertex Set (MFVS) [4] algorithms can be used to break the cycles and vertices in the MFVS solution should also be included in $\mathcal{N}$ to make sure every removed record is inferable.

The recovery is simple in our technique: given that the circular dependencies are resolved in the summarization, all removed records can be regenerated by iteratively evaluating Horn rules in $\mathcal{H}$ until no record is added to the database. Counterexamples should be excluded to keep data consistency.

Association rules adopt a limited number (usually one [14] or two [21]) of universally quantified variables, and the patterns are only expressed by the co-occurrences of entities. Thus, general correlations represented by more variables and existential quantifiers are not captured. In the above example, the only inducible association pattern by LLC is the following (or the reverse):

$$type(X, man) \leftarrow gender(X, male) \quad (10)$$

The triples in relation **gender** (or **type**) are not removable. Thus, the conciseness, coverage, and semantic completeness are low in LLC, even though part of the schematic overview has been correctly induced from the data. It is possible to hardcode various semantics into different association structures, such as varying the variables from the subject to the object or even the relation. However, the structures rely on human input and are often tedious to enumerate.

3.3   First-Order Horn Rule Mining

The most critical component in our technique is the induction of first-order Horn rules. Logic rule mining has been extensively studied in the Inductive Logic Programming (ILP) community [5]. Both top-down and bottom-up methodologies have been proposed and optimized for over three decades. The bottom-up strategy regards facts as specific-most rules and merges correlated ones to generalize [19]. The top-down strategy operates in the inverse direction, where rules are constructed from general to specific by imposing new restrictions on candidates [23,27]. Top-down mining techniques are easier to understand and optimize and are employed in more knowledge-based applications.

Our technique also follows the top-down methodology, and the specialization is refined to improve performance. In previous works, such as FOIL [23] and AMIE [9], candidate rules are specialized by simply appending new atoms to the body of Horn rules. The specialization in the pattern semantics is not well-organized because some newly imposed conditions are repeatedly applied to the candidates, and the number of applicable predicates in each step of specialization is exponential to the maximum arity of the relations if inducing on relational databases. In our approach, a candidate rule is extended in a smaller step size which corresponds to the equivalence between a column and another or a constant value. For example, Rule (8) is constructed in the following order:

$$gender(?, ?) \leftarrow$$
$$gender(X, ?) \leftarrow father(X, ?)$$
$$gender(X, male) \leftarrow father(X, ?)$$

The benefit of this modification is three-fold: 1) The extension operations are feasible to relations of arbitrary arities without increasing the difficulty of enumerating applicable predicates to the body. The number of applicable extensions is polynomial to the rule length and the arity of relations. 2) The small-step exploration employs existentially quantified variables with lower cost compared to current logic rule mining techniques, no mention of the association ones. 3) The specialization maximizes the reusability of intermediate results and is better cooperated with caching techniques in relational databases, such as materialization. The reason is that the specialization by each newly imposed condition is updated and stored only once during the induction. Together with pruning [27] and parallelization techniques [8,26], the performance of logic rule mining will no longer be the stopping reason for RDF KB summarization.

Searching for the best logic rule is accomplished with the beam search, similar to the FOIL system, except that an RDF KB does not provide negative examples. Therefore, the Closed World Assumption (CWA) is adopted in our technique to enumerate the negative examples if necessary. The quality of a Horn rule r is measured by the reduction of overall size:

$$\delta(r) = |\mathcal{E}_r^+| - |\mathcal{E}_r^-| - |r| \tag{11}$$

Table 3. Dataset overview

Datasets	Short	#Rel.	#Entity	#Triple	#Label
Elti	E	10	47	318	–
Family.simple	Fs	4	82	322	1
Dunur	D	17	26	466	–
DBpedia.factbook	DBf	2	335	880	Default
Family.medium	Fm	9	142	1242	1
Student Loan	S	9	1031	6317	–
UMLS	U	46	135	6664	Default
WN18	WN	18	41K	193K	Default
NELL	N	1083	44K	278K	821
FB15K	FB	1345	15K	607K	Default

4 Evaluation

This section evaluates our technique and answers the following research questions:

Q1 To what extent are RDF KBs summarized by first-order Horn rules?
Q2 How and why does our technique outperform state-of-the-art methods?
Q3 How fast does our technique induce logic rules?

Datasets. We use ten open-access datasets, without deliberate selection, from various domains, including relational databases, fragments of popular knowledge graphs that are widely used as benchmarks, and two synthetic datasets. Table 3 shows statistics of these datasets.[1] "E", "D", and "S" are relational databases themselves, and the others are converted to the corresponding relational form. Given that KGIST requires entity labels in databases, we assign a default label to datasets where the label information is unavailable. Datasets tested in LLC are outdated and no longer accessible thus are not used in our tests. The datasets are not extremely large because FOIL and KGIST are not implemented in a parallel manner, and we compare the speed in a single thread mode to demonstrate the impact of the small-step specialization operations. More importantly, the datasets are sufficient to emphasize the superiority of our technique.

Rivals and Settings. We compare our approach against four state-of-the-art techniques: FOIL, LLC, AMIE, and KGIST. The summarization quality is compared mainly against LLC. KGIST and AMIE are also compared for summarization, as KGIST is devoted to the same purpose via a graph-based approach, and AMIE can be slightly modified, for a fair comparison, to summarize KGs by selecting the rules useful for reducing the overall size. FOIL and AMIE are chosen as the competitors for speed comparison, as both of them induce first-order

---

[1] E, D, S are available at: https://relational.fit.cvut.cz/; Fm, Fs are synthetic, and the generators are available with the project source code.

Horn rules and are the most similar to ours. However, neither the source code nor compiled tool is available for LLC. Therefore, we reimplemented the algorithm according to the instructions in [14]. The latest version of AMIE, AMIE3, is used in the experiment, and Partial Completeness Assumption (PCA) is employed in AMIE. Our technique is implemented in Java 11 and is open-source on GitHub[2] All tests were carried out in a single thread on Deepin Linux (kernel: 5.10.36-amd64-desktop) with Ryzen 3600 and 128 GB RAM. The beamwidth for our technique is 5.

Metrics. The quality of summarization is quantitatively reported by the summarization ratio (θ), pattern/rule complexity ($|r|$) and connectivity (ρ), and knowledge coverage (τ). The summarization ratio is defined as:

$$\theta(\mathcal{D}) = (\|\mathcal{H}\| + |\mathcal{N}| + |\mathcal{C}|)/|\mathcal{D}| \tag{12}$$

where $\|\mathcal{H}\|$, $|\mathcal{N}|$, $|\mathcal{C}|$, and $|\mathcal{D}|$ in LLC, AMIE, and our technique follow Definition 1. The components in KGIST are measured by the length of bit strings. The connectivity is the connection density in relations and reflects the completeness of exhausting hidden semantics in a knowledge base:

$$\rho(\mathcal{H}) = |\{(p,q)|p, q \in \mathcal{P}, p, q \text{ appear in the same rule } r \in \mathcal{H}\}|/|\mathcal{P}|^2 \tag{13}$$

In our technique, the converted "type" or "label" relations are counted as one single relation, as is calculated in other techniques. The knowledge coverage is the ratio of all inferable (not necessarily removable) triples over the entire set:

$$\tau(\mathcal{D}) = |\{e \in \mathcal{D}|\mathcal{D} \models_\mathcal{H} e\}|/|\mathcal{D}| \tag{14}$$

4.1   Summarization with Horn Rules

The results in this section answer Q1: The summarization and compression ratios of our technique are up to 40%; Circular entailments frequently appear in the summarization but are easy to resolve.

Figure 4 shows summarization statistics of our technique on the datasets. Θ refers to the compression ratio measured by input/output files in Bytes. The bars in three different colors in Fig. 4a add up to the total summarization ratio, and it is shown that more than 60% contents are replaced by logic rules in the datasets. Compared to the number of remaining triples, the sizes of rules and counterexamples are negligible. The reason is that there are usually clear topics and themes in modern knowledge bases, and within the topics, some relations extend details of complex concepts. Moreover, necessary redundancies are included for high completeness of domain knowledge, as most facts are automatically extracted from the open-source text and checked by human. For example, the followings are some rules induced from the datasets:

$$part\_of(X, Y) \leftarrow has\_part(Y, X) \tag{15}$$
$$uncle(X, Y) \leftarrow brother(X, Z), aunt(Z, Y) \tag{16}$$
$$aunt(X, Y) \leftarrow sister(X, Z), uncle(Z, Y) \tag{17}$$

---

[2] https://github.com/TramsWang/SInC.

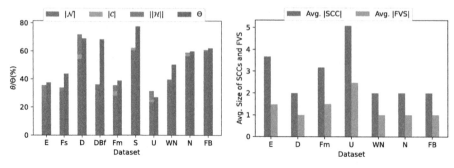

(a) Summarization/Compression Ratios   (b) Sizes of SCCs and MFVS Solutions

Fig. 4. Summarization detail

Relation `aunt` and `uncle` in dataset "Fm" can be mutually defined by each other with some auxillary relations. Many relations, such as `part_of` and `has_part` in "WN", are symmetric, and this is a common circumstance in modern KGs. Figure 4b shows the evidence by counting the sizes of Strongly Connected Components (SCCs) in the graph that represents inference dependencies of triples. The average sizes of SCCs in large-scale KGs, such as "FB", "WN", and "N", are approximately 2, which testifies the above analysis. Moreover, from the figure, we can conclude that the cycles are not large in the datasets and can be efficiently solved by MFVS algorithms, even in a greedy manner.

Figure 4a also shows that our technique successfully applies to relational data-bases. Moreover, the summarization ratio is close to the file compression ratio. It is proper to define the size of a triple as one. θ and Θ have an apparent difference in "DBf" because the following induced rule eliminates entities after triples are removed: $sameAs(X, X) \leftarrow$, and extra information for the entities should be recorded for a complete recovery. However, the information is not included in Definition 1, as the above case is rare in practice.

4.2 Quality of Summarizations

In this section, we compare our technique against the state-of-the-art tools: LLC, AMIE, and KGIST, and answer Q2: Our technique induces more expressive logic rules than the state-of-the-art; Rules in our technique cover more triples, reflect more comprehensive semantics, and are more representative.

Figure 5 shows the overall summarization ratios of the techniques. "E", "D", and "S" are not compared as the competitors cannot handle relational databases. Our technique outperforms the others in almost all datasets. Some of the ratios by LLC are larger than 100% because many rules are induced but not used to replace triples. For example, the following two rules are induced from "DBf" by

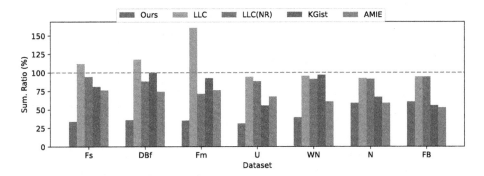

Fig. 5. Summarization comparison

Table 4. Blocked rules (%) in LLC

Dataset	Fs	DBf	Fm	U	WN	N	FB
Blocked rules (%)	52.63	48.39	68.46	60.00	44.23	42.90	45.00

LLC:

$$spokenIn(X, Russia) \land type(X, default) \leftarrow spokenIn(X, Kazakhstan) \quad (18)$$

$$spokenIn(X, Russia) \land type(X, default) \leftarrow spokenIn(X, Uzbekistan) \quad (19)$$

But Rule (19) is blocked from replacing the head triples if Rule (18) is applied. According to Table 4, about half of the rules in LLC are blocked due to the above reason.

However, excluding the size of rules (shown as "LLC (NR)" in Fig. 5) does not change the fact that LLC is not competitive to logic-rule-based techniques. The main reason is that association patterns are applicable to only a small part of triples in the datasets. For example, Rule (18) is the most frequently used in "DBf", and it replaces only 18 triples, the proportion of which is only 2.05%, in the dataset. Figure 6a compares the overall coverage of all techniques. The association patterns induced by LLC cover only about 20% triples in a KG. The low coverage is further explained by Fig. 6b. The figure shows that the number of itemsets, i.e., potential association patterns, exponentially decreases with increasing size of the itemset. More importantly, the number is much smaller than the matching arguments, represented by variables. Therefore, the association patterns are not representative as first-order logic rules are.

Figure 6c compares the connectivity (see Eq. (13)) of induced patterns. Given that the connectivity varies a lot in datasets, for a clear illustration, we compare the connectivity of other techniques to ours. Therefore, the red line at value 1.0 denotes the connectivity of our technique, and the others are the relative values. In most cases, our technique induces patterns that correlate the most relations, thus reflecting the semantics more comprehensively in the data. Although LLC combines more relations in "DBf" and "WN", the average numbers of triples

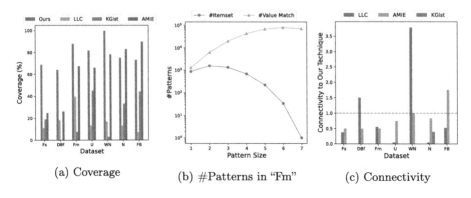

(a) Coverage   (b) #Patterns in "Fm"   (c) Connectivity

Fig. 6. Pattern comparison

inferable by the rules induced from the two datasets are 2.07 and 11.17, while the numbers for our technique are 124 and 8358.22. Hence, the coverage of our technique remains extensive even though the connectivity is occasionally low.

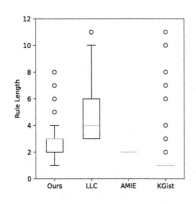

Fig. 7. Rule Lengths on NELL

Figure 7 compares the length of patterns in the dataset "N" according to the measure proposed in Sect. 3.1. The reason why LLC induces longer patterns is that the association patterns consist mainly of entities, each of which is size 1 in the new length measure, while variables represent the matching between arguments with much less cost.

The comparisons with other state-of-the-art techniques also approve that logic-rule-based approaches generalize better than graph-pattern-based ones, thus producing more concise summaries. The summarization ratios for our technique and AMIE are smaller than LLC and KGist. The knowledge coverage is also significantly more extensive than the graph-pattern-based approaches. Most of the rules in KGist are at length one because the patterns it describes usually involve a single relation and the direction, and this is also the reason for almost-zero connectivity in KGist.

Our technique summarizes better than AMIE because rules induced in our technique are longer and contain existentially quantified UVs. For example, the following rule is simple but out of reach of AMIE, because it contains a UV:

$$gender(X, female) \leftarrow mother(X, ?)$$

Moreover, the rule evaluation metric adopted in AMIE is based on PCA, which assumes the functionality of relations in knowledge bases. However, the PCA in

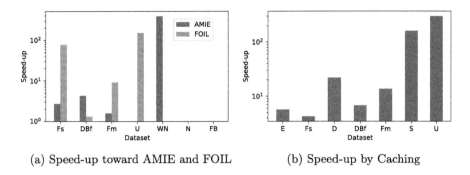

(a) Speed-up toward AMIE and FOIL   (b) Speed-up by Caching

Fig. 8. Rule mining speed comparison

AMIE is not suitable for the summarization purpose. For example, our technique covers triples in relation **produces** with only 10 counterexamples, while AMIE does with 97.

4.3   Rule Mining Speed

The results in this section answer Q3: The speed of our technique is up to two orders faster than FOIL and AMIE, and the speed-up is mainly due to the small-step specialization operations together with caching.

Both AMIE and FOIL induce first-order Horn rules and are the most similar techniques to ours. AMIE restricts the length and applicable variables in the rules, and it runs in multi-threads. The maximum length and the number of threads in AMIE should be set to 5 and 1 to compare the performance under approximately equal expressiveness. However, AMIE frequently ends up with errors under the above setting. The adopted parameters for maximum length and threads are 4 and 3, respectively. Therefore, the actual speed-up is larger than the recorded numbers in Fig. 8a. In the figure, the missing numbers are because of program failures due to program errors or memory issues in FOIL and AMIE.

The results show that our technique performs one to two orders faster than AMIE and FOIL. Although AMIE adopts an estimation metric for heuristically selecting promising specializations of rules, it tends to repeatedly cover triples by different rules. No more than 10% rules produced by AMIE are used in the summarization. Although AMIE employs an in-memory database with combinatorial indices, the caching is not fully explored due to the types of terms it appends to the rules. For example, Rules (21) and (22) are two extensions of Rule (20) in AMIE. The condition $grandfather.0 = father.0$ has been repeatedly imposed on the base rule during the extension.

$$grandfather(X, Y) \leftarrow \qquad (20)$$
$$grandfather(X, Y) \leftarrow father(X, Y) \qquad (21)$$
$$grandfather(X, Y) \leftarrow father(X, Z) \qquad (22)$$

FOIL finds the best description for relations under the metric "Information Gain". FOIL does not over-explore the search space of Horn rules as AMIE does, but the tables are repeatedly joined, as FOIL does not cache the intermediate result of candidate rules during the construction. Figure 8b shows the speed-up by caching intermediate results, and this explains most of the difference between FOIL and our technique. Moreover, Fig. 8b also shows that the speed-up by caching is more significant in larger datasets.

5   Conclusion

This paper proposes a novel summarization technique on RDF KBs by inducing first-order Horn rules. Horn rules significantly extend the coverage, completeness, and conciseness due to extensive exploration of variables compared to the association and graph-structure patterns. The small-step specialization operations also improve the performance of rule induction by maximizing the reusability of cached contents. As shown in the experiments, our technique summarizes KBs to less than 40% of the original size, covers more than 70% triples, and is up to two orders faster than the rivals. Our technique not only produces a concise and faithful summary of RDF KBs but is also applicable to relational databases. Therefore, the new technique is practical for a broader range of knowledge-based applications.

References

1. Ahmed, M.: Data summarization: a survey. Knowl. Inf. Syst. **58**(2), 249–273 (2019)
2. Belth, C., Zheng, X., Vreeken, J., Koutra, D.: What is normal, what is strange, and what is missing in a knowledge graph: Unified characterization via inductive summarization. In: The Web Conference (WWW) (2020)
3. Čebirić, Š, et al.: Summarizing semantic graphs: a survey. VLDB J. **28**(3), 295–327 (2019)
4. Chen, J., Liu, Y., Lu, S., O'sullivan, B., Razgon, I.: A fixed-parameter algorithm for the directed feedback vertex set problem. In: Proceedings of the Fortieth Annual ACM Symposium on Theory of Computing, pp. 177–186 (2008)
5. Cropper, A., Dumancic, S., Muggleton, S.H.: Turning 30: New ideas in inductive logic programming. In: IJCAI (2020)
6. Dudáš, M., Svátek, V., Mynarz, J.: Dataset summary visualization with LODSight. In: Gandon, F., Guéret, C., Villata, S., Breslin, J., Faron-Zucker, C., Zimmermann, A. (eds.) ESWC 2015. LNCS, vol. 9341, pp. 36–40. Springer, Cham (2015). https://doi.org/10.1007/978-3-319-25639-9_7
7. Fan, W., Li, J., Wang, X., Wu, Y.: Query preserving graph compression. In: Proceedings of the 2012 ACM SIGMOD International Conference on Management of Data, pp. 157–168 (2012)
8. Fonseca, N.A., Srinivasan, A., Silva, F., Camacho, R.: Parallel ILP for distributed-memory architectures. Mach. Learn. **74**(3), 257–279 (2009)

9. Galárraga, L., Teflioudi, C., Hose, K., Suchanek, F.M.: Fast rule mining in onto-logical knowledge bases with AMIE+. VLDB J. **24**(6), 707–730 (2015)

10. Gunaratna, K., Thirunarayan, K., Sheth, A.: Faces: diversity-aware entity sum-marization using incremental hierarchical conceptual clustering. In: Twenty-Ninth AAAI Conference on Artificial Intelligence (2015)

11. Hammer, P.L., Kogan, A.: Quasi-acyclic propositional horn knowledge bases: opti-mal compression. IEEE Trans. Knowl. Data Eng. **7**(5), 751–762 (1995)

12. Han, J., Pei, J., Yin, Y.: Mining frequent patterns without candidate generation. ACM SIGMOD Rec. **29**(2), 1–12 (2000)

13. Hose, K., Schenkel, R.: Towards benefit-based RDF source selection for SPARQL queries. In: Proceedings of the 4th International Workshop on Semantic Web Infor-mation Management, pp. 1–8 (2012)

14. Joshi, A.K., Hitzler, P., Dong, G.: Logical Linked data compression. In: Cimiano, P., Corcho, O., Presutti, V., Hollink, L., Rudolph, S. (eds.) ESWC 2013. LNCS, vol. 7882, pp. 170–184. Springer, Heidelberg (2013). https://doi.org/10.1007/978-3-642-38288-8_12

15. Kushk, A., Kochut, K.: Esdl: Entity summarization with deep learning. In: The 10th International Joint Conference on Knowledge Graphs, pp. 186–190 (2021)

16. Luo, Y., Fletcher, G.H., Hidders, J., Wu, Y., De Bra, P.: External memory k-bisimulation reduction of big graphs. In: Proceedings of the 22nd ACM Interna-tional Conference on Information & Knowledge Management, pp. 919–928 (2013)

17. Meier, M.: Towards rule-based minimization of RDF graphs under constraints. In: Calvanese, D., Lausen, G. (eds.) RR 2008. LNCS, vol. 5341, pp. 89–103. Springer, Heidelberg (2008). https://doi.org/10.1007/978-3-540-88737-9_8

18. Motta, E., et al.: A novel approach to visualizing and navigating ontologies. In: Aroyo, L., et al. (eds.) ISWC 2011. LNCS, vol. 7031, pp. 470–486. Springer, Hei-delberg (2011). https://doi.org/10.1007/978-3-642-25073-6_30

19. Muggleton, S.H., Lin, D., Pahlavi, N., Tamaddoni-Nezhad, A.: Meta-interpretive learning: application to grammatical inference. Mach. Learn. **94**(1), 25–49 (2014)

20. Palmonari, M., Rula, A., Porrini, R., Maurino, A., Spahiu, B., Ferme, V.: ABSTAT: linked data summaries with ABstraction and STATistics. In: Gandon, F., Guéret, C., Villata, S., Breslin, J., Faron-Zucker, C., Zimmermann, A. (eds.) ESWC 2015. LNCS, vol. 9341, pp. 128–132. Springer, Cham (2015). https://doi.org/10.1007/978-3-319-25639-9_25

21. Pan, J.Z., Pérez, J.M.G., Ren, Y., Wu, H., Wang, H., Zhu, M.: Graph pattern based rdf data compression. In: Supnithi, T., Yamaguchi, T., Pan, J.Z., Wuwongse, V., Buranarach, M. (eds.) JIST 2014. LNCS, vol. 8943, pp. 239–256. Springer, Cham (2015). https://doi.org/10.1007/978-3-319-15615-6_18

22. Pires, C.E., Sousa, P., Kedad, Z., Salgado, A.C.: Summarizing ontology-based schemas in pdms. In: 2010 IEEE 26th International Conference on Data Engi-neering Workshops (ICDEW 2010), pp. 239–244. IEEE (2010)

23. Quinlan, J.R.: Learning logical definitions from relations. Mach. Learn. **5**(3), 239–266 (1990)

24. Raedt, L.D., Kersting, K.: Statistical relational learning. In: Sammut, C., Webb, G.I. (eds.) Encyclopedia of Machine Learning, pp. 916–924. Springer (2010). https://doi.org/10.1007/978-0-387-30164-8_786

25. Sanders, P., Schulz, C.: High quality graph partitioning. Graph Partition. Graph Cluster. **588**(1), 1–17 (2012)

26. Srinivasan, A., Faruquie, T.A., Joshi, S.: Data and task parallelism in ILP using mapreduce. Mach. Learn. **86**(1), 141–168 (2012)

27. Zeng, Q., Patel, J.M., Page, D.: Quickfoil: scalable inductive logic programming. Proc. VLDB Endow. **8**(3), 197–208 (2014)
28. Zneika, M., Lucchese, C., Vodislav, D., Kotzinos, D.: Summarizing linked data RDF graphs using approximate graph pattern mining. In: EDBT 2016., pp. 684–685 (2016)

Social Network Analysis

A Heterogeneous Propagation Graph Model for Rumor Detection Under the Relationship Among Multiple Propagation Subtrees

Guoyi Li[1,2], Jingyuan Hu[1,2], Yulei Wu[3], Xiaodan Zhang[1,2(✉)], Wei Zhou[1,2], and Honglei Lyu[1,2]

[1] Institute of Information Engineering, Chinese Academy of Sciences, Beijing, China
{liguoyi,hujingyuan,zhangxiaodan,zhouwei,lvhonglei}@iie.ac.cn
[2] School of Cyber Security, University of Chinese Academy of Sciences, Beijing, China
[3] Department of Computer Science, University of Exeter, Exeter, UK
Y.L.Wu@exeter.ac.uk

Abstract. Pervasive rumors in social networks have significantly harmed society due to their seditious and misleading effects. Existing rumor detection studies only consider practical features from a propagation tree, but ignore the important differences and potential relationships of subtrees under the same propagation tree. To address this limitation, we propose a novel heterogeneous propagation graph model to capture the relevance among different propagation subtrees, named Multi-subtree Heterogeneous Propagation Graph Attention Network (MHGAT). Specifically, we implicitly fuse potential relationships among propagation subtrees using the following three methods: 1) We leverage the structural logic of a tree to construct different types of propagation subtrees in order to distinguish the differences among multiple propagation subtrees; 2) We construct a heterogeneous propagation graph based on such differences, and design edge weights of the graph according to the similarity of propagation subtrees; 3) We design a propagation subtree interaction scheme to enhance local and global information exchange, and finally, get the high-level representation of rumors. Extensive experimental results on three real-world datasets show that our model outperforms the most advanced method.

Keywords: Rumor detection · Heterogeneous graph · Propagation subtrees · Local and global relations · Message passing

1 Introduction

Due to the popularity of Twitter, Facebook and other social media in recent years, a growing number of rumor generating methods have emerged. Taking the COVID-19 pandemic as an example, there were growing concerns about

© The Author(s), under exclusive license to Springer Nature Switzerland AG 2023
M.-R. Amini et al. (Eds.): ECML PKDD 2022, LNAI 13714, pp. 207–223, 2023.
https://doi.org/10.1007/978-3-031-26390-3_13

the spread of misinformation about the pandemic, known as the "information epidemic" [27]. Social media have been widely used to facilitate the spread of misinformation. These issues are even more pressing in that atmosphere since the information flowing through social media is directly related to human health and safety. It is therefore of paramount importance to effectively identify rumors.

Most existing efforts mainly focus on using linguistic features from text to detect rumors, ranging from deceptive clues to writing styles. For example, Li et al. [12] combined user information and text features to train an LSTM to capture their potential associations. Other algorithms such as the Bayesian network were applied to compute text-similarity of microblogs [11]. This kind of rumor detection methods was mainly to capture text features of rumors, which is vulnerable to the negative influence of forged text because the language used in social media is highly informal, ungrammatical, and dynamic.

To address the above issue, studies of rumor propagation structures have been carried out. For instance, Kumar et al. [10] proposed a new way to represent social-media conversations as propagation trees and used Tree LSTM models to capture conversation features. Ma et al. [19] proposed recursive neural models based on a bottom-up and a top-down tree-structured neural networks, to learn discriminative features from tweet's content by following their non-sequential propagation structures. Since temporal structural characteristics only concern the sequence of spreading rumors but ignore the consequence of rumor spreading, these approaches have significant limitations in terms of effectiveness. The structure of rumor dispersion also reflects important features of rumor spreading.

To consider such crucial features, researchers have started to apply graph convolution methods to detect rumors. Yu et al. [25] used GCN to realize the fusion of rumors in the propagation tree, the user information, and text features of retweets. Choi et al. [3] proposed a dynamic GCN to construct a time graph and utilized the characteristics of tweets published in adjacent times to strengthen the structural features of rumor propagation.

While the above methods have shown effectiveness of introducing the graph structure of data into a model, these approaches face two major shortcomings which make the rumor representation vulnerable to the local structural relationships and the characteristics of adjacent nodes. **First, existing studies only consider the aggregated information of each tweet and its neighbour, but ignore the important correlation of all retweets in the same propagation subtree. Second, the graphical structure of data ignores the potential impact among different propagation subtrees**.

To facilitate the understanding, Fig. 1 exemplifies the propagation structure of a (rumor) tweet "Says Bill O'Reilly wrote a post claiming that the coronavirus was created as a bioweapon by the Chinese government." In the first case, tweets x_1 and x_2 have the same characteristics ["article", "criminal acts"]; they have a certain correlation but no real connection. In fact, x_2 negates the basis of x_1. In the second case, x_1 and x_{11} incline that s is true and has a positive impact on the s. Even though x_{12} deems s was wrong and had more common characteristics with s, it can only affect s along the x_1, while x_1 and x_{11} prefer s is false and cannot well incorporate the features of the deeper retweets. The above two

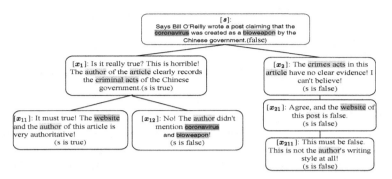

Fig. 1. An example of a false rumor.

situations are common in rumor detection, and their cumulative impact may lead to unexpected errors. Therefore, in this paper we propose to enhance the effect of rumor detection by constructing a local representation.

The starting point of our approach is an observation: tweets in a similar propagation location show certain relevance (such as $[x_1, x_{11}, x_{12}]$, $[x_2, x_{21}, x_{211}]$ in Fig. 1). Thus, we propose a new way of message passing to obtain the high-level representation of rumor propagation: (1) According to the structural logic of a tree and the spatial relationship among nodes, we model the different propagation subtrees of the tree where the nodes are located. We construct a heterogeneous propagation graph model with the weights of edges designed according to the propagation subtrees' similarity. (2) We initialize each node to integrate the relative temporal information carried by the parent tweet and the source post information, and apply structure-aware self-attention to propagation subtrees. (3) We design a two-layer attention mechanism to realize the interaction among propagation subtrees.

The main contribution of this paper can be summarized as follows:

- We propose a novel MHGAT model, which applies the propagation subtree as the computing unit to construct the heterogeneous propagation graph. It improves the performance of rumor detection by distinguishing the differences of local structures on the propagation tree.
- The model utilizes the heterogeneous propagation graph to guide the direction of message transmission. Moreover, the interaction between local information and global information is constructed to obtain the high-level rumor representation.
- The model fuses the parent tweet text features with the corresponding time information and the source text feature in appropriate places, to make the representation of the local structure more accurate.
- We conduct extensive experiments using three public real-world datasets. Experimental results show that our model significantly outperforms the state-of-the-art models in rumor classification and early detection tasks.

2   Related Work

Rumor detection aims to detect whether a tweet is a rumor according to the relevant information of the tweet published on the social media platform, such as text content and propagation mode.

Content-Based Classification Methods: Content-based classification methods [7,24] generally detect rumors based on linguistic clues such as writing style [20], bag-of-words [4], temporal characteristics [17], etc. However, these methods relying only on the text content to detect rumors, ignore the correlation between tweets, and its accumulative effect on a large number of tweets can affect the performance of detection.

Propagation-Based Classification Methods: Recent studies can be divided into two groups: Attention-based and GCN-based models. Attention-based models primarily utilize the attention mechanism to focus on pairs or sequences of posts with some inherent order [8,10,15,19]. Several recent works applied the transformer to enhance the representation learning for responsive tweets [8,15]. The difference lies in that Khoo et al. [8] defined time delay (the time interval between the tweet and retweet) as the intrinsic order, while Ma et al. [15] applied the topological order of the propagation tree as the inherent order. However, these methods are susceptible to the negative impact of unrelated tweets and require more time cost for detection. GCN-based models enhance the tweet representation by aggregating the features of related retweets [1,25]. For example, BiGCN [1] applied graph convolution to strengthen root features and learn local structure information. To better weight different types of neighbor nodes, in recent years several studies have applied heterogeneous graph model Graph Attention Network (GAT) that combines attention mechanism and GCN for rumour detection [13,26]. For example, Lin et al. [13] represented the propagation tree as an undirected interaction graph and utilized GAT aggregating information from parent and sibling nodes, taking the average representation as rumor representation that makes it difficult to distinguish the global structure of the rumor.

However, the above methods treat a tree's substructures as independent units, ignoring their differences and potential global associations. Our model will take advantage of the propagation tree structure and the heterogeneous graph model to construct the interaction between local and global information in order to enhance the representation of tweets in rumor detection.

3   Multi-subtree Heterogeneous Propagation Graph Attention Network Model

This section details the proposed MHGAT algorithm as shown in Fig. 2. Our algorithm can be divided into four parts. First, we construct the heterogeneous propagation graph to refine different subtrees (substructures) of the ordinary

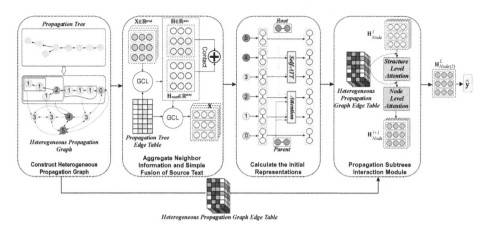

Fig. 2. Our proposed rumor detection model.

Fig. 3. Propagation tree of a false rumor.

propagation tree (Sect. 3.1). Second, to obtain a direct local representation, we utilize GCN to aggregate neighbor's features of the ordinary propagation tree (Sect. 3.2). Third, we get the initial representation of different subtrees in the heterogeneous propagation graph (Sect. 3.3). Finally, we design a heterogeneous graph convolution algorithm to realize the interaction between local and global information to enhance the rumor representation (Sect. 3.4).

Formally, let each node represent a tweet. The source node denotes the source tweet, and the children nodes are retweets that have responded to it directly. First, based on the retweet and the response relationships, we construct the origin event tree for a rumor c_i. In each training period, a propagation subtree has the probability p, of being discarded to reduce overfitting [1]. The probability of subtree pruning is positively correlated with the depth of the tree: $P_{drop} \propto dep(root\_of\_subtree)$. We denote the event tree after being discarded as $\langle V, E \rangle$ (see Fig. 3).

3.1 Construct Heterogeneous Propagation Graph

The heterogeneous propagation graph $\langle V', E' \rangle$ is designed to distinguish the differences of propagation subtrees better and address the two limitations mentioned above. This process is implemented with a general tree structure data processing method Depth-First-Search [21]. Our heterogeneous propagation graph

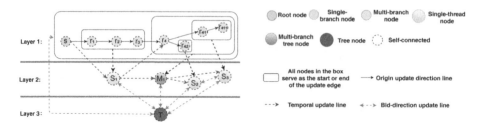

Fig. 4. The heterogeneous propagation graph is constructed by classifying the nodes of the original propagation tree in the first layer and building the nodes and edges of propagation subtrees in the last two layers. In addition, each node has a self-connected edge.

includes six types of structural nodes as shown in Fig. 4: (1) Root node ($Node_{(0)}$): Source tweet. (2) Single-branch node ($Node_{(1)}$): Leaf or the node with the single child except for the root node. (3) Multi-branch node ($Node_{(2)}$): The node with multiple branches except for the root node. (4) Single-thread node ($Node_{(3)}$): the node representing the single propagation thread without other branches. (5) Multi-branch tree node ($Node_{(4)}$): a propagation subtree with multiple branches. (6) Tree node ($Node_{(5)}$): a complete tree. Intuitively, we exemplify a false rumor claim and illustrate its propagation on twitter in Fig. 3. We observe that a group of tweets in the single-chain from r_1, $[r_1, r_2, r_3]$ tend to a point of view or a content, and construct their local representation S_1 in Fig. 3 to enhance features like ["composite", "edited", "modified"]. Moreover, we refer M_1 to the representation of $\{r_4, r_{41}, r_{411}, r_{42}\}$ which contains a stronger collection of different opinions about one content arising from the multi-branch node r_4. Essentially, multi-branch nodes have a broader direct impact influence than single-branch nodes. Finally, we refer to the tree node as the global representation to enhance rumor representation by realizing the interaction between the local information in the first five structure nodes and the global information.

In addition to the connection of nodes within the propagation subtree, we have added two effective connecting edges between $Node_{(3)}$ and $Node_{(4)}$:

(1) Considering that the nodes of a thread in the propagation tree have the corresponding time relationship (like $s \rightarrow r_1$ in Fig. 3), we now extend this feature to propagation subtrees in the heterogeneous propagation graph (such as $s \rightarrow S_1$ in Fig. 4). We define: when u and v are propagation subtrees of the type $Node_{(3)}$ or $Node_{(4)}$, r_i and r_j are retweet nodes in the propagation subtrees u and v respectively, where $i \neq j$, $u \neq v$. If r_i connects to r_j, u has a directed edge to v.

(2) Considering that two retweets forwarding the same tweet (the parent of r_i and r_j is the same node) may have similar characteristics (for example r_{42} and $\{r_{41}, r_{411}\}$ have common features ["source", "author"] in Fig. 3), and the propagation subtrees (u, v) are also related (S_2 and S_3 in Fig. 4), we define: when u and v are propagation subtrees of the type $Node_{(3)}$ or $Node_{(4)}$, r_i

and r_j are retweet nodes in propagation subtrees u and v respectively, where $i \neq j$. If $\text{Father}(r_i) = \text{Father}(r_j)$, u has a undirected edge to v.

Normalization: Considering the large difference in the number of nodes of the same type connecting different nodes, it may have an adverse impact on model learning. We normalize the weights of the edges of the starting nodes of the same type. Among the neighbors pointing to node u, the node set of type i is marked as $N_{(u)}^i$, and the set size is marked as $\mathbf{Num}(N_{(u)}^i)$. The edge regularization weight from any $v \in N_{(u)}^i$ to u is normalized to: $(\mathbf{Num}(N_{(u)}^i))^{-1}$. Thus, we get the normalized adjacency matrix $\tilde{\mathbf{A}}$ of heterogeneous propagation graph $\langle V', E' \rangle$.

3.2 Aggregate Neighbour Information and Simple Fusion of Root Features

This module aims to strengthen the representation of nodes in the propagation tree $\langle V, E \rangle$ by aggregating adjacent nodes and the source tweet. Graph convolution is an essential operation for aggregating neighbor information to extract local features. In addition, the source tweet can enhance the effect of rumor texts on retweets. As for nodes, let $\mathbf{A} \in \mathbb{R}^{n \times n}$ denote the normalized adjacency matrix, and $\mathbf{X} \in \mathbb{R}^{n \times d}$ represent the input signals of nodes of the propagation tree $\langle V, E \rangle$. First, we aggregate neighbour's features from node embedding $\mathbf{X}$:

$$\mathbf{H} = \mathbf{ReLU}(\mathbf{A}\mathbf{X}W_0). \tag{1}$$

Second, the aggregated features are fused with the root,

$$\mathbf{H}' = concat(\mathbf{H}, \mathbf{H_{root}}). \tag{2}$$

Last, we perform another layer of graph convolution to get a high-level representation of the node:

$$\tilde{\mathbf{X}} = \mathbf{ReLU}(\mathbf{A}\mathbf{H}'W_1), \tag{3}$$

where $\mathbf{H}$, $\tilde{\mathbf{X}} \in \mathbb{R}^{n \times d}$ are the hidden feature matrices computed by the Graph Conventional Layer (GCL), $W_0 \in \mathbb{R}^{d \times c}, W_1 \in \mathbb{R}^{(c+c) \times d}$. W_0, W_1 are the filter parameter matrices of graph convolution layer, and $\mathbf{H_{root}}$ represents the root representation after first-layer graph convolution. $\tilde{\mathbf{X}}$ is the node representation after two layers of graph convolutional layers.

3.3 Calculate the Initial Representation

We apply the attention mechanism to fuse parent node and source text feature (root node) to enhance the representation of propagation subtrees in heterogeneous propagation graph $\langle V', E' \rangle$, which can fuse the corresponding time and the source text information. For the root node ($Node_{(0)}$) and the tree node ($Node_{(5)}$): the node is initialized to the representation of the processed root embedding: $\mathbf{H}_{(Node_{(0)})} = \mathbf{H}_{(Node_{(5)})} = \mathbf{X}_{root}$. For single-branch nodes and multi-branch nodes, we fuse the source text feature and the parent tweet text feature, and these embeddings are calculated as:

$$\mathbf{H}_{(Node_{(1)\sim(2)})} = \mathbf{ATTN}(\tilde{\mathbf{X}}_{(Node_{(1)\sim(2)})}, \mathbf{H}_{pr}), \tag{4}$$

where

$$\mathbf{H}_{pr} = concat(\mathbf{H}_{parent}, \mathbf{H}_{root}). \tag{5}$$

where $\mathbf{ATTN}$ is a function $f : \mathbf{X}_{key} \times \varphi \to \mathbf{X}_{val}$, which maps the feature vector $\mathbf{X}_{key}$ and candidate feature vector set φ to the weighted sum of elements in $\mathbf{X}_{val}$ [22].

For the single-thread node ($Node_{(3)}$) and the multi-branch tree node ($Node_{(4)}$), these two types of nodes represent point sets, and we utilize attention mechanism to fuse the point sets into one representation:

$$\mathbf{H}_{(node_{(3)\sim(4)})} = \mathbf{Self\text{-}ATT}(\tilde{\mathbf{X}}_{(Node_{(3)\sim(4)})}), \tag{6}$$

where $\mathbf{Self\text{-}ATT}(.)$ includes the fusion process of self-attention and attention fusion [22]. Moreover, the gated mechanism is applied to strengthen the root features to get a high-level representation:

$$\alpha = \sigma(W_r \mathbf{H}_{(node_{(3)\sim(4)})} + W_{root} \tilde{\mathbf{X}}_{root} + b), \tag{7}$$

$$\mathbf{H}_{(Node_{(3)\sim(4)})} = \alpha \times \mathbf{H}_{(node_{(3)\sim(4)})} + (1 - \alpha)\tilde{\mathbf{X}}_{root}, \tag{8}$$

where $\sigma(\cdot) = \frac{1}{1+exp(\cdot)}$ is sigmoid activation function, and $W_r, W_{root} \in \mathbb{R}^{d \times 1}$, $b \in \mathbb{R}$ are parameters of the fusion gate.

Weight Introduction: In addition to the regularized weights that eliminate quantitative differences, since these new potential links may introduce noise where not all neighbors are equal in contributing important information for the aggregation when modelling the propagation subtrees, we shall calculate the weight of links between propagation subtrees in heterogeneous propagation graph $\langle V', E' \rangle$. To this end, we first use the cosine similarity $s(u, v) = h_u \cdot h_v^T / (|h_u| \cdot |h_v|)$ between nodes u and v to measure their similarity, where h is the embedding of the node. To properly define node's similarity, we introduce an asymmetric regularization term to balance the difference of the sum of similarity on every neighbor node:

$$\mathbf{R}_u(s(u, v)) = s(u, v) / \sum_t^n s(u, t), \tag{9}$$

where n is the set of u neighbor nodes. Combining the topology and attribute information, the similarity between u and v is

$$w(u, v) = W_{\tau_u \tau_v}(b(u, v) + \beta \cdot \mathbf{R}_u(h_u \cdot h_v^T / (|h_u| \cdot |h_v|)), \tag{10}$$

where β is a parameter to make a tradeoff between network topology and attributes, and $W_{\tau_u \tau_v}$ represents the trainable similarity relationship between propagation subtree type τ_u and τ_v. $b(u,v)$ is a network topology term: (1) If τ_u, $\tau_v \in [Node_{(0\sim2)}]$, $b(u,v) = 0$, which regards the points are the same in the topology. (2) If $\tau_u \in [Node_{(0\sim2)}]$, $\tau_v \in [Node_{(3\sim5)}]$, $b(u,v) = (-1)^{\delta(u,v)}\gamma_{\tau_u\tau_v}$, $\delta(u,v) = 1$ where γ is a trainable parameter, if u is a point in propagation subtree v, $\delta(u,v) = 0$. (3) If τ_u, $\tau_v \in [Node_{(3\sim4)}]$, $b(u,v) = n_u n_v/2e$, n_u represents the number of points in propagation subtree u.

Therefore, for propagation subtrees, let $\tilde{\mathbf{A}}'$ represent the matrix $\tilde{\mathbf{A}}$ with weights introduced, and $\tilde{\mathbf{A}}'_s \in \mathbb{R}^{|n| \times |n_s|}$ denote the submatrix of $\tilde{\mathbf{A}}'$, whose rows represent all the nodes and columns denote their neighboring nodes with the type s.

3.4   Propagation Subtree Interaction Module

This module is designed to realize the interaction between local and global structral features in the heterogeneous propagation graph. In other words, tree nodes aggregate local structural information in each iteration while other structural nodes aggregate local and global structural information. It consists of two attention layers to aggregate various types of subtrees. First, we calculate the structure-level attention scores based on the node embedding h_u and the propagation subtree type embedding h_s:

$$\alpha_s = softmax(\textbf{LeakyRelu}(w_s^T[h_u||h_s])), \tag{11}$$

$$s = \sum_{v' \in N_u} \tilde{\mathbf{A}}'_{uu'}h_{u'}, \tag{12}$$

where h_s is the sum of neighbouring node features, and $h_{u'}$ refers to the embedding of nodes $u' \in N_u$ with the same propagation subtree type s.

Then, as for the node-level attention part, given a specific node v with the structure type s and its neighboring node $u' \in N_u$ with the structure type s', we compute the node-level attention scores based on the node embeddings h_u and $h_{u'}$ with the structure-level attention weight α_s for the node u:

$$v_{uu'} = softmax(LeakyRelu(w_{node}^T \cdot \alpha_s[h_u||h_{u'}])), \tag{13}$$

where w_{node}^T is the attention vector. Then, we merge structure-level and node-level attention into heterogeneous propagation graph convolution.

$$\mathbf{H}^{(l+1)} = \sigma(\sum_{s \in Node_{(*)}} I_s \cdot \mathbf{H}_s^{(l)} \cdot W_s^{(l)}). \tag{14}$$

Here, I_s represents the attention matrix, whose element in the uth row u'^{th} column is $v_{uu'}$.

Finally, after going through an L times propagation subtree interaction process, the label of the event $\tilde{S}$ is calculated as:

$$\tilde{y} = softmax(FC(\mathbf{H}_{Node_{(5)}}^{L})), \tag{15}$$

where $\tilde{y} \in \mathbb{R}^{1 \times C}$ is a vector of probabilities for all the classes used to predict the label of the rumor.

4   Experiments

4.1   Datasets

Almost all prevalent datasets for experimental evaluation in the field of rumor detection come from two source platforms: Twitter and Sina Weibo. We evaluate the proposed model on three real-world datasets: *Twitter15* [18], *Twitter16* [18] and *Weibo* [14]. In all the three datasets, nodes refer to source tweets and retweets, edges represent response relationships, and features are the extracted top-5000 words in terms of the TF-IDF values. The Twitter15 and Twitter16 datasets contain four different labels, namely "false rumor" (FR), "non-rumor" (NR), "unverified" (UR), and "true rumor" (TR). Moreover, the Weibo dataset only contains binary labels, i.e., "true rumor" and "false rumor". Details of the three datasets are shown in Table 1.

4.2   Baselines and Evaluations Metrics

We compare our proposed model with the following baseline and state-of-the-art models. **ClaHi-GAT** [13]: An undirected interaction graph model utilizes GAT to capture interactions between posts with responsive parent-child or sibling relationships. **BiGCN** [1]: A bottom-up and a top-down tree-structured fusion model based on GCN for rumor detection. **PLAN** [8]: A transformer-based rumour detection model that can capture the interaction between any pair of tweets, even irrelevant ones. **RvNN** [19]: A bottom-up and a top-down tree-structured model based on recursive neural networks for rumor detection on Twitter. **SVM-TK** [18]: A SVM model uses Tree kernel to capture the propagation structure. **SVM-TS** [17]: A linear SVM classifier that uses content features to build a time-series model. **DTC** [2]: A decision tree-based model that utilizes a combination of news characteristics.

Table 1. Details of the datasets

Statistic	Twitter15	Twitter16	Weibo
# of source tweets	1490	818	4664
# of posts	331,612	204,820	3,805,656
# of users	276,663	173,487	2,746,818
# True rumors	374	205	2351
# False rumors	370	205	2313
# Unverified rumors	374	203	0
# Non-rumors	372	205	0

Table 2. Experimental results on *Weibo* dataset.

Metric	Class	DTC	SVM-TS	SVM-TK	RvNN	PLAN	BiGCN	ClaHi-GAT	MHGAT
Acc.	–	0.767	0.756	0.786	0.794	0.831	0.863	0.852	**0.914**
Prec.	F	0.735	0.732	0.916	0.833	0.823	0.971	0.953	**0.978**
	T	0.685	0.714	0.613	0.727	0.885	0.775	0.754	**0.841**
Rec.	F	0.763	0.804	0.819	0.783	0.841	0.717	0.739	**0.853**
	T	0.786	0.821	0.753	0.833	0.766	0.971	0.952	**0.985**
F_1	F	0.749	0.774	0.864	0.812	0.832	0.824	0.861	**0.868**
	T	0.732	0.717	0.773	0.808	0.821	0.862	0.842	**0.897**

Table 3. Experimental results on *Twitter15* and *Twitter16*.

Method	Twitter15					Twitter16				
	Acc.	N	F	T	U	Acc.	N	F	T	U
		F_1	F_1	F_1	F_1		F_1	F_1	F_1	F_1
DTC	0.625	0.716	0.519	0.642	0.523	0.607	0.652	0.432	0.573	0.739
SVM-TS	0.581	0.394	0.520	0.463	0.549	0.645	0.546	0.638	0.654	0.668
SVM-TK	0.705	0.619	0.756	0.485	0.835	0.732	0.814	0.713	0.745	0.801
RvNN	0.759	0.714	0.765	0.814	0.714	0.722	0.628	0.712	0.833	0.714
PLAN	0.795	0.784	0.810	0.793	0.802	0.825	0.846	0.803	0.774	0.832
BiGCN	0.814	0.772	0.827	0.830	0.786	0.816	0.751	0.839	0.904	0.781
ClaHi-GAT	0.823	0.805	0.843	0.894	0.807	0.838	0.763	0.864	0.892	0.816
MHGAT	**0.862**	**0.836**	**0.872**	**0.925**	**0.823**	**0.874**	**0.836**	**0.896**	**0.912**	**0.852**

For a fair comparison, we adopt the same evaluation metrics that have already been widely used in existing work [5,6]. Thus, for the Weibo dataset, we evaluate the Accuracy (Acc.), Precision (Prec.), Recall (Rec.) and F_1 measure (F_1) on each class. For the two Twitter datasets, we evaluate the Accuracy (Acc.) and F_1 on each class.

4.3 Data Processing and Experiments Setup

To be more realistic, we randomly select 15% of the instances as the development dataset that the model has not seen at all, and split the remaining instances into training and test datasets at a ratio of 4:1 in all datasets; this similar to the settings in existing studies [16,26]. In order to reduce the randomness, we repeat the experiments fifty times and take the average value as the result. We optimize the model using the Adam algorithm [9]. The dimension of each node's hidden feature vector is 128. The number of head K of self-attention is set to 8. The dropping

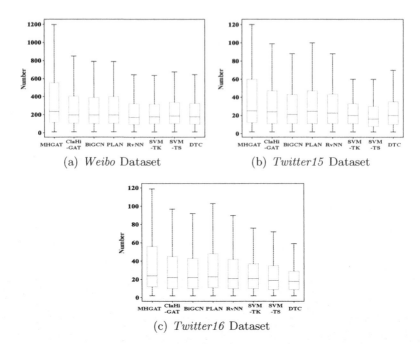

(a) *Weibo* Dataset

(b) *Twitter15* Dataset

(c) *Twitter16* Dataset

Fig. 5. Comparison of the number of correctly detected rumor data (the vertical axis), where each rumor data is a propagation tree consisting of one source tweet and a number of retweets. The horizontal axis represents the proposed model and the various baselines.

rate in Subtree Drop is 0.1 for all three datasets. The training process is iterated upon 150 epochs and early stopping [23] is applied when the validation loss stops decreasing by 10 epochs.

4.4   Results and Analysis

Tables 2 and 3 show the performance of the proposed method and all comparison methods on Weibo and Twitter datasets. Compared with the content-based methods like DTC, SVM-TS, the propagation-based methods considering the propagation structure's characteristics, are generally more effective. The success rate of PLAN is higher than that of SVM-TK and RVNN that focus on propagation characteristics, because the potential relevance of all posts is considered in PLAN, but it tends to cause noises weakening the topological structure of the propagation tree. BiGCN and ClaHi-GAT pay more attention to the topology of propagation trees and aggregate the local characteristics of the propagation tree. The former demonstrates the effectiveness of incorporating the structure of dispersion and the source text features enhancement into rumor detection, while the latter shows the effectiveness of considering potentially associated tweets based on topological structures. However, these two methods can only take the averageof all local representations as rumor representation, ignoring the differences among local representations and the impact of the global structure.

MHGAT considers the influence of the dispersion and the sequence structure of rumor propagation, the difference among local structures, and the interaction between local and global information. In addition, it strengthens the rumor representation by incorporating the source text feature and the parent text feature where appropriate. Thus, MHGAT outperforms all the baselines and state-of-the-art methods on all three datasets, especially in the large-scale *Weibo* dataset.

In order to further illustrate the detection performance of the model, we compare the number of correctly detected rumor data by different methods as shown in Fig. 5. By comparing box sizes and the upper and lower bounds, we found that methods (PLAN, BiGCN, ClaHi-GAT, MHGAT) that consider the local propagation structure and the potential correlation of posts tend to work better with most data than the other methods. Clearly, MHGAT has a wider upper and lower limit and can cover a broader range of data than the other methods. It proves that our method does not need a large amount of complex data to learn and can cope with the high-flow hot spot rumor, showing its outstanding performance in a more complex real-world scenario.

4.5   Ablation Study

To analyze the effect of each module of MHGAT, we conduct a series of ablation studies on different parts of the model. The ablation study is conducted in the following order: **w/o SBN**: Removing single-branch subtree nodes (SBN) and the related edges, and utilizing the remaining information on the graph for rumor detection. **w/o MBN**: Removing multi-branch subtree nodes (MBN) and the related edges, and utilizing the remaining information on the graph for rumor detection. **w/o STN**: Removing single-thread subtree nodes (STN) and the related edges, and utilizing the remaining information on the graph for rumor detection. **w/o MBTN**: Removing multi-branch subtree nodes (MBTN) and the related edges, and utilizing the remaining information on the graph for rumor detection. **w/o TN**: Removing the tree node (TN) and the related edges, and taking the mean representation of all nodes in the heterogeneous propagation graph as the final representation of the rumour for rumor detection.

We can observe the effect of removing all kinds of propagation subtrees covering local information in Table 4, which proves the universality of propagation subtrees and the necessity of classifying differences among local structures. Specifically, removing STN has the most significant impact on the results, and the accuracy on the Weibo, Twitter15 and Twitter16 datasets has dropped by 7.2%, 4.7% and 4.7%, respectively. This result is predictable. The information carried by the SBN is fragmented, whereas the information carried by the STN is able to cover the local relevance better and still has a better effect without SBN. Furthermore, there is a decrease in the accuracy rate without TN, but it is still higher than the baselines and the other variants of the ablation study due to the interaction among local subtrees in the interaction process, confirming the importance of local information interaction and the effect of the interaction

Table 4. The ablation study results on the *Weibo*, *Twitter15* and *Twitter16* datasets.

Models	Weibo accuracy	Twitter15 accuracy	Twitter16 accuracy
MHGAT	0.914	0.862	0.874
w/o SBN	0.853	0.829	0.813
w/o MBN	0.871	0.837	0.849
w/o STN	0.842	0.815	0.827
w/o MBTN	0.883	0.841	0.845
w/o TN	0.889	0.847	0.853

between local and global information. Since the proposed method is integrated with the source text feature and the parent text feature, it is necessary to analyze the effects of each component. As shown in Fig. 6, we compare the results of the complete model and its variants and find that the complete model is better than the ones without the fusion of source text feature or parent text feature. This shows incorporating the source text feature and the corresponding time information of the parent node in appropriate places can improve the performance of our model.

Moreover, when the model introduces implicit links between subtrees, not all neighbors can contribute important information to the aggregation. Thus we introduce weights for subtree aggregation. As shown in Fig. 6, the model with added weight is better than the model without weight, which proves that the weight we designed reasonably solves the noise problem introduced by implicit links and further enhances the effect of our rumor detection model.

4.6 Early Detection

One of the most crucial tasks in rumor detection is the early detection of rumors. In the early rumor detection task, we compare different detection methods at elapsed time checkpoints. As shown in Fig. 7, from the performance of our method and the baseline method on different time delays in the Twitter

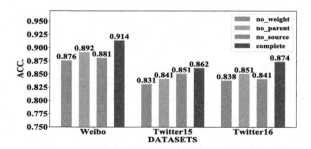

Fig. 6. Comparison of MHGAT and its variants.

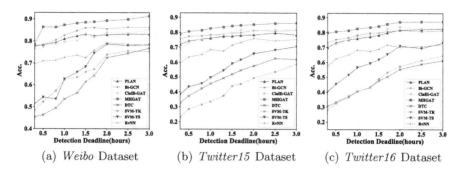

(a) *Weibo* Dataset   (b) *Twitter15* Dataset   (c) *Twitter16* Dataset

Fig. 7. Results of early rumor detection.

and Weibo datasets, it can be seen that our method achieves higher accuracy very quick as soon as the initial broadcast of the source and can still maintain higher accuracy as the time delay goes up. It is worth noting that some baselines decrease slightly when the time delay increases. This is because as the rumor is propagated, more similar structural and semantic information shows, and more noises are introduced simultaneously. The results show that our model is more suitable for a complex real-world case and has a better stability.

5 Conclusions

This paper proposed a novel Multi-subtree Heterogeneous Propagation Graph Attention Network, which is used for social media rumor detection. This method refined propagation subtrees of the rumor propagation tree, strengthened the interaction between local and global structure information, and improved the ability to learn high-level rumor representation, hence achieving the best performance. Extensive experiments proved the superiority of the proposed method. However, one of the existing obstacles of rumor detection is the performance degradation caused by data uncertainty. To address this issue, in future we will study how to use uncertainty estimation to explain the model's performance in rumor propagation.

References

1. Bian, T., et al.: Rumor detection on social media with bi-directional graph convolutional networks. In: Proceedings of the AAAI Conference on Artificial Intelligence, vol. 34, pp. 549–556 (2020)
2. Castillo, C., Mendoza, M., Poblete, B.: Information credibility on twitter. In: Proceedings of the 20th International Conference on World Wide Web, pp. 675–684 (2011)
3. Choi, J., Ko, T., Choi, Y., Byun, H., Kim, C.k.: Dynamic graph convolutional networks with attention mechanism for rumor detection on social media. Plos One **16**(8), e0256039 (2021)

4. Enayet, O., El-Beltagy, S.R.: Niletmrg at semeval-2017 task 8: determining rumour and veracity support for rumours on twitter. In: Proceedings of the 11th International Workshop on Semantic Evaluation (SemEval-2017), pp. 470–474 (2017)
5. Fuller, C.M., Biros, D.P., Wilson, R.L.: Decision support for determining veracity via linguistic-based cues. Decis. Supp. Syst. **46**(3), 695–703 (2009)
6. Giudice, K.D.: Crowdsourcing credibility: the impact of audience feedback on web page credibility. Proc. Am. Soc. Inf. Sci. Technol. **47**(1), 1–9 (2010)
7. Jin, Z., Cao, J., Zhang, Y., Luo, J.: News verification by exploiting conflicting social viewpoints in microblogs. In: Proceedings of the AAAI Conference on Artificial Intelligence, vol. 30 (2016)
8. Khoo, L.M.S., Chieu, H.L., Qian, Z., Jiang, J.: Interpretable rumor detection in microblogs by attending to user interactions. In: Proceedings of the AAAI Conference on Artificial Intelligence, vol. 34, pp. 8783–8790 (2020)
9. Kingma, D.P., Ba, J.: Adam: a method for stochastic optimization. arXiv preprint arXiv:1412.6980 (2014)
10. Kumar, S., Carley, K.M.: Tree lstms with convolution units to predict stance and rumor veracity in social media conversations. In: Proceedings of the 57th Annual Meeting of the Association for Computational Linguistics, pp. 5047–5058 (2019)
11. Li, C., Liu, F., Li, P.: Text similarity computation model for identifying rumor based on bayesian network in microblog. Int. Arab J. Inf. Technol. **17**(5), 731–741 (2020)
12. Li, Q., Zhang, Q., Si, L.: Rumor detection by exploiting user credibility information, attention and multi-task learning. In: Proceedings of the 57th Annual Meeting of the Association for Computational Linguistics, pp. 1173–1179 (2019)
13. Lin, H., Ma, J., Cheng, M., Yang, Z., Chen, L., Chen, G.: Rumor detection on twitter with claim-guided hierarchical graph attention networks. arXiv preprint arXiv:2110.04522 (2021)
14. Liu, X., Nourbakhsh, A., Li, Q., Fang, R., Shah, S.: Real-time rumor debunking on twitter. In: Proceedings of the 24th ACM International on Conference on Information and Knowledge Management, pp. 1867–1870 (2015)
15. Ma, J., Gao, W.: Debunking rumors on twitter with tree transformer. In: ACL (2020)
16. Ma, J., et al.: Detecting rumors from microblogs with recurrent neural networks (2016)
17. Ma, J., Gao, W., Wei, Z., Lu, Y., Wong, K.F.: Detect rumors using time series of social context information on microblogging websites. In: Proceedings of the 24th ACM International on Conference on Information and Knowledge Management, pp. 1751–1754 (2015)
18. Ma, J., Gao, W., Wong, K.F.: Detect rumors in microblog posts using propagation structure via kernel learning. Association for Computational Linguistics (2017)
19. Ma, J., Gao, W., Wong, K.F.: Rumor detection on twitter with tree-structured recursive neural networks. Association for Computational Linguistics (2018)
20. Potthast, M., Kiesel, J., Reinartz, K., Bevendorff, J., Stein, B.: A stylometric inquiry into hyperpartisan and fake news. arXiv preprint arXiv:1702.05638 (2017)
21. Tarjan, R.: Depth-first search and linear graph algorithms. SIAM J. Comput. **1**(2), 146–160 (1972)
22. Vaswani, A., et al.: Attention is all you need. Adv. Neural Inf. Process. Syst. **30**, 5998–6008 (2017)
23. Yao, Y., Rosasco, L., Caponnetto, A.: On early stopping in gradient descent learning. Constr. Approx. **26**(2), 289–315 (2007)

24. Yu, F., Liu, Q., Wu, S., Wang, L., Tan, T., et al.: A convolutional approach for misinformation identification. In: IJCAI, pp. 3901–3907 (2017)
25. Yu, K., Jiang, H., Li, T., Han, S., Wu, X.: Data fusion oriented graph convolution network model for rumor detection. IEEE Trans. Netw. Serv. Manag. **17**(4), 2171–2181 (2020)
26. Yuan, C., Ma, Q., Zhou, W., Han, J., Hu, S.: Jointly embedding the local and global relations of heterogeneous graph for rumor detection. In: 2019 IEEE International Conference on Data Mining (ICDM), pp. 796–805. IEEE (2019)
27. Zarocostas, J.: How to fight an infodemic. The Lancet **395**(10225), 676 (2020)

DeMis: Data-Efficient Misinformation Detection Using Reinforcement Learning

Kornraphop Kawintiranon$^{(\boxtimes)}$ and Lisa Singh

Georgetown University, Washington, DC, USA
{kk1155,lisa.singh}@georgetown.edu

Abstract. Deep learning approaches are state-of-the-art for many natural language processing tasks, including misinformation detection. To train deep learning algorithms effectively, a large amount of training data is essential. Unfortunately, while unlabeled data are abundant, manually-labeled data are lacking for misinformation detection. In this paper, we propose DeMis, a novel reinforcement learning (RL) framework to detect misinformation on Twitter in a resource-constrained environment, i.e. limited labeled data. The main novelties result from (1) using reinforcement learning to identify high-quality weak labels to use with manually-labeled data to jointly train a classifier, and (2) using fact-checked claims to construct weak labels from unlabeled tweets. We empirically show the strength of this approach over the current state of the art and demonstrate its effectiveness in a low-resourced environment, outperforming other models by up to 8% (F1 score). We also find that our method is more robust to heavily imbalanced data. Finally, we publish a package containing code, trained models, and labeled data sets.

Keywords: Reinforcement learning · Misinformation detection

1 Introduction

Social media sites allow users to share different types of online content. Unfortunately, there is no requirement that the content be true. As a result, we are seeing varying levels of accuracy in shared content. False information (fake information, misinformation, and disinformation) detection is not a new problem, and a significant amount of research has emerged (see [1,7] for surveys). Most research studies focus on detecting the spread of fake news by news sources [16,17], e.g. CNN and Washington Post. Some researchers have also worked on utilizing fact-checked information to verify the truth of social media content generated by users [4,19]. While this previous research can effectively identify false information on Twitter, in practice, the methods either requires a large amount of training data for each false claim or myth being detected, or expect balanced training data.

To mitigate these challenges, we propose a novel reinforcement learning (RL) framework for detecting misinformation on Twitter in a constrained environment, i.e. where data labels are limited and imbalanced. Our approach, DeMis,

M.-R. Amini et al. (Eds.): ECML PKDD 2022, LNAI 13714, pp. 224–240, 2023.
https://doi.org/10.1007/978-3-031-26390-3_14

uses fact-checking articles (FC-articles) as background knowledge. The framework requires a small number of FC-articles related to the target myth theme. Then it weakly labels the unlabeled tweets given the chosen FC-articles. We design the RL mechanism to select high-quality tweets. These weak-labeled tweets are then used to help train the detector. While the joint training of classifier and selector [21] is often used to maximize the model performance, we partially train the classifier before jointly training the classifier and selector. This guides the classifier to gain knowledge about the manually-labeled data prior to learning from the weak and manually labeled data together.

Our Contributions Are as Follows: (1) We propose a novel data-efficient RL framework in which state, action and reward are exclusively designed for misinformation detection. (2) We propose an approach (DeMis) to incorporate FC-articles as expert knowledge as a form of weak supervision. (3) We integrate multiple learning paradigms (reinforcement learning, multi-source joint learning, neural learning) into a framework for identifying misinformation. (4) We compare our model to multiple classic, neural, and reinforcement models and show that our model generally performs better. (5) We demonstrate the effectiveness of our framework when the training data is heavily imbalanced. (6) We release a package for misinformation detection using reinforcement learning, including the code, trained models and data sets.[1]

2   Related Works

Misinformation detection is an active area of research (see [1] for a recent survey). Because fake information can be produced by bots or humans, our work and review focuses on post-level misinformation instead of user-level and reinforcement learning approaches for generating additional training data.

Misinformation Detection: Research on misinformation detection typically falls into two categories based on types of information used to train a classifier [1], content-based and social context-based. Content-based approaches use information extracted from the content of posts such as text, images, and videos. Social context-based approaches use human-content interaction data such as retweets, replies, and likes. While using both types of information achieves slightly better results [10,13,26], because of the additional cost of data collection and the need for timely identification of misinformation, we focus on content-based methods.

Many studies use the lexical and syntactic features extracted from textual data [2,14]. Jin et al. [5] convert the detection problem into a text matching problem. They classify misinformation tweets based on the similarity scores between input tweets and the original verified-false posts. Their best algorithm is BM25 with an accuracy of 0.799. Recently, deep learning models have been shown to be state of the art for misinformation detection [1]. Wang et al. [20] propose EANN, a model that uses convolution neural networks (CNN) to learn latent semantic

---

[1] https://github.com/GU-DataLab/misinformation-detection-DeMis.

text representations and use it along with image data to train a classification layer. Their models are evaluated on Twitter and Weibo data that have both text and images, achieving F1 scores of 0.719 and 0.829, respectively. A CNN model with an attention mechanism has also been proposed [24], improving the state of the art by 9 and 12% on the same data sets. These data sets are balanced and pseudo-labeled using keywords. Hossain et al. [4] introduced *COVIDLIES*, a manually-labeled Twitter data set about COVID-19 misinformation. It consists of 86 myths and 6761 tweets. Their approach has two sub-tasks including related-myth retrieval and stance detection. Using a BERT-based sentence similarity algorithm [25], they achieve the best Hit@k of 60.8 to 96.9 for different k values on the related-myth retrieval task but they obtain an F1 score of only 50.2 on the stance detection task because the data are imbalanced. Recently, Vo et al. [19] proposed a framework to search for fact-checking articles given a tweet, using a large amount of labeled training data (over 10K tweets and 2K FC-articles).

Data-Efficient and Reinforcement Learning. Generally, a large amount of labeled data is required to train a reasonably accurate neural network (NN) model. Weak supervision aims to reduce human effort by automatically generating labels given unlabeled data. The quality of labels then heavily relies on the labeling algorithms [23]. An automatic data annotator based on the sources of news articles was proposed in [3]. Each tweet containing at least one URL to a news article was labeled true or false based on trustworthy or untrustworthy sources. Reinforcement learning (RL) techniques [18] have been adopted in many classification tasks to learn a high-quality data selector [21,23]. A model with a RL-based selector in [22] achieves an average F1 score of 0.692 on the Twitter click-bait classification task. Yoon and colleagues [23] propose a RL-based algorithm that quantifies the quality of labeled data. Their experimental results show that removing low-quality data from the training process improves the overall model performance on several classification tasks with accuracy scores ranging from 0.448 to 0.903. Mosallanezhad et al. [11] propose RL-based domain-adaptive learning which learns domain-invariant features and utilizes auxiliary information for fake news detection. Recently, WeFEND [21] was proposed for fake news detection on WeChat. The model trains a weak-labeling annotator using private user reports attached to each news article then selects the high-quality samples using a reinforced selector for training. The model obtains an F1 score of 0.81 on balanced WeChat data. Conceptually, we take a similar approach, building a model using reinforcement learning to identify weak labels. However, our annotator and joint learning paradigm are different.

3   Background and Problem Definition

Misinformation has many definitions. One common feature of these definitions is that misinformation must contain a piece of false information. Kumar et al. [7] define misinformation as false information spread without the intent to deceive, while others [26] define it as any false or inaccurate information regardless of

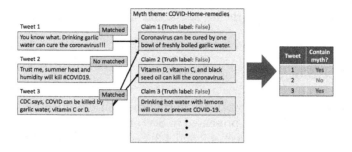

Fig. 1. Examples of misinformation tweets and supporting evidence.

intention. In this paper, we follow the later and refer to a misinformation tweet as a tweet containing a piece of myth-related information. A *myth* is a false claim verified by trustworthy fact-checkers. This task is different from fake news detection which focuses on detecting a news article published by a news outlet that is verifiably false, and rumor detection which aims to determine if a story or online post is a rumor or non-rumor regardless of its veracity [8]. Figure 1 demonstrates how tweets are determined to be misinformation. For example, a claim saying "boiled garlic water could kill the coronavirus" is false. A tweet containing such information (even if it is being refuted) is classified as misinformation conversation, regardless of the user stance. In other words, our goal is to identify that misinformation is being discussed on social media, not the intent or the position of the poster. A tweet that does not is labeled as true information.

Generally, fact checkers provide a set of *FC-article* that each contain a claim, truth label, and fact. A *claim* is a truth-verifiable statement that may be true, false, partially true or have insufficient information to determine whether or not it is true. A *truth label* is the factual state of the related claim at a particular time. It is manually verified by experts in the relevant areas. Different fact checkers have different rating schemes. For example, PolitiFact claims are usually rated using six level of falseness. The *fact* is the supporting information that provides context about the claim and explains details about why a particular truth label is assigned. Different claims of FC-articles may be in the same myth theme as shown in Fig. 1. In this paper, the goal is to predict whether a tweet contains the same piece of misinformation as in the claims of interest. We only use claims and truth labels verified as false since our goal is to detect misinformation discussion.

More formally, the problem we investigate is content-based misinformation. Let $\mathbf{M}$ represent a set of myths and $\mathbf{C}$ represent a set of claims from FC-articles. Suppose we are given a set of target claims $\bar{C}_p$ that are related to the pre-defined myth theme M_p. Our task is to determine a class label y_r for a tweet t_r from Twitter data $\mathbf{T}$ using claim information $(\bar{c}_{pq} \in \bar{C}_p)$ related to M_p. If t_r contains misinformation, $(y_r = 1)$, otherwise, $(y_r = 0)$. We assume that claims across myths in $\mathbf{M}$ are non-overlapping, $\bigcap_{p=1}^{|\mathbf{M}|} \bar{C}_p = \emptyset$. For example, claims $\bar{C}_1$ under the myth theme M_1 about a specific weather condition killing coronavirus, and claims $\bar{C}_2$ under the myth theme M_2 about COVID home-remedies, are not overlapped ($\bar{C}_1 \cap \bar{C}_2 = \emptyset$).

4   Methodology

We propose DeMis, a framework for misinformation detection on Twitter. An overview of the framework is presented in Sect. 4.1. The main components of the framework are presented in Sect. 4.2 and 4.3. Section 4.4 presents the integration of all the components.

4.1   Overview of DeMis

The overview of the framework is shown in Fig. 2. We begin by extracting claims $\mathbf{C}$ and target claims $\bar{C}_p$ related to the myth themes of interest from existing FC-articles. Each theme of interest M_p has a small number of manually-labeled tweets. We refer to these tweets as *strong-labeled* tweets. The automatic annotator (Sect. 4.2) uses a sentence similarity algorithm to calculate similarity scores between all claims $\mathbf{C}$ and unlabeled tweets in $\mathbf{T}$. The scores are used to generate labels for the unlabeled tweets using our proposed labeling function. We refer to tweets with labels generated by the automatic annotator as *weak-labeled* tweets. Once the reinforced selector (Sect. 4.3) chooses high-quality weak-labeled tweets, they are combined with the strong-labeled tweets for training the misinformation detector. The samples that are selected by the reinforced selector are referred to as *selected* tweets. The reward is computed based on the model performance and used to update the selector for the next iteration. The updated selector selects high-quality weak-labeled tweets to train the detector until the detection classifier converges. The misinformation detector $D_n(\cdot; \theta_n)$ is a transformer-based model with a neural network on top as a classifier layer, where θ_n denotes its parameters. We now present the details.

4.2   Automatic Annotation Based on Claims

We propose an unsupervised approach for automatically labeling tweets.[2] There are two main components: sentence similarity ranking and labeling. First, among all claims C, there are claims $\bar{C}_p \subset C$ belonging to the target myth theme M_p that we are interested in. We calculate the similarity scores between each tweet t_r and all claims $c_q \in C$. For each tweet, we obtain a list of all claims L_r ranked by the similarity scores. If at least one of the target claims $\bar{c}_{pq} \in \bar{C}_p$ appears in the top K of the list, then the tweet is labeled as positive (about misinformation). Otherwise, the label is negative (not about misinformation). Any similarity score is reasonable. Given that we are using short texts, we use a sentence transformer [15] in our empirical evaluation. We convert a claim and a tweet into vectors and compute the final similarity score using cosine similarity.

---

[2] We use the term *unsupervised* because we do not use any labeled data at this stage.

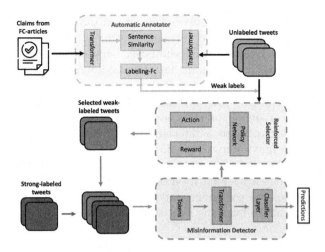

Fig. 2. The architecture of our proposed misinformation detection framework.

4.3 Data Selection via Reinforcement Learning

The goal of the data selection component is to select high-quality weak-labeled samples that improve the detector performance. We propose a performance-driven data selector that uses the policy-gradient reinforcement learning mechanism called the *reinforced selector*. It takes weak-labeled data as input, selects high-quality samples, and then sends them to the classifier to use during training. The reward is computed based on the model performance and used to update the policy network. Because the reward is computed after the data selection process is finished, the policy update is delayed. This is inefficient. To obtain rewards and train the policy network more efficiently, we split the input data $\mathcal{X} = \{x_1, ..., x_n\}$ into N bags $\mathcal{B} = \{B^1, ..., B^N\}$. Each bag B^k contains a sequence of unlabeled samples $\{x_1^k, x_2^k, ..., x_{|B^k|}^k\}$. Each bag is fed into the reinforced selector. For each sample in the bag, the reinforced selector decides on an *action* to retain or remove. The action of the current sample x_i^k is based on the current *state* vector and all the actions of previous samples in the current bag $\{x_1^k, x_2^k, ..., x_{i-1}^k\}$. The *reward* is computed based on the change in performance of the misinformation detector. The remainder of this subsection presents the details of the main components of the RL mechanism: *state*, *action*, *reward* and *optimization*.

State. s_i^k represents the state vector of sample x_i^k. The action a_i^k is decided based on the current and selected samples in the same bag, B^k. The state vector s_i^k consists of two major components, including the representation of the current sample and the average representation of selected samples. We consider quality and diversity for a representation of a sample. For the quality of the sample, we consider a prediction output from the misinformation detector and a small number of elements from the sentence similarity algorithm (Sect. 4.2). For the current sample, these elements include: (i) the highest similarity score between the current sample and all claims C, (ii) the K-th highest similarity score,

(iii) the highest similarity score between the current sample and the target claims $\bar{C}_p$, (iv) the subtraction of (i) and (iii), and (v) the subtraction of (iii) and (ii). For diversity, we calculate the cosine similarity between the current sample and all selected samples in the bag, and then the maximum similarity score is used as the representation of the diversity of the current sample among the selected samples. The weak label of the current sample is also included in the representation vector as a signal for the class distribution. The final current state representation vector contains eight elements: 1) the output probability from the detector, 2) the maximum cosine similarity score between the current sample and the selected samples, 3) the weak label of the current sample, and five elements from the sentence similarity described above. Once we have the current representation vector, we concatenate it with the average of previously selected representation vectors to form the final state vector s_i^k.

Action. An action value of the reinforced selector for any sample is either 1 representing an action to *retain*, or 0 representing an action to *remove* the sample from the training set. We train a policy network $P(\cdot; \theta_s)$ to determine action values, where θ_s indicates its parameters. The policy network is a neural network of two fully-connected layers with the sigmoid (σ) and ReLU activation functions and is defined as $P(s_i^k; \theta_s) = \sigma(W_2 \cdot ReLU(W_1 \cdot s_i^k))$, where W_1 and W_2 are randomly initialized weights. The network outputs the probability of the *retain* action p_i^k for the sample x_i^k given the corresponding state vector s_i^k. Next, the policy $\pi_{\theta_s}(s_i^k, a_i^k)$ determines the action a_i^k by sampling using the output probability p_i^k as follows $\pi_{\theta_s}(s_i^k, a_i^k) = a_i^k p_i^k + (1 - a_i^k)(1 - p_i^k)$.

Reward. As previously mentioned, we use the performance changes of the misinformation detector $D_n(\cdot; \theta_n)$ as the reward function. To determine the initial baseline performance F_{base}, we train the detector on the strong-labeled training set and evaluate it on the validation set. For the k-th bag, the reinforced selector chooses high-quality samples. They are used to re-train the detector, then the performance F_k for the k-th bag is obtained by evaluating the re-trained detector on the validation set. Formally, the reward R_k for the k-th bag is the subtraction of F_{base} and F_k as shown in the equation $R_k = F_{base} - F_k$.

Optimization. The goal is to maximize the expected total reward for each bag B^k. However, the magnitude of reward R_k is undoubtedly small because a performance change ranges from zero to one. Therefore, we use the summation of reward R_k weighted by policy values $\pi_{\theta_s}(s_i^k, a_i^k)$ from every sample in the bag $\{x_i^k\}_{i=1}^{|B^k|}$. Finally, the objective function for the k-th bag is defined as: $J(\theta_s) = \sum_{i=1}^{|B^k|} \pi_{\theta_s}(s_i^k, a_i^k) R_k$, and its derivative function is: $\nabla_\theta J(\theta_s) = \mathbb{E}_{\theta_s}[\sum_{i=1}^{|B^k|} R_k \nabla_{\theta_s} \log \pi_{\theta_s}(s_i^k, a_i^k)]$.

Since we are using policy-gradient reinforcement learning [18], we update the policy network using the gradient ascend: $\theta_s \leftarrow \theta_s + \alpha \sum_{i=1}^{|B^k|} R_k \nabla_{\theta_s} \log \pi_{\theta_s}(s_i^k, a_i^k)$, where α is the learning rate.

Algorithm 1: The Overall Training Process of DeMis

Input : Misinformation detector $D_n(\cdot; \theta_n)$, policy network $P(\cdot; \theta_s)$ of reinforced selector with random weights, strong-labeled data $\mathcal{D}$

1. Pre-train the detector $D_n(\cdot; \theta_n)$ to predict misinformation using the strong-labeled training data $\mathcal{D}_t$.
2. Pre-train the policy network $P(\cdot; \theta_s)$ by running Algorithm 2 with the misinformation detector $D_n(\cdot; \theta_n)$ fixed.
3. Re-initialize the parameters of the detector $D_n(\cdot; \theta_n)$ with random weights.
4. Warm up the detector $D_n(\cdot; \theta_n)$ by training for $\mathcal{L}$ epochs.
5. Jointly train $D_n(\cdot; \theta_n)$ and $P(\cdot; \theta_s)$ using Algorithm 2 until convergence.

Output: The trained models $D_n(\cdot; \theta_n)$ and $P(\cdot; \theta_s)$.

4.4 Model Training

The overall training process is described in Algorithm 1. First, we randomly initialize weights of the misinformation detector and policy network of the reinforced selector. The detection classifier $D_n(\cdot; \theta_n)$ is a neural network model: $p(y|x; \theta_n) = Softmax(W_{L2}(\tanh(W_{L1}x_t + b_1)) + b_2)$, where $p(y|x; \theta_n)$ represents the output probability of being class y given input x from the linear classifier, x represents a contextual representation vector of tweet t from the pre-trained language model (BERTweet) after the dropout layer, W_{Li} is a weight vector at layer i randomly initialized, and b_i is a bias vector at layer i where $i \in \{1, 2\}$. The weights of the classifier are updated using the cross-entropy loss function. We use the softmax function to normalize the values of the output vector from the classifier in order to obtain a probability score for each class.

Second, we get the baseline performance F_{base} by training the detector using the strong-labeled training data $\mathcal{D}_t$ and evaluating it on the validation set $\mathcal{D}_v$. Next, because the joint-training technique can result in a detector over-fitting the small data set, we re-initialize the weights of the detector model and train it for $\mathcal{L}$ epochs instead of training it until convergence (Algorithm 1, step 4–5). This makes the detector under-fit, leaving some room for joint-training. Finally, we jointly train the detector and reinforced selector together until convergence.

Algorithm 2 explains how to train the detector and reinforced selector jointly. The detector provides the mechanism to compute the reward based on its evaluation performance. The selector uses the reward to refine its ability to select high-quality samples that potentially enhance the detector performance. To improve the training stability we update the target policy network slowly: $\theta'_s = \tau \theta_s + (1 - \tau)\theta'_s$.

Algorithm 2: Learning Algorithm of Reinforced Selector

Input : Strong-labeled training data $\mathcal{D}_t$. N bags of weak-labeled training data $\mathcal{B} = \{B^1, ..., B^N\}$. A misinformation detector $D_n(\cdot; \theta_n)$ and a policy network $P(\cdot; \theta_s)$. Epoch number L.

Initialize the target networks as: $\theta'_n \leftarrow \theta_n$ and $\theta'_s \leftarrow \theta_s$

for *epoch* $\ell \leftarrow 1$ **to** L **do**

   Shuffle $\mathcal{B}$ to get a sequence of bags $\{B^1, B^2, ..., B^N\}$ **foreach** *bag* $B^k \in \mathcal{B}$ **do**

      /* We omit superscript k for clarity   */

      Sample actions for each data sample in B with θ'_s:

      $A = \{a_1, ..., a_{|B|}\}$, $a_i \sim \pi_{\theta'_s}(s_i, a_i)$

      Train the detector $D_n(\cdot; \theta_n)$ using selected samples based on actions A and update weights θ_n

      Compute delayed reward R_k

      Update the parameters θ_s of reinforced selector:

      $\theta_s \leftarrow \theta_s + \alpha \sum_{i=1}^{|B|} R_k \nabla_{\theta_s} \log \pi_{\theta_s}(s_i, a_i)$

   end

   Update the weights of target policy network: $\theta'_s = \tau \theta_s + (1 - \tau)\theta'_s$

   Train the target detector using the selected samples from the target selector then update weights θ'_n

   Reset the weights of detector: $\theta_n \leftarrow \theta'_n$

end

Output: The trained models $D_n(\cdot; \theta_n)$ and $P(\cdot; \theta_s)$.

---

5 Experimental Design

5.1 Data Collection

Our empirical evaluation uses one large unlabeled and three manually-labeled Twitter data sets: COVIDLIES [4], COMYTH-W and COMYTH-H. These data sets have different characteristics in terms of myth diversity and training data imbalance. The sizes of positive samples in a training set range from only 40 to 200. In COVIDLIES, misinformation tweets contain claims belonging to multiple myth themes (high-diversity) and have class-imbalances (high-imbalance). In COMYTH-W and H, misinformation tweets contain claims belonging to one myth theme (low-diversity), COVID-weather and COVID-home-remedies, respectively. While COMYTH-W is a balanced data set (low-imbalance), COMYTH-H is not (high-imbalance). Table 1 presents the statistics of these data.

Unlabeled Twitter Data. Our research team collected English tweets related to COVID-19 using hashtags and keywords through the Twitter Streaming API. Between March 2020 and August 2020, we collected over 20 million tweets, not including quotes and retweets. These unlabeled tweets were used to train all models that require unlabeled data.[3]

COVILDLIES. This data set, shared by Hossain et al. [4], contains 62 claims, along with 6591 tweet-claim pairs. Each tweet has at least one related claim and an annotated stance of the tweet content towards the claims (agree, disagree, no stance). We follow the labeling approach of the original paper [4] and label a tweet as misinformation if and only if the tweet contains a stance. A tweet with no stance is labeled as no misinformation. Among 62 claims, only four claims (of different themes) have more than 100 tweets containing a stance towards them, indicating high diversity. There are 811 annotated tweets, 136 containing misinformation and 675 regular tweets.

COMYTH. To conduct experiments on data sets with specific myth themes, we created a data set of COVID-myth-related tweets and claims from a random sample of tweets. We focus on two myth themes, weather and home-remedies. Our data were labeled using Amazon Mechanical Turk (MTurk). The labeling choices were yes, no, and unsure. Each tweet has three annotations from three different MTurk workers. We compute inter-annotator agreement scores to assess the quality of our labeled data. The task-based and worker-based metrics are recommended by the MTurk official site[4], given their annotating mechanism. All scores range from 85% up to 97%, indicating the high inter-rater reliability for these data sets. The majority voting among three annotators is used to determine a label for each tweet (containing related myths or not). Finally, there are 930 labeled tweets for the weather theme (COMYTH-W), of which 459 tweets contain weather myths. For the home-remedies theme (COMYTH-H), there are 779 labeled tweets, of which 101 tweets contain home-remedies myths. To build a data set of COVID-related claims, we collected claims from PolitiFact, FactCheck.org and Snopes. Our research team manually identified 3 COVID-weather-related claims and 12 COVID-home-remedies-related claims as target claims for our framework.

5.2 Data Preparation

Data sets are split into train, validation and hold out sets with an approximate ratio of 5/2/3. Each tweet is preprocessed by replacing mentions with *@USER* and links with *HTTPURL*. To build weak-labeled data sets, we run our weak annotator as described in Sect. 4.2 on the unlabeled data set and sample 10K tweets for each class (myth/not-myth).

5.3 Baselines

Our baseline models are categorized into four algorithm groups. The first group contains classic machine learning models, including Naive Bayes (NB),

---

[3] Our unlabeled tweets do not overlap with any of our labeled data.

[4] https://docs.aws.amazon.com/AWSMechTurk/latest/AWSMturkAPI/
ApiReference_HITReviewPolicies.html. Amazon Mechanical Turk - HIT Review Policies.

Table 1. Data set details.

Data Set	Myth theme	Split	# Tweets	# Myth	# Not-myth	Myth ratio
COVIDLIES	COVID-mixed	Train	380	64	316	∼17%
		Val	163	27	136	
		Test	268	45	223	
COMYTH-W	COVID-weather	Train	436	213	223	∼50%
		Val	187	96	91	
		Test	307	150	157	
COMYTH-H	COVID-home-remedies	Train	365	48	317	∼13%
		Val	156	26	130	
		Test	258	27	231	

Myth theme indicates whether the data set is for a specific myth or mixed myth themes. Myth ratio indicates the ratio of misinformation.

k-Nearest-Neighbor (kNN), Logistic Regression (LR), Support Vector Machine (SVM), Decision Tree (DT), Random Forest (RF) and Elastic Net (EN). We adopt the implementations in [6] because their approaches are shown to be highly accurate for detecting low-quality textual content on Twitter. Their feature sets include simple counting properties in a tweet content (Count), Bag-of-Words (BoW) and Term-Frequency-Inverse Document-Frequency (TF-IDF). All the models are trained using different combinations of these features. The second baseline group contains neural network models, including a vanilla neural network (NN) and a convolution neural network (CNN). We follow the setup used in EANN [20]. The third group consists of transformer-based models. We use RoBERTa (RB), BERTweet (BT) and BERTweet-covid (BTC). RoBERTa is an optimized version of BERT. BERTweet is RoBERTa trained on Twitter data, and BERTweet-covid is BERTweet additionally trained on COVID-related tweets. The classification layer is a single layer neural network. The last group contains RL-based models including DVRL [23] and WeFEND [21].

We use DVRL to select high-quality weak-labeled samples. We run the model to estimate the quality of our weak-labeled data. We combine the top v percent of weak-labeled data, sorted by the quality scores with the strong-labeled data, where $v \in \{10, 20, ..., 100\}$ as used in the original paper. In addition, we also use smaller values for $v \in \{0.5, 1, ..., 5\}$ in order to have a more complete stability and sensitivity analysis. Using $v = 100$ means we combine all weak-labeled data with the strong-labeled data for training. For each combination, we train the same misinformation detector as used in our model (Sect. 4.1) and report the best results based on F1 scores. We implement the WeFEND framework as described in the original paper since the code was not available. Because the original framework uses user reports to generate weak labels but there is no such report publicly available for Twitter, we modify the framework by substituting the weak label annotation part with our weak label annotator to investigate its potential to use public accessible expert knowledge (FC-articles). The rest of the framework remains the same.

Table 2. Experimental results. The best results are bolded.

Type	Algorithm	COVIDLIES (multiple myth themes)				COMYTH-W (one myth theme)				COMYTH-H (one myth theme)			
		Acc	Pr	Re	F1	Acc	Pr	Re	F1	Acc	Pr	Re	F1
Classic ML	Count [6]	0.7724	0.3095	0.2889	0.2989	0.6645	0.6158	0.8333	0.7082	0.7985	0.2093	0.3333	0.2571
	+BoW [6]	0.8396	0.5200	0.5778	0.5474	0.9414	0.9400	0.9400	0.9400	0.9031	0.5500	0.4074	0.4681
	+TFIDF [6]	0.8545	0.5682	0.5556	0.5618	0.9479	0.9467	0.9467	0.9467	0.9070	0.5600	0.5185	0.5385
DL	NN [6]	0.8408	0.5484	0.2963	0.3845	0.8795	0.7565	0.7862	0.7672	0.9160	0.6149	0.5309	0.5696
	CNN [20]	0.3340	0.1508	0.7183	0.2446	0.7492	0.7275	0.9003	0.7816	0.2313	0.1295	0.5565	0.1953
Transformer	RB [9]	**0.8756**	**0.6550**	0.5481	0.5964	0.9739	**0.9755**	0.9711	0.9733	0.9328	0.6541	0.7901	0.7132
	BT [12]	0.8557	0.5891	0.5185	0.5450	0.9511	0.9272	0.9778	0.9515	0.9160	0.5936	0.6790	0.6248
	BTC [12]	0.8595	0.5733	0.6370	0.6035	0.9631	0.9531	0.9733	0.9627	0.9367	0.6995	0.7037	0.6990
RL	DVRL [23]	0.8333	0.5204	0.6444	0.5667	0.9577	0.9369	0.9800	0.9578	0.9057	0.5752	0.7654	0.6323
	WeFEND [21]	0.4378	0.1991	0.6765	0.2553	0.9338	0.9323	0.9346	0.9328	0.6460	0.4733	0.7509	0.4995
	DeMis (ours)	0.8483	0.5644	**0.7226**	**0.6210**	**0.9750**	0.9638	**0.9894**	**0.9762**	**0.9406**	**0.7353**	**0.8991**	**0.7887**
Compare DeMis	vs. best scores	-0.0273	-0.0906	+0.0043	+0.0175	+0.0011	-.0117	+0.0094	+0.0029	+0.0039	+0.0358	+0.1090	+0.0755
	vs. best model	-0.0113	-0.0089	+0.0856	+0.0175	+0.0011	-.0117	+0.0183	+0.0029	+0.0078	+0.0812	+0.1090	+0.0755

5.4   Evaluations and Hyperparameter Tuning

We evaluate all models using accuracy, precision, recall and F1 scores based on positive class (misinformation). We evaluate all models on the test set three times with different random seeds to determine the stability of the results. The average results are reported. For our classic ML models, we conduct a sensitivity analysis using a grid-search on influential parameters. The best parameters varied by classifiers, data sets, and feature sets. For neural network and transformer-based models, we use different learning rates (1e–4, 1e–5, 2e–5, 3e–5, 1e–6). We report the best results based on F1 scores from the parameter tuning step. We present results for the learning rate of 1e–5 for DeMis and use a learning rate for target network τ of 0.001.

6   Results and Analysis

Table 2 shows the experimental results on the test sets, averaged over three runs. The models from four different categories are evaluated on all data sets. The variances of results from different models are not significantly different. Our proposed model outperforms the best baselines F1 scores by ~2%, ~1% and ~8% on COVIDLIES, COMYTH-W and COMYTH-H, respectively. The last two rows of the table show the comparison of DeMis result with the best scores in the same column, and with the second best models based on F1 score.

6.1   Experimental Results

We hypothesize that the most complicated data set is COVIDLIES because of the high diversity of the myth themes and the data imbalance. The baseline models have F1 scores ranging from 0.2446 (CNN) to 0.6035 (BERTweet-covid). Our proposed model outperforms the baselines with an F1 score of 0.6210, slightly better than BERTweet-covid. The difficulty of this data set is two-fold. First, with 136 positive training samples for different myth themes, there are only 10 to 42 samples for each myth theme. This is insufficient for training deep learning models; therefore, the transformer models (RoBERTa and BERTweet-covid) and two of the classic models (Count+Bow and Count+TFIDF) perform better than the deep learning models. The second complexity is the mix of multiple myth themes, each having different contexts, signal words, and writing styles. These signals from different myth themes can mislead the classifiers, resulting in inefficient learning of the positive class. For example, in a batch size of 32, there are likely samples from at least two myth themes. If their characteristics are completely different, then the loss computed using the error from the samples in the batch could be misleading, resulting in under-fitting. While our models perform comparably to the state-of-the-art ones on this high diversity and imbalanced data (COVIDLIES), our model performs better on data sets containing one myth and possible imbalances.

We anticipate that the least complicated data set for this task is COMYTH-W since it contains one myth theme and is balanced data. On this data set, the baseline models perform reasonably with F1 scores ranging from 0.7082 (a classic model with Count features) to 0.9733 (RoBERTa). The notably high F1 score from RoBERTa shows that the data set is uncomplicated for the misinformation detection task and implies marginal room for improvement. Our model performs comparably with RoBERTa, having an F1 score of 0.9762.

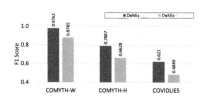

Fig. 3. The model performance of DeMis with and without RL (DeMis−).

We anticipate that the COMYTH-H data set is the second most complicated because it contains one myth theme but has a similar level of imbalance as COVIDLIES (the myth ratios of both data sets are around 10%, see Table 1). The baseline models have F1 scores ranging from 0.1953 (CNN-based model) to 0.7132 (RoBERTa), indicating that this data set is moderately complex for the task. We see that the lowest and highest F1 scores of baseline models on COMYTH-H are much lower than COVIDLIES (0.19/0.71 vs. 0.70/0.97) due to class imbalance and the nature of the myth themes. While there are only three claims related to COVID-weather, there are 12 claims about COVID-home-remedies, leading to a more diverse set of topics about home-remedies, i.e. higher content (vocabulary) diversity. Our model significantly outperforms all baselines with an F1 score of 0.7887 on COMYTH-H, an approximate 8% improvement over RoBERTa (second best).

To better understand the characteristics of the misclassified samples, we look at their distribution. We find that from 20 misclassified samples by RoBERTa and 14 misclassified samples by our model, 12 samples are the same. Our model corrects six false positives and two false negatives that the RoBERTa model misclassifies, but we have two additional false negatives, meaning that our model tends to error on the side of false negative, not false positives.

To investigate the advantage of the reinforced selector, we train our DeMis without RL by substituting it with a random selector (DeMis−). It randomly selects samples instead of selecting only high-quality samples. The results are shown in Fig. 3. On COMYTH-W, the F1 score (yellow) of DeMis without RL is 10% lower than DeMis with RL. Similarly, F1 scores are substantially higher for DeMis with RL on the other two data sets. We observe that recall scores stay the same between DeMis with and without RL because the model without RL still learns good positive examples from the strong-labeled samples. However, the precision scores drop significantly, producing more false positives when low-quality samples are selected. These empirical results suggest that incorporating RL is beneficial for improving the data selection process.

6.2 Robustness of Model

We further investigate the robustness of our model on two imbalanced data sets, COMYTH-H and COVIDLIES. We compare our model with RoBERTa since it is the second-best performer on COMYTH-H and performs comparably to BERTweet-covid on COVIDLIES. A random oversampling algorithm is used to balance the class distribution of these two data sets. We train RoBERTa on these

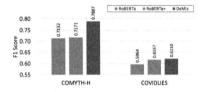

Fig. 4. The model performance of RoBERTa, RoBERTa+, and DeMis.

balanced data sets separately and report the results (RoBERTa+). We see that making the data sets more balanced for RoBERTa slightly increases the F1 scores by 0.39% and 2.03% on COMYTH-H and COVIDLIES, respectively. Without any data modification, our model that used imbalanced training data outperforms RoBERTa+ by 7.16% and 0.43%, further highlighting our model's robustness to data imbalances.

We also investigate the robustness of our model when smaller sizes of training data are provided. We randomly sample training data of sizes 200 and 300 while keeping the same level of imbalance. Figure 5 shows the F1 scores of the top performers. Our model outperforms other baselines on smaller sizes of all training data sets. However, we see that smaller sizes of data lead to larger performance deterioration on both imbalanced data sets (COMYTH-H and COVIDLIES) by all the models. In other words, when there are less than 300 training samples, the models underfit the data.

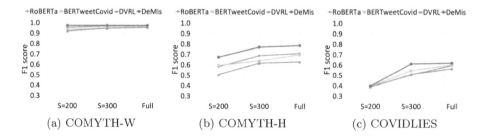

Fig. 5. The performance of top models on different sizes of training data.

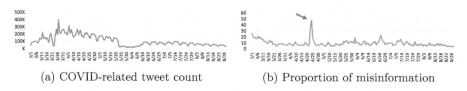

(a) COVID-related tweet count   (b) Proportion of misinformation

Fig. 6. Daily tweet counts and proportion of COVID-weather per 10,000 tweets.

6.3 Analysis on Big Data

We conduct a small case study to better understand the prevalence of misinformation on Twitter, we run our model on data containing Covid-related hashtags (Sect. 5.1) to predict levels of misinformation conversation. We find over 20K misinformation tweets about COVID-weather between March 1 to August 31, 2020. Figure 6 illustrates the daily number of tweets and the diffusion of misinformation on Twitter related to COVID-weather by DeMis. Misinformation conversation was spreading before March and reached its peak on April 24th (red arrow), the day after the White House promoted new lab results suggesting heat and sunlight slow coronavirus on April 23rd[5]. This small analysis highlights the level of misinformation on a public health related data stream and demonstrates the role prominent leaders play in spreading and/or reinforcing it.

7 Conclusions

This paper proposes DeMis, a novel RL-based framework for misinformation detection that requires only a small amount of labeled training data. We design a novel RL mechanism, inspired by policy-gradient reinforcement learning, that provides high-quality data selection, improving our overall detection performance. We evaluate models on three data sets, and show that they outperforms other baselines by up to 8% (F1 score). Our approach is particularly strong in the presence of class imbalances and comparable to other models when there is high diversity in the myth themes. Finally, we release a resource package to support the community to studying misinformation.

---

[5] News on Washington Posts.

Acknowledgement. This research was funded by National Science Foundation awards #1934925 and #1934494, and the Massive Data Institute (MDI) and McCourt Institute at Georgetown University. We would like to thank our funders, the MDI staff, and the Georgetown DataLab for their support.

References

1. Guo, B., Ding, Y., Yao, L., Liang, Y., Yu, Z.: The future of false information detection on social media: new perspectives and trends. ACM Comput. Surv. **53**(4), 1–36 (2020)
2. Haber, J., et al.: Lies and presidential debates: How political misinformation spread across media streams during the 2020 election. Harv. Kennedy School Misinform. Rev. (2021)
3. Helmstetter, S., Paulheim, H.: Weakly supervised learning for fake news detection on twitter. In: ASONAM (2018)
4. Hossain, T., Logan IV, R.L., Ugarte, A., Matsubara, Y., Young, S., Singh, S.: COVIDLies: detecting COVID-19 misinformation on social media. In: Workshop on NLP for COVID 2019 (Part 2) at EMNLP (2020)
5. Jin, Z., Cao, J., Guo, H., Zhang, Y., Wang, Y., Luo, J.: Detection and analysis of 2016 us presidential election related rumors on twitter. In: SBP-BRiMS (2017)
6. Kawintiranon, K., Singh, L., Budak, C.: Traditional and context-specific spam detection in low resource settings. Mach. Learn. **111**, 2515–2536 (2022)
7. Kumar, S., Shah, N.: False Information on Web and Social Media: A Survey. CRC Press, Boca Raton (2018)
8. Li, Q., Zhang, Q., Si, L., Liu, Y.: Rumor detection on social media: datasets, methods and opportunities. In: NLP4IF Workshop at EMNLP (2019)
9. Liu, Y., et al.: RoBERTa: a robustly optimized BERT pretraining approach. arXiv preprint (2019)
10. Min, E., et al.: Divide-and-conquer: Post-user interaction network for fake news detection on social media. In: WWW (2022)
11. Mosallanezhad, A., Karami, M., Shu, K., Mancenido, M.V., Liu, H.: Domain adaptive fake news detection via reinforcement learning. In: WWW (2022)
12. Nguyen, D.Q., Vu, T., Nguyen, A.T.: BERTweet: a pre-trained language model for english tweets. In: EMNLP: System Demonstrations (2020)
13. Nielsen, D.S., McConville, R.: Mumin: a large-scale multilingual multimodal fact-checked misinformation social network dataset. In: SIGIR (2022)
14. Pérez-Rosas, V., Kleinberg, B., Lefevre, A., Mihalcea, R.: Automatic detection of fake news. In: COLING (2018)
15. Reimers, N., Gurevych, I.: Sentence-BERT: sentence Embeddings using Siamese BERT-Networks. In: EMNLP (2019)
16. Singh, L., et al.: A first look at Covid-19 information and misinformation sharing on twitter. arXiv preprint (2020)
17. Singh, L., Bode, L., Budak, C., Kawintiranon, K., Padden, C., Vraga, E.: Understanding high-and low-quality URL sharing on covid-19 twitter streams. J. Comput. Social Sci. **3**(2), 343–366 (2020)
18. Sutton, R.S., Barto, A.G.: RL: An Introduction. MIT Press, London (2018)
19. Vo, N., Lee, K.: Where are the facts? searching for fact-checked information to alleviate the spread of fake news. In: EMNLP (2020)
20. Wang, Y., et al.: Event adversarial neural networks for multi-modal fake news detection. In: KDD (2018)

21. Wang, Y., et al.: Weak supervision for fake news detection via reinforcement learning. In: AAAI (2020)
22. Wu, J., Li, L., Wang, W.Y.: Reinforced co-training. In: NAACL (2018)
23. Yoon, J., Arik, S., Pfister, T.: Data valuation using reinforcement learning. In: ICML (2020)
24. Yu, F., Liu, Q., Wu, S., Wang, L., Tan, T.: Attention-based convolutional approach for misinformation identification from massive and noisy microblog posts. Comput. Secur. **83**, 106–121 (2019)
25. Zhang, T., Kishore, V., Wu, F., Weinberger, K.Q., Artzi, Y.: BERTScore: evaluating text generation with bert. In: ICLR (2020)
26. Zhou, X., Zafarani, R.: A survey of fake news: fundamental theories, detection methods, and opportunities. ACM Comput. Surv. **53**(5), 1–40 (2020)

The Burden of Being a Bridge: Analysing Subjective Well-Being of Twitter Users During the COVID-19 Pandemic

Ninghan Chen[1], Xihui Chen[2], Zhiqiang Zhong[1], and Jun Pang[1,2(✉)]

[1] Faculty of Science, Technology and Medicine, University of Luxembourg,
Esch-sur-Alzette, Luxembourg
{ninghan.chen,zhiqiang.zhong,jun.pang}@uni.lu
[2] Interdisciplinary Centre for Security, Reliability and Trust,
University of Luxembourg, Esch-sur-Alzette, Luxembourg
xihui.chen@uni.lu

Abstract. The outbreak of the COVID-19 pandemic triggers *infodemic* over online social media, which significantly impacts public health around the world, both physically and psychologically. In this paper, we study the impact of the pandemic on the mental health of influential social media users, whose sharing behaviours significantly promote the diffusion of COVID-19 related information. Specifically, we focus on subjective well-being (SWB), and analyse whether SWB changes have a relationship with their *bridging performance* in information diffusion, which measures the speed and wideness gain of information transmission due to their sharing. We accurately capture users' bridging performance by proposing a new measurement. Benefiting from deep-learning natural language processing models, we quantify social media users' SWB from their textual posts. With the data collected from Twitter for almost two years, we reveal the greater mental suffering of influential users during the COVID-19 pandemic. Through comprehensive hierarchical multiple regression analysis, we are the first to discover the strong relationship between social users' SWB and their bridging performance.

Keywords: Subjective well-being · COVID-19 · Information diffusion

1 Introduction

Since its outbreak, COVID-19 has become an unprecedented global health crisis and incited a worldwide *infodemic*. The term "infodemic" outlines the perils of misinformation during disease outbreaks mainly on social media [7,15]. Apart from accelerating virus transmission by distracting social reactions, the infodemic increases cases of psychological diseases such as anxiety, phobia and depression during the pandemic [10]. As a result, the infodemic impairs the UN's

This work is supported by Luxembourg National Research Fund via grants DRVIVEN (PRIDE17/12252781), Spsquared (PRIDE15/10621687), and HETERS (CORE/C21/IS/16281848).

sustainable development goals (SDGs), especially SDG3 which aims to promote mental health and well-being.

To combat infodemic, both governments and healthcare bodies have launched a series of social media campaigns to diffuse trustworthy information. To amplify the speed and wideness of information spread, users with a large number of followers are invited to help share messages [1,33]. Healthcare professionals and social activists also voluntarily and actively participate in relaying information they deem as useful with their social media accounts. All these people actually play a bridging role on social media delivering information to the public, although their *bridging performance* differs. We use bridging performance as an analogy to estimate how efficient and wide information can spread across social media due to the sharing of a user.

Subjective well-being (SWB), one important indicator of SDG3, evaluates individuals' cognitive (e.g., life satisfaction) and affective (i.e., positive and negative) perceptions of their lives [19]. Since the onset of the COVID-19 pandemic, the decrease of SWB has been unanimously recognised across the world. With studies for various sub-populations [12,17], many factors have been discovered correlating to SWB changes such as professions, immigration status and gender. In this paper, we concentrate on influential social media users who play the bridging role in diffusing COVID-19 information, and study the impact of the pandemic on their SWB. We further examine whether their active participation in diffusing COVID-19 information is a predictor of the SWB changes. To the best of our knowledge, we are the first to study the mental health of this specific group of people during the pandemic.

We identify two main challenges to overcome before conducting our analysis. First, there are no measurements available that can accurately quantify users' real bridging performance in diffusing COVID-19 related information. The measurements, widely used in crisis communications and online marketing, rely on social connections, and have been found insufficient in capturing users' actual bridging performance, especially in such a global health crisis [27]. For instance, although some healthcare professionals are not super tweeters with thousands of followers, their professional endorsement significantly promotes the popularity of the posts they retweeted [27]. The second challenge is the access to the SWB levels of a large number of social media users whose bridging performance is simultaneously available.

In this paper, we take advantage of the information outbreak on social media incurred by the COVID-19 pandemic and the advances of artificial intelligence to address the two challenges. For the first challenge, we propose a new bridging performance measurement based on *information cascades* [29] which abstract both information spread processes and social connections. To address the second challenge, we leverage the success of deep learning in Natural Language Processing (NLP) and estimate users' SWB by referring to the sentiments expressed in their textual posts. In spite of the inherent biases, the power of social media posts has been shown in recent studies [19] for robust extraction of well-being with supervised data-driven methods. In this paper, instead of manually constructed features, we use the state-of-the-art transformer-based text embedding to automatically learn the representative features of textual posts.

Our Contributions. We collect data from Twitter generated from *the Greater Region of Luxembourg* (GR). GR is a cross-border region centred around Luxembourg and composed of adjacent regions of Belgium, Germany and France. One important reason to select this region is its intense inter-connections of international residents from various cultures, which is unique as a global financial centre. Moreover, they well represent the first batch of countries administering COVID vaccines. Our collection spans from October 2019 to the end of 2021 for over 2 years, including 3 months before the outbreak of the COVID-19 pandemic. Our contributions are summarised as follows:

- We propose a new measurement to capture the actual bridging performance of individual users in diffusing COVID-19 related information. Compared to existing social connection-based measurements, it is directly derived from information diffusion history. Through manual analysis of the collected dataset, our measurement allows for identifying the accounts of influential health professionals and volunteers that are missed before in addition to super tweeters.
- Through deep learning-based text embedding methods, we implement a classification model which can accurately extract the sentiments expressed in social media messages. With the sentiments of posts, we quantitatively estimate individual users' SWB, and confirm the greater suffering of influential users in their SWB during the pandemic.
- Through the hierarchical multiple regression model, we reveal that users' SWB has a strong negative relationship with their bridging performance in COVID-19 information diffusion, but weak relationship with their social connections.

Our research provides policy makers with an effective method to identify influential users in the fight against infodemic. Moreover, we contribute to the realisation of SDG3 by highlighting the necessity to pay special attention to the mental well-being of people who actively participate in transmitting information in health crises like COVID-19.

2 Related Work

2.1 Measuring Bridging Performance

A considerable amount of literature has been published quantifying users' bridging performance based on social connections to identify amplifiers in social media. We can divide the measurements into two types. The first type of measurements implicitly assume that influential users are likely to hold certain topology properties on social networks such as large degrees, strong betweenness centrality or community centrality [14]. The second type of measurements assume that influential users tend to be more likely reachable from other users through random walks. PageRank [25] and its variant TwitterRank [30] among the representative benchmarks of this type of measurements. PageRank is calculated only with network structures while TwitterRank additionally takes into account topic similarities between users. All the two categories of measurements have been widely applied in practice, from public health crisis communication [23] to online marketing [21]. However, recent studies pointed out that they may not truly capture users' actual bridging performance in information diffusion during a specific public healthy crisis [23]. Although new measurements are proposed by extending existing ones with fusion indicators, their poor efficiency prevents them from being applied to real-world large-scale networks like Twitter and Facebook [18].

2.2 Subjective Well-Being Extraction

Subjective well-being is used to measure how people subjectively rate their lives both in the present and in the near future [9]. Many methods have been used to assess subjective well-being, from traditional self-reporting methods [8] to the recent ones exploiting social media [32]. Studies have cross-validated SWB extracted from social media data with the Gallup-Sharecare Well-Being Index survey,[1] a classic reference used to investigate public SWB, and found that SWB extracted from social media is a reliable indicator of SWB [19]. Twitter-based studies usually calculate SWB as the overall scores of positive or negative emotions (i.e., sentiment or valence) [19]. Sentiment analysis has developed from the original lexicon-based approaches [3] to the data-driven ones which ensure better performance [19]. We adopt the recent advances of the latter approaches, and make use of the pre-trained XLM-RoBERTa [24], a variant of RoBERTa [22], to automatically learn the linguistic representation of textual posts. As a deep learning model, RoBERTa and its variants have been shown to overwhelm traditional machine learning models in capturing the linguistic patterns of multilingual texts [2].

3   The GR-Ego Twitter Dataset

In this section, we describe how we build our Twitter dataset, referred to as *GR-ego*. In addition to its large number of active users, we have another two considerations to select Twitter as our data source. First, the geographical addresses of posters are attached with tweets and thus can be used to locate users. Second, tweet status indicates whether a tweet is retweeted. If a tweet is retweeted, the corresponding original tweet ID is provided. Together with the time stamps, we can track the diffusion process of an original tweet. Our GR-ego dataset consists of two components: (i) the social network of GR users recording their following relations; (ii) the tweets posted or retweeted by GR users during the pandemic. We follow three sequential steps to collect our GR-ego dataset. Table 1 summarises its main statistics.

---

[1] https://www.gallup.com/175196/gallup-healthways-index-methodology.aspx.

Table 1. Statistics of the GR-ego dataset.

Social network	#node	5,808,938
	#edge	12,511,698
	Average degree	2.15
Timeline tweets	#user	14,756
	#tweet before COVID	5,661,949
	#tweet during COVID	18,523,099
	#tweet per user before COVID	388.44
	#tweet per user during COVID	1255.29

Step 1. Meta Data Collection. Our purpose of this step is to collect seed users in GR who actively participate in COVID-19 discussions. Instead of directly searching by COVID-19 related keywords, we make use of a publicly available dataset of COVID-19 related tweets for the purpose of efficiency [4]. Restricted by Twitter's privacy policies, this dataset only consists of tweet IDs. We extract the tweet IDs posted between October 22nd, 2019, which is about three months before the start of the COVID-19 pandemic, and December 31st, 2021. Then with these IDS, we download the corresponding tweet content. On Twitter, geographical information, i.e., the locations of tweet posters and original users if tweets are re-tweeted, is either maintained by Twitter users, or provided directly by their positioning devices. We stick to the device-input positions, and only use user-maintained ones when such positions are unavailable. Due to the ambiguity of user-maintained positions, we leverage the geocoding APIs, Geopy and ArcGis Geocoding to regularise them into machine-parsable locations. With regularised locations, we filter the crawled tweets and only retain those from GR. In total, we obtain 128,310 tweets from 8,872 GR users.

Step 2. Social Network Construction. In this step, we search GR users from the seed users and construct the GR-ego social network. We adopt an iterative approach to gradually enrich the social network. For each seed user, we obtain his/her followers and only retain those who have a mutual following relation with the seed user, because such users are more likely to reside in GR [6]. We then extract new users' locations from their profile data and regularise them. Only users from GR are added to the social network as new nodes. New edges are added if there exist users in the network with following relationships with the newly added users. After the first round, we continue going through the newly added users by adding their mutually followed friends that do not exist in the current social network. This process continues until no new users can be added. Our collection takes 5 iterations before termination. In the end, we take the largest weakly connected component as the *GR-ego social network*.

Step 3. COVID-19 Related Timeline Tweets Crawling. In this step, we collect tweets originally posted or re-tweeted by the users in our dataset. These tweets will be used to extract users' SWB. Thus, the collected tweets are *not limited* to

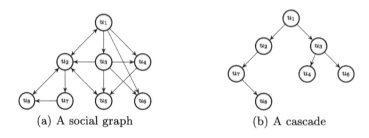

(a) A social graph   (b) A cascade

Fig. 1. Example of a cascade.

those relevant to the COVID-19 pandemic. Due to the constraints of Twitter, it is not tractable to download all the users' past tweets. We select a sufficiently large number of representative users who actively participated in retweeting COVID-19 related messages, and then crawl their history tweets. In detail, we choose 14,756 users who (re)tweeted at least three COVID-19 related messages. With the newly released Twitter API which allows for downloading 500 tweets of any given month for each user, we collect $37,281,824$ tweets spanning between October 22nd, 2019 and December 31st, 2021. This period also contains the last three months before the pandemic is officially claimed. We release the IDs of our collected tweets via Github.[2]

4 Data Processing

4.1 Cascade Computation

A cascade records the process of the diffusion of a message. It stores all activated users and the time when they are activated. In our dataset, a user is activated in diffusing a message when he/she retweets the message. In this paper, we adopt the widely accepted *cascade tree* to represent the cascade of a message [5,6,29].

The first user who posted the message is regarded as the root of the cascade tree. Users who retweeted the message, but received no further retweeting comprise the leaf nodes. Note that a tweet with the quotation to another tweet is also considered as a retweet of the quoted message. An edge from u to u' is added to the cascade if u' follows u and u' re-tweeted the message after u, indicating u activated u'. If many of the users who u' follows ever retweeted the message, meaning u' may be activated by any of them, we select the one who lastly retweeted as the parent node of u'. Figure 1(b) shows a cascade of the social network in Fig. 1(a). In this example, user u_4 can be activated by the messages retweeted by either u_1 or u_3. Since u_3 retweeted after u_1, we add the edge from u_3 to u_4 indicating that the retweeting of u_3 activated u_4.

We denote the root node of a cascade C by $r(C)$. We call a path that connects the root and a leaf node a *cascade path*, which is actually a sequence of nodes ordered by their activation time. For instance, (u_1, u_3, u_4) is a cascade path in our example indicating that the diffusion of a message started from u_1 and

---

[2] https://github.com/NinghanC/SWB4Twitter.

reached u_4 in the end through u_3. In this paper, we represent a cascade tree as a set of cascade paths. For instance, the cascade in Fig. 1(b) is represented by the following set $\{(u_1, u_2, u_7, u_8), (u_1, u_3, u_4), (u_1, u_3, u_6)\}$.

For our study, we follow the method in [20] to construct tweet cascades. Recall that when a tweet's status is '*Retweeted*', the ID number of the original tweet is also recorded. We first create a set of original tweets with all the ones labelled in our meta data as '*Original*'. Second, for each original tweet, we collect the IDs of users that have retweeted the message. At last, we generate the cascade for every original tweet based on the following relations in our GR-ego social network and their retweeting time stamps. We eliminate cascades with only two users where messages are just retweeted once. In total, 614,926 cascades are built and the average size of these cascades is 7.13.

4.2   Sentiment Analysis

Previous works [34] leverage user-provided mood (e.g., angry, excited) or status to extract users' sentiment (i.e., positive or negative) and use them to approximately estimate affective subjective well-being. However, such information is not available on Twitter. We refer to the sentiments expressed in textual posts to extract users' SWB. In this paper, we treat sentiment extraction as a tri-polarity sentiment analysis for short texts, and classify a tweet as *negative, neutral* or *positive*. In order to deal with the multilingualism of our dataset, we benefit from the advantages of deep learning in sentiment analysis [2], and build an end-to-end deep learning model to conduct the classification. Our model is composed of three components. The first component uses a pre-trained multilingual language model, i.e., XLM-RoBERTa [24], to calculate the representation of tweets. The representations are then sent to the second component, a fully-connected ReLU layer with dropout. The last component is a linear layer added on the top of the second component's outputs with sigmoid as the activation function. We use cross-entropy as the loss function and optimise it with the Adam optimiser.

Model Training and Testing. We train our model on the *SemEval-2017 Task 4A* dataset [26], which has been used for sentiment analysis on COVID-19 related messages [11]. The dataset contains 49,686 messages which are annotated with one of the three labels, i.e., positive, negative and neutral. We shuffle the dataset and take the first 80% for training and the rest 20% for testing. We assign other training parameters following the common principles in existing works. We run 10 epochs with the maximum string length as 128 and dropout ratio as 0.5. When tested with macro-average F1 score and accuracy metrics, we achieve an accuracy of 70.09% and macro-average F1 score of 71.31%.

Despite its effectiveness on classifying *SemEval-2017 Task 4A* data, in order to check whether such performance will persist on our GR-ego dataset, we construct a new testing dataset. This dataset consists of 500 messages, 100 for each of the top 5 most popular languages. We hire two annotators to manually label the selected tweets and the annotated labels are consistent between them with

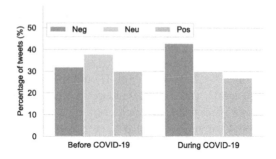

Fig. 2. Sentiment distribution of users' timeline tweets.

Cohen's Kappa coefficient $k = 0.93$. When applied on this new manually annotated dataset, our trained model achieves a similar accuracy of about 87%.

Analysing our GR-ego Dataset. Before applying our sentiment classification
model on our GR-ego dataset, we clean tweet contents by removing all URLs,
and mentioned usernames. Figure 2 summarises the statistics obtained from user
timeline tweets before and during the pandemic. The numbers of users' timeline
tweets are consistent with previous studies. For instance, users tend to become
more negative after the outbreak of the COVID-19 pandemic [12,17].

5   Bridging Performance of Users in Information Diffusion

We devote this section to addressing the first challenge regarding identifying
users that play the bridging role in transmitting COVID-19 related information.

5.1   Measuring User Bridging Performance

We evaluate users' overall performance in the diffusion of observed COVID-19
related tweets. As a user can participate in diffusing a number of tweets, we first
focus on her/his importance in the diffusion of one single tweet and then combine
her/his importance in all tweets into one single measurement. We consider a
user *more important* in diffusing a tweet when his/her retweeting behaviour
activates more users, or leads to a given number of activated users with fewer
subsequent retweets. In other words, a more important user promotes wider
acceptance of the information or accelerates its propagation. Given a cascade
path $S = (u_1, u_2, \ldots, u_n)$, we use $S^*(u_i)$ ($1 \le i < n$) to denote the subsequence
composed of the nodes after u_i (including u_i), i.e., $(u_i, u_{i+1}, \ldots, u_n)$. For any
u that does not exist in S, we have $S^*(u) = \varepsilon$ where ε represents an empty
sequence and its length $|\varepsilon| = 0$.

Definition 1 (Cascade bridging value). *Given a cascade tree C and a user u ($u \neq r(C)$), the cascade bridging value of u in C is calculated as:*

$$\alpha_C(u) = \left(\sum_{S \in C} \frac{|S^*(u)|}{|S|} \right) / |C|.$$

Note that our purpose is to evaluate the importance of users as transmitters of messages. Therefore, the concept of cascade bridging value is not applicable to the root user, i.e., the message originator.

Example 1. In Fig. 1(b), u_3 participated in two cascade paths, i.e., $S_1 = (u_1, u_3, u_4)$ and $S_2 = (u_1, u_3, u_6)$. Thus, $S_1^* = (u_3, u_4)$ and $S_2^* = (u_3, u_6)$. We then have $\alpha_C(u_3) = \frac{2/3 + 2/3}{3} \approx 0.44$.

In Definition 1, we do not simply use the proportion of users activated by a user in a cascade to evaluate her/his bridging performance. This is because it only captures the number of activated users and ignores the speed of the diffusion. Taking u_2 in Fig. 1(b) as an example, according to our definition, $\alpha_C(u_2) = 0.25$ which is smaller than $\alpha_C(u_3)$. This is due to the fact that u_2 activated two users through two retweets while u_3 only used one. However, if we only consider the proportion of activated users, the values of these two users will be the same.

With a user's bridging value calculated in each cascade, we define *user bridging magnitude* to evaluate her/his overall importance in the diffusion of a given set of observed messages. Intuitively, we first add up the bridging values of a user in all his/her participated cascades and then normalise the sum by the maximum number of cascades participated by a user. This method captures not only the bridging value of a user in each participated cascade, but also the number of cascades she/he participated in. This indicates that, a user who is more active in sharing COVID-19 related information is considered more important in information diffusion.

Definition 2 (User bridging magnitude (UBM)). *Let $\mathcal{C}$ be a set of cascades on a social network and $\mathcal{U}$ be the set of users that participate in at least one cascade in $\mathcal{C}$. A user u's user bridging magnitude (UBM) is calculated as:*

$$\omega_{\mathcal{C}}(u) = \frac{\sum_{C \in \mathcal{C}} \alpha_C(u)}{\max_{u' \in \mathcal{U}} |\{C \in \mathcal{C} | \alpha_C(u') > 0\}|}.$$

With this measurement, we can compare the UBM values of any two given users, and learn which one plays a more important role in information diffusion.

5.2   Validation of UBM

Experimental Results. We compare the effectiveness of our UBM to five widely used topology-based measurements in the literature, i.e., in-degree, PageRank [25], TwitterRank [30], betweenness centrality [14] and community centrality [14]. We randomly split the set of cascades into two subsets. The first

Table 2. Comparison of bridging performance with benchmarks.

	in-degree	PageRank	TwitterRank	Betweenness centrality	Community centrality	UBM
Avg. #activated user/minute	0.042	0.057	0.064	0.043	0.056	0.104
Avg. #activated users	13.99	16.84	17.68	15.54	17.00	23.81
%impacted user	32.17	52.54	57.44	43.44	56.54	71.66

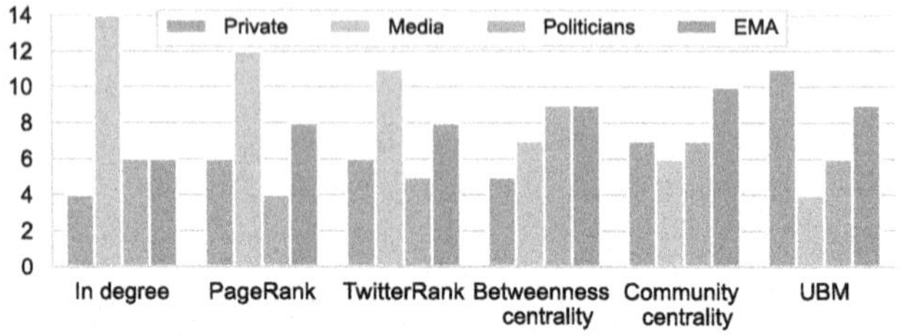

Fig. 3. Profile distribution of the top 30 accounts with highest bridging performance

subset accounts for 80% of the cascades and is used to calculate the bridging performance of all users. Then we select the top 20% users with the highest bridging performance in every adopted measurement and use the other subset to compare their actual influences in information diffusion. We adopt three measurements to quantitatively assess the effectiveness of UBM and the benchmarks. We use the *average number of activated users per minute* to evaluate the efficiency of the information diffusion. The more users activated in a minute, the faster information can be spread when it is shared by the influential users. The *average number of activated users* counts the users who received the information after the retweeting behaviour of an identified influential user. It is meant to evaluate the expected wideness of the spread once an influential user retweets a message. The *percentage of impacted users* gives the proportion of users that have ever received a message due to the sharing behaviours of identified influential users. This measurement is to compare the overall accumulated influence of all the selected influential users. We show the results of UBM and other benchmark measurements in Table 2. We can observe that it takes less time on average for the influential users identified according to UBM to activate an additional user, with 0.104 users activated a minute due to their retweets. With 23.81 users activated, UBM allows for finding the users whose retweeting action can reach more than 35% users than those identified by the benchmarks. In the end, the top 20% influential users identified by UBM spread their shared information to 71% users in our dataset, which overwhelms that of the best benchmark by about 15%. From the above analysis in terms of the three measurements, we can see that our UBM can successfully identify influential users whose sharing on social media manages to promote both the wideness and the speed of the diffusion of COVID-19 related information.

Manual Analysis. In order to understand the profiles of the calculated influential users by the measurements, we select the top 30 users with the highest
bridging performance of each measurement. We identify four types of user profiles: *private, media, politicians* and *emergency management agencies* (EMA).
Figure 3 shows the distributions of their profiles. We can observe that the distributions vary due to the different semantics of social connections captured by
the measurements. For instance, due to the large numbers of followers, Twitter
accounts managed by traditional media are favoured by in-degree. This obviously
underestimates the importance of accounts such as those of EMAs in publishing
pandemic updates. With reachability and importance in connecting users and
communities considered, more accounts of politicians and EMAs stand out. The
proportion of private accounts also starts to increase. When UBM is applied,
the percentage of private accounts becomes dominant. A closer check discovers
that 10 out of the 11 private accounts belong to health professionals and celebri-
ties. This is consistent with the literature [16] which highlights the importance
of health professionals and individuals in broadcasting useful messages about
preventive measures and healthcare suggestions in the pandemic.

6   Impact of COVID-19 on the SWB of Influential Users

6.1   Measuring SWB

We extend the definition proposed in [34] to measure the level of subjective well-
being of users based on the sentiment expressed in their past tweets. Specifically,
we extend it from bi-polarity labels, i.e., negative and positive affection, to tri-
polarity with neutral sentiment by multiplying a scaling factor to simulate the
trustworthiness of the bi-polarity SWB.

Definition 3 (Social media Subjective well-being value (SWB)). *We
use $N_p(u)$, $N_{neg}(u)$ and $N_{neu}(u)$ to denote the number of positive, negative
and neutral posts of a user u, respectively. The subjective well-being value of
u, denoted by $swb(u)$, is calculated as:*

$$\frac{N_p(u) - N_{neg}(u)}{N_p(u) + N_{neg}(u)} \cdot \left(\frac{N_p(u) + N_{neg}(u)}{N_p(u) + N_{neg}(u) + N_{neu}(u)} \right)^{\frac{1}{2}}.$$

If all messages are neutral, then $swb(u)$ is 0.

Discussion. Note that i) consistent with [34], we focus on affective SWB (i.e.,
positive and negative) in this paper, while ignoring its cognitive dimension;
ii) users' SWB is evaluated based on their original messages: originally posted
tweets and quotations; iii) for tweets with quotations to other messages, only the
texts are considered without the quoted messages. As retweets may not explicitly
include users' subjective opinions, we exclude them from the SWB calculation.

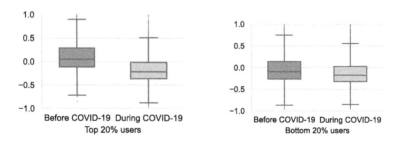

Fig. 4. SWB changes after the outbreak of the pandemic.

6.2 Analysing SWB Changes of Influential Users

With the proposed SWB measurement, we study how users' subjective well-being changes due to the outbreak of the COVID-19 pandemic. We calculate the UBM values of the users in our collected dataset and order them descendingly. Then we select the top 20% users as well as the bottom 20% users and compare the two groups' responses to the pandemic. For each group, we calculate users' SWBs according to their posts before the pandemic and after the pandemic to capture the changes. Note that we only consider the users with more than 5 posts in each time period. In Fig. 4, we show the SWB distributions of the two user groups. On average, the users with high UBM have positive SWB of 0.11 before the pandemic while the users with low UBM are negative. *The SWB of both user groups decreases after the pandemic but the SWB of the top 20% users drops more significantly.* Specifically, their SWB falls by 0.33, which is two times as much as that of the bottom 20% users. The lowest value of the top 20% users' SWB slightly decreases after the pandemic, while the lowest value of the bottom 20% of users does not change significantly. Note that the minimum values here do not include outliers that lie outside the box whiskers. This indicates that the top 20% users become even more negative than the bottom 20% users, in terms of mean and minimum values. To sum up, the pandemic causes more negative mental impacts on the social media users who play a more important bridging role in transmitting COVID-19 related information.

6.3 Relation Between SWB and Bridging Performance

We conduct the first attempt to study if a user's bridging performance has a relationship with the SWB changes of the users actively participating the diffusion of COVID-19 related information. In addition to UBM and the five benchmark measurements used in Sect. 5.2, we consider two additional variables: *out-degree* and *activity*. Out-degree is used to check whether the number of accounts a user follows correlates with SWB changes. The activity variable evaluates how active a user is engaged in the online discourse and is quantified by the number of messages he/she posted. In order to isolate the impacts of these variables, we adopt the method of *hierarchical multiple regression* [28]. The intuitive idea is to check whether the variables of interest can explain the SWB variance after accounting for some variables.

To check the validity of applying hierarchical multiple regression, we conduct first-line tests to ensure a sufficiently large sample size and independence between variables. We identify the variables corresponding to community centrality and TwitterRank fail to satisfy the multi-collinearity requirement. We thus ignore them in our analysis. The ratio of the number of variables to the sample size is 1:2108, which is well below the requirement of 1:15 [28]. This indicates the sample size is adequate. We iteratively input the variables into the model with three stages. The results are shown in Table 3. In the first stage, we input the variables related to network structures, i.e., in-degree, out-degree, Pagerank and Betweenness centrality. The combination of the variables can explain 4.30% of the SWB variance ($F = 4.379, p < 0.05$). Note that an F-value of greater than 4 indicates the linear equation can explain the relation between SWB and the variables. This demonstrates that there exists a positive relationship between the topology-based variables and SWB, but this relationship is rather weak. A closer check on the t-values show that out-degree is irrelevant to SWB and the rest three variables are weakly related. In the second stage, we add the variable of activity to the model. After controlling all the variables of the first stage, we

Table 3. Hierarchical multiple regression model examining variance in SWB explained by independent variables, $*p < 0.05$; $* * p < 0.001$

Variable	B	SEB	b	t	R	R^2	ΔR^2
Stage 1					−0.207	0.043	0.043
In-degree	0.234	0.103	0.160	2.272*			
Out-degree	0.861	0.680	0.054	1.267			
Pagerank	3.081	0.148	0.180	2.082*			
Betweenness centrality	−3.287	0.728	−1.453	−4.515**			
Stage 2					−0.312	0.097	0.054
In-degree	0.228	0.102	0.158	2.239*			
Out-degree	0.075	0.080	0.050	0.945			
Pagerank	0.307	0.150	0.180	2.049*			
Betweenness centrality	−3.268	0.723	−1.390	−4.520**			
Activity	0.861	0.123	0.037	0.716			
Stage 3					−0.579	0.335	**0.238**
In-degree	0.158	0.123	0.107	1.125*			
Out-degree	0.516	0.45	0.050	1.147			
Pagerank	0.191	0.143	0.168	1.338*			
Betweenness centrality	−1.105	0.541	−0.509	−2.066**			
Activity	0.067	0.133	0.053	0.508			
UBM	−2.254	−0.196	−1.797	**−11.469****			

observe that user activity does not significantly contribute to the model with t-value of 0.716. This suggests that user activity is not a predictor of SWB. In the third stage, we introduce UBM to the model. The addition of UBM, with the variables in the previous two stages controlled, reduces the R value from –0.219 to –0.579. UBM contributes significantly to the overall model with $F = 147.82$ ($p < 0.001$) and increases the predicted SWB variance by 23.8%. Together with the t-value of –11.469 ($p < 0.001$), we can see there exists a strong negative relation between UBM and SWB, and UBM is a strong predictor for SWB.

Discussion. The results illustrate that UBM is strongly related to SWB, while in-degree, Pagerank and betweenness centrality are weakly related. This difference further shows that UBM can more accurately capture users' behaviour changes after the outbreak of the pandemic while topology features remain similar to those before the pandemic. This may be explained by the recent studies [17] that once considered as a change in life after the pandemic outbreak, this extra bridging responsibility in diffusing COVID-19 related messages is likely to associate with lower life satisfaction.

7   Conclusion and Limitation

In this paper, we concentrated on the social media users whose sharing behaviours significantly promote the popularity of COVID-19 related messages. By proposing a new measurement for bridging performance, we identified these influential users. With our collected Twitter data of an international region, we successfully show the influential users suffer from more decrease in their subjective well-being compared to those with smaller bridging performance. We then conducted the first research to reveal the strong relationship between a user's bridging performance in COVID-19 information diffusion and his/her SWB. Our research provides a cautious reference to public health bodies that some users can be mobilised to help spread health information, but special attention should be paid to their psychological health.

This paper has a few limitations that deserve further discussion. First, we only focused on the affective dimension of subjective well-being while noticing its multi-dimensional nature. This allows us to follow previous SWB studies to convert the calculation of SWB to sentiment analysis, but does not comprehensively evaluate users' cognitive well-being, such as life satisfaction. In our following research, we will attempt to leverage more advanced AI models to investigate cognitive aspects such as *happy* and *angry*. Second, extracting SWB from users' online disclosure inevitably incurs bias compared to social surveys although it supports analysis of an unprecedented large number of users. Third, socio-demographic information of users is not taken into account in this paper. It is known that SWB varies among different socio-demographic groups, and such variation may have an impact on the results of the hierarchical multiple regression [19]. Currently deep learning based models exist for socio-demographic inference. In our future work, we will use the models to extract users' socio-demographic information such as age, gender, income and political orientation to ascertain whether the regression results will change due to the variations of socio-demographic information. Last, we notice that the region we targeted at may introduce additional bias in our results. As a continuous work, we will extend our study to a region of multiple European countries and cross-validate our findings with other published results in social science.

Ethical Considerations. This work is based completely on public data and does not contain private information of individuals. Our dataset is built in accordance with the FAIR data principles [31] and Twitter Developer Agreement and Policy and related policies. Meanwhile, there have been a significant amount of studies on measuring users' subjective well-being through social media data. It has become a consensus that following the terms of service of social media networks is adequate to respect users' privacy in research [13]. To conclude, we have no ethical violation in the collection and interpretation of data in our study.

References

1. Banerjee, D., Meena, K.S.: COVID-19 as an "Infodemic" in public health: Critical role of the social media. Front. Public Health **9**, 231–238 (2021)
2. Barbieri, F., Camacho-Collados, J., Espinosa Anke, L., Neves, L.: TweetEval: unified benchmark and comparative evaluation for tweet classification. In: Proceedings of 2020 Conference on Empirical Methods in Natural Language Processing (EMNLP), pp. 1644–1650. Association for Computational Linguistics (2020)
3. Bradley, M.M., Lang, P.J.: Affective norms for English words (ANEW): Instruction manual and affective ratings. Tech. rep., the Centre for Research in Psychophysiology, University of Florida (1999)
4. Chen, E., Lerman, K., Ferrara, E.: Tracking social media discourse about the COVID-19 pandemic: development of a public coronavirus Twitter data set. JMIR Public Health Surveill. **6**(2), e19273 (2020)
5. Chen, N., Chen, X., Zhong, Z., Pang, J.: From #jobsearch to #mask: improving COVID-19 cascade prediction with spillover effects. In: Proceedings of 2021 International Conference on Advances in Social Networks Analysis and Mining(ASONAM), pp. 455–462. ACM (2021)
6. Chen, N., Chen, X., Zhong, Z., Pang, J.: Exploring spillover effects for COVID-19 cascade prediction. Entropy **24**(2) (2022)
7. Cinelli, M., et al.: The COVID-19 social media infodemic. Sci. Rep. **10**(1), 1–10 (2020)
8. Diener, E., Emmons, R.A., Larsen, R.J., Griffin, S.: The satisfaction with life scale. J. Pers. Assess. **49**(1), 71–75 (1985)
9. Diener, E., Oishi, S., Lucas, R.E.: Personality, culture, and subjective well-being: Emotional and cognitive evaluations of life. Annu. Rev. Psychol. **54**(1), 403–425 (2003)
10. Dubey, S., et al.: Psychosocial impact of COVID-19. Diab. Metabol. Synd. Clin. Res. Rev. **14**(5), 779–788 (2020)
11. Duong, V., Luo, J., Pham, P., Yang, T., Wang, Y.: The ivory tower lost: how college students respond differently than the general public to the COVID-19 pandemic. In: Proceedings 2020 IEEE/ACM International Conference on Advances in Social Networks Analysis and Mining (ASONAM), pp. 126–130. IEEE (2020)

12. Engel de Abreu, P.M., Neumann, S., Wealer, C., Abreu, N., Coutinho Macedo, E., Kirsch, C.: Subjective well-being of adolescents in Luxembourg, Germany, and Brazil during the COVID-19 pandemic. J. Adolesc. Health **69**(2), 211–218 (2021)

13. Fernando, S., López, J.A.D., Şerban, O., Gómez-Romero, J., Molina-Solana, M., Guo, Y.: Towards a large-scale Twitter observatory for political events. Futur. Gener. Comput. Syst. **110**, 976–983 (2020)

14. Freeman, L.C.: Centrality in social networks conceptual clarification. Soc. Netw. **1**(3), 215–239 (1978)

15. Guarino, S., Pierri, F., Giovanni, M.D., Celestini, A.: Information disorders during the COVID-19 infodemic: the case of Italian Facebook. Online Soc. Netw. Media **22**, 100124 (2021)

16. Hernandez, R.G., Hagen, L., Walker, K., O'Leary, H., Lengacher, C.: The COVID-19 vaccine social media infodemic: healthcare providers' missed dose in addressing misinformation and vaccine hesitancy. Hum. Vacc. Immunother. **17**(9), 2962–2964 (2021)

17. Hu, Z., Lin, X., Kaminga, A.C., Xu, H.: Impact of the COVID-19 epidemic on lifestyle behaviors and their association with subjective well-being among the general population in mainland China: Cross-sectional study. J. Med. Internet Res. **22**(8), e21176 (2020)

18. Huang, S., Lv, T., Zhang, X., Yang, Y., Zheng, W., Wen, C.: Identifying node role in social network based on multiple indicators. PLoS ONE **9**(8), e103733 (2014)

19. Jaidka, K., Giorgi, S., Schwartz, H.A., Kern, M.L., Ungar, L.H., Eichstaedt, J.C.: Estimating geographic subjective well-being from Twitter: a comparison of dictionary and data-driven language methods. Proc. Natl. Acad. Sci. **117**(19), 10165–10171 (2020)

20. Kupavskii, A., et al.: Prediction of retweet cascade size over time. In: Proc. 2012 International Conference on Information and Knowledge Management (CIKM), pp. 2335–2338. ACM (2012)

21. Li, Y.M., Lai, C.Y., Chen, C.W.: Discovering influencers for marketing in the blogosphere. Inf. Sci. **181**(23), 5143–5157 (2011)

22. Liu, Y., et al.: RoBERTa: A robustly optimized BERT pretraining approach. In: ICLR (2019)

23. Mirbabaie, M., Bunker, D., Stieglitz, S., Marx, J., Ehnis, C.: Social media in times of crisis: learning from Hurricane Harvey for the coronavirus disease 2019 pandemic response. J. Inf. Technol. **35**(3), 195–213 (2020)

24. Ou, X., Li, H.: Ynu_oxz @ haspeede 2 and AMI : XLM-RoBERTa with ordered neurons LSTM for classification task at EVALITA 2020. In: Proceedings of 2020 Evaluation Campaign of Natural Language Processing and Speech Tools for Italian (EVALITA), vol. 2765 (2020)

25. Page, L., Brin, S., Motwani, R., Winograd, T.: The PageRank citation ranking: Bringing order to the web. Tech. rep, Stanford InfoLab (1999)

26. Rosenthal, S., Farra, N., Nakov, P.: Semeval-2017 task 4: Sentiment analysis in Twitter. In: Proceedings of 2017 International Workshop on Semantic Evaluation (SemEval), pp. 502–518 (2017)

27. Struweg, I.: A twitter social network analysis: the South African health insurance bill case. In: Hattingh, M., Matthee, M., Smuts, H., Pappas, I., Dwivedi, Y.K., Mäntymäki, M. (eds.) I3E 2020. LNCS, vol. 12067, pp. 120–132. Springer, Cham (2020). https://doi.org/10.1007/978-3-030-45002-1_11

28. Tabachnick, B.G., Fidell, L.S., Ullman, J.B.: Using Multivariate Statistics. Pearson Education (2007)

29. Wang, Y., Shen, H., Liu, S., Gao, J., Cheng, X.: Cascade dynamics modeling with attention-based recurrent neural network. In: Proceedings of 2017 International Joint Conference on Artificial Intelligence (IJCAI), pp. 2985–2991. IJCAI (2017)

30. Weng, J., Lim, E.P., Jiang, J., He, Q.: Twitterrank: finding topic-sensitive influential twitters. In: Proceedings of 2010 ACM International Conference on Web Search and Data Mining (WSDM), pp. 261–270 (2010)

31. Wilkinson, M.D., et al.: The fair guiding principles for scientific data management and stewardship. Sci. Data **3**(1), 1–9 (2016)

32. Yang, C., Srinivasan, P.: Life satisfaction and the pursuit of happiness on Twitter. PLoS ONE **11**(3), e0150881 (2016)

33. Zarocostas, J.: How to fight an infodemic. Lancet **395**(10225), 676 (2020)

34. Zhou, X., Jin, S., Zafarani, R.: Sentiment paradoxes in social networks: why your friends are more positive than you? In: Proceedings of 2020 International Conference on Web and Social Media (ICWSM), pp. 798–807. AAAI Press (2020)

SkipCas: Information Diffusion Prediction Model Based on Skip-Gram

Dedong Ren and Yong Liu[✉]

School of Computer Science and Technology, Heilongjiang University, Harbin, China
2201840@s.hlju.edu.cn, liuyong123456@hlju.edu.cn

Abstract. The development of social network platforms such as Twitter and Weibo has accelerated the generation and transmission of information. Predicting the growth size of the information cascade is widely used in the fields of preventing rumor spread, viral marketing, recommendation system and so on. However, most of the existing methods either cannot fully capture the structural representation of the cascade graph, or cannot effectively utilize the dynamic changes of information diffusion, which often leads to poor prediction results. Therefore, in this paper, we propose a novel deep learning model called SkipCas to predict the growth size of the information cascade. First, we use the diffusion path and time effect at each diffusion time in the cascade graph to obtain the dynamic process of the information diffusion. Second, we put the sequence of biased random walk sampling into the skip-gram model to obtain the structural representation of the cascade graph. Finally, we combine the dynamic diffusion process and the structural representation to predict the growth size of the information cascade. Extensive experiments on two real datasets show that our model SkipCas significantly improves the prediction accuracy compared with the state-of-the-art models.

Keywords: Information cascade · Cascade size prediction · Structural information · Random walk

1 Introduction

Online social networking platforms such as Twitter, Weibo and Facebook have become the main sources of information in people's daily life. Being able to accurately predict the size of information diffusion after a certain period has attracted widespread attention in the academic community, which plays a critical role in suppressing rumors information diffusion, improving content recommendation and other many down-stream applications [1,2].

Many approaches have been proposed for predicting information diffusion. It mainly falls into three categories: 1) Feature-based approaches: They mainly focus on identifying and incorporating hand-crafted features for cascade prediction, such as temporal features [3,4], structural features [5,6], and content features [7,8], etc. Their performance depends on extracted features, which are difficult to generalize to new domains. 2) Generative approaches: The popularity

M.-R. Amini et al. (Eds.): ECML PKDD 2022, LNAI 13714, pp. 258–273, 2023.
https://doi.org/10.1007/978-3-031-26390-3_16

of information cascades over time is considered as a dynamic time series fitting problem [9], leading to the development of certain macroscopic distributions or stochastic processes based on various strong assumptions. These approaches rely heavily on the designed self-excited mechanisms and intensity functions [10,11]. This usually has a huge gap with the real world, resulting in poor predictive power. 3) Deep learning-based approaches: In recent years, researchers leverage various deep learning techniques to capture the temporal and sequential processes of information diffusion. For example, DeepCas [12], Topo-LSTM [13], and DeepCon+Str [14] model the network topology for information diffusion prediction; DeepHawkes [15] and RNN-based CRPP [16] model the temporal information for information diffusion prediction.

Despite obvious improvements in modeling cascade diffusion, existing deep learning methods still face several key challenges: 1) The dynamics of information diffusion are not effectively utilized in existing methods. 2) The structural representation of the cascade network are critical for accurately predicting information cascades. However, most methods fail to fully obtain the structural representation, resulting in unsatisfactory prediction results.

To address the above challenges, we propose a novel information cascade prediction model called SkipCas, which attempts to capture the dynamic diffusion process of the information cascade and obtain the structural representation of the cascade network. To capture the dynamic diffusion process, we put the diffusion path at each diffusion time in the cascade graph into GRU to obtain path representations, weight path representations with diffusion time, and then pool all path representations. To obtain the structural representation of the cascade network, we represent the cascade graph as a set of biased random walk paths and fed them into the skip-gram model to obtain node representations, and then pool all node representations. Finally, we integrate the dynamic diffusion process with the structural representation to predict the growth size of the information cascade. Our main contributions can be summarized as follows:

1) We propose a novel deep learning model called SkipCas for information growth size prediction.
2) We encode the diffusion path at each diffusion time in the cascade graph, which can well preserve the dynamic diffusion process of information diffusion.
3) We leverage the skip-gram model to capture the network structure and obtain the structural representation of the cascade graph.
4) Extensive experiments on several real-world cascade datasets show that SkipCas can significantly improve the cascade size prediction performance compared with the state-of-the-art approaches.

2   Related Works

2.1   Cascades Prediction

The existing methods on information cascade prediction fall into the following three categories:

Feature-based approaches extract various hand-crafted features from the original data, usually including information temporal features [3,4], cascade structural features [5,6], content features [7,8] and user features [17], and then predict its popularity through various machine learning models. However, their performance relies heavily on the relevant features extracted by hand, and may not be directly applied when they are not in a specific environment, thus the feature-based approaches are not easy to generalize.

Generative approaches typically treat the growing size of the information cascade as a cumulative stochastic process [18], modeling it as a parametric model and then estimating the parameters for each event by maximizing the probability of the event occurring at the observed time. [19] divided the observed popularity into multiple stages at equal-sized time intervals, modeled them using multiple linear regression and auto-regression, respectively. In addition to the simple regression-based model, they also used different point processes, such as Poisson [20,21] and Hawkes processes [10,22]. However, as mentioned in [1], the Poisson process is too simple to capture the diffusion patterns, and Hawkes usually overestimates their popularity, probably due to their underlying self-excitation mechanism. In contrast, SkipCas enables incorporates both structural and temporal information.

Deep learning-based approaches are inspired by deep neural networks and have achieved significant performance improvements in many applications. Deep-Cas [12] is the first deep learning-based information cascade prediction model, which learns the representation of cascade graphs in an end-to-end manner. DeepHawkes [15] inherits the high interpretability of the Hawkes process and has the high predictive ability of deep learning methods. CasCN [23] samples the cascade graph as cascade subgraphs and employs a dynamic multi-directional convolutional network to learn the structural information of the cascade graph. VaCas [24] extends the deterministic cascade embedding with random node representation and diffuse uncertainty, enabling more robust cascade prediction. In addition, methods such as CYAN-RNN [25], Topo-LSTM [13], and SNIDSA [26] extract the full path of diffusion from sequential observations of information infections, using recurrent neural networks and attention mechanisms to model information growth and predict diffusion size. However, they lack better learning ability in cascading structural information and dynamics modeling, due to the bias of sampling methods and the inefficiency of local structure embedding.

2.2   Graph Representation

Learning node embeddings in graphs aims to learn low-dimensional latent representations of nodes in the networks, and the learned feature representations can be used as features for various graph-based tasks, such as classification, clustering, link prediction, and visualization. Word2vec [27] is an unsupervised learning technique that given a word can guess its surrounding context. Inspired by it, the DeepWalk [28] algorithm first introduced a word vector training model to the network. To capture the diversity of network structures, node2vec [29] generated biased second-order random walk, rather than uniform ones. In addition, inspired by Convolutional Neural Networks, GCN [30] has also been developed to learn representations of nodes in graphs from neighboring node representations, such as GraphSage [31] and DiffPool [32]. We fuse dynamic diffusion processes to predict the cascade growth size based on the skip-gram model.

3 Preliminaries

In this section, we will formally define the cascade prediction problem.

Definition 1. Social Graph. Given a snapshot of a social network graph $G = (V, E)$, where V is the set of vertices of the social graph and $E \subset V \times V$ is the set of edges. A vertex can be a user of a social platform or a paper in the network of academic papers, and an edge represents the relationship between two nodes, such as retweeting or citing.

Definition 2. Cascade Graph. Suppose there are M messages in the social network, for the i-th message we use the cascade graph C_i to represent. Each cascade graph C_i corresponds to an evolution sequence, we use the cascade $g_i(t_j) = \left\{ V_i^{t_j}, E_i^{t_j}, t_j \right\}$ to represent the diffusion process of the cascade graph C_i within time t_j, where $V_i^{t_j}$ denotes the users participating in the cascade within time t_j, $E_i^{t_j}$ denotes the feedback relationship between users in $V_i^{t_j}$ (e.g., retweeting or citation), t_j is the time between retweets of the original post. The diffusion process of the cascade graph is shown in Fig. 1, i.e., $g_i(t_0) = \{\{A\}, \{\oslash\}, t_0\}$, $g_i(t_1) = \{\{A, B\}, \{(A, B)\}, t_1\}$, ... , and so on.

Definition 3. Growth Size. In this paper, the growth size of the cascade is defined as the number of retweets or citations of a message or paper. Specifically, given a cascade C_i, within the observation time window T, our research task is to predict the growth size $\triangle S_i$ of C_i at the fixed time interval $\triangle t$, e.g., $\triangle S_i = |V_i^{T+\triangle t}| - |V_i^T|$.

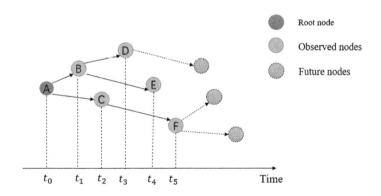

Fig. 1. Diffusion cascade graph of a certain message.

4   Model

The framework of our proposed SkipCas model takes the cascade as input and predicts the growth size $\triangle S_i$ of the cascade graph C_i as output. The model is shown in Fig. 2. SkipCas consists of four main components: 1) Diffusion path coding: the diffusion paths are coded by recurrent neural networks according to the observed cascade diffusion order; 2) Time effect: the encoded diffusion paths combine with temporal effects to further extract the cascade representation; 3) Structural modeling: the sequence of random walk sampling is used to obtain the structural representation of the cascade graph through the skip-gram; 4) Prediction: the cascaded representation with time effect and the structural representation are fed into the multilayer perceptron for cascade size prediction.

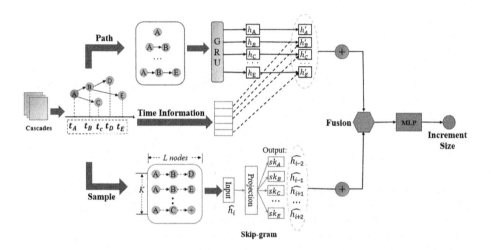

Fig. 2. Framework of SkipCas model.

4.1   Diffusion Path Encoding

Users participating in cascading diffusion will not only be affected by users who have just occurred retweeting behavior, but also by previous users; similarly, previous participants will also influence their direct retweeters and indirect retweeters. As shown in Fig. 1, user A published a message, user B retweeted the message from user A, and D retweeted the message of user B, then the retweet path of this message is A → B → D, user A still has influence on the delivery of the message. This illustrates that each user in the cascade may have an impact on the whole information transfer that follows it. Therefore, we encode the entire cascaded diffusion path.

We use the Gated Recursive Unit (GRU) to encode the entire diffusion path. Specifically, each user in the diffusion path is first represented by a one-hot vector, and then according to the order of the diffusion path, the k-th in the diffusion path, denoted as $x_k \in R^d$, is fed to the GRU unit. The hidden state $h_k = GRU(x_k, h_{k-1})$ is updated after the update operation on it, where the output $h_k \in R^H$, the input $x_k \in R^d$, h_{k-1} represents the hidden state before the update, d is the dimension size of the user, and H is the dimension size of the hidden state. The update formula of GRU is as follows:

The reset gate $r_k \in R^H$ is calculated by

$$r_k = \sigma(W_r x_k + U_r h_{k-1} + b_r). \tag{1}$$

The update gate $z_k \in R^H$ is calculated by

$$z_k = \sigma(W_z x_k + U_z h_{k-1} + b_z). \tag{2}$$

The actual activation of hidden state h_k is calculated by

$$h_k = z_k \cdot h_{k-1} + (1 - z_k) \cdot tanh(W_h x_k + U_h h_{k-1} + b_h), \tag{3}$$

where $\sigma(\cdot)$ is the sigmoid activation function, $W_r \in R^{H \times d}$, $W_z \in R^{H \times d}$, $W_h \in R^{H \times d}$, $U_r \in R^{H \times H}$, $U_h \in R^{H \times H}$, $U_z \in R^{H \times H}$ and $b_r \in R^H$, $b_z \in R^H$, $b_h \in R^H$ are independent trainable parameters.

4.2 Time Effect

The time effect is a common phenomenon of cascading information diffusion and plays an important role in cascading prediction. For example, a post on Weibo is usually frequently retweeted in the first period after it is published, and the number of retweets decreases with time.

Suppose a cascade C_i whose duration after generation is t, then it is easy to know how long the time interval between its generation and each retweet or citation. Then we can get the time interval of each user's retweet in the cascade graph, e.g., $\left\{ t'_v = t^r_v - t_0 \mid 0 \leq t'_v \leq t, v \in V_i^{t_j} \right\}$, where t^r_v is the time when user v retweets the message, and t_0 is the original posting time of the post.

In order to learn the effect of time on the cascade, we employ the following time decay effect. Supposing the time window of the observed cascade is $[0, T]$, we divide the time window into l equal-sized time intervals as $\{[t_0, t_1), ..., (t_{l-1}, t_l]\}$, where $t_0 = 0$, $t_l = T$. It can assign a corresponding interval to each diffusion time, thus we can compute the corresponding time interval β of the time decay effect for a retweet at time t:

$$\beta = \lfloor \frac{t'_v}{T/l} \rfloor \quad (0 \leq t'_v \leq t). \tag{4}$$

The function of the time decay effect is:

$$\lambda_\beta = \frac{1}{1 + \frac{t'_v}{t_0}}. \tag{5}$$

Then we add the time decay effect to the obtained cascaded hidden state h_t, and further obtain

$$h'_t = \lambda_\beta h_t. \tag{6}$$

Summation to obtain the representation vector for the cascade C_i:

$$h'(C_i) = \sum_{t=1}^{T} h'_t. \tag{7}$$

4.3   Structural Modeling

The future size of the cascade depends heavily on who is the information "propagator", i.e., the nodes in the current cascade graph. Therefore, a straightforward way to represent a graph is to treat it as a bag of nodes. However, this approach ignores the structural information in the cascade graph, which is important in predicting diffusion. The biased random walk considers the breadth-first and depth-first sampling strategies, which can better capture the structural information of the cascade graph. Therefore, we represent the cascade graph C_i as a set of cascade paths sampled through multiple biased random walk processes. For each random walk process, we first sample the starting node with the following probability:

$$p(u) = \frac{deg_{C_i}(u) + \alpha}{\sum_{u \in V_{C_i}} (deg_G(u) + \alpha)}, \tag{8}$$

where α is the smoother, deg_{C_i} is the out-degree of node u in cascade C_i, and $deg_G(u)$ is the degree of u in the global graph G, V_{C_i} is the set of nodes in cascaded C_i. Then, after the starting node, the neighboring nodes are sampled with the following probability:

$$p(u \in N_{C_i}(v) \mid v) = \frac{deg_{C_i}(u) + \alpha}{\sum_{u \in N_{C_i}(v)} (deg_G(u) + \alpha)}, \tag{9}$$

where $N_{C_i}(v)$ represents the set of neighbors of v in the cascade graph C_i.

The number and length of random walk sampling sequences play a key role in determining the representation of the cascade graph. Therefore, in order to better perform the sampling process, we set two parameters L and K, where K represents the number of sequences sampled, and L represents the length of each sequence. We fix L and K as constants, the specific settings will be explained in the next section of the experiment. Sampling of a sequence stops when we reach a predefined length L or when we reach a node without any outgoing neighbors. If the length of the one sequence is less than L, the sequence is filled with a special node '+'. This process of sampling sequences continues until K sequences are sampled.

The skip-gram model was originally proposed in [28] and has been applied to deal with word representations in natural language. It aims to classify as many words as possible based on another word in the same sentence. Specifically, the representation of each given word is the input, and the model uses logistic regression to predict the words within a certain distance before and after the input word in the sentence. Similarly, we use the sequence of nodes obtained by random walk as input, and after the logarithmic function mapping of the projection layer, we get the embedding vector of each node. Suppose $N_{C_i}(v)$ is the neighborhood list of node v generated by the neighborhood sampling strategy, and the embedding representation is denoted as $\hat{H} = \left\{\hat{h}_1, \hat{h}_2, ..., \hat{h}_n\right\}$, where n is the number of nodes. The following objective can be optimized by the skip-gram model to maximize the log-probability of the node neighborhood $N_{C_i}(v)$ for all $v \in V_{C_i}$ as follows:

$$\max_{\hat{H}} \prod_{v \in V_{C_i}} P(N_{C_i}(v) \mid \hat{h}_v). \tag{10}$$

According to the conditional independence assumption, we get:

$$P(N_{C_i}(v) \mid \hat{h}_v) = \prod_{p \in N_{C_i}(v)} P(\hat{h}_p \mid \hat{h}_v). \tag{11}$$

According to the feature space symmetry assumption in Node2vec. We assume that the source node and the neighbor nodes have symmetric effects with each other in the embedding space, the conditional likelihood function for each source-neighbor node pair can be modeled using a softmax function parameterized by the dot product of its features:

$$P(\hat{h}_q \mid \hat{h}_v) = \frac{exp(\hat{h}_p \cdot \hat{h}_v)}{\sum_{q \in V_{C_i}} exp(\hat{h}_q \cdot \hat{h}_v)}. \tag{12}$$

With the above assumptions, the final objective function can be simplified to:

$$\min_{\hat{H}} S^{loss} = \sum_{v \in V_{C_i}} (\log \sum_{q \in V_{C_i}} exp(\hat{h}_q \cdot \hat{h}_v) - \sum_{p \in N_{C_i}(v)} (\hat{h}_p \cdot \hat{h}_v)). \tag{13}$$

4.4   Prediction

We integrate the minimization of the squared loss between the predicted growth size and the ground truth, where a multilayer perceptron is used as the prediction, the formula is as follows:

$$\min_{\theta} O^{loss} = \sum_{i=1}^{M} (\log \Delta S_i - \log \Delta \tilde{S}_i)^2. \tag{14}$$

$$\Delta S_i = MLP(h'(C_i) \oplus \sum_{v \in V_{C_i}} \hat{h}_v). \tag{15}$$

where θ denotes the trainable parameters of the MLP, ΔS_i denotes the predicted growth size for cascade C_i, and $\Delta \tilde{S}_i$ denotes the ground truth.

5   Experiments

In this section, we describe the details of the experiments performed on real-world datasets and the analysis of the results between our proposed model and baseline methods.

5.1   Datasets

We evaluate the effectiveness of the proposed model in two information cascade prediction scenarios and compare it with previous work using publicly available datasets, i.e., Weibo and APS. The statistics of the dataset are shown in Table 1.

Sina Weibo is a public dataset provided by [15], where each tweet and its retweets can form a retweet cascade. We follow a similar experimental setup to [15] with observation time windows of length $T = 1\,h$, $2\,h$ and $3\,h$. Due to the effect of circadian rhythms, we focus on tweets posted between 8 am and 6 pm. We randomly select 70% for training, 15% for validation, and the remaining 15% for testing.

American Physical Society (APS) [20] contains scientific papers published by APS journals. Each paper and its citations in the APS dataset form a citation cascade, and the growth size of the cascade is the number of citations. We only use papers published between 1893 and 1989, so that each paper has at least 20 years to develop its cascade. For the length T of the observation time window, we choose T = 5 years, 7 years and 9 years. Similarly, the first 70% of the data is used for training, 15% for validation, and 15% for testing.

Table 1. Statistics of datasets

	Dataset	Weibo			APS		
Number of Cascades	All	119,311			207,685		
Number of Nodes	All	325,380			616,014		
Number of Edges	All	8,466,858			4,710,547		
T		1 h	2 h	3 h	5 years	7 years	9 years
Cascades	Trian	25,515	29,515	31,780	16,299	21,171	24,658
	val	5,386	6,324	6,810	3,582	4,507	5,254
	Test	5,386	6,324	6,810	3,475	4,589	5,279

5.2   Baselines

We compare the proposed model with some state-of-the-art cascade prediction methods, including:

Feature-Based: Recent studies have shown that structural features, temporal features, and other features (e.g., content features) are useful for information cascade prediction. We select several features commonly used in cascade graphs (e.g., the number of nodes, the number of edges, average degree, edge density) and predicted the size of the cascade through Feature-linear and Feature-Deep.

Node2vec [29]: It is the representative of node embedding methods. We perform random walks on the cascade graph and generate an embedding vector for each node. Then the embeddings of all nodes in the cascade graph are fed into the MLP for prediction.

DeepCas [12]: The first deep learning architecture for information cascade prediction, which represents the cascade graph as a set of random walk paths via random walks, and uses GRU and attention mechanism to model and predict cascade sizes in an end-to-end manner.

Topo-LSTM [13]: It uses a directed acyclic graph as the diffusion topology, the LSTM is used to model the relationship between nodes in the graph. The hidden state and cell of each node at a given time depends on the hidden state and cell of each previous node that was infected before that time instant.

DeepHawkes [15]: It integrates the high predictive power of deep learning into the interpretable factors of the Hawkes process for cascading size prediction. Bridging the gap between predicting and understanding information cascades.

CasCN [23]: It samples the cascade graph as a sequence of sub-cascade graphs, learns the local structure of each sub-cascade by graph convolution, and then captures the evolution of the cascade structure using LSTM.

DeepCon+Str [14]: It learns the embeddings of the cascade as a whole. It first constructs higher-order graphs based on content and structural similarity to learn the low-dimensional representation of each cascade graph, and then makes cascade predictions through a semi-supervised language model.

5.3 Experimental Settings

The models mentioned above involve several hyper-parameters. For example, the L2 coefficient in Feature-linear is chosen to be 0.05. For Feature-deep, the parameters are similar to deep learning-based approaches. For the sampling sequence of the cascade graph, we set $K = 200$ paths and the length of each path $L = 10$. For Node2vec, we follow the work in [29].

For DeepCas, DeepHawkes, Topo-LSTM, CasCN, DeepCon+Str and our model SkipCas all follow the settings of [12], where the user embedding dimension size is 50, the hidden layer of each GRU is 32 units, and the hidden dimensions of the two-layer MLP are 32 and 16, respectively. The learning rate is 0.005, the batch size is set to 32, and the smoother α is set to 0.01.

5.4 Evaluation Metric

Following the existing work, we adopt mean squared log-transformed error (MSLE) to evaluate the accuracy of predictions on the test set, which is widely used in cascaded prediction evaluation. MSLE is defined as:

$$MSLE = \frac{1}{M} \sum_{i=1}^{M} (\log \Delta S_i - \log \Delta \tilde{S}_i)^2, \qquad (16)$$

where M is the total number of messages, ΔS_i denotes the predicted growth size for cascade C_i, and $\Delta \tilde{S}_i$ denotes the ground truth.

Table 2. Overall performance comparison of information cascades prediction among different methods.

Datasets	Weibo			APS		
Metric	MSLE					
T	1 h	2 h	3 h	5 years	7 years	9 years
Features-deep	3.682	3.361	3.296	1.593	1.514	1.465
Features-linear	3.501	3.435	3.324	1.582	1.508	1.456
Node2vec	3.795	3.523	3.513	2.278	2.003	1.982
DeepCas	3.649	3.250	3.056	1.629	1.538	1.467
Topo-LSTM	2.772	2.643	2.423	1.511	1.483	1.462
DeepHawkes	2.501	2.384	2.275	1.286	1.236	1.162
CasCN	2.348	2.243	2.066	1.455	1.353	1.222
DeepCon+Str	2.670	2.391	2.377	1.468	1.382	1.327
SkipCas	**2.251**	**2.103**	**1.890**	**1.163**	**1.086**	**1.045**

5.5   Experimental Results

We compare the performance of the proposed model with several baseline methods on the Weibo and APS datasets, and the results are shown in Table 2. Experimental results show that the SkipCas model performs relatively well on information cascade prediction for both datasets. It not only outperforms traditional methods, but also state-of-the-art deep learning methods, with a statistically significant drop in MSLE. We plot the training process of SkipCas on the Weibo and APS datasets as shown in Fig. 3. It can be seen that the SkipCas loss gradually converges to a lower result.

The performance gap between Feature-deep and Feature-linear is very small, and Feature-linear outperforms Feature-deep on the APS dataset. This means that deep learning does not always perform better than traditional prediction methods if there is a representative set of information cascading features. However, the performance of these methods depends heavily on the relevant features extracted by hand, and it is difficult to generalize to other domains.

For the embedding method, Node2vec performs poorly on both datasets. It only uses the nodes in the graph to represent the network and ignores other structural and content information in the cascade.

DeepCas shows better performance than feature-based methods on the Weibo dataset, but it is inferior to feature-based methods on the APS dataset, which again shows that deep learning methods are not necessarily better than feature-based methods. However, it still performs worse than other deep learning-based methods because it ignores temporal features and topology of cascaded graphs; similarly, Topo-LSTM lacks temporal features and cannot extract enough information from the cascade, so that its performance is slightly worse compared to our model. DeepHawkes does not consider the topological information of the cascade, and its performance depends on the time series modeling ability. Although CasCN utilizes the structure and time information of the cascade network at the same time, its performance is not the best due to its weak ability to learn structural information. DeepCon+Str utilizes the similarity of cascade graph structure and content to obtain the embedding of the whole cascade graph, but it does not consider the time factor, which affects the prediction performance.

Among these baselines, SkipCas has the best performance and achieves good results on both datasets because it fully investigates the dynamic diffusion process and structural representation of information cascades.

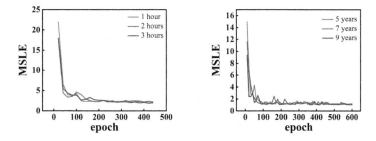

Fig. 3. Convergence of SkipCas on Weibo and APS datasets.

5.6   Ablation Study

To better investigate the effectiveness of each component of SkipCas, we propose four variants. Table 3 summarizes the performance comparison between the models and variants.

SkipCas-LSTM: This method uses LSTM to replace the GRU of the proposed model. Similar to GRU, the LSTM variant models the cascading information through extra gating units.

SkipCas-Time: This method does not consider the time effect of the cascade graph, and is to test the necessity of the time effect in the proposed model.

SkipCas-Path: This method uses a cascade sequence of random walk samples instead of diffusion paths.

SkipCas-Skipgarm: This method does not consider the skip-gram component of the proposed model and only uses GRU and temporal features for prediction, which is to test the importance of the structure of the cascade graph.

From Table 3, we can see that compared with other variants, the prediction error of the original model SkipCas has a certain reduction. Although the error of SkipCas-LSTM is not different from the original model, it can still show that our choice of recurrent neural network is correct; by comparing SkipCas-Time, we find that ignoring the time effect leads to a significant increase in prediction error, which indicates that the time effect is essential in cascading predictions. Similarly, the prediction performance of SkipCas-Path is also decreased significantly, which indicated that the diffusion path could better reflect the change process of the cascade graph. In addition, compared with the original model, the prediction effect of SkipCas-Skipgram is significantly reduced, which fully shows that the structural information of the cascade graph is very important in cascade prediction.

In summary, the time effect of the cascade and the structural information of the cascade are important for future cascade prediction, and our experiments also demonstrate the validity and necessity of the individual components of the proposed model, which essentially improve the performance of the information cascade prediction.

Table 3. Performance comparison between SkipCas and its variants.

Datasets	Weibo			APS		
Metric	MSLE					
T	1 h	2 h	3 h	5 years	7 years	9 years
SkipCas-LSTM	2.301	2.194	1.958	1.325	1.166	1.088
SkipCas-Time	2.523	2.438	2.321	1.582	1.458	1.356
SkipCas-Path	2.332	2.286	2.147	1.465	1.364	1.229
SkipCas-Skipgram	2.495	2.423	2.348	1.529	1.328	1.267
SkipCas	**2.251**	**2.103**	**1.890**	**1.163**	**1.086**	**1.045**

5.7 Parameter Analysis

The observation time window T is an important parameter of the model. As shown in Fig. 4, we can observe that the value of MSLE decreases continuously with increasing observation time on the Weibo dataset, and the prediction error improves by 16% for 3 h compared to 1 h; similarly, the same effect is observed on the APS citation dataset, where the prediction performance continues to improve with the increase of observation years, and the prediction error improves by 10.1% for 9 years compared to 5 years. This shows that as the observation time window T increases, the more information we can observe, the easier it is to make more accurate predictions, which is also a natural result of the increase in training data.

For the time interval l, we choose the datasets with Weibo of 2 h and APS of 7 years for analysis. It can be seen from Fig. 5 (left) that with the increase of the time interval, the prediction performance of the model gradually improves, but when the time interval exceeds 8, the performance starts to decrease again. Therefore, the experiment in this paper adopts the time interval $l = 8$.

For the user embedding dimension size d, we also choose the datasets with Weibo of 2 h and APS of 7 years. The experimental results are shown in Fig. 5 (right). With the increase of dimension size d, the prediction performance of the model improves. When d is 50, the minimum value of MSLE indicates that the prediction effect is the best at this time. However, when the user dimension size exceeds 50, the prediction performance does not improve but decreases. Therefore, in this paper, the user embedding dimension size d is 50.

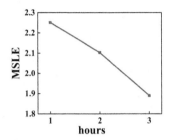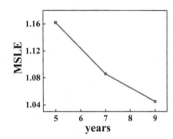

Fig. 4. The effect of observation window on the performance of Weibo (left) and APS (right) datasets.

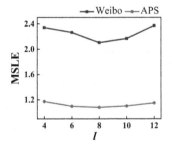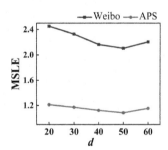

Fig. 5. The effect of time interval l (left) and user embedding dimension size d (right) on datasets performance.

6   Conclusion

In this paper, we propose a novel information cascade prediction model called SkipCas. Our model encodes the diffusion path at each diffusion time in the cascade graph to obtain the dynamic process of information diffusion, uses the sequence of random walk sampling to obtain the structural representation of the cascade graph through skip-gram, and finally predicts the growth size of the information cascade by combining the diffusion process and the structural representation. The experimental results on two real datasets show that SkipCas significantly improves the cascade prediction performance. As for future works, we plan to incorporate relevant message features such as text content to improve prediction performance and explore more effective methods to further mine the structural information between the cascades.

Acknowledgements. This work was supported by the National Natural Science Foundation of China (No. 61972135), the Natural Science Foundation of Heilongjiang Province in China (No. LH2020F043), and the Foundation of Graduate Innovative Research of Heilongjiang University in China (No. YJSCX2022-236HLJU).

References

1. Gao, X., Cao, Z., Li, S., Yao, B., Chen, G., Tang, S.: Taxonomy and evaluation for microblog popularity prediction. In: TKDD, pp. 1–40 (2019)
2. Zhou, F., Xu, X., Trajcevski, G., Zhang, K.: A Survey of information cascade analysis: models, predictions, and recent advances. ACM Comput Surv. **54**(2), 1–36 (2021)
3. Szabo, G., Huberman, B.A.: Predicting the popularity of online content. Commun. ACM **53**(8), 80–88 (2010)
4. Pinto, H., Almeida, J.M., Gonçalves, M.A.: Using early view patterns to predict the popularity of youtube videos. In: WSDM, pp. 365–374 (2013)
5. Bao, P., Shen, H., Huang, J., Cheng, X.: Popularity prediction in microblogging network: a case study on sina weibo. In: WWW, pp. 177–178 (2013)
6. Weng, L., Menczer, F., Ahn, Y.: Predicting successful memes using network and community structure. In: ICWSM (2014)
7. Tsur, O., Rappoport, A.: What's in a hashtag? Content based prediction of the spread of Ideas in microblogging communities. In: WSDM,, pp. 643–652 (2012)
8. Ma, Z., Sun, A., Cong, G.: On predicting the popularity of newly emerging hashtags in Twitter. Assoc. Inf. Sci. Technol. **64**(7), 1399–1410 (2013)
9. Bao, Z., Liu, Y., Zhang, Z., Liu, H., Cheng, J.: Predicting popularity via a generative model with adaptive peeking window. Phys. A **522**, 54–68 (2019)
10. Zhao, Q., Erdogdu, M.A., He, H.Y., Rajaraman, A., Leskovec, J.: SEISMIC: a self-Exciting point process model for predicting tweet popularity. In: SIGKDD, pp. 1513–1522 (2015)
11. Rizoiu, M., Xie, L., Sanner, S., Cebrian, M., Yu, H., Hentenryck, P.V.: Expecting to be HIP: Hawkes intensity processes for social media popularity. In: WWW, pp. 735–744 (2017)
12. Li, C., Ma, J., Guo, X., Mei, Q.: DeepCas: an end-to-end predictor of information cascades. In: WWW, pp. 577–586 (2017)
13. Wang, J., Zheng, V.W., Liu, Z., Chang, K.C.: Topological recurrent neural network for diffusion prediction. In: ICDM, pp. 475–484 (2017)
14. Feng, X., Zhao, Q., Liu, Z.: Prediction of information cascades via content and structure proximity preserved graph level embedding. Inf. Sci. **560**, 424–440 (2021)
15. Cao, Q., Shen, H., Cen, K., Ouyang, W.R., Cheng, X.: DeepHawkes: bridging the gap between prediction and understanding of information cascades. In: CIKM, pp. 1149–158 (2017)
16. Saha, A., Samanta, B., Ganguly, N., De, A.: CRPP: competing recurrent point process for modeling visibility dynamics in information diffusion. In: CIKM, pp. 537–546 (2018)

17. Cui, P., Jin, S., Yu, L., Wang, F., Zhu, W., Yang, S.: Cascading outbreak prediction in networks: a data-driven approach. In: SIGKDD, pp. 901–909 (2013)
18. Yu, L., Cui, P., Wang, F., Song, C., Yang, S.: From micro to macro: uncovering and predicting information cascading process with behavioral dynamics. In: ICDM, pp. 559–568 (2015)
19. Pinto, H., Almeida, J.M., Gonçalves, M.A.: Using early view patterns to predict the popularity of youtube videos. In: WSDM, pp. 365–374 (2013)
20. Shen, H., Wang, D., Song, C., Barabasi, A.L.: Modeling and predicting popularity dynamics via reinforced poisson processes. In: AAAI, pp. 291–297 (2014)
21. Iwata, T., Shah, A., Ghahramani, Z.: Discovering latent influence in online social activities via shared cascade poisson processes. In: SIGKDD, pp. 266–274 (2013)
22. Zaman, T., Fox, E.B., Bradlow, E.T.: A Bayesian approach for predicting the popularity of tweets. Ann. Appl. Stat. **8**(3), 1583–1611 (2014)
23. Chen, X., Zhou, F., Zhang, K., Trajcevski, G., Zhong, T.: Information diffusion prediction via recurrent cascades convolution. In: ICDE, pp. 770–781 (2019)
24. Zhou, F., Xu, X., Zhang, K., Trajcevski, G., Zhong, T.: Variational information diffusion for probabilistic cascades prediction. In: INFOCOM, pp. 1618–1627, (2020)
25. Wang, Y., Shen, H., Liu, S., Gao, J., Cheng, X.: Cascade dynamics modeling with attention-based recurrent neural network. In: IJCAI, pp. 2985–2991 (2017)
26. Wang, Z., Chen, C., Li, W.: A sequential neural information diffusion model with structure attention. In: CIKM, pp. 1795–1798 (2018)
27. Mikolov, T., Sutskever, I., Chen, K., Corrado, G.S.: Distributed representations of words and phrases and their compositionality. In: NIPS, pp. 3111–3119 (2013)
28. Perozzi, B., Al-Rfou, R., Skiena, S.: DeepWalk: online learning of social representations. In: SIGKDD, pp. 701–710 (2014)
29. Grover, A., Leskovec, J.: Node2vec: scalable feature learning for networks. In: SIGKDD, pp. 855–864 (2016)
30. Kipf, T., Welling, M.: Semi-supervised classification with graph convolutional networks. In: ICLR, (2017)
31. Hamilton, W.L., Ying, Z., Leskovec, J.: Inductive representation learning on large graphs. In: NIPS, pp. 1025–1035 (2017)
32. Ying, R., You, J., Morris, C., Ren, X., Hamilton, W.L., Leskovec, J.: Hierarchical graph representation learning with differentiable pooling. In: NeurIPS, pp. 4800–4810 (2018)

Probing Spurious Correlations in Popular Event-Based Rumor Detection Benchmarks

Jiaying Wu[(✉)] and Bryan Hooi

School of Computing, National University of Singapore, Singapore, Singapore
jiayingwu@u.nus.edu, bhooi@comp.nus.edu.sg

Abstract. As social media becomes a hotbed for the spread of misinformation, the crucial task of rumor detection has witnessed promising advances fostered by open-source benchmark datasets. Despite being widely used, we find that these datasets suffer from spurious correlations, which are ignored by existing studies and lead to severe overestimation of existing rumor detection performance. The spurious correlations stem from three causes: (1) *event-based* data collection and labeling schemes assign the same veracity label to multiple highly similar posts from the same underlying event; (2) merging multiple data sources spuriously relates source identities to veracity labels; and (3) labeling bias. In this paper, we closely investigate three of the most popular rumor detection benchmark datasets (i.e., Twitter15, Twitter16 and PHEME), and propose *event-separated rumor detection* to eliminate spurious cues. Under the event-separated setting, we observe that the accuracy of existing state-of-the-art models drops significantly by over 40%, becoming only comparable to a simple neural classifier. To better address this task, we propose Publisher Style Aggregation (PSA), a generalizable approach that aggregates publisher posting records to learn writing style and veracity stance. Extensive experiments demonstrate that our method outperforms existing baselines in terms of effectiveness, efficiency and generalizability.

Keywords: Rumor detection · Spurious correlations · Benchmarks · Text mining · Social network

1 Introduction

In the battle against escalating online misinformation, recent years have witnessed growing interest in automatic rumor detection on social media. Numerous real-world rumor detection datasets including TWITTER15 [13], TWITTER16 [13] and PHEME [15] have emerged as valuable resources fueling continuous development in this field. From early feature engineering models [1] to recent content-based [5,20] and propagation-based [6,29] methods, the ever-evolving approaches have achieved promising advances.

However, most existing methods ignore the *spurious attribute-label correlations* induced by dataset construction pitfalls, which arise from dataset-related

© The Author(s), under exclusive license to Springer Nature Switzerland AG 2023
M.-R. Amini et al. (Eds.): ECML PKDD 2022, LNAI 13714, pp. 274–290, 2023.
https://doi.org/10.1007/978-3-031-26390-3_17

artifacts instead of relationships generalizable to practical real-world settings. The commonly adopted *event-based* data collection framework first fact-checks newsworthy events, and then automatically scrapes a large number of highly similar microblogs (e.g., tweets) containing the same event keywords. Some benchmark datasets also merge data samples from multiple existing sources to balance their class distribution. These factors consequently induce event-label and source-label correlations, which may not hold in practice.

Negligence of spurious cues can lead to unfair over-predictions that limit model generalization and adaptability. Similar issues have been identified in several natural language processing tasks including sentiment classification [17], argument reasoning comprehension [18] and fact verification [35], but the task of social media rumor detection remains underexplored.

Hence, in this paper, we make an effort to thoroughly investigate the causes of spurious correlations in existing rumor detection benchmark datasets, and take solid steps to counteract their impact. Specifically, we identify three causes of spurious correlations: (1) event-based data collection and labeling strategies associate event keywords with veracity labels; (2) merging data sources for label balancing creates correlations between source-specific propagation patterns and microblog veracity; and (3) event-level annotation strategies give rise to labeling bias. Under the post-level data splitting scheme commonly adopted by existing approaches, the prevalence of spurious cues can bring about numerous shared spurious correlations between the training and test data. For instance, the training and test data might even contain identical microblog texts, leading to data leakage (see examples in Fig. 1). If left unchecked, these correlations can lead to severe overestimation of model performance.

To offset the impact of spurious cues, we study a more practical task, namely *event-separated rumor detection*, where the test data contains microblogs from a set of events unseen during training. Without prior knowledge of event-specific cues in the test set, we empirically demonstrate stark performance deterioration of existing approaches, e.g. state-of-the-art rumor detection accuracy plummets from 90.2% to 44.3% on TWITTER16 [13], one of the most widely adopted datasets.

Striving for reliable rumor detection, we propose Publisher Style Aggregation (PSA), a novel method inspired by human fact-checking logic (i.e. reading through a user's homepage to determine user stance and credibility). Specifically, our approach (1) encodes the textual features of source posts and user comments; (2) learns publisher-specific features based on multiple microblog instances produced by each source post publisher; and (3) augments each local microblog representation with its corresponding global publisher representation.

We evaluate PSA on the event-separated rumor detection task using 3 real-world benchmark datasets and compare it against 8 state-of-the-art baselines. Extensive experiments show that PSA outperforms its best competitors by a significant margin, respectively boosting test accuracy and F1 score by 14.18% and 15.26% on average across all 3 datasets. Furthermore, we empirically demonstrate the efficiency and generalizability of PSA via two experimental objectives, namely early rumor detection and cross-dataset rumor detection. Our code is publicly available at: https://github.com/jiayingwu19/PSA.

Dataset	Tweet Content	Freq.
Twitter15	we are running out of chocolate, warns world's largest chocolate manufacturer URL URL	5
	lego letter from the 1970s still offers a powerful message to parents 40 years later URL URL	4
Twitter16	steve jobs was adopted. his biological father was abdulfattah jandali, a syrian muslim.	12
	this is how people in china are riding escalators after a horrific accident URL URL	5

Steve Jobs was adopted. His biological father was Abdulfattah Jandali, a Syrian Muslim.
9:38 AM · Sep 28, 2015

Steve Jobs was adopted. His biological father was Abdulfattah Jandali, a Syrian Muslim.
9:37 AM · Oct 22, 2015

...

Steve Jobs was adopted. His biological father was Abdulfattah Jandali, a Syrian Muslim.
1:12 AM · Nov 25, 2015

Fig. 1. Automated event-based scraping results in numerous duplicate microblog texts in benchmark datasets Twitter15 and Twitter16, causing data leakage under random splitting. The highlighted words are event keywords obtained from the Snopes fact-checking website, in line with the datasets' data collection scheme (see Sect. 3.1).

2 Related Work

Social Media Rumor Detection. Real-world rumor detection datasets, with microblog posts and propagation patterns retrieved from social media platforms such as Twitter [12,13,15] and Weibo [4], form the bedrock of rapidly evolving approaches.

Recent advances in automated rumor detection typically adopt neural network based frameworks. *Content-based* approaches utilize microblog and comment features. For instance, [4,9] respectively employ Recurrent Neural Networks and Convolutional Neural Networks to model the variations of text and user representations over time. Hierarchical attention networks [10] and pretrained language models [11] have also proven effective. Another line of work leverages *propagation-based* information diffusion patterns to encode information flow along user interaction edges. Some inject structural awareness into recursive neural networks [20] and multi-head attention [21–23], while others achieve success with Graph Neural Networks [6,40].

Closely related to our topic is fake news detection. [38] trains an event discriminator to overlook domain-specific knowledge under the multi-modal setting, [39] formulates domain-agnostic fake news detection as a continual learning problem, [37] studies the case with limited labeling budget, and [7,36] take advantage of auxiliary user descriptions and large-scale user corpus.

Existing methods either overlook the publisher-microblog relationship or require external knowledge (e.g. images and additional user description). In contrast, we seek to achieve generalizable rumor detection by capturing rumor-indicative publisher characteristics based on aggregation of multiple microblog data samples.

Investigation of Spurious Correlations. Despite the promising performance of deep learning models, reliance on dataset-related cues has been observed in a wide range of tasks including text classification [17], natural language inference [32] and visual question answering [30]. In fact-checking scenarios, language models can capture underlying identities of news sites [19], and rumor instances can possess time-sensitive characteristics [11]. Spurious artifacts lead to model failure on out-of-domain test instances, as empirically observed by [23, 29, 41].

However, systematic investigation into social media rumor detection remains unexplored. We bridge this gap by discussing three types of spurious correlations specific to this topic, and provide a solution to offset the impact of spurious correlations (i.e. event-separated rumor detection).

3   Spurious Correlations in Event-Based Datasets

3.1   Event-Based Data Collection

In this subsection, we outline the event-based data collection scheme adopted by benchmark datasets.

Newsworthy Event Selection: Newsworthy events serve as vital information sources, from which rumors and non-rumors arise and diffuse on social media. Existing studies either collect events from leading fact-checking websites (e.g., Snopes, Emergent, and PolitiFact) [4, 12, 14], or obtain candidate events identified by professionals [15].

Keyword-Based Microblog Retrieval: To facilititate mass collection, existing datasets typically adopt automated *event-based* data collection strategies, i.e. for each event, (1) extract keywords from its claim; (2) scrape microblogs via keyword-based search; and (3) select influential microblogs. Event keywords are mostly *neutral* (e.g. places, people or objects), carrying little or no stance.

Microblog Labeling Scheme: Existing rumor detection datasets conduct fact-checking at either event-level [4, 12, 14] or post-level [15]. While *event-level* labeling assigns all source posts under an event with the same event-level fact-checking label, *post-level* labeling annotates every source post independently. Although both event- and post-level annotations are performed by trained professionals, the former is more vulnerable to data selection pitfalls, on which we elaborate in Sect. 3.2.

3.2   Possible Causes of Spurious Correlations

We investigate three of the most popular event-based rumor detection benchmark datasets containing source posts, propagation structures and conversation threads, namely TWITTER15 [13], TWITTER16 [13] and PHEME [15], and summarize the dataset statistics in Table 1. As TWITTER15 and TWITTER16 both consist of class-balanced tweets with abundant interactions, we also sample class-balanced tweets involving at least 10 users in PHEME for a fair comparison.

Intra-Event Textual Similarity: Under each event, the automated keyword-based microblog retrieval framework collects a large number of highly similar keyword-sharing samples with the same label, even obtaining identical microblog texts (Fig. 1). Consequently, the correlations between event keywords and class labels result in strong textual cues that generalize poorly beyond the current event.

Table 1. Dataset statistics.

Dataset	TWITTER15	TWITTER16	PHEME
Labeling scheme	Event-level	Event-level	Post-level
# of Events	298	182	9
# of Source Posts	1490	818	973
# of Non-Rumors	374	205	245
# of False Rumors	370	205	241
# of True Rumors	372	207	244
# of Unverified Rumors	374	201	243
# of Distinct Users	480,984	289,675	12,905
# of User Interactions	622,927	362,713	21,169

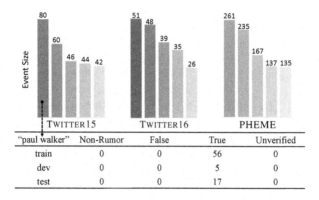

"paul walker"	Non-Rumor	False	True	Unverified
train	0	0	56	0
dev	0	0	5	0
test	0	0	17	0

Fig. 2. The size of largest events in three datasets; "paul walker" directly correlates with "True" label in the post-level random splits adopted by the SOTA method [22].

Under the post-level data splitting scheme adopted by existing works, these cues would scatter into different splits, creating shared correlations between the training and test data that do not hold in the real world. We illustrate such correlations via TWITTER15's largest event about the death of Paul Walker (Fig. 2). All 80 microblogs reporting this event are assigned the "True" label, among which 78 contain the keywords "paul walker". These event-specific keywords produce a strong correlation between "paul walker" and the "True" label. Under post-level random splitting, these samples spread across different data splits, creating undesirable textual similarity between the training and test data. As shown in Fig. 2, the datasets are dominated by such large-size events. Specifically, the top-5 largest events cover 96.09% of data samples in PHEME, while large-size events (containing more than 5 keyword-sharing tweets) cover more than 70% of samples in TWITTER15 and TWITTER16. Large event sizes lead to the prevalence of event-specific keyword-label correlations, further exacerbating the problem.

Merge of Data Sources: For label balancing purposes, TWITTER15 and TWITTER16 merge tweets from multiple sources including [4,12,16], and scrape additional news events from verified media accounts. While the events covered by

	# of source posts	Source	Avg. # of Interactions / Post	Avg. Time Range (hours)
Non-Rumor	205			
False	205	news_accounts	623.11	867.17
True	78 129	IJCAI	439.59	1214.65
	29	PLOS_ONE	308.12	184.95
Unverified	172	snopes	337.65	566.76

Fig. 3. Source-label correlations in TWITTER16 suggest underlying spurious properties. IJCAI and PLOS_ONE refer to [4, 16], respectively.

Fig. 4. One mislabeled instance from TWITTER15, where the source tweet, inconsistent with the Snopes claim, is wrongly assigned the same event-level label.

different sources do not overlap, direct 1-1 correlations between data sources and labels can possibly induce spurious correlations between data source features and the labels. As demonstrated in Fig. 3, the user interaction count (comments and reposts) and interaction time range of tweets from each source form distinctive patterns. For instance, all tweets from PLOS_ONE [16] are "True", spread very quickly and tend to arouse less interactions. These source-specific propagation patterns could possibly be exploited by graph- or temporal-based models, which we empirically demonstrate in Sect. 6.5 (Table 4).

Labeling Bias: While automated event-based data retrieval and event-level annotations allow for easier construction of large-scale benchmark datasets, the lack of post-level scrutiny induces vulnerability to labeling bias. For instance, as shown in Fig. 4, Snopes marked a MH17-related event claim as "False" due to image fabrication. However, in view of highly similar keywords "Malaysian Airlines" and "photo", the data collection framework retrieved an MH17-relevant tweet linking to an authentic photo by Reuters and mistakenly labeled it as "False" in TWITTER15. Such mislabelings exacerbate our previously mentioned problem of intra-event textual similarity, making the resulting keyword-label correlations stronger but more deceptive. To make the best use of valuable data resources, we suggest that future approaches incorporate techniques that are robust to label noise (e.g. noise-tolerant training [34]).

4   Event-Separated Rumor Detection

4.1   Problem Formulation

Social media rumor detection aims to learn a classification model that is able to detect and fact-check rumors. Let $\mathcal{T} = \{T_1, T_2, \ldots, T_N\}$ be a rumor detection dataset of size N, and $\mathcal{Y} = \{y_1, y_2, \ldots, y_N\}$ be the corresponding ground-truth labels, with $y_i \in \mathcal{C} = \{1, \ldots, C\}$. Each microblog instance T_i consists of a source post publisher u_i, a source post p_i, and related comments $c_i^1, \ldots, c_i^k$. p_i (c_i^j) has a corresponding textual feature vector $\mathbf{r}_i$ ($\mathbf{r}_i^j$). Denote the event behind microblog instance T_i as e_i. Consequently, training data $\mathcal{T}_{tr}$ and test data $\mathcal{T}_{te}$ respectively contain events $\mathcal{E}_{tr} = \{e_i | T_i \in \mathcal{T}_{tr}\}$ and $\mathcal{E}_{te} = \{e_i | T_i \in \mathcal{T}_{te}\}$.

Most existing approaches ignore the underlying microblog-event relationship and adopt event-mixed post-level data splits, resulting in significant overlap between $\mathcal{E}_{tr}$ and $\mathcal{E}_{te}$. However, prior knowledge of test data is not always guaranteed in practice (e.g. the model's performance gains from duplicate tweets in the training and test data are unlikely to generalize), and previous assumptions can lead to performance overestimation caused by intra-event textual similarity (see Sect. 3.2).

In order to eliminate these confounding event-specific correlations, we propose to study a more practical problem, namely **event-separated rumor detection**, where $\mathcal{E}_{tr} \cap \mathcal{E}_{te} = \varnothing$. This task is challenging due to the underlying event distribution shift, and it thereby provides a means to evaluate debiased rumor detection performance.

4.2   Existing Approaches

We compare the event-mixed and event-separated rumor detection performance of representative approaches on TWITTER15, TWITTER16 and PHEME datasets to investigate the impact of event-specific spurious correlations.

Propagation-Based:(1) **TD-RvNN** [20]: a recursive neural model that encodes long-distance user interactions with gated recurrent units. (2) **GLAN** [21]: a global-local attentive model based on microblog-user heterogeneous graphs. (3) **BiGCN** [6]: a GCN-based model that encodes augmented bidirectional rumor propagation patterns. (4) **SMAN** [22]: a multi-head attention model that enhances training with user credibility modeling.

Content-Based: (1) **BERT** [24]: a deep language model that encodes bidirectional context with Transformer blocks. (2) **XLNet** [27]: a generalized autoregressive approach that trains on all possible permutations of word factorizations. (3) **RoBERTa** [25]: a refined BERT-like approach that adopts dynamic masking for varied masking patterns over different epochs. (4) **DistilBERT** [26]: a distilled model that maintains 97% of BERT's performance with 60% less parameters.

Data Splitting. For all three datasets, we sample 10% instances for validation, then split the remainder 3:1 into training and test sets. Specifically, we obtain the event-separated splits based on the *publicly available event IDs* released in TWITTER15 [13], TWITTER16 [13] and PHEME [15], respectively.

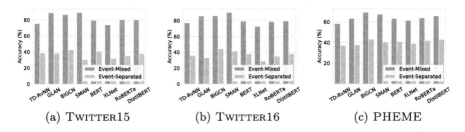

(a) TWITTER15   (b) TWITTER16   (c) PHEME

Fig. 5. Existing rumor detection approaches fail to generalize across events. Comparing between event-mixed and event-separated settings, mean accuracy based on 20 different runs of each approach demonstrates drastic performance deterioration. (Over 40% drop on TWITTER15 & 16, and over 20% on PHEME.)

4.3   SOTA Models' Performance Is Heavily Overestimated

Figure 5 reveals a stark contrast between event-mixed and event-separated rumor detection performance. More specifically, test accuracy plummets from 74.0%–89.2% to 30.5%–42.8% on TWITTER15, from 72.7%–90.2% to 28.8%–44.3% on TWITTER16, and from 58.3%–69.2% to 37.3%–43.4% on PHEME. Furthermore, despite the consistency of best event-separated performance across all three datasets, all models achieve significantly higher event-mixed performance on TWITTER15 and TWITTER16 than on PHEME, where the former adopts event-level labeling and the latter adopts post-level labeling (see Sect. 3.1). This gap is in line with our hypothesis that direct event-label correlations induce additional bias.

Results imply the heavy reliance of existing methods on spurious event-specific correlations. Despite performing well under the event-mixed setting, these models cannot generalize to unseen events, resulting in poor real-world adaptability.

5   Proposed Method

To tackle the challenges of event-separated rumor detection, we propose Publisher Style Aggregation (PSA), a novel approach that learns generalizable publisher characteristics based on each publisher's aggregated posts, as illustrated in Fig. 6.

5.1   Consistency of Publisher Style

Source post publishers are highly influential users who produce claims towards newsworthy events. Therefore, each publisher's unique credibility stance and writing style can exhibit distinctive traits that help determine the veracity of their statements. For a more intuitive view, we illustrate the TWITTER15 publisher tendency towards each class in Fig. 7. Specifically, for publisher u, we define u's tendency score under class c as (# microblogs posted by u under class

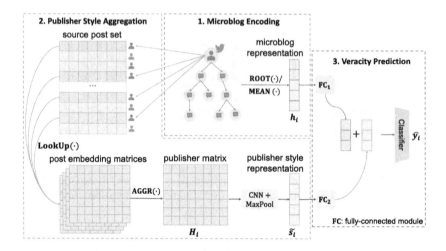

Fig. 6. Overview of our proposed PSA framework.

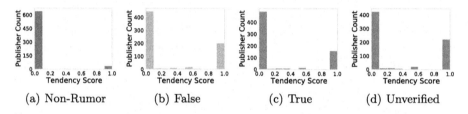

(a) Non-Rumor   (b) False   (c) True   (d) Unverified

Fig. 7. Publishers tend to post tweets of the same credibility type, as exemplified by TWITTER15 publisher behavior patterns.

$c)/(\#$ microblogs posted by u). Figure 7 shows that most publishers have scores either approaching 0 or approaching 1 towards one particular class, i.e., most publishers tend to post microblogs under a single veracity label, which verifies our hypothesis of publisher style consistency.

5.2   Content-Based Microblog Encoding

We first propose two simple neural classifiers, namely RootText and MeanText, to study the importance of source post and comment features in social media rumor detection. In each dataset, all source posts and comments constitute a vocabulary of size $|V|$. Following [6,13], we formulate each source post feature $\mathbf{r}_i \in \mathbb{R}^{|V|}$ and its related comment features $\mathbf{r}_i^j \in \mathbb{R}^{|V|}$ as the sum of all one-hot word vectors within the corresponding source post or comment.

RootText: Source posts contain the claims to be fact-checked. Therefore, we propose to encode each microblog instance T_i solely based on its source post textual feature $\mathbf{r}_i$, i.e., $\mathbf{h}_i := \mathbf{r}_i$.

MeanText: We also propose to consider the user comments of source posts for more robust credibility measurement. Here, we adopt mean-pooling to condense source post and comment features into microblog representations:

$$\mathbf{h}_i := \frac{\mathbf{r}_i + \sum_{j=1}^{k}\mathbf{r}_i^{j}}{k+1}. \tag{1}$$

We obtain the encoding $\mathbf{h}_i \in \mathbb{R}^{|V|}$ of microblog T_i based on either RootText or MeanText, and extract high-level features $\tilde{\mathbf{h}}_i \in \mathbb{R}^{n}$ via a two-layer fully-connected neural network with the ReLU activation function. Then, we employ dropout to prevent overfitting before passing $\tilde{\mathbf{h}}_i$ through the final fully-connected layer with output dimensionality $|\mathcal{C}|$ for veracity prediction.

5.3   Publisher Style Aggregation

As shown in Sect. 5.1, the writing stance and credibility of highly influential source post publishers remain relatively stable in a fixed timeframe. Inspired by this observation, we further propose Publisher Style Aggregation (PSA), a generalizable method that jointly leverages multiple microblog instances produced by each publisher and extracts distinctive publisher traits to enhance local features learned within each individual microblog. More specifically, PSA (1) looks up a set of microblog instances produced by each publisher, (2) learns publisher style representations based on these source posts' aggregated textual features, and (3) augments the representation of each microblog (i.e. $\tilde{\mathbf{h}}_i$ learned via RootText / MeanText) with its corresponding publisher style representation $\tilde{\mathbf{s}}_i$.

Publisher Style Modeling: Assume that publisher u_i has produced $m_i \geq 1$ microblog instances, with the corresponding source posts denoted as $\mathcal{P}(u_i) = \{p_k | u_k = u_i, k = 1, \ldots, N\}$; note that only accessible data are used during training. We treat the j-th post $p_i^{j} \in \mathcal{P}(u_i)$ as a word token sequence with maximum length L. Then, we construct an embedding matrix $\mathbf{W}_i^{j} \in \mathbb{R}^{L \times d}$ for p_i^{j} based on trainable d-dimensional word embeddings. We aggregate all post embedding matrices $\{\mathbf{W}_i^{j}\}_{j=1}^{m_i}$ of u_i, and obtain the corresponding publisher matrix $\mathbf{H}_i \in \mathbb{R}^{L \times d}$ as follows:

$$\mathbf{H}_i = \mathsf{AGGR}(\{\mathbf{W}_i^{j}\}_{j=1}^{m_i}), \tag{2}$$

where the AGGR operator can be either MEAN or SUM.

To capture high-level publisher characteristics, we apply convolution on each $\mathbf{H}_i$ to extract latent publisher style features. Specifically, we use three convolutional layers with different window sizes to learn features with varied granularity. Each layer consists of F filters, and each filter outputs a feature map $\mathbf{f}_* = [f_*^{1}, f_*^{2}, \ldots, f_*^{L-k+1}]$, with

$$f_*^{j} = \mathsf{ReLU}\left(\mathbf{W}_f \cdot \mathbf{H}_i[j : j + k - 1] + b\right), \tag{3}$$

where $\mathbf{W}_f \in \mathbb{R}^{k \times d}$ the convolution kernel, k the window size and $b \in \mathbb{R}$ a bias term. We perform max-pooling to extract the most prominent value of each $\mathbf{f}_*$, and stack these values to form a style feature vector $\mathbf{s} \in \mathbb{R}^F$. Then, we concatenate the $\mathbf{s}_*$ produced by each of the three CNN layers to obtain the publisher style representation $\tilde{\mathbf{s}}_i \in \mathbb{R}^{3F}$:

$$\tilde{\mathbf{s}}_i = \mathsf{Concat}[\mathbf{s}_1; \mathbf{s}_2; \mathbf{s}_3]. \tag{4}$$

Microblog Veracity Prediction: We augment microblog representation $\tilde{\mathbf{h}}_i \in \mathbb{R}^n$ with the corresponding publisher style representation $\tilde{\mathbf{s}}_i$. Finally, we utilize a fully connected layer to predict the microblog veracity label $\hat{\mathbf{y}}_i$:

$$\hat{\mathbf{y}}_i = \mathsf{Softmax}(\mathbf{W}_2^\mathsf{T}(\tilde{\mathbf{h}}_i + \mathbf{W}_1^\mathsf{T}\tilde{\mathbf{s}}_i)), \tag{5}$$

where transformations $\mathbf{W}_1 \in \mathbb{R}^{3F \times n}$ and $\mathbf{W}_2 \in \mathbb{R}^{n \times |\mathcal{C}|}$. We also apply dropout before the final layer to prevent overfitting.

Model parameters are optimized by minimising the cross-entropy loss between $\hat{\mathbf{y}}_i$ and ground truth y_i.

6   Experiments

In this section, we review our experiments for answering the following questions:

Q1 (Model Performance): Does PSA outperform the existing baselines on event-separated rumor detection?

Q2 (Early Rumor Detection): Does PSA work well under temporal rumor detection deadlines?

Q3 (Model Generalization): Is PSA effective under cross-dataset settings?

6.1   Experimental Setup

We implement our proposed PSA model and its variants based on PyTorch 1.6.0 with CUDA 10.2, and train them on a server running Ubuntu 18.04 with NVIDIA RTX 2080Ti GPU and Intel(R) Xeon(R) CPU E5-2690 v4 @ 2.60GHz. We adopt an Adam optimizer with $(\beta_1, \beta_2) = (0.9, 0.999)$, learning rate of 10^{-4} (0.005), and weight decay 10^{-5} (10^{-4}) for TWITTER15/16 (PHEME). We obtain microblog representations via a 2-layer neural network with layer sizes of 128 and 64. We utilize the 300-dimensional word vectors from [21] to form publisher matrices, employ three CNN layers with the same filter number $F = 100$ but different window sizes $k \in \{3, 4, 5\}$, concatenate their outputs and use a fully-connected layer to extract publisher style representations with the size of 64. We implement AGGR in Eq. 2 as both SUM and MEAN, and report average performance over 20 different runs on the event-separated data splits presented in Sect. 4.2.

Table 2. PSA significantly improves event-separated rumor detection accuracy (%) and Macro F1 Score (%) (S stands for SUM and M for MEAN; averaged over 20 runs).

Method	Twitter15		Twitter16		PHEME	
	Acc.	F1	Acc.	F1	Acc.	F1
TD-RvNN [20]	38.62 ± 1.85	36.40 ± 2.38	36.15 ± 1.90	35.66 ± 1.89	37.30 ± 2.54	34.17 ± 2.56
GLAN [21]	38.56 ± 3.38	35.52 ± 5.31	33.13 ± 4.54	27.93 ± 5.53	38.10 ± 2.85	34.60 ± 3.04
BiGCN [6]	42.83 ± 2.27	38.17 ± 3.04	44.28 ± 3.39	42.31 ± 3.77	43.36 ± 1.71	37.93 ± 2.16
SMAN [22]	30.52 ± 2.62	28.80 ± 3.30	41.42 ± 2.65	40.62 ± 2.95	40.74 ± 1.36	36.02 ± 1.62
BERT [24]	40.95 ± 4.80	37.47 ± 7.56	37.89 ± 6.68	34.76 ± 9.22	40.90 ± 3.22	36.33 ± 4.02
XLNet [27]	32.05 ± 6.78	26.00 ± 9.20	28.82 ± 4.08	20.59 ± 7.06	39.14 ± 5.14	34.35 ± 6.65
RoBERTa [25]	35.30 ± 5.65	28.41 ± 7.93	34.84 ± 6.69	27.90 ± 9.74	42.25 ± 4.47	37.67 ± 5.70
DistilBERT [26]	38.09 ± 4.48	33.74 ± 5.21	38.02 ± 4.24	33.98 ± 5.56	43.33 ± 3.62	38.37 ± 4.16
RootText(RT)	33.80 ± 2.74	30.56 ± 2.92	30.54 ± 1.72	28.87 ± 2.32	42.75 ± 1.25	38.74 ± 1.47
MeanText(MT)	48.68 ± 1.80	47.18 ± 1.63	45.19 ± 1.86	44.23 ± 1.72	32.57 ± 1.70	30.48 ± 1.51
RT+PSA(S)	47.85 ± 5.64	45.26 ± 5.13	57.88 ± 3.16	55.30 ± 4.64	43.52 ± 0.93	37.81 ± 1.22
RT+PSA(M)	45.67 ± 0.82	38.55 ± 0.89	47.28 ± 2.87	42.74 ± 4.15	$\mathbf{46.30 \pm 1.28}$	$\mathbf{41.57 \pm 1.49}$
MT+PSA(S)	$\mathbf{61.83 \pm 1.43}$	$\mathbf{58.75 \pm 2.08}$	$\mathbf{64.89 \pm 1.75}$	$\mathbf{64.31 \pm 1.70}$	37.43 ± 1.05	32.36 ± 1.25
MT+PSA(M)	54.73 ± 1.04	50.85 ± 1.56	60.16 ± 2.76	58.16 ± 3.13	40.19 ± 1.20	35.98 ± 1.45

6.2   Q1. Model Performance

We compare PSA (base classifier: RootText/MeanText) with existing approaches in Table 2.

Importance of Textual Features: We observe that MeanText outperforms existing methods on Twitter15&16, while RootText only achieves 0.6% lower accuracy than the best baseline on PHEME. This implies severe degradation of overparameterized models when the spurious attribute-label correlations (i.e. event-specific cues) in the training data do not apply to the test data, in line with prior work [41]. Comparing between RootText and MeanText, we also observe that the former performs better on PHEME but otherwise on Twitter15 and Twitter16. Different labeling schemes may account for such differences; as PHEME labels each microblog independently, the source posts would contain the most distinctive features. While the source post content is not as distinctive in Twitter15 and Twitter16, both datasets exhibit more complicated propagation patterns (see Table 1). Therefore, adopting MeanText to aggregate comment features proves more effective in these cases.

Effectiveness of PSA: Our proposed PSA approach, with AGGR implemented as either SUM or MEAN, significantly enhances the base classifiers RootText and MeanText. The best PSA combinations outperform the best baseline by a large margin; they boost event-separated rumor detection accuracy by 19.00% on Twitter15, 20.61% on Twitter16, and 2.94% on PHEME. Unlike existing methods, PSA explicitly aggregates publisher style features across microblog instances from multiple events, thereby enhancing the model's capability to learn

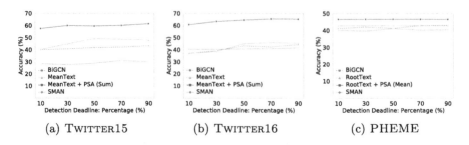

(a) TWITTER15   (b) TWITTER16   (c) PHEME

Fig. 8. Early rumor detection accuracy (%) of PSA models against the best propagation-based baselines for varying detection deadlines on TWITTER15, TWITTER16 and PHEME under the event-separated setting (averaged over 20 runs).

Table 3. Cross-dataset rumor detection accuracy (%) and Macro F1 Score (%) of PSA between T15 (TWITTER15) and T16 (TWITTER16), compared with the best propagation- and content-based baseline methods (averaged over 20 runs).

Train	Test	Method	Acc.	F1
T15	T16	DistilBERT	40.70 ± 4.78	39.98 ± 5.02
		BiGCN	36.80 ± 3.79	35.76 ± 3.92
		MeanText	46.55 ± 1.33	44.11 ± 1.71
		+ PSA (Sum)	$\mathbf{63.99 \pm 1.53}$	$\mathbf{61.95 \pm 1.86}$
T16	T15	DistilBERT	41.66 ± 4.95	37.07 ± 6.06
		BiGCN	44.39 ± 1.89	42.41 ± 2.23
		MeanText	48.00 ± 1.70	44.83 ± 1.63
		+ PSA (Sum)	$\mathbf{60.82 \pm 1.47}$	$\mathbf{57.97 \pm 2.07}$

event-invariant features. As a result, PSA is able to capture stance and style pertaining to distinctive publisher characteristics, leading to substantial performance improvements.

6.3   Q2. Early Rumor Detection

Accurate and timely misinformation detection is of vital importance. Given only partial propagation information, we compare the best PSA combinations and their corresponding base classifiers with their best propagation-based competitors. Figure 8 shows the consistent superiority of PSA over baseline methods at all detection deadlines. Even with only the earliest 10% of comments, PSA achieves 57.53% accuracy on TWITTER15, 60.65% on TWITTER16, and 46.30% on PHEME. Note that RootText (+PSA) models maintain stable performance across all deadlines, as they provide instant predictions solely based on source posts. The results demonstrate that augmenting rumor detection models with publisher style representations achieves both efficiency and effectiveness.

Table 4. Classification accuracy (%) based on user interaction count and interaction time range suggests potential source-specific correlations.

Feature	TWITTER15	TWITTER16	PHEME
A: Interaction Count	43.03	43.32	23.15
B: Time Range	37.09	44.92	27.31
A + B	**44.21**	**55.61**	26.85

6.4   Q3. Cross-Dataset Rumor Detection

To study the generalization ability of PSA, we conduct cross-dataset experiments on TWITTER15 and TWITTER16, where the model is trained on one dataset and tested on the other. For a fair comparison, we utilize the same event-separated data splits adopted in Sects. 6.2 and 6.3. If overlapping events exist between the training set from dataset A and the test set from dataset B, we remove all instances related to these events in the training set, and replace them with the same number of non-overlapping instances randomly sampled from A's test set.

The cross-dataset setting is inherently more challenging, as the training and test events stem from different timeframes, which can create temporal concept shifts. However, Table 3 shows that PSA continues to excel and enhances the base classifier (our MeanText method) by 17.44% on TWITTER15 and 12.82% on TWITTER16, which further demonstrates PSA's generality to unseen events.

6.5   Discussion: Source-Specific Spurious Cues

In Table 4, we empirically show the potential impact of source-specific spurious cues (Sect. 3.2). Under the event-separated setting, we construct a simple Random Forest classifier based on each microblog's user interaction count and the time range covered by these interactions. Surprisingly, the classifier outperforms existing baseline methods on both TWITTER15 and TWITTER16, and achieves comparable performance even with only one feature. In contrast, the single-source PHEME remains unaffected. Although neither our proposed approaches nor existing methods exploit these features, we nevertheless suggest the integration of debiasing techniques in future graph- and temporal-based models.

7   Conclusion

In this paper, we systematically analyze how event-based data collection schemes create event- and source-specific spurious correlations in social media rumor detection benchmark datasets. We study the task of event-separated rumor detection to remove event-specific correlations, and empirically demonstrate severe limitations on existing methods' generalization ability. To better address this task, we propose PSA to augment microblog representations with aggregated publisher style features. Extensive experiments on three real-world datasets show substantial improvement on cross-event, cross-dataset and early rumor detection.

For future work, we suggest (1) event-separated rumor detection performance as a major evaluation metric; (2) same-source samples and post-level expert annotations in dataset construction; and (3) integration of causal reasoning and robust learning techniques in model design, in the hope that our findings could motivate and measure further progress in this field.

Acknowledgements. This work was supported in part by NUS ODPRT Grant R252-000-A81-133.

References

1. Castillo, C., Mendoza, M., Poblete, B.: Information credibility on twitter. In: WWW (2011)
2. Ma, J., Gao, W., Wei, Z., Lu, Y., and Wong, K.-F.: Detect rumors using time series of social context information on microblogging websites. In: CIKM (2015)
3. Kwon, S., Cha, M., Jung, K., Chen, W., Wang, Y.: Prominent features of rumor propagation in online social media. In: ICDM (2013)
4. Ma, J., Gao, W., Mitra, P., Kwon, S., Jansen, B.J., Wong, K.-F., Cha, M.: Detecting rumors from microblogs with recurrent neural networks. In: IJCAI (2016)
5. Zhang, J., Dong, B., Yu, P.S.: FakeDetector: effective fake news detection with deep diffusive neural network. In: ICDE (2020)
6. Bian, T., et al.: Rumor detection on social media with bi-directional graph convolutional networks. In: AAAI (2020)
7. Nguyen, V.-H., Sugiyama, K., Nakov, P., Kan, M.-Y.: FANG: leveraging social context for fake news detection using graph representation. In: CIKM (2020)
8. Shu, K., Wang, S., Liu, H.: Beyond news contents: the role of social context for fake news detection. In: WSDM (2019)
9. Liu, Y. Wu, Y.-F.: Early detection of fake news on social media through propagation path classification with recurrent and convolutional networks. In: AAAI (2018)
10. Shu, K., Cui, L., Wang, S., Lee, D., Liu, H.: DEFEND: explainable fake news detection. In: KDD (2019)
11. Pelrine, K., Danovitch, J., Rabbany, R.: The surprising performance of simple baselines for misinformation detection. In: WWW (2021)
12. Liu, X., Nourbakhsh, A., Li, Q., Fang, R., Shah, S.: Real-time rumor debunking on twitter. In: CIKM (2015)
13. Ma, J., Gao, W., Wong, K.-F.: Detect rumors in microblog posts using propagation structure via kernel learning. In: ACL (2017)
14. Shu, K., Mahudeswaran, D., Wang, S., Lee, D., Liu, H.: FakeNewsNet: A Data Repository with News Content, Social Context and Dynamic Information for Studying Fake News on Social Media. arXiv preprint arXiv:1809.01286 (2018)
15. Kochkina, E., Liakata, M., Zubiaga, A.: All-in-one: multi-task learning for rumour verification. In: COLING (2018)
16. Zubiaga, A., Liakata, M., Procter, R., Wong, S.H.G., Tolmie, P.: Analysing How People Orient to and Spread Rumours in Social Media by Looking at Conversational Threads. Public Library of Science, PLOS ONE (2016)

17. Wang, Z., Culotta, A.: Identifying spurious correlations for robust text classification. In: Findings of EMNLP (2020)
18. Niven, T., Kao, H.-Y.: Probing neural network comprehension of natural language arguments. In: ACL (2019)
19. Zhou, X., Elfardy, H., Christodoulopoulos, C., Butler, T., Bansal, M.: Hidden biases in unreliable news detection datasets. In: EACL (2021)
20. Ma, J., Gao, W., Wong, K.-F.: Rumor detection on twitter with tree-structured recursive neural networks. In: ACL (2018)
21. Yuan, C., Ma, Q., Zhou, W., Han, J., Hu, S.: Jointly embedding the local and global relations of heterogeneous graph for rumor detection. In: ICDM (2019)
22. Yuan, C., Ma, Q., Zhou, W., Han, J., Hu, S.: Early detection of fake news by utilizing the credibility of news, publishers, and users based on weakly supervised learning. In: COLING (2020)
23. Khoo, L.M.S., Chieu, H.L., Qian, Z., Jiang, J.: Interpretable rumor detection in microblogs by attending to user interactions. In: AAAI (2020)
24. Devlin, J., Chang, M.-W., Lee, K., Toutanova, K.: BERT: Pre-training of deep bidirectional transformers for language understanding. In: NAACL (2019)
25. Liu, Y., et al.: RoBERTa: A Robustly Optimized BERT Pretraining Approach. arXiv preprint arXiv: 1907.11692 (2019)
26. Sanh, V., Debut, L., Chaumond, J., Wolf, T.: DistilBERT, a distilled version of BERT: smaller, faster, cheaper and lighter. arXiv preprint arXiv: 1910.01108 (2019)
27. Yang, Z., Dai, Z., Yang, Y., Carbonell, J., Salakhutdinov, R.R. Le, Q.V.: XLNet: Generalized Autoregressive Pretraining for Language Understanding. In: Advances in Neural Information Processing Systems (2019)
28. Wolf, T., Debut, L., et al.: Transformers: state-of-the-art natural language processing. In: EMNLP (2020)
29. Huang, Y.-H., Liu, T.-W., Lee, S.-R., Alvarado, C., Henrique, F. Chen, Y.-S.: Conquering cross-source failure for news credibility: learning generalizable representations beyond content embedding. In: WWW (2020)
30. Agarwal, V., Shetty, R., Fritz, M.: Towards causal vqa: revealing and reducing spurious correlations by invariant and covariant semantic editing. In: CVPR (2020)
31. Geirhos, R., Rubisch, P., Michaelis, C., Bethge, M., Wichmann, F.A., Brendel W.: ImageNet-trained CNNs are biased towards texture; increasing shape bias improves accuracy and robustness. In: ICLR (2019)
32. McCoy, T., Pavlick, E., Linzen, T.: Right for the wrong reasons: diagnosing syntactic heuristics in natural language inference. In: ACL (2019)
33. Srivastava, M., Hashimoto, T.B., Liang, P.: Robustness to Spurious Correlations via Human Annotations. In: ICML (2020)
34. Li, J., Wong, Y., Zhao, Q., Kankanhalli, M.S.: Learning to Learn from noisy labeled data. In: CVPR (2019)
35. Schuster, T., Shah, D., Yeo, Y.J.S., Filizzola, D., Santus, E., Barzilay, R.: Towards debiasing fact verification models. In: EMNLP-IJCNLP (2019)
36. Dou, Y., Shu, K., Xia, C., Yu, P.S. Sun, L.: User preference-aware fake news detection. In: SIGIR (2021)
37. Silva, A., Luo, L., Karunasekera, S., Leckie, C.: Embracing Domain differences in fake news: cross-domain fake news detection using multi-modal data. In: AAAI (2021)
38. Wang, Y., et al.: EANN: event adversarial neural networks for multi-modal fake news detection. In: KDD (2018)

39. Han, Y., Karunasekera, S., Leckie, C.: Graph neural networks with continual learning for fake news detection from social media. arXiv preprint arXiv: 2007.03316 (2020)
40. Huang, Q., Yu, J., Wu, J., Wang, B.: Heterogeneous Graph Attention Networks for Early Detection of Rumors on Twitter. arXiv preprint arXiv:2006.05866 (2020)
41. Sagawa, S., Raghunathan, A., Koh, P. W., Liang, P.: An investigation of why overparameterization exacerbates spurious correlations. In: ICML (2020)

Graph Neural Networks

Self-supervised Graph Learning
with Segmented Graph Channels

Hang Gao[1,2], Jiangmeng Li[1,2], and Changwen Zheng[2(✉)]

[1] University of Chinese Academy of Sciences, Zhongguancun East Road. 80,
Haidian District, Beijing 100081, China
[2] Science and Technology on Integrated Infomation System Laboratory,
Institute of Software Chinese Academy of Sciences, Zhongguancun South Fourth
Street. 4, Haidian District, Beijing 100083, China
{Hang,Jiangmeng}@iscas.ac.cn, changwen@iscas.ac.cn
https://www.ucas.ac.cn/, http://www.iscas.cn/

Abstract. Self-supervised graph learning adopts self-defined signals as
supervision to learn representations. This learning paradigm solves the
critical problem of utilizing unlabeled graph data. Conventional self-
supervised graph learning methods rely on graph data augmentation to
generate different views of the input data as self-defined signals. However, the views generated by such an approach contain amounts of identical node features, which leads to the learning of redundant information.
To this end, we propose *Self-Supervised Graph Learning with Segmented
Graph Channels* (SGL-SGC) to address the issue. SGL-SGC divides the
input graph data across the feature dimensions as Segmented Graph Channels (SGCs). By combining SGCs with data augmentation, SGL-SGC can
generate views that vastly reduce the redundant information. We further
design a feature-level weight-sensitive loss to jointly accelerate optimization and avoid the model falling into a local optimum. Empirically, the
experiments on multiple benchmark datasets demonstrate that SGL-SGC
outperforms the state-of-the-art methods in contrastive graph learning
tasks. Ablation studies verify the effectiveness and efficiency of different
parts of SGL-SGC.

Keywords: Graph neural network · Self-supervised learning ·
Unsupervised learning · Contrastive learning · Node classification

1 Introduction

Graph representation learning (GRL) aims to learn effective representations of
graph-structured data. Such representations play an important role in a variety
of real-world applications, including knowledge graphs [33], molecules [5], social
networks [12], physical processes [19], and codes [1]. Recently, Graph Neural
Networks (GNNs) emerged as a powerful approach to conducting graph representation learning. Various GNNs, including Graph Convolutional Networks

---

Supplementary Information The online version contains supplementary material
available at https://doi.org/10.1007/978-3-031-26390-3_18.

(GCN) [12], Graph Attention Networks (GAT) [27], and Graph Isomorphism Networks (GIN) [31], achieve eye-catching success in graph representation learning. These approaches require labeled graph data for training. However, labeling graph data is a rather challenging task as it requires large amounts of onerous work, particularly with large-scale graphs.

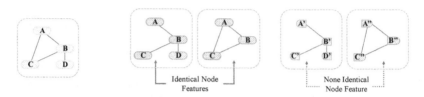

(a) Original input graph data

(b) Views generated with data augmentation alone

(c) Views generated with SGCs and data augmentation

Fig. 1. Examples of views generated using different methods. View generated with graph data augmentation contain identical node features. SGCs help eliminate them

To reduce the dependence on labeled data, recent research efforts are dedicated to developing self-supervised learning for GNNs. In computer vision (CV), self-supervised learning utilizing unlabeled data has already made significant progress [4,7,10]. Viewing its success in CV, some researchers combine self-supervised learning with graph learning and propose a variety of powerful self-supervised graph learning (SGL) methods [8,17,28]. SGL methods rely on views, i.e., human-defined data transformations that preserve the invariance of intrinsic properties of graph data, as training signals to conduct representation learning [34]. Previous works leverage the mutual information maximization principle (InfoMax) [15] and obtain graph representations by maximizing the mutual information between representations of different views. However, the InfoMax principle can be risky. It only encourages the maximization of mutual information while this mutual information may contain redundancy. Based on the information bottleneck principle [25,26,30] points out that when the task-related information contained in the views is not damaged, the redundant mutual information between views should be minimized. To minimize such redundancy, the choice of views is critical [30,34].

In recent years, researchers proposed various view generation methods in graph self-supervised learning, including node dropping, edge perturbation, attribute masking, and subgraph [35]. These methods can be summarized as graph data augmentation that generates different views by making minor changes to the graph data without damaging the task-related information of the graph. We analyze the graph data augmentation methods and propose that they can be expressed as perturbing the original graph data with a specific form of noise. Views generated with such a mechanism contain a large number of identical node features, which will lead to learning redundant information. Considering Fig. 1,

Fig. 1(b) demonstrates the different views of an input graph (Fig. 1(a)) generated with data augmentation. The data augmentation drop nodes and delete edges but leaves the features of node features A, B, and C unmodified, leaving identical node features (marked with red) between the views. However, we cannot reduce such redundant information with more perturbation, i.e., adding more noise. Otherwise, the task-related information of the original graph may be corrupted or even completely changed. The conventional graph view generation methods show limitations here.

To address such limitations, we look to the view generation methods in CV for inspiration. Self-supervised methods generate different views by splitting input image data across channels, e.g., an RGB image can be split into three views for R, G, and B channels [24]. The advantage of this view generation method is that there is no identical feature between different views. Furthermore, it does not introduce more noises. Since each channel provides a relatively condensed and expressive view, this method allows the neural network to pay more attention to task-related semantic information instead of redundant information. We believe that a similar approach can also be applied to graph learning. Suppose we regard each node in the graph as a pixel on the picture and artificially divide the node features into different channels. In that case, we can generate "channels" on the graph, which we denote as Segmented Graph Channels (SGCs).

With SGCs, we propose the Self-Supervised Graph Learning with Segmented Graph Channel (SGL-SGC) to enhance graph representation learning. We combine SGCs with conventional graph data augmentation methods to generate views for self-supervised learning. Figure 1(c) gives an example. Due to combining two different view generation methods, our method can generate amounts of views without introducing more noise. We design an objective function named feature-level weight-sensitive loss to train the encoders with these views. This loss function helps reduce the computational burden while avoiding the model falling into a local optimum. Furthermore, it can assign different weights to different samples according to their importance, further enhancing the learning capability of SGL-SGC.

We summarize our contributions as follows:

- We propose a novel view generate method based on segmented graph channels to generate views with less redundant information. These views strengthen the ability of our proposed method to perform representation learning.
- We design a feature-level weight-sensitive loss as an objective function for training the encoders with the generated views. Feature-level weight-sensitive loss reduces computational burden while avoiding the model falling into a local optimum. Furthermore, our loss function emphasizes the samples with more importance.
- We conduct experiments to compare our method with state-of-the-art graph self-supervised learning approaches on benchmark datasets, and the results prove the superiority of our method.

2   Related Works

This section reviews some representative works on graph learning and self-supervised graph learning, as they are related to this article.

Graph Neural Networks (GNNs). GNNs learn the representation of the graph nodes through aggregating the neighboring information. The learned representations can then be applied to different downstream tasks. Varieties of GNN frameworks have been raised since the concept of GNNs was proposed. Graph Convolutional Networks (GCNs) [12] extend convolution neural networks to graph data. As a widely used GNN, GCN adopts convolution operation to aggregate the features from a node's graph neighborhood. Graph Attention Networks (GATs) [27] introduce attention mechanisms into graph learning. GATs measure the importance of the neighboring features before aggregating them. By comparing the GNNs with the WL test, [31] proposes that GNNs are most powerful as the WL test in distinguishing graphs and proposed Graph Isomorphism Networks (GIN). Our proposed SGL-SGC adopts GCNs as the basic encoder.

Self-supervised Learning. Self-supervised learning, which aims to learn data representations without labels, is a thriving learning approach with multiple applications. Contrastive Predictive Coding (CPC) [16] proposes a self-supervised framework that contrasts predictive features with original features. CMC [24] conducts self-supervised learning by contrasting different views of an image. Similar frameworks were later applied to graph learning. This self-supervised learning approach successfully improves the utilization of unlabeled data. Given the success of these approaches, [2,14] conduct theoretical analysis on the reason behind them. [6,11,15] elaborates on the objectives of self-supervised learning from the perspective of information theory.

With the proposal of GNNs, neural networks based on self-supervised graph learning have become a research hotspot. [35] propose a framework that adopts graph data augmentation to generate different views and maximize the agreement between different representations of different views. [32] propose an approach that adopts the EM algorithm to enhance the representation learning of local and global structures. Our method focuses on reducing the redundancy of information in the learned representations.

3   Methods

This section introduces our proposed Self-supervised Graph Learning with Segmented Graph Channels (SGL-SGC). The architecture of SGL-SGC is illustrated in Fig. 2. SGL-SGC adopts a novel view generator to acquire more independent views than conventional unsupervised graph learning methods. We utilize multiple encoders for representation learning to process these views and a feature-level weight-sensitive loss function for fast and effective training.

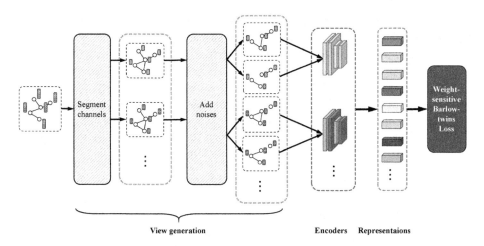

Fig. 2. The structure of SGL-SGC. SGL-SGC can be divided into three parts, including view generation, encoders, and loss function. In conventional self-supervised graph learning, the view generation part usually only consists of data augmentation operations. We, on the other hand, adopt a two-phase method, including segmenting channels and adding noise. SGL-SGC generates multiple expressive views with less redundancy. We also use a feature-level weight-sensitive loss to train the encoders to learn better representations of these views.

3.1 Preliminary

We first recap some preliminary concepts and notations for further exposition. In graph learning, the input attribute graphs can be denoted as $G = (V, E)$, where V is a node set and E is an edge set. V have attributes $\{X_v \in \mathbb{R}^F | v \in V\}$. For each node v, its neighbors are denoted as $\mathcal{N}_v$.

Learning Graph Representations. Given a set of graphs G_i, $i = 1, 2, ..., n$, in some universe $\mathcal{G}$, our objective is to learn the latent representation z_i. z_i preserves the network structures and node attributes of G_i. It can be further used for downstream tasks such as label predicting. Typically, the graph data is fed into graph neural networks (GNNs) to acquire z_i:

$$z_i = GNNs(G_i). \tag{1}$$

Graph Neural Networks. As described earlier, GNNs developed multiple variants, yet their structures still share large similarity. For a graph $G = (V, E)$, a graph neural network layer can be expressed as:

$$h_v^{(k+1)} = combine^{(k)}\left(h_v^k, aggregate^{(k)}(h_u^k, \forall u \in \mathcal{N}_v)\right),$$

where $h^{(k+1)}$ is the representation of node v, acquired by passing the initial node features of v through k layers of graph neural networks. $update(\cdot)$ and *aggregate*(·) are trainable functions. The graph representation z can be obtained by pooling the node representations of the last layer:

$$z = pool(h_v^k, v \in V), \tag{2}$$

Mutual Information Theory. Graph contrastive learning, one of the most popular self-supervised graph learning approaches, defines its learning objective as maximizing the mutual information between the graph and its representation, which is known as the mutual information maximization principle:

$$\max_f I(G; f(G)), \ where \ G \sim \mathbb{P}_\mathcal{G}. \tag{3}$$

$I(\cdot)$ denotes the mutual information between variables. In general, graph contrastive learning achieves mutual information maximization by maximizing the mutual information between different views generated with data augmentation [9,34,35]. Such learning objectives can be expressed as follows:

$$\max_{f_1,f_2} I\Big(f_1(V_1); f_2(V_2)\Big), \ where \ V_1, V_2 \ are \ different \ views \ of \ G, \ G \sim \mathbb{P}_\mathcal{G} \tag{4}$$

$f_1(\cdot)$ and $f_2(\cdot)$ are encoders corresponding to each view. In some methods, the encoders share the same set of parameters. We follow the same learning objective as graph contrastive learning.

3.2   Segmented Graph Channels

We follow [35] and categorize the data augmentation approaches for view generation into four different kinds. **Node dropping** drops a certain amount of nodes along with the edges linked to them. **Edge perturbation** changes the connectivity of the graph by deleting or adding some edges. **Attribute masking** masks are part of the node features. **Subgraph sampling** samples a subgraph from the original graph.

These augmentation methods can be summarized as changing the graph structures or node features. They can be seen as imposing some noise signal S on the original graph data. Depending on the specific content, S could lead to node dropping, edge perturbation, some parts of the features being masked, and making the influenced graph a subgraph of the original one.

Definition 1. *(Graph Data Augmentation with Noise). For a graph G, $q(G, S)$ denote a graph data augmentation of G, where S is a noise signal and $q(\cdot)$ denote the function modifying G according to S. S can be randomly generated or created according to specific rules.*

As we follow the learning objective of graph contrastive learning, with Definition 1 and Eq. 4, we define our learning objective as:

$$\max_{f_1,f_2} I\Big(f_1\big(q(G, S_1)\big); f_2\big(q(G, S_2)\big)\Big), \ where \ G \sim \mathbb{P}_\mathcal{G}. \tag{5}$$

S_1 and S_2 are different noise signals corresponding to different views. e.g., S_1 and S_2 could represent the nodes and edges to be dropped, and $q(\cdot)$ could be the operation that drops them. The mechanism of data augmentation results in that there will still be a large amount of identical node features between $q(G, S_1)$ and $q(G, S_2)$. We denote the optimal choices for the noise signals as S_1^* and S_2^*. Following [23], we propose that:

$$(S_1^*, S_2^*) = \arg\min_{S_1, S_2} I\Big(q(G, S_1); q(G, S_2)\Big)$$
$$s.t.\ I(q(G, S_1); Y) = I(q(G, S_2); Y) = I(G; Y), \tag{6}$$
$$where\ (G, Y) \sim \mathbb{P}_{\mathcal{G} \times \mathcal{Y}}.$$

Ideally, the values of S_1 and S_2 should be chosen to minimize the redundant mutual information between views, which means more modifications will be made to the input graph, e.g., more nodes dropped or edges deleted. Such modifications will lead to an increase in input noises, which will inevitably lead to the corruption of the original input graph. When the graphs get corrupted and changed, they may represent different things. e.g., one node feature of a graph might represent "movie". After the graph is changed, the new node feature might be the same as those representing "paper". Thus, the changed graph has a different set of labels. We denote such labels as Y'. We measure the mutual information between Y and Y' with the following theorem:

Theorem 1. *The mutual information between the original graph label Y and distorted label Y' decreases as the amount of information of input noises S increases.*

Please see Appendix B for proof. Unfortunately, in the task of self-supervised learning, Y' is not available. We have to conduct the training under the assumption that Y is almost the same as Y'. However, suppose we rely on increasing the input noises S to decrease the mutual information between views. In that case, the mutual information between Y' and Y will drop significantly, making the learning meaningless. On the other hand, If we do not increase the input noises that much, there are bound to be identical node features between different views, which will lead to learning redundant information. An alternative is required.

Inspired by contrastive learning algorithms in the computer vision domain [24], we propose the concept of the Segmented Graph Channel (SGC) to generate different views.

Definition 2. *(Segmented Graph Channel). For graph $G = (V, E)$ with attributed node features $\{X_v \in \mathbb{R}^F | v \in V\}$, we denote SGCs of G as C, $C = (V', E)$, V' is a node set that is the same as V except for attributed node features $\{X_{v'} \in \mathbb{R}^{F'} | v' \in V'\}$. $X_{v'}$ is a feature vector that consists of part of the data extracted out of X_v, $F' \le F$. The feature vectors of graph G are split into different parts to get different SGCs. During the generation of $X_{v'}$, the extraction location on each node feature is the same.*

Different from graph augmentations, SGCs generate new views without damaging any graph information. However, SGCs alone can't serve as different views because of their lack of deformation of the graph structure. We combine the SGCs with the data augmentation method and propose our new learning objective:

$$\max_{f_1, f_2} I\Big(f_1\big(q(C_1, S_1)\big); f_2\big(q(C_2, S_2)\big)\Big), \tag{7}$$

where C_1 and C_2 are two SGCs of G. We can completely eliminated identical node features between $q(C_1, S_1)$ and $q(C_2, S_2)$ as we extract different parts of node features. The edge features can be augmented using conventional approaches. Thus, our new learning objective can effectively reduce the redundant mutual information between views compared to conventional data augmentation methods. Furthermore, we achieve such a goal without further introducing noise.

3.3   Network Structure

The network structure of SGL-SGC is demonstrated in Fig. 2. We first adopt SGCs and data augmentation with noise to generate multiple views of the original graph. We will combine each SGC with multiple noise signals to generate different views. We use the same encoder to process the views generated with the same SGC. Parameters are not shared between these encoders.

With k different SGCs $\boldsymbol{C} = \{C_i\}_{i=1}^k$ and m different noise signals $\boldsymbol{S} = \{S_j\}_{j=1}^m$, we could get $k * m$ different views, these views will go through the corresponding encoders to get the representations. For single input graph $G = (V, E)$ with n nodes, we will acquire $k * m * n$ different representations for all nodes. We denote the output representation as $\boldsymbol{z}_{vij}$. $\boldsymbol{z}_{vij}$ can be formulated as:

$$\boldsymbol{z}_{v,i,j} = f_i\big(q(C_i, S_j)\big). \tag{8}$$

These representations will be processed with a loss function. In our task, we want to make use of all of them. Nevertheless, the conventional contrastive loss will require contrasting negative samples with all of the $k * m * n$ different representations for a single input graph, which will cost too much computing resources. Moreover, these representations will contribute differently to training. We want to emphasize those that contribute more.

3.4   Feature-Level Weight-Sensitive Loss

In order to solve the problems mentioned above, we need a loss function that can process the representations in a negative-sample-free way. Inspired by [36], we adopt a feature level learning objective so as to avoid calculating a large amount of negative samples. Given graph G with n nodes, k segmented graph channels, and m augmenters, we will acquire representations $\{\boldsymbol{z}_{1,1,1}, ..., \boldsymbol{z}_{v,i,j}, ..., \boldsymbol{z}_{n,k,m}\}$. Then, we define matrix set $\boldsymbol{M} = \{M_{1,1,2}, ..., M_{v,h,h'}, ... M_{n,k*m-1,k*m}\}$.

The subscript v refers to different nodes, and h, h' refer to different views. $M_{v,h,h'}$ can be formulated as follows:

$$M_{v,h,h'} = z_{v,i,j}^T \times z_{v,i',j'}, \ i \neq i' \ or \ j \neq j' \tag{9}$$

$z_{v,i,j}$ and $z_{v,i',j'}$ are representations of the same node but under different views. Our loss function can be formulated as:

$$\mathcal{L}_{fl} = \sum_{h'}\sum_{h}\sum_{v} \left(OnDiag(M_{v,h,h'}) + \lambda \ OffDiag(M_{v,h,h'}) \right), \ h \neq h' \tag{10}$$

where λ is a hyperparameter that trades off the importance of two terms, $OnDiag(M_{v,h,h'})$ and $OffDiag(M_{v,h,h'})$, we define them as follows:

$$OnDiag(M) = \sum_{a}(1 - m_{a,a})^2,$$

$$OffDiag(M) = \sum_{a}\sum_{a\neq b}(m_{a,b})^2. \tag{11}$$

where $m_{a,a}$ is the element on the diagonal of matrix M, $m_{a,b}$ represents the rest elements. Subscripts a and b are the coordinates. $OnDiag(M)$ implements the optimization objective we described in Eq. 7 at the feature level. As we built our optimization objective without the usage of negative samples, we adopt $OffDiag(M)$ to prevent trivial solutions from optimization.

Following [20], we consider the samples that are further away from the optimal goals more crucial. We adopt weight factors $\omega_{v,h,h'}$ to measure the importance of the representations that are used to calculate $M_{v,h,h'}$. $\omega_{v,h,h'}$ can be denoted as:

$$\omega_{v,h,h'} = \left((OnDiag(M_{v,h,h'}) + \lambda \ OffDiag(M_{v,h,h'})) - O \right)^\tau, \tag{12}$$

where O is the optimal value of the sum of the first two terms, in our task, $O = 0$. τ is a hyperparameter that controls the effect of ω. We use $\omega_{v,h,h'}$ to help emphasize the more crucial samples. Substituting Eq. 12 into Eq. 10, we have:

$$\mathcal{L}_{wsfl} = \sum_{h'}\sum_{h}\sum_{v} \omega_{v,h,h'} \left(OnDiag(M_{v,h,h'}) + \lambda \ OffDiag(M_{v,h,h'}) \right), \ h \neq h' \tag{13}$$

The new loss function is weight-sensitive, which emphasizes representations that are considered more crucial.

4   Experiments

This section demonstrates the effectiveness of our proposed SGL-SGC by conducting extensive experiments on various benchmark datasets.

4.1   Comparison with the State-of-the-Art Methods

Datasets. We select five widely used graph datasets, including three citation network datasets: Cora, Citeseer, and PubMed [3,21], and two relationship datasets: Amazon-Computers and Amazon-Photo [37]. We download all the datasets with DGL APIs, which can be found at https://www.dgl.ai/. For the experimental protocol, we follow [9,37], and adopt the same train/validation/test splits. We report the mean classification accuracy with standard deviation over ten runs of training.

Baselines. For baselines, we select supervised, semi-supervised and unsupervised graph learning approaches. The supervised approaches include GCN [12] and GAT [27]. The semi-supervised approaches include CG^3 [29]. The unsupervised graph learning approaches include Deepwalk [18], GAE [13], DGI [28], MVGRL [9], GCA [37], and InfoGCL [30].

Table 1. Classification accuracy of compared methods on Cora, CiteSeer, PubMed, Amazon Computers, and Amazon Photos. According to different learning strategies, the records are divided into two groups. The records that are not associated with standard deviations due to the reason that they are directly taken from [22], which did not report their standard deviations. **Bold** denotes the best records.

Methods	Cora	CiteSeer	PubMed	Amazon Computers	Amazon Photo
Supervised & Semi-Supervised Approaches					
GCN	81.5	70.3	79.0	87.0 ± 0.3	92.6 ± 0.4
GAT	83.0 ± 0.7	72.5 ± 0.7	79.0 ± 0.3	86.5 ± 0.5	92.4 ± 0.2
CG^3	83.4 ± 0.7	73.6 ± 0.8	80.2 ± 0.8	79.9 ± 0.6	89.4 ± 0.5
Unsupervised Approaches					
DeepWalk	70.7 ± 0.6	51.4 ± 0.5	74.3 ± 0.9	85.7 ± 0.1	89.4 ± 0.1
GAE	71.5 ± 0.4	65.8 ± 0.4	72.1 ± 0.5	85.3 ± 0.2	91.6 ± 0.1
DGI	83.8 ± 0.5	72.0 ± 0.6	77.9 ± 0.3	84.0 ± 0.5	91.6 ± 0.2
MVGRL	83.2 ± 0.6	72.9 ± 0.3	79.8 ± 0.6	87.5 ± 0.1	91.7 ± 0.1
GCA	82.1 ± 0.4	71.7 ± 0.2	78.9 ± 0.7	87.9 ± 0.3	92.5 ± 0.2
InfoGCL	83.5 ± 0.3	73.5 ± 0.4	79.1 ± 0.2	–	–
SGL-N	83.1 ± 0.7	73.2 ± 0.5	79.5 ± 0.2	85.6 ± 0.3	91.6 ± 0.2
SGL-SGC	$\mathbf{84.2 \pm 0.5}$	$\mathbf{74.0 \pm 0.3}$	$\mathbf{80.8 \pm 0.4}$	$\mathbf{88.7 \pm 0.2}$	$\mathbf{93.1 \pm 0.3}$

Evaluation Protocol. For evaluation protocol, we follow [28] and pre-train the model on all the nodes in the graph without supervision. Then, we freeze the parameters and feed the acquired node representations into a logistic regression model for label prediction. We only use nodes from the training set to train the logistic regression model, and we report the classification accuracy on testing sets.

We adopt the Adam optimizer with an initial learning rate of 10^{-3} for model training. For view generation, we used a total of three SGCs. Each SGC is followed by two data augmentations with different noise signals. SGL-SGC generates six views in total. We adopt three different encoders, each corresponding to an SGC. Each encoder consists of a 2-layer GCN with a hidden dimension of 512. Their outputs are concatenated together for downstream tasks. For Cora, Citeseer, and PubMed datasets, the pre-training epochs were 100, 20, and 100. For Amazon-Computers and Amazon-Photo, the pre-training epochs were set as 60. The hyperparameter that controls the effect of ω is set to 0.2. All of our experiments were conducted on an Nvidia RTX 5000. For the ablation study, we built a network with the same structure as SGL-SGL except for the SGCs. We remove them and generate the same amount of views as SGL-SGC with conventional data augmentations. The new network is named SGL-N.

Results. The classification results are reported in Table 1 . We highlight the highest records in bold. As we can see from the table, SGL-SGC outperforms all the other methods across all datasets. The results demonstrate our method's potential to outperform supervised, semi-supervised, and unsupervised methods on various datasets. We attribute this potential to the fact that SGL-SGC can generate views that contain less redundant information. Moreover, we design a feature-level weights-sensitive loss function that can be used to train the encoders better to learn from these views.

Another observable phenomenon is that SGL-N can only achieve comparable results to other methods, while SGL-SGC outperforms it. This outcome proves that only utilizing six different views generated with graph data augmentation does not help produce better performances. Furthermore, it proves the necessity of our proposed SGCs in helping increase the performance of self-supervised representation learning.

4.2 Comparison of Computing Resource Consumption

To analyze the computational resource overhead of our method, we conduct a set of comparative experiments. For comparison, We built a graph contrastive learning framework utilizing conventional InfoNCE loss instead of feature-level weight-sensitive loss, named I-GCL. I-GCL adopts two-layer GCNs as elemental encoders, the same as SGL-SGC. We use the same augmenter for each framework.

The results are demonstrated in Table 2. It shows that SGL-SGC costs much less memory than I-GCL under six views, which proves our proposed feature-level weight-sensitive loss can vastly reduce computing costs. Another interesting phenomenon is that SGL-SGC with six views still costs less memory than I-GCL with two views. Such records prove that, in our task, SGL-SGC does not cost more computing resources than conventional graph contrastive methods. We can also see from the table that when the amount of views increases, the memory cost of I-GCL rises by 47%. On the other hand, the memory cost of SGL-SGC only rises by 16%. Such results suggest that increasing the number of segmented graph channels does not significantly increase the computational cost when using a feature-level weight-sensitive loss. Furthermore, the time cost of SGL-SGC is also less than those of I-GCL.

Table 2. Memory costs and time costs of different methods under different conditions. Since the network width of each layer is the same, a single column of hidden dimensions is used to represent them. Views and SGCs represent the number of graph views and SGCs we used. We conduct all the experiments on an Nvidia RTX 5000.

Methods	Hidden dimension	Views	SGCs	Memory costs (GB)	Time costs per Epoch (second)
I-GCL	512	2	1	2.82	0.13
I-GCL	512	6	3	5.35	0.16
SGL-SGC	512	2	1	1.78	0.08
SGL-SGC	512	6	3	2.12	0.09
I-GCL	256	2	1	1.96	0.10
I-GCL	256	6	3	3.89	0.11
SGL-SGC	256	2	1	1.34	0.06
SGL-SGC	256	6	3	1.73	0.08

4.3   Evaluation of the Weight Factors

In this part, we further evaluate the weight factor ω that we introduced in our proposed loss function. We modify the value of hyperparameter τ that controls the effectiveness of ω and observe how the performance changes. We perform such experiments on multiple datasets. The results are shown in Fig. 3.

As we can see, the performance peaks when the value of τ is 0.2 on all three datasets. As τ decreases, ω will hold less influence on training. It is observable that the performance of SGL-SGC drops when τ decreases from 0.2 to 0, which indicates that the influence of ω does improve the representation learning ability of SGL-SGC. We believe ω help emphasize the samples that contribute more to training, thus improving the overall performance. Another observable phenomenon is that when τ takes a larger value than 0.2, the performance of SGL-SGC also drops. This phenomenon shows that the effect of ω cannot be expanded indefinitely, and it is necessary to use the hyperparameter τ to limit it.

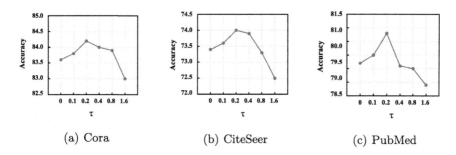

(a) Cora   (b) CiteSeer   (c) PubMed

Fig. 3. Results of SGL-SGC's performance on Cora, CiteSeer, and PubMed datasets with different values of τ. The ordinates in the figure represent different accuracy rates, while the abscissas represent different values of τ .

4.4   Representation Capability Analysis

For further analysis of the representation capability of our proposed method, we visualize the outputs to make an intuitive observation. For comparison, we adopt a deformation of SGL-SGC with InfoNCE loss and only two different views without SGCs. The new framework is named IN-GCL.

Figure 4 demonstrates the results. We can see that the untrained encoder output features without much distinguishability. We can observe many vertical lines running through multiple blocks of different labels. These lines represent similar representations, indicating there are common features shared between different classes. The output of SGL-SGC under ten epochs of training shows some interesting developments. The vertical lines of each block become clearer than in the previous column, and there is still not much distinguishability. For

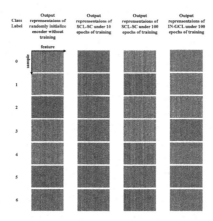

Fig. 4. Visualized output representations of different frameworks. Each horizontal line in each small block represents an output representation. In each small block, the horizontal axis represents different dimensions of the output representation, and the vertical axis represents different samples. All representations are grouped by category.

the outputs of SGL-SGC after 100 epochs of training, we can see much difference between each class. We believe that SGL-SGC will first increase the independence of each dimension of the feature during the training process, making the output more expressive. After that, the distinction between the various classes appears, indicating that the encoders have learned meaningful information.

The last column of Fig. 4 shows the output of IN-GCL without feature-level weight-sensitive loss and SGCs. It can be seen that after the same 100 rounds of training, SGL-SGC can learn more discriminative features than IN-GCL, which demonstrate the superiority of our method.

5   Conclusions

This paper proposed a self-supervised graph learning method with segmented graph channels. We enhance the conventional view generation with segmented graph channels to reduce the redundant mutual information between multiple views while avoiding introducing more noises. We also proposed a feature-level weight-sensitive loss as our training objective. This loss function can emphasize samples with more contribution to training and reduce consumption of computing resources. We conducted multiple experiments to prove the superiority of our proposed method.

Acknowledgements. This work is supported by the National Key Research and Development Program of China (No. 2019YFB1405100).

References

1. Allamanis, M., Brockschmidt, M., Khademi, M.: Learning to represent programs with graphs. In: International Conference on Learning Representations (2018)
2. Arora, S., Khandeparkar, H., Khodak, M., Plevrakis, O., Saunshi, N.: A theoretical analysis of contrastive unsupervised representation learning. In: 36th International Conference on Machine Learning, ICML 2019, pp. 9904–9923. International Machine Learning Society (IMLS) (2019)
3. Bojchevski, A., Günnemann, S.: Deep gaussian embedding of graphs: Unsupervised inductive learning via ranking. In: International Conference on Learning Representations (2018)
4. Chen, T., Kornblith, S., Norouzi, M., Hinton, G.: A simple framework for contrastive learning of visual representations. In: International conference on machine learning, pp. 1597–1607. PMLR (2020)
5. Duvenaud, D.K., et al.: Convolutional networks on graphs for learning molecular fingerprints. In: Advances in neural information processing systems, vol. 28 (2015)
6. Federici, M., Dutta, A., Forré, P., Kushman, N., Akata, Z.: Learning robust representations via multi-view information bottleneck. In: International Conference on Learning Representations (2019)
7. Grill, J.B., et al.: Bootstrap your own latent-a new approach to self-supervised learning. Adv. Neural. Inf. Process. Syst. **33**, 21271–21284 (2020)

8. Grover, A., Zweig, A., Ermon, S.: Graphite: Iterative generative modeling of graphs. In: International conference on machine learning, pp. 2434–2444. PMLR (2019)
9. Hassani, K., Khasahmadi, A.H.: Contrastive multi-view representation learning on graphs. In: International Conference on Machine Learning, pp. 4116–4126. PMLR (2020)
10. He, K., Fan, H., Wu, Y., Xie, S., Girshick, R.: Momentum contrast for unsupervised visual representation learning. In: Proceedings of the IEEE/CVF Conference on Computer Vision and Pattern Recognition, pp. 9729–9738 (2020)
11. Hjelm, R.D., et al.: Learning deep representations by mutual information estimation and maximization. In: International Conference on Learning Representations (2018)
12. Kipf, T.N., Welling, M.: Semi-supervised classification with graph convolutional networks. arXiv preprint arXiv:1609.02907 (2016)
13. Kipf, T.N., Welling, M.: Variational graph auto-encoders. In: Bayesian Deep Learning Workshop@NIPS (2016)
14. Lee, J.D., Lei, Q., Saunshi, N., Zhuo, J.: Predicting what you already know helps: Provable self-supervised learning. In: Advances in Neural Information Processing Systems, vol. 34 (2021)
15. Linsker, R.: Self-organization in a perceptual network. Computer **21**(3), 105–117 (1988)
16. Oord, A.v.d., Li, Y., Vinyals, O.: Representation learning with contrastive predictive coding. arXiv preprint arXiv:1807.03748 (2018)
17. Peng, Z., et al.: Graph representation learning via graphical mutual information maximization. In: Proceedings of The Web Conference 2020, pp. 259–270 (2020)
18. Perozzi, B., Al-Rfou, R., Skiena, S.: Deepwalk: Online learning of social representations. In: Proceedings of the 20th ACM SIGKDD international conference on Knowledge discovery and data mining, pp. 701–710 (2014)
19. Sanchez-Gonzalez, A., et al.: Graph networks as learnable physics engines for inference and control. In: International Conference on Machine Learning, pp. 4470–4479. PMLR (2018)
20. Schroff, F., Kalenichenko, D., Philbin, J.: Facenet: A unified embedding for face recognition and clustering. In: Proceedings of the IEEE Conference on Computer Vision and Pattern Recognition, pp. 815–823 (2015)
21. Sen, P., Namata, G., Bilgic, M., Getoor, L., Galligher, B., Eliassi-Rad, T.: Collective classification in network data. AI Mag. **29**(3), 93–93 (2008)
22. Sun, F.Y., Hoffman, J., Verma, V., Tang, J.: Infograph: Unsupervised and semi-supervised graph-level representation learning via mutual information maximization. In: International Conference on Learning Representations (2020)
23. Suresh, S., Li, P., Hao, C., Neville, J.: Adversarial graph augmentation to improve graph contrastive learning. In: Advances in Neural Information Processing Systems, vol. 34 (2021)
24. Tian, Y., Krishnan, D., Isola, P.: Contrastive multiview coding. In: Vedaldi, A., Bischof, H., Brox, T., Frahm, J.-M. (eds.) ECCV 2020. LNCS, vol. 12356, pp. 776–794. Springer, Cham (2020). https://doi.org/10.1007/978-3-030-58621-8_45
25. Tishby, N., Pereira, F.C., Bialek, W.: The information bottleneck method. arXiv preprint physics/0004057 (2000)
26. Tishby, N., Zaslavsky, N.: Deep learning and the information bottleneck principle. In: 2015 IEEE Information Theory Workshop (itw), pp. 1–5. IEEE (2015)
27. Veličković, et al.: Graph attention networks. arXiv preprint arXiv:1710.10903 (2017)

28. Velickovic, P., Fedus, W., Hamilton, W.L., Liò, P., Bengio, Y., Hjelm, R.D.: Deep graph infomax. ICLR (Poster) **2**(3), 4 (2019)
29. Wan, S., Pan, S., Yang, J., Gong, C.: Contrastive and generative graph convolutional networks for graph-based semi-supervised learning. In: Proceedings of the AAAI Conference on Artificial Intelligence. vol. 35, pp. 10049–10057 (2021)
30. Xu, D., Cheng, W., Luo, D., Chen, H., Zhang, X.: Infogcl: Information-aware graph contrastive learning. In: Advances in Neural Information Processing Systems, vol. 34 (2021)
31. Xu, K., Hu, W., Leskovec, J., Jegelka, S.: How powerful are graph neural networks? In: International Conference on Learning Representations (2018)
32. Xu, M., Wang, H., Ni, B., Guo, H., Tang, J.: Self-supervised graph-level representation learning with local and global structure. In: International Conference on Machine Learning, pp. 11548–11558. PMLR (2021)
33. Xu, X., Feng, W., Jiang, Y., Xie, X., Sun, Z., Deng, Z.H.: Dynamically pruned message passing networks for large-scale knowledge graph reasoning. In: International Conference on Learning Representations (2019)
34. Yang, L., Zhang, L., Yang, W.: Graph adversarial self-supervised learning. In: Advances in Neural Information Processing Systems, vol. 34 (2021)
35. You, Y., Chen, T., Sui, Y., Chen, T., Wang, Z., Shen, Y.: Graph contrastive learning with augmentations. Adv. Neural. Inf. Process. Syst. **33**, 5812–5823 (2020)
36. Zbontar, J., Jing, L., Misra, I., LeCun, Y., Deny, S.: Barlow twins: Self-supervised learning via redundancy reduction. In: International Conference on Machine Learning, pp. 12310–12320. PMLR (2021)
37. Zhu, Y., Xu, Y., Yu, F., Liu, Q., Wu, S., Wang, L.: Graph contrastive learning with adaptive augmentation. In: Proceedings of the Web Conference 2021, pp. 2069–2080 (2021)

TopoAttn-Nets: Topological Attention in Graph Representation Learning

Yuzhou Chen[1,4]([✉]), Elena Sizikova[2], and Yulia R. Gel[3,5]

[1] Department of Electrical and Computer Engineering, Princeton University, Princeton, USA
yc0774@princeton.edu
[2] Center for Data Science, New York University, New York, USA
es5223@nyu.edu
[3] Department of Mathematical Sciences, University of Texas at Dallas, Dallas, USA
ygl@utdallas.edu
[4] Lawrence Berkeley National Laboratory, Berkeley, USA
[5] National Science Foundation, Virginia, USA

Abstract. Topological characteristics of graphs, that is, properties that are invariant under continuous transformations, have recently emerged as a new alternative form of graph descriptors which tend boost performance of graph neural networks (GNNs) on a wide range of graph learning tasks, from node classification to link prediction. Furthermore, GNNs coupled with such topological information tend to be more robust to attacks and perturbations. However, all prevailing topological methods for GNNs consider a scenario of a fixed learning approach and do not allow for distinguishing between topological noise and topological signatures of the graph which might be the most valuable for the current learning task. To exploit the inherent task-specific topological graph descriptors, we propose a new versatile framework known as Topological Attention Neural Networks (TopoAttn-Nets) (Our code is available at https://github.com/TopoAttn-Nets/TopoAttn-Nets.git). As the first meta-representation of topological knowledge, TopoAttn-Nets employs the attention operation on both local and global data properties and offers their geometric augmentation. We derive theoretical guarantees of the proposed topological learning framework and evaluate TopoAttn-Nets in conjunction with graph classification. TopoAttn-Nets delivers the highest accuracy, outperforming 26 state-of-the-art classifiers on benchmark datasets.

Keywords: Meta-representation · Topological signatures · Representation learning · Graph classification

1 Introduction

Accurately classifying graphs by inferring their geometric and topological properties has recently witnessed an ever increasing interest in many data science

---

Supplementary Information The online version contains supplementary material available at https://doi.org/10.1007/978-3-031-26390-3_19.

M.-R. Amini et al. (Eds.): ECML PKDD 2022, LNAI 13714, pp. 309–325, 2023.
https://doi.org/10.1007/978-3-031-26390-3_19

applications [6,10,33]. In particular, an emerging sub-field of geometric deep learning (GDL) aims to generalize the concept of deep learning (DL) to data in non-Euclidean spaces by bridging the gap between graph theory and deep neural networks [3]. In turn, many recent studies indicate that integration of topological descriptors, i.e., systematic shape characteristics, into graph learning often results in noticeable performance gains in such tasks as graph classification, link prediction, and anomaly detection [6,12,18,33,46,47]. Furthermore, incorporating the topological signatures into GDL enhances robustness of graph learning to perturbations and attacks. This phenomenon can be explained by important complementary information and deeper insight into the intrinsic graph organizational structure provided by topological data summaries, as compared to conventional non-topological descriptors. Here we aim to further advance topological approaches to graph learning by offering a systematic and versatile framework for extracting the essential *task-specific shape* information.

In particular, topological data analysis (TDA) offers rigorous mathematical tools to explore structural shape properties of the graph-structured data [4,9,14]. Here by shape we broadly understand data properties which are invariant under continuous transformations such as stretching, bending, and twisting. Persistence homology (PH) is a methodology under the TDA framework that analyzes evolution of various patterns in a graph $\mathcal{G}$ as we vary certain user-selected (dis)similarity threshold (i.e., a scale). As such, we can say that PH studies the observed graph $\mathcal{G}$ at multiple resolutions or evaluates its structural properties through multiple lenses. All extracted shape patterns can be then summarized in a form of multi-set in $\mathbb{R}$, known as a persistence diagram (PD). PDs record a type of the topological patterns we detect as well as how long we observe each topological feature as a function of the scale parameter. We are particularly interested in topological features with a longer lifespan, since such features tend to contain valuable information about hidden mechanisms behind graph organization and as such, play a more important role in graph learning. Features with a longer lifespan are said to persist. In turn, features with shorter lifespans are likely to be attributed to topological noise. However, there exists a number of interlinked fundamental challenges on the way of successful integration of topological information into graph learning. The first key problem is how to distinguish important topological features from topological noise [8,9,15]. Second, since PDs are point multi-sets, there exists no straightforward approach to combine the extracted topological summaries in a form of PDs with DL models, as DL often requires input data in vector form. As such, there are multiple approaches to make PDs compatible with DL inputs [1,18,24]. One of the most popular PD representations allowing for construction of a fully trainable topological layer is adaptively kernelization of PDs. However, existing kernel representations of PDs assume that influence of persistent features on the learning process is *fixed*. Furthermore, typically only a single PD is computed from the graph $\mathcal{G}$–, either upon extracting topological features directly from $\mathcal{G}$, referred to as the *topological domain*, or from the spectral signatures of $\mathcal{G}$, (e.g., Heat Kernel Signatures (HKS) with a single (fixed) diffusion parameter t), referred to as the *spectral domain*. As such, the current kernel representations of PDs do not allow for distinguishing topological graph characteristics which are *the most valuable for the current learning task*, from topological noise.

New Topological Meta-Representation Paradigm. We propose a new flexible and unified framework, TopoAttn-Nets, for meta-representation topological signatures of the graph $\mathcal{G}$ extracted from its PDs. That is, we instill topological signatures from different domains and embed them into meta-representation with attention mechanism which shows an end-to-end learning approach that in turn can be used to learn multiple persistence representations. Furthermore, inspired by the recent meta-learning mechanisms in deep neural networks [20], we combine all kernel-based representation of PDs in various domains into a joint aggregated attention layer, where attention mechanism is used to explicitly encode the structural information of $\mathcal{G}$ from a global perspective. The resulting TopoAttn-Nets represents a trainable, task-specific framework to extract the most informative topological signatures of graph $\mathcal{G}$ from multiple domains in an efficient and provably stable manner.

Contributions. Contrary to all conventional TDA methods in DL where a given task is tackled using a *fixed* learning approach, this paper aims to enhance the topological learning algorithm itself, thereby being the first step toward the paradigm of *topological meta-learning*. The ultimate idea of TopoAttn-Nets is to systematically integrate joint topological features, persistence-based information from multiple domains, and PD transform learning. Specifically, compared to all previous approaches for topological features/kernels/layers, our meta-representation: (1) is not restricted to a particular type of input data and a fixed parametrization map of topological summaries, (2) is more robust to perturbations, (3) allows for learning relationships among topological signatures by providing their geometric augmentation. As a part of the new topological meta-representation, the attention mechanism learns to focus on the most essential topological characteristics of the data and learning algorithms. This is particularly important for web-based data, e.g., usage graphs from social media or other web sources, that exhibit variation at different scales. Capturing both finer scale and larger scale variations using a fixed learning model is challenging. In contrast, TopoAttn-Nets offers a representation that captures both local and global properties, and as a result, improved tractability and generalization performance. Our extensive numerical results indicate that TopoAttn-Nets is competitive in graph classification in comparison to the state of the art: it outperforms 26 top methods in accuracy and is more robust under graph perturbations.

2   Related Work

Kernels for Graph Classification. Traditionally, one of the most popular graph classification tools over the past two decades were graph kernel approaches. There is a wide variety of graph kernel frameworks, including marginalized kernel [21], shortest-path kernel [2], graphlet kernel [35], Weisfeiler-Lehman graph kernel [34], and Weisfeiler-Lehman hash graph kernel [29]. These more classical graph-based kernels only consider generating graph level features through aggregating node representations. While powerful and expressive, the existing kernel-based techniques suffer from limited ability to capture similarities among higher order graph properties of local neighborhoods which in contrast can be inferred from topological structures. To address this limitation, we propose a new flexible topological meta-representation neural network model which coupled with attention mechanism, enables the graph-based learning framework to systematically incorporate higher order graph information both at the local and global levels.

Neural Networks for Graph Classification. There generally exist three neural network-based approaches for graph classification: (i) GNN architectures that encode both local graph structure and features of nodes [22, 26, 28, 39, 41], (ii) stable vectorizations of PDs within GNNs [1, 46] or embedding multiple graph filtrations [19], and (iii) kernelization of topological information within GNNs [19, 24, 45, 47]. In contrast, our approach is built upon meta-representation of *multiple* kernelized PDs, that is, choice of topological meta-knowledge to meta-learn. Armed with the proposed meta-representation machinery, we can then exploit the relations between tasks or domains, and learning algorithms.

3   Background on Persistent Homology

Let $\mathcal{G} = (\mathcal{V}, \mathcal{E})$ be the observed graph, where $\mathcal{V}$ denotes the set of nodes, $\mathcal{E}$ denotes the set of edges, and $e_{uv} \in \mathcal{E}$ denoting an edge between nodes $u, v \in \mathcal{V}$. The fundamental postulate is to view $\mathcal{G}$ as a sample from some metric space $\mathbb{M}$ whose intrinsic topological structure has been lost due to sampling. Our goal is then to regain knowledge on the lost structural properties of $\mathbb{M}$ via characterizing shape of the observed graph $\mathcal{G}$. The key approach here is to first associate $\mathcal{G}$ with some filtration of $\mathcal{G}$: let $\mathcal{G}_1 \subseteq \mathcal{G}_2 \subseteq \ldots \subseteq \mathcal{G}_k = \mathcal{G}$ be a nested sequence of subgraphs, and let $\mathcal{C}_i$ be the simplicial complex induced by the subgraph $\mathcal{G}_i$ (e.g., clique complex). Then, the nested sequence of these simplicial complexes $\mathcal{C}_1 \subseteq \mathcal{C}_2 \subseteq \ldots \subseteq \mathcal{C}_k$ is called *a filtration* of $\mathcal{G}$. We then can track lifespan of shape characteristics of $\mathcal{G}$ throughout this nested sequence of simplicial complexes. Such shape features include connected components, loops, cavities, and more generally k-dimensional holes. We detect them by means of *a homology*, an algebraic topological invariant. To define the lifespan of a topological feature, we say that the feature is born at $\mathcal{G}_b$ if it does not come from $\mathcal{G}_{b-1}$, and it dies at $\mathcal{G}_d$ ($d \geq b$) if the feature disappears entering $\mathcal{G}_d$ [5]. Hence, its corresponding lifespan, or *persistence* is $d - b$. The resulting persistent homology can be then coded as a multi-set $\mathcal{D}$ of points in $\mathbb{R}^2$, called a PD, with x and y coordinates being the birth and death of each topological feature, respectively. Since $d \geq b$, all points in $\mathcal{D}$ are in the half-space on or above $y = x$. The multiplicity of a point $(b, d) \in \Omega = \{(x, y) \in \mathbb{R}^2 : y > x\}$ is the number of k-dimensional topological features that are born at b and die at d, while points at the diagonal $\Delta = \{(b, b) | b \in \mathbb{R}\}$ have infinite multiplicities. Finally, there exist multiple approaches to construct a filtration of $\mathcal{G}$ [9]. One common method is to use a descriptor function (usually conveys domain information) $f : \mathcal{V} \to \mathbb{R}$ and a sequence of real numbers $a_1 <$

$a_2 < \cdots < a_k$, one can define a nested sequence of subgraphs with $\mathcal{G}_i = (\mathcal{V}_i, \mathcal{E}_i)$ where $\mathcal{V}_i = \{v \in \mathcal{V} | f(v) \le a_i\}$ and $\mathcal{G}_i$ is the induced subgraph of $\mathcal{G}$ by $\mathcal{V}_i$, i.e., $\mathcal{E}_i = \{e_{uv} \in \mathcal{E} | u, v \in \mathcal{V}_i\}$. Similarly, for a weighted graph $\mathcal{G} = (\mathcal{V}, \mathcal{E}, w)$ and a sequence of real numbers $a'_1 < a'_2 < \cdots < a'_s$, one can use the weights to define $\mathcal{G}_j = (\mathcal{V}_j, \mathcal{E}_j)$ with $\mathcal{E}_j = \{e_{uv} \in \mathcal{E} | w_{uv} \le a'_j\}$ and $\mathcal{V}_j = \{v \in \mathcal{V} | e_{uv} \in \mathcal{E}_j\}$.

4 Learnable Topological Meta-Representation for Deep Attention Networks

4.1 Persistence Meta-Representation

In spirit of recent approaches to learnable PD vectorizations [6,18,24], we define an individual representation function s of $\mathcal{D}$ as a composite function of three point transformations in $\mathbb{R}^2$: $s = k \circ \tau_\eta \circ \rho : \Theta \to \{f : \Omega \cup \Delta \to \mathbb{R}\}$, where k is a parametrized functional (e.g., the Gaussian kernel) such that $k(x, -\infty) = 0$, $\rho : \mathbb{R}^2 \to \mathbb{R}^2$ is a linear birth-lifetime coordinate transform such that $\rho(x, y) = (x, y - x)$, τ_η is a rationally stretched birth-lifetime, or spike point transform $\tau_\eta : \mathbb{R} \times [0, \infty] \to \mathbb{R} \times (\mathbb{R} \cup \{-\infty\})$, $\eta > 0$, and Θ is a parameter space. Representation of s as a composite function allows us to study PD parametrization over $\mathbb{R}^2$ and, hence, enables a more tractable mathematical formalism and application of a broader range of weighting functions to distinguish topological features in terms of their contribution to the learning task.

Based on the PH framework, we can obtain a set of different representation of topological signatures for the same input graph $\mathcal{G}$ by (i) considering different choices of simplicial complexes, (ii) using different filtering functions, and (iii) defining $\mathcal{G}$ on different domains. Our idea is to harness complementary information from multiple PDs and their learnable representations, hence, capitalizing on the concepts of *meta-analysis*. In particular, here we focus on representation learning of persistence diagrams with respect to two domains. Armed with the set of learnable representations $s = \{s_1, s_2, \ldots, s_Q\}$ and a collection of PDs $\mathcal{D} = \{\mathcal{D}_1, \mathcal{D}_2, \ldots, \mathcal{D}_Q\}$, we propose an aggregated, i.e., a meta-representation, of multiple PDs. We first assign each dimension $i \in \{1, \cdots, Q\}$ a 2-dimensional base representation $s_i(x, y)$ (where (x, y) belongs to $\mathcal{D}_i$) and construct an aggregated representation with n-th order as: $\mathfrak{s}_{agg_n}(x, y) = \underset{1 \le i_1 < i_2 < \cdots < i_n \le Q}{\mathcal{I}} \left[\omega_{i_1, \ldots, i_n} \times \phi(s_{i_1}, \ldots, s_{i_n}) \right]$, where Q is the dimension of the input space; $\mathcal{I}[\cdot]$ refers to the aggregation scheme such as sum and average; function $\phi(\cdot)$ takes multiple base representations as input and outputs to a new representation – *meta-representation*; ω is a weight controlling the effect of corresponding *meta-representation*. For the sake of notation, we omit indices of the base representation as $s_i(\cdot)$. In particular, when $n = Q$, the Q-th order *meta-representation* can be written of the form: $\mathfrak{s}_{agg_Q}(x, y) = \phi(s_1, \ldots, s_Q)$.

To extract topological signatures from a graph $\mathcal{G}$, we can compute persistent homology directly from the observed graph $\mathcal{G}$ and from spectral descriptors of $\mathcal{G}$. The resulting persistence-based summaries contain complementary information and can be plugged into compatible learning representations via different kernel types.

Option 1. *Spectral domain:* Following [6, 32], we compute $\mathcal{D}$ for graph by replacing original filtration with HKS for fixed diffusion parameter t as the feature function. Given a real-valued function $h(\cdot;\cdot) : \mathbb{R}_+^2 \to \mathbb{R}$, set $h(t;\lambda_k) = e^{-t\lambda_k}$. The HKS $p(\cdot;\cdot) : \mathbb{R}^2 \to \mathbb{R}$ is defined as $p_t(x,y) = \sum_{i=0}^\infty e^{-\lambda_i t}\varphi_i(x)\varphi_i(y)$, where λ_i and φ_i are the i-th eigenvalue and the i-th eigenfunction of the Laplace-Beltrami operator, respectively. HKS on a graph $\mathcal{G}$ can be represented as $p : v \to \sum_{i=1}^n e^{-\lambda_i t}\varphi_i^2(v)$, where v is a node of $\mathcal{G}$, λ_i and φ_i are eigenvalues and eigenvectors of the normalized graph Laplacian. Since the heat kernel can be viewed as a *low-pass filter*, HKS contains information mainly from low frequencies (and hence higher frequencies are suppressed by increasing t). To capture all the low and high frequencies in $\mathcal{G}$, we use a *meta-representation* to include multiple PDs extracted from HKS with various diffusion parameters.

Option 2. *Topological domain:* To make the model invariant to changes in position and orientation, rotation has been shown to significantly increase classification and segmentation performance [13, 23]. The key operation in the topological domain is to produce transformed training samples of PDs and feed them to the DL model. For a persistence diagram $\mathcal{D}$, rotation augmentations are done by rotating the points on the x- and y-coordinates by θ degrees. In machine learning terminology, these coordinates can be referred to as features. This allows us to characterize the $\mathcal{D}$ generated by each data point as a compact feature vector. The application of rotation augmentation to $\mathcal{D}$ allows us to encode importance of different topological summaries in a vector representation.

Learnable PD Representation in the Topological Domain. Let $R_\theta : \mathbb{R}^2 \to \mathbb{R}^2$ be a *rotational* operator for rotation by an angle θ. Applying R_θ to $\mathcal{D}$ results in $R_\theta(\mathcal{D}) = \mathcal{D}_\theta = \{(\cos(\theta)x + \sin(\theta)y, \cos(\theta)y - \sin(\theta)x) \in \mathbb{R}^2 | (x,y) \in \mathcal{D}\}$. A PD can be rotated by multiple angles $\boldsymbol{\theta} = \{\theta_1, \ldots, \theta_\aleph\}$, $\aleph \geq 2$, where either θ_i is sampled from the uniform distribution $U(0,\pi)$ or $\boldsymbol{\theta}$ is a deterministic sequence of angles. Number of rotated angles $\aleph$ is user-specified to meet computational constraints. This rotational procedure provides a set of candidate latent features for meta-learning [7, 20].

What are Advantages of the PD Random Rotation? Random rotation of a PD achieves two interlinked goals: (i) improves the extraction of prominent topological information from PDs and (ii) enhances learning the ring of algebraic functions on PDs. On (i), topological features near the diagonal Δ exhibit a higher level of uncertainty but may still contain useful information for classification tasks [27]. Indeed, since we cannot explicitly define how close a feature ought to be to Δ in order to be viewed as topological noise, we aim to extract the signal out of such features under uncertainty. Note that since $\theta \sim U(0,\pi)$, the range of topological feature lifespan in the rotation image $R_\theta(\mathcal{D})$ is $(y - x, -y + x)$. As a result, features with a shorter lifespan in the original unrotated space are stretched in the rotated space and may have a longer lifespan. That is, intuitively, while we still give a higher weight to more persistent features in the original space, upon rotation with a random angle θ, we attempt to assign topological features whose original lifespan may be shorter due, e.g., to various uncertainties, a chance to contribute to the topological learning. Since $\mathbb{E}[\theta] = \pi/2$ and, hence, the expected rotated lifespan of each topological feature translates to its mean point in the unrotated space, and vice versa, we still incorporate the conventional lifespan characterization of PD. As a result, we extract more signal out of all available topological features than the standard TDA tools (i.e., in (i)), while the attention mechanism mitigates the impact of including the potential topological noise. On (ii), random sampling of θ in the rotation operator R_θ allows us to enrich the set of elements of the affine coordinate ring (i.e., functions on the coordinatized PD space), thereby improving learning of the associated algebraic variety under uncertainties. Such random rotation may be also viewed as a semi-parametric bootstrap of lifespans of each topological feature. To infer potential long-range and periodic relations in the *rotational* transformation of PDs, we propose the generalized locally periodic (GLP) kernel for rotated PDs.

Definition 1. *Let $p_i, l_i, \mu_i, \alpha_i \in \mathbb{R}$, $i = 1, 2$. Then the generalized locally periodic (GLP) kernel is nonnegative function $\mathbb{R}^2 \to \mathbb{R}_+$ is defined as:*

$$
\begin{aligned}
k_{GLP}(x, y) = \sigma^2 e^{\left\{ -2\sin^2\left(\frac{\pi(x-\alpha_1)^2}{p_1} \right) - \frac{(x-\mu_1)^2}{2l_1^2} \right\}} \\
\times e^{\left\{ -2\sin^2\left(\frac{\pi(y-\alpha_2)^2}{p_2} \right) - \frac{(y-\mu_2)^2}{2l_2^2} \right\}}.
\end{aligned}
\tag{1}
$$

The advantages of the generalized locally periodic (GLP) kernel are as follows: (i) compared to the Gaussian kernel, it is more appropriate to adopt a periodic kernel that can reflect the similarities between different PDs and (ii) strict periodicity is too rigid (i.e., the purely periodic kernel) since variance exists.

Lemma 1. *The GLP kernel $k_{GLP}(x, y)$ is (a) Lipschitz continuous on $\mathbb{R}^2$, and (b) positive semidefinite.*

Proof. See Appendix A.1.

Furthermore, here we extend the rationally stretched birth-lifetime transform of [18] and consider a generalized spike transform:

$$
\tau_\eta^m(x, y) = \begin{cases} (x, y), & y \in [\eta, \infty), \\ (x, \frac{m}{m-1}\eta - \frac{1}{m-1}\frac{\eta^m}{y^{m-1}}), & y \in (0, \eta), \\ (x, -\infty), & y = 0, \end{cases}
\tag{2}
$$

where $m \in \mathbb{Z}, m \geq 2$.

Lemma 2. *Let $m \in \mathbb{Z}$ and $m \geq 2$, then τ_η is continuous on $\mathbb{R} \times \mathbb{R}_+$ and belongs to a class $\mathcal{C}^1$ of continuously differentiable functions on $\mathbb{R} \times \mathbb{R}_+$.*

Proof. See Appendix A.2.

Armed with Lemmas 1 and 2, we now show the key result needed to derive stability of $s_{ROT} = k_{GLP} \circ \tau_\eta^m \circ \rho$.

Lemma 3. $\lim_{y \to 0} \left| \left(k_{GLP} \circ \tau_\eta \right)'_y \right| < C$ for $\mathbb{R} \times [0, \epsilon)$, $C > 0$.

Proof. See Appendix A.3.

Lemma 3 implies that $k_{GLP} \circ \tau_\eta$ is Lipschitz continuous and, hence, we can derive stability of rotationally transformed PD representations.

Corollary 1 (Stability of Rotationally Transformed PD Representations). *Following the rotational operator procedure, let $\mathcal{D}_{\theta_1}$ and $\mathcal{D}_{\theta_2}$ be two rotated persistence diagrams by two angles (i.e., θ_1, θ_2) and let $s_{ROT} = k_{GLP} \circ \tau_\eta^m \circ \rho$ where τ_η^m is defined by (2) and $\rho : \Omega \cup \Delta \to \mathbb{R} \times \mathbb{R}_{\geq 0}$. Then $\left| s_{ROT}(\mathcal{D}_{\theta_1}) - s_{ROT}(\mathcal{D}_{\theta_2}) \right| \leq C W_1^q(\mathcal{D}_{\theta_1}, \mathcal{D}_{\theta_2})$, where $C > 0$ and*

$$W_1^q(\mathcal{D}_{\theta_1}, \mathcal{D}_{\theta_2}) = \inf_{\gamma} \left(\sum_{x \in \mathcal{D}_{\theta_1}} \|x - \gamma(x)\|_q \right)$$

is 1-Wasserstein distance with $q \in \mathbb{Z}^+$, γ ranging over all bijections between $\mathcal{D}_{\theta_1} \cup \Delta$ and $\mathcal{D}_{\theta_2} \cup \Delta$, and $\|z\|_\infty = \max_i |z_i|$.

Proof. See Appendix A.4.

Learnable PD Representation in the Spectral Domain. Based on the multi-scale property of the heat kernel, for small values of t, the function $p_i(t)$ is mainly determined by small neighborhoods of node i, and heat diffuses to larger and larger neighborhoods as t increases. This means $p_i(t)$ can capture both local and global information from the view point of node i when varying t. Let $\mathcal{D}_t$ be a PD obtained from graph $\mathcal{G}$ by using the multiscale heat kernel $p(t)$ with diffusion parameter t. Similar to [17, 24, 32], we consider the Gaussian-based kernel as a representation for PD, but we utilize a higher-order Gaussian kernel which can be beneficial for better distinguishing topological signals from topological noise [36].

Definition 2. *Let $\boldsymbol{\mu} = (\mu_1, \mu_2)^\top \in \mathbb{R}^2$, $\boldsymbol{\sigma} = (\sigma_1, \sigma_2) \in \mathbb{R}^2$, and $\boldsymbol{\rho} = (\rho_1, \rho_2) \in \mathbb{R}_+^2$. We define the higher-order Gaussian (HOG) kernel through the following equation:*

$$k_{HOG}(x, y) = e^{\left(-\left(\frac{(x-\mu_1)^2}{\sigma_1^2} \right)^{\rho_1} - \left(\frac{(y-\mu_2)^2}{\sigma_2^2} \right)^{\rho_2} \right)}. \tag{3}$$

Note that $k_{HOG}(x, y)$ belongs to class $\mathcal{C}^\infty(\mathbb{R}^2)$ and is Lipschitz continuous on $\mathbb{R}^2$.

Similar to s_{ROT}, we derive the following theoretical properties on the learnable PD representation in the spectral domain, i.e., Lipschitz continuity in Lemma 4 and stability of the PD representation using the HOG kernel.

Lemma 4. $\lim_{y \to 0} \left| \left(k_{HOG} \circ \tau_\eta \right)'_y \right| < C$ for $\mathbb{R} \times [0, \epsilon)$, $C > 0$.

Proof. See Appendix A.5.

Corollary 2 (Stability of PD Representations in the Spectral Domain). *Let $\mathcal{D}_{t_1}$ and $\mathcal{D}_{t_2}$ be two persistence diagrams over two diffusion parameters (e.g., t_1, t_2) and let $s_{TOP} = k_{HOG} \circ \tau_\eta^m \circ \rho$, where τ_η^m is defined by (2) and $\rho : \Omega \cup \Delta \to \mathbb{R} \times \mathbb{R}_{\geq 0}$. Then $\left| s_{TOP}(\mathcal{D}_{t_1}) - s_{TOP}(\mathcal{D}_{t_2}) \right| \leq CW_1^q(\mathcal{D}_{t_1}, \mathcal{D}_{t_2})$.*

Proof. See Appendix A.6.

Persistence-Based Weight Mechanism. Recall that points $d = (x, y) \in \mathcal{D}$ with a longer persistence $(y - x)$ are likelier to contain intrinsic structural information on the graph $\mathcal{G}$, while points with shorter persistence tend to be topological noise [?]. As such, assigning a higher weight to more persistent points in $\mathcal{D}$ tends to improve classification performance. Here we consider a weighting function $\mathcal{F}(x, y) = \arctan\big(C((y - x))^\zeta\big)$, where $C > 0$ and $\zeta \in \mathbb{Z}^+$.

Theorem 1 (Stability of the Weighted Kernel Embedding). *Let $\mathcal{D}_1$ and $\mathcal{D}_2$ be two persistence diagrams. Let $h(x, y) = \mathcal{F}(x, y)s(x, y)$, where $\mathcal{F}(x, y) = \arctan\big(C((y - x))^\zeta\big)$, $C > 0$ and $\zeta \in \mathbb{Z}^+$, and $s : \Omega \bigcup \Delta \to \mathbb{R}$ where s is either s_{ROT} (1) or s_{TOP} (2). Then, for $\zeta = 1$,*

$$\left\| \sum_{(x,y) \in \mathcal{D}_1} h(x, y) - \sum_{(x',y') \in \mathcal{D}_2} h(x', y') \right\| \leq CW_1^q(\mathcal{D}_1, \mathcal{D}_2).$$

Proof. See Appendix A.7.

4.2 Aggregated Attention Layer

We now proceed to construction of TopoAttn-Nets. First, note that HKS at lower and higher values of t capture high- and low-frequency information, respectively. Since higher frequencies are more sensitive to changes of t than lower frequencies, in a bid to capture the global and local information of input graph $\mathcal{G}$, we propose a new model, TopoAttn-Nets, that can learn relationships between spectral and geometric information, including mixing feature representations of different frequencies and transformations. As discussed earlier, aggregated representations in machine learning constitute a powerful architecture allowing for automatic combination of multi-source information. Contrary to [18, 24, 32], all key constituents in the proposed TopoAttn-Nets framework – kernel locations, kernel lengths, kernel scales, and the stretched parameter (i.e., parameters defined for a *meta-representation*) are learnable during training. For any domain, we use $\mathbf{D}_{\vartheta_i} = \{\mathcal{D}_{\vartheta_i}^1, \cdots, \mathcal{D}_{\vartheta_i}^N\}$, where $i = \{1, 2, \cdots, I\}$ and N is the number of PDs, to represent a set of PDs over HKS diffusion scale t_i (i.e., $\vartheta_i \leftarrow t_i$) or rotation angle θ_i (i.e., $\vartheta_i \leftarrow \theta_i$). Finally, the TopoAttn-Nets can be formulated as:

$$H^{(l+1)} = \begin{cases} \oplus_i \sigma(\alpha_i s(\mathbf{D}_{\vartheta_i}) \cdot \Theta_i^{(l)}), & \text{1st-order} \\ \oplus_{\substack{i \neq j \\ i < j}} \sigma(\alpha_{ij}[s(\mathbf{D}_{\vartheta_i}); s(\mathbf{D}_{\vartheta_j})]\Theta_{ij}^{(l)}), & \text{2nd-order} \end{cases} \quad (4)$$

where $\oplus$ denotes concatenation of vectors, $H^{(l+1)}$ is the first-order feature vector, $\Theta_i^{(l)}$ and $\Theta_{ij}^{(l)}$ are trainable weights in the layer, and $\sigma(\cdot)$ is the activation function, e.g., $\mathrm{ReLU}(\cdot) = \max(0, \cdot)$. Notice that function $s(\cdot)$ is either $s_{ROT}(\cdot)$ or $s_{TOP}(\cdot)$, which depends on the type of $\mathcal{D}_{\vartheta_i}$. To make learnable weights comparable across different components, we normalize them by a softmax operation. That is, (i) 1st-order: $\alpha_i = \exp(\omega_i) / \sum_i \exp(\omega_i)$, where $\omega_i = \mathrm{diag}(\mathfrak{F}(\mathcal{D}_{\vartheta_i}^1), \cdots, \mathfrak{F}(\mathcal{D}_{\vartheta_i}^N))$; (ii) 2nd-order: $\alpha_{ij} = \exp(\omega_{ij}) / \sum_j \exp(\omega_{ij})$, where $\omega_{ij} = \mathrm{diag}\left(\sum_{\kappa=1}^{2} \mathfrak{F}(\mathcal{D}_{\vartheta_i}^\kappa), \cdots, \sum_{\kappa=N-1}^{N} \mathfrak{F}(\mathcal{D}_{\vartheta_i}^\kappa)\right)$ and $\mathfrak{F}(\mathcal{D}_{\vartheta_i}^\kappa) = \left(\mathcal{F}(x_1, y_1)_{\vartheta_i}^\kappa, \mathcal{F}(x_2, y_2)_{\vartheta_i}^\kappa, \cdots, \mathcal{F}(x_m, y_m)_{\vartheta_i}^\kappa\right)$ (where $\mathfrak{F}(\mathcal{D}_{\vartheta_i}^\kappa)$ is the arctangent function for k-th PD $\mathcal{D}_{\vartheta_i}^\kappa$ and $\mathcal{F}(x, y) = \arctan\left(C((y - x))^\zeta\right)$ (every point $(x, y) \in \mathcal{D}_{\vartheta_i}^\kappa, C > 0, \zeta \in \mathbb{Z}^+)$). Here m is the number of points in $\mathcal{D}_{\vartheta_i}^\kappa$. The relative architectures of the feature vectors based on HKS at various diffusion parameters with the A-th order and rotation by different angles with the B-th order can be written as $H_{\mathrm{hks}_A}^{(l+1)}$ and $H_{\mathrm{rot}_B}^{(l+1)}$, respectively (where $A, B \in \{1, 2\}$). We can now rewrite the output $Z^{l+1} = \{H_{\mathrm{hks}_A}^{(l+1)}, H_{\mathrm{rot}_B}^{(l+1)}\}$ of the TopoAttn-Nets using column-wise concatenation as $Z^{(l+1)} = \oplus_j H_j^{(l+1)}$, where $j \in \{\mathrm{hks}_A, \mathrm{rot}_B\}$.

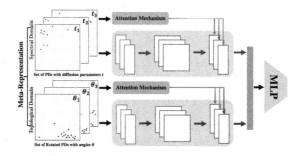

Fig. 1. Architecture of TopoAttn-Nets. A detailed description is given in Appendix C.

5   Experiments

For graph classification, we validate our method on the following standard graph benchmarks: (i) biological frameworks MUTAG and PTC, where nodes represent mutable and carcinogenic molecules, (ii) internet movie collaborations IMDB-B and IMDB-M, where nodes are actors/actresses and edges are common movie occurrences, and (iii) Reddit (an online aggregation and discussion website) discussion threads REDDIT-5K and REDDIT-12K, where nodes are Reddit users and edges are direct replies in the discussion threads. Each dataset includes multiple graphs of each class, and we aim to classify graph classes. For all graphs, we use the split setting of [18], that is, a 90/10 random training/test split. Furthermore, we perform a one-sided two-sample t-test between the best result and

Table 1. Performance summary (accuracy with standard deviation) on the graph classification tasks.

Method	MUTAG	PTC	IMDB-B	IMDB-M	REDDIT-5K	REDDIT-12K
GK [35]	83.5 (0.6)	59.2 (0.5)	65.9 (0.3)	43.9 (0.4)	41.0 (0.2)	31.8 (0.1)
RetGK [44]	90.3 (1.1)	62.5 (1.6)	71.9 (1.0)	47.7 (0.3)	56.1 (0.5)	**48.7 (0.2)**
DGK [42]	87.4 (2.7)	60.1 (2.5)	67.0 (0.6)	44.6 (0.4)	41.3 (0.2)	32.2 (0.1)
RF [17]	89.0 (3.8)	61.5 (2.7)	71.5 (0.8)	50.7 (0.7)	50.9 (0.3)	42.7 (0.3)
WL [34]	84.4 (1.5)	55.4 (1.5)	70.8 (0.5)	49.8 (0.5)	51.2 (0.3)	32.6 (0.3)
Deep-WL [42]	82.9 (2.7)	60.1 (2.5)	–	–	–	–
WWL [37]	87.3 (1.5)	66.3 (1.2)	–	–	–	–
P-WL [33]	86.3 (1.4)	63.1 (1.7)	72.8 (0.5)	–	–	–
P-WL-C [33]	90.5 (1.3)	64.0 (0.8)	73.2 (0.8)	–	–	–
P-WL-UC [33]	85.2 (0.3)	63.5 (1.6)	73.0 (1.0)	–	–	–
PF [25]	85.6 (1.7)	62.4 (1.8)	71.2 (1.0)	48.6 (0.7)	56.2 (1.1)	47.6 (0.5)
WKPI [45]	88.3 (2.6)	68.1 (2.4)	75.1 (1.1)	49.5 (0.4)	59.5 (0.6)	48.4 (0.5)
TopoGNN [19]	–	–	72.0 (2.3)	–	–	–
TopoGNN$_{(stat)}$ [19]	–	–	72.8 (5.4)	–	–	–
sPBoW [47]	-	-	-	-	45.6 (5.4)	31.6 (2.8)
PI$_{(NN)}$ [18]	89.8 (2.5)	63.5 (2.6)	71.2 (2.5)	48.8 (2.8)	46.7 (0.5)	35.1 (0.5)
Essential$_{(NN)}$ [18]	90.0 (1.7)	63.0 (2.3)	73.5 (2.0)	52.0 (1.8)	54.5 (0.6)	44.5 (0.4)
DGCNN [43]	85.8 (5.5)	58.6 (7.1)	70.0 (0.8)	47.8 (3.4)	48.7 (4.5)	–
GAT [38]	87.4 (5.3)	63.7 (8.2)	72.3 (5.1)	50.1 (3.6)	57.2 (2.2)	–
GraphSAGE [16]	85.7 (4.7)	63.9 (7.7)	72.3 (5.3)	50.9 (2.2)	-	–
CapsGNN [40]	86.7 (6.9)	66.0 (5.9)	71.7 (3.4)	48.5 (4.1)	52.9 (2.2)	–
PSCN [30]	89.0 (4.4)	62.3 (5.7)	71.0 (2.3)	45.2 (2.8)	49.1 (0.7)	41.3 (0.4)
GIN [41]	90.0 (8.8)	66.6 (6.9)	75.1 (5.1)	52.3 (2.8)	57.5 (1.5)	–
GCN [22]	85.6 (5.8)	64.2 (4.3)	74.0 (3.4)	51.9 (3.8)	56.7 (1.7)	–
PersLay [6]	89.8 (1.5)	-	71.2 (2.5)	48.8 (1.0)	55.6 (1.1)	47.7 (0.9)
FC [31]	87.3 (0.7)	65.1 (3.9)	73.8 (0.4)	46.8 (0.4)	52.4 (0.4)	–
TopoAttn-Nets (ours)	***92.4 (1.5)	68.3 (5.1)	75.2 (2.1)	***54.2 (0.6)	59.5 (0.5)	45.0 (0.5)

the best performance achieved by the runner-up, where *, **, *** denote significant, statistically significant, highly statistically significant results, respectively. The statistics of data we used in the Experiments section are summarized in Appendix B, Table 1.

Baselines. For graph classification, we perform an expansive evaluation the performance of TopoAttn-Nets with respect to the 26 most recent state-of-the-art (SOA) approaches: (i) graph kernel-based approaches: graphlet kernel (GK) [35], deep graphlet kernel (DGK) [42], Weisfeiler-Lehman kernel (WL) [34], deep variant of subtree features (Deep-WL) [42], graph-feature + random forest approach (RF) [17], Wasserstein Weisfeiler-Lehman (WWL) [37], probability-based graph kernel (RetGK) [44], persistent Weisfeiler-Lehman kernels (P-WL, P-WL-C, P-WL-UC) [33], and Persistence Fisher kernel (PF) [25]; (ii) topological information in kernel-based methods: Stable Persistence Bag of Words (sPBoW) [47], weighted-kernel for persistence images (WKPI) [45], and Filtration Curves (FC) [31]; (iii) graph neural networks: PATCHYSAN (PSCN) [30], Graph Convolutional Network (GCN) [22], Graph attention networks (GAT) [38], GraphSAGE [16], Deep Graph CNN (DGCNN) [43], Graph Isomorphism Network (GIN) [41], and Capsule Graph Neural Network (CapsGNN) [40]; (iv) topological-based deep neural networks: persistence images (PI) combined with a convolutional neural network [17,18], essential features

(Essential) combined with a convolutional neural network [18], GNN augmented with global graph persistence yielded from multiple filtrations (TopoGNN) [19], and the generic neural network layer for persistence diagrams (PersLay) [6].

Parameters Setting. In our experiments, We adopt the Adam optimizer for our TopoAttn-Nets model training with an initial learning rate $lr = 1 \times 10^{-3}$. We fix the number of training epoch to 500 for all datasets. We train the model using early stopping with a window size of 200. To prevent over-fitting, we use 1×10^{-4} L_2 regularization on the weights, and dropout input and hidden layers. To analyze behavior of HKS, i.e., $p(t, x) = \sum_{i=0}^{\infty} e^{-\lambda_i t} \varphi_i(x)^2$ (where λ_i and $\varphi_i(x)$ are the i-th eigenvalue and the i-th eigenfunction of the Laplace-Beltrami operator, respectively) under different time values t and to capture all of the information contained in the heat kernel, we set $t = \{0.1, 1, 5, 10, 50, 100, 150, 200, 1000\}$. We then conduct a random combination

Table 2. Analysis of kernel hyperparameters, attention mechanism, numbers of PDs as input, and rotation angles. Classification accuracy (st. dev.) on IMDB-B.

Kernel	k_{HOG}	
	$\rho = 1.0$	$\rho = 2.0$
	72.0 (3.4)	**75.2 (2.1)**
Framework	Attention mechanism	
	W/o Attn	With Attn
	74.0 (3.7)	**76.2 (2.1)**
	The number of PDs	
	1 PD	3 PDs
	71.0 (2.2)	**75.2 (2.1)**
Rotation	Rotation angles	
	$\theta = 45°$	$\theta = 90°$
	71.1 (1.1)	**71.2 (4.6)**

method to determine the best combination of local and global information. For topological signature rotation, the rotation could be implemented by infinite angles among the range $[0°, 180°]$. To avoid repetition and redundancy, we rotate topological signatures at the set of angles θ, i.e., $\theta = [0°, 30°, 45°, 60°, 90°, 120°, 135°, 150°, 180°]$ and find an optimal combination of topological information through random combination method. Since points near the diagonal in the persistence diagram $\mathcal{D}$ have shorter lifetimes (i.e., $y - x$) and are considered "topological noise", we determine the number of persistent pairs for model training through $argsort(f(\mathcal{D}))[-num\_pairs :]$, where $f(\mathcal{D}) = (y_1 - x_1, y_2 - x_2, \cdots, y_m - x_m)$ and num_pairs is the minimum number of persistent pairs in PDs $\{\mathcal{D}_1, \cdots, \mathcal{D}_N\}$ for each graph in the dataset.

Graph Classification. Table 1 reports results of mean accuracy and standard deviation across all models tested. The proposed model outperforms 26 SOAs on 5 benchmark datasets, except for REDDIT-12K. Compared to baseline methods, which extract PDs from only one domain, TopoAttn-Nets combines multi-frequencies and topological information across different domains in a single framework. RetGK outperforms our proposed model on the REDDIT-12K dataset may be due to REDDIT-12K has the weakest structural information, i.e., with very few links per node on average (its average density $\approx 2 \times 10^{-6}$ which is too sparse to deliver sufficient information on higher-order topological properties). In addition, for attributed graphs (i.e., MUTAG and PTC), TopoAttn-Nets still outperforms GCN-based approaches which use additional node features/labels, because kernel-based *meta-representation* equipped with neural network architecture can extract aggregated information from different scales that greatly benefits graph classification tasks.

Ablation Study. To better evaluate the performance of TopoAttn-Nets, we conduct a comprehensive ablation study on IMDB-B (see Table 2) by testing (i) kernel hyperparameters, (ii) attention mechanism (Attn), (iii) the number of PDs as input to our TopoAttn-Nets model, and (iv) rotation angle. The performances of TopoAttn-Nets with different (kernel) hyperparameters indicate that kernel hyperparameters enable control the effect of persistence, i.e., extracting meaningful information via a good approximation of the kernel. The comparison between with and without attention mechanism shows that adding attention mechanism can help capture importance of different PDs. Examining the results of different PDs as input, we can observe that a large improvement brought by applying multiple PDs to the input of TopoAttn-Nets. Comparison among different rotation angles underscores contribution of rotations to variability.

Table 3. Learned attention weights α_{hks} and α_{rot} of TopoAttn-Nets for *multi-frequency* and *topological* features.

Dataset	Learned value	
Attention weights	α_{hks}	α_{rot}
IMDB-B	**0.53**	0.47
IMDB-M	**0.60**	0.40
REDDIT-5K	0.42	**0.58**
REDDIT-12K	0.32	**0.68**

Sensitivity and Robustness. We evaluate robustness of TopoAttn-Nets w.r.t. adversarial attacks on REDDIT-5K. Here we consider graph structural perturbations of [48]) and present a comparison against two runner-ups which are the closest competitors of TopoAttn-Nets, namely, WKPI [45] and GIN [41]. Table 4 indicates that TopoAttn-Nets outperforms SOAs both in terms of accuracy and standard deviation under all attacks. Hence, TopoAttn-Nets may be viewed as the most reliable and accurate alternative under perturbations.

Table 4. Classification accuracy (st. dev.) under adversarial attack on REDDIT-5K.

Method	Perturbation rate			
	0%	5%	10%	15%
WKPI [45]	59.5 (0.6)	51.3 (3.3)	50.5 (2.2)	50.0 (2.0)
GIN [41]	57.5 (1.5)	51.2 (3.5)	49.0 (1.5)	47.7 (1.6)
TopoAttn-Nets	**59.5 (0.5)**	**51.9 (2.9)**	**51.2 (2.3)**	**50.1 (1.4)**

Relative Importance of Features. Table 3 reports the TopoAttn-Nets learned attention weights. Interestingly, we notice the attention weight of the *multi-frequency* feature is larger than that of *topological* feature for smaller graphs (i.e., biological and internet movie collaboration graphs). That is, the attention component reveals the relative importance of intrinsic finer- or coarser-grain variability in the data shape. For example, in learning tasks for sparser graphs, local variability often tends to be the key factor. Table 3 shows that indeed topological features addressing finer-grain shape properties of very sparse REDDIT-5K and REDDIT-12K, with average diameters of 11.96 and 10.91 and densities of 0.90 and 1.79, respectively, tend to be more valuable for classification. This also implies that importance of multi-frequency or topological information might depend more on the graph size rather than the specific type of data.

Computational Costs. Complexity of computing distances among PDs is $\mathcal{O}(m^3)$, where m is the number of points. All experiments are compiled and tested on a Tesla V100-SXM2-16GB GPU. Table 5 reports average running time to generate PDs and mean training time per epoch of TopoAttn-Nets on IMDB-B and REDDIT-5K, respectively.

Table 5. Complexity of TopoAttn-Nets: average time (in sec) to generate PD and training time per epoch.

Dataset	Avg. points in PD	Avg. time taken	
		PD generation	Train per epoch
IMDB-B	84.51	6×10^{-3}	1.15
REDDIT-5K	521.35	5×10^{-1}	0.53

6   Conclusion

We have developed a new flexible framework for meta-representation of persistence information in graphs, which may be viewed as the first step toward topological meta-learning on graphs. We have derived stability guarantees of the proposed approach and assessed its robustness to perturbations. The exhaustive experimental validation has indicated high competitiveness of the proposed meta-representation ideas in respect to the benchmarks. Future research include multiple directions. First, we will explore few shot concepts for topological meta-learning on graphs. Second, we will investigate utility of topological meta-representation for link prediction. Third, we will explore the proposed meta-representation and attention ideas in conjunction with multiparameter persistence [11] and local topological algorithms [10, 46].

Acknowledgements. This work is sponsored by the National Science Foundation under award numbers ECCS 2039701, INTERN supplement for ECCS 1824716, DMS 1925346 and the Department of the Navy, Office of Naval Research under ONR award number N00014-21-1-2530. Part of this material is also based upon work supported by (while serving at) the National Science Foundation. Any opinions, findings, and conclusions or recommendations expressed in this material are those of the author(s) and do not necessarily reflect the views of the National Science Foundation and/or the Office of Naval Research. The authors are grateful to Baris Coskunuzer for insightful discussions.

References

1. Adams, H., et al.: Persistence images: a stable vector representation of persistent homology. JMLR **18**(1), 218–252 (2017)
2. Borgwardt, K.M., Kriegel, H.P.: Shortest-path kernels on graphs. In: ICDM (2005)
3. Bronstein, M.M., Bruna, J., LeCun, Y., Szlam, A., Vandergheynst, P.: Geometric deep learning: going beyond euclidean data. IEEE Signal Process. Mag. 34(4), 18–42 (2017)
4. Carlsson, G.: Topology and data. BAMS **46**(2), 255–308 (2009)
5. Carlsson, G.: Topological pattern recognition for point cloud data. Acta Numerica **23**, 289–368 (2014)
6. Carrière, M., Chazal, F., Ike, Y., Lacombe, T., Royer, M., Umeda, Y.: Perslay: A simple and versatile neural network layer for persistence diagrams. In: AISTATS (2020)
7. Charles, C.K., Taylor, C., Keller, J.: Meta-analysis: From data characterisation for meta-learning to meta-regression. In: PKDD Workshop on data mining, decision support, meta-learning and ILP (2000)
8. Chazal, F., Fasy, B., Lecci, F., Michel, B., Rinaldo, A., Wasserman, L.: Robust topological inference: Distance to a measure and kernel distance. J. Mach. Learn. Res. **18**, 1–40 (2017)
9. Chazal, F., Michel, B.: An introduction to topological data analysis: fundamental and practical aspects for data scientists. Front. Artif. Intell. **4** 667963 (2021)
10. Chen, Y., Coskunuzer, B., Gel, Y.: Topological relational learning on graphs. In: NeurIPS. vol. 34, pp. 27029–27042 (2021)
11. Chen, Y., Segovia-Dominguez, I., Coskunuzer, B., Gel, Y.: TAMP-S2GCNets: coupling time-aware multipersistence knowledge representation with spatio-supra graph convolutional networks for time-series forecasting. In: ICLR (2022)
12. Chen, Y., Segovia-Dominguez, I., Gel, Y.R.: Z-GCNETs: Time zigzags at graph convolutional networks for time series forecasting. In: ICML (2021)
13. Dieleman, S., Willett, K.W., Dambre, J.: Rotation-invariant convolutional neural networks for galaxy morphology prediction. MNRAS **450**(2), 1441–1459 (2015)
14. Edelsbrunner, H., Letscher, D., Zomorodian, A.: Topological persistence and simplification. Discrete Comput. Geom. **28**, 511–533 (2002)
15. Fasy, B.T., Lecci, F., Rinaldo, A., Wasserman, L., Balakrishnan, S., Singh, A.: Confidence sets for persistence diagrams. AoS **42**(6), 2301–2339 (2014)
16. Hamilton, W., Ying, Z., Leskovec, J.: Inductive representation learning on large graphs. In: NeurIPS, pp. 1024–1034 (2017)
17. Hofer, C., Kwitt, R., Niethammer, M., Uhl, A.: Deep learning with topological signatures. In: NeurIPS, pp. 1634–1644 (2017)
18. Hofer, C.D., Kwitt, R., Niethammer, M.: Learning representations of persistence barcodes. J. Mach. Learn. Res. **20**(126), 1–45 (2019)
19. Horn, M., De Brouwer, E., Moor, M., Moreau, Y., Rieck, B., Borgwardt, K.: Topological graph neural networks. In: ICLR (2022)
20. Hospedales, T., Antoniou, A., Micaelli, P., Storkey, A.: Meta-learning in neural networks: A survey. arXiv:2004.05439 (2020)

21. Kashima, H., Tsuda, K., Inokuchi, A.: Marginalized kernels between labeled graphs. In: ICML, pp. 321–328 (2003)
22. Kipf, T.N., Welling, M.: Semi-supervised classification with graph convolutional networks. In: ICLR (2017)
23. Krizhevsky, A., Sutskever, I., Hinton, G.E.: Imagenet classification with deep convolutional neural networks. In: NeurIPS, pp. 1097–1105 (2012)
24. Kusano, G., Fukumizu, K., Hiraoka, Y.: Kernel method for persistence diagrams via kernel embedding and weight factor. J. Mach. Learn. Res. 18(1), 6947–6987 (2017)
25. Le, T., Yamada, M.: Persistence fisher kernel: A Riemannian manifold kernel for persistence diagrams. In: NeurIPS, pp. 10007–10018 (2018)
26. Levie, R., Monti, F., Bresson, X., Bronstein, M.M.: Cayleynets: Graph convolutional neural networks with complex rational spectral filters. IEEE Signal Process. Mag. 67(1), 97–109 (2018)
27. Maroulas, V., Mike, J.L., Oballe, C.: Nonparametric estimation of probability density functions of random persistence diagrams. J. Mach. Learn. Res. 20(151), 1–49 (2019)
28. Monti, F., Boscaini, D., Masci, J., Rodola, E., Svoboda, J., Bronstein, M.M.: Geometric deep learning on graphs and manifolds using mixture model CNNs. In: CVPR, pp. 5115–5124 (2017)
29. Morris, C., Kriege, N.M., Kersting, K., Mutzel, P.: Faster kernels for graphs with continuous attributes via hashing. In: IEEE ICDM, pp. 1095–1100 (2016)
30. Niepert, M., Ahmed, M., Kutzkov, K.: Learning convolutional neural networks for graphs. In: ICML. pp. 2014–2023 (2016)
31. O'Bray, L., Rieck, B., Borgwardt, K.: Filtration curves for graph representation. In: ACM SIGKDD, pp. 1267–1275 (2021)
32. Reininghaus, J., Huber, S., Bauer, U., Kwitt, R.: A stable multi-scale kernel for topological machine learning. In: CVPR, pp. 4741–4748 (2015)
33. Rieck, B., Bock, C., Borgwardt, K.: A persistent Weisfeiler-Lehman procedure for graph classification. In: ICML, pp. 5448–5458 (2019)
34. Shervashidze, N., Schweitzer, P., Van Leeuwen, E.J., Mehlhorn, K., Borgwardt, K.M.: Weisfeiler-Lehman graph kernels. J. Mach. Learn. Res. 12(77), 2539–2561 (2011)
35. Shervashidze, N., Vishwanathan, S., Petri, T., Mehlhorn, K., Borgwardt, K.: Efficient graphlet kernels for large graph comparison. In: AISTATS, pp. 488–495 (2009)
36. Tashev, I., Acero, A.: Statistical modeling of the speech signal. In: IWAENC (2010)
37. Togninalli, M., Ghisu, E., Llinares-López, F., Rieck, B., Borgwardt, K.: Wasserstein Weisfeiler-Lehman graph kernels. In: NeurIPS, pp. 6436–6446 (2019)
38. Veličković, P., Cucurull, G., Casanova, A., Romero, A., Lio, P., Bengio, Y.: Graph attention networks. In: ICLR (2018)
39. Verma, S., Zhang, Z.L.: Hunt for the unique, stable, sparse and fast feature learning on graphs. In: NeurIPS, pp. 88–98 (2017)
40. Xinyi, Z., Chen, L.: Capsule graph neural network. In: ICLR (2018)
41. Xu, K., Hu, W., Leskovec, J., Jegelka, S.: How powerful are graph neural networks? In: ICLR (2019)
42. Yanardag, P., Vishwanathan, S.: Deep graph kernels. In: ACM SIGKDD, pp. 1365–1374 (2015)
43. Zhang, M., Cui, Z., Neumann, M., Chen, Y.: An end-to-end deep learning architecture for graph classification. In: AAAI (2018)
44. Zhang, Z., Wang, M., Xiang, Y., Huang, Y., Nehorai, A.: Retgk: Graph kernels based on return probabilities of random walks. In: NeurIPS, pp. 3964–3974 (2018)

45. Zhao, Q., Wang, Y.: Learning metrics for persistence-based summaries and applications for graph classification. In: NeurIPS, pp. 9855–9866 (2019)
46. Zhao, Q., Ye, Z., Chen, C., Wang, Y.: Persistence enhanced graph neural network. In: AISTATS, pp. 2896–2906 (2020)
47. Zieliński, B., Lipiński, M., Juda, M., Zeppelzauer, M., Dłotko, P.: Persistence bag-of-words for topological data analysis. In: IJCAI, pp. 4489–4495 (2019)
48. Zügner, D., Akbarnejad, A., Günnemann, S.: Adversarial attacks on neural networks for graph data. In: ACM SIGKDD, pp. 2847–2856 (2018)

SEA: Graph Shell Attention in Graph Neural Networks

Christian M. M. Frey[1,2(✉)] (ID), Yunpu Ma[2] (ID), and Matthias Schubert[2] (ID)

[1] Christian-Albrecht University of Kiel, Kiel, Germany
cfr@informatik.uni-kiel.de
[2] Ludwig Maximilian University of Munich, Munich, Germany
{ma,schubert}@dbs.ifi.lmu.de

Abstract. A common problem in *Graph Neural Networks* (GNNs) is known as *over-smoothing*. By increasing the number of iterations within the message-passing of GNNs, the nodes' representations of the input graph align and become indiscernible. The latest models employing attention mechanisms with *Graph Transformer Layers* (GTLs) are still restricted to the layer-wise computational workflow of a GNN that are not beyond preventing such effects. In our work, we relax the GNN architecture by means of implementing a routing heuristic. Specifically, the nodes' representations are routed to dedicated experts. Each expert calculates the representations according to their respective GNN workflow. The definitions of distinguishable GNNs result from k-localized views starting from the central node. We call this procedure *Graph $\boldsymbol{Sh}$ell $\boldsymbol{A}$ttention* (SEA), where experts process different subgraphs in a transformer-motivated fashion. Intuitively, by increasing the number of experts, the models gain in expressiveness such that a node's representation is solely based on nodes that are located within the receptive field of an expert. We evaluate our architecture on various benchmark datasets showing competitive results while drastically reducing the number of parameters compared to state-of-the-art models.

1 Introduction

Graph Neural Networks (GNNs) have been proven to be an important tool in a variety of real-world applications building on top of graph data [22]. These range from predictions in social networks over property predictions in molecular graph structures to content recommendations in online platforms. From a machine learning perspective, we can categorize them into various theoretical problems that are known as *node classification, graph classification/regression* - encompassing binary decisions or modeling a continuous-valued function -, and *relation prediction*. In our work, we propose a novel framework and show its applicability on graph-level classification and regression, as well as on node-level classification tasks.

The high-level intuition behind GNNs is that by increasing the number of iterations $l = 1, \ldots, L$, a node's representation contains, and therefore relies

© The Author(s), under exclusive license to Springer Nature Switzerland AG 2023
M.-R. Amini et al. (Eds.): ECML PKDD 2022, LNAI 13714, pp. 326–343, 2023.
https://doi.org/10.1007/978-3-031-26390-3_20

successively more on its k-hop neighborhood. However, a well-known issue with the vanilla GNN architecture refers to a problem called *over-smoothing* [23]. In simple words, the information flow in GNNs between two nodes $u, v \in \mathcal{V}$, where $\mathcal{V}$ denotes a set of nodes, is proportional to the reachability of node v on a k-step random walk starting from u. By increasing the layers within the GNN architecture, the information flow of every node approaches the stationary distribution of random walks over the graph [7]. As a consequence, the localized information flow is getting lost, i.e., increasing the number of iterations within the message-passing of GNN results in representations for all the nodes in the input graph that align and become indiscernible [15]. One strategy for increasing a GNN's effectiveness is adding an attention mechanism. An adaption of the *Transformer* model [19] on graph data has been introduced as *Graph Transformer Layer* (GTL) [3]. Generally, multi-headed attention shows competitive results whenever we have prior knowledge to indicate that some neighbors might be more informative than others. Our framework further improves the representational capacity by adding an expert heuristic into the GTL architecture. More specifically, to compute a node's representation, a routing module first decides upon an expert that is responsible for a node's computation. The experts differ in how their k-hop localized neighborhood is processed and they capture individually various depths of GNNs/GTLs. We refer to different substructures that experts process as *Graph Shells*. As each expert attends to a specific subgraph of the input graph, we introduce the concept of *Graph **Shell** **Attention*** (SEA). Hence, whereas a vanilla GNN might suffer from over-smoothing the nodes' representations, we introduce additional degrees of freedom in our architecture to simultaneously capture short- and long-term dependencies being processed by respective experts. In summary, our contributions are as follows:

- Integration of expert-routing into graph neural nets;
- Novel Graph Shell Attention (SEA) models capturing short- and long-term dependencies, simultaneously;
- Experiments showing a reduction in the number of model parameters compared to SOTA models;

2   Related Work

In recent years, the AI community proposed various forms of (self-)attention mechanisms in numerous domains. *Attention* itself refers to a mechanism in neural networks where a model learns to make predictions by selectively attending to a given set of data. The success of applying attention heuristics was further boosted by introducing the *Transformer* model [19]. It relies on scaled dot-product attention, i.e., given a query matrix Q, a key matrix K, and a value matrix V, the output is a weighted sum of the value vectors, where the dot-product of the query with corresponding keys determines the weight that is assigned to each value.

Transformer architectures have also been successfully applied to graph data. The work by Dwivedi et al. [3] evaluates transformer-based GNNs. They conclude that the attention mechanism in Transformers applied on graph data should only aggregate the information from local neighborhoods, ensuring graph sparsity. As in *Natural Language Processing* (NLP), where a positional encoding is applied, they propose to use Laplacian eigenvectors as the positional encodings for further improvements. In their results, they outperform baseline GNNs on the graph representation task. A similar work [13] proposes a full Laplacian spectrum to learn the position of each node within a graph. Yun et al. [25] proposed *Graph Transformer Networks* (GTN) that are capable of learning on heterogeneous graphs. The target is to transform a given heterogeneous input graph into a meta-path-based graph and apply a convolution operation afterwards. Hence, the focus of their attention framework is on interpreting generated meta-paths. Another transformer-based architecture that has been introduced by Hu et al. [9] is *Heterogeneous Graph Transformer* (HGT). Notably, their architecture can capture graph dynamics w.r.t the information flow in heterogeneous graphs. Specifically, they take the relative temporal positional encoding into account based on differences of temporal information given for the central node and the message-passing nodes. By including the temporal information, Zhou et al. [26] built a transformer-based generative model for generating temporal graphs by directly learning from the dynamic information in networks. The work of Ngyuen et al. [14] proposes another idea for positional encoding. The authors of this work introduced a graph transformer for arbitrary homogeneous graphs with a coordinate embedding-based positional encoding scheme. In [24], the authors introduced a transformer motivated architecture where various encodings are aggregated to compute the hidden representations. They propose graph structural encodings subsuming a spatial encoding, an edge encoding, and a centrality encoding. Furthermore, a work exploring the effectiveness of large-scale pre-trained GNN models is proposed by the *GROVER* model [16]. The authors include an additional GNN operating in the attention sublayer to produce vectors for Q, K, and V. Moreover, they apply single long-range residual connections and two branches of feedforward networks to produce node and edge representations separately. In a self-supervised fashion, they first pre-train their model on 10 million unlabeled molecules before using the resulting node representations in downstream tasks. Typically, all the models are built in a way such that the same parameters are used for all inputs. To gain more expressiveness, the motivation of the mixture of experts (MoE) heuristic [18] is to apply different parameters w.r.t the input data. Recently, Google proposed *Switch Transformer* [5], enabling training above a trillion parameter networks but keeping the computational cost in the inference step constant. We provide an approach how a similar routing mechanism can be integrated in GNNs.

3   Preliminaries

3.1   Notation

Let $\mathcal{G} = (\mathcal{V}, \mathcal{E})$ be an undirected graph where $\mathcal{V}$ denotes a set of nodes and $\mathcal{E}$ denotes a set of edges connecting nodes. We define $N_k(u)$ to be the k-hop neighborhood of a node $u \in \mathcal{V}$, i.e., $N_k(u) = \{v \in \mathcal{V} : d_\mathcal{G}(u,v) \leq k\}$, where $d_G(u,v)$ denotes the hop-distance between u and v on $\mathcal{G}$. For $N_1(u)$ we will simply write $N(u)$ and omit the index k. The induced subgraph by including the k-hop neighbors starting from node u is denoted by $\mathcal{G}_u^k$. Moreover, in the following we will use a real-valued representation vector $h_u \in \mathbb{R}^d$ for a node u, where d denotes the embedding dimensionality.

3.2   Recap: Graph Transformer Layer

As formalized in [3], a *Graph Transformer Layer* (GTL) update for layer $l \in [1..L]$ including edge features is defined as:

$$\hat{h}_u^{l+1} = O_h^l \mathbin{\Big\|}_{i=1}^{H} \left(\sum_{v \in N(u)} w_{uv}^{i,l} V^{i,l} h_v^l \right), \tag{1}$$

$$\hat{e}_{uv}^{l+1} = O_e^l \mathbin{\Big\|}_{i=1}^{H} (\hat{w}_{uv}^{i,l}), \text{where}, \tag{2}$$

$$w_{uv}^{i,l} = \text{softmax}_v(\hat{w}_{uv}^{i,l}), \tag{3}$$

$$\hat{w}_{uv}^{i,l} = \left(\frac{Q^{i,l} h_u^l \cdot K^{i,l} h_v^l}{\sqrt{d_i}} \right) \cdot E^{i,l} e_{uv}^l, \tag{4}$$

where $Q^{i,l}, K^{i,l}, V^{i,l}, E^{i,l} \in \mathbb{R}^{d_i \times d}$, and $O_h^l, O_e^l \in \mathbb{R}^{d \times d}$. The operator $\|$ denotes the concatenation of attention heads $i = 1, \ldots, H$. Subsequently, the outputs $\hat{h}_u^{l+1}$ and $\hat{e}_{uv}^{l+1}$ are passed to feedforward networks and succeeded by residual connections and normalization layers yielding the representations h_u^{l+1} and e_{uv}^{l+1}.

A graph's embedding $h_\mathcal{G}$ is derived by a permutation-invariant readout function w.r.t. the nodes in $\mathcal{G}$:

$$h_\mathcal{G} = readout(\{h_u | u \in \mathcal{V}\}) \tag{5}$$

A common heuristic for the readout function is to choose a function READOUT$(\cdot) \in \{mean(\cdot), sum(\cdot), max(\cdot)\}$.

4   Methodology

In this section, we introduce our *Graph **S**hell **A**ttention* (SEA) architecture for graph data. SEA builds on top of the message-passing paradigm of *Graph Neural Networks* (GNNs) while integrating an expert heuristic.

4.1   Graph Shells Models

In our approach, we implement *Graph Transformer Layers* (GTLs) [3] and extend our framework by a set of *experts*. A routing layer decides which expert is most relevant for computing a node's representation. An expert's calculation for

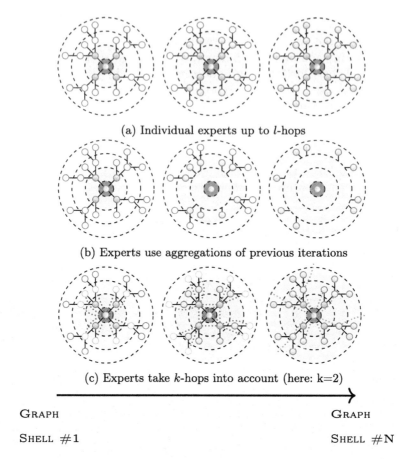

(a) Individual experts up to l-hops

(b) Experts use aggregations of previous iterations

(c) Experts take k-hops into account (here: k=2)

GRAPH   GRAPH

SHELL #1   SHELL #N

Fig. 1. Three variants of SEA models; for each model, the respective fields of 3 experts are shown from left to right.

a node representation differs in how k-hop neighbors are stored and processed within GTLs.

Generally, starting from a central node, *Graph Shells* refer to subgraphs that include only nodes that have at maximum a k-hop distance (k-neighborhood). Formally, the i-th expert comprises the information given in the i-th neighborhood $N_i(u) = \{v \in \mathcal{V} : d_G(u, v) \leq i\}$, where $u \in \mathcal{V}$ denotes the central node. We refer to the subgraph $\mathcal{G}_u^i$ as the expert's *receptive field*. Notably, increasing the number of iterations within GTLs/GNNs correlates with the number of experts being used. In the following, we introduce three variants on how experts process graph shells:

• **SEA-GTL.** The first graph shell model exploits the vanilla architecture of GTLs for which shells are defined by the standard graph neural net construction. For a maximal number of L iterations, we define a set $\{E_i(u)\}_{i=1}^{N}$ of $N = L$ experts. The embeddings after the l-th iteration are fed to the l-th expert, i.e., according to Eq. 1, the information of nodes in $\mathcal{G}_u^l$ for a central node u have been processed. Figure 1a illustrates this model. From left to right, the information of nodes being reachable by more hops is processed. Experts processing information in early iterations refer to short-term dependencies, whereas experts processing more hops yield information of long-term dependencies.

• **SEA-AGGREGATED.** For the computation of the hidden representation h_u^{l+1} for node u on layer $l + 1$, the second model employs an aggregated value from the preceding iteration. Following Eq. 1, the aggregation function (sum) in GLT considers all 1-hop neighbors $N_1(u)$. For *SEA*-AGGREGATED, we propagate the aggregated value back to all of u's 1-hop neighbors. For a node $v \in N_1(u)$, the values received by v are processed according to an aggregation function AGG $\in \{mean(\cdot), sum(\cdot), max(\cdot)\}$. Formally:

$$h_u^{l+1} = \text{AGG}^l(\{h_v^{l+1} : v \in N(u)\}) \tag{6}$$

Figure 1b illustrates this graph shell model. In the first iteration, there are no preceding layers, hence, the first expert processes the information in the same way as in the first model. In succeeding iterations, the aggregated representations are first sent to neighboring nodes, which in turn process the incoming representations. These aggregated values define the input for the current iteration. Full-colored shells illustrate aggregated values from previous iterations.

• **SEA-K-HOP.** For this model we relax the aggregate function defined in Eq. 1. Given a graph $\mathcal{G}$, we also consider k-hop linkages in the graph connecting a node u with all entities having a maximum distance of $d_{\mathcal{G}}(u, v) = k$. The relaxation of Eq. 1 is formalized as:

$$\hat{h}_u^{l+1} = O_h^l \, \overset{H}{\underset{i=1}{||}} \, (\sum_{v \in N_k(u)} \overset{\text{attention scores Eq. 3}}{w_{uv}^{i,l}} V^{i,l} \underset{\substack{\text{embds. of nodes} \\ \text{in k-hop dist.}}}{h_v^l}), \tag{7}$$

where $N_k(u)$ denotes the k-hop neighborhood set. This approach allows for processing each $N_1(u), \ldots, N_k(u)$ by own submodules, i.e., for each k-hop neighbors we use respective feedforward networks to compute Q, K, V in GTLs. Notably, Eq. 7 can be interpreted as a generalization of the vanilla architecture, which is given by setting $k = 1$. Figure 1c shows the k-hop graph shell model with $k = 2$.

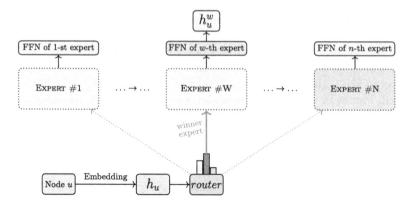

Fig. 2. Routing mechanism to N experts

4.2   SEA: Routing Mechanism

By endowing our models with experts referring to various graph shells, we gain variable expressiveness for short- and long-term dependencies. Originally introduced for language modeling and machine translation, Shazeer et al. [18] proposed a Mixture-of-Experts (MoE) layer. A routing module decides to which expert the attention is steered. We use a *single expert* strategy [5].

The general idea relies on a routing mechanism for a node u's representation to determine the best expert from a set $\{E_i(u)\}_{i=1}^{N}$ of N experts processing graph shells as described in the previous Sect. 4.1. The router module consists of a linear transformation whose output is normalized via softmaxing. The probability of choosing the i-th expert for node u is defined as:

$$p_i(u) = \frac{\exp(r(u)_i)}{\sum_j^N \exp(r(u)_j)}, \quad r(u) = h_u^T W_r + b_r, \tag{8}$$

where $r(\cdot)$ denotes the routing operation with $W_r \in \mathbb{R}^{d \times N}$ being the routing's learnable weight matrix, and b_r denotes a bias term. The idea is to select the winner expert $E_w(\cdot)$ that is the most representative for a node's representation, i.e., where $w = \underset{i=1,\dots,N}{\arg\max}\ p_i(u)$[1]. A node's representation calculated by taking the winner's graph shells into account is then used as input for the expert's individual linear transformation:

$$h_u^w = E_w(u)^T W_w + b_w, \tag{9}$$

---

[1] In DL libraries, the $\arg\max(\cdot)$ operation implicitly calls $\arg\max(\cdot)$ forwarding the maximum of the input. Hence, it is differentiable w.r.t to the values yielded by the max op., not to the indices.

where $W_w \in \mathbb{R}^{d \times d}$ denotes the weight matrix of expert $E_w(\cdot)$, b_w denotes the bias term. The node's representation according to expert $E_w(\cdot)$, is denoted by h_u^w. Figure 2 shows how the routing is integrated into our architecture.

4.3   Shells vs. Over-smoothing

Over-smoothing in GNNs is a well-known issue [23] and exacerbates the problem when we build deeper graph neural net models. Applying the same number of iterations for each node inhibits the simultaneous expressiveness of short- and long-term dependencies. We gain expressiveness by routing each node representation towards dedicated experts processing only nodes in their k-localized receptive field.

Let $\mathcal{G} = (\mathcal{V}, \mathcal{E})$ be an undirected graph. Following the proof scheme of [15], let $A = (\mathbb{1}_{(i,j) \in \mathcal{E}})_{i,j \in [N] := \{1, \ldots, N\}} \in \mathbb{R}^{N \times N}$ be the adjacency matrix and $D := \mathrm{diag}(\deg(i)_{i \in [N]}) \in \mathbb{R}^{N \times N}$ be the degree matrix of $\mathcal{G}$ where $\deg(i) := |\{j \in V \mid (i,j) \in \mathcal{E}\}|$ is the degree of node i. Let $\tilde{A} := A + I_N$, $\tilde{D} := D + I_N$ be the adjacent and the degree matrix of graph $\mathcal{G}$ augmented with self-loops, where I_N denotes the identity matrix of size N. The augmented normalized Laplacian of $\mathcal{G}$ is defined by $\tilde{\Delta} := I_N - \tilde{D}^{-\frac{1}{2}} \tilde{A} \tilde{D}^{-\frac{1}{2}}$ and set $P := I_N - \tilde{\Delta}$. Let $L, C \in \mathbb{N}_+$ be the layer and channel sizes, respectively. W.l.o.g, for weights $W_l \in \mathbb{R}^{C \times C} (l \in [L] := \{1, \ldots, L\})$, we define a GCN associated with $\mathcal{G}$ by $f = f_L \circ \ldots \circ f_1$ where $f_l : \mathbb{R}^{N \times C} \to \mathbb{R}^{N \times C}$ is defined by $f_l(X) = \sigma(P X W_l)$, where $\sigma(\cdot)$ denotes the ReLU activation function. For $M \leq N$, let U be a M-dimensional subspace of $\mathbb{R}^N$. Furthermore, we define a subspace $\mathcal{M}$ of $\mathbb{R}^{N \times C}$ by $\mathcal{M} = U \otimes \mathbb{R}^C = \{\sum_{m=1}^M e_m \otimes w_m \mid w_m \in \mathbb{R}^C\}$, where $(e_m)_{m \in [M]}$ is the orthonormal basis of U. For an input $X \in \mathbb{R}^{N \times C}$, the distance between X and $\mathcal{M}$ is denoted by $d_\mathcal{M} = \inf\{\|X - Y\|_F \mid Y \in \mathcal{M}\}$.

Considering $\mathcal{G}$ as M connected components, i.e. $V = V_1 \cup \ldots \cup V_m$, where an indicator vector of the m-th connected component is denoted by $u_m = (\mathbb{1}_{\{n \in V_m\}})_{n \in [N]} \in R^N$. The authors of [15] investigated the asymptotic behavior of the output X^L of the GCN when $L \to \infty$:

Proposition 1. *Let $\lambda_1 \leq \ldots \leq \lambda_N$ be the eigenvalue of P sorted in ascending order. Then, we have $-1 < \lambda_1, \lambda_{N-M} < 1$, and $\lambda_{N-M+1} = \ldots = \lambda_N = 1$. In particular, we have $\lambda = \max_{n=1, \ldots N-M} |\lambda_n| < 1$. Further, $e_m = \tilde{D}^{\frac{1}{2}} u_m$ for $m \in [M]$ are the basis of the eigenspace associated with the eigenvalue 1.*

Table 1. Summary dataset statistics

Domain	Dataset	#Graphs	Task
Chemistry	ZINC	12K	Graph Regression
	OGBG-MOLHIV	41K	Graph Classification
Mathematical Modeling	PATTERN	14K	Node Classification

Let $s = \sup_{l \in \mathbb{N}_+} s_l$ with s_l denoting the maximum singular value of W_l, the major theorem and their implications for GCNs is stated as follows:

Theorem 1. *For any initial value $X^{(0)}$, the output of l-th layer $X^{(l)}$ satisfies $d_{\mathcal{M}}(X^{(l)}) \leq (s\lambda)^l d_{\mathcal{M}}(X^{(0)})$. In particular, $d_{\mathcal{M}}(X^{(l)})$ exponentially converges to 0 when $s\lambda < 1$.*

Proofs of Proposition 1 and Theorem 1 are formulated in [15].

Intuitively, the representations X align subsequently with the subspace $\mathcal{M}$, where the distance between both converges to zero. Therefore, it can also be interpreted as information loss of graph neural nets in the limit of infinite layers.

The theoretical justification for the routing mechanism applied in our SEA models comes to light when we exploit the monotonous behavior of the exponential decay where the initial distance $d_{\mathcal{M}}(X^{(0)})$ is treated as a constant value. The architecture includes the experts in a cascading manner, where the routing mechanism allows to point to each of the $(d_{\mathcal{M}}(f_l(X)))_{l=1,...,L}$, separately. From Theorem 1, we get:

$$d_{\mathcal{M}}(X^{(L)}) \leq (s\lambda)^L d_{\mathcal{M}}(X^{(0)}) \leq (s\lambda)^{L-1} d_{\mathcal{M}}(X^{(0)})$$
$$\leq \ldots \leq (s\lambda)^1 d_{\mathcal{M}}(X^{(0)}),$$

where each inequality is supported by the output of the l-th expert, separately:

L-th expert:	$d_{\mathcal{M}}(X^{(L)})$	$\leq (s\lambda)^L d_{\mathcal{M}}(X^{(0)})$
L-1-th expert:	$d_{\mathcal{M}}(X^{(L-1)})$	$\leq (s\lambda)^{L-1} d_{\mathcal{M}}(X^{(0)})$
$\ldots$	$\ldots$	$\leq \ldots$
1-st expert:	$d_{\mathcal{M}}(X^{(1)})$	$\leq (s\lambda)^1 d_{\mathcal{M}}(X^{(0)})$

Hence, our architecture does not suffer from overs-smoothing the same way as standard GNNs, as each captures a different distance $d_{\mathcal{M}}$ compared to using a GNN where a pre-defined number of layer updates is applied for all nodes equally and potentially leading to an over-smoothed representation.

5   Evaluation

5.1   Experimental Setting

Datasets

ZINC [10] is one of the most popular real-world molecular dataset consisting of 250K graphs. A subset consisting of 10K train, 1K validation, and 1K test graphs is used in the literature as benchmark [4].

We also evaluate our models on **ogbg-molhiv** [8]. Each graph within the dataset represents a molecule, where nodes are atoms and edges are chemical bonds.

A benchmark dataset generated by the *Stochastic Block Model* (SBM) [1] is **PATTERN**. The graphs within this dataset do not have explicit edge features. The benchmark datasets are summarized in Table 1.

Implementation Details

Our implementation uses PyTorch, Deep Graph Library (DGL) [21], and OGB [8]. The models are trained on an NVIDIA GeForce RTX 2080 Ti.[2]

Model Configuration

We use the Adam optimizer [11] with an initial learning rate $\in \{1e\text{-}3, 1e\text{-}4\}$. We apply the same learning rate decay strategy for all models that half the learning rate if the validation loss does not improve over a fixed number of 5 epochs. We tune the pairing (#heads, hidden dimension) $\in \{(4, 32), (8, 56), (8, 64))\}$ and use READOUT $\in \{sum\}$ as function for inference on the whole graph information. *Batch Normalization* and *Layer Normalization* are disabled, whereas residual connections are activated per default in GTLs. For dropout, we tuned the value to be $\in \{0, 0.01, 0.05, 0.07, 0.1\}$ and a weight decay $\in \{5e\text{-}5, 5e\text{-}7\}$. For the number of graph shells, i.e., number of experts being used, we report values $\in \{4, 6, 8, 10, 12\}$. As aggregation function we use AGG $\in \{mean\}$ for Eq. 6. As laplacian encoding, the 8 smallest eigenvectors are used.

5.2   Prediction Tasks

In the following series of experiments, we investigate the performance of the *Graph Shell Attention* mechanism on graph-level prediction tasks for the datasets ogbg-molhiv [8] and ZINC [10], and a node-level classification task on PATTERN [1]. We use commonly used metrics for the prediction tasks as they are used in [4], i.e., mean absolute error (MAE) for ZINC, the ROC-AUC score on ogbg-molhiv, and the accuracy on PATTERN.

Competitors. We evaluate our architectures against state-of-the-art GNN models achieving competitive results. Our report subsumes the vanilla GCN [12], GAT [20] that includes additional attention heuristics, or more recent GNN architectures building on top of Transformer-enhanced models like SAN [13] and Graphormer [24]. Moreover, we include GIN [23] that is more discriminative towards graph structures compared to GCN [12], GraphSage [6], and DGN [2] being more discriminative than standard GNNs w.r.t the Weisfeiler-Lehman 1-WL test.

---

[2] Code: https://github.com/christianmaxmike/SEA-GNN.

Table 2. Comparison to state-of-the-art; results are partially taken from [4,13]; color coding (gold/silver/bronze)

Model	ZINC #params.	MAE
GCN [12]	505K	0.367
GIN [23]	509K	0.526
GAT [20]	531K	0.384
SAN [13]	508K	**0.139**
Graphormer-SLIM [24]	489K	**0.122**
Vanilla GTL	83K	0.227
SEA-GTL	347K	0.212
SEA-AGGREGATED	112K	0.215
SEA-2-HOP	**430K**	**0.159**
SEA-2-HOP-AUG	709K	0.189

(a) ZINC [10]

Model	OGBG-MOLHIV #params.	%ROC-AUC
GCN-GRAPHNORM [12]	526K	76.06
GIN-VN [23]	3.3M	77.80
DGN [2]	114K	79.05
Graphormer-FLAG [24]	47.0M	**80.51**
Vanilla GTL	386K	78.06
SEA-GTL	347K	79.53
SEA-AGGREGATED	**133K**	**80.18**
SEA-2-HOP	511K	**80.01**
SEA-2-HOP-AUG	594K	79.08

(b) ogbg-molhiv [8]

Model	PATTERN #params.	% ACC
GCN [12]	500K	71.892
GIN [23]	100K	85.590
GAT [20]	526K	78.271
GraphSage [6]	101K	50.516
SAN [13]	454K	**86.581**
Vanilla GTL	82K	84.691
SEA-GTL	132K	85.006
SEA-AGGREGATED	69K	57.557
SEA-2-HOP	**48K**	**86.768**
SEA-2-HOP-AUG	152K	**86.673**

(c) PATTERN [1]

Results. Tables 2a, b, and c summarize the performances of our SEA models compared to baselines on ZINC, ogbg-molhiv, and PATTERN. *Vanilla GTL* shows the results of our implementation of the GNN model including Graph Transformer Layers [3]. *SEA*-2-HOP includes the 2-hop connection within the input graph, whereas *SEA*-2-HOP-AUG process the input data the same way as the 2-HOP heuristic, but uses additional feedforward networks for computing Q, K, V values for the 2-hop neighbors.

For PATTERN, we observe the best result using the *SEA*-2-HOP model, beating all other competitors. On the other hand, distributing an aggregated value to neighboring nodes according to *SEA*-AGGREGATED yields a too coarse view for graphs following the SBM and loses local graph structure.

In the sense of *Green AI* [17] that focuses on reducing the computational cost to encourage a reduction in resources spent, our architecture reaches state-of-the-art performance on ogbg-molhiv while drastically reducing the number of parameters being trained. Comparing *SEA*-AGGREGATED to the best result reported for *Graphormer* [24], our model economizes on **99.71%** of the number of parameters while still reaching competitive results.

The results on ZINC enforces the argument of using individual experts compared to vanilla GTLs, where the best result is reported for *SEA*-2-HOP.

5.3   Number of Shells

Next, we examine the performance w.r.t the number of experts. Notably, increasing the number of experts correlated with the number of *Graph Shells* which are taken into account. Table 3 summarizes the results where all other hyperparameters are frozen, and we only have a variable size in the number of experts. We train each model for 500 epochs and report the best-observed metrics on the test datasets. We apply an early stopping heuristic, where we stop the learning procedure if we have not observed any improvements w.r.t the evaluation metrics or if the learning rate scheduler reaches a minimal value which we set to 10^{-6}. Each evaluation on the test data is conducted after 5 epochs, and the early stopping is effective after 10 consecutive evaluations on the test data with no improvements. First, note that increasing the number of experts also increases the model's parameters linearly. This is due to additional routings and linear layer being defined for each expert separately. Secondly, we report also the average running time in seconds [s] on the training data for each epoch. By construction, the running time correlates with the number of parameters that have to be trained. The number of parameters differs from one dataset to another with the same settings due to a different number of nodes and edges within the datasets and slightly differs if biases are used or not. Note that we observe better results of *SEA*-AGGREGATED by decreasing the embedding size from 64 to 32, which also applies for the PATTERN dataset in general. The increase of parameters of the augmented 2-hop architecture *SEA*-2-HOP-AUG is due to the additional feedforward layers being used for the k-hop neighbors to compute the inputs Q, K, V in the graph transformer layer. Notably, we also observe that similar settings apply for datasets where the structure is an important feature

Table 3. Influence of the number of experts applied on various SEA models; best configurations are highlighted in green

Model	#experts	ZINC			OGBG-MOLHIV			PATTERN		
		#params	MAE	time/epoch	#params	%ROC-AUC	time/epoch	#params	% ACC	time/epoch
SEA-GTL	4	183K	0.385	13.60	182K	79.24	49.21	48K	78.975	58.14
	6	266K	0.368	20.93	263K	78.24	68.67	69K	82.117	82.46
	8	349K	0.212	26.24	345K	79.53	84.35	90K	82.983	108.41
	10	433K	0.264	31.63	428K	79.35	107.11	111K	84.041	133.73
	12	516K	0.249	38.26	511K	79.18	122.99	132K	85.006	168.47
SEA-AGGREGATED	4	49K	0.257	31.24	48K	77.87	60.98	48K	57.490	99.10
	6	70K	0.308	44.61	69K	79.21	86.26	69K	57.557	106.79
	8	91K	0.249	57.89	90K	77.19	86.93	90K	54.385	131.57
	10	112K	0.215	73.49	111K	77.48	102.40	111K	57.221	173.74
	12	133K	0.225	87.08	132K	80.18	124.08	132K	57.270	206.73
SEA-2-HOP	4	182K	0.309	14.28	180K	76.30	43.51	48K	86.768	94.04
	6	265K	0.213	20.13	263K	77.27	59.82	69K	86.706	138.10
	8	347K	0.185	24.91	345K	76.61	79.56	90K	86.707	178.64
	10	430K	0.159	32.68	428K	78.38	95.69	111K	86.680	232.91
	12	513K	0.188	38.73	511K	80.01	112.93	132K	86.699	269.71
SEA-2-HOP-AUG	4	248K	0.444	16.86	248K	77.21	48.65	65K	84.889	124.96
	6	363K	0.350	24.84	363K	75.19	70.05	94K	85.141	203.38
	8	478K	0.285	31.48	476K	76.55	90.78	123K	86.660	270.85
	10	594K	0.205	39.25	594K	79.08	109.91	152K	86.673	363.58
	12	709K	0.189	46.51	707K	77.52	133.48	181K	86.614	421.46

Table 4. Influence of parameter k for the *SEA*-K-HOP model; best configuration for each model is highlighted in green

Model	#exp.	k	ZINC #prms	MAE	OGBG-MOLHIV #prms	%ROC-AUC	PATTERN #prms	%ACC
SEA-K-HOP	6	2	265K	0.213	263K	**77.27**	69K	**86.768**
		3	266K	**0.191**	263K	76.15	69K	86.728
		4	266K	0.316	263K	73.48	69K	86.727
	10	2	430K	**0.159**	428K	**78.38**	111K	86.680
		3	433K	0.171	428K	74.67	111K	**86.765**
		4	433K	0.239	428K	73.72	111K	86.725

of the graph, like in molecules (ZINC + ogbg-molhiv). In contrast to that is the behavior on graphs following the stochastic block model (PATTERN). On the latter one, the best performance could be observed by including k-hop information, whereas an aggregation yields too simplified features to be competitive. For the real-world molecules (ZINC + ogbg-molhiv) datasets, we observe that more experts boost the performance for the various SEA extensions.

5.4 Stretching Locality in SEA-K-HOP

Lastly, we investigate the influence of the parameter k for the *SEA*-K-HOP model. Generally, by increasing the parameter k, the model diverges to the full model being also examined for the *SAN* architecture explained in [13]. In short, the full setting takes edges into account that is given by the input data and also sends information over non-existent edges, i.e., the argumentation is on a full graph setting. In our model, we smooth the transition from edges being given in the input data to the full setting that naturally arises when k, the number of hops, is set to a sufficiently high number. Table 4 summarizes the results for the non-augmented model, i.e., no extra linear layers are used for each k-hop neighborhood. The number of parameters stays the same by increasing k.

5.5 Distribution of Experts

We evaluate the distributions of the experts being chosen to compute the nodes' representations in the following. We set the number of experts to 8. Figure 3 summarizes the relative frequencies of the experts being chosen on the datasets ZINC, ogbg-molhiv, and PATTERN. Generally, the performance of the shell attention heuristic degenerates whenever we observe *expert collapsing*. In the extreme case, just one expert expresses the mass of all nodes, and the capability to distribute nodes' representations over several experts is not leveraged. To overcome *expert collapsing*, we can use a heuristic where in the early stages of

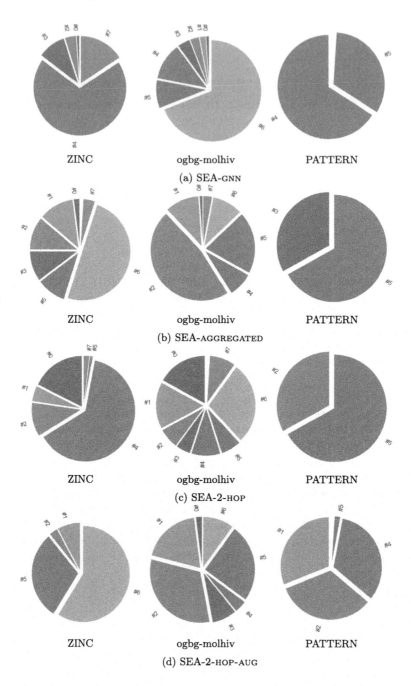

Fig. 3. Distribution of 8 experts for models *SEA*-GNN, *SEA*-AGGREGATED, *SEA*-2-HOP, and *SEA*-2-HOP-AUG for datasets ZINC, ogbg-molhiv and PATTERN. Relative frequencies are shown for values ≥ 1%. Numbers attached to the slices refer to the respective experts.

the learning procedure, an additional epsilon parameter ϵ introduces randomness. Like a decaying greedy policy in Reinforcement Learning (RL), we choose a random expert with probability ϵ and choose the expert with the highest probability according to the routing layer with a probability of $1 - \epsilon$. The epsilon value slowly decays over time. This ensures that all experts' expressiveness is being explored to find the best matching one w.r.t to a node u and prevents getting stuck in a local optimum. The figure shows the distribution of experts that are relevant for the computation of the nodes' representations. For illustrative purposes, values below 1% are omitted. Generally, nodes are more widely distributed over all experts in the molecular datasets - ZINC and ogbg-molhiv - for all models compared to PATTERN following a stochastic block model. Therefore, various experts are capable of capturing individual topological characteristics of molecules better than vanilla graph neural networks for which over-smoothing might potentially occur. We also observe that the mass is distributed to only a subset of the available experts for the PATTERN dataset. Hence, the specific number of iterations is more expressive for nodes within graph structures following SBM.

6   Conclusion

We introduced the theoretical foundation for integrating an expert heuristic within transformer-based graph neural networks. This opens a fruitful direction for future works that go beyond successive message-passing to develop even more powerful architectures in graph learning. We provide an engineered solution that allows selecting the most representative experts for nodes in the input graph. For that, our model exploits the idea of a routing layer steering the nodes' representations towards the individual expressiveness of dedicated experts. As experts process different subgraphs starting from a central node, we introduce the terminology of *Graph Shell Attention* (SEA), where experts solely process nodes that are in their respective receptive field. Therefore, we gain expressiveness by capturing varying short- and long-term dependencies expressed by individual experts. In a thorough experimental study, we show on real-world benchmark datasets that the gained expressiveness yields competitive performance compared to state-of-the-art results while being more economically. Additionally, we report experiments that stress the number of graph shells that are taken into account.

Acknowledgements. This work has been funded by the German Federal Ministry of Education and Research (BMBF) under Grant No. 01IS18050C (MLWin) and Grant No. 01IS18036A (MCML). The authors of this work take full responsibilities for its content.

References

1. Abbe, E.: Community detection and stochastic block models. Found. Trends Commun. Inf. Theory (2018). https://doi.org/10.1561/0100000067

2. Beaini, D., Passaro, S., Létourneau, V., Hamilton, W.L., Corso, G., Liò, P.: Directional graph networks. CoRR (2020). https://arxiv.org/abs/2010.02863
3. Dwivedi, V.P., Bresson, X.: A generalization of transformer networks to graphs (2021)
4. Dwivedi, V.P., Joshi, C.K., Laurent, T., Bengio, Y., Bresson, X.: Benchmarking graph neural networks. arXiv preprint arXiv:2003.00982 (2020)
5. Fedus, W., Zoph, B., Shazeer, N.: Switch transformers: scaling to trillion parameter models with simple and efficient sparsity. CoRR (2021). https://arxiv.org/abs/2101.03961
6. Hamilton, W., Ying, Z., Leskovec, J.: Inductive representation learning on large graphs. In: Guyon, I., et al. (eds.) Advances in Neural Information Processing Systems (2017). https://proceedings.neurips.cc/paper/2017/file/5dd9db5e033da9c6fb5ba83c7a7ebea9-Paper.pdf
7. Hoory, S., Linial, N., Wigderson, A.: Expander graphs and their applications. Bull. Amer. Math. Soc. (2006). https://doi.org/10.1090/s0273-0979-06-01126-8
8. Hu, W., et al.: Open graph benchmark: Datasets for machine learning on graphs. arXiv preprint arXiv:2005.00687 (2020)
9. Hu, Z., Dong, Y., Wang, K., Sun, Y.: Heterogeneous graph transformer. In: Proceedings of the Web Conference 2020, WWW 2020 (2020). https://doi.org/10.1145/3366423.3380027
10. Irwin, J.J., Sterling, T., Mysinger, M.M., Bolstad, E.S., Coleman, R.G.: Zinc: a free tool to discover chemistry for biology. J. Chem. Inf. Model. **52**(7), 1757–1768 (2012)
11. Kingma, D.P., Ba, J.: Adam: a method for stochastic optimization. In: International Conference on Learning Representations (ICLR) (2015)
12. Kipf, T.N., Welling, M.: Semi-supervised classification with graph convolutional networks. In: Proceedings of the 5th International Conference on Learning Representations, ICLR 2017 (2017). https://openreview.net/forum?id=SJU4ayYgl
13. Kreuzer, D., Beaini, D., Hamilton, W.L., Létourneau, V., Tossou, P.: Rethinking graph transformers with spectral attention (2021)
14. Nguyen, D.Q., Nguyen, T.D., Phung, D.: Universal self-attention network for graph classification. arXiv preprint arXiv:1909.11855 (2019)
15. Oono, K., Suzuki, T.: Graph neural networks exponentially lose expressive power for node classification. arXiv: Learning (2020)
16. Rong, Y., et al.: Self-supervised graph transformer on large-scale molecular data. In: Larochelle, H., Ranzato, M., Hadsell, R., Balcan, M.F., Lin, H. (eds.) Advances in Neural Information Processing Systems (2020). https://proceedings.neurips.cc/paper/2020/file/94aef38441efa3380a3bed3faf1f9d5d-Paper.pdf
17. Schwartz, R., Dodge, J., Smith, N.A., Etzioni, O.: Green AI. Commun. ACM (2020). https://doi.org/10.1145/3381831
18. Shazeer, N., et al.: Outrageously large neural networks: the sparsely-gated mixture-of-experts layer. arXiv preprint arXiv:1701.06538 (2017)
19. Vaswani, A., et al.: Attention is all you need. In: Proceedings of the 31st International Conference on Neural Information Processing Systems, NIPS 2017 (2017)
20. Veličković, P., Cucurull, G., Casanova, A., Romero, A., Liò, P., Bengio, Y.: Graph attention networks. In: 6th International Conference on Learning Representations (2017)
21. Wang, M., et al.: Deep graph library: a graph-centric, highly-performant package for graph neural networks. arXiv preprint arXiv:1909.01315 (2019)

22. Wu, Z., Pan, S., Chen, F., Long, G., Zhang, C., Yu, P.S.: A comprehensive survey on graph neural networks. IEEE Trans. Neural Netw. Learn. Syst. (2021). https://doi.org/10.1109/TNNLS.2020.2978386

23. Xu, K., Hu, W., Leskovec, J., Jegelka, S.: How powerful are graph neural networks? CoRR (2018). http://arxiv.org/abs/1810.00826

24. Ying, C., et al.: Do transformers really perform bad for graph representation? arXiv preprint arXiv:2106.05234 (2021)

25. Yun, S., Jeong, M., Kim, R., Kang, J., Kim, H.J.: Graph transformer networks. In: Wallach, H., Larochelle, H., Beygelzimer, A., d' Alché-Buc, F., Fox, E., Garnett, R. (eds.) Advances in Neural Information Processing Systems (2019). https://proceedings.neurips.cc/paper/2019/file/9d63484abb477c97640154d40595a3bb-Paper.pdf

26. Zhou, D., Zheng, L., Han, J., He, J.: A Data-Driven Graph Generative Model for Temporal Interaction Networks (2020). https://doi.org/10.1145/3394486.3403082

Edge but not Least: Cross-View Graph Pooling

Xiaowei Zhou[1,3], Jie Yin[2(✉)], and Ivor W. Tsang[1,4]

[1] Australian Artificial Intelligence Institute (AAII), University of Technology Sydney, Ultimo, NSW 2007, Australia
`Xiaowei.Zhou@student.uts.edu.au, ivor.tsang@uts.edu.au`
[2] Discipline of Business Analytics, The University of Sydney, Camperdown, NSW 2006, Australia
`jie.yin@sydney.edu.au`
[3] Data61, CSIRO, Eveleigh, NSW 2015, Australia
[4] Center for Frontier AI Research A*STAR, Singapore, Singapore

Abstract. Graph neural networks have emerged as a powerful representation learning model for undertaking various graph prediction tasks. Various graph pooling methods have been developed to coarsen an input graph into a succinct graph-level representation through aggregating node embeddings obtained via graph convolution. However, because most graph pooling methods are heavily node-centric, they fail to fully leverage the crucial information contained in graph structure. This paper presents a cross-view graph pooling method (Co-Pooling) that explicitly exploits crucial graph substructures for learning graph representations. Co-Pooling is designed to fuse the pooled representations from both node view and edge view. Through cross-view interaction, edge-view pooling and node-view pooling mutually reinforce each other to learn informative graph representations. Extensive experiments on one synthetic and 15 real-world graph datasets validate the effectiveness of our Co-Pooling method. Our results and analysis show that (1) our method is able to yield promising results over graphs with various types of node attributes, and (2) our method can achieve superior performance over state-of-the-art pooling methods on graph classification and regression tasks.

Keywords: Graph pooling · Graph representation learning

1 Introduction

With widespread digitization occurring in various domains, a significant portion of data takes the form of graphs, such as social networks, chemical molecular graphs, and financial transaction networks. As such, learning effective graph representations plays a crucial role in a variety of tasks, such as drug discovery, molecule property prediction, and traffic forecast, etc. Recently, graph neural networks (GNNs) have emerged as state-of-the-art models for graph representation learning, including graph convolutional network (GCN) [9], graph attention network (GAT) [20], graph isomorphism network (GIN) [22], and Graph-SAGE [8]. Most of these GNN models rely on message passing to learn the

M.-R. Amini et al. (Eds.): ECML PKDD 2022, LNAI 13714, pp. 344–359, 2023.
https://doi.org/10.1007/978-3-031-26390-3_21

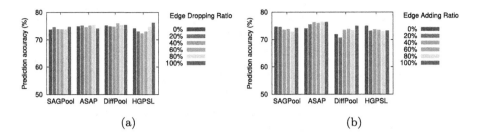

Fig. 1. Classification accuracy on PROTEINS with different edge ratios. Accuracy does not significantly drop when different ratios of edges are (a) randomly dropped from original graphs, or (b) randomly added from no-edge graphs.

embedding of each node by aggregating and transforming the features of its neighbouring nodes. To obtain the representation of the entire graph, node embeddings are aggregated via a readout function or graph pooling methods [15,24–26]. Graph pooling methods coarsen an input graph into a compact vector-based representation for the entire graph, which is then used for graph prediction tasks, such as graph classification or graph regression.

To learn expressive graph representations, various graph pooling methods have been proposed in recent years. Sampling based methods (*e.g.*, SAGPool [10], ASAP [15], HGPSL [26], and gPool [6]) calculate an importance score for each node and then select the top K important nodes to generate an induced subgraph. For example, SAGPool [10] selects nodes by learning importance scores via a self-attention mechanism. HGPSL [26] samples important nodes and uses an additional structure learning mechanism to learn new graph connectivity for the sampled nodes. Clustering based methods, like differentiable graph pooling (DiffPool) [24], learn an assignment matrix to cluster nodes into several supernodes level by level. Then, a hierarchy of the induced subgraphs can be generated for representing the whole graph. Nonetheless, we argue that the existing pooling methods focus primarily on aggregating node-level information, so they fail to exploit key graph substructures for learning graph-level representations. The loss of information present in graph structure would hinder message passing in subsequent layers and consequently jeopardize the graph representation expressiveness.

To verify our argument, we select four state-of-the-art pooling methods: SAGPool [10], ASAP [15], DiffPool [24], and HGPSL [26], and analyze the influence of changing graph topological structure on the graph classification accuracy. We use PROTEINS, a macromolecule dataset containing rich structural information, as a case study, where we change graph topological structure by randomly dropping or adding edges with different ratios. As shown in Fig. 1, we find that the random edge manipulation does not cause a significant drop in the graph classification accuracy. Surprisingly, when there are no edges at all, *i.e.*, dropping 100% edges in Fig. 1(a) and adding 0% edges (no edges) in Fig. 1(b), the classification accuracy still retains at the same level as other edge ratios. In particular, for HGPSL (which implicitly uses edge information), the classification accuracy

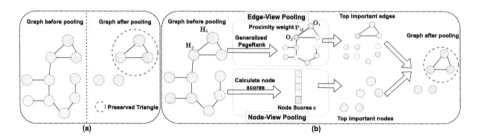

Fig. 2. (a) Illustration of substructure (triangle) preserved by Co-Pooling. (b) Overview of our proposed Co-Pooling framework. Co-Pooling is composed of two complementary components – edge-view pooling and node-view pooling – that reinforce each other to better learn informative graph-level representations.

is the highest when all edges are removed. Our empirical studies indicate that current graph pooling methods are heavily node-centric and lack the ability to fully leverage the crucial information contained in graph structure.

To fill this research gap, we propose a novel cross-view graph pooling method (Co-Pooling) that explicitly exploits graph substructures for learning graph-level representations. Our main motivations are twofold. First, we would like to capture crucial graph substructures through explicitly pruning unimportant edges in the graph. Key substructures, such as functional groups (*e.g.*, triangles, rings) in molecular networks, or cliques in protein-protein interaction networks and social networks, have been widely recognized as a crucial source for graph prediction tasks [13]. For example, in molecular chemistry, certain patterns of atoms (*e.g.*, triangles) are considered highly indicative for predicting certain molecule's properties [17]. As illustrated in Fig. 2(a), we need to preserve circular connectivity among three nodes in order to capture a triangle substructure. The crux is that, if graph pooling operates in a node-centric way or on the pairwise adjacency matrix, such higher-order, triangle circular connectivity cannot be properly preserved. Thus, we propose to preserve key substructures through learning higher-order proximity weights, which are then used to prune unimportant edges for substructure extraction. Second, apart from structural information, real-world graphs often have various types of node properties, such as one-hot attributes, real-valued attributes, or even no attributes (see Table 2). Hence, it is highly desirable for our pooling method to fuse useful information from both edge and node views and to make the best of node-level attributes when available.

Co-Pooling is composed of two key components: *edge-view pooling* and *node-view pooling*. Figure 2(b) shows the overview of Co-Pooling. Edge-view pooling aims to preserve crucial graph substructures, which are informative for subsequent graph prediction tasks. This is achieved by capturing high-order structural and attribute proximity via generalized PageRank and then pruning the edges with lower proximity weights. For node-view pooling, an importance score is calculated for each node, and top-ranked important nodes are selected for pooling. The learning of graph pooling from the edge and node views mutually reinforces each other through exchanging proximity weights and the selected important nodes. The final pooled graph is obtained by fusing graph representations learned from these two views. Through cross-view interaction, Co-Pooling enables edge-view pooling and node-view pooling to complement each other towards learning expressive graph representations. Our contributions are summarised as follows:

- We investigate the ineffectiveness of the existing node-centric graph pooling methods in fully leveraging graph structure.
- We propose a novel graph pooling method (Co-Pooling) to learn graph representations by fusing the pooled graph from both node view and edge view. Co-Pooling has the ability to preserve crucial graph substructures and to handle different types of graphs (node-labeled/attributed/plain graphs).
- We validate the effectiveness of Co-Pooling in graph classification and regression tasks across one synthetic and 15 real-world graph datasets, demonstrating its competitive performance over state-of-the-art pooling methods.

2   Related Work

Graph pooling is a key component of GNNs for learning a vector representation of an input graph. The existing graph pooling methods can be divided into two categories: *sampling based pooling* and *clustering based pooling*.

Sampling based pooling methods generate a smaller induced graph by selecting the top important nodes according to certain importance scores of nodes. SortPooling [27] ranks the nodes according to node embeddings learned from graph convolution and stacks the embeddings of selected nodes as graph representation. SAGPool [10] uses a self-attention mechanism to calculate an importance score for each node and then chooses top-ranked nodes to induce the pooled graph. Ranjan et al. [15] propose adaptive structure aware pooling (ASAP) that updates node embeddings by aggregating the features of nodes in a local region and then calculates a fitness score for each node to select the top-K nodes. Gao et al. [7] propose neighborhood information gain as a criterion to select top important nodes and then construct a coarsened graph from selected nodes. The above mentioned methods, however, do not fully leverage the crucial information contained in graph structure in the pooling process. HGPSL [26] takes one step forward to learn new connections between the selected nodes, but fails to capture crucial substructures contained in the original graph.

Clustering based pooling methods learn an assignment matrix to cluster nodes into super-nodes. DiffPool [24] learns a differentiable soft cluster assignment, which is used to group nodes into several clusters in the subsequent layer. HaarPooling [21] relies on compressive Haar transform filters to generate the induced graph of smaller size. HAP [11] uses master-orthogonal attention to learn a soft assignment to cluster nodes. SUGAR [18] samples several subgraphs from the input graph and clusters the top important subgraphs into super-nodes via reinforcement learning. However, this method is highly dependent on the sampling strategy used to obtain useful subgraph candidates.

Most of current graph pooling methods operate on a single node view; they are unable to fully leverage crucial graph structure. Although preliminary attempts (*e.g.*, EdgePool [3] and EdgeCut [5]) have been made to pool the input graph from an edge view, these methods simply rely on local connectivity to calculate pairwise edge scores. In contrast, our edge-view pooling leverages higher-order structural and attribute proximity to measure the importance of edges, which is fed to further guide the selection of important nodes for node-view pooling. To the best of our knowledge, we are the first to propose a cross-view graph pooling method, which enables pooling to fuse useful information from both edge and node views towards learning informative graph representations.

3   Methodology

In this section, we first introduce preliminaries and notations, and then present the details of our proposed cross-view graph pooling method.

3.1   Preliminaries and Notations

Suppose we are given m input graphs $\mathbb{G} = \{G^{(0)}, G^{(1)}, \cdots, G^{(m)}\}$ and their corresponding targets $\mathbb{Y} = \{y^{(0)}, y^{(1)}, \cdots, y^{(m)}\}$. For graph classification, $y^{(i)}$ is a discrete class label; for graph regression, $y^{(i)}$ is a continuous target variable $y^{(i)} \in \mathbb{R}$. An arbitrary graph $G^{(g)}$ is represented as $(\mathcal{V}^{(g)}, \mathcal{E}^{(g)}, \mathbf{X}^{(g)})$. For simplicity, $(\mathcal{V}^{(g)}, \mathcal{E}^{(g)}, \mathbf{X}^{(g)})$ is also noted as $(\mathcal{V}, \mathcal{E}, \mathbf{X})$, where $\mathcal{V}$ is the node set and $\mathcal{E}$ is the edge set. $\mathbf{X} \in \mathbb{R}^{n \times d}$ denotes the node attribute matrix, where $n = |\mathcal{V}|$ is the number of nodes and d is the dimension of node attributes; $\mathbf{A}$ is the adjacent matrix. If there is an edge between node i and j, $\mathbf{A}_{ij} = 1$; otherwise $\mathbf{A}_{ij} = 0$. $\hat{\mathbf{A}} = \mathbf{A} + \mathbf{I}$ denotes the adjacent matrix with self-loops.

In this work, we use graph convolution network (GCN) as our backbone to learn representations for graphs. The graph convolution operation is defined as:

$$\mathbf{H} = \hat{\mathbf{D}}^{-1/2} \hat{\mathbf{A}} \hat{\mathbf{D}}^{-1/2} \mathbf{X} \mathbf{\Theta} \tag{1}$$

where $\mathbf{H} \in \mathbb{R}^{n \times f}$ is node embedding after convolution, f is the dimension of node embedding; $\hat{\mathbf{D}}_{ii} = \sum_{j=0} \hat{\mathbf{A}}_{ij}$ is diagonal degree matrix; $\mathbf{\Theta}$ is a learnable parameter.

After node embeddings are learned, graph pooling aims to generate a vector representation for the whole graph. To facilitate downstream graph prediction tasks, the learned graph representation is expected to preserve the information conveyed by both graph structure and node attributes.

3.2   Cross-View Graph Pooling: Co-Pooling

The key idea of Co-Pooling is to preserve crucial "signals" that are beneficial to downstream graph prediction tasks. Unlike previous studies that dominantly focus on node-level information, we take both node and edge views to preserve crucial substructures reflected by graph structure and node attributes. To this end, we propose to perform graph pooling from both edge and node views.

As shown in Fig. 2(b), our proposed Co-Pooling framework consists of two complementary components: edge-view pooling and node-view pooling. Edge-view pooling prunes unimportant edges to capture meaningful substructures (*e.g.*, triangles). Node-view pooling, on the other hand, further selects top-ranked important nodes. Through cross-view interaction, edge-view pooling and node-view pooling reinforce each other to induce informative graph representations.

Edge-View Pooling. The key objective of edge-view pooling is to preserve crucial substructures contained in the original graph. Extracting useful substructures requires to incorporate high-order structural and node attribute information. Thus, we propose to use generalized PageRank (GPR) [2] to jointly optimize node attribute and topological information extraction.

To be specific, we first update node embeddings via GPR to capture the information from multi-hop neighbours. As shown in Eq. (2), node embeddings are updated by multiplying a GPR weight β_t at each step t. When $t = 0$, we have $\mathbf{H}^0 = \mathbf{H}$; when $t > 0$, we have $\mathbf{H}^t = \hat{\mathbf{D}}^{-1/2}\hat{\mathbf{A}}\hat{\mathbf{D}}^{-1/2}\mathbf{H}^{t-1}$. Through GPR, node embeddings propagate T steps, and the GPR weight β_t is learned at each step. Thus, the contribution of each propagation step towards node embeddings can be learned adaptively. The GPR operation of T steps helps incorporate the information from multi-hop neighbours to learn expressive node embeddings.

$$\mathbf{O} = \sum_{t=0}^{T} \beta_t \mathbf{H}^t. \tag{2}$$

After updating node embeddings via GPR, we calculate pairwise proximity weights that reflect high-order structural and attribute proximity between nodes. This process can be illustrated using Eq. (3), where $\mathbf{O}_i$ and $\mathbf{O}_j$ are the updated embeddings of node i and node j by GPR. We first transform node embeddings $\mathbf{O}_i$ and $\mathbf{O}_j$ via a linear transformation parameterized with $\mathbf{W}$. Then, the transformed embeddings are concatenated and fed to another linear transformation with learnable parameters $\mathbf{a}$. Finally, the proximity weight $\mathbf{P}_{ij}$ between node i and node j is obtained via a Sigmoid function. To preserve the original adjacency of graphs, we multiply the proximity weight with the adjacent matrix $\mathbf{A}$.

$$\mathbf{P}_{ij} = \sigma(\mathbf{a}^T[\mathbf{WO}_i\|\mathbf{WO}_j]) \odot \mathbf{A}_{ij}, \tag{3}$$

where $\mathbf{P}_{ij}$ is the proximity weight between node i and node j; σ is Sigmoid function; $\|$ represents the concatenation operation; $\mathbf{a}$ and $\mathbf{W}$ are learnable parameters; $\odot$ represents matrix element-wise multiplication; $\mathbf{A}_{ij} = 1$ *or* 0 indicates whether or not there is an edge connecting node i and node j.

According to the proximity weight $\mathbf{P}_{ij}$ of each node pair, we can obtain the proximity matrix $\mathbf{P}$ for all node pairs. For undirected graphs, we average the proximity weights at symmetric positions by $\mathbf{P}_{\text{sym}} = (\hat{\mathbf{P}} + \hat{\mathbf{P}}^T)/2$.

For a specific prediction task, the edges constituting discriminative substructures are expected to have higher proximity weights. Conversely, less important edges would have lower proximity weights. Thus, we prune unimportant edges with low proximity weights during edge-view pooling. For a given edge preserving ratio γ, we have the pruned proximity matrix $\mathbf{P}_{\text{prune}} = \text{Top}_\gamma(\mathbf{P}_{\text{sym}})$, where $\text{Top}_\gamma()$ is the operation that preserves the top γ percentage of edges with high proximity weights. Accordingly, we update the adjacent matrix to reflect the removal of edges. The pruned proximity matrix signifies certain crucial substructures preserved by pruning unimportant edges. The pruned proximity matrix is further fed to node-view pooling to guide the selection of important nodes.

Node-View Pooling. For node-view pooling, the aim is to select the top K important nodes for coarsening the input graph. To better exploit the connectivity between nodes, we take the pruned proximity matrix from edge-view pooling to compute an importance score for each node, given by

$$\mathbf{s} = \hat{\mathbf{D}}_{\text{prune}}^{-1/2} \hat{\mathbf{P}}_{\text{prune}} \hat{\mathbf{D}}_{\text{prune}}^{-1/2} \mathbf{H}\mathbf{1}^T, \tag{4}$$

where $\mathbf{s}$ is the score vector for all nodes; $\hat{\mathbf{D}}_{\text{prune}}$ is the diagonal degree matrix of $\hat{\mathbf{P}}_{\text{prune}}$, $\hat{\mathbf{P}}_{\text{prune}} = \mathbf{P}_{\text{prune}} + \mathbf{I}$; and $\mathbf{1}^T$ is a vector with all entries as one.

Based on node importance scores, we select the top $K = \lceil n \times \epsilon \rceil$ nodes, where ϵ is the node pooling ratio. For selected nodes, we can obtain their indices and corresponding node embeddings.

Edge-Node View Interaction. To enable edge-view pooling and node-view pooling to reinforce each other, Co-Pooling exchanges the pruned proximity matrix and the indices of selected nodes, which serve as the mediator for the interaction between two views.

For node-view pooling, the pruned proximity matrix from edge-view pooling is used to calculate the important score for each node. The pruned proximity matrix better reflects higher-order structural and attribute proximity between nodes, thus providing a better measure than the original adjacent matrix to quantify the importance of nodes contained in certain substructures. After obtaining the node scores, we select the top-K important nodes as the pooled graph, *i.e.*, $\mathbf{H}(\text{indices}, :)$, where $(\text{indices}, :)$ indicates the index selection operation.

For edge-view pooling, the indices of selected nodes obtained from node-view pooling are used to aggregate node embeddings from neighborhoods based on the pruned proximity matrix. The pooled representation from edge-view pooling is obtained as $\mathbf{P}_{\text{prune}}(\text{indices}, :)\mathbf{H}$. This enables to extract meaningful substructures centered on important nodes and perform neighborhood aggregation over important edges only.

Lastly, the pooled representations from node-view pooling and edge-view pooling are fused to form the final graph representation as:

$$\mathbf{Z} = \mathbf{W}[\mathbf{P}_{\text{prune}}(\text{indices}, :)\mathbf{H}\|\mathbf{H}(\text{indices}, :)] \tag{5}$$

where $\mathbf{W}$ is a learnable parameter of linear transformation; $\|$ indicates the concatenation operator; and $\mathbf{Z}$ is the graph representation after pooling. Through edge-node view interaction, Co-Pooling enables edge-view pooling and node-view pooling to complement each other for learning informative graph representations.

4   Experiments

We evaluate the performance of Co-Pooling on three graph prediction tasks, including substructure counting, graph classification, and graph regression. For substructure counting (Sect. 4.1), we empirically assess the performance of Co-Pooling in preserving important substructures. For graph classification, we compare Co-Pooling against several state-of-the-art pooling methods under two settings: *attribute-complete graphs* (Sect. 4.2) and *attribute-incomplete graphs* (Sect. 4.3). Furthermore, we compare Co-Pooling against baseline pooling methods on graph regression (Sect. 4.5). The source code of Co-Pooling is available at: https://github.com/zhouxiaowei1120/Co-Pooling.

Baselines. As our focus is upon designing new graph pooling methods, we compare Co-Pooling with five state-of-the-art graph pooling methods rather than specially designed GNNs for graph classification. These baseline methods include: SAGPool [10], ASAP [15], DiffPool [24], HGPSL [26], and EdgePool [3]. When training DiffPool, we use the auxiliary link prediction loss function with entropy regularization as in the original paper. For comparison, all graph pooling methods are built on top of the same GCN architecture for downstream tasks.

4.1   Substructure Counting on Random Graphs

To verify the capacity of Co-Pooling in preserving graph substructures, we consider a substructure counting task, with the aim to count the number of triangles contained in random graphs.

Dataset. For substructure counting task, we use the synthetic **Syn-triangle** dataset [1], consisting of 5,000 Erdös-Renyi random graphs. Each graph contains 10 nodes and $p = 0.3$ is the probability that an edge exists. Akin to [1], we use 30%-20%-50% graphs as training-validation-test sets.

Experimental Setup. For training a regression model on Syn-triangle, we use GCN as the backbone and inject two pooling layers before an MLP layer. Adam optimizer with learning rate decay is used to train the model. The optimization stops if the validation loss does not decrease after 50 epochs. Following [1], we use L2 loss and set the initial learning rate and weight decay as 0.02 and 0.001, respectively. The GPR operation step T is set as 3. We train the regression model with four different random seeds. The results are measured by normalized mean square error (MSE) on the test set (*i.e.*, MSE divided by the variance of the ground truth counts of the triangles over all graphs in the test set).

Table 1. Normalized MSE for substructure counting on Syn-triangle.

Methods	Syn-triangle
SAGPool	0.849 ± 0.061
ASAP	0.701 ± 0.140
DiffPool	0.762 ± 0.194
HGPSL	0.878 ± 0.079
EdgePool	0.704 ± 0.009
Co-Pooling	$\mathbf{0.448 \pm 0.046}$

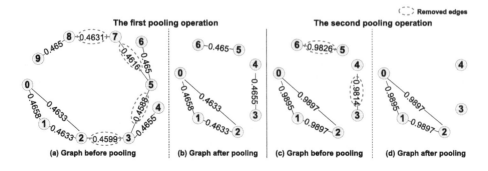

Fig. 3. Illustration of the pooled graphs by Co-Pooling.

Results. The results on triangle counting are given in Table 1. As can be seen, Co-Pooling outperforms all baseline methods, yielding markedly smaller errors than the second best performer ASAP. This empirically verifies that Co-Pooling is able to preserve crucial substructures during the pooling process.

Figure 3 gives an example to illustrate two pooling operations of Co-Pooling. For a given graph, the proximity weights of edges and the pooled graphs are shown in the figure. During the first pooling operation, four edges with lower proximity weights (marked in red dashed ellipse) are removed, and afterwards, nodes 7, 8, and 9 with lower importance scores are further removed to generate the pooled graph. It is clear to find that the triangle substructure is preserved after the first pooling layer (see Fig. 3(b)). A similar process can be observed during the second pooling operation, where the triangle substructure is also preserved in the pooled graph (see Fig. 3(d)), which is highly indicative for triangle counting.

4.2 Graph Classification on Attribute-Complete Graphs

Datasets. We undertake graph classification on a total of 13 benchmark graph datasets with various attribute properties, including three *attributed graph* datasets with real-valued node attributes, five *labeled graph* datasets with only one-hot node attributes, and five *plain graph* datasets without node attributes. The detailed statistics about these datasets are listed in Table 2.

- **BZR-A** [19] is a dataset of chemical compounds for classifying biological activities. The node attributes are 3D coordinates of compound structures.
- **AIDS-A** [16] is composed of graphs representing molecular compounds. It contains two classes of graphs, which are against HIV or not.
- **FRANKENSTEIN** [14] consists of molecules as mutagens or non-mutagens for binary classification. The node attributes are 780-dimensional MNIST image vectors of pixel intensities, representing chemical atom symbols.
- **D&D** [12] and **PROTEINS** [12] include macromolecules as graph datasets in bioinformatics, which are for enzyme and non-enzyme classification task.

Table 2. Details of graph datasets for graph classification evaluation.

Dataset	# Graphs	# Classes	Avg. $|V|$	Avg. $|E|$	Node Attributes	Type
BZR-A	405	2	35.75	38.36	Real-valued	*Attributed*
AIDS-A	2,000	2	15.69	16.20	Real-valued	*Attributed*
FRANKENSTEIN	4,337	2	16.90	17.88	Real-valued	*Attributed*
PROTEINS	1,113	2	39.06	72.82	Node label	*Labeled*
D&D	1,178	2	284.32	715.66	Node label	*Labeled*
NCI1	4,110	2	29.87	32.30	Node label	*Labeled*
NCI109	4,127	2	29.68	32.13	Node label	*Labeled*
MSRC_21	563	20	77.52	198.32	Node label	*Labeled*
COLLAB	5,000	3	74.49	2457.78	None	*Plain*
IMDB-B	1,000	2	19.77	96.53	None	*Plain*
IMDB-M	1,500	3	13.00	65.94	None	*Plain*
REDDIT-B	2,000	2	429.63	497.75	None	*Plain*
REDDIT-M	11,929	11	391.41	456.89	None	*Plain*

- **NCI1** [12] and **NCI109** [12] contain chemical compounds as small molecules, which are used for anticancer activity classification task.
- **MSRC_21** [12] is a graph dataset constructed by semantic images. Each image is represented as a conditional Markov random field graph. Nodes in a graph represent the segmented superpixels in an image. If the segmented superpixels are adjacent, the corresponding nodes are connected. Each node is assigned a semantic label as node attribute.
- **COLLAB** [23] is a collection of scientific collaboration graphs, where the task is to classify the graphs into different research fields.
- **IMDB-B** [23] and **IMDB-M** [23] are two datasets for classifying graphs into movie genres. Each graph is an ego-network for each actor/actress.
- **REDDIT-B** and **REDDIT-M** [23] are two datasets generated from online discussions. Each graph represents a discussion thread where nodes indicate different users. If one user responds to another one, there is an edge between them. The task is to classify which section each discussion belongs to.

Baselines. Apart from other graph pooling baselines, we also compare with two ablated variants of Co-Pooling: Co-Pooling w/o GPR that removes generalized PageRank and Co-Pooling w/o NV that removes node-view pooling.

Experimental Setup. For all datasets, we use the same GNN architecture for a fair comparison. The GNN consists of three GCN layers, two pooling layers (constructed by different pooling methods), and three linear transformation layers. A softmax layer is then connected after the last linear transformation layer. Note that, the input to the first linear transformation layer is the concatenated features after each pooling layer.

Akin to prior work [24], we perform 10-fold cross-validation to train the GNN model. We randomly partition each dataset into training, validation, and test sets using a 80%-10%-10% split. We use Adam optimizer with early stopping;

Table 3. Graph classification accuracy on 13 graph datasets.

Dataset	SAGPool	ASAP	DiffPool	HGPSL	EdgePool	Co-Pooling (w/o GPR)	Co-Pooling (w/o NV)	Co-Pooling
BZR-A	82.95 ± 4.91	83.70 ± 6.00	83.93 ± 4.41	83.23 ± 6.51	83.43 ± 6.00	81.00 ± 5.82	81.69 ± 5.80	85.67 ± 5.29
AIDS-A	98.85 ± 0.78	99.00 ± 0.74	99.40 ± 0.58	99.10 ± 0.66	99.05 ± 0.69	98.85 ± 0.71	98.90 ± 0.58	99.45 ± 0.42
FRANK	60.94 ± 2.90	66.73 ± 2.76	65.08 ± 1.50	62.19 ± 1.74	62.99 ± 2.21	64.01 ± 1.70	67.00 ± 2.37	64.15 ± 1.34
D&D	76.91 ± 3.42	77.84 ± 3.41	78.01 ± 2.70	77.33 ± 4.22	76.66 ± 2.05	75.81 ± 3.81	77.00 ± 5.04	77.85 ± 2.21
PROTEINS	73.68 ± 4.63	74.85 ± 5.18	75.11 ± 2.95	74.13 ± 4.12	77.01 ± 5.41	73.68 ± 2.33	76.28 ± 5.09	76.19 ± 4.13
NCI1	71.51 ± 4.51	76.59 ± 1.71	74.14 ± 1.43	73.48 ± 2.42	78.39 ± 2.43	77.25 ± 2.11	79.15 ± 2.04	78.66 ± 1.48
NCI109	69.69 ± 3.27	74.73 ± 3.48	72.04 ± 1.43	72.30 ± 2.18	77.01 ± 2.39	75.60 ± 1.46	78.07 ± 1.77	77.08 ± 2.03
MSRC_21	90.22 ± 2.82	90.41 ± 3.91	90.41 ± 3.58	88.97 ± 4.78	90.05 ± 3.02	91.64 ± 2.79	91.29 ± 3.70	92.54 ± 2.63
COLLAB	70.58 ± 2.31	72.84 ± 1.84	72.18 ± 1.68	74.20 ± 2.72	-	74.82 ± 2.10	68.90 ± 5.59	77.30 ± 2.29
IMDB-B	60.90 ± 2.34	65.50 ± 2.80	58.27 ± 5.92	62.50 ± 3.50	60.30 ± 5.08	70.40 ± 3.85	70.80 ± 3.60	72.10 ± 4.44
IMDB-M	39.80 ± 3.39	45.93 ± 4.03	40.00 ± 4.52	40.53 ± 4.88	44.27 ± 4.50	47.60 ± 4.55	44.80 ± 3.94	49.07 ± 3.28
REDDIT-B	83.55 ± 4.53	-	84.61 ± 2.42	-	88.35 ± 2.31	88.90 ± 2.00	88.00 ± 4.69	89.35 ± 1.25
REDDIT-M	40.56 ± 3.30	-	41.21 ± 1.96	-	-	46.84 ± 2.26	49.02 ± 1.56	46.85 ± 2.62

"-" means the results can not be obtained in an acceptable time, $i.e.$ 24 h.

the optimization stops if the validation loss does not improve after 50 epochs. The maximum epoch number is set as 300. Following [10], we use grid search to obtain optimal hyperparameters for each method. The ranges of different hyperparameters are set as follows: learning rate in $\{0.005, 0.0005, 0.001\}$, weight decay in $\{0.0001\ 0.001\}$, node pooling ratio in $\{0.5, 0.25\}$, hidden size in $\{128, 64\}$, dropout ratio in $\{0, 0.5\}$, and edge preserving ratio γ in $\{0.3, 0.6, 1.0\}$. Akin to [2], step T of GPR is set to 10. To implement the convolution operation on plain graphs without node attributes, we follow the implementation of DiffPool to pad each node with a constant vector, $i.e.$ an all-one vector of 10 dimensions.

Comparison with State-of-the-Art. Table 3 shows graph classification accuracy of all methods averaged over 10-fold cross-validation on 13 datasets. For a fair comparison, all baseline methods and our method are trained using the same training strategy. As we can observe, among all methods, our Co-Pooling method achieves the best performance on all datasets except on D&D and PRO-TEIN, where Co-Pooling achieves the second best performance. In particular, Co-Pooling significantly improves the best baseline method by 6.6%, 3.14%, 7.81%, 2.13%, and 1.74% on IMDB-B, IMDB-M, REDDIT-M, MSRC_21, and BZR-A, respectively. This proves the effectiveness of Co-Pooling in predicting different types of graphs with various attribute properties. It is worth noting that Co-Pooling achieves the best performance on all five datasets without node attributes. This shows the superiority of our method to complement node-view pooling with edge-view pooling, when node attributes are not informative.

When comparing different variants of our method, Co-Pooling consistently outperforms Co-Pooling w/o GPR on all datasets. This shows the importance of using generalized PageRank to capture higher-order structural information. Co-Pooling yields higher accuracy than Co-Pooling w/o NV on most (8/13) of the datasets. This demonstrates the effectiveness of Co-Pooling in combining two complementary views. In particular, its performance gains on *attributed graphs* with real-valued node attributes are more significant than those on *labeled graphs*

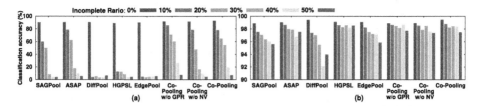

Fig. 4. Graph classification accuracy on (a) *labeled graph* dataset (MSRC_21) and (b)*attributed graph* dataset (AIDS-A) under attribute-incomplete settings.

with one-hot attributes. This is because real-valued node attributes provide more accurate information to select important nodes for node-view pooling as opposed to one-hot attributes. This in turn reinforces edge-view pooling more effectively for learning the final graph representations.

4.3   Graph Classification on Attribute-Incomplete Graphs

Next, we compare the performance of our method and all baselines on attribute-incomplete graphs. For attribute-incomplete graphs, a portion of nodes has completely missing attributes. This set of experiments are used to evaluate the effectiveness of our method in real-world scenarios, where attribute information for some nodes is inaccessible due to privacy or legal constraints.

Experimental Setup. We perform experiments on *attributed graph* dataset (AIDS-A) and *labeled graph* dataset (MSRC_21) as a case study. For each graph from the two datasets, we randomly select different ratios of nodes and remove their original node attributes, while keeping the rest of nodes unchanged. We define the ratio of nodes with all their attributes removed as the incomplete ratio. For example, if we remove all attributes for 10% of nodes, the incomplete ratio is 10%. The resulting attribute-incomplete graph datasets are randomly divided into training set (80%), validation set (10%), and test set (10%). We train the GNN model with different pooling methods on training set. The GNN model architecture and the best hyperparameters are the same as in Sect. 4.2. We report graph classification accuracy averaged over 10-fold cross-validation.

Comparison with State-of-the-Art. Figure 4 (a) compares the classification accuracy of all methods on attribute-incomplete MSRC_21 datasets. For all baseline methods, the classification accuracy drops significantly as the incomplete ratio increases from 0% to 50%. In contrast, the accuracy of Co-Pooling and its variants decreases at a much lower rate. Especially for DiffPool, HGPSL, and EdgePool, the classification accuracy drops down to 3.73%, 12.33%, and 4.61%, respectively, although only 10% nodes have their attributes missing. With the 10% incomplete ratio, Co-Pooling and its variants can still achieve at least 77.93% accuracy. Compared with the best baseline method ASAP, Co-Pooling achieves an average of 8.62% accuracy increase on attribute-incomplete MSRC_21 datasets with different incomplete ratios (from 0% to 50%).

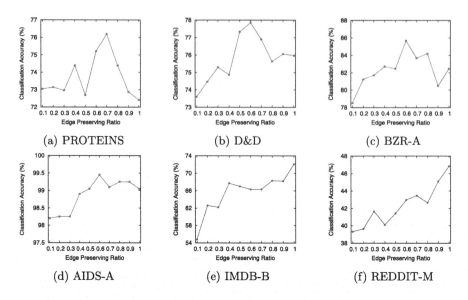

Fig. 5. Graph classification accuracy with different edge preserving ratios (γ).

Figure 4(b) compares graph classification accuracy of all methods on attribute-incomplete AIDS-A datasets. We can see that, for SAGPool, DiffPool, and EdgePool, the classification accuracy drops by 3.25%, 5.45%, and 3.2%, respectively, when the incomplete ratio increases from 0% to 50%. On the contrary, the accuracy of Co-Pooling and its variants drops by around 1.15% only. Our methods beat ASAP with all incomplete ratios. Compared with HGPSL, our methods achieve better performance with 0%, 10%, 20%, and 40% incomplete ratios, with higher average accuracy on all attribute-incomplete AIDS-A datasets.

The comparisons on attribute-incomplete graph datasets demonstrate the effectiveness of our method in handling graphs with missing node attributes. This further testifies the complementary advantage of our method by fusing node-view and edge-view pooling, especially when node attributes are less informative.

4.4   Parameter Sensitivity

The Co-Pooling method has the edge preserving ratio (γ) as an important parameter to determine the percentages of edges preserved during edge-view pooling. To investigate the effect of the edge preserving ratio (*i.e.*, γ) on the graph classification accuracy of Co-Pooling, we conduct empirical studies on six representative graph datasets, including two *labeled graph* datasets, two *attributed graph* datasets, and two *plain graph* datasets. On each dataset, we train the GNN model with an edge preserving ratio ranging from 10% to 100%. All other hyperparameters are set as the best values obtained in Sect. 4.2. We also use the same GNN model architecture and training strategy as in Sect. 4.2. We report the average classification accuracy on 10-fold cross-validation.

Figure 5 plots the change in classification accuracy with respect to γ on the six datasets. On the two *labeled graph* datasets (PROTEINS and D&D), we find that keeping all edges ($\gamma = 1.0$) is not the best choice for graph classification. As shown in Fig. 5 (a) and (b), Co-Pooling achieves the highest classification accuracy when $\gamma = 0.7$ on PROTEIN, and $\gamma = 0.6$ on D&D, respectively. A similar phenomenon can also be observed on the two *attributed graphs* (BZR-A and AIDS-A). As shown in Fig. 5 (c) and (d), Co-Pooling yields the best performance when γ is set to 0.6 on the two datasets. The results on the four datasets indicate that not all edges are useful for graph classification when graphs have informative node attributes. Again, this confirms the effectiveness of our method in preserving crucial edge information through edge-view pooling and using this knowledge to further guide node-view pooling. On the other hand, on the two *plain graph* datasets (IMDB-B and REDDIT-M), keeping all edges renders the highest classification accuracy. As shown in Fig. 5 (e) and (f), Co-Pooling achieves the best performance on both graphs when keeping all edges ($\gamma = 1.0$). This is what we have expected, because when graphs have no node attributes, preserving all graph structures would best benefit graph classification.

4.5   Graph Regression

Lastly, we carry out experiments to evaluate the efficacy of our method on the graph regression task. We compare Co-Pooling with the same state-of-the-art pooling methods on the following two graph datasets:

- **ZINC** [12] contains 250,000 molecules. The task is to regress the properties of molecules. We focus on predicting one specific graph property, contained solubility. Following the setting in [4], we use 10,000 graphs from ZINC for training, 1,000 graphs for validation, and 1,000 graphs for testing.
- **QM9** [12] is a graph dataset consisting of 13,000 molecules with 19 regression targets. We focus on regressing dipole moment μ, one of 19 molecular properties. All 13,000 molecules are randomly divided into training-validation-test sets using a 80%-10%-10% split.

For training a regression model on each dataset, we use the same GNN architecture and training strategy as in Sect. 4.1. Following [4], we use L1 loss to train each model. The initial learning rate and weight decay are set as 0.001 and 0.0001, respectively. We train regression models with four different random seeds and report the average mean absolute error (MAE) on the test set.

We compare our Co-Pooling method with SAGPool, ASAP, DiffPool, HGPSL, and EdgePool on the two datasets. As shown in Table 4, Co-Pooling consistently yields lower error than other baseline methods. Particularly, Co-Pooling outperforms DiffPool and HGPSL by a large margin on both datasets. These results indicate that our method effectively learns a better graph-level representation by fusing edge-view pooling and node-view pooling, leading to competitive performance on graph regression tasks as well.

Table 4. MAE results of graph regression on ZINC and QM9. Lower is better.

Methods	ZINC	QM9
GCN+SAGPool	0.378 ± 0.031	0.545 ± 0.010
GCN+ASAP	0.372 ± 0.026	0.500 ± 0.017
GCN+DiffPool	1.641 ± 0.026	1.331 ± 0.014
GCN+HGPSL	1.326 ± 0.096	1.035 ± 0.049
GCN+EdgePool	0.382 ± 0.030	0.489 ± 0.022
GCN+Co-Pooling (ours)	0.340 ± 0.036	0.439 ± 0.009

5   Conclusion

We proposed a new graph pooling method (Co-Pooling) for learning graph-level representations. We argued that most of current graph pooling methods are highly node-centric and fail to leverage crucial graph substructures, which are in fact beneficial to various prediction tasks. Our proposed Co-Pooling method fuses the pooled graph information from two views. From the edge view, generalized PageRank is used to aggregate valuable structural information from multi-hop neighbours. The proximity weights between node pairs are then calculated to prune less important edges. From the node view, the node importance scores are computed through the proximity matrix to select the top important nodes. Through cross-view interaction, edge-view pooling and node-view pooling complement each other to effectively learn informative graph representations. Extensive experiments on 16 graph datasets demonstrate the superior performance of Co-Pooling on both graph classification and regression tasks.

Acknowledgements. Xiaowei Zhou is supported by a Data61 PhD Scholarship from CSIRO. Ivor W. Tsang is supported by the Center for Frontier AI research, A*STAR, and ARC under grants DP200101328. This work is partially supported by the USYD-Data61 Collaborative Research Project grant.

References

1. Chen, Z., Chen, L., Villar, S., Bruna, J.: Can graph neural networks count substructures? NeurIPS **33**, 10383–10395 (2020)
2. Chien, E., Peng, J., Li, P., Milenkovic, O.: Adaptive universal generalized pagerank graph neural network. In: ICLR (2021)
3. Diehl, F.: Edge contraction pooling for graph neural networks. arXiv preprint arXiv:1905.10990 (2019)
4. Dwivedi, V.P., Joshi, C.K., Laurent, T., Bengio, Y., Bresson, X.: Benchmarking graph neural networks. arXiv preprint arXiv:2003.00982 (2020)
5. Galland, A.: Graph pooling by edge cut (2021)
6. Gao, H., Ji, S.: Graph u-nets. In: ICML, pp. 2083–2092. PMLR (2019)

7. Gao, X., Dai, W., Li, C., Xiong, H., Frossard, P.: iPool-information-based pooling in hierarchical graph neural networks. IEEE TNNLS **33**, 1–13 (2021)
8. Hamilton, W.L., Ying, R., Leskovec, J.: Inductive representation learning on large graphs. In: NIPS, pp. 1025–1035 (2017)
9. Kipf, T.N., Welling, M.: Semi-supervised classification with graph convolutional networks. In: ICLR (2017)
10. Lee, J., Lee, I., Kang, J.: Self-attention graph pooling. In: ICML, pp. 3734–3743 (2019)
11. Liu, N., Jian, S., Li, D., Zhang, Y., Lai, Z., Xu, H.: Hierarchical adaptive pooling by capturing high-order dependency for graph representation learning. IEEE TKDE (2021)
12. Morris, C., Kriege, N., Bause, F., Kersting, K., Mutzel, P., Neumann, M.: Tudataset: a collection of benchmark datasets for learning with graphs. arXiv:2007.08663 (2020)
13. Milo, R., Shen-Orr, S., Itzkovitz, S., Kashtan, N., Chklovskii, D., Alon, U.: Network motifs: simple building blocks of complex networks. Science **298**, 824–827 (2002)
14. Orsini, F., Frasconi, P., De Raedt, L.: Graph invariant kernels. In: IJCAI, pp. 3756–3762 (2015)
15. Ranjan, E., Sanyal, S., Talukdar, P.: ASAP: adaptive structure aware pooling for learning hierarchical graph representations. In: AAAI, pp. 5470–5477 (2020)
16. Riesen, K., Bunke, H.: IAM graph database repository for graph based pattern recognition and machine learning. In: da Vitoria Lobo, N., et al. (eds.) SSPR /SPR 2008. LNCS, vol. 5342, pp. 287–297. Springer, Heidelberg (2008). https:// doi.org/10.1007/978-3-540-89689-0_33
17. Shang, J., et al.: Assembling molecular sierpiński triangle fractals. Nat. Chem. **7**(5), 389–393 (2015)
18. Sun, Q., et al.: Sugar: subgraph neural network with reinforcement pooling and self-supervised mutual information mechanism. In: Proceedings of the Web Conference 2021, pp. 2081–2091 (2021)
19. Sutherland, J.J., O'brien, L.A., Weaver, D.F.: Spline-fitting with a genetic algorithm: a method for developing classification structure-activity relationships. J. Chem. Inf. Comput. Sci. **43**(6), 1906–1915 (2003)
20. Veličković, P., Cucurull, G., Casanova, A., Romero, A., Lio, P., Bengio, Y.: Graph attention networks. In: ICLR (2018)
21. Wang, Y.G., Li, M., Ma, Z., Montufar, G., Zhuang, X., Fan, Y.: Haar graph pooling. In: ICML, pp. 9952–9962 (2020)
22. Xu, K., Hu, W., Leskovec, J., Jegelka, S.: How powerful are graph neural networks? In: ICLR (2019)
23. Yanardag, P., Vishwanathan, S.: Deep graph kernels. In: SIGKDD, pp. 1365–1374 (2015)
24. Ying, R., You, J., Morris, C., Ren, X., Hamilton, W.L., Leskovec, J.: Hierarchical graph representation learning with differentiable pooling. In: NIPS, pp. 4805–4815 (2018)
25. Yuan, H., Ji, S.: Structpool: structured graph pooling via conditional random fields. In: ICLR (2020)
26. Zhang, Z., et al.: Hierarchical graph pooling with structure learning. arXiv preprint arXiv:1911.05954 (2019)
27. Zhang, M., Cui, Z., Neumann, M., Chen, Y.: An end-to-end deep learning architecture for graph classification. In: AAAI, pp. 4438–4445 (2018)

GNN Transformation Framework for Improving Efficiency and Scalability

Seiji Maekawa[1(✉)], Yuya Sasaki[1], George Fletcher[2], and Makoto Onizuka[1]

[1] Osaka University, 1–5 Yamadaoka, Suita, Osaka, Japan
{maekawa.seiji,sasaki,onizuka}@ist.osaka-u.ac.jp
[2] Eindhoven University of Technology, P.O. Box 513, 5600 MB Eindhoven,
The Netherlands
g.h.l.fletcher@tue.nl

Abstract. We propose a framework that automatically transforms non-scalable GNNs into precomputation-based GNNs which are efficient and scalable for large-scale graphs. The advantages of our framework are two-fold; 1) it transforms various non-scalable GNNs to scale well to large-scale graphs by separating local feature aggregation from weight learning in their graph convolution, 2) it efficiently executes precomputation on GPU for large-scale graphs by decomposing their edges into small disjoint and balanced sets. Through extensive experiments with large-scale graphs, we demonstrate that the transformed GNNs run faster in training time than existing GNNs while achieving competitive accuracy to the state-of-the-art GNNs. Consequently, our transformation framework provides simple and efficient baselines for future research on scalable GNNs.

Keywords: Graph Neural Networks · Large-scale graphs · Classification

1 Introduction

Graph is a ubiquitous structure that occurs in many domains, such as Web and social networks. As a powerful approach for analyzing graphs, Graph Neural Networks (GNNs) have gained wide research interest [25,30]. Many GNNs have been proposed for node classification and representation learning including GCN [15], which is the most popular GNN variant. Most existing GNNs adopt graph convolution that performs three tasks; 1) feature aggregation, 2) learnable weight multiplication, and 3) activation function application (e.g., ReLU, a non-linear function). By stacking multiple graph convolutional layers, they propagate node features over the given graph topology. However, these existing GNNs cannot be efficiently trained on large-scale graphs since the GNNs need to perform three tasks in graph convolution every time learnable weights are updated. In addition, large-scale graphs cannot be put on GPU memory for efficient matrix operations. As a result, graph convolution is not efficient and scalable for large-scale graphs.

A major approach to apply GNNs to large-scale graphs is to separate feature aggregation from graph convolution so that GNNs can precompute aggregated features [8,18,24]. These methods are called *precomputation-based* GNNs. In detail, they remove non-linearity, i.e., activation functions, from graph convolution so that feature aggregation is separated from weight learning. Thanks to the independence of feature aggregation and weight learning, precomputation-based GNNs are efficient in learning steps by precomputing feature aggregation before training learnable weights.

Though some existing works tackle the scalability problem of GNNs as discussed above, most widely studied GNNs are not scalable to large-scale graphs for the following two reasons. First, existing studies on precomputation-based GNNs [8,18,24] focus on introducing several specific GNN architectures that are manually designed. So, it is laborsome to apply the same precomputation idea to other GNNs. An interesting observation is that they share the common motivation: precomputation of feature aggregation is indispensable for high scalability. To our best knowledge, there are no works that study a general framework that transforms non-scalable GNNs to scalable precomputation-based GNNs. Second, existing precomputation schemes are not scalable because they need to put complete graphs (e.g., graphs with one billion edges [12]) on GPU memory. Since the size of large graphs typically exceeds the memory size of general GPU, existing works precompute feature aggregation on CPU.

To tackle the above issues, we address two research questions: **Q1**: *Can we design a general procedure that transforms non-scalable GNNs to efficient and scalable precomputation-based GNNs while keeping their classification performance?* and **Q2**: *Can we efficiently execute the precomputation on GPU?* There are two technical challenges which must be overcome to answer our questions. First, we need to automatically transform non-scalable GNNs to precomputation-based GNNs. We should develop a common transformation procedure that can be applied to various non-scalable GNNs while preserving their expressive power. Second, we need to decompose large graphs into small groups each of which can be handled efficiently with GPU. Typically, graph decomposition suffers from an imbalance problem since node degree distributions usually follow power law distributions [19]. Hence, we should divide graphs into balanced groups and select an appropriate group size so that precomputation time is optimized.

In this paper, we propose a framework[1] that automatically transforms non-scalable GNNs into precomputation-based GNNs with a scalable precomputation schema. As for the first challenge, we develop a new transformation procedure, called Linear Convolution (LC) transformation, which can be applied to various non-scalable GNNs so that transformed GNNs work efficiently and scale well to large-scale graphs. Our transformation procedure removes non-linear functions from graph convolution, but incorporates non-linear functions into weight learning. This idea is derived from our hypothesis that it is not crucial to incorporate non-linearity into graph convolutional layers but into weight learning for prediction. Since our transformation preserves the major functionality of graph convolution and a similar expressive power to original GNNs, the transformed GNNs can achieve competitive prediction performance to the original ones while improving their scalability. As for the second challenge, we develop a block-

---

[1] Our codebase is available on (https://github.com/seijimaekawa/LCtransformation).

wise precomputation scheme which optimally decomposes large-scale graphs into small and balanced blocks each of which can fit into GPU memory. We introduce a simple decomposition approach to ensure that blocks are balanced and give minimization formulas that decide the optimal block size under limited GPU memory.

Through extensive experiments, we validate that our transformation procedure and optimized block-wise precomputation scheme are quite effective. First, we show that our LC transformation procedure transforms non-scalable GNNs to efficient and scalable precomputation-based GNNs while keeping their node classification accuracy. Second, we show that our precomputation scheme is more efficient than that of existing precomputation-based GNNs. In summary, our transformation procedure provides simple and efficient baselines for future research on scalable GNNs by shining a spotlight on existing non-scalable methods.

The rest of this paper is organized as follows. We describe notations and fundamental techniques for our method in Sect. 2. Section 3 proposes our framework. We give the purpose and results of experiments in Sect. 4. Section 5 describes the details of related work. Finally, we conclude this paper in Sect. 6.

2   Preliminaries

An *undirected attributed graph with class labels* is a triple $G = (A, X, C)$ where $A \in \{0, 1\}^{n \times n}$ is an adjacency matrix, $X \in \mathbb{R}^{n \times d}$ is an attribute matrix assigning attributes to nodes, and a class matrix $C \in \{0, 1\}^{n \times y}$ contains class information of each node, and n, d, y are the numbers of nodes, attributes and classes, respectively. If there is an edge between nodes i and j, A_{ij} and A_{ji} are set to one. We define the degree matrix $D = \text{diag}(D_1, \ldots, D_n) \in \mathbb{R}^{n \times n}$ as a diagonal matrix, where D_i expresses the degree of node i. We also define an identity matrix $I = \text{diag}(1, \ldots, 1) \in \mathbb{R}^{n \times n}$ and an adjacency matrix extended with self-loops $\tilde{A} = A + I$. We define node embeddings $H \in \mathbb{R}^{n \times h}$, where h is the dimension of a hidden layer. We summarize notation and their definitions in Table 1.

2.1   Graph Convolutional Networks

Multi-layer GCN is a standard GCN model which was proposed in [15]. GCNs learn a feature representation for the feature of each node over layers. For the k-th graph convolutional layer, we denote the input node representations of all nodes by the matrix $H^{(k-1)}$ and the output node representations by $H^{(k)}$. The initial node representations are set to the input features, i.e., $H^{(0)} = X$. Let S denote the normalized adjacency matrix

$$S = \tilde{D}^{-\frac{1}{2}} \tilde{A} \tilde{D}^{-\frac{1}{2}}. \tag{1}$$

This normalized adjacency matrix is commonly used as a graph filter for graph convolution. The graph filter is known as a low-pass filter that filters out noise

Table 1. Notation and definitions

n	Number of nodes
d	Dimension of features
y	Number of classes
h	Dimension of hidden layer
K	Number of hidden layers
$A \in \mathbb{R}^{n \times n}$	Adjacency matrix
$\tilde{A} \in \mathbb{R}^{N \times N}$	Extended adjacency matrix
$S \in \mathbb{R}^{n \times n}$	Normalized adjacency matrix
$X \in \mathbb{R}^{n \times d}$	Feature matrix
$C \in \mathbb{R}^{n \times y}$	Class matrix
$D \in \mathbb{R}^{n \times n}$	Degree matrix
$H \in \mathbb{R}^{n \times h}$	Node embeddings
$W_1 \in \mathbb{R}^{d \times h}, W_2, \ldots, W_{K-1} \in \mathbb{R}^{h \times h}, W_K \in \mathbb{R}^{h \times y}$	Weight matrices
$Y \in \mathbb{R}^{n \times y}$	Predicted label matrix

in node features [15]. For each layer, GCN propagates the embedding of a node to its neighbors as follows:

$$H^{(k)} = \sigma(S H^{(k-1)} W_k), \tag{2}$$

where W_k denotes the weight matrix of the k-th layer and σ denotes a non-linear function, e.g., ReLU. In the output layer, K-layer GCN outputs a predicted label matrix $Y \in \mathbb{R}^{n \times y}$ as:

$$Y = \text{softmax}(S H^{(K-1)} W_K), \tag{3}$$

where $\text{softmax}(P)_{ij} = \frac{\exp(P_{ij})}{\sum_{j=1}^{y} \exp(P_{ij})}$ for a matrix P. The number of layers is typically set to $K = 2$ [15].

2.2 Precomputation-Based GNNs

Several precomputation-based GNNs have been proposed recently [8,18,24]. Their fundamental and common idea is to remove non-linear functions between each layer in order to precompute feature aggregation. We explain Simplifying Graph Convolution (SGC for short) [24] which is the simplest precomputation-based GNN. Thanks to the removal, K-layer GCN can be rewritten as follows by unfolding the recursive structure:

$$Y = \text{softmax}(S \ldots S X W_1 \ldots W_K). \tag{4}$$

The repeated multiplication with the normalized adjacency matrix S can be simplified into a K-th power matrix S^K and the multiple weight matrices can be reparameterized into a single matrix $W = W_1 \ldots W_K$. The output becomes

$$Y = \text{softmax}(S^K X W). \tag{5}$$

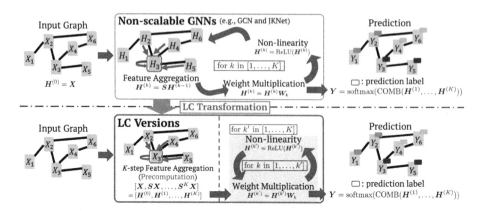

Fig. 1. Example of LC transformation. Upper part: non-scalable GNNs operate K-layer graph convolution combining feature aggregation, weight multiplication, and activation function application (ReLU). This example corresponds to K-layer GCN if COMB outputs only $\boldsymbol{H}^K$. Lower part: LC transformation separates feature aggregation and weight learning while keeping the similar architectures with the original GNNs. LC versions avoid recomputing feature aggregation whenever learnable weights are updated at each learning step. (Color figure online)

By separating graph feature aggregation and weight learning, SGC precomputes $\boldsymbol{S}^K \boldsymbol{X}$ before learning $\boldsymbol{W}$. The other methods also follow the same idea: separating feature aggregation and weight learning and precomputing feature aggregation.

3   GNN Transformation Framework

We propose a general framework that automatically transforms non-scalable GNNs to efficient and scalable precomputation-based GNNs and efficiently executes precomputation of feature aggregation on GPU. We first introduce a transformation procedure that automatically rewrites the formulations of non-scalable GNNs so that the transformed GNNs run efficiently and scale well to large-scale graphs (Sect. 3.1). We also describe a limitation of our transformation, namely, that it does not support GNNs that require dynamical changes of graph filters during weight learning. Our transformation procedure is applicable not only to GCN [15] but also to the state-of-the-art GNNs, such as JKNet [27], H2GCN [32] and GPRGNN [7]. Next, we introduce a block-wise precomputation scheme that efficiently computes feature aggregation for large-scale graphs (Sect. 3.2). The core idea is to decompose an adjacency matrix and feature matrix into disjoint and balanced blocks each of which can be handled on GPU. Also, we formulate and solve an optimization problem that decides the optimal size of blocks. Note that this scheme is a general approach since it can be applied to existing precomputation-based GNNs [8,18,24].

3.1   Linear Convolution Transformation

LC transformation is the first concrete procedure that transforms non-scalable GNNs to efficient and scalable precomputation-based GNNs, which have a similar functionality to the input GNNs. We call the output the *LC version* of the input GNN. LC transformation is motivated by the effectiveness of SGC and Multi-Layer Perceptron (MLP). SGC preserves the major benefit of graph convolution with efficient training by precomputing feature aggregation, but it degrades the accuracy due to the lack of non-linearity [7]. Beside, MLP outperforms linear regression in classification task by using non-linear functions but does not capture the structures of graphs. LC version of GNN leverages both the strengths of SGC and MLP by precomputing feature aggregation and then learning weights with non-linearity.

Figure 1 demonstrates an example of LC transformation by comparing it with non-scalable GNNs. Intuitively, LC transformation separates feature aggregation from graph convolution that performs 1) feature aggregation, 2) weight multiplication, and 3) activation function application (e.g., ReLU, a non-linear function). Notice that a normalized adjacency matrix S is adjacent to the feature matrix X in the formulation of LC versions (see the left part of the red box of the figure). So, we can precompute $S^k X$ in the same way as SGC [24]. Thanks to the separation, LC versions can avoid computing feature aggregation whenever learnable weights are updated at each learning step (see the right part of the red box of the figure). Hence, LC versions efficiently work and scale well to large-scale graphs.

Discussion. We discuss why LC versions work from two aspects, feature aggregation and weight learning. As in the discussion on the spectral analysis [24], feature aggregation acts as a low-pass filter that produces smooth features over the graph, which is the major benefit of graph convolution. In this sense, LC versions are expected to have the same functionality as the input GNNs since LC transformation preserves feature aggregation within multi-hops. As for weight learning, LC versions have a similar learning capability to their original GNNs since they have a similar model architecture of multi-layer neural networks. As a result, LC versions can achieve a similar prediction performance to their original GNNs while scaling to large-scale graphs.

Procedure. Next, we describe the procedure of LC transformation, which removes non-linear functions from graph convolution, but incorporates non-linear functions into weight learning. We first give the definition of LC transformation below:

Definition 1 (LC transformation). *Given a non-scalable GNN algorithm, LC transformation iteratively applies a function f_{LC} to the formulation of the input GNN since non-scalable GNNs have multiple graph convolutional layers. f_{LC} commutes matrix multiplication of S and a non-linear function σ as follows:*

$$f_{LC} : g_2(S\sigma(g_1(X))) \xrightarrow[f_{LC}]{} g_2(\sigma(Sg_1(X))), \qquad (6)$$

where g_1 and g_2 indicate any functions that input and output matrices. The iteration continues until the formulation does not change. LC transformation outputs a precomputation-based GNN having the transformed formulation, i.e., the LC version of the input GNN.

To intuitively explain the details, we use JKNet [27] as an example, which is a widely used GNN. The formulation of JKNet (GCN-based) is as follows:

$$H = \text{COMB}_{k=1}^{K}(S\sigma(S\sigma(\ldots(SXW_1)\ldots)W_{k-1})W_k), \qquad (7)$$

where COMB expresses a skip connection between different layers, such as concatenation of intermediate representations or max pooling. By applying a softmax function to feature representations H, JKNet outputs a prediction result Y, i.e., $Y = \text{softmax}(H)$. We apply f_{LC} to it in order to transform the formulation of an input GNN. To this end, we assign $g_1(X) = S\sigma(\ldots(SXW_1)\ldots)W_{k-1}$ and $g_2(S\sigma(g_1(X))) = \text{COMB}_{k=1}^{K}(S\sigma(g_1(X))W_k)$. By utilizing f_{LC}, g_1, and g_2, we transform Eq. (7) as follows:

$$H \xrightarrow{f_{LC}} \text{COMB}_{k=1}^{K}(\sigma(S^2\sigma(\ldots(SXW_1)\ldots)W_{k-1})W_k). \qquad (8)$$

Then, we iteratively apply f_{LC} to the formulation by appropriately assigning g_1 and g_2 for each iteration. Finally, we obtain the formulation of the LC version of the input GNN, H^{LC}, as follows:

$$H^{LC} = \text{COMB}_{k=1}^{K}(\sigma(\sigma(\ldots(S^kXW_1)\ldots)W_{k-1})W_k). \qquad (9)$$

Then, in the same way as the input GNN, the LC version outputs a predicted label matrix $Y = \text{softmax}(H^{LC})$.

The LC transformation procedure is applicable not only to JKNet but also to general non-scalable GNNs including APPNP [16], MixHop [1], H2GCN [32], and GPRGNN [7]. We give another example of applying LC transformation in Appendix A.

Limitation. Precomputation-based GNNs can use multiple graph filters such as an exact 1-hop away adjacency matrix and Personalized PageRank diffusion matrix [16]. Those GNNs do not dynamically control the propagation of features during weight learning, since they use constant graph filters in order to precompute feature aggregation. Since our framework also leverages a precomputation scheme, it cannot support those existing GNNs [21,22,26] which dynamically sample edges or modify the importance of edges during weight learning. For example, Dropedge [21] randomly reduces a certain number of edges at each iteration. A possible future research direction is that we simulate random edge reduction by utilizing the deviations of feature aggregation.

3.2   Efficient Precomputation

Existing precomputation-based GNNs need to use CPUs to compute feature aggregations for large-scale graphs since they do not fit on GPU memory. This CPU computation has large cost and a deteriorating effect on efficiency.

To tackle this problem, we propose a simple yet efficient block-wise precomputation scheme and provide a formulation for optimal decomposition for our block-wise precomputation scheme. The core idea is to decompose the edge set of a given graph into disjoint and balanced groups, while existing approaches [31] decompose the node set into groups, i.e., row/column wise decomposition. Our scheme is inspired by edge partitioning [9,17], which aims to decompose a graph into groups having similar numbers of edges such that communication costs for graph operations are minimized in distributed environments. Our scheme consists of three steps. First, it decomposes an adjacency matrix and feature matrix into small disjoint blocks each of which can be put on GPU memory. Second, the scheme computes block-wise matrix operations for the disjoint blocks on GPU. Third, it aggregates the results of the block-wise matrix operations and obtains the whole matrix operation result.

Precomputation on GPU. There are two matrix operations to be precomputed, adjacency matrix normalization and feature aggregation. First, we describe the computation of adjacency matrix normalization shown by Eq. (1). Since an adjacency matrix is typically sparse, we utilize adjacency list $(i, j) \in \mathcal{E}$, where $\tilde{A}_{ij} = 1$. To obtain small blocks each of which can be loaded on GPU memory, we decompose $\mathcal{E}$ into disjoint sets that include similar numbers of edges, $\mathcal{E}^{(1)} \cup \cdots \cup \mathcal{E}^{(a)}$, where a is a number of sets and $\mathcal{E}^{(p)} \cap \mathcal{E}^{(q)} = \emptyset$ if $p \neq q$. Note that the sizes of the sets $\mathcal{E}^{(1)}, \ldots, \mathcal{E}^{(a)}$ are balanced. Then, we decompose $\tilde{A} = \tilde{A}^{(1)} + \cdots + \tilde{A}^{(a)}$, where $\tilde{A}^{(1)}, \ldots, \tilde{A}^{(a)} \in \mathbb{R}^{n \times n}$ and $\tilde{A}_{ij}^{(l)} = 1$ if $(i, j) \in \mathcal{E}^{(l)}$. Then, we can rewrite Eq. (1) as follows:

$$S = \tilde{D}^{-\frac{1}{2}} \tilde{A} \tilde{D}^{-\frac{1}{2}} = \sum_{l=1}^{a} \tilde{D}^{-\frac{1}{2}} \tilde{A}^{(l)} \tilde{D}^{-\frac{1}{2}}. \tag{10}$$

By appropriately selecting the number of blocks a, $\tilde{D}^{-\frac{1}{2}} \tilde{A}^{(l)} \tilde{D}^{-\frac{1}{2}}$ can be executed on GPU. We sum the results of the block-wise matrix computations. This summation can be efficiently computed on CPU by disjoint union of edge lists since $\mathcal{E}^{(l)}$, i.e., $\tilde{A}^{(l)}$, is disjoint each other. Since our decomposition is agnostic on nodes, the decomposed blocks can be easily balanced while row/column(node)-wise decomposition approaches suffer from an imbalance problem. Further discussion on Limitations follows below in this subsection.

Next, we introduce a block-wise computation for feature aggregation on GPU. Algorithm 1 describes the procedure of the computation. To obtain small blocks of a normalized adjacency matrix S, we decompose it into $S^{(1)}, \ldots, S^{(b)} \in \mathbb{R}^{n \times n}$ where b is a number of blocks (line 2). Similarly to the decomposition of A, each corresponding edge list is disjoint and includes similar numbers of edges. Also, in order to obtain small blocks of a feature matrix X, we decompose it into $X^{(1)}, \ldots, X^{(c)}$, where c is a number of blocks (line 5). Since we assume that X is a dense matrix, we adopt column-wise decomposition, i.e., $X = \text{concat}(X^{(1)}, \ldots, X^{(c)})$. Then, we compute matrix multiplication $S^{(j)} X^{(i)}$ for each pair on GPU (line 9). We aggregate $S^{(j)}$ by summation (line 10) and aggregate X_{tmp} by concatenation (lines 11–14). X_{prev} is updated by the aggregated features X_{conc} (line 16). We repeat this aggregation K times (lines 4–16).

Algorithm 1 Block-wise feature aggregation.

Require: normalized adjacency matrix S, feature matrix X, number of layers K
Ensure: aggregated feature list SX_list
 1: $SX\_list = []$
 2: $S^{(1)}, S^{(2)}, \ldots, S^{(b)} = \mathrm{split}(S)$ $\triangleright$ disjoint edge sets
 3: $X_{prev} = X$
 4: **for** $k = 1$ to K **do**
 5:     $X^{(1)}, X^{(2)}, \ldots, X^{(c)} = \mathrm{split}(X_{prev})$
 6:     **for** $i = 1$ to c **do**
 7:        $X_{tmp} = [0]^{n \times \lceil d/c \rceil}$ $\triangleright$ same size to $X^{(i)}$
 8:        **for** $j = 1$ to b **do**
 9:           $Z_{tmp} = S^{(j)} X^{(i)}$ $\triangleright$ on GPU
10:           $X_{tmp} = X_{tmp} + Z_{tmp}$ $\triangleright$ on GPU
11:        **if** $i == 1$ **then**
12:           $X_{conc} = X_{tmp}$ $\triangleright$ on CPU
13:        **else**
14:           $X_{conc} = \mathrm{concat}(X_{conc}, X_{tmp})$ $\triangleright$ on CPU
15:     $SX\_list.\mathrm{append}(X_{conc})$
16:     $X_{prev} = X_{conc}$

Optimal Graph Decomposition. We discuss an optimal decomposition for our block-wise precomputation scheme. We have two requirements to decompose large matrices into disjoint blocks. First, each matrix operation for disjoint blocks can be executed on GPU. Second, the number of disjoint blocks is as small as possible to reduce the number of block-wise matrix operations. To simplify the discussion, we assume that the running time of a matrix operation on GPU is the same regardless of the matrix size.

As for the block-wise adjacency matrix normalization, we minimize a number of disjoint blocks, a. We formulate the minimization as follows:

$$\min(a), \text{ subject to } \frac{\alpha_A B_A + \alpha_S B_S}{a} + \alpha_D B_D \leq B_{\mathrm{GPU}}, \tag{11}$$

where $\alpha_A, \alpha_S, \alpha_D$ indicate coefficients for executing matrix operations regarding A, S, D, respectively, and $B_A, B_S, B_D, B_{\mathrm{GPU}}$ indicate the volume of an adjacency matrix, the volume of a normalized adjacency matrix, the volume of a degree matrix, and the available volume of a GPU, respectively. As for block-wise feature aggregation, we minimize the number of pairs of disjoint blocks, bc. We formulate the minimization as follows:

$$\min_{b,c}(bc), \text{ subject to } \frac{\beta_S B_S}{b} + \frac{\beta_X B_X}{c} \leq B_{\mathrm{GPU}}, \tag{12}$$

Table 2. Summary of datasets.

Dataset	Nodes	Edges	Features	Classes
Flickr	89,250	899,756	500	7
Reddit	232,965	11,606,919	602	41
arxiv	169,343	1,166,243	128	40
papers100M	111,059,956	1,615,685,872	128	172

where β_S, β_X indicate coefficients for executing matrix operations regarding S, X, respectively, and B_X indicates the volume of a feature matrix. Note that $\alpha_A, \alpha_S, \alpha_D, \beta_S$, and β_X depend on execution environments[2].

Next, we discuss optimization regarding Eq. (11) and (12). As for Eq. (11), it is trivial to find the minimum number of blocks a since there are no other parameters. As for Eq. (12), an exhaustive search is applicable since the number of combinations of b and c (natural numbers) is not large. Consequently, these optimization problems can be easily solved.

Limitation. Our precomputation scheme focuses on feature aggregation on a whole graph. This indicates that our scheme is not suitable for node-wise operations since it may decompose the edge set of the same node into different groups. However, accelerating feature aggregation on a whole graph is still crucial since many graph neural networks [8,15,18,24] adopt it.

4   Experiments

We design our experiments to answer the following questions; **Q1**: Can our LC transformation improve the efficiency and scalability of GNNs? **Q2**: Can our block-wise precomputation scheme accelerate precomputation?

Dataset. We use four commonly used datasets, Flickr [29], Reddit [10], ogbn-arxiv (arxiv for short), and ogbn-papers100M (papers100M for short) [12]. Table 2 provides the summary of the datasets. The sizes of the datasets range from 9K nodes to 110M.

In the Flickr dataset, nodes represent images uploaded to Flickr. If two images share common properties such as same geographic location, same gallery, comments by the same users, there is an edge between the nodes. Node features represent the 500-dimensional bag-of-words associated with the image (node). As for node labels, the authors of [29] scan over 81 tags of each image and manually merged them to 7 classes. In the Reddit dataset, nodes represent posts. If the same user left comments on two posts, then there is an edge between the two posts. Node features are the embedding of the contents of the posts. The labels of nodes indicate

---

[2] In real environments, users can measure $\alpha_A, \alpha_S, \alpha_D, \beta_S$, and β_X by monitoring the memory usage on small graphs, even if users do not know the details of their own environments.

communities which the nodes belong to. In the ogbn-arxiv dataset, nodes represent ARXIV papers and edges indicate that one paper cites another one. Node features represent 128-dimensional feature vectors obtained by averaging the embeddings of words in titles and abstracts. Node labels indicate subject areas of ARXIV CS papers[3]. In the ogbn-papers100M (papers100M) dataset, its graph structure and node features are constructed in the same way as ogbn-arxiv. Among its nodes, approximately 1.5 million nodes are labeled with one of ARXIV's subject areas. As in [28], Flickr and Reddit are under the inductive setting. ogbn-arxiv and ogbn-papers100M are under the transductive setting.

Baseline. We compare three types of existing methods as baselines; non-scalable GNNs, precomputation-based GNNs, and sampling-based GNNs which are scalable but inefficient (we discuss the details in Sect. 5). As for non-scalable GNNs, we use GCN[4] [15], JKNet[5] [27], and GPRGNN[6] [7]. As for precomputation-based GNNs, we use SGC[7] [24] and FSGNN[8] [18]. As for sampling-based GNNs, we use ShaDow-GNN[9] [28]. FSGNN and ShaDow-GNN are the state-of-the-art precomputing-based and sampling-based GNNs, respectively. We note that we use our block-wise precomputation to the precomputation-based GNNs instead of using its original CPU computation for a fair comparison.

Setup. We tune hyperparameters on each dataset by Optuna [2] and use Adam optimizer [14]. We adopt mini-batch training for precomputation-based GNNs, sampling-based GNNs, and LC-versions to deal with large-scale graphs[10]. As for ShaDow-GNN, we use the best hyperparameter sets provided by the authors and adopt GAT [22] as a backbone model since ShaDow-GAT achieves the best accuracy in most cases reported in the paper. We measure training time on a NVIDIA Tesla V100S GPU (32 GB) and Intel(R) Xeon(R) Gold 5220R CPUs (378 GB).

4.1   Effectiveness of LC Transformation (Q1)

Table 3 shows the test accuracy of LC versions and the baselines. LC versions (GCN_LC, JKNet_LC, and GPRGNN_LC) achieve comparable test accuracy with their original GNNs (GCN, JKNet, and GPRGNN) for all datasets. Next, Table 4 shows the training time of LC versions and the baselines. The LC versions run faster than their original GNNs. Note that LC versions tend to stop earlier than non-scalable GNNs since LC versions train their models more times due

---

[3] https://arxiv.org/archive/cs.

[4] https://github.com/tkipf/pygcn.

[5] Since official codes of JKNet from the authors are not provided, we simply implement JKNet based on the implementation of GCN.

[6] https://github.com/jianhao2016/GPRGNN.

[7] https://github.com/Tiiiger/SGC.

[8] https://github.com/sunilkmaurya/FSGNN.

[9] https://github.com/facebookresearch/shaDow_GNN.

[10] We will provide hyperparameter search space and the best parameters to reproduce experiments on our codebase that will be publicly available on acceptance.

Table 3. Comparison on test accuracy. We report the average values (standard deviation) over 5 runs.

	Flickr	Reddit	arxiv
GCN	0.525(0.003)	0.945(0.000)	0.702(0.005)
JKNet	0.526(0.004)	0.941(0.006)	0.712(0.001)
GPRGNN	0.494(0.006)	0.918(0.012)	0.694(0.006)
SGC	0.494(0.037)	0.948(0.001)	0.692(0.004)
FSGNN	0.513(0.001)	0.964(0.001)	0.722(0.003)
ShaDow-GAT	0.531(0.003)	0.947(0.003)	0.716(0.004)
GCN_LC	0.515(0.003)	0.947(0.001)	0.710(0.001)
JKNet_LC	0.517(0.004)	0.951(0.000)	0.710(0.003)
GPRGNN_LC	0.513(0.001)	0.961(0.000)	0.720(0.004)

Table 4. Comparison on training time (per epoch/total). Note that total training time includes precomputation time for SGC, FSGNN, ShaDow-GAT, GCN_LC, JKNet_LC, and GPRGNN_LC. We report the average values over 5 runs.

	Flickr	Reddit	arxiv
GCN	64.62 [ms]/129.24 [s]	654.70 [ms]/1309.40 [s]	210.81 [ms]/421.63 [s]
JKNet	170.43 [ms]/253.25 [s]	1428.51 [ms]/2552.45 [s]	529.05 [ms]/1058.10 [s]
GPRGNN	272.86 [ms]/539.48 [s]	1456.01 [ms]/2806.62 [s]	523.08 [ms]/961.76 [s]
SGC	51.18 [ms]/30.31 [s]	141.68 [ms]/285.43 [s]	50.27 [ms]/42.23 [s]
FSGNN	346.97 [ms]/133.63 [s]	1066.66 [ms]/1793.91 [s]	284.73 [ms]/382.67 [s]
ShaDow-GAT	120.85e3 [ms]/3634.65 [s]	376.42e3 [ms]/11321.09 [s]	163.67e3 [ms]/4913.29 [s]
GCN_LC	56.75 [ms]/49.85 [s]	165.73 [ms]/212.16 [s]	62.59 [ms]/120.60 [s]
JKNet_LC	144.78 [ms]/78.24 [s]	430.41 [ms]/865.71 [s]	138.52 [ms]/277.63 [s]
GPRGNN_LC	287.54 [ms]/164.88 [s]	818.13 [ms]/1645.49 [s]	219.66 [ms]/204.56 [s]

to mini-batch training. For example, in Flickr data LC versions more efficiently train than non-scalable GNNs even if they have similar training time per epoch. These results indicate that our framework transforms non-scalable GNNs to efficient precomputation-based GNNs with the comparable classification accuracy to the original GNNs.

Comparison on Large-Scale Graph. Table 5 shows the performance comparison on papers100M having more than 100 million nodes and one billion edges. Non-scalable GNNs (GCN, JKNet, and GPRGNN) cannot work on papers100M since the whole graph cannot be put on GPU memory. GPRGNN_LC achieves comparable accuracy (approximate one percent difference) with FSGNN, which is the state-of-the-art precomputation-based GNN while GPRGNN_LC runs faster than FSGNN. Though ShaDow-GAT achieves the highest accuracy, it requires more than 10× total training time than other models. This is because it needs to operate graph convolutions on many enclosing subgraphs extracted from the whole graph. SGC obtains lower accuracy than GCN_LC. This result

Table 5. Results on papers100M. We show test/validation accuracy (standard deviation) and training time (per epoch/total). Total training time includes precomputation time. OOM indicates that the execution is out of memory.

	Test accuracy	Val accuracy	Time (epoch/total)
GCN	OOM	OOM	OOM
JKNet	OOM	OOM	OOM
GPRGNN	OOM	OOM	OOM
SGC	0.623(0.007)	0.667(0.002)	425.15 [ms]/2211.23 [s]
FSGNN	0.665(0.003)	0.706(0.001)	3550.82 [ms]/8612.48 [s]
ShaDow-GAT	0.666(0.003)	0.703(0.001)	2948.50e3 [ms]/92264.76 [s]
GCN_LC	0.647(0.006)	0.688(0.002)	611.90 [ms]/2477.55 [s]
JKNet_LC	0.641(0.003)	0.689(0.004)	1488.80 [ms]/3396.69 [s]
GPRGNN_LC	0.658(0.002)	0.696(0.001)	2749.27 [ms]/7410.47 [s]

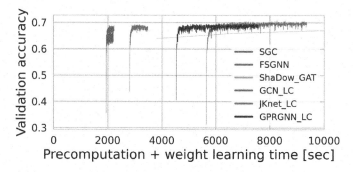

Fig. 2. Validation accuracy over training time (precomputation and weight learning time) on papers100M. Plots indicate epochs. LC versions (GCN_LC, JKNet_LC, and GPRGNN_LC) are faster than FSGNN and ShaDow-GAT while achieving competitive accuracy.

validates that non-linearity contributes to weight learning for better classification.

In order to analyze the results on papers100M in details, we show the validation accuracy at each epoch over total training time in Fig. 2. Note that total training time consists of precomputation and weight learning time. We observe that GCN_LC, JKNet_LC, and GPRGNN_LC are plotted in the upper left corner of the figure. This observation indicates that they require less total training time than FSGNN and ShaDow-GAT. The LC versions achieve competitive performance with them. Through these experiments, we demonstrate that LC versions are efficient and scalable for large-scale graphs.

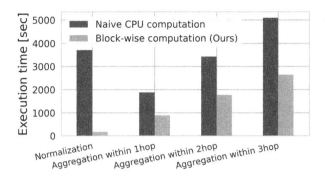

Fig. 3. Precomputation time comparison between a naive CPU computation and our block-wise computation.

4.2 Precomputation Efficiency (Q2)

To validate the efficiency of our block-wise precomputation, we compare it with naive CPU computation adopted by existing works [12,18]. We use a large-scale graph, papers100M, which requires a 67 GB normalized adjacency matrix and a 57 GB feature matrix. For adjacency matrix normalization, we set the number of disjoint blocks of an adjacency matrix to $a = 3$, which satisfies Eq. (11). Also, for feature aggregation we set numbers of disjoint blocks of a normalized adjacency matrix and feature matrix to $b = 10, c = 16$, respectively, which satisfy Eq. (12).

Figure 3 shows the precomputation time for normalization and feature aggregation on CPU and GPU. The result demonstrates that our block-wise precomputation is $20\times$ faster than CPU computation for normalization. Also, the result indicates that our precomputation is up to twice faster than CPU computation for feature aggregation. Hence, we conclude that our precomputation is more efficient than CPU computation on a single machine.

5 Related Work

Relationship Between Non-scalable GNNs and LC Versions. We discuss the background of non-scalable GNNs and their LC versions. Graph convolution is motivated by the 1-dim Weisfeiler-Lehman (WL-1) algorithm [23] which is used to test graph isomorphism; two graphs are called isomorphic if they are topologically identical. WL-1 iteratively aggregates the labels of nodes and their neighbors, and hashes the aggregated labels into unique labels. The algorithm decides whether two graphs are isomorphic or not by using the labels of nodes at some iteration. Non-scalable GNNs such as GCN [15] replace the hash function of WL-1 with a graph convolutional layer which consists of feature aggregation, weight multiplication, and non-linear function application. As for LC versions, they replace the hash function of WL-1 with feature aggregation. These observations indicate that WL-1 is analogous to feature aggregation of LC versions, similarly to graph convolution of non-scalable GNNs.

Sampling-Based GNNs. Sampling-based GNNs [5, 6, 10, 28, 29] avoid keeping a whole graph on GPU by computing node representations from enclosing subgraphs of the input graph. The major drawback of the sampling-based GNNs is that they need costly training time since they need to operate graph convolutions on many enclosing subgraphs extracted from the input graph.

GNNs Dynamically Modifying the Importance of Edges. As we discussed in Sect. 3.1, our transformation cannot support GNNs which dynamically control the propagation of features during weight learning. An example of such GNNs is GAT [22], which learns attention parameters controlling the importance of edges for each iteration. Another example is GIN [26] learns a parameter controlling a weight between self features and features from neighbors. One possible direction is that we first determine the parameters by training on a subset of an input graph, then fix them in order to precompute feature aggregation.

Distributed Matrix Operations. Matrix operations can be parallelized for distributed computing [3, 4]. For example, the authors of [11] proposed Mars which is an approach for hiding the programming complexity of MapReduce on GPU. Also, MR-Graph [20] is a customizable and unified framework for GPU-based MapReduce. It allows its users to implement their applications more flexibly. As for distributed graph neural network training, DistDGL [31] has proposed mini-batch training on graphs, which scales beyond a single machine. It suffers from an imbalance problem since it uses a typical graph clustering algorithm METIS [13] to partition large-scale graphs into subgraphs, while our scheme can partition an edge set into balanced subsets. For further scale up of graphs, it would be important to combine distributed computing and our block-wise precomputation for graphs.

6   Conclusion

We presented a framework that automatically transforms non-scalable GNNs to efficient and scalable precomputation-based GNNs. There are two major characteristics of our framework: 1) it supports a novel transformation procedure that transforms non-scalable GNNs to efficient and scalable precomputation-based GNNs having a similar functionality to the original GNNs, 2) the precomputation of the transformed GNNs can be efficiently executed by our block-wise precomputation scheme that decomposes large-scale graphs into disjoint and balanced blocks each of which can be handled on GPU memory. Through our experiments, we demonstrated that the transformed GNNs run more efficiently than their original GNNs and can be scaled to graphs with millions of nodes and billions of edges. Due to the strong performance of LC versions, we argue that LC versions will be beneficial as baseline comparisons for future research on scalable GNNs.

Acknowledgement. This work was supported by JSPS KAKENHI Grant Numbers JP20H00583 and JST PRESTO Grant Number JPMJPR21C5.

A LC Version of GPRGNN

We show an example of LC transformation for the state-of-the-art GNN model, GPRGNN [7]. We give the formulation of GPRGNN as follows:

$$\boldsymbol{H} = \sum_{k=0}^{K} \gamma_k \boldsymbol{S}^k (\sigma(\ldots \sigma(\boldsymbol{X}\boldsymbol{W}_1)\ldots)\boldsymbol{W}_T), \tag{13}$$

where γ_k is an attention parameter learning the importance of k-th layer and T is the number of layers for Multi-layer perceptrons. Note that $\boldsymbol{S}^k$ cannot be efficiently precomputed since the number of non-zero elements significantly increases when $k \geq 2$ for large-scale graphs. By iteratively applying f_{LC} to Eq. (13), we obtain the formulation of its LC version as follows:

$$\boldsymbol{H}^{LC} = \sum_{k=0}^{K} \gamma_k (\sigma(\ldots \sigma(\boldsymbol{S}^k \boldsymbol{X}\boldsymbol{W}_1)\ldots)\boldsymbol{W}_T). \tag{14}$$

$\boldsymbol{S}^k \boldsymbol{X}$ can be precomputed since it does not need to be updated when learnable weights $\boldsymbol{W}_1 \ldots \boldsymbol{W}_T$ and a parameter γ are updated.

References

1. Abu-El-Haija, S., et al.: MixHop: higher-order graph convolutional architectures via sparsified neighborhood mixing. In: ICML (2019)
2. Akiba, T., Sano, S., Yanase, T., Ohta, T., Koyama, M.: Optuna: a next-generation hyperparameter optimization framework. In: KDD (2019)
3. Awaysheh, F.M., Alazab, M., Garg, S., Niyato, D., Verikoukis, C.: Big data resource management & networks: taxonomy, survey, and future directions. IEEE Commun. Surv. Tutor. **23**(4), 2098–2130 (2021)
4. Boehm, M., et al.: SystemML: declarative machine learning on spark. PVLDB **9**(13), 1425–1436 (2016)
5. Chen, J., Ma, T., Xiao, C.: FastGCN: fast learning with graph convolutional networks via importance sampling. arXiv preprint (2018)
6. Chiang, W.L., Liu, X., Si, S., Li, Y., Bengio, S., Hsieh, C.J.: Cluster-GCN: an efficient algorithm for training deep and large graph convolutional networks. In: KDD (2019)
7. Chien, E., Peng, J., Li, P., Milenkovic, O.: Adaptive universal generalized pagerank graph neural network. In: ICLR (2021). https://openreview.net/forum?id=n6jl7fLxrP
8. Frasca, F., Rossi, E., Eynard, D., Chamberlain, B., Bronstein, M., Monti, F.: Sign: scalable inception graph neural networks. In: ICML 2020 Workshop on Graph Representation Learning and Beyond (2020)
9. Gonzalez, J.E., Low, Y., Gu, H., Bickson, D., Guestrin, C.: Powergraph: distributed graph-parallel computation on natural graphs. In: 10th USENIX Symposium on Operating Systems Design and Implementation (OSDI 2012), pp. 17–30 (2012)
10. Hamilton, W.L., Ying, R., Leskovec, J.: Inductive representation learning on large graphs. In: NeurIPS (2017)
11. He, B., Fang, W., Luo, Q., Govindaraju, N.K., Wang, T.: Mars: a mapreduce framework on graphics processors. In: PACT (2008)
12. Hu, W., et al.: Open graph benchmark: datasets for machine learning on graphs. arXiv preprint (2020)

13. Karypis, G., Kumar, V.: A fast and high quality multilevel scheme for partitioning irregular graphs. SIAM J. Sci. Comput. **20**(1), 359–392 (1998)
14. Kingma, D.P., Ba, J.: Adam: a method for stochastic optimization. arXiv preprint (2014)
15. Kipf, T.N., Welling, M.: Semi-supervised classification with graph convolutional networks. In: ICLR (2017)
16. Klicpera, J., Bojchevski, A., Günnemann, S.: Predict then propagate: graph neural networks meet personalized pagerank. In: ICLR (2019)
17. Low, Y., Gonzalez, J., Kyrola, A., Bickson, D., Guestrin, C., Hellerstein, J.M.: Distributed GraphLab: a framework for machine learning in the cloud. PVLDB **5**(8), 716–727 (2012)
18. Maurya, S.K., Liu, X., Murata, T.: Improving graph neural networks with simple architecture design. arXiv preprint (2021)
19. Newman, M.E.: Networks: An Introduction. Oxford University Press, Oxford (2010)
20. Qiao, Z., Liang, S., Jiang, H., Fu, S.: A customizable mapreduce framework for complex data-intensive workflows on GPUs. In: IPCCC (2015)
21. Rong, Y., Huang, W., Xu, T., Huang, J.: DropEdge: towards deep graph convolutional networks on node classification. arXiv preprint (2019)
22. Veličković, P., Cucurull, G., Casanova, A., Romero, A., Lio, P., Bengio, Y.: Graph attention networks. arXiv preprint (2017)
23. Weisfeiler, B., Lehmann, A.A.: A reduction of a graph to a canonical form and an algebra arising during this reduction. Nauchno-Technicheskaya Informatsia **2**(9), 12–16 (1968)
24. Wu, F., Souza, A., Zhang, T., Fifty, C., Yu, T., Weinberger, K.: Simplifying graph convolutional networks. In: ICML (2019)
25. Wu, Z., Pan, S., Chen, F., Long, G., Zhang, C., Philip, S.Y.: A comprehensive survey on graph neural networks. IEEE Trans. Neural Netw. Learn. Syst. **32**(1), 4–24 (2020)
26. Xu, K., Hu, W., Leskovec, J., Jegelka, S.: How powerful are graph neural networks? arXiv preprint arXiv:1810.00826 (2018)
27. Xu, K., Li, C., Tian, Y., Sonobe, T., Kawarabayashi, K.I., Jegelka, S.: Representation learning on graphs with jumping knowledge networks. In: ICML (2018)
28. Zeng, H., et al.: Deep graph neural networks with shallow subgraph samplers. CoRR abs/2012.01380 (2020). https://arxiv.org/abs/2012.01380
29. Zeng, H., Zhou, H., Srivastava, A., Kannan, R., Prasanna, V.: GraphSAINT: graph sampling based inductive learning method. In: International Conference on Learning Representations (2020). https://openreview.net/forum?id=BJe8pkHFwS
30. Zhang, Z., Cui, P., Zhu, W.: Deep learning on graphs: a survey. IEEE TKDE **34**(1), 249–270 (2020)
31. Zheng, D., et al.: DistDGL: distributed graph neural network training for billion-scale graphs. In: 2020 IEEE/ACM 10th Workshop on Irregular Applications: Architectures and Algorithms (IA3), pp. 36–44. IEEE (2020)
32. Zhu, J., Yan, Y., Zhao, L., Heimann, M., Akoglu, L., Koutra, D.: Beyond homophily in graph neural networks: current limitations and effective designs. In: NeurIPS, vol. 33 (2020)

Masked Graph Auto-Encoder Constrained Graph Pooling

Chuang Liu[1], Yibing Zhan[2], Xueqi Ma[3], Dapeng Tao[4], Bo Du[1],
and Wenbin Hu[1(✉)]

[1] School of Computer Science, Wuhan University, Wuhan, China
{chuangliu,dubo,hwb}@whu.edu.cn
[2] JD Explore Academy, Beijing, China
zhanyibing@jd.com
[3] School of Software, Tsinghua University, Beijing, China
xueqima@s.upc.edu.cn
[4] Yunnan University, Kunming, China
dptao@ynu.edu.cn

Abstract. The node drop pooling is a significant type of graph pooling that is required for learning graph-level representations. However, existing node drop pooling models still suffer from the information loss problem, impairing their effectiveness in graph classification. To mitigate the detrimental effect of the information loss, we propose a novel and flexible technique called <u>M</u>asked <u>G</u>raph <u>A</u>uto-encoder constrained <u>P</u>ooling (MGAP), which enables vanilla node drop pooling methods to retain sufficient effective graph information from both node-attribute and network-topology perspectives. Specifically, MGAP reconstructs the original node attributes of the graph using a graph convolutional network and the node degree of the graph (i.e., structural information) using a feedforward neural network with exponential neurons from the pooled (masked) graphs generated by the vanilla node drop pooling models. Notably, MGAP is a plug-and-play technique that can be directly adopted in the current node drop pooling methods. To evaluate the effectiveness of MGAP, we conduct extensive experiments on eleven real-world datasets by applying MGAP to three commonly-used methods, i.e., TopKPool, SAGPool, and GSAPool. The experimental results reveal that MGAP has the capacity to consistently improve the performance of all the three node drop pooling models in the graph classification task.

Keywords: Graph Neural Nnetworks · Graph pooling · Graph auto-encoder · Graph classification

1 Introduction

Graph Neural Networks (GNNs) have demonstrated their significant effectiveness in a variety of graph classification tasks [2,6], including molecular property

---

C. Liu—This work has been done when Chuang Liu was an intern at JD Explore Academy.

M.-R. Amini et al. (Eds.): ECML PKDD 2022, LNAI 13714, pp. 377–393, 2023.
https://doi.org/10.1007/978-3-031-26390-3_23

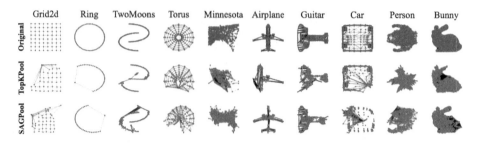

Fig. 1. Graph reconstruction with two typical node drop pooling operators. The point locations in the figure above represent the node attributes.

prediction [1], cancer diagnosis [26], and brain-data analysis [17]. In contrast to node-level tasks (e.g., the node classification), which mainly leverage the graph convolutional network (GCN) to generate node representations for downstream tasks [13], graph classification requires obtaining holistic graph-level representations. Therefore, for graph classification, the pooling mechanism is an essential component that condenses the input graph with GCN-learned node representations into a single vector or a coarser graph with a smaller size.

Early adopted graph pooling techniques such as average and maximum pooling disregard node correlations, hence restricting overall performance [5,40]. Subsequently, graph pooling utilizes hierarchical architecture to model the node correlations [20,34] and can be roughly classified into node clustering pooling and node drop pooling. Node clustering pooling requires clustering nodes into new nodes, which is time-and space-consuming [3,37,38]. In comparison, node drop pooling preserves only representative nodes by assessing their importance, and is hence more efficient and suitable for large-scale networks [7,14,39].

Although efficient and effective, the current node drop pooling methods are affected by information loss, resulting in suboptimal graph-level representations and unsatisfactory performance in the graph classification task. To substantiate the above idea, we conduct experiments on graph reconstruction to directly quantify the amount of retained information after pooling. Specifically, we employ two node drop pooling algorithms (i.e., TopKPool [7] and SAGPool [14]) on 10 synthetic point cloud graphs. The experimental settings are consistent with those proposed in prior research [1,3]. Figure 1 depicts the reconstructed results of the point cloud's original attributes (i.e., coordinates) from its pooled graph, which is generated by the node drop pooling operators. As shown in Fig. 1, node drop pooling approaches frequently fail to recover the original graph signal, indicating that they discard part of critical graph information, which explains their inferior performance in graph classification.

We provide an intuitive explanation for the aforementioned phenomenon as follows. Indeed, nodes connected in a graph typically share similar attributes [22], and their similarity rises further after message propagation using GNNs (such as GCN [13] or GAT [33]). Node drop pooling methods, such as TopKPool and SAGPool, generate node scores based on the node attributes, resulting in

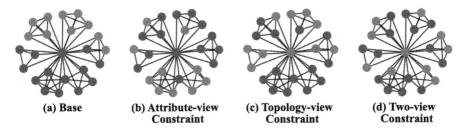

(a) Base   (b) Attribute-view   (c) Topology-view   (d) Two-view
                      Constraint   Constraint   Constraint

Fig. 2. Illustration of the constraint mechanism. Visualization of node selection results with and without (**Base**) constraints. The reserved nodes are highlighted in Red. (Color figure online)

a high potential for most selected nodes to share similar attributes or to be connected. Consequently, node drop pooling models may be stuck in significant local structures, thus selecting redundant nodes and ignoring significant nodes from other substructures. To empirically validate our analysis, we test the SAGPool model [14] on a real-world dataset (i.e., IMDB-BINARY). The experimental conditions are identical to those for the graph classification problem, which is described in detail in Sect. 4.1. In this example study, 40% of the nodes that are selected as significant nodes by the first pooling layer are marked in red. As shown in Fig. 2 (a), SAGPool (i.e., **Base**) is more likely to select nodes concentrated in the same area, confirming our hypothesis. As a result, existing node drop pooling methods may overlook critical information in other parts of a graph, causing loss of critical information and the lower performance in the graph classification task.

To address the limitations of existing node drop pooling methods, we design a masked graph auto-encoder constrained strategy called Masked Graph Auto-encoder constrained Pooling (MGAP), which mitigates the information-loss impact associated with graph pooling. Specifically, we incorporate a graph auto-encoder layer with two decoders into graph pooling models in order to impose implicit restrictions on the pooled graphs from two perspectives. Firstly, from the node-attribute perspective, we apply GCN layers to the embeddings of pooled nodes to reconstruct the original node attributes (Sect. 3.1), aiming to prevent the pooled graph from losing excessive critical attribute information. Secondly, from the network-topology perspective, we adopt a feedforward neural network to rebuild the node degree (Sect. 3.2), which enables the node drop pooling models to reserve more important nodes with regard to the topology aspect. As illustrated in Fig. 2 (b) and (c), the selected nodes are distributed across different substructures or cover the fundamental nodes in the graph, demonstrating that the proposed attribute- and topology-view constraints enable models to reserve significant nodes from the attribute and topology aspects (retaining more attribute and topological information), respectively. Additionally, Fig. 2 (d) illustrates the effect of combining constraints from the two views. Subsequently, we describe how to integrate MGAP with the present architecture for node drop pooling (Sect. 3.3). To further demonstrate the practical efficiency of our MGAP, we provide an in-depth analysis of time efficiency (Sect. 3.4).

Furthermore, we extensively examine MGAP across three backbone models and eleven benchmark datasets, which vary in content domains and dataset sizes. The experimental results demonstrate that MGAP generally and consistently improves the performance of the current node drop pooling models (e.g., TopKPool, SAGPool, and GSAPool). Our contributions are summarized as follows.

- We propose MGAP to alleviate the information-loss effect in graph pooling from the perspectives of attribute space and topology space.
- We demonstrate that MGAP is a plug-and-play and easy-to-compute module, which can be combined with node drop pooling methods to enhance their performance in the graph classification task. Furthermore, MGAP maintains controllable time and memory complexities.
- We conduct extensive experiments using three typical node drop pooling methods with and without MGAP in the graph classification task across eleven real-world datasets. The experimental results comprehensively demonstrate the effectiveness of MGAP.

2   Preliminaries and Related Works

2.1   Notations

Let $G = (\mathcal{V}, \mathcal{E})$ denote a graph with the node set $\mathcal{V}$ and edge set $\mathcal{E}$. The node attributes are denoted by $\boldsymbol{X} \in \mathbb{R}^{n \times d}$, where n is the number of nodes and d is the dimension of node attributes. The graph topology is represented by an adjacency matrix $\boldsymbol{A} \in \{0, 1\}^{n \times n}$.

2.2   Problem Statement

Definition 1 (Graph Classification). *The task of graph classification is to learn a mapping function f :*

$$f : \mathcal{G} \to \mathcal{Y}, \tag{1}$$

where $\mathcal{G} = \{G_1, G_2, \ldots, G_t\}$ is the set of input graphs, $\mathcal{Y} = \{y_1, y_2, \ldots, y_t\}$ is the set of labels associated with the graphs, and t is the number of graphs.

2.3   Graph Convolutional Networks

Recently, numerous studies have been conducted based on Graph Convolutional Networks, which generalize convolutional operation in graph data. The basic idea behind such methods as Graph Convolutional Network (GCN) [13], GraphSAGE [10], Graph Attention Network (GAT) [33], and Graph Isomorphism Network (GIN) [36], is to update the embedding of each node with messages from its neighbor nodes. Formally, the above message-passing mechanism in the l-th layer can be formalized as follows:

$$h_v^{(l+1)} = \text{COM}^{(l)} \left(\left\{ h_v^{(l)}, \text{AGG}^{(l)} \left(\left\{ h_{v'}^{(l)} : v' \in \mathcal{N}_v \right\} \right) \right\} \right),$$ (2)

where $\mathcal{N}_v$ is the set of neighbors of node v. $h_v^{(l)} \in \mathbb{R}^c$ is the representation vector for node v in the l-layer, where c is the dimension of the node embeddings. AGG and COM refer to the aggregation and combination functions, respectively. The methods mentioned above have achieved excellent performance in the node classification and link prediction tasks. However, an additional pooling operation is required to obtain a representation of the entire graph for downstream graph-level tasks (for example, the graph classification).

2.4   Graph Pooling

Definition 2 (Graph Pooling). *Let a graph pooling operator be defined as any function* POOL *that maps a graph G to a new pooled graph $G' = (\mathcal{V}', \mathcal{E}')$:*

$$G' = \text{POOL}(G),$$ (3)

where $|\mathcal{V}'| < |\mathcal{V}|$ and $|\mathcal{V}|$ is the number of nodes[1]. The generic goal of graph pooling is to reduce the number of nodes in a graph while preserving its semantic information.

Graph pooling, which plays a crucial role in representing the entire graph, could be roughly divided into global pooling and hierarchical pooling. Global pooling performs global sum/average/max-pooling [5] or more sophisticated operations [36,40] on all node attributes to produce graph-level representations, which disregard the topology of graphs. Contrarily, hierarchical pooling models are later proposed considering the graph topology, which could be classified into node clustering pooling and node drop pooling. **1) Node clustering pooling** considers the graph pooling problem as a node clustering problem to map the nodes into a set of clusters [3,20,37,38], which is limited by time and memory complexities caused by the dense soft-assignment matrix computation. Additionally, as discussed in previous studies [9,23], clustering-enforcing regularization that enforces clustering is typically ineffective. **2) Node drop pooling** exploits learnable scoring functions to eliminate nodes with relatively lower significance scores [7,8,14,16,39,41]. While the node drop pooling is more economical and suitable to large-scale networks than node clustering pooling, it suffers from an inevitable information loss. For a detailed description of graph pooling, please refer to the recent review [19].

---

[1] In some very specific cases, there exists $|\mathcal{V}'| \geq |\mathcal{V}|$, causing the graph to be up scaled by pooling.

2.5   Graph Auto-Encoder

Recent years have seen a surge of interest in studying the framework of auto-encoder for graph embedding. The non-probabilistic graph auto-encoder model (GAE) [12] consists of a GCN encoder, integrating the graph topology and node attributes, and a nonlinear inner product decoder, reconstructing the adjacency matrix. Formally, the auto-encoder can be summarized as:

$$\hat{A} = \sigma(ZZ^{\top}), \text{ with } \quad Z = \text{GCN}(X, A), \tag{4}$$

where $\hat{A} \in \{0,1\}^{n \times n}$ is the reconstructed adjacency matrix, and σ is the logistic sigmoid function. $Z \in \mathbb{R}^{n \times c}$ is the node embedding matrix, where c is the dimension of the node embeddings.

Instead of reconstructing the graph topology in GAE, some methods attempt to design a decoder to reconstruct the node attributes [15,25] or both the topology and attributes [31]. However, these methods are unsuitable for large-scale graphs. Therefore, some methods [27,28] have introduced general frameworks to scale GAE to large-scale graphs. Unlike the above methods, which perform auto-encoder in the Euclidean space, some recent studies [21,24] have attempted to encode and decode graphs in the hyperbolic space. Furthermore, Salha et al. [30] extended the GAE frameworks to address link prediction in directed graphs using gravity-inspired decoder scheme. Due to the space limitation, some other GAE methods, such as linear GAE [29], permutation-invariant GAE [35], and adaptive GAE [18], are not presented here in detail. Compared with the above studies, our manuscript heuristically exploits auto-encoder to constrain the pooled graphs of node drop pooling methods.

3   MGAP: Masked Graph Auto-Encoder Constrained Pooling

The whole structure of the proposed MGAP is illustrated in Fig. 3, which contains two parts: the constraint from the perspective of attribute (Sect. 3.1) and the constraint from the perspective of topology (Sect. 3.2). Additionally, we discuss how to integrate the present node drop pooling methods with MGAP (Sect. 3.3). Finally, we conduct an extensive investigation of complexity (Sect. 3.4).

3.1   Constraint in Attribute Space

The node attribute in a graph is essential for graph representation learning because each node attribute depicts partial characteristics of the graph. However, as illustrated in Fig. 1, current node drop pooling methods tend to discard a large amount of node attribute information, which may cause the decreased performance in the graph classification task. Therefore, we suggest compensating for this information loss through the use of an auto-encoder system. Specifically,

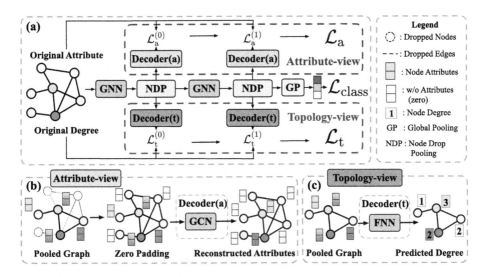

Fig. 3. The illustration of the proposed MGAP, which includes two parts: the attribute-view constrained module and the topology-view constrained module.

given the hidden representations of partially pooled nodes, we attempt to reconstruct the original node attributes of all nodes in the graph. In the proposed approach, there are three Components: **(C1)** the node drop pooling encoder, **(C2)** the designed decoder in attribute space, and **(C3)** the reconstruction target, the details of which are introduced as below.

(C1) Node Drop Pooling Encoder. Instead of following the standard GAE approaches [12], which adopt the well-established GNN models shown in Eq. (2) as an encoder, we employ one-layer GCN with graph pooling (GNN and NDP in Fig. 3 (a)) as the encoder. The objective of the encoder is to learn the embeddings of each node and to select which partial nodes to discard (mask). The embeddings of these masked nodes will not be observed by the decoder. As shown in Fig. 3 (a), the encoder first performs message propagation on the graphs to generate node embeddings using Eq. (2), and then generates coarsened graphs using node drop pooling methods. We used the SAGPool model [14] to describe the process of node drop pooling encoder, which consists of three disjoint parts:

1) Generating Scores. SAGPool predicts the significance scores for each node by using graph convolution as follows:

$$S^{(l)} = \text{GCN}(Z^{(l)}, A^{(l)}) \in \mathbb{R}^{n^{(l)} \times 1}, \tag{5}$$

where $S^{(l)} \in \mathbb{R}^{n^{(l)} \times 1}$ is the score matrix for the nodes, $A^{(l)} \in \{0, 1\}^{n^{(l)} \times n^{(l)}}$ is the adjacency matrix of the coarsened graph in the layer l, and $n^{(l)}$ is the number of reserved nodes in the coarsened graph.

2) Selecting Nodes. Subsequently, SAGPool selects the nodes with the top-k significance scores as follows:

$$\mathrm{idx}^{(l)} = \mathrm{TOP}_k(\boldsymbol{S}^{(l)}), \tag{6}$$

where TOP_k ranks values and returns the indices of the largest k values in $\boldsymbol{S}^{(l)}$, and $\mathrm{idx}^{(l)}$ indicates the reserved node indices for new graphs.

3) Coarsening Graphs. With the selected nodes, a new graph coarsened from the original one is obtained by learning new attribute and adjacency matrices:

$$
\begin{aligned}
\boldsymbol{Z}^{(l+1)} &= \boldsymbol{Z}^{(l)}_{\mathrm{idx}^{(l)}} \odot \boldsymbol{S}^{(l)}_{\mathrm{idx}^{(l)}} \in \mathbb{R}^{n^{(l+1)} \times 1}; \\
\boldsymbol{A}^{(l+1)} &= \boldsymbol{A}^{(l)}_{(\mathrm{idx}^{(l)},\mathrm{idx}^{(l)})} \in \{0,1\}^{n^{(l+1)} \times n^{(l+1)}},
\end{aligned} \tag{7}
$$

where $\cdot_{\mathrm{idx}}$ is an indexing operation, $\boldsymbol{Z}^{(l)}_{\mathrm{idx}^{(l)}}$ is the row-wise indexed embedding matrix, $\odot$ is the broadcast elementwise product, and $\boldsymbol{A}^{(l)}_{(\mathrm{idx}^{(l)},\mathrm{idx}^{(l)})}$ is the row-wise and column-wise indexed adjacency matrix. $\boldsymbol{Z}^{(l+1)}$ and $\boldsymbol{A}^{(l+1)}$ are the new attribute and corresponding adjacency matrices, respectively.

(C2) Decoder in Attribute Space. Unlike traditional GAE, MGAP, with the embeddings of pooled nodes, aims to recover the original attributes of graphs containing pooled nodes and masked (dropped) nodes. In particular, following [7], we first initialize an empty attribute matrix $\hat{\boldsymbol{X}}_0 \in \mathbb{R}^{n \times c}$ for the new graph. Subsequently, we insert the pooled node embeddings $\boldsymbol{Z}^{(l)} \in \mathbb{R}^{n^{(l)} \times c}$ into $\hat{\boldsymbol{X}}_0$ to obtain a new embedding matrix $\hat{\boldsymbol{X}}^{(l)} \in \mathbb{R}^{n \times c}$ (the zero padding operation in Fig. 3 (b)). The other row vectors (embeddings of the dropped nodes) remain zero. Then, we adopt graph convolution as the decoder, as introduced in Eq. (2), on the new node embedding matrix $\hat{\boldsymbol{X}}$ and the original adjacency matrix $\boldsymbol{A}^{(0)} = \boldsymbol{A} \in \mathbb{R}^{n \times n}$ to reconstruct the node attributes:

$$\psi_{\mathrm{a}}\left(\hat{\boldsymbol{X}}^{(l)}\right) = \mathrm{GCN}\left(\hat{\boldsymbol{X}}^{(l)}, \boldsymbol{A}^{(0)}\right) \in \mathbb{R}^{n \times d}. \tag{8}$$

(C3) Reconstruction Target. The constraint in attribute space aims to improve the power of pooling methods to preserve node information; that is, the learned embeddings of reserved nodes can recover the original attributes of the whole graph. Therefore, we directly measure the Euclidean distance between the reconstructed attribute matrix $\psi_{\mathrm{a}}(\hat{\boldsymbol{X}}^{(l)})$ and the original input attribute matrix $\boldsymbol{X} \in \mathbb{R}^{n \times d}$, and consider it as the loss function, which is formalized as follows:

$$\mathcal{L}_{\mathrm{a}}^{(l)} = \left\| \boldsymbol{X} - \psi_{\mathrm{a}}(\hat{\boldsymbol{X}}^{(l)}) \right\|_F^2, \tag{9}$$

where $\mathcal{L}_{\mathrm{a}}^{(l)}$ is the attribute-view constrained loss in the layer l, which enables pooling models to reserve additional important nodes from the perspective of node attributes, and $\| \cdot \|_F$ is the Frobenius norm.

3.2 Constraint in Topology Space

In addition to the attribute information discussed previously (Sect. 3.1), topological information in a graph is essential in graph representation learning. Therefore, it is logical and critical to ensure that the pooled nodes can reassemble the network topology. We propose to reconstruct the node degree, motivated by NWR-GAE [32], with the goal of capturing topological information. Specifically, our solution consists of three **Components**: **(C1)** the node drop pooling encoder, which is the same as the encoder in attribute-view constraint and will not be given any further details here, **(C2)** the designed decoder in topology space, and **(C3)** the reconstruction target, all of which are described in detail below.

(C2) Decoder in Topology Space. The decoders in existing graph auto-encoders are designed to drive the embeddings of the linked nodes similar, which appears away from our motivation that enables models to capture the topological information. Therefore, we suggest reconstructing the node degree, which is a typical graph topological feature that reflects the receptive field of a node. Specifically, given the embedding of the pooled nodes, we adopt an FNN layer with an activation function ReLU$(\cdot)$, which makes the predicted value non-negative, to reconstruct the node degree in the l-th layer:

$$\psi_t \left(\boldsymbol{Z}^{(l)} \right) = \text{ReLU} \left(\text{FNN} \left(\boldsymbol{Z}^{(l)} \right) \right) \in \mathbb{R}^{n^{(l)} \times 1}. \tag{10}$$

(C3) Reconstruction Target. We measure the Euclidean distance between the truth degree $\boldsymbol{D}^{(l)} \in \mathbb{R}^{n^{(l)} \times 1}$ and the predict degree $\psi_t(\boldsymbol{Z}^{(l)})$, which is formalized as follows:

$$\mathcal{L}_t^{(l)} = \left\| \boldsymbol{D}^{(l)} - \psi_t \left(\boldsymbol{Z}^{(l)} \right) \right\|_F^2, \tag{11}$$

where $\mathcal{L}_t^{(l)}$ is the topology-view constrained loss in the layer l. With this loss, a fraction of important nodes from the perspective of the topology can be reserved.

3.3 Node Drop Pooling Framework with MGAP

Figure 3 (a) illustrates in detail how to apply MGAP to the node drop pooling framework. Concretely, we view a GCN layer followed by a node drop pooling layer, such as TopKPool or SAGPool, as a pooling function unit and name it GCN-Pool layer for convenience. A GCN-Pool layer takes a graph as input and outputs a pooled graph that is represented by an embedding matrix and a new adjacency matrix. The two decoders ψ_a (Decoder (a) in Fig. 3) and ψ_t (Decoder (t) in Fig. 3) are trained to simultaneously reconstruct the original node attributes and network topology, which generates two losses, $\mathcal{L}_a$ and $\mathcal{L}_t$. The pooled graph is then fed into the next GCN-Pool layer and, simultaneously, a readout module, in which the node embeddings are added up as the graph embedding in this layer. Finally, the graph embeddings in all layers are added up to the final graph representation, that is taken as the input of an Multi-layer Perceptron (MLP) classifier for predicting the label of the original graph. Classification error is defined by the cross-entropy loss $\mathcal{L}_{\text{class}}$.

By combining the classification loss $\mathcal{L}_{\text{class}}$ and two constrained losses in Eq. (9) and (11), we obtained the total loss:

$$\mathcal{L}_{\text{total}} = \mathcal{L}_{\text{class}} + \lambda_a \mathcal{L}_a + \lambda_t \mathcal{L}_t, \tag{12}$$

where λ_a and λ_t are the trade-off weight parameters, and $\mathcal{L}_a$ and $\mathcal{L}_t$ are the average attribute- and topology-view constrained losses of all layers, respectively. Notably, the graph classification task is performed by GCN-Pool layer in the proposed framework, and the decoders were only used for constraining the pooled nodes and their embeddings in GCN-Pool. Thus, the decoders are used only in the training phase. After obtaining the trained model, we apply GCN-Pool to perform graph classification in the test set without decoders.

3.4 Complexity Analysis

Our proposed MGAP is highly efficient because the major operations involved in it are only GCN and FNN, as shown in Fig. 3 (b) and (c), respectively. Theoretically, the time complexity of GCN layer is $\mathcal{O}\left(L\|A\|_0 d + Lnd^2\right)$, where L is the number of layers, n is the number of nodes, and $\|A\|_0$ is the number of nonzeros in the adjacency matrix A. The time complexity of FNN layer is $\mathcal{O}(1)$. The time complexity for calculating $\mathcal{L}_a$ by Eq. 9 is $\mathcal{O}(nd)$ and calculating $\mathcal{L}_t$ by Eq. 11 is $\mathcal{O}(n')$, where n' is the number of the pooled nodes. Thus, the total time complexity of the proposed method is $\mathcal{O}\left(L\|A\|_0 d + Lnd^2\right)$, which is on par with the neighborhood aggregation operation in node drop pooling methods.

4 Experiments

In this section, we study the effectiveness of MGAP for graph classification. Specifically, we would like to answer the following questions:

Q1. How often and how much does MGAP improve the performance of the base node drop pooling methods? (Sect. 4.2)

Q2. Does each component of MGAP contribute to the improvements in performance? (Sect. 4.3)

Q3. How much extra computation time and memory does MGAP incur?(Sect. 4.4)

Q4. How would the parameters affect the performance? (Sect. 4.5)

4.1 Experimental Settings

Datasets. To answer **Q1**, we use 11 publicly available and well-known benchmark datasets, including bioinformatics datasets (D&D, PROTEINS, and ENZYMES), molecule datasets (NCI1, NCI109, PTC-MR, MUTAG, MUTA-GENICITY, and FRANKENSTEIN), and social network datasets (IMDB-BINARY and IMDB-MULTI). The above 11 real-world datasets vary in content domains and dataset sizes, and the dataset statistics are summarized in Table 1.

Table 1. Statistics and properties of benchmark datasets (TUdatasets[a]).

	Datasets	# Graphs	# Classes	Avg. # Nodes	Avg. # Edges
Bioinformatics	D&D	1,178	2	284.32	715.66
	PROTEINS	1,113	2	39.06	72.82
	ENZYMES	600	6	32.63	124.20
Molecules	NCI1	4,110	2	29.87	32.30
	NCI109	4,127	2	29.68	32.13
	PTC-MR	344	2	14.30	14.69
	MUTAG	188	2	17.93	19.79
	MUTAGENICITY	4,337	2	30.32	30.77
	FRANKENSTEIN	4,337	2	16.90	17.88
Social Networks	IMDB-BINARY	1,000	2	19.77	96.53
	IMDB-MULTI	1,500	3	13.00	65.94

[a]TUDatasets: https://chrsmrrs.github.io/datasets/docs/datasets/

Backbone Models. We select three representative node drop pooling models as the backbone: **TopKPool** [7]. This method selects the top k nodes based on the scores generated by a learnable function that only considers node attributes. **SAGPool** [14]. This method selects the important nodes with higher scores that are generated by a graph convolution layer, which involves node attributes and network topology. Particularly, this method has two variants: *1) SAGPool (G)* is a global node drop pooling method that drops unimportant nodes at one time at the end of the architecture. *2) SAGPool (H)* is a hierarchical node drop pooling method that sequentially drops unimportant nodes with multiple graph convolution layers. We use SAGPool (H) in this study. **GSAPool** [39]. This method predicts scores from two perspectives: *1)* using an MLP layer to capture the significant node attributes and *2)* using a GNN layer to capture the significant network topology. Subsequently, the model linearly combines the two scores mentioned above.

Implementation Details. For a fair comparison, we adopt the same settings on all datasets and models. Specifically, we evaluate the model performance with a 10-fold cross validation setting, and the dataset split is based on the conventionally used training/test splits [1,36]. Each convolution layer consists of 128 hidden neurons, and the pooling ratio in each pooling layer is set as 0.5, i.e., removing 50% of nodes per graph after a pooling operation. We employ Adam [11] to optimize the parameters with learning rate as $5e^{-4}$ and weight decay as $1e^{-4}$, and adopt early stopping to control the training epochs based on validation loss with patience set as 50. We then report the average performance on the test sets, by performing overall experiments 100 times with different seeds from 42 to 51.

Table 2. MGAP performance across three backbone models and 11 datasets in the graph classification task. The reported results are mean and standard deviation over 100 different runs.

| | Molecules | | | | | |
	NCI1	NCI109	MUTAG	PTC-MR	MUTAGE.	FRANK.
SAGPool	$71.71_{\pm0.75}$	$70.70_{\pm0.95}$	$72.56_{\pm3.09}$	$56.41_{\pm1.63}$	$74.27_{\pm1.04}$	$58.74_{\pm0.61}$
+ MGAP	$\mathbf{73.02}_{\pm1.00}$	$\mathbf{71.95}_{\pm0.63}$	$\mathbf{73.72}_{\pm1.88}$	$\mathbf{58.80}_{\pm2.29}$	$\mathbf{74.99}_{\pm1.22}$	$\mathbf{59.06}_{\pm0.81}$
Gain	1.83% ↑	1.76% ↑	1.60% ↑	4.23% ↑	0.97% ↑	0.54% ↑
TopKPool	$71.90_{\pm1.22}$	$70.69_{\pm1.00}$	$71.83_{\pm1.66}$	$57.15_{\pm3.14}$	$75.10_{\pm0.94}$	$58.84_{\pm0.80}$
+ MGAP	$\mathbf{72.83}_{\pm1.24}$	$\mathbf{72.35}_{\pm1.03}$	$\mathbf{73.06}_{\pm2.77}$	$\mathbf{58.35}_{\pm2.70}$	$\mathbf{76.46}_{\pm1.05}$	$\mathbf{58.96}_{\pm0.53}$
Gain	1.29% ↑	2.35% ↑	1.71% ↑	2.10% ↑	1.81% ↑	0.0% ↑
GSAPool	$73.70_{\pm0.89}$	$71.83_{\pm1.65}$	$\mathbf{72.56}_{\pm2.41}$	$56.10_{\pm1.83}$	$76.65_{\pm1.12}$	$59.11_{\pm0.69}$
+ MGAP	$\mathbf{75.36}_{\pm1.68}$	$\mathbf{74.10}_{\pm1.95}$	$72.44_{\pm3.42}$	$\mathbf{57.59}_{\pm2.81}$	$\mathbf{77.94}_{\pm1.03}$	$\mathbf{59.57}_{\pm0.32}$
Gain	2.25% ↑	3.16% ↑	0.16% ↓	2.66% ↑	1.68% ↑	0.78% ↑

| | Bioinformatics | | | Social Networks | | Average |
	D&D	PROT.	ENZYM.	IMDB-B	IMDB-M	
SAGPool	$74.21_{\pm1.23}$	$72.65_{\pm1.26}$	$47.42_{\pm1.54}$	$70.71_{\pm1.36}$	$48.43_{\pm0.81}$	65.25
+ MGAP	$\mathbf{76.20}_{\pm0.73}$	$\mathbf{74.53}_{\pm1.04}$	$\mathbf{48.93}_{\pm3.69}$	$\mathbf{72.88}_{\pm1.22}$	$\mathbf{49.71}_{\pm0.80}$	$\mathbf{66.70}$
Gain	2.68% ↑	2.58% ↑	3.18% ↑	3.07% ↑	2.64% ↑	2.22% ↑
TopKPool	$73.71_{\pm1.04}$	$72.81_{\pm0.74}$	$46.02_{\pm3.53}$	$70.96_{\pm1.15}$	$48.97_{\pm0.60}$	65.27
+MGAP	$\mathbf{75.97}_{\pm0.96}$	$\mathbf{73.95}_{\pm1.23}$	$\mathbf{49.58}_{\pm2.01}$	$\mathbf{72.05}_{\pm0.50}$	$\mathbf{49.52}_{\pm0.79}$	$\mathbf{66.64}$
Gain	3.06% ↑	1.56% ↑	7.73% ↑	1.53% ↑	1.12% ↑	2.10% ↑
GSAPool	$74.19_{\pm1.32}$	$73.20_{\pm1.11}$	$49.08_{\pm2.02}$	$71.06_{\pm1.15}$	$49.03_{\pm0.50}$	66.00
+ MGAP	$\mathbf{76.24}_{\pm1.03}$	$\mathbf{73.49}_{\pm1.39}$	$\mathbf{55.05}_{\pm2.86}$	$\mathbf{72.34}_{\pm1.26}$	$\mathbf{49.68}_{\pm0.61}$	$\mathbf{67.59}$
Gain	2.76% ↑	0.40% ↑	12.16% ↑	1.80% ↑	1.32% ↑	2.40% ↑

Hyper-parameter tuning. As described in Sect. 3.3, two hyper-parameters λ_a and λ_t are used in our MGAP, which serve as trade-off weights in the loss function. We utilize a grid search to tune the above two hyper-parameters with a search space $\{1, 1e^{-1}, 1e^{-2}\}$.

Environments. 1) Software. All models are implemented with Python 3.7, PyTorch 1.9.0 or above (which further requires CUDA 10.2 or above), and PyTorch-Geometric 1.7.3 or above. **2) Hardware.** Each experiment was run on a single GPU (NVIDIA V100 with a 16 GB memory size), and the experiments were run on the server at any given time[2].

---

[2] The source code is available at https://github.com/liucoo/mgap.

Table 3. Ablation study results. **Bold:** the best performance per backbone model and dataset. <u>Underline:</u> the second best performance per backbone model and dataset.

	PTC-MR			IMDB-BINARY		
	SAGPool	TopKPool	GSAPool	SAGPool	TopKPool	GSAPool
Base	$56.41_{\pm1.6}$	$57.15_{\pm3.1}$	$56.10_{\pm1.8}$	$70.71_{\pm1.4}$	$70.96_{\pm1.2}$	$71.06_{\pm1.2}$
MGAP	$\mathbf{58.80_{\pm2.2}}$	$\mathbf{58.35_{\pm2.7}}$	$\mathbf{57.59_{\pm2.8}}$	$\mathbf{72.88_{\pm1.2}}$	$\mathbf{72.05_{\pm0.5}}$	$\mathbf{72.34_{\pm1.3}}$
w/o attr-const	$\underline{58.44_{\pm1.9}}$	$57.21_{\pm3.5}$	$56.21_{\pm3.1}$	$\underline{72.43_{\pm0.9}}$	$\underline{71.87_{\pm0.9}}$	$\underline{71.89_{\pm0.9}}$
w/o topo-const	$58.35_{\pm2.6}$	$\underline{57.76_{\pm1.6}}$	$\underline{56.56_{\pm2.0}}$	$71.72_{\pm1.1}$	$71.29_{\pm0.85}$	$71.50_{\pm0.9}$

4.2   Overall Results

To answer **Q1**, we conduct extensive experiments for graph classification on 11 datasets using three backbone models. The accuracy results of all models summarized in Table 2 are averaged over **100 runs** with random weight initializations (10 different seeds through the 10-fold cross validation). We highlight the best performance **in bold** per backbone model and dataset. In Table 2, we report the improvement achieved by MGAP on each backbone model and each dataset. We obtain the following findings. **1)** Evidently, MGAP consistently improves the accuracy of all three node drop pooling models on all datasets in most cases, sometimes by large margins. **2)** Specifically, MGAP achieves improvements over three node drop pooling models (averaged across datasets): 2.22% (SAGPool), 2.10% (TopKPool), and 2.40% (GSAPool). **3)** MGAP obtains more significant enhancement on bioinformatics datasets and increases the accuracy by up to 12.16%. Intuitively, this may be because the information loss, caused by the condensation of selected nodes into the local structure, makes a greater impact on bioinformatics datasets. In summary, the above results indicate that MGAP is a general framework for improving the performance of base node drop pooling methods.

4.3   Ablation Study

To answer **Q2**, we conduct ablation studies on the dataset PTC-MR (social domain) and IMDB-BINARY (biochemical domain) using the SAGPool model. For convenience, we name the methods without attribute-view and topology-view constraints as **w/o attr-const** and **w/o topo-const**, respectively. Note that except the selected component, the rest remain the same as the complete model. From Table 3, we obtain that all variants with some components removed exhibit clear performance drops compared to the complete model, indicating that each component contributes to the improvements. Furthermore, MGAP without the topology-view constraint performs poorly on the IMDB-BINARY dataset, thereby demonstrating the significance of the proposed topology-view constraint for datasets in social domain, where network topology plays an important role.

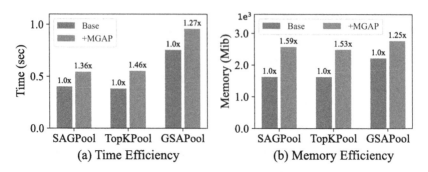

Fig. 4. Memory and time efficiency of MGAP compared with three backbone models. (a) The reported values are the average per-epoch training time on all 11 datasets. (b) The reported values are the average GPU memory usage on all 11 datasets.

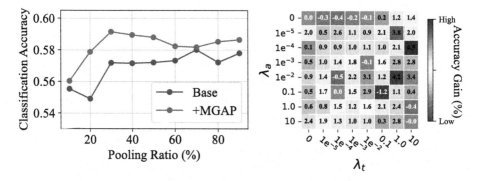

Fig. 5. Parameter analysis on the PTC-MR dataset. *Left.* Model performance varying with the pooling ratio. *Right.* Parameter sensitivity of trade-off weights λ_a and λ_t.

4.4 Efficiency Analysis

To answer **Q3**, we compare the time and memory efficiency of MGAP with that of three backbone models. **1) Time Efficiency**. Fig. 4 (a) illustrates the average per-epoch training time on all 11 datasets. We fix the training epochs to 10 with 10 different random seeds. It is observed that the additional time consumption keeps relatively low. **2) Memory Efficiency**. The experimental settings are the same as those in measuring the time efficiency. Figure 4 (b) shows that our MGAP is efficient in terms of memory. The above results confirm that our MGAP is practically efficient.

4.5 Parameter Analysis

To answer **Q4**, we investigate the sensitivity of the parameters of two types on the PTC-MR dataset using the SAGPool model. **1) Inherent Parameter Sensitivity**. We study how the graph pooling ratio would affect the graph classification performance. As shown in the left part of Fig. 5, SAGPool equipped with MGAP (+MGAP) performs better in all cases, suggesting that the proposed method enable node drop pooling methods to select the nodes that are essential for graph-level representation learning regardless of the pooling ratio. **2) Introduced Parameter Sensitivity.** We investigate the effects of two new parameters, λ_a and λ_t, which serve as the trade-off weights in the loss function. In this parameter sensitivity study, both parameters are searched within the range of $\{10, 1, 1e^{-1}, 1e^{-2}, 1e^{-3}, 1e^{-4}, 1e^{-5}, 0\}$. Note that the search space is only $\{1, 1e^{-1}, 1e^{-2}\}$ in the graph classification experiments. As shown in the right part of Fig. 5, the method performs best when λ_a is $1e^{-4}$ and λ_t is 10, demonstrating the importance of combining attribute and topology constraints of the pooled graphs. The results of the above two experiments further validate the robustness and effectiveness of the proposed MGAP.

5   Conclusion and Future Work

Conclusion. In this study, we empirically verify the information-loss problem of current node drop pooling models and propose MGAP, a novel plug-in and easy-to-compute module, to solve this problem from the perspectives of attribute space and topology space. Through extensive experiments, we demonstrate that MGAP generally improves common node drop pooling methods across various benchmark datasets in the graph classification task.

Future Work. For future directions, **1)** choose various formulas of GNNs, such as attention mechanism [33], as a decoder, in addition to GCN, for reconstructing the original attributes of nodes. **2)** Consider other topological features, such as triangle count, local clustering score, eigenvector centrality, and betweenness, in addition to node degrees. **3)** Further design other evaluation criteria for topological information loss, such as some criteria studied in graph coarsening algorithms [4]. **4)** Explore the effects of MGAP on other tasks such as graph reconstruction, graph compression, and node classification, and further design more reasonable constraints for these tasks.

Acknowledgements. This work was supported in part by the Natural Science Foundation of China (Nos. 61976162, 82174230, 62002090), Artificial Intelligence Innovation Project of Wuhan Science and Technology Bureau (No.2022010702040070), Science and Technology Major Project of Hubei Province (Next Generation AI Technologies) (No. 2019AEA170), and Joint Fund for Translational Medicine and Interdisciplinary Research of Zhongnan Hospital of Wuhan University (No. ZNJC202016).

References

1. Baek, J., Kang, M., Hwang, S.J.: Accurate learning of graph representations with graph multiset pooling. In: ICLR (2021)
2. Bai, L., Jiao, Y., Cui, L., Hancock, E.R.: Learning aligned-spatial graph convolutional networks for graph classification. In: ECML-PKDD, pp. 464–482 (2019)

3. Bianchi, F.M., Grattarola, D., Alippi, C.: Spectral clustering with graph neural networks for graph pooling. In: ICML. vol. 119, pp. 874–883 (2020)

4. Cai, C., Wang, D., Wang, Y.: Graph coarsening with neural networks. In: ICLR (2021)

5. Duvenaud, D., et al.: Convolutional networks on graphs for learning molecular fingerprints. In: NeurIPS, pp. 2224–2232 (2015)

6. Errica, F., Podda, M., Bacciu, D., Micheli, A.: A fair comparison of graph neural networks for graph classification. In: ICLR (2020)

7. Gao, H., Ji, S.: Graph u-nets. In: ICML, pp. 2083–2092 (2019)

8. Gao, X., Dai, W., Li, C., Xiong, H., Frossard, P.: ipool-information-based pooling in hierarchical graph neural networks. IEEE Trans. Neural Netw. Learn. Syst. **33**(9), 5032–5044 (2021)

9. Grattarola, D., Zambon, D., Bianchi, F.M., Alippi, C.: Understanding pooling in graph neural networks. arXiv:2110.05292 (2021)

10. Hamilton, W., Ying, Z., Leskovec, J.: Inductive representation learning on large graphs. In: NeurIPS, vol. 30 (2017)

11. Kingma, D.P., Ba, J.: Adam: A method for stochastic optimization. arXiv:1412.6980 (2014)

12. Kipf, T.N., Welling, M.: Variational graph auto-encoders. In: NeurIPS Workshop on Bayesian Deep Learning (2016)

13. Kipf, T.N., Welling, M.: Semi-supervised classification with graph convolutional networks. In: ICLR (2017)

14. Lee, J., Lee, I., Kang, J.: Self-attention graph pooling. In: ICML, pp. 3734–3743 (2019)

15. Li, J., Li, J., Liu, Y., Yu, J., Li, Y., Cheng, H.: Deconvolutional networks on graph data. In: NeurIPS (2021)

16. Li, M., Chen, S., Zhang, Y., Tsang, I.: Graph cross networks with vertex infomax pooling. In: NeurIPS, vol. 33, pp. 14093–14105 (2020)

17. Li, X., et al.: Braingnn: Interpretable brain graph neural network for fmri analysis. Med. Image Anal. **73**, 102233 (2021)

18. Li, X., Zhang, H., Zhang, R.: Adaptive graph auto-encoder for general data clustering. IEEE Trans. Patt. Anal. Mach. Intell. **44**(12), 9725–9732 (2021)

19. Liu, C., Zhan, Y., Li, C., Du, B., Wu, J., Hu, W., Liu, T., Tao, D.: Graph pooling for graph neural networks: Progress, challenges, and opportunities. arXiv:2204.07321 (2022)

20. Ma, Y., Wang, S., Aggarwal, C.C., Tang, J.: Graph convolutional networks with eigenpooling. In: SIGKDD, pp. 723–731 (2019)

21. Mathieu, E., Le Lan, C., Maddison, C.J., Tomioka, R., Teh, Y.W.: Continuous hierarchical representations with poincaré variational auto-encoders. In: NeurIPS (2019)

22. McPherson, M., Smith-Lovin, L., Cook, J.M.: Birds of a feather: Homophily in social networks. Ann. Rev. sociol. **27**(1), 415–444 (2001)

23. Mesquita, D., Souza, A., Kaski, S.: Rethinking pooling in graph neural networks. In: NeurIPS. vol. 33, pp. 2220–2231 (2020)

24. Park, J., Cho, J., Chang, H.J., Choi, J.Y.: Unsupervised hyperbolic representation learning via message passing auto-encoders. In: CVPR, pp. 5516–5526 (2021)

25. Park, J., Lee, M., Chang, H.J., Lee, K., Choi, J.Y.: Symmetric graph convolutional autoencoder for unsupervised graph representation learning. In: ICCV, pp. 6519–6528 (2019)

26. Rhee, S., Seo, S., Kim, S.: Hybrid approach of relation network and localized graph convolutional filtering for breast cancer subtype classification. In: IJCAI, pp. 3527–3534 (2018)
27. Salha, G., Hennequin, R., Remy, J.B., Moussallam, M., Vazirgiannis, M.: Fastgae: scalable graph autoencoders with stochastic subgraph decoding. Neural Netw. **142**, 1–19 (2021)
28. Salha, G., Hennequin, R., Tran, V.A., Vazirgiannis, M.: A degeneracy framework for scalable graph autoencoders. In: IJCAI, pp. 3353–3359 (2019)
29. Salha, G., Hennequin, R., Vazirgiannis, M.: Simple and effective graph autoencoders with one-hop linear models. In: ECML-PKDD, pp. 319–334 (2020)
30. Salha, G., Limnios, S., Hennequin, R., Tran, V.A., Vazirgiannis, M.: Gravity-inspired graph autoencoders for directed link prediction. In: CIKM, pp. 589–598 (2019)
31. Sun, D., Li, D., Ding, Z., Zhang, X., Tang, J.: Dual-decoder graph autoencoder for unsupervised graph representation learning. Knowl.-Based Syst. **234**, 107564 (2021)
32. Tang, M., Li, P., Yang, C.: Graph auto-encoder via neighborhood wasserstein reconstruction. In: ICLR (2022)
33. Veličković, P., Cucurull, G., Casanova, A., Romero, A., Liò, P., Bengio, Y.: Graph attention networks. In: ICLR (2018)
34. Wang, Z., Ji, S.: Second-order pooling for graph neural networks. IEEE Trans. Pattern Anal. Mach. Intell. (2020, early access). https://doi.org/10.1109/TPAMI.2020.2999032
35. Winter, R., Noé, F., Clevert, D.A.: Permutation-invariant variational autoencoder for graph-level representation learning. In: NeurIPS, vol. 34 (2021)
36. Xu, K., Hu, W., Leskovec, J., Jegelka, S.: How powerful are graph neural networks? In: ICLR (2019)
37. Ying, Z., You, J., Morris, C., Ren, X., Hamilton, W., Leskovec, J.: Hierarchical graph representation learning with differentiable pooling. In: NeurIPS. pp. 4805–4815 (2018)
38. Yuan, H., Ji, S.: Structpool: Structured graph pooling via conditional random fields. In: ICLR (2020)
39. Zhang, L., et al.: Structure-feature based graph self-adaptive pooling. In: WWW, pp. 3098–3104 (2020)
40. Zhang, M., Cui, Z., Neumann, M., Chen, Y.: An end-to-end deep learning architecture for graph classification. In: AAAI (2018)
41. Zhang, Z., et al.: Hierarchical multi-view graph pooling with structure learning. IEEE Trans. Knowl. Data Eng. **34**(1), 545–559 (2021)

Supervised Graph Contrastive Learning for Few-Shot Node Classification

Zhen Tan[1(✉)], Kaize Ding[1], Ruocheng Guo[2], and Huan Liu[1]

[1] Arizona State University, Tempe, AZ, USA
{ztan36,kding9,huanliu}@asu.edu
[2] Bytedance AI Lab, London, UK

Abstract. Graphs present in many real-world applications, such as financial fraud detection, commercial recommendation, and social network analysis. But given the high cost of graph annotation or labeling, we face a severe graph label-scarcity problem, i.e., a graph might have a few labeled nodes. One example of such a problem is the so-called *few-shot node classification*. A predominant approach to this problem resorts to *episodic meta-learning*. In this work, we challenge the status quo by asking a fundamental question whether meta-learning is a must for few-shot node classification tasks. We propose a new and simple framework under the standard few-shot node classification setting as an alternative to meta-learning to learn an effective graph encoder. The framework consists of supervised graph contrastive learning with novel mechanisms for data augmentation, subgraph encoding, and multi-scale contrast on graphs. Extensive experiments on three benchmark datasets (CoraFull, Reddit, Ogbn) show that the new framework significantly outperforms state-of-the-art meta-learning based methods.

Keywords: Few-shot learning · Graph Neural Networks · Graph contrastive learning

1 Introduction

Graphs are ubiquitous in many real-world applications. Graph Neural Networks (GNNs) [20,30,35] have been applied to model a myriad of network-based systems in various domains, such as social networks [13], citation networks [18], and knowledge graphs [24]. Despite these breakthroughs, it has been noticed that conventional GNNs fail to make accurate predictions when the labels are scarcely available [6,7,40]. One such problem is so-called *few-shot node classification*. It consists of two disjoint phases: In the first phase (train), classes with substantial labeled nodes (i.e., *base classes*) are available to learn a GNN model; and in the second phase (test), the GNN classifies nodes of unseen or *novel classes* with few labeled nodes. A few-shot node classification task is called N-*way K-shot node classification* if a node is classified into N classes and each class contains only a few K labeled nodes in the test phase.

M.-R. Amini et al. (Eds.): ECML PKDD 2022, LNAI 13714, pp. 394–411, 2023.
https://doi.org/10.1007/978-3-031-26390-3_24

This shortage of labeled training data for the novel classes poses a great challenge to learning effective GNNs. A prevailing paradigm to tackle this problem is *episodic meta-learning* [7,12,23,33,41]; its representative algorithms are Matching Network [31], MAML [10], and Prototypical Network [25]. Episodic meta-learning is inspired by how humans learn unseen classes with few samples via utilizing previously learned prior knowledge. During the training phase, it generates numerous meta-train tasks (or episodes) by emulating the test tasks, following the same N-way K-shot node classification structure. An example of episodic meta-learning is shown in Fig. 1. In each episode, K labeled nodes are randomly sampled from N base classes, forming a *support set*, to train the GNN model while emulating the N-way K-shot node classification in the test phase. The GNN then predicts labels for an emulated *query set* of nodes randomly sampled from the same classes as the support set. A Cross-Entropy Loss is used for backpropagation to update the GNN. Current research [7,9,12,33,41] has shown that via numerous such episodic emulations across different sampled meta-tasks on bases classes, the trained encoder can extract transferable meta-knowledge to fast adapt to unseen novel classes.

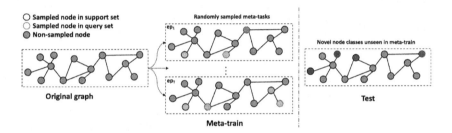

Fig. 1. Episodic meta-learning for few-shot node classification (ep$_i$ is the ith episode). Colors indicate different classes. Specially, grey nodes mean those nodes are not sampled. Different types of nodes indicate if nodes are from a support set or a query set. (Color figure online)

These episodic meta-learning based methods entail the following steps: (1) random sampling in each episode for meta-train in order to acquire the topological knowledge crucial for learning representative node embeddings. Since only a small portion of the nodes and classes are randomly selected per episode, the topological knowledge learned with those nodes is piecemeal and insufficient to train expressive GNN encoder, especially if the selected nodes share little correlation. (2) To boost accuracy, those meta-learning based methods have to rely on a large number of episodes. In other words, a large number of samples from the original graph is required for meta-train to capture the topological knowledge that can be transferred for use in the test phase. Consequently, these meta-learning algorithms can take time to converge with their emulation-based meta-train. The two problems are closely related. The *piecemeal knowledge* from emulation-based meta-learning unique for few-shot node classification entails the need for *large numbers of episodes* to acquire better topological knowledge. In this paper, we investigate if an alternative approach to episodic meta-learning can be developed so that the two problems can be addressed from their root causes for better performance of few-shot node classification. We posit that the key to few-shot node classification is to learn a generalizable GNN encoder that can produce discriminative representation on novel classes by learning transferable topological patterns implied in bases classes. If we can address the piecemeal knowledge problem, we can better use the few labeled nodes from novel classes to fine-tune another simple classifier (e.g., Logistic Regression, SVM, shallow MLP, etc.) to predict labels for other unlabeled nodes.

As an alternative to episodic meta-learning, we propose a novel approach for few-shot node classification, supervised graph contrastive learning [11,19]. Graph contrastive learning is proven effective in training powerful GNNs [14,38,42]. Multiple views of the original graph are first created through predefined transformations [8,14,38,42] (e.g., randomly dropping edges, randomly perturbing node attributes). Then a contrastive loss is applied to maximize feature consistency under differently augmented views. However, none of those existing methods accommodate the unique characteristics of few-shot node classification. In this paper, we propose a novel graph contrastive learning method especially designed for few-shot node classification. We will present technical details on how the new supervised contrastive learning can avoid the two problems with the episodic meta-learning after a formal problem statement is given.

Contributions. Our contributions include: (1) we are the first to investigate an alternative to the prevailing meta-learning paradigm for graph few-shot learning; (2) we propose a supervised graph contrastive learning method tailored for few-shot node classification by developing novel mechanisms for data augmentation, subgraph encoding, and multi-scale contrast on a graph; and (3) we conduct systematic experiments to assess the proposed framework in comparison with the existing meta-learning based graph few-shot learning methods and representative graph contrastive learning methods in terms of accuracy and efficiency.

2   Problem Formulation

In this paper, we focus on few-shot node classification on a single graph. Formally, given an attributed network $G = (\mathcal{V}, \mathcal{E}, \mathbf{X}) = (\mathbf{A}, \mathbf{X})$, where $\mathcal{V}$, $\mathcal{E}$, $\mathbf{A}$ and $\mathbf{X}$ denote the nodes, edges, adjacency matrix and node attributes, respectively. The few-shot node classification problem assumes the existence of a series of homogeneous node classification tasks $\mathcal{T} = \{\mathcal{D}^i\}_{i=1}^I = \{(\mathbf{A}_{\mathcal{C}^i}, \mathbf{X}_{\mathcal{C}^i})\}_{i=1}^I$, where $\mathcal{D}^i$ denotes the given dataset of a task, I denotes the number of such tasks, $\mathbf{X}_{\mathcal{C}^i}$ denotes the attributes of nodes whose labels belong to the label space $\mathcal{C}^i$, and $\mathbf{A}_{\mathcal{C}^i}$ similarly. Following the literature [7,12,23,33,41], we call the classes available during training as base classes, $\mathcal{C}_{base}$, and the classes for target test phase as novel classes, $\mathcal{C}_{novel}$, $\mathcal{C}_{base} \cap \mathcal{C}_{novel} = \varnothing$. Conventionally, there are substantial gold-labeled nodes for $\mathcal{C}_{base}$, but few labeled nodes for novel classes $\mathcal{C}_{novel}$. We can formally define the problem of few-shot node classification as follows:

Definition 1. ***Few-shot Node Classification:*** *Given an attributed graph $G =$ (A, X) with a divided node label space $C = \{C_{base}, C_{novel}\}$, we have substantial labeled nodes from C_{base}, and few-shot labeled nodes (support set S) for C_{novel}. The task is to predict the labels for unlabeled nodes (query set Q) from C_{novel}.*

3   Methodology

In graph representation learning, usually a GNN encoder g_θ is employed to model the high dimensional graph knowledge and project the node attributes to a low dimensional latent space. The classifier f_ψ is then applied to the latent node representations for node classification. The essence of few-shot node classification is to learn a encoder g_θ that can transfer the topological and semantic knowledge learned from substantial data of base classes to generate discriminative embeddings for nodes from novel classes with limited supervisory information.

As a prevailing paradigm, meta-learning is adopted [7,33,41] to jointly learn g_θ and f_ψ by episodically optimizing the Cross-Entropy Loss (CEL) on sampled meta-tasks. However, optimizing CEL over sampled piecemeal graph structures will engender node embeddings excessively discriminative against the current sampled nodes, rendering them sub-optimal for nodes classification in the test phase where the nodes are sampled arbitrarily from unseen novel classes. To mitigate such issues, those meta-learning based methods rely on a large number of episodes to train the model on numerous differently sampled meta-tasks to learn more transferable node embeddings, which makes the training process highly unscalable, especially for large graphs.

As a remedy, in this paper, we propose a decoupled method to learn the graph encoder g_θ and the final node classifier f_ψ separately. We put forward a supervised graph contrastive learning to firstly pretrain the GNN encoder g_θ to generate more transferable node embeddings. With such high-quality node embeddings, we can fine-tune a simple linear classifier to perform the final few-shot node classification. As shown in Fig. 2, our framework consists of the following key components:

- An augmentation function $T(\cdot)$ that transforms a sampled centric node into a correlated view. We propose a node connectivity based augmentation mechanism to sample nodes that are highly correlated to the centric nodes to form a subgraph as its augmented view (Sect. 3.1).
- A GNN encoder g_θ that encodes the subgraphs rather than the whole graph like all the existing meta-learning methods do. In such a manner, our model will consume much less time to converge (Sect. 3.2).
- A contrastive mechanism that enables the encoder g_θ to discriminates embeddings between differently augmented views by capturing structural patterns across base classes and extrapolating such knowledge onto unseen novel classes (Sect. 3.3).
- A linear classifier f_ψ fine-tuned by a few labeled nodes from novel classes and is tasked to predict labels for those unlabeled nodes (Sect. 3.4).

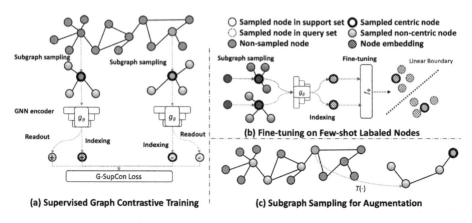

Fig. 2. Colors indicate different classes. Specially, grey nodes mean those nodes are not sampled. Different types of nodes indicate if nodes are from a support set or a query set, or sampled as centric nodes. Ombré nodes indicate the nodes are sampled non-centric nodes for the subgraphs. Crisscross nodes are node embeddings from the GNN encoder. (a) Supervised graph contrastive training framework. (b) Fine-tuning on few-shot labeled nodes from novel classes. (c) Node connectivity based subgraph sampling strategy samples nodes that are strongly connected to the centric nodes. (Color figure online)

3.1 Data Augmentation

Given a graph $\mathcal{G} = (\mathcal{V}, \mathcal{E}, \mathbf{X}) = (\mathbf{A}, \mathbf{X})$, following the conventions in contrastive learning methods [4,15], a transformation $T(\cdot)$ is used to generate a new view $\mathbf{x}'_j$ for a node j ($\forall j \in \{1, ..., M\}$, M is the number of nodes) with node attributes $\mathbf{X}$ and the adjacency matrix $\mathbf{A}$:

$$\mathbf{x}'_j = T(\mathbf{A}, \mathbf{X}, j) \tag{1}$$

There are multiple transformations for graph data, such as node masking or feature perturbation. However, such methods can introduce extra noise that impairs the learned node representations. In this paper, we propose a node connectivity based subgraph sampling strategy as the data augmentation mechanism. Connectivity score is a family of metrics that measures the connection strength between a pair of nodes in a graph without using the node attributes [3,18]. Notably, since only the adjacency matrix is needed for calculation, such augmentation can be pre-computed before training:

$$\mathbf{S} = Connect(\mathcal{V}, \mathcal{E}) = Connect(\mathbf{A}), \tag{2}$$

where each column $\mathbf{s}_j$ of $\mathbf{S}$ contains the scores between node j and all nodes in the graph. We set the score between a node and itself as a constant value $\gamma = 0.3$ for better performance. Intuitively, nodes that share more semantic similarities tend to have more correlations. However, such nodes may not be geologically close to each other. Node connectivity can capture the correlation between nodes by considering both the global and local graph structures [3,17]. As shown in Fig. 2(c), for any given node, we treat it as the centric node and sample other nodes that have the highest connectivity scores with the centric node to build a contextualized subgraph as its augmented view. Now the transformation for data augmentation can be defined as:

$$
\begin{aligned}
\mathbf{x}'_j &= T(\mathbf{A}, \mathbf{X}, j) \\
&= \mathbf{X}[top\_rank(\mathbf{S}[j,:], \alpha)] \\
&= \mathbf{X}[top\_rank(Connect(\mathbf{A})[j,:], \alpha)] \\
&= (\mathbf{A}'_j, \mathbf{X}'_j) = \mathcal{G}_s(j)
\end{aligned}
\tag{3}
$$

where top_rank is a function that returns the indices of the top α values, α is a hyperparameter that controls the augmented subgraph size $(\alpha + 1)$, and $\mathcal{G}_s(j)$ is the augmented subgraph for node j with adjcency matrix $\mathbf{A}'_j$ and sampled node attributes $\mathbf{X}'_j$. In this paper, we present two methods to calculate the node connectivity scores: Node Algebraic Distance (NAD) [3] and Personalized PageRank (PPR) [17].

Node Algebraic Distance (NAD). Following [3], we first randomly assign a random value u_i $(0 < u_i < 1)$ to any node i in the graph to form a vector $\mathbf{u} \in \mathbb{R}^M$. Then, we iteratively updates the value of a node by aggregating its neighboring weighted values: At the t-th iteration, for node i we have:

$$
\hat{u}_i^t = \sum_j \mathbf{A}(i,j) u_j^{t-1} / \sum_j \mathbf{A}(i,j),
\tag{4}
$$

$$
\mathbf{u}^t = (1 - \eta)\mathbf{u}^{t-1} + \eta \hat{\mathbf{u}}^t,
\tag{5}
$$

where η is a parameter set to 0.5. After a few iterations, the difference between the values of node i and j indicates the coupling between them. The smaller difference stands for a stronger connection. The final score matrix is:

$$
\mathbf{S} = \{\mathbf{S}(i,j)\}_{i,j=1}^M = \frac{1}{|u_i - u_j| + \epsilon}
\tag{6}
$$

where ϵ is a parameter set to 0.01. We column-normalize $\mathbf{S}$ and then set the value of $\mathbf{S}(i,i), (\forall i \in M)$ to γ.

Personalized PageRank (PPR). For PPR we have:

$$
\mathbf{S} = \phi \cdot (\mathbf{I} - (1 - \phi) \cdot \mathbf{AD}^{-1})
\tag{7}
$$

where ϕ is a parameter usually set as 0.15, and $\mathbf{D}$ is a diagonal matrix with: $\mathbf{D}(i,i) = \sum_j \mathbf{A}(i,j)$. We column-normalize the scores $\mathbf{S}$ to make the scores on the diagonal equal to $\gamma = 0.3$ as in NAD.

3.2   Subgraph Encoder

With the pre-computed connectivity score matrix, for each node j, we sample the top α nodes with the highest connectivity scores to construct a subgraph $\mathcal{G}_s(j) = (\mathbf{A}'_j, \mathbf{X}'_j)$. The larger α is, the richer context is given in the subgraph (We set $\alpha = 19$ for memory limitation). Then, as shown in Fig. 2(a), we feed the resulting subgraph into a GNN encoder g_θ. In particular, as the size of the subgraphs is a fixed small number (20 in our case) as opposed to the original graph, compared to the existing methods where the encoder encodes the whole graph, our method reduces the dimension for neighborhood aggregation to a much smaller magnitude and makes it highly scalable and faster to converge:

$$\mathbf{Z}'_j = g_\theta(\mathbf{x}'_j) = g_\theta(\mathbf{A}'_j, \mathbf{X}'_j), \tag{8}$$

where $\mathbf{Z}'_j \in \mathbb{R}^{(\alpha+1) \times F}$ and F is the embedding size. We normalize the latent features for better performance. To enable the contrastive learning across different scales (e.g. nodes and subgraphs), we propose a readout function $R(\cdot)$ which maps the representation of all nodes in the subgraph, $\mathbf{Z}'_j$, to a vector $\mathbf{z}'_j$ as the summarized subgraph representation. Intuitively, $\mathbf{z}'_j$ is a weighted average of the representations of all nodes in the subgraph. The weights are their normalized connectivity scores $\hat{\mathbf{s}}_j \in \mathbb{R}^\alpha$ to node j. We set the connectivity score of a node j to itself as a constant value γ (as defined in Sect. 3.1) before normalization:

$$\mathbf{z}'_j = R(\mathbf{Z}', \hat{\mathbf{s}}_j) = \sigma(\hat{\mathbf{s}}_j^T \cdot \mathbf{Z}') \tag{9}$$

where σ is the sigmoid function, and $\mathbf{z}'_j \in \mathbb{R}^F$ is the subgraph representation. The centric node embedding $\mathbf{z}_j \in \mathbb{R}^F$ can be directly indexed from $\mathbf{Z}'_j$.

3.3   Multi-scale Graph Contrastive Learning with Augmented Views

Since the augmented subgraphs contain topological information crucial for learning expressive encoder, we propose to use multi-scale contrastive learning that combines three categories of contrastive pairs to enforce the model to learn from both individual attributes and contextualized knowledge from sampled views: the contrast between (1) nodes and nodes ($\mathbf{z}_i$ & $\mathbf{z}_j$), (2) subgraphs and subgraphs ($\mathbf{z}'_i$ & $\mathbf{z}'_j$), and (3) nodes and subgraphs ($\mathbf{z}_i$ & $\mathbf{z}'_j$), where i, j are arbitrary node indices. For any node, the nodes in the same class together with their corresponding subgraphs are viewed as positives, and all the rest are viewed as negatives. Now, given the batch size B, we define a *duo-viewed* batch, consisting of the representations of both nodes and their corresponding augmented subgraphs:

$$\{(\mathbf{h}_b, y_b)\}_b^{2B} = \{(\mathbf{z}'_b, y_b), (\mathbf{z}_b, y_b)\}_b^B. \tag{10}$$

Loss Function. To better suit the setting of few-shot node classification, we adapt the Supervised Contrastive loss (SupCon) from [19] to pretrain the GNN encoders. We term the proposed loss G-SupCon for convenience. Compared to unsupervised contrastive loss (e.g. Deep InfoMax (DIM) [1], SimCLR [4], Margin Loss [18] etc.), and Cross-Entropy Loss (CEL), G-SupCon utilizes the ground-truth label to sample mini-batches to help to better align the representation of nodes in the same class more closely and push nodes from different classes further apart. Such learning patterns can be easier to transfer to unseen novel classes to generate highly discriminative representations. To ensure the balance in training data, we sample $B/|C_{tr}|$ nodes per class as centric nodes in each mini-batch for training, where $|C_{tr}|$ is the number of classes for pretraining (i.e., base classes). We term it Balanced Sampling (BS) for convenience. Then, the loss function is defined as:

$$\mathcal{L} = \sum_{b \in B} \frac{-1}{|P(b)|} \sum_{p \in P(b)} \log \frac{\exp\left(\mathbf{h}_b \cdot \mathbf{h}_p / \tau\right)}{\sum_{a \in A(b)} \exp\left(\mathbf{h}_b \cdot \mathbf{h}_a / \tau\right)}, \tag{11}$$

where $A(b)$ is the set of indices from 1 to $2B$ excluding b, and $P(b)$ is the set of indices of all positives in a duo-viewed batch excluding b. This contrastive loss includes all the three categories of contrastive pairs mentioned before. Additionally, $\tau \in \mathbb{R}^+$ is a scalar temperature parameter defined as $\tau = \beta/\sqrt{degree(\mathcal{G})}$, where $degree(\mathcal{G})$ is the average degree of the graph, and β is a hyperparameter. The temperature parameter controls the sensitivity of the trained model to the hard negative samples. As nodes in different classes also share some correlation with the centric nodes, especially for more complex graphs with higher average degrees, we do not want them to be fully separated apart in the representation space. So, we add the degree as a penalty for the temperature parameter to guide the separateness.

Under this fully-supervised setting, we pretrain our GNN encoder with all the base node labels and fix it during fine-tuning.

3.4   Linear Classifier Fine-Tuning

As indicated in Fig. 2(b), with a pretrained GNN encoder, when fine-tuning on a target few-shot node classification dataset $\mathcal{D}^i$ on novel classes, we tune a separate linear classifier f_ψ, (e.g. logistic regression, SVM, a linear layer, etc.) with the few labeled nodes in the support set $\mathcal{S}^i$, and task it to predict the labels for nodes in the query set $\mathcal{Q}^i$. The representations of nodes in $\mathcal{D}^i$ are obtained through the same procedure: treat each node as centric node to retrieve a contextualized subgraph, and then feed the sampled subgraphs to the pretrained GNN encoder, and the centric node embedding can be directly indexed and used to fine-tune the classifier f_ψ by optimizing a naive Cross-Entropy Loss.

4   Experiments

In this section, we design experiments to evaluate the proposed framework by comparing with three categories of methods for few-shot node classification: (1) naive supervised pretraining using GNNs, (2) state-of-the-art meta-learning based methods, and (3) by going one step further to compare with contrastive learning methods. For the third category, to our best knowledge, we are the first to implement state-of-the-art self-supervised graph contrastive methods to few-shot node classification and compare them with the proposed graph supervised contrastive learning.

4.1 Experimental Settings

Table 1. Statistics of the commonly used datasets

	# Nodes	# Edges	# Features	# Labels	Base	Novel
CoraFull	19,793	126,842	8,710	70	42	28
Reddit	232,965	11,606,919	602	41	24	17
Ogbn-arxiv	169,343	1,166,243	128	40	24	16

Evaluation Datasets. We conduct our experiments on three widely used graph few-shot learning benchmark datasets where a sufficient number of node classes are available for sampling few-shot node classification tasks: CoraFull [2], Reddit [13], and Ogbn-arxiv [35]. Their statistics are given in Table 1.

Baseline Methods. In this work, we compare our framework with the following 3 categories of methods.

- Naive supervised pretraining. We use GCN [20] as a naive encoder and pretrain it with all nodes from the base classes, and following convention, we fine-tune a single linear layer as the classifier for each few-shot node classification task. Also, we implement an initialization strategy, TFT [5] for the classifier by setting its weight as a matrix consisting of concatenated prototype vectors of novel classes.
- State-of-the-art meta-learning methods for few-shot node classification on a single graph: MAML based Meta-GNN [41], Matching Network [31] based AMM-GNN [33], and Prototypical Network based GPN [7]. We do not include methods like [34,36] in the baselines because they require extra auxiliary graph data, nor methods like [21,23] because they have similar performance with the chosen baselines according to their original papers.
- State-of-the-art self-supervised pretraining methods on a single graph: MVGRL [14], SUBG-Con [18], GraphCL [38], and GCA [42]. These methods pretrain a GNN encoder with nodes from base classes without using the labels. The classifiers are then fine-tuned on novel support nodes and their accuracy is reported on predicting labels for novel query nodes.

Evaluation Protocol: To make fair comparison, all the scores reported are accuracy values averaged over 10 random seeds and all the baselines share the same splits of base classes and novel classes as shown in Table 1.

Implementation Details: In Table 1, the specific data splits are listed for each dataset. For a fair comparison, we adopt the same encoder for all compared methods. Specifically, the graph encoder g_θ consists of one GCN layer [20] with PReLU activation. The effect of the encoder architecture are further explored in Sect. 4.4. We choose logistic regression as the linear classifier for fine-tuning. The encoder is trained with Adam optimizers whose learning rates are set to be 0.001 initially with a weight decay of 0.0005. And the coefficients for computing running averages of gradient and square are set to be $\beta_1 = 0.9$, $\beta_2 = 0.999$. The default values of batch size B and graph temperature parameter β are set to 500 and 1.0, respectively. For the baseline methods, we use the default parameters provided in their implementations.

4.2   Overall Evaluation

We present the comparative results between our framework and three categories of baseline methods described earlier. It is worth mentioning that, this work is the first to investigate the necessity of episodic meta-learning for Few-shot Node Classification (FNC) problems. For a fair comparison, all methods share the same GCN encoder architecture as the proposed framework. Also, when experimented on each dataset, they share the same random seeds for data split, leading to identical evaluation data. The results are shown in Table 2. We summarize our findings next.

Necessity of employing episodic meta-learning style methods for graph few-shot learning. First and foremost, we find that almost all the contrastive pre-training based methods outperform the existing meta-learning based FNC algorithms. Even the most straightforward one, TFT, which only leverages a simple initialization to the separate classifier, can produce comparable scores to the best meta-learning based FNC method, GPN. We have shown that through appropriate pretraining, including self-supervised and supervised training, adding a simple linear classifier can outperform the existing meta-learning based framework by a significant margin.

Effectiveness of our framework to learn discriminative node embeddings for FNC problems. Compared with other existing meta-learning based methods, our framework outperforms them by a large margin under different settings. We attribute this to the following facets: (1) The effectiveness of the proposed node-connectivity-based sampling strategy that can provide highly correlated context-specific information for the centric node by considering both global and local information. Besides, NAD and PPR can provide similar outcomes. NAD can perform better on graphs with a higher average degree; and (2) The supervised contrastive learning loss function G-SupCon can utilize label information to further enforce the GNN encoder to generate more discriminative representations by minimizing the distances among nodes from the same classes while segregating nodes from different classes in the representation space.

Robustness to various N-way K-shot settings. Similar to the meta-learning based FNC methods, the performance of contrastive pretraining based methods also degrades when the K decreases or N increases. However, from the results

Table 2. Comparative results: three datasets under different N-way K-shot settings

Methods	Loss	CoraFull (%)			Reddit (%)			Ogbn-arxiv (%)		
		10-way 5-shot	5-way 5-shot	3-way 1-shot	10-way 5-shot	5-way 5-shot	3-way 1-shot	10-way 5-shot	5-way 5-shot	3-way 1-shot
GCN	CEL	37.25	45.68	43.23	36.28	44.34	39.62	30.83	38.40	35.41
TFT	CEL	63.50	70.18	66.41	59.75	69.80	58.32	47.68	62.25	60.48
Meta-GNN	CEL	55.23	66.25	60.25	48.62	65.50	55.78	41.20	58.67	55.68
AMM-GNN	CEL	60.80	70.52	65.27	53.28	67.20	55.84	44.33	61.02	58.64
GPN	CEL	62.02	73.40	67.07	59.20	69.31	60.20	50.58	64.12	62.20
GraphCL	SimCLR	79.68	87.35	84.80	82.51	87.34	84.16	57.30	67.34	62.27
MVGRL	DIM	81.34	88.40	85.03	84.78	89.65	86.29	57.82	68.24	63.36
SUBG-Con	Margin Loss	80.71	88.02	86.26	83.70	89.55	85.57	56.29	66.36	62.10
GCA	SimCLR	82.43	90.52	87.89	85.61	90.96	87.26	59.03	69.53	65.49
Ours (NAD)	G-SupCon	**86.13**	**93.65**	89.42	88.52	93.36	91.70	**62.24**	73.25	71.63
Ours (PPR)	G-SupCon	85.34	93.62	**90.56**	**89.10**	**94.12**	**92.31**	61.85	**73.65**	**72.90**

shown in Table 2, we find that those methods are more robust to settings with decreasing K and increasing N. This means that the encoder can better extrapolate to novel classes by generating more discriminative node representations. To a large extent, the degradation lies in the less accurate classifier due to fewer training nodes. Learning to better measure the classifier under scenarios with extremely scarce support nodes (very small K) is also worth further research.

4.3   Further Experiments

To further evaluate our framework, we conduct more experiments next. The default setting for the following experiments is 5-way 5-shot, where we set $\beta = 1.0$, $B = 500$, and use a single GCN layer as the encoder.

Efficiency Study. As discussed in Sect. 3.2, our method is scalable and has much less convergence time because we construct a GNN encoder for the sampled subgraph rather than the whole graph. Also, Sect. 4.2 shows that performance is not sacrificed for achieving scalability. On the contrary, due to the effective sampling strategy, only fine-grained data are fed into the encoder, resulting in a considerable boost in accuracy. We show the actual time consumption for a single run of training of our model and typical meta-learning and pretraining baselines. The experiment is conducted on a single RTX 3090 GPU. In Table 3, we show that our method can achieve excellent performance and consume much less training time. Note that here we only consider the training time, excluding the time for computing the node connectivity scores **S** which can be pre-calculated.

Table 3. Training time of GPN, GCA, and Ours (proposed method)

Methods	CoraFull	Reddit	Ogbn-arxiv
GPN	1352 s	3248 s	2562 s
GCA	584 s	3651 s	3237 s
Ours	**10 s**	**21 s**	**54 s**

Representation Clustering. This experiment is designed to demonstrate the high-quality representation generated by the encoder in the proposed framework. We show the clustering results of the representation of the query nodes in 5 randomly sampled novel classes in Table 4 and visualize the embedding in Fig. 3 (without fine-tuning on support set). It can be observed that an encoder trained by the proposed pretraining strategy possesses a stronger extrapolation ability to generate highly discriminative boundaries for unseen novel classes.

Table 4. Performance on novel classes query node embedding clustering, reported in Normalized Mutual Information (**NMI**) and Adjusted Rand Index (**ARI**)

Methods	CoraFull		Reddit		Ogbn-arxiv	
	NMI	ARI	NMI	ARI	NMI	ARI
GPN	0.5134	0.4327	0.3690	0.3115	0.2905	0.2235
GCA	0.7531	0.7351	0.7824	0.7756	0.3786	0.3219
Ours	**0.8567**	**0.8229**	**0.8890**	**0.8720**	**0.5280**	**0.4485**

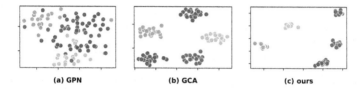

Fig. 3. t-SNE embedding visualization on CoraFull data: (a) GPN (b) GCA (C) Ours

4.4   Ablation and Parameter Analysis

In this section we study how sensitive the proposed framework is to the design choice of its components. In particular, we consider the architecture of encoder g_θ, graph temperature parameter β, and batch size B.

Analysis on Encoder Architecture. Our framework is independent of the encoder architecture. We evaluate our framework with three widely-used GNNs: GCN [20], GAT [30], and GIN [35], as shown in Table 5. The difference between encoder architectures is insignificant, so we choose GCN as the default encoder.

Analysis on Loss Function. To better demonstrate the effectiveness of our G-SubCon loss, we design an experiment to compare different loss functions. The SimCLR loss does not consider the label information, so the Balance Sampling (BS) strategy we proposed in Sect. 3.3 is not feasible for it. But in order to explicitly show the influence from the loss function, we also show the result of SimCLR with the same sampled data splits from BS as our G-SubCon. We list the results from all the candidate loss functions under both settings in Table 6. It can be observable that all the losses suffer from the class imbalance issue, and the simple BS scheme can improve performance. Also, our proposed G-SupCon outperforms all others even with identical data splits. In short, both the BS scheme and the G-SupCon loss function are effective in terms of accuracy on few-shot node classification tasks.

Analysis on Graph Temperature Parameter and Batch Size. As presented in Fig. 4(a), we test the sensitivity of our framework regarding the graph temperature parameter β and batch size B. Observably, our framework is not that sensitive to these two hyperparameters. The best value of β is 1.0 and we

Table 5. Results of our model with different encoders

Encoder	CoraFull (%)	Reddit (%)	Ogbn-arxiv (%)
GCN	**93.62**	**94.12**	73.65
GAT	92.05	94.04	73.46
GIN	93.25	93.86	**74.10**

set it as default. Generally speaking, the larger batch size can produce higher scores because more contrast can be made in each batch. We choose 500 as the default batch size for computational efficiency and its decent performance.

Table 6. Results of our model trained with different loss functions

Loss function	BS	CoraFull (%)	Reddit (%)	Ogbn-arxiv (%)
Cross Entropy	No	70.18	69.80	62.25
Cross Entropy	Yes	73.86	74.08	63.67
SimCLR	No	87.54	89.60	66.34
SimCLR	Yes	89.48	92.88	69.98
G-SupCon	**Yes**	**93.62**	**94.12**	**73.65**

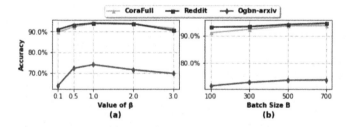

Fig. 4. (a) Accuracy vs Graph temperature parameter β (b) Accuracy vs Batch size B on the three datasets

5   Related Work

Few-shot Node Classification. Graph Neural Networks (GNNs) [20,30,35] are a family of deep neural network models for graph-structured data, which exploits recurrent neighborhood aggregation to preserve the graph structure information and transform the node attributes simultaneously. Recently, increasing attention has been paid to few-shot node classification problems, episodic meta-learning [10] has become the most dominant paradigm. It trains the GNN encoders by explicitly emulating the test environment for few-shot learning [10], where the encoders are expected to gain the adaptability to extrapolate onto new domains. Meta-GNN [41] applies MAML [10] to learn directions for optimization with limited labels. AMM-GNN [33] deploys Matching Network [31] to learn transferable metric among different meta-tasks. GPN [7] adopts Prototypical Networks [25] to make the classification based on the distance between the node feature and the prototypes. MetaTNE [21] and RALE [23] also use episodic meta-learning to enhance the adaptability of the learned GNN encoder and achieve similar results. Furthermore, HAG-Meta [27] extends the problem to incremental learning setting. Recently, people in the image domain argue that the reason for the fast adaptation in the existing works lies in feature reuse rather than those complicated mate-learning algorithms [5,28]. In other words, with a carefully pretrained encoder, decent performance can be obtained through direct fine-tuning a simple classifier on the target domain. Since then, various pretraining strategies [22,28] have been put forward to tackle the few-shot image classification problem. However, no research has been done in the graph domain with its crucial distinction from images that nodes in a graph are not i.i.d. data. Their interactive relationships are reflected by both the topological and semantic information. Our work here is the first attempt to bridge the gap by developing a novel graph supervised contrastive learning for few-shot node classification.

Graph Contrastive Learning. Contrastive learning has become popular representation learning paradigm in image [4], text [32], and graph [14,42] domains. Starting from the self-supervised setting, contrastive learning methods are proved to learn discriminative representation by contrasting a predefined distance between positive and negative samples. Usually, those samples are augmented through some heuristic transformations from original data. Specifically, in the graph domain, the transformations can be categorized into the following types: (1) graph structure based augmentation, e.g., randomly drop edges or nodes [14,42], randomly sample subgraphs [18]. (2) graph feature based augmentation, e.g., randomly mask or perturb attributes of nodes or edges [29]. [42] further improves those augmentations by adding masks according to feature importance. Some works [26,37,39] try to explore different ways to automatically generate augmented views. Recently, for image [19] and text [11] domains, people notice that by injecting label information to the contrastive loss to compact or enlarge the distances of augmentations of instances within the same or different classes, supervised contrastive loss outperforms the original unsupervised version and even cross-entropy loss in the setting of transfer learning, by providing highly discriminative representation learned from texts or images. However, no existing work has focused on its extrapolation ability for graphs, especially, under an extremer few-shot situation.

6   Conclusion, Limitations and Outlook

In this work, we question the fundamental question whether episodic meta-learning is necessary for few-shot node classification. To answer the question, we propose a graph supervised contrastive learning tailored for the few-shot node classification problem and demonstrate its superb adaptability to extrapolate onto novel classes by fine-tuning a simple linear classifier. Through extensive experiments on benchmark datasets, we demonstrate that our framework can surpass episodic meta-learning methods for few-shot node classification in terms of both accuracy and efficiency. Therefore, this work offers the answer: episodic meta-learning is not a must for few-shot node classification.

Due to limited space, limitations of our work need to be acknowledged. (1) *Limited strategy consideration.* To pretrain the GNN encoder on base classes, we only consider naive supervised training and graph contrastive training. There are many other pretraining strategy that worth further investigation (e.g. [16]). (2) *Lack of theoretical justification.* Our work mainly presents empirical studies which may throw up many questions in need of further theoretical justification, for instance, to what magnitude contrastive pretraining surpasses meta-learning and the reason behind it.

We hope our work will shed new light on few-shot node classification tasks. There are many promising directions worth further research. For example, when base classes also have very limited labeled nodes or even no label at all, from Table 4.2, we can see that self-supervised pretraining can also help improve the performance. So it would be interesting to investigate the pretraining strategy under semi-supervised or unsupervised settings. In addition, since the pretraining phase may involve extra noise through data sampling or augmentation, methods to calibrate the learned embedding or refine the obtained prototypes are also potential directions for research.

Acknowledgments. This work is partially supported by Army Research Office (ARO) W911NF2110030 and Army Research Lab (ARL) W911NF2020124.

References

1. Bachman, P., Hjelm, R.D., Buchwalter, W.: Learning representations by maximizing mutual information across views. arXiv preprint arXiv:1906.00910 (2019)
2. Bojchevski, A., Günnemann, S.: Deep Gaussian embedding of graphs: unsupervised inductive learning via ranking. arXiv preprint arXiv:1707.03815 (2017)
3. Chen, J., Safro, I.: A measure of the connection strengths between graph vertices with applications. arXiv preprint arXiv:0909.4275 (2009)
4. Chen, T., Kornblith, S., Norouzi, M., Hinton, G.: A simple framework for contrastive learning of visual representations. In: ICML (2020)
5. Dhillon, G.S., Chaudhari, P., Ravichandran, A., Soatto, S.: A baseline for few-shot image classification. In: ICLR (2019)
6. Ding, K., Wang, J., Caverlee, J., Liu, H.: Meta propagation networks for graph few-shot semi-supervised learning. In: AAAI (2022)

7. Ding, K., Wang, J., Li, J., Shu, K., Liu, C., Liu, H.: Graph prototypical networks for few-shot learning on attributed networks. In: CIKM (2020)
8. Ding, K., Xu, Z., Tong, H., Liu, H.: Data augmentation for deep graph learning: a survey. arXiv preprint arXiv:2202.08235 (2022)
9. Ding, K., Zhou, Q., Tong, H., Liu, H.: Few-shot network anomaly detection via cross-network meta-learning. In: TheWebConf (2021)
10. Finn, C., Abbeel, P., Levine, S.: Model-agnostic meta-learning for fast adaptation of deep networks. In: ICML (2017)
11. Gunel, B., Du, J., Conneau, A., Stoyanov, V.: Supervised contrastive learning for pre-trained language model fine-tuning. In: ICLR (2020)
12. Guo, Z., et al.: Few-shot graph learning for molecular property prediction. In: WWW (2021)
13. Hamilton, W.L., Ying, R., Leskovec, J.: Inductive representation learning on large graphs. In: NeurIPS (2017)
14. Hassani, K., Khasahmadi, A.H.: Contrastive multi-view representation learning on graphs. In: ICML (2020)
15. He, K., Fan, H., Wu, Y., Xie, S., Girshick, R.: Momentum contrast for unsupervised visual representation learning. In: CVPR (2020)
16. Hu, W., et al.: Strategies for pre-training graph neural networks. In: ICLR (2020)
17. Jeh, G., Widom, J.: Scaling personalized web search. In: WWW (2003)
18. Jiao, Y., Xiong, Y., Zhang, J., Zhang, Y., Zhang, T., Zhu, Y.: Sub-graph contrast for scalable self-supervised graph representation learning. In: ICDM (2020)
19. Khosla, P., et al.: Supervised contrastive learning. In: NeurIPS (2020)
20. Kipf, T.N., Welling, M.: Semi-supervised classification with graph convolutional networks. arXiv preprint arXiv:1609.02907 (2016)
21. Lan, L., Wang, P., Du, X., Song, K., Tao, J., Guan, X.: Node classification on graphs with few-shot novel labels via meta transformed network embedding. In: NeurIPS (2020)
22. Liu, C., et al.: Learning a few-shot embedding model with contrastive learning. In: AAAI (2021)
23. Liu, Z., Fang, Y., Liu, C., Hoi, S.C.: Relative and absolute location embedding for few-shot node classification on graph. In: AAAI (2021)
24. Park, N., Kan, A., Dong, X.L., Zhao, T., Faloutsos, C.: Estimating node importance in knowledge graphs using graph neural networks. In: KDD (2019)
25. Snell, J., Swersky, K., Zemel, R.S.: Prototypical networks for few-shot learning (2017)
26. Suresh, S., Li, P., Hao, C., Neville, J.: Adversarial graph augmentation to improve graph contrastive learning. In: NeurIPS (2021)
27. Tan, Z., Ding, K., Guo, R., Liu, H.: Graph few-shot class-incremental learning. In: WSDM (2022)
28. Tian, Y., Wang, Y., Krishnan, D., Tenenbaum, J.B., Isola, P.: Rethinking few-shot image classification: a good embedding is all you need? In: Vedaldi, A., Bischof, H., Brox, T., Frahm, J.-M. (eds.) ECCV 2020. LNCS, vol. 12359, pp. 266–282. Springer, Cham (2020). https://doi.org/10.1007/978-3-030-58568-6_16
29. Tong, Z., Liang, Y., Ding, H., Dai, Y., Li, X., Wang, C.: Directed graph contrastive learning. In: NeurIPS (2021)
30. Veličković, P., Cucurull, G., Casanova, A., Romero, A., Lio, P., Bengio, Y.: Graph attention networks. arXiv preprint arXiv:1710.10903 (2017)
31. Vinyals, O., Blundell, C., Lillicrap, T., Wierstra, D., et al.: Matching networks for one shot learning. In: NeurIPS (2016)

32. Wang, D., Ding, N., Li, P., Zheng, H.: Cline: contrastive learning with semantic negative examples for natural language understanding. In: ACL (2021)
33. Wang, N., Luo, M., Ding, K., Zhang, L., Li, J., Zheng, Q.: Graph few-shot learning with attribute matching. In: CIKM (2020)
34. Wen, Z., Fang, Y., Liu, Z.: Meta-inductive node classification across graphs. In: SIGIR (2021)
35. Xu, K., Hu, W., Leskovec, J., Jegelka, S.: How powerful are graph neural networks? arXiv preprint arXiv:1810.00826 (2018)
36. Yao, H., et al.: Graph few-shot learning via knowledge transfer. In: AAAI (2020)
37. You, Y., Chen, T., Shen, Y., Wang, Z.: Graph contrastive learning automated. In: ICML (2021)
38. You, Y., Chen, T., Sui, Y., Chen, T., Wang, Z., Shen, Y.: Graph contrastive learning with augmentations. In: NeurIPS (2020)
39. You, Y., Chen, T., Wang, Z., Shen, Y.: Bringing your own view: graph contrastive learning without prefabricated data augmentations. In: WSDM (2022)
40. Zhang, C., et al.: Few-shot learning on graphs: a survey. In: IJCAI (2022)
41. Zhou, F., Cao, C., Zhang, K., Trajcevski, G., Zhong, T., Geng, J.: Meta-GNN: on few-shot node classification in graph meta-learning. In: CIKM (2019)
42. Zhu, Y., Xu, Y., Yu, F., Liu, Q., Wu, S., Wang, L.: Graph contrastive learning with adaptive augmentation. In: WWW (2021)

A Piece-Wise Polynomial Filtering Approach for Graph Neural Networks

Vijay Lingam$^{(\boxtimes)}$, Manan Sharma, Chanakya Ekbote, Rahul Ragesh,
Arun Iyer, and Sundararajan Sellamanickam

Microsoft Research India, Bengaluru, India
vijaylingam0810@gmail.com,
{t-cekbote,rahulragesh,ariy,ssrajan}@microsoft.com

Abstract. Graph Neural Networks (GNNs) exploit signals from node features and the input graph topology to improve node classification task performance. Recently proposed GNNs work across a variety of homophilic and heterophilic graphs. Among these, models relying on polynomial graph filters have shown promise. We observe that polynomial filter models need to learn a reasonably high degree polynomials without facing any oversmoothing effects. We find that existing methods, due to their designs, either have limited efficacy or can be enhanced further. We present a spectral method to learn a bank of filters using a piece-wise polynomial approach, where each filter acts on a different subsets of the eigen spectrum. The approach requires eigendecomposition only for a few eigenvalues at extremes (i.e., low and high ends of the spectrum) and offers flexibility to learn sharper and complex shaped frequency responses with low-degree polynomials. We theoretically and empirically show that our proposed model learns a better filter, thereby improving classification accuracy. Our model achieves performance gains of up to ~6% over the state-of-the-art (SOTA) models while being only ~2x slower than the recent spectral approaches on graphs of sizes up to ~169K nodes.

Keywords: Graph Neural Networks · Representation learning · Polynomial filtering

1 Introduction

We are interested in the problem of classifying nodes in a graph where a graph with features for all nodes, and labels for a few nodes are made available for learning. Inference is done using the learned model for the remaining nodes (*aka* transductive setting). Graph Neural Networks (GNNs) perform well on such problems [1]. Most GNNs predict a node's label by aggregating information from its neighbours in a certain way, making them dependent on some correlation

---

V. Lingam, M. Sharma and C. Ekbote—Equal contribution. Work done while author was at Microsoft Research India.

M.-R. Amini et al. (Eds.): ECML PKDD 2022, LNAI 13714, pp. 412–452, 2023.
https://doi.org/10.1007/978-3-031-26390-3_25

between the structure and the node labels[1]. For example, in the simplest case, GNNs work well when the node and its neighbours share similar labels. However, the performance can be poor if this criterion is not satisfied. Recently, several modeling approaches have been proposed to build/learn robust GNN models. Some modify the aggregation mechanism [3–5], while others propose to estimate and leverage the label-label compatibility matrix as a prior [6].

More recent approaches have tackled this problem from a graph filter learning perspective [7,8,32,36–38]. With eigenvalues having frequency interpretations [26], one or more filters (i.e., a bank of filters) that selectively accentuates and suppresses various spectral components of graph signals are learned using task-specific available information. The filtering operation enables learning better node representation which translates to improved classification accuracy.

Designing effective graph filters is a challenging problem, and most recent methods [8,10,36,37] suggest interesting ways to learn polynomial filters having finite impulse response (FIR) characteristics. These models are efficient and attractive, as they make use of local neighborhood (i.e., using sparse adjacency matrix repeatedly) and do not require to pre-compute eigendecomposition, which is expensive (when done over the entire spectrum, i.e., for all eigenpairs). Though these models are able to learn better filters and give good performance gains, they are still unable to learn richer and complex frequency responses, which require higher-order polynomials. One key reason for their inability to learn effective high-order polynomials is that they only *mitigate* the over-smoothing problem. This aspect of the problem becomes clear when we analyze a general class of FIR filters (GFIR) and find that the over-smoothing problem exists for the whole class, of which simplified GCN [16], GPR-GNN [8] and several other models are special cases. We also find that while constraining the model space of GFIR (e.g., [8]) helps to mitigate over-smoothing, it is still unable to learn complex-shaped and sharper frequency responses. Considering this background, our interest lies in learning a bank of effective filters in spectral domain to model complex shaped frequency responses, as needed for graphs with diverse label correlations. Our contributions are:

1. We propose a novel piece-wise polynomial filtering approach to learn a filter bank tuned for the task at hand. Since full eigendecomposition is expensive, we present an efficient method that makes use of only a few extremal eigenpairs and leverages GPR-GNN to learn multiple filters. (While computing the extremal eigenpairs does lead to an increased computational cost, we show in A.9 that such a cost is indeed managable, i.e. the model is only ∼2x slower than recent spectral SOTA methods.)

2. We analyze, theoretically and experimentally, the shortcomings of a general class of FIR (GFIR) filters. We show that the proposed piece-wise polynomial

---

[1] Characterizing the correlation between the graph structure and node features/labels is an active area of research. Several metrics have been proposed including edge homophily [5,13], node homophily [4], class homophily [27]. All these metrics show that standard GNNs perform well when the graphs and node labels are positively correlated.

GNN (PP-GNN) solution is more expressive and is capable of modeling richer and complex frequency responses.
3. We conduct a comprehensive experimental study to compare PP-GNN with a wide range of methods ($\sim$20), covering both spatial and spectral convolution based methods on nearly a dozen datasets. Experimental results show that PP-GNN performs significantly better, achieving up to $\sim$6% gains on several datasets.

2 Related Work

Graph Neural Networks (GNNs) have become increasingly popular models for semi-supervised classification with graphs. [11] set the stage for early GNN models, which was then followed by various modifications [1,2,9,12] and improvements along with several different directions such as improved aggregation and attention mechanisms [2,3,9], efficient implementation of spectral convolution [12,16], incorporating random walk information [13–15], addressing over-smoothing [10,13–15,28–30], etc.

Another line of research explored the question of where GNNs help. The key understanding is that the performance of GNN is dependent on the correlation of the graphs with the node labels. Several approaches [5,13,31] considered edge homophily and proposed a robust GNN model by aggregating information from several higher-order hops. [3] also considered edge homophily and mitigated the issue by learning robust attention models. [4] talks about node homophily and proposes to aggregate information from neighbours in the graph and neighbours inferred from the latent space. [6] proposes to estimate label-label compatibility matrix and uses it as a prior to update posterior belief on the labels.

Recent approaches motivated by the developments in graph signal processing [25], focus on learning graph filters with filter functions that operate on the eigenvalues of the graph directly or indirectly, adapting the frequency response of graph filters for the desired task. [7] models the filter function as an attention mechanism on the edges, which learns the difference in the proportion of low-pass and high-pass frequency signals. [8] proposes a polynomial filter on the eigenvalues that directly adapts the graph for the desired task. [32] decompose the graph into low-pass and high-pass frequencies, and define a framelet based convolutional model. [37] propose to learn graph filters using Bernstein approximation of arbitrary filtering function. [36] suggest to learn adaptive graph filters for different feature channels and frequencies by stacking multiple layers. Our work is closely related to these lines of exploration. All these works still need high-degree polynomials when sharper frequency responses are needed; however, though improved performance is observed and over-smoothing is mitigated, further improvements seem possible. Another class of Infinite Impulse Response (IIR) filters have been proposed to learn complex filter responses. ARMA [38] achieves this by using autoregressive moving average, but empirically have been found to have limited effectiveness. Implementing precise ARMA filters for graphs is a challenging problem and has high computation costs. [38] proposes several approximations to mitigate the issues, but these come with limited efficacy. In our work, we propose to learn a filter function as a sum of polynomials over different subsets of the eigenvalues (in essence, a bank of filters) by operating directly in the spectral domain, enabling design of effective filters to model task-specific complex frequency responses with compute trade-offs.

3   Problem Setup and Motivation

We focus on the problem of semi-supervised node classification on a simple graph $\mathcal{G} = (\mathcal{V}, \mathcal{E})$, where $\mathcal{V}$ is the set of vertices and $\mathcal{E}$ is the set of edges. Let $\mathbf{A} \in \{0,1\}^{n \times n}$ be the adjacency matrix associated with $\mathcal{G}$, where $n = |\mathcal{V}|$ is the number of nodes. Let $\mathcal{Y}$ be the set of all possible class labels. Let $\mathbf{X} \in \mathbb{R}^{n \times d}$ be the d-dimensional feature matrix for all the nodes in the graph. Given a training set of nodes $\mathcal{D} \subset \mathcal{V}$ whose labels are known, along with $\mathbf{A}$ and $\mathbf{X}$, our goal is to predict the labels of the remaining nodes. Let $\mathbf{A_I} = \mathbf{A} + \mathbf{I}$ where $\mathbf{I}$ is the identity matrix. Let $\mathbf{D_{A_I}}$ be the degree matrix of $\mathbf{A_I}$ and $\widetilde{\mathbf{A}} = \mathbf{D_{A_I}^{-1/2}} \mathbf{A_I} \mathbf{D_{A_I}^{-1/2}}$. Let $\widetilde{\mathbf{A}} = \mathbf{U \Lambda U}^T$ be the eigendecomposition. The spectral convolution of $\mathbf{X}$ on the graph $\mathbf{A}$ can be defined via the reference operator $\widetilde{\mathbf{A}}$ and a general Finite Impulse Response (FIR) filter [39], parameterized by $\mathbf{\Theta}$ as:

$$\mathbf{Z} = \sum_{j=1}^{k} \widetilde{\mathbf{A}}^j \mathbf{X} \mathbf{\Theta}_j \tag{1}$$

The term, $\widetilde{\mathbf{A}}^j \mathbf{X}$ uniformly converges to a stationary value as the value of j increases, making the node features indistinguishable (often referred to as the problem of over-smoothing), thereby reducing the importance of the corresponding term for the task at hand. We formalize the argument via commenting on the Dirichlet energy of the higher-order terms [40]. Dirichlet energy reveals the embedding smoothness with the weighted node pair distance. A smaller value is highly related to over-smoothing [41]. Under some conditions, the upper bound of Dirichlet energy of higher terms is theoretically proved to converge to 0 in the limit of infinite layers. In other words, all nodes converge to a trivial fixed point in the embedding space and hence do not contribute to the discriminative signals. This is formalized as follows:

Proposition 3.1: The upper bound of Dirichlet energy for the higher-order terms in the general FIR model exponentially decreases to 0 with the order, k. Formally, with $\mathbf{S}$ as any graph shift operator (in our case, the normalized adjacency), and $\mathbf{\Theta}_k$ be the set of parameters, indexed by k:

$$E(\mathbf{S}^k \mathbf{X} \mathbf{\Theta}_k) \leq (1 - \lambda)^{2k} s_{\mathbf{\Theta}_k} E(\mathbf{X})$$

where, λ is the positive eigenvalue of the graph Laplacian Δ that is closest to 0; $s_{\mathbf{\Theta}_k}$ is the largest singular value of $\mathbf{\Theta}_k$. We relegate the proof of the corollary as well as the formal definition of a few terms in Sect. A.3 of the supplementary material.

The family of general FIR filters is ubiquitous and gives rise to various other filter families (e.g. polynomial) simply by placing constraint on the form of parameterization. We experiment with placing simple constraints on the bare GFIR model in Sect. A.7 of supplementary and observe that while constraining helps improving the performance, it does not help in learning complex responses. It is not difficult to see that the models of [8,12], etc. are just instantiations of the GFIR family. Particularly, by restricting $\Theta_j = \alpha_j \mathbf{I}$, we recover the linear model (without MLP) of [8], which can now be interpreted as the polynomial filter function h operating on the eigenvalues, in the Fourier domain [8,25] as,

$$\mathbf{Z} = \sum_{j=1}^{k} \alpha_j \widetilde{\mathbf{A}}^j \mathbf{X} = \mathbf{U} h(\mathbf{\Lambda}) \mathbf{U}^T \mathbf{X} \tag{2}$$

with $h : \mathbb{R} \rightarrow \mathbb{R}$ is defined as $h(\lambda; \boldsymbol{\alpha}) = \sum_{i=1}^{k} \alpha_i \lambda^i$ where α_i's are coefficients of the polynomial, k is the order of the polynomial and λ is any eigenvalue from $\mathbf{\Lambda}$. $h()$ is applied element-wise across $\mathbf{\Lambda}$ in Eq. 2. In this process, the filter function is essentially adapting the graph for the desired task at hand.

It is well-known that polynomial filters can approximate any graph filter [25, 26]. Since polynomial filters are a class of the GFIR filter family, they inherit the same problem of over-smoothing as the order of the polynomial becomes higher. [8] show that they achieve the diminishing of the contribution of higher-order terms by showing that their coefficients converge to zero during training. While this mitigates the over-smoothing problem, use of lower-order polynomials results in an imprecise approximation when the dataset requires a complex spectral filter for obtaining a superior performance, which we will show is the case for certain datasets (See Fig. 1 and supplementary's A.6). Empirical results demonstrating the key points discussed in this section: a) smoothening of the higher-order terms (can be found in Fig. 5a of the supplementary material) and b) their effect on the test performance on a few datasets (can be found in Fig. 5b and 5c of supplementary material). These problems indicate the need for a method that can approximate arbitrarily complex filters better and at the same time mitigate the effects of over-smoothing.

4   Proposed Approach

We propose to learn a bank of polynomial filters with each filter operating on different parts of the spectrum, taking task-specific requirements into account. We show that our proposed filter design can approximate the latent optimal graph filter better than a single polynomial, and the resultant class of learnable filters is richer.

4.1   Piece-Wise Polynomial Filters

We start with the expression (2) for node embedding rewritten with an MLP network transforming input features, $\mathbf{X}$:

$$\mathbf{Z} = \sum_{i=1}^{n} h(\lambda_i) \mathbf{u}_i \mathbf{u}_i^T \mathbf{Z}_x(\mathbf{X}; \boldsymbol{\Theta}) \tag{3}$$

where $\mathbf{u}_i$ is the eigenvector corresponding to the eigenvalue, λ_i, and $\mathbf{Z}_x(\mathbf{X}; \boldsymbol{\Theta})$ is an MLP network with parameters $\boldsymbol{\Theta}$. Our goal is to learn a filtering function, $h(\lambda)$ jointly with MLP network, using which we compute the node embedding, $\mathbf{Z}$. We model $h(\lambda)$ as a piece-wise polynomial or spline function where each polynomial is of a lower degree (e.g., a cubic polynomial). We partition the spectrum in $[-1, 1]$ (or $[0, 2]$ as needed) into contiguous intervals and approximate the desired frequency response by fitting a low degree polynomial in each interval. This process helps us to learn a more complex shaped frequency response as needed for the task. Let $\mathcal{S} = \{\sigma_1, \sigma_2, \ldots, \sigma_m\}$ denote a partition of the spectrum, containing m contiguous intervals and $h_{i,k_i}(\lambda; \alpha_i)$ denote a k_i-degree polynomial filter function defined over the interval σ_i (and 0 elsewhere) with polynomial coefficients α_i. We define piece-wise polynomial GNN (PP-GNN) filter function as:

$$h(\lambda) = \sum_{\sigma_i \in \mathcal{S}} h_{i,k_i}(\lambda; \alpha_i) \tag{4}$$

and learn a smooth filter function by imposing additional constraints to maintain continuity between polynomials of contiguous intervals at different endpoints (*aka* knots). This class of filter functions is rich, and its complexity is controlled by choosing intervals (i.e., endpoints and number of partitions) and polynomial degrees. Given the filter function, we compute the PP-GNN node embedding matrix as:

$$\mathbf{Z} = \sum_{\sigma_i \in \mathcal{S}} \mathbf{U}_i h(\lambda_{\sigma_i}) \mathbf{U}_i^T \mathbf{Z}_x(\mathbf{X}; \boldsymbol{\Theta}) \tag{5}$$

where $\mathbf{U}_i$ is a matrix with columns as eigenvectors corresponding to eigenvalues that lie in σ_i and $h(\lambda_\sigma)$ is the diagonal matrix with diagonals containing the h_i evaluated at the eigenvalues lying in σ_i. Thus, the node embedding, $\mathbf{Z}$, is computed as a sum of outputs from a bank of polynomial filters with each filter operating over a spectral interval, σ_i.

4.2 Practical and Implementation Considerations

The filter function (5) requires computing full eigendecomposition of $\widetilde{\mathbf{A}}$ and is expensive, therefore, not scalable for very large graphs. We address this problem by performing eigendecomposition only for a few extreme values (i.e., at low and high ends of the spectrum) for sparse matrices, for which efficient algorithms exist [42] with corresponding off-the-shelf implementations. The primary motivation is that many recent works including GPR-GNN investigated the problem of designing robust graph neural networks that work well across homophilic and heterophilic graphs, and, they found that graph filters that amplify or attenuate low and high-frequency components of signals (i.e., low-pass and high-pass filters) are critical to improving performance on several benchmark datasets. However, there is still one question: *how do we extract signals from the remaining (middle) portion of the spectrum, and that too efficiently?* We answer this question as follows. Using the observation that the GPR-GNN method learns a graph filter but operates on the entire spectrum by sharing the filter coefficients across the spectrum, our proposal is to use an efficient variant of (4) as:

$$\tilde{h}(\lambda) = \eta_l \sum_{\sigma_i \in \mathcal{S}^l} h_i^{(l)}(\lambda; \gamma_i^{(l)}) + \eta_h \sum_{\sigma_i \in \mathcal{S}^h} h_i^{(h)}(\lambda; \gamma_i^{(h)}) + \eta_{gpr} h_{gpr}(\lambda; \gamma) \tag{6}$$

where $\mathcal{S}^l$ consists of partitions over low-frequency components, $\mathcal{S}^h$ consists of partitions over high-frequency components, the first and second terms fit piece-wise polynomials[2] in low/high-frequency regions, as indicated through super-scripts. We refer PP-GNN models using only filters corresponding to the first and second terms alone in (6) as PP-GNN (Low) and PP-GNN (High), respectively. We extract any useful information from other frequencies in the middle region by adding the GPR-GNN filter function, $h_{gpr}(\lambda; \gamma)$ (the final term in 6), which is computationally efficient. Since $h_{gpr}(\lambda; \gamma)$ is a special case of (4) and the terms in (6) are additive, it is easy to see that (6) is same as (4) with a modified set of polynomial coefficients. Furthermore, we can control the contributions from each term by setting or optimizing over hyperparameters, η_l, η_h and η_{gpr}. Thus, the proposed model offers richer capability and flexibility to learn complex frequency response and balance computation costs over GPR-GNN. Please see Sect. A.3 for implementation details.

Model Training. Like GPR-GNN, we apply SOFTMAX activation function on (5) and use the standard cross-entropy loss function to learn the sets of polynomial coefficients (γ) and classifier model parameters ($\boldsymbol{\Theta}$) using labeled data. To ensure smoothness of the learned filter functions, we add a regularization term that penalizes squared differences between the function values of polynomials of contiguous intervals at each other's interval end-points. More details can be found in the supplementary material (A.3).

Discussion. In our model (4), we alleviate the over-smoothing problem using low-order polynomials, and learning complex and sharper frequency responses is feasible as we approximate higher-order polynomial functions effectively using several low-order piece-wise polynomials. However, this comes with eigende-composition compute cost for a few (k) extreme eigenvalues, but is control-lable by choosing k in an affordable way[3]. We observe this cost is (one time) pre-training cost and can be amortized over multiple rounds of model train-ing required for the optimization of hyperparameters. Also, we need to com-pute each filter specific embedding with non-local eigen-graphs (via the opera-tions, $\mathbf{U}_i H_i(\gamma_i) \mathbf{U}_i^T \mathbf{Z}_x(\mathbf{X}; \boldsymbol{\Theta})$); thus, we lose (spatial) local neighborhood prop-erty of conventional methods like GPR-GNN. We compute node embeddings afresh whenever the model parameters are updated, thereby incurring an addi-tional cost (over GPR-GNN) of $O(nkL)$ where k and L denote the number of

---

[2] For brevity, we dropped the polynomial degree dependency.

[3] Most algorithms for this task utilize Lanczos' iteration, convergence bounds of which depends on the input matrix' spectrum [33,34], which although have superlinear convergence, but are observed to be efficient in practice.

selected low/high eigenvalues and classes, respectively. We conduct a comprehensive experimental study to assess the time taken by our method, compare against other state-of-the-art methods and present our findings in the experiment section.

4.3   Analysis

This section is arranged as follows: (a) Theorem 1 establishes superior capabilities of our model in approximating arbitrary filters than a standard polynomial filter; (b) Theorem 2 demonstrates the new space of filters that our model learns from, each region of which induces a controllable, strong bias towards certain parts of the spectrum while at the same time has dimension of the same order as the corresponding polynomial family.

Theorem 1. *For any frequency response h^*, and an integer $K \in \mathbb{N}$, let $\tilde{h} := h + h_f$, with h_f having a continuous support over a subset of the spectrum, σ_f. Assume that h and h_f are parameterized by independent K and K'-order polynomials, p and p_f, respectively, with $K' \leq K$. Then there exists $\tilde{h}$, such that $\min \|\tilde{h} - h^*\|_2 \leq \min \|h - h^*\|_2$, where the minimum is taken over the polynomial parameterizations. Moreover, for multiple polynomial adaptive filters $h_{f_1}, h_{f_2}, ..., h_{f_m}$ parameterized by independent K'-degree polynomials with $K' \leq K$ but having disjoint, contiguous supports, the same inequality holds for $\tilde{h} = h + \sum_{i=1}^m h_{f_i}$.*

For a detailed proof please refer to A.3 of the supplementary. We also conducted an experiment to illustrate the main conclusion of the above theorem in Sect. A.2 of the supplementary material.

Next, we note that since an actual waveform is not observed in practice and instead, we estimate it by optimizing over the observed labels via learning a graph filter, we theoretically show that the family of filters that we learn is a strict superset of the polynomial filter family. The same result holds for the families of the resulting adapted graphs.

Theorem 2. *Define $\mathbb{H} := \{h(\cdot) \mid \forall$ possible K-degree polynomial parameterizations of $h\}$ to be the set of all K-degree polynomial filters, whose arguments are $n \times n$ diagonal matrices, such that a filter response over some Λ is given by $h(\Lambda)$ for $h(\cdot) \in \mathbb{H}$. Similarly $\mathbb{H}' := \{\tilde{h}(\cdot) \mid \forall$ possible polynomial parameterizations of $\tilde{h}\}$ is set of all filters learnable via PP-GNN , with $\tilde{h} = h + h_{f_1} + h_{f_2}$, where h is parameterized by a K-degree polynomial supported over entire spectrum, h_{f_1} and h_{f_2} are localized adaptive filters parameterized by independent K'-degree polynomials which only act on top and bottom t diagonal elements respectively, with $t < n/2$ and $K' \leq K$; then $\mathbb{H}$ and $\mathbb{H}'$ form a vector space, with $\mathbb{H} \subset \mathbb{H}'$. Also, $\frac{dim(\mathbb{H}')}{dim(\mathbb{H})} = \frac{K+2K'+3}{K+1}$.*

Corollary 1. *The corresponding adapted graph families $\mathbb{G} := \{\mathbf{U}h(\cdot)\mathbf{U}^T \mid \forall h(\cdot) \in \mathbb{H}\}$ and $\mathbb{G}' := \{\mathbf{U}\tilde{h}(\cdot)\mathbf{U}^T \mid \forall \tilde{h}(\cdot) \in \mathbb{H}'\}$ for any unitary matrix $\mathbf{U}$ form a vector space, with $\mathbb{G} \subset \mathbb{G}'$ and $\frac{dim(\mathbb{G}')}{dim(\mathbb{G})} = \frac{K+2K'+3}{K+1}$.*

The above theorem can be trivially extended to an arbitrary number of adaptive filters with arbitrary support. The presence of each adaptive filter induces a bias in the model towards learning a bank of filters that operate only on the corresponding support. Since the number of filters and their support sizes are hyperparameters, tuning them offers control and flexibility to model richer frequency responses over the entire spectrum. Thus, our model learns from a more diverse space of filters and the corresponding adapted graphs. The result also implies that our model learns from a space of filters that is only $O(1)$-fold greater than that of polynomial filters[4]. Note that learning from this diverse region is feasible. This observation comes from the proofs of Theorem 4.2 and Corollary 4.2.1 (A.3 and A.3 in supplementary). Using the localized adaptive filters without any filter with the entire spectrum as support results in learning a set of adapted graphs, $\hat{\mathbb{G}}$. This set is disjoint from $\mathbb{G}$, with $\mathbb{G}' = \mathbb{G} \oplus \hat{\mathbb{G}}$. We conduct various ablative studies where we demonstrate the effectiveness of learning from $\hat{\mathbb{G}}$ and $\mathbb{G}'$.

Our model formulation is a generalization of the formulation by [8], and we show in Sect. A.3 of the supplementary material by extending their analysis to our model that it still inherits their property of mitigating oversmoothing effects when high degree polynomial is used. Our experiments show that we are able to obtain superior performance without needing the higher-order polynomials.

4.4 Comparison Against Other Filtering Methods

General FIR filter are a generalization of the polynomial filter family and thus a precursor to the models based on the latter. As per the study conducted in Sect. A.7 of the supplementary, constraining the model is required to obtain better performance. Restricting to polynomial filters can be seen as having an implicit regularization on the same and we also empirically observe that such a restriction (restricting to polynomial filters) gives much better performance than constraining GFIR (see Sect. 5.1 and A.7) by simpler regularization methods such as L2 and/or dropout. We have also shown in Theorem 2 that PP-GNN increases the space of graph filters (over GPR-GNN) and we observe in 5.1 that this increase in graph space results in an increased performance, over other polynomial filter methods. Thus, it requires a careful balance of the constraints imposed on the filter family, while also appropriately increasing the graph space to obtain better performance. A comprehensive study of this balance is beyond the scope of this work and we leave that as future work. Below, we first show the different ways of constraining the space (via polynomial filters) and compare them against PP-GNN.

Polynomial filters are a class of filters constructed and evaluated from polynomials. These filters can be constructed via multiple bases (for instance monomial, Bernstein) in the polynomial vector space. APPNP, GPR-GNN, and BERNNET are all instances of polynomial graph filters defined in different bases

---

[4] We leave the formal bias-variance analysis for adapted graph families as future work.

and with different constraints. Below, we illustrate the differences between these three methods and also discuss the shortcomings of each of them.

APPNP: One of the early works, APPNP [10], can be interpreted as a fixed polynomial graph filter that works with monomial basis. The polynomial coefficients correspond to Personalised PageRank (PPR) [43]. The node embeddings are learnt by APPNP as described in A.8. The main shortcoming of this method is the assumption that the optimal coefficients for the polynomial filter (for all tasks) are PPR coefficients, which need not necessarily be the case.

GPR-GNN: GPR-GNN builds on APPNP by overcoming this shortcoming by making the coefficients γ_k (see A.8) learnable. [8] identified that negative coefficients allows the model to exploit high frequency signals required for better performance on heterophilic graphs. GPR-GNN, like APPNP, uses the monomial basis. The node embeddings are learnt by GPR-GNN as described in A.8. While this method is an improvement over APPNP, adapting an arbitrary filter response which requires a high-order polynomial is difficult due to the oversmoothing problem. GPR-GNN mitigates oversmoothing by showing that the higher order terms' coefficients uniformly converge to zero during training. Mitigating the oversmoothing problem limits the complexity of the filter learnt, and therefore making GPR-GNN ineffective at learning complex frequency responses.

BernNet: While oversmoothing is one shortcoming of GPR-GNN, BERN-NET identified another shortcoming that GPR-GNN and other polynomial filtering based methods can result in ill-posed solutions and face optimization issues (converging to saddle points) by not constraining the filter response to non-negative values. [37] proposed a model that learns a non-negative frequency response, a constraint that can be easily enforced by modifying the learning problem from learning the coefficients of the monomial basis functions to learning the coefficients of the Bernstein basis functions, since the latter are non-negative in their standard domain. [37] argue that constraining coefficients to take on non-negative values is required for stability and interpretability of the learned filters and is the main reason for performance improvements. The node embeddings are learnt as described in A.8. Note that in the expression referenced, $\theta_k(\forall k)$ are learnable coefficients and are constrained to non-negative values. We first replace $\frac{1}{2^K}\binom{K}{r}\sum_{p=0}^{q}\binom{K-r}{q-p}\binom{r}{p}(-1)^p$ with α_{rq} and then subsequently replace $\sum_{r=0}^{K}\theta_r\alpha_{rq}$ with w_q. Such an exercise was done to show that the filter defined by BERNNET does indeed fall into the class of polynomial filters. We tabulate the important attributes of each of the polynomial filters described above in Table 11 of the supplementary material.

All of these approaches run into the oversmoothing issue with an increase in the degree of the polynomial filter (A.1 of supplementary). PP-GNN, owing to its piece-wise definition, can model more complex shaped responses better without the need to increase the degree. Our proposed model only requires extremal eigendecomposition (i.e. computing only the extreme eigenpairs), for which there exists efficient algorithms to compute [44, 45]. Further, as mentioned earlier, this is a one time pre-training cost, that can be amortized over training multiple models for hyper-parameter tuning. We illustrate this through a comprehensive empirical study in Sect. A.9 of the supplementary material. In the next section, we experimentally show the benefits of PP-GNN.

5   Experiments

We conduct extensive experiments to demonstrate the effectiveness and competitiveness of the proposed method over standard baselines and state-of-the-art (SOTA) GNN methods. We conduct ablative studies to demonstrate the usefulness of different filters and the number of eigenpairs required in PP-GNN. We also compare the quality of the embeddings learned and the time to train different models. We first describe our experimental setup along with baselines and information on hyper-parameter tuning.

We evaluate our model on several real-world heterophilic and homophilic datasets. We resort detailed descriptions of dataset statistics, preprocessing steps, and baselines to the Appendix (A.4). We report the mean and standard deviation of test accuracy over splits to compare model performance.

5.1   PP-GNN Versus SOTA Models

Table 1. Results on a few heterophilic and homophilic datasets. GFIR-1 corresponds to unconstrained setting. GFIR-2 corresponds to constrained setting. For a more detailed comparison and description please refer to Appendix A.5

	Squirrel	Chameleon	Cora	Computer	Photos
GFIR-1	36.50 ± 1.12	51.71 ± 3.11	87.93 ± 0.90	78.39 ± 1.09	89.26 ± 1.00
GFIR-2	41.12 ± 1.17	61.27 ± 2.42	87.46 ± 1.26	79.57 ± 2.12	89.38 ± 1.03
FAGCN [7]	42.59 ± 0.79	55.22 ± 3.19	$\underline{88.21 \pm 1.37}$	82.16 ± 1.48	90.91 ± 1.11
APPNP [10]	39.15 ± 1.88	47.79 ± 2.35	88.13 ± 1.53	82.03 ± 2.04	$\underline{91.68 \pm 0.62}$
LGC [22]	44.26 ± 1.49	61.14 ± 2.07	88.02 ± 1.44	83.44 ± 1.77	91.56 ± 0.74
GPRGNN [8]	46.31 ± 2.46	62.59 ± 2.04	87.77 ± 1.31	82.38 ± 1.60	91.43 ± 0.89
AdaGNN [36]	$\underline{53.50 \pm 0.96}$	$\underline{65.45 \pm 1.17}$	86.72 ± 1.29	81.27 ± 2.10	89.93 ± 1.22
BernNET [37]	52.56 ± 1.69	62.02 ± 2.28	88.13 ± 1.41	$\underline{83.69 \pm 1.99}$	91.61 ± 0.51
ARMA [38]	47.37 ± 1.63	60.24 ± 2.19	87.37 ± 1.14	78.55 ± 2.62	90.26 ± 0.48
UFG [32]	42.06 ± 1.55	56.29 ± 1.58	87.93 ± 1.52	80.01 ± 1.78	90.20 ± 1.41
PP-GNN	$\mathbf{59.15 \pm 1.91}$	$\mathbf{69.10 \pm 1.37}$	$\mathbf{89.52 \pm 0.85}$	$\mathbf{85.23 \pm 1.36}$	$\mathbf{92.89 \pm 0.37}$

Heterophilic Datasets. We perform comprehensive experiments to show the effectiveness of PP-GNN on several Heterophilic graphs and tabulate the results in Table 5 in the Appendix (A.5). Datasets like Texas, Wisconsin, and Cornell contain graphs with high levels of Heterophily and *rich node features*. Standard non-graph baselines like LR and MLP perform competitively or better on these datasets compared to many spatial and spectral-based methods. PP-GNN offers significant lifts in performance with gains of up to $\sim 6\%$. The node features in datasets like Chameleon and Squirrel are not adequately discriminative, and significant improvements are possible via convolutions, as we compare non-graph and graph-based methods in Table 5. Spatial GNN methods, in general, offer improvements over non-graph counterparts. In specific, methods like GCN, which also have a spectral connotation, show better performance on these datasets. We observe from the Table that Spectral methods offer additional improvements over models like GCN. The difference in performance among spectral methods majorly comes from their ability to learn better frequency responses of graph filters. Our proposed model shows significant lifts over all the baselines with gains up to ∼6% and ∼4% on the Squirrel and Chameleon datasets. These improvements empirically support the efficacy of PP-GNN's filter design.

Homophilic Datasets. The input graphs for these datasets contain informative signals, and one can expect competitive task performance from even basic spatial-convolution based methods as observed in Table 7 present in Appendix (A.5). We can see that spatial models are among the top performers for several Homophilic datasets. Existing spectral methods marginally improve over spatial methods on a few datasets. Not surprisingly, our PP-GNN model with effective filter design can exploit additional discriminatory signals from an already rich informative source of signals. PP-GNN offers additional gains up to 1.3% over other baselines.

Due to space constraints we have shown a small subset of our results in Table 1. For a more detailed comparison please refer to the Appendix A.5 section, where we compare against more SOTA methods and on other datasets as well.

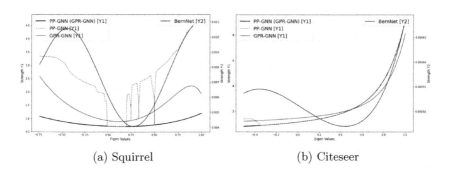

(a) Squirrel   (b) Citeseer

Fig. 1. Learned filter responses of PP-GNN, GPR-GNN, and BERNNET.

5.2   PP-GNN Model Investigation

We conducted several experimental studies to understand and illustrate how the PP-GNN model works. Our studies include: (a) how does the frequency response of PP-GNN look like?, (b) what happens when we learn only individual sub-filter banks (e.g., PP-GNN (Low), PP-GNN (Low + GPR-GNN)? and (c) does PP-GNN learn better embeddings?

Frequency Response. In Fig. 1a and 1b, we show the learned frequency responses (i.e., $h(\lambda)$) of the overall PP-GNN model, GPR-GNN component of PP-GNN (PP-GNN (GPR-GNN)), stand-alone (GPR-GNN) model and BERN-NET model on the Squirrel and Citeseer datasets. For Squirrel (a heterophilic dataset), we can observe that while GPR-GNN and BERNNET learns the importance of low and high-frequency signals, it is unable to capture their relative strengths/importance adequately, and this happens due to the restriction of learning a single polynomial globally. PP-GNN learns sharper and richer responses at different parts of the spectrum, thereby improving classification accuracy. For Citeseer (a homophilic dataset) we can observe that all the models in comparison learn a smooth polynomial, GPR-GNN is not able to capture the complex transition that can be seen at the lower end of the spectrum, while BERNNET is doing it some degree. This inability to capture the complex transition leads to a lower classification accuracy. A similar trend can be found on two other datasets in A.6.

Quality of Learned Embeddings: We qualitatively assess the difference in the learned embedding of PP-GNN, GPR-GNN and BERNNET. Towards this, we generated t-SNE plots of the learned node embeddings and visually inspected them. From Fig. 2a, 2b and 2c, we observe that PP-GNN discovers more discriminative features resulting in discernible clusters on the Squirrel dataset compared to GPR-GNN and BERNNET, enabling PP-GNN to achieve significantly improved performance.

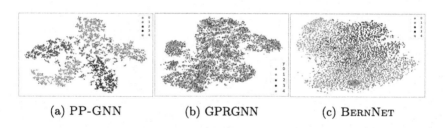

(a) PP-GNN   (b) GPRGNN   (c) BERNNET

Fig. 2. t-SNE plots of learned embeddings on the Squirrel dataset

5.3   Additional Experiments

We summarize a list of experiments that can be found in the supplementary material. Studies on varying the number of eigenvectors used by PP-GNN and the importance of MLP can be found in Sect. A.6. Analysis on the effect of varying the order of GPR-GNN's polynomial on performance is presented in A.1. Experimental details for PP-GNN with boundary regularization is in A.3.

6 Conclusion

Several recently proposed methods attempt to build robust models for diverse graphs exhibiting different correlations between graph and node labels. We build on the filter-based approach of GPR-GNN which can be extended further with Generalized FIR models. This work proposed an effective polynomial filter bank design using a piece-wise polynomial filtering approach. We combine GPR-GNN with additional polynomials resulting in a bank of filters that adapt to low and high-end spectrums using multiple polynomial filters. While our method makes an unconventional choice of extremal eigendecomposition, it does help to get improved performance, albeit with some additional but manageable cost. Our experiments demonstrate that the proposed approach can learn effective filter functions that improve node classification accuracy significantly across diverse graphs. While our work shows merit, it is still founded upon the polynomial formulation, and even though piecewise polynomial filters are more expressive than conventional polynomial filters, they still retain the properties of the polynomial filters locally. Hence, there is still room for even more expressive filter formulations that are well motivated, and we leave their exploration as future work.

A Appendix

The appendix is structured as follows. In Sect. A.1, we present additional evidence of the limitations of GPR-GNN. In Sect. A.2, we show a representative experiment that motivates Sect. 4. In Sect. A.3, we provide proofs for theorems, propositions and corollaries defined in Sect. 3 and 4.3. In Sect. A.4, we provide more details regarding the baselines, datasets, their respective splits and additional implementation details including hyper-parameter ranges. We also provide details and results of additional experiments. In Sect. A.5, we provide detailed comparison of PP-GNN against numerous SOTA models. In Sect. A.7, we provide details on GFIR and compare our proposed model against it. Section A.8 provides additional information on differences between our model and other polynomial filtering methods. In Sect. A.9, we provide a comprehensive timing analysis.

A.1 Motivation

Node Feature Indistinguishably Plots. In the main paper (Fig. 5), we plot the average of pairwise distances between node features for the Cora dataset, after computing $\widetilde{\mathbf{A}}^j X$ for increasing j values, and showed that the mean pairwise node feature distance decreases as j increases. We observe that this is consistent across three more datasets: Citeseer, Chameleon and Squirrel. This is observed in Fig. 3.

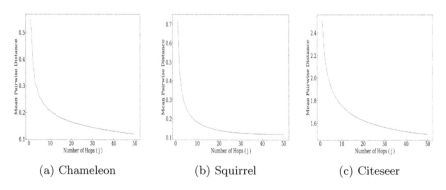

Fig. 3. Average of pairwise distances between node features, after computing $\widetilde{\mathbf{A}}^j X$, for increasing j values

We also observed the mean of the variance of each dimension of node features, after computing $\widetilde{\mathbf{A}}^j X$, for increasing j values. We observe that this mean does indeed reduce as the number of hops increase. We also observe that the variance of each dimension of node features reduces for Cora, Squirrel and Chameleon as the number of hops increase; however, we don't observe such an explicit phenomenon for Citeseer. See Fig. 4.

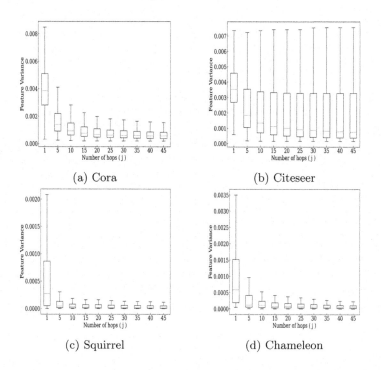

Fig. 4. Variance of each dimension of node features

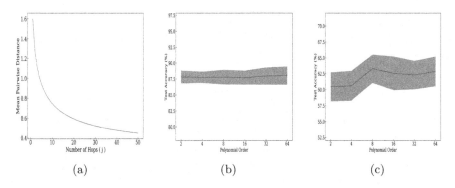

(a)   (b)   (c)

Fig. 5. In (a), we plot the average of pairwise distances between node features for the Cora dataset, after computing $\widetilde{\mathbf{A}}^j X$ for increasing j values. X-axis represents the various powers j and the Y-axis represents the average of pairwise distances between node features. In (b) and (c), we plot the test accuracies of the model in [8] for increasing order of polynomials for Cora and Chameleon dataset respectively. X-axis represents the order of the polynomial and Y-axis represents the test accuracy achieved for that order.

Effect of Varying the Order of the GPR-GNN Polynomial. In the main paper (Fig. 5a), we plot the test accuracies of the GPR-GNN model while increasing the order of the polynomials for the Cora and Chameleon dataset, respectively. We observe that on increasing the polynomial order, the accuracies do not increase any further. We can show a similar phenomenon on two other datasets, Squirrel and Citeseer, in Fig. 6.

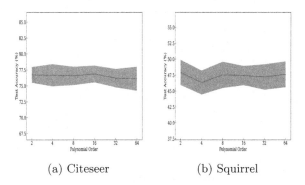

(a) Citeseer   (b) Squirrel

Fig. 6. Accuracy of the GPR-GNN model on inceasing the order of the polynomial

In Sect. 3 of the main paper, we claim that due to the over-smoothing effect, even on increasing the order of the polynomial, there is no improvement in the test accuracy. Moreover, in Fig. 1 we can see that our model can learn a complicated filter polynomial while GPR-GNN cannot. This section shows that even on increasing the order of the GPR-GNNpolynomial, neither does the test accuracy increase nor does the waveform become as complicated as PP-GNN. See Fig. 7.

A.2   Fictitious Polynomial

In Sect. 4, we claim that having multiple disjoint low order polynomials can approximate a complicated waveform more effectively than a single higher-order polynomial. To demonstrate, we create a representative experiment that shows this phenomenon by creating a fictitious complicated polynomial and try to fit it using a single unconstrained polynomial (representative of GPR-GNN), a single constrained polynomial (indicative of BERNNET, where the coefficients should be non-negative) and a disjoint piece-wise polynomial (indicative of PP-GNN). We setup a least square optimization problem to obtain the coefficients for these different polynomial variants. We evaluate and plot these polynomials in Fig. 8. To quantify the effectiveness of different polynomial variants, we compute the approximation error (RMSE) with respect to the optimal waveform. We observe that piece-wise polynomials achieve much lower RMSE (1.5053) compared to constrained polynomial (3.9659) and unconstrained polynomial (3.3854)

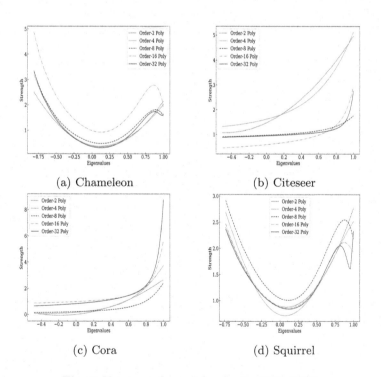

(a) Chameleon   (b) Citeseer

(c) Cora   (d) Squirrel

Fig. 7. Varying polynomial order in GPR-GNN

A.3   Proposed Approach

Details Regarding Boundary Regularization. To induce smoothness in the learned filters, we add a regularization term that penalizes squared differences between function values of polynomials at knots (endpoints of contiguous bins). Our regularization term looks as follows:

$$\sum_{i=1}^{m-1} \exp^{-(\sigma_i^{max} - \sigma_{i+1}^{min})^2} (h_i(\sigma_i^{max}) - h_{i+1}(\sigma_{i+1}^{min}))^2 \tag{7}$$

In Eq. 7, σ_i^{max} and σ_i^{min} refer to the maximum and minimum eigenvalues in σ_i (Refer to Sect. 4). This regularization term is added to the Cross-Entropy loss. We perform experiments with this model and report the performance in Table 2. We observe that we are able to reach similar performance even without the presence of this regularization term. Therefore, majority of the results reported in our Main paper are without this regularization term.

Implementing the Filter in Practice. We provide more details to explain the filtering operation. An Equation similar to Eq. 2 can be derived for our model by substituting Eq. 6 from the paper into Eq. 5. On substitution, we get:

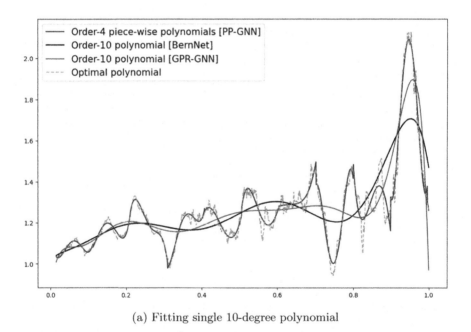

(a) Fitting single 10-degree polynomial

Fig. 8. To demonstrate the effectiveness of adaptive polynomial filter, we try to approximate a complex waveform (green dashed line) via (a) 10 disjoint adaptive polynomial filters of order 4 (colored blue) (b) a single constrained order 10 polynomial (colored red), (c) an unconstrained order 10 polynomial (colored purple). The corresponding RMSE values are: (a) 1.5053, (b) 3.9659, (c) 3.3854 (Color figure online)

Table 2. Results with and without boundary regularization

Test Acc	Computer	Chameleon	Citeseer	Cora	Squirrel
PPGNN (with reg)	83.53 (1.67)	67.92 (2.05)	76.85 (2)	88.19 (1.19)	55.42 (2.1)
PPGNN	85.23 (1.36)	69.10 (1.37)	78.25 (1.76)	89.52 (0.85)	59.15 (1.91)

$$Z = \eta_l \sum_{\sigma_i \in S^l} \sum_{j=1}^{k_1} \gamma_{ij}^{(l)} U_{\sigma_i}^{(l)} \left(\Lambda_{\sigma_i}^{(l)} \right)^j U_{\sigma_i}^{(l)T} Z_0(X; \theta)$$

$$+ \eta_h \sum_{\sigma_i \in S^h} \sum_{j=1}^{k_2} \gamma_{ij}^{(h)} U_{\sigma_i}^{(h)} \left(\Lambda_{\sigma_i}^{(h)} \right)^j U_{\sigma_i}^{(h)T} Z_0(X; \theta)$$

$$+ \eta_{gpr} \sum_{j=1}^{k_3} \gamma_j \widetilde{A}^j Z_0(X; \theta) \quad (8)$$

where

$$\widetilde{A}_i^{(l)} = U_{\sigma_i}^{(l)} \Lambda_{\sigma_i}^{(l)} U_{\sigma_i}^{(l)T}, \sigma_i \in S^l$$

$$\widetilde{A}_i^{(h)} = U_{\sigma_i}^{(h)} \Lambda_{\sigma_i}^{(h)} U_{\sigma_i}^{(h)T}, \sigma_i \in S^h$$

$U_{(\sigma_i)}^{(l)}$, $\Lambda_{(\sigma_i)}^{(l)}$ are the matrices containing eigenvectors and eigenvalues corresponding to the partition σ_i of the low frequency components and $U_{(\sigma_i)}^{(h)}$, $\Lambda_{(\sigma_i)}^{(h)}$ are the matrices containing eigenvectors and eigenvalues corresponding to the partition σ_i of the high frequency components and $\widetilde{A}$ (See Eq. 1 in the main paper) where $U_{(\sigma_i)} \in \mathbb{R}^{n \times |\sigma_i|}$ and $\Lambda_{(\sigma_i)} \in \mathbb{R}^{|\sigma_i| \times |\sigma_i|}$ with latter being a diagonal matrix.

The way we have implemented the filter is that we pre-compute the top and bottom eigenvalues/vectors of $\widetilde{A}$ and use them to compute partition specific node embeddings. Note that Eq. 8 can be rewritten as:

$$Z = \eta_l \sum_{\sigma_i \in S^l} U_{\sigma_i}^{(l)} H_{\sigma_i}^{(l)} U_{\sigma_i}^{(l)T} Z_0(X; \theta)$$

$$+ \eta_h \sum_{\sigma_i \in S^h} U_{\sigma_i}^{(h)} H_{\sigma_i}^{(h)} U_{\sigma_i}^{(h)T} Z_0(X; \theta)$$

$$+ \eta_{gpr} \sum_{j=1}^{k_3} \gamma_j \widetilde{A}^j Z_0(X; \theta)$$

(9)

where $H_{\sigma_i}^{(l)} = \sum_{j=1}^{k_1} \gamma_{ij}^{(l)} \left(\Lambda_{\sigma_i}^{(l)} \right)^j$ and $H_{\sigma_i}^{(h)} = \sum_{j=1}^{k_2} \gamma_{ij}^{(h)} \left(\Lambda_{\sigma_i}^{(h)} \right)^j$ form the effective low and high frequency component filters. Thus, a weighted combination of low and high frequency component based embeddings (i.e., the first and second term in Equation above) and the GPR-GNN term based embedding (i.e., third term) is computed. We implement our model based on Eq. 9.

Note that the following discussion is just for illustration purpose and we do not explicitly calculate the newer terms introduced here: We can also interpret the GPR-GNN term in terms of piece-wise polynomial filters defined on a mutually exclusive partition of the spectrum, with a difference that the coefficients and the order of the polynomial are shared across all partitions:

$$
\eta_{gpr} \sum_{j=1}^{k_3} \gamma_j \widetilde{A}^j Z_0(X;\theta) = \eta_{gpr} \Big(\sum_{\sigma_i \in S^l} U_{\sigma_i}^{(l)} H_{\sigma_i}^{(gpr)} U_{\sigma_i}^{(l)T}
$$
$$
+ \sum_{\sigma_i \in S^h} U_{\sigma_i}^{(h)} H_{\sigma_i}^{(gpr)} U_{\sigma_i}^{(h)T}
$$
$$
+ \sum_{\sigma_i \in S^{mid}} U_{\sigma_i}^{(mid)} H_{\sigma_i}^{(gpr)} U_{\sigma_i}^{(mid)T} \Big) Z_0(X;\theta) \quad (10)
$$

where $S^{mid} := S - (S^l \cup S^h)$, and $H_{\sigma_i}^{(gpr)} = \sum_{j=1}^{k_3} \gamma_j \left(\Lambda_{\sigma_i}^{(mid)} \right)^j$. Hence, we can club the respective terms of the partitions and obtain the final embeddings as:

$$
Z = \sum_{\sigma_i \in S^l} U_{\sigma_i}^{(l)} (\eta_l H_{\sigma_i}^{(l)} + \eta_{gpr} H_{\sigma_i}^{(gpr)}) U_{\sigma_i}^{(l)T} Z_0(X;\theta)
$$
$$
+ \sum_{\sigma_i \in S^h} U_{\sigma_i}^{(h)} (\eta_h H_{\sigma_i}^{(h)} + \eta_{gpr} H_{\sigma_i}^{(gpr)}) U_{\sigma_i}^{(h)T} Z_0(X;\theta) \quad (11)
$$
$$
+ \eta_{gpr} \sum_{\sigma_i \in S^{mid}} U_{\sigma_i}^{(mid)} H_{\sigma_i}^{(gpr)} U_{\sigma_i}^{(mid)T} Z_0(X;\theta)
$$

From Eq. 11, it is clear that PP-GNN also adapts the responses from the middle parts of the spectrum, albeit by a single polynomial. One can also interpret each term of Eq. 11 as an effective polynomial filter acting only on the corresponding part of the spectrum, with each effective polynomial filter can be influenced by a shared polynomial filter.

Notation Used. Vectors are denoted by lower case bold Roman letters such as $\mathbf{x}$, and all vectors are assumed to be column vectors. In the paper, h with any sub/super-script refers to a frequency response, which is also considered to be a vector. A superscript T denotes the transpose of a matrix or vector; Matrices are denoted by bold Roman upper case letters, such as $\mathbf{M}$. A field is represented by $\mathbb{K}$; sets of real and complex numbers are denoted by $\mathbb{R}$ and $\mathbb{C}$ respectively. $\mathbb{K}[x_1, \ldots, x_n]$ denotes a multivariate polynomial ring over the field $\mathbb{K}$, in indeterminates $x_1, \ldots, x_n$. Set of $n \times n$ square matrices with entries from some set $\mathbb{S}$ are denoted by $\mathbb{M}_n(\mathbb{S})$. Moore-Penrose pseudoinverse of a matrix $\mathbf{A}$ is denoted by $\mathbf{A}^\dagger$. Eigenvalues of a matrix are denoted by λ, with $\lambda_1, \lambda_2, \ldots$ denoting a decreasing order when the eigenvalues are real. A matrix $\mathbf{\Lambda}$ denotes a diagonal matrix of eigenvalues. Set of all eigenvalues, i.e., spectrum, of a matrix is denoted by $\sigma_{\mathbf{A}}$ or simply σ when the context is clear. L_p norms are denoted by $\| \cdot \|_p$. Frobenius norm over matrices is denoted by $\| \cdot \|_F$. Norms without a subscript default to L_2 norms for vector arguments and Frobenius norm for matrices. $\oplus$ denotes a direct sum. For maps f_i defined from the vector spaces

$V_1, \cdots, V_m$, with a map of the form $f : V \mapsto W$, with $V = V_1 \oplus V_2 \oplus \cdots \oplus V_m$, the phrase "$f : V \mapsto W$ by mapping $f(\mathbf{v_i})$ to $f_i(g(\mathbf{v_i}))$" means that f maps a vector $\mathbf{v} = \mathbf{v_1} + \ldots + \mathbf{v_m}$ with $\mathbf{v}_i \in V_i$ to $f_1(g(\mathbf{v_1})) + \ldots + f_m(g(\mathbf{v_m}))$.

Proof of Theorem 1

Theorem. *For any desired frequency response h^*, and an integer $K \in \mathbb{N}$, let $\tilde{h} := h + h_f$, with h_f having a continuous support over a subset of the spectrum, σ_f. Assuming h and h_f to be parameterized by independent K and K'-order polynomials p and p_f respectively, with $K' \le K$, then there exists $\tilde{h}$, such that $\min \|\tilde{h} - h^*\|_2 \le \min \|h - h^*\|_2$, where the minimum is taken over the polynomial parameterizations. Moreover, for multiple polynomial adaptive filters $h_{f_1}, h_{f_2}, ..., h_{f_m}$ parameterized by independent K'-degree polynomials with $K' \le K$ but having disjoint, contiguous supports, the same inequality holds for $\tilde{h} = h + \sum_{i=1}^m h_{f_i}$.*

Proof. We make the following simplifying assumptions:

1. $|\sigma_{f_i}| > K$, $\forall i \in [m]$, i.e., that is all support sizes are lower bounded by K (and hence K')
2. All eigenvalues of the reference matrix are distinct

For methods that use a single polynomial filter, the polynomial graph filter, $h_K(\mathbf{\Lambda}) = diag(\mathbf{V}\gamma)$ where γ is a vector of coefficients (i.e., γ parameterizes h), with eigenvalues sorted in descending order in components, and $\mathbf{V}$ is a Vandermonde matrix:

$$\mathbf{V} = \begin{bmatrix} 1 & \lambda_1 & \lambda_1^2 & \cdots & \lambda_1^K \\ 1 & \lambda_2 & \lambda_2^2 & \cdots & \lambda_2^K \\ \vdots & \vdots & \vdots & \ddots & \vdots \\ 1 & \lambda_n & \lambda_n^2 & \cdots & \lambda_n^K \end{bmatrix}$$

And to approximate a frequency response h^*, we have the following objective:

$$\min \|h - h^*\|_2^2 := \min_{\gamma} \|diag(h^*) - diag(\mathbf{V}\gamma)\|_F^2$$

$$= \min_{\gamma} \|h^* - \mathbf{V}\gamma\|_2^2$$

$$= \min_{\gamma} \|\mathbf{e}_p(\gamma)\|_2^2$$

where $\|\|_F$ and $\|\|_2$ are the Frobenius and L_2 norms respectively. Due to the assumptions, the system of equations $h^* = \mathbf{V}\gamma$ is well-defined and has a unique minimizer, $\gamma^* = \mathbf{V}^\dagger h^*$, and thus $\|\mathbf{e}_p(\gamma^*)\| = \min_\gamma \|\mathbf{e}_p(\gamma)\|$. Next we break this error vector as:

$$\mathbf{e}_p(\gamma^*) := h^* - \mathbf{V}\gamma^*$$

$$= \sum_{i=1}^m (h_i^* - \mathbf{V}_i \gamma^*) + (h_L^* - \mathbf{V}_L \gamma^*)$$

$$:= \sum_{i=1}^m \mathbf{e}_{p_i}^* + \mathbf{e}_{p_L}^*$$

where $e_{p_i}^* := (h_i^* - V_i\gamma^*)$ with similar definition for e_{p_L}; h_i^* is a vector whose value at components corresponding to the set $\sigma(h_{f_i})$ is same as that of h^* and rest are zero. Similarly, V_i^* is a matrix whose rows corresponding to the set $\sigma(h_{f_i})$ are same as that of V with other rows being zero. Also, $V_L = V - \sum_{i=0}^m V_i$ and $h_L^* = h^* - \sum_{i=0}^m h_i^*$. Note that as a result of this construction, $[e_{p_i}^*] \cup e_{p_L}^*$ is a linearly independent set since the supports $[\sigma(h_{f_i})]$ form a disjoint set (note the theorem statement). We split the proof in two cases:

Case 1: $K' = K$. We now analyze the case where we have m polynomial adaptive filters added, all having an order of K, where the objective is $\min \|\tilde{h} - h^*\|$, which can be written as:

$$\min_{\gamma,[\gamma_i]} \|diag(h^*) - diag\left(V\gamma + \sum_{i=0}^m V_i\gamma_i\right)\|_F^2$$

$$= \min_{\gamma,[\gamma_i]} \|h^* - V\gamma - \sum_{i=0}^m V_i\gamma_i\|_2^2$$

$$= \min_{\gamma,[\gamma_i]} \|e_g(\gamma, [\gamma_i])\|_2^2$$

Before characterizing the above system, we break a general error vector as:

$$e_g(\gamma, [\gamma_i]) := h^* - V\gamma - \sum_{i=0}^m V_i\gamma_i$$

$$= \sum_{i=1}^m (h_i^* - V_i(\gamma + \gamma_i)) + (h_L^* - V_L\gamma)$$

$$:= \sum_{i=1}^m e_{g_i} + e_{g_L}$$

where $e_{g_i} := (h_i^* - V_i(\gamma + \gamma_i))$ with similar definition for e_{g_L}. Clearly, the systems of equations, $e_{g_i} = 0, \ \forall i$ and $e_{g_L} = 0$ are well-defined due to the assumptions 1 and 2. Since all the systems of equations have independent argument, unlike in the polynomial filter case where the optimization is constrained over a single variable; one can now resort to individual minimization of squared norms of e_{g_i} which results in a minimum squared norm of e_g. Thus, we can set:

$$\gamma = V_L^\dagger h_L^* = \gamma_g^* \qquad \gamma_i = V_i^\dagger h_i^* - V_L^\dagger h_L^* = \gamma_i^*, \ \forall i \in [m]$$

to minimize squared norms of e_{g_i} and e_{g_L}. Note that $[e_{g_i}] \cup e_{g_L}$ is a linearly independent set since the supports $[\sigma(h_{f_i})]$ form a disjoint set and by the above construction, this is also an orthogonal set, and hence we have $\|e_g\|^2 = \sum_{i=1}^m \|e_{g_i}\|^2 + \|e_{g_L}\|^2$, and hence the above assignment implies:

$$\|e_g(\gamma_g^*, [\gamma_i^*])\| = \min_{\gamma,[\gamma_i]} \|e_g(\gamma_g, [\gamma_i])\| := \min \|\tilde{h} - h^*\|_2$$

Hence, it follows that, $\min_x \|h_i^* - \mathbf{V}_i x\|^2 = \|\mathbf{e}_{g_i}^*\|^2 \leq \|\mathbf{e}_{p_i}^*\|^2 = \|h_i^* - \mathbf{V}_i \gamma^*\|^2$ and $\min_x \|h_L^* - \mathbf{V}_L x\|^2 = \|\mathbf{e}_{g_L}^*\|^2 \leq \|\mathbf{e}_{p_L}^*\|^2 = \|h_L^* - \mathbf{V}_L \gamma^*\|^2$. Hence,

$$\sum_{i=1}^{m} \|\mathbf{e}_{g_i}^*\|^2 + \|\mathbf{e}_{g_L}^*\|^2 \leq \sum_{i=1}^{m} \|\mathbf{e}_{p_i}^*\|^2 + \|\mathbf{e}_{p_L}^*\|^2$$

$$\min \|\tilde{h} - h^*\| \leq \min \|h - h^*\|$$

Case 2: $K' < K$. We demonstrate the inequality showing the existence of an $\tilde{h}$ that achieves a better approximation error. By definition, the minimum error too will be bounded above by this error. For this, we fix γ, the parameterization of h as $\gamma = \mathbf{V}^\dagger h^* = \gamma_p^*$ (say). Note that $\gamma_{p*} = \arg\min_\gamma \|\mathbf{e}_p(\gamma)\|$. Now our objective function becomes

$$\mathbf{e}_g(\gamma_p^*, [\gamma_i]) := h^* - \mathbf{V}\gamma_p^* - \sum_{i=0}^{m} \mathbf{V}_i' \gamma_i$$

$$= \sum_{i=1}^{m} (h_i^* - \mathbf{V}_i \gamma_p^* + \mathbf{V}_i' \gamma_i) + (h_L^* - \mathbf{V}_L \gamma_p^*)$$

$$= \sum_{i=1}^{m} \mathbf{e}_{g_i}' + \mathbf{e}_{g_L}'$$

where $h_i^*, h_L^*, \mathbf{V}_i, \mathbf{V}_L$ have same definitions as that in case 1 and $\mathbf{V}_i'$ is a matrix containing first $K' + 1$ columns of $\mathbf{V}_i$ as its columns (and hence has full column rank), and, $\gamma_i \in \mathbb{R}^{K'+1}$. By this construction, we have

$$\|\mathbf{e}_g(\gamma_p^*, \mathbf{0})\| = \min_\gamma \|\mathbf{e}_p(\gamma)\| = \|\mathbf{e}_p(\gamma_p^*)\|$$

Our optimization objective becomes $\min_{[\gamma_i]} \|\mathbf{e}_g(\gamma_p^*, [\gamma_i])\|$, which is easy since the problem is well-posed by assumption 1 and 2. The unique minimizer of this is obtained by setting

$$\gamma_i = \mathbf{V}_i'^\dagger (h_i^* - \mathbf{V}_i \gamma_p^*) = \gamma_i^* \text{ (say)} \ \forall i \in [m]$$

Now,

$$\|\mathbf{e}_g(\gamma_p^*, [\gamma_i^*])\| = \min_{[\gamma_i]} \|\mathbf{e}_g(\gamma_p^*, [\gamma_i])\| \leq \|\mathbf{e}_g(\gamma_p^*, \mathbf{0})\|$$

and,

$$\|\mathbf{e}_g(\gamma_p^*, \mathbf{0})\| = \min_\gamma \|\mathbf{e}_p(\gamma)\| = \min \|h - h^*\|$$

By the definition of minima, $\min_{\gamma, [\gamma_i]} \|\mathbf{e}_g(\gamma, [\gamma_i])\| \leq \min_{[\gamma_i]} \|\mathbf{e}_g(\gamma_p^*, [\gamma_i])\|$, and by the definition, $\min \|\tilde{h} - h^*\| = \min_{\gamma, [\gamma_i]} \|\mathbf{e}_g(\gamma, [\gamma_i])\|$, we have:

$$\min \|\tilde{h} - h^*\| \leq \min \|h - h^*\|$$

$\square$

Proof of Theorem 2

Theorem. *Define* $\mathbb{H} := \{h(\cdot) \mid \forall \text{ possible } K\text{-degree polynomial parameterizations}$ *of* $h\}$ *to be the set of all* K-*degree polynomial filters, whose arguments are* $n \times n$ *diagonal matrices, such that a filter response over some* $\mathbf{\Lambda}$ *is given by* $h(\mathbf{\Lambda})$ *for* $h(\cdot) \in \mathbb{H}$. *Similarly* $\mathbb{H}' := \{\tilde{h}(\cdot) \mid \forall \text{ possible polynomial parameterizations of } \tilde{h}\}$ *is set of all filters learn-able via* PP-GNN, *with* $\tilde{h} = h + h_{f_1} + h_{f_2}$, *where* h *is parameterized by a* K-*degree polynomial supported over entire spectrum,* h_{f_1} *and* h_{f_2} *are adaptive filters parameterized by independent* K'-*degree polynomials which only act on top and bottom* t *diagonal elements respectively, with* $t < n/2$ *and* $K' \leq K$; *then* $\mathbb{H}$ *and* $\mathbb{H}'$ *form a vector space, with* $\mathbb{H} \subset \mathbb{H}'$. *Also,* $\frac{dim(\mathbb{H}')}{dim(\mathbb{H})} = \frac{K+2K'+3}{K+1}$.

Proof. We start by constructing the abstract spaces on top of the polynomial vector space. Consider the set of all the univariate polynomials having degree at most K in the vector space over the ring $\mathbb{K}_n^x := \mathbb{K}[x_1, \ldots, x_n]$ where $\mathbb{K}$ is the field of real numbers. Partition this set into n subsets, say $V_1, \ldots, V_n$, such that for $i \in [n]$, V_i contains all polynomials of degree up to K in x_i. It is easy to see that $V_1, \ldots, V_n$ are subspaces of $\mathbb{K}[x_1, \ldots, x_n]$. Define $V = V_1 \oplus V_2 \oplus \cdots \oplus V_n$ where $\oplus$ denotes a direct sum. Define the matrix $D_i[c]$ whose $(i,i)^{\text{th}}$ entry is c and all the other entries are zero. For $i \in [n]$, define linear maps $\phi_i : V_i \to \mathbb{M}_n(\mathbb{K}_n^x)$ by $f(x_i) \mapsto D_i[f(x_i)]$. $\text{Im}(\phi_i)$ forms a vector space of all diagonal matrices, whose (i,i) entry is the an element of V_i. Generate a linear map $\phi : V \to \mathbb{M}_n(\mathbb{K}_n^x)$ by mapping $\phi(f(x_i))$ to $\phi_i(f(x_i))$ for all $i \in [n]$ as the components of the direct sum present in its argument. Note that ϕ_i for $i \in [n]$ are injective maps, making ϕ an injective map. This implies that $\mathbb{H} \subset \text{Im}(\phi)$ is a subspace with basis $\mathcal{B}_h := \{\phi(x_1^0 + \cdots + x_n^0), \phi(x_1 + \cdots + x_n), \ldots, \phi(x_1^K + \cdots + x_n^K)\}$, making $\dim(\mathbb{H}) = K + 1$. Similarly we have, $\mathbb{H}' \subset \text{Im}(\phi)$, a subspace with basis $\mathcal{B}_{h'} := \mathcal{B}_h \bigcup \{\phi(x_1^0 + \cdots + x_t^0 + 0 + \cdots + 0), \phi(x_1 + \cdots + x_t + 0 + \cdots + 0), \ldots, \phi(x_1^{K'} + \cdots + x_t^{K'} + 0 + \cdots + 0)\} \bigcup \{\phi(0 + \cdots + 0 + x_{n-t+1}^0 + \cdots + x_n^0), \phi(0 + \cdots + 0 + x_{n-t+1} + \cdots + x_n), \ldots, \phi(0 + \cdots + 0 + x_{n-t+1}^{K'} + \cdots + x_n^{K'})\}$ where x_i^0 and 0 are the corresponding multiplicative and additive identities of $\mathbb{K}_n^x$, implying $\mathbb{H} \subset \mathbb{H}'$ and $\dim(\mathbb{H}') = K + 2K' + 3$. $\square$

Proof of Corollary 1

Corollary. *The corresponding adapted graph families* $\mathbb{G} := \{\mathbf{U}h(\cdot)\mathbf{U}^T \mid \forall h(\cdot) \in \mathbb{H}\}$ *and* $\mathbb{G}' := \{\mathbf{U}\tilde{h}(\cdot)\mathbf{U}^T \mid \forall \tilde{h}(\cdot) \in \mathbb{H}'\}$ *for any unitary matrix* $\mathbf{U}$ *form a vector space, with* $\mathbb{G} \subset \mathbb{G}'$ *and* $\frac{dim(\mathbb{G}')}{dim(\mathbb{G})} = \frac{K+2K'+3}{K+1}$.

Proof. Consider the injective linear maps $f_1, f_2 : \mathbb{M}_n(\mathbb{K}_n^x) \to \mathbb{M}_n(\mathbb{K}_n^x)$ as $f_1(\mathbf{A}) = \mathbf{U}^T\mathbf{A}$ and $f_2(\mathbf{A}) = \mathbf{A}\mathbf{U}$. Define $f_3 : \mathbb{H} \to \mathbb{M}_n(\mathbb{K}_n^x)$ and $f_4 : \mathbb{H}' \to \mathbb{M}_n(\mathbb{K}_n^x)$ as $f_3(\mathbf{A}) = (f_1 \circ f_2)(\mathbf{A})$ for $\mathbf{A} \in \mathbb{H}$ and $f_4(\mathbf{A}) = (f_1 \circ f_2)(\mathbf{A})$ for $\mathbf{A} \in \mathbb{H}'$. Since $\mathbf{U}$ is given to be a unitary matrix, f_3 and f_4 are monomorphisms. Using this with the result from Theorem 4.2, $\mathbb{H} \subset \mathbb{H}'$, we have $\mathbb{G} \subset \mathbb{G}'$. $\square$

PPGNN Mitigates Oversmoothing. For showing that our model mitigates oversmoothing for the higher orders, we extend a few results by [8].

Lemma 1. *Assume that the nodes in an undirected and connected graph G have one of C labels. Then, for k large enough, we have,*

$$\mathbf{H}^k_{:j} = \beta_j \boldsymbol{\pi} + o_k(1)$$

$$(\mathbf{H}^k_{\sigma_i})_{:j} = \begin{cases} \beta_j \boldsymbol{\pi} + o_k(1), & \text{if } \pm 1 \in \sigma_i \\ \mathbf{0}, & \text{otherwise} \end{cases}$$

for $j \in [C]$. Here $\boldsymbol{\pi}_i = \dfrac{\sqrt{\tilde{D}_{ii}}}{\sqrt{\sum_{v \in V} \tilde{D}_{vv}}}$ and $\beta^T = \boldsymbol{\pi}^T \mathbf{H}_0$.

Proof. The first equality is a standard result. For the second equality, note that all $\mathbf{S}_{\sigma_i}$ have nullspace of dimension $n - |\sigma_i|$, and rest eigenvalues have their absolute values ≤ 1. By definition, $\hat{\mathbf{A}}$ is a doubly stochastic matrix, the stationary distribution for $\mathbf{S}_{\sigma_i}$ can only be reached if it contains an eigenvalue of absolute value 1. (easily seen that the largest eigenvalue of $\hat{\mathbf{A}}$ is 1). □

Thus, whenever the label prediction is dominated by higher order $\mathbf{H}^k_{()}$, all nodes have a representation proportional to $\tau\beta$, giving same label prediction for each node.

Definition 1 *(Oversmoothing).* *If oversmoothing occurs in PPGNN for K sufficiently large, we have $\mathbf{Z}_{:j} = c_1\beta_j\boldsymbol{\pi}, \forall j \in [C]$ for some $c_1 > 0$ if $\tau_k > 0$ and $\mathbf{Z}_{:j} = -c_1\beta_j\boldsymbol{\pi}, \forall j \in [C]$ for some $c_1 > 0$ if $\tau_k < 0$.*

Following lemma is the extended from the corresponding lemma of [8].

Lemma 2. *Let $L = \sum_{i \in \mathcal{T}} L_i = \sum_{i \in \mathcal{T}} -\log(\langle \mathbf{P}^T_{i:}, \mathbf{Y}^T_{i:}\rangle)$ be the cross-entropy loss and $\mathcal{T}$ be the training set. The gradient of τ_k for k large enough is $\frac{\partial L}{\partial \tau_k} = \sum_{i \in \mathcal{T}} \boldsymbol{\pi}_i \langle \mathbf{P}_{i:} - \mathbf{Y}_{i:}, \beta \rangle + o_k(1)$.*

Now the main result follows in same way as [8] from the above lemmas:

Theorem 3 *(Extension of Theorem 4.2 of [8]).* *If the training set contains nodes from each of C classes, then PP-GNN can always avoid over-smoothing. That is, for a large enough k and for a parameter associated with a k-order term, $\tau \in [\gamma_i] \cup [\gamma_i^{(h)}] \cup [\gamma_i^{(l)}], i \in [K] \cup \{0\}$, we have:*

$$\frac{\partial L}{\partial \tau} = \begin{cases} \sum_{i \in \mathcal{T}} \boldsymbol{\pi}_i \left(\max_{j \in [C]} \beta_j - \beta_{\mathbf{1}[\mathbf{Y}_{i:}]}\right) + o_k(1), \tau > 0 \\ \sum_{i \in \mathcal{T}} \boldsymbol{\pi}_i \left(\min_{j \in [C]} \beta_j - \beta_{\mathbf{1}[\mathbf{Y}_{i:}]}\right) + o_k(1), \tau < 0 \end{cases}$$

where, $\boldsymbol{\pi}_i = \dfrac{\sqrt{\tilde{D}_{ii}}}{\sqrt{\sum_{v \in V} \tilde{D}_{vv}}}$ and $\beta^T = \boldsymbol{\pi}^T \mathbf{H}_0$. This implies that all parameters, τ and their gradients $\frac{\partial L}{\partial \tau}$ are of same sign for sufficiently high orders. Since the gradients are bounded, higher order parameters τ will approach to 0 until we escape oversmoothing.

A.4   Experiments

Datasets. We evaluate on multiple benchmark datasets to show the effectiveness of our approach. Detailed statistics of the datasets used are provided in Table 3. We borrowed **Texas, Cornell, Wisconsin** from WebKB[5], where nodes represent web pages and edges denote hyperlinks between them. **Actor** is a co-occurrence network borrowed from [17], where nodes correspond to an actor, and edge represents the co-occurrence on the same Wikipedia page. **Chameleon, Squirrel** are borrowed from [18]. Nodes correspond to web pages and edges capture mutual links between pages. For all benchmark datasets, we use feature vectors, class labels from [3]. For datasets in (Texas, Wisconsin, Cornell, Chameleon, Squirrel, Actor), we use 10 random splits (48%/32%/20% of nodes for train/validation/test set) from [4]. We borrowed **Cora, Citeseer,** and **Pubmed** datasets and the corresponding train/val/test set splits from [4]. The remaining datasets were borrowed from [3]. We follow the same dataset setup mentioned in [3] to create 10 random splits for each of these datasets. We also experiment with two slightly larger datasets **Flickr** [20] and **OGBN-arXiv** [21]. We use the publicly available splits for these datasets.

Table 3. Dataset Statistics.

Properties	Texas	Wisconsin	Actor	Squirrel	Chameleon	Cornell	Flickr	Cora-Full
Homophily Level	0.11	0.21	0.22	0.22	0.23	0.30	0.32	0.59
#Nodes	183	251	7600	5201	2277	183	89250	19793
#Edges	492	750	37256	222134	38328	478	989006	83214
#Features	1703	1703	932	2089	500	1703	500	500
#Classes	5	5	5	5	5	5	7	70
#Train	87	120	3648	2496	1092	87	446625	1395
#Val	59	80	2432	1664	729	59	22312	2049
#Test	37	51	1520	1041	456	37	22313	16349

Properties	OGBN-arXiv	Wiki-CS	Citeseer	Pubmed	Cora	Computer	Photos
Homophily Level	0.63	0.68	0.74	0.80	0.81	0.81	0.85
#Nodes	169343	11701	3327	19717	2708	13752	7650
#Edges	1335586	302220	12431	108365	13264	259613	126731
#Features	128	300	3703	500	1433	767	745
#Classes	40	10	6	3	7	10	8
#Train	90941	580	1596	9463	1192	200	160
#Val	29799	1769	1065	6310	796	300	240
#Test	48603	5487	666	3944	497	13252	7250

Hyperparameter Tuning. We provide the methods in comparison along with the hyper-parameters ranges for each model. For all the baseline models, we sweep the common hyper-parameters in the same ranges. Learning rate is swept over $[0.001, 0.003, 0.005, 0.008, 0.01]$, dropout over $[0.2, 0.3, 0.4, 0.5, 0.6, 0.7, 0.8]$, weight decay over $[1e-4, 5e-4, 1e-3, 5e-3, 1e-2, 5e-2, 1e-1]$, and hidden dimensions over $[16, 32, 64]$. For model-specific hyper-parameters, we tune over author prescribed ranges. We use undirected graphs with symmetric normalization for all graph networks in comparison. For all models, we report test

---

[5] http://www.cs.cmu.edu/afs/cs.cmu.edu/project/theo-11/www/wwkb.

accuracy for the configuration that achieves the highest validation accuracy. We report standard deviation wherever applicable.

The hyper-parameter search space as described above, was prescribed in [46] where the authors have shown that with thorough tuning, GCN(s) thoroughly outperform several existing SOTA models. Training and observing the results of all the models on this hyper-parameter search space, while also maintaining a consistent number of Optuna Trials (per SOTA model, per dataset, per split), is a convincing indicator of the fact that our proposed method does hold merit, and is not attributed to a statistical anomaly.

LR and MLP: We trained Logistic Regression classifier and Multi Layer Perceptron on the given node features. For MLP, we limit the number of hidden layers to one.

GCN: We use the GCN implementation provided by the authors of [8].

SGCN: SGCN [16] is a spectral method that models a low pass filter and uses a linear classifier. The number of layers in SGCN is treated as a hyper-parameter and swept over $[1, 2]$.

SuperGAT: SUPERGAT [3] is an improved graph attention model designed to also work with noisy graphs. SUPERGAT employs a link-prediction based self-supervised task to learn attention on edges. As suggested by the authors, on datasets with homophily levels lower than 0.2 we use SUPERGAT$_{SD}$. For other datasets, we use SUPERGAT$_{MX}$. We rely on authors code for our experiments.

Geom-GCN: GEOM-GCN [4] proposes a geometric aggregation scheme that can capture structural information of nodes in neighborhoods and also capture long range dependencies. We quote author reported numbers for Geom-GCN. We could not run Geom-GCN on other benchmark datasets because of the unavailability of a pre-processing function that is not publicly available.

H$_2$GCN: H$_2$GCN [5] proposes an architecture, specially for heterophilic settings, that incorporates three design choices: i) ego and neighbor-embedding separation, higher-order neighborhoods, and combining intermediate representations. We quote author reported numbers where available, and sweep over author prescribed hyper-parameters for reporting results on the rest datasets. We rely on author's code for our experiments.

FAGCN: FAGCN [7] adaptively aggregates different low-frequency and high-frequency signals from neighbors belonging to same and different classes to learn better node representations. We rely on author's code for our experiments.

APPNP: APPNP [10] is an improved message propagation scheme derived from personalized PageRank. APPNP's addition of probability of teleporting back to root node permits it to use more propagation steps without oversmoothing. We use GPR-GNN's implementation of APPNP for our experiments.

LinearGCN (LGC): LINEARGCN (LGC) [22] is a spectrally grounded GCN that adapts the entire eigen spectrum of the graph to obtain better node feature representations.

GPR-GNN: GPR-GNN [8] adaptively learns weights to jointly optimize node representations and the level of information to be extracted from graph topology. We rely on author's code for our experiments.

TDGNN: TDGNN [31] is a tree decomposition method which mitigates feature smoothening and disentangles neighbourhoods in different layers. We rely on author's code for our experiments.

ARMA: ARMA [38] is a spectral method that uses K stacks of $ARMA_1$ filters in order to create an $ARMA_K$ filter (an ARMA filter of order K). Since [38] do not specify a hyperparameter range in their work, following are the ranges we have followed: GCS stacks (S): $[1, 2, 3, 4, 5, 6, 7, 8, 9, 10]$, stacks' depth$(T)$: $[1, 2, 3, 4, 5, 6, 7, 8, 9, 10]$. However we only select configurations such that the number of learnable parameters are less than or equal to those in PP-GNN. The input to the ARMAConv layer are the node features and the output is the number of classes. This output is then passed through a softmax layer. We use the implementation from the official PyTorch Geometric Library[6]

BernNet: BernNet [37] is a method that approximates any filter over the normalised Laplacian spectrum of a graph, by a Kth Order Bernstein Polynomial Approximation. We use the model specific hyper-parameters prescribed by the authors of the paper. We vary the Propagation Layer Learning Rate as follows: $[0.001, 0.002, 0.01, 0.05]$. We also vary the Propagation Layer Dropout as follows: $[0.2, 0.3, 0.4, 0.5, 0.6, 0.7, 0.8]$. We rely on the authors code for our experiments.

AdaGNN: AdaGNN [36] is a method that captures the different importance's for varying frequency components for node representation learning. We use the model specific hyper-parameters prescribed by the authors of the paper. The No. of Layers hyper-parameter is varied as follows: $[2, 4, 8, 16, 32, 128]$. We rely on the authors code for our experiments.

UFG: UFG [32] decompose the graph into low-pass and high-pass frequencies, and define a framelet based convolutional model. We use the model specific hyper-parameters as prescribed by the authors of the paper. We rely on the authors code for our experiments.

GFIR - Unconstrained Setting: In this setting, we do not impose any regularization constraints such as dropout and L2 regularization.

GFIR - Constrained Setting: In this setting, we impose dropouts as well as L2 regularization on the GFIR model. Both dropouts and L2 regularization were applied on the H_k's (the learnable filter matrices from the above equation).

Links to the authors' codebases can be found in Table 4.

---

[6] https://pytorch-geometric.readthedocs.io/en/latest/_modules/torch_geometric/nn/conv/arma_conv.html#ARMAConv.

Table 4. Links to the codebases of certain baselines.

Method	Code links	Commit ID
GCN	https://github.com/jianhao2016/GPRGNN	dc246833865af87ae5d4e965d189608f8832ddac
SuperGAT	https://github.com/dongkwan-kim/SuperGAT	2d3f44acbb10af5850aa17a3903dea955a29d2e2
H2GCN	https://github.com/GemsLab/H2GCN	08011c5199426e1c49b80ee2944d338dfd55e2b5
FAGCN	https://github.com/bdy9527/FAGCN	23bb10f6bf0b1d2e5874140cd4b266c60a7c63f3
APPNP	https://github.com/jianhao2016/GPRGNN	dc246833865af87ae5d4e965d189608f8832ddac
GPRGNN	https://github.com/jianhao2016/GPRGNN	dc246833865af87ae5d4e965d189608f8832ddac
TDGNN	https://github.com/YuWVandy/TDGNN	505b1af90255aace255744ec81a7033a5d682b90
BernNet	https://github.com/ivam-he/BernNet	7b9c1652dbe43730f52d647957761bf6d3f17425
AdaGNN	https://github.com/yushundong/AdaGNN	f178d3144921c8845027234cac68a7f0dd057fe2
UFG	https://github.com/YuGuangWang/UFG	229acd89b33f4f4e1bab2c0d92fb93d146127fd1

Implementation Details. In this subsection, we present several important points that are useful for practical implementation of our proposed method and other experiments related details. Our approach is based on adaptation of a few eigen graphs constructed using eigen components. Following [1], we use a symmetric normalized version ($\tilde{\mathbf{A}}$) of adjacency matrix $\mathbf{A}$ with self-loops: $\tilde{\mathbf{A}} = \tilde{D}^{-\frac{1}{2}}(\mathbf{A} + \mathbf{I})\tilde{D}^{-\frac{1}{2}}$ where $\tilde{D}_{ii} = 1 + D_{ii}$, $D_{ii} = \sum_j A_{ij}$ and $\tilde{D}_{ij} = 0, i \neq j$. We work with eigen matrix and eigen values of $\tilde{\mathbf{A}}$.

To reduce the learnable hyper-parameters, we separately partition the low-end and high-end eigen values into several contiguous bins and use shared filter parameters for each of these bins. The number of bins, which can be interpreted as number of filters, is swept in the range $[2, 4, 5]$. The orders of the polynomial filters are swept in the range $[2, 4, 6]$. The number of EVD components are swept in the range $[256, 1024]$. In our experiments, we set $\eta_l = \eta_h$ and we vary the η_l parameter in range $(0, 1)$ and $\eta_{gpr} = 1 - \eta_l$. The range of η_l, η_h and η_{gpr} is kept between $(0, 1)$ in order to have a bounded and weighted contribution of every term. Since previous works either use directly or use a variant of the GPR term, these ranges make it feasible to carry out an analysis of how the term contributes to the learning of the representations and also to compare it with the contribution of the terms of the proposed model, as keeping parameters between $(0, 1)$, adding to 1 provides room for a weighted contribution of each term.

For optimization, we use the Adam optimizer [19]. We set early stopping to 200 and the maximum number of epochs to 1000. We utilize learning rate with decay, with decay factor set to 0.99 and decay frequency set to 50. All our experiments were performed on a machine with Intel Xeon 2.60 GHz processor, 112 GB Ram, Nvidia Tesla P-100 GPU with 16 GB of memory, Python 3.6, and PyTorch 1.9.0 [24]. We used Optuna [23] and set the number of trials to 20 to optimize the hyperparameter search for PP-GNN. For other baseline models, we set the number of trials to 100.

Note: Several baselines report elevated results on some of our benchmark datasets. This difference is because of the difference in splits. We use the splits from [4]. Baselines including BERNNET, GPR-GNN evaluate on random splits with 60/20/20 distribution for train/val/test labels.

A.5   PP-GNN v/s SOTA Models (Extension)

Comparison Against Other Baslines. We compare our method against three category of methods: (a) standard LR (Logistic Regression) and MLP (Multi-Layer Perceptron), (b) traditional and spatial convolution-based GNN models including GCN, SGCN, SUPERGAT, TDGNN, H_2GCN, and GEOM-GCN, and (c) recent spectral convolution-based methods (with emphasis on graph filters) such as GPR-GNN, FAGCN, APPNP, LGC, ARMA, ADAGNN, BERNNET, GFIR, and UFG. Our tabular results are organized as per this grouping, along with references. In some cases, the grouping of spatial convolution-based methods is somewhat overlapping with spectral filtering based-methods since spectral interpretations are available for the former. The results for Heterophilic graphs are tabulated in Table 5, and Homophilic graphs are reported in Table 7.

Large Datasets. We also observe gains on moderately large datasets like Flickr and perform competitively on the OGBN-arXiv dataset. Please note that our latter numbers are slightly inferior for baselines like GCN compared to the leaderboard[7] numbers. These differences are because we turn off the optimization tricks like Batch Normalization.

A.6   Additional Experiments

Role of Different Filter-Banks. Recall that the PP-GNN model is a filter-bank model comprising several polynomial filters operating at different parts of the spectrum. We evaluate PP-GNN's performance by learning each group of filters alone (e.g., PP-GNN(Low), PP-GNN(High)) and report results on several datasets in Table 6. We see that the Heterophilic datasets (like Squirrel and Chameleon) largely benefit from high-frequency signals. In contrast, Homophilic datasets (Cora and Citeseer) exhibit a reverse trend. Incorporating the GPR-GNN filter as part of the PP-GNN filter helps to get improved performance over individual filters (PP-GNN(Low) or PP-GNN (High)) and shows considerable improvements over a wide variety of datasets.

Adaptable Frequency Responses. In Fig. 1 of the main paper, we observe that PP-GNN learns a complicated frequency response for a heterophilic dataset (Squirrel) and a simpler frequency response for a homophilic dataset (Citeseer). We observe that this trend follows for two other datasets Chameleon (heterophilic) and Computer (homophilic). See Fig. 9.

---

[7] https://tinyurl.com/oarxiv.

Table 5. Results on Heterophilic Datasets. '*' indicates that the results were borrowed from the corresponding papers. Bold indicates the best performing model; underline for second-best.

	Texas	Wisconsin	Squirrel	Chameleon	Cornell
LR	81.35 (6.33)	84.12 (4.25)	34.73 (1.39)	45.68 (2.52)	83.24 (5.64)
MLP	81.24 (6.35)	84.43 (5.36)	35.38 (1.38)	51.64 (1.89)	83.78 (5.80)
SGCN [16]	62.43 (4.43)	55.69 (3.53)	45.72 (1.55)	60.77 (2.11)	62.43 (4.90)
GCN [1]	61.62 (6.14)	58.82 (4.89)	47.78 (2.13)	62.83 (1.52)	62.97 (5.41)
SuperGAT [3]	61.08 (4.97)	56.47 (3.90)	31.84 (1.26)	43.22 (1.71)	57.30 (8.53)
Geom-GCN [4]	67.57*	64.12*	38.14*	60.90*	60.81*
H2GCN [5]	84.86 (6.77)*	86.67 (4.69)*	37.90 (2.02)*	58.40 (2.77)	82.16 (4.80)*
TDGNN [31]	83.00 (4.50)*	85.57 (3.78)*	43.84 (2.16)	55.20 (2.30)	82.92 (6.61)*
GFIR-1[a]	73.24 (6.91)	77.84 (3.21)	36.50 (1.12)	51.71 (3.11)	72.43 (7.62)
GFIR-2[b]	74.59 (4.45)	79.41 (3.10)	41.12 (1.17)	61.27 (2.42)	74.05 (7.77)
FAGCN [7]	82.43 (6.89)	82.94 (7.95)	42.59 (0.79)	55.22 (3.19)	79.19 (9.79)
APPNP [10]	81.89 (5.85)	85.49 (4.45)	39.15 (1.88)	47.79 (2.35)	81.89 (6.25)
LGC [22]	80.20 (4.28)	81.89 (5.98)	44.26 (1.49)	61.14 (2.07)	74.59 (3.42)
GPR-GNN [8]	81.35 (5.32)	82.55 (6.23)	46.31 (2.46)	62.59 (2.04)	78.11 (6.55)
AdaGNN [36]	71.08 (8.55)	77.70 (4.91)	53.50 (0.96)	65.45 (1.17)	71.08 (8.36)
BernNET [37]	83.24 (6.47)	84.90 (4.53)	52.56 (1.69)	62.02 (2.28)	80.27 (5.41)
ARMA [38]	79.46 (3.65)	82.75 (3.56)	47.37 (1.63)	60.24 (2.19)	80.27 (7.76)
UFG-ConvR [32]	66.22 (7.46)	68.63 (4.98)	42.06 (1.55)	56.29 (1.58)	69.19 (6.40)
PP-GNN	**89.73 (4.90)**	**88.24 (3.33)**	**59.15 (1.91)**	**69.10 (1.37)**	82.43 (4.27)

[a] GFIR-1: unconstrained setting
[b] GFIR-2: constrained setting

Table 6. Performance of different filters

Test Acc	Squirrel	Chameleon	Citeseer	Cora
PP-GNN (Low)	45.75 (1.69)	56.73 (4.03)	76.23 (1.54)	88.03 (0.79)
PP-GNN (High)	58.70 (1.60)	**69.19 (1.88)**	55.50 (6.38)	73.76 (2.03)
PP-GNN (GPR[a]+Low)	50.96 (1.26)	63.71 (2.69)	78.07 (1.71)	**89.56 (0.93)**
PP-GNN (GPR+High)	60.39 (0.91)	67.83 (2.30)	**78.30 (1.60)**	89.42 (0.97)
GPR-GNN	42.06 (1.55)	56.29 (1.58)	76.74 (1.33)	87.93 (1.52)
PP-GNN	**59.15 (1.91)**	69.10 (1.37)	78.25 (1.76)	89.52 (0.85)

[a] GPR here refers to GPR-GNN

Effect of Number of Eigenvalues/Vectors (EVs). Since the number of EVs to adapt might not be known apriori, we conducted a study to assess the effect of using different number of EVs on test performance. We report results on a few representative datasets. From Fig. 10, we see that Homophilic datasets can benefit by adapting as small as 32 eigen components. Heterophilic datasets achieve peak performance by adapting (~250–500) number of eigen components. These results indicate that the number of EVs required to get competitive/superior performance is typically small, therefore, computationally feasible and affordable.

Does the MLP Even Matter? In PP-GNN there is a two layered MLP (that transforms the input node features) followed by a single graph filtering layer similar to GPR-GNN.

To understand MLP's significance, we ran an additional experiment, where we have used a single linear layer, instead of the two layered MLP. We continue to observe competitive (with respect to our original PP-GNN model) performance, across most datasets. The results can be found in Table 9. Also, the two layer MLP is not the significant contributor towards performance. This can also be seen by comparing GPR-GNN's performance with that of LGC's. LGC can be interpreted as a linear version of GPR-GNN, and achieves comparable performance as GPR-GNN.

Table 9 seems to suggest that PP-GNN (Linear) is competitive to PP-GNN (Original).

We can infer that adding MLP may give marginal improvements over the linear version. This phenomenon is also observed in GPR-GNN. To illustrate this, we can compare GPR-GNN with LGC (linear version of GPR-GNN). We can observe in Table 9 that GPR-GNN and LGC are comparable in performance.

A.7 Comparison Against General FIR Filters

Instead of using a polynomial filter, we can use a general FIR filter (GFIR) which is described by the following equation (Table 8):

$$Z = \sum_{k=0}^{K} S^k X H_k$$

Table 7. Results on homophilic datasets.

	Cora-Full	Wiki-CS	Citeseer	Pubmed	Cora	Computer	Photos
LR	39.10 (0.43)	72.28 (0.59)	72.22 (1.54)	87.00 (0.40)	73.94 (2.47)	64.92 (2.59)	77.57 (2.29)
MLP	43.03 (0.82)	73.74 (0.71)	73.83 (1.73)	87.77 (0.27)	77.06 (2.16)	64.96 (3.57)	76.96 (2.46)
SGCN	61.31 (0.78)	78.30 (0.75)	76.77 (1.52)	88.48 (0.45)	86.96 (0.78)	80.65 (2.78)	89.99 (0.69)
GCN	59.63 (0.86)	77.64 (0.49)	76.47 (1.33)	88.41 (0.46)	87.36 (0.91)	82.50 (1.23)	90.67 (0.68)
SuperGAT	57.75 (0.97)	77.92 (0.82)	76.58 (1.59)	87.19 (0.50)	86.75 (1.24)	83.04 (1.02)	90.31 (1.22)
Geom-GCN	NA	NA	77.99*	90.05*	85.27*	NA	NA
H2GCN	57.83 (1.47)	OOM	77.07 (1.64)*	89.59 (0.33)*	87.81 (1.35)*	OOM	91.17 (0.89)
TDGNN	OOM	79.58 (0.51)	76.64 (1.54)*	89.22 (0.41)*	88.26 (1.32)*	84.52 (0.92)	92.54 (0.28)
GFIR (unconstrained)	60.87 (0.78)	79.15 (0.65)	75.83 (1.94)	88.47 (0.45)	87.93 (0.90)	78.39 (1.09)	89.26 (1.00)
GFIR (constrained)	60.92 (0.80)	79.15 (0.63)	76.24 (1.43)	88.47 (0.39)	87.46 (1.26)	79.57 (2.12)	89.38 (1.03)
FAGCN	60.07 (1.43)	79.23 (0.66)	76.80 (1.63)	89.04 (0.50)	88.21 (1.37)	82.16 (1.48)	90.91 (1.11)
APPNP	60.83 (0.55)	79.13 (0.50)	76.86 (1.51)	89.57 (0.53)	88.13 (1.53)	82.03 (2.04)	91.68 (0.62)
LGC	**61.84 (0.90)**	79.82 (0.49)	76.96 (1.73)	88.78 (0.51)	88.02 (1.44)	83.44 (1.77)	91.56 (0.74)
GPR-GNN	61.37 (0.96)	79.68 (0.50)	76.84 (1.69)	89.08 (0.39)	87.77 (1.31)	82.38 (1.60)	91.43 (0.89)
AdaGNN	59.57 (1.18)	77.87 (4.95)	74.94 (0.91)	89.33 (0..57)	86.72 (1.29)	81.27 (2.10)	89.93 (1.22)
BernNET	60.77 (0.92)	79.75 (0.52)	77.01 (1.43)	89.03 (0.55)	88.13 (1.41)	83.69 (1.99)	91.61 (0.51)
ARMA	60.23 (1.21)	78.94 (0.32)	78.15 (0.74)	88.73 (0.52)	87.37 (1.14)	78.55 (2.62)	90.26 (0.48)
UFG-ConvR	60.98 (0.82)	78.56 (0.43)	76.74 (1.33)	85.68 (0.62)	87.93 (1.52)	80.01 (1.78)	90.20 (1.41)
PP-GNN	61.42 (0.79)	**80.04 (0.43)**	**78.25 (1.76)**	**89.71 (0.32)**	**89.52 (0.85)**	**85.23 (1.36)**	**92.89 (0.37)**

Table 8. Results on large datasets.

	LR	MLP	GCN	SGCN	SuperGAT	H2GCN	TDGNN	PP-GNN
Flickr	46.51	46.93	53.40	50.75	53.47	OOM	OOM	**55.30**
OGBN-arXiv	52.53	54.96	**69.37**	68.51	55.1*	OOM	OOM	<u>69.28</u>

Table 9. Comparision of linear GPR-GNN and linear PP-GNN with respect to other pertinent baselines.

	Computers	Chameleon	Citeseer	Cora	Squirrel	Texas	Wisconsin
GPR-GNN	82.38 (1.60)	62.59 (2.04)	76.84 (1.69)	87.77 (1.31)	46.31 (2.46)	81.35 (5.32)	82.55 (6.23)
LGC	83.44 (1.77)	61.14 (2.07)	76.96 (1.73)	88.02 (1.44)	44.26 (1.49)	80.20 (4.28)	81.89 (5.98)
PP-GNN (Original)	85.23 (1.36)	69.10 (1.37)	78.25 (1.76)	89.52 (0.85)	59.15 (1.91)	89.73 (4.90)	88.24 (3.33)
PP-GNN (Linear)	84.27 (1.19)	67.88 (1.62)	77.86 (1.74)	88.43 (0.69)	55.11 (1.72)	85.58 (4.70)	86.24 (3.23)

Table 10. Comparing PP-GNN and GPR-GNN against the GFIR filter models.

Train Acc/Test Acc	Computers	Chameleon	Citeseer	Cora	Squirrel	Texas	Wisconsin
GFIR (Unconstrained)	78.39 (1.09)	51.71 (3.11)	75.83 (1.94)	87.93 (0.90)	36.50 (1.12)	73.24 (6.91)	77.84 (3.21)
GFIR (Constrained)	79.57 (2.12)	61.27 (2.42)	76.24 (1.43)	87.46 (1.26)	41.12 (1.17)	74.59 (4.45)	79.41 (3.10)
GPR-GNN	82.38 (1.60)	62.59 (2.04)	76.84 (1.69)	87.77 (1.31)	46.31 (2.46)	81.35 (5.32)	82.55 (6.23)
PP-GNN	**85.23** (1.36)	**67.74** (2.31)	**78.25** (1.76)	**89.52** (0.85)	**56.86** (1.20)	**89.73** (4.90)	**88.24** (3.33)

Table 11. Different polynomial filtering based methods. Note that the coefficients of APPNP are fixed (not learnable) PPR coefficients ($\gamma_k \, \forall \, k$) and the coefficients of GPR-GNN ($\gamma_k \, \forall k$) and BERNNET ($\theta_k \, \forall k$) are learnable.

Method	Polynomial basis	Filter response	Constraints
APPNP	Monomial	$h(\lambda) = \sum_{k=0}^{K} \gamma_k \lambda^k$	$\gamma_k = \alpha(1-\alpha)^k; \gamma_K = (1-\alpha)^K; \alpha$ is a hyper-parameter
GPR-GNN	Monomial	$h(\lambda) = \sum_{k=0}^{K} \gamma_k \lambda^k$	γ_k are unconstrained
BERNNET	Bernstein	$h(\lambda) = \sum_{k=0}^{K} \frac{\theta_k}{2^k} \binom{K}{k} (2-\lambda)^{K-k} \lambda^k$	$\theta_k \geq 0$

where S is the graph shift operator (which in our case is $\widetilde{A}$), X is the node feature matrix and H_k's are learnable filter matrices. One can see GCN, SGCN, GPR-GNN as special cases of this GFIR filter, which constrain the H_k in different ways.

We first demonstrate that constraint on the GFIR filter is necessary for getting improvement in performance, particularly on heterophilic datasets. Towards this, we build two versions of GFIR: one with regularization (constrained), and the other without regularization (unconstrained). We ensure that the number of trainable parameters in these models are comparable to those used in PP-GNN. We provide further details of the versions of the GFIR models below and report the results in Table 10 below:

– **Unconstrained Setting**: In this setting, we do not impose any regularization constraints such as dropout and L2 regularization.

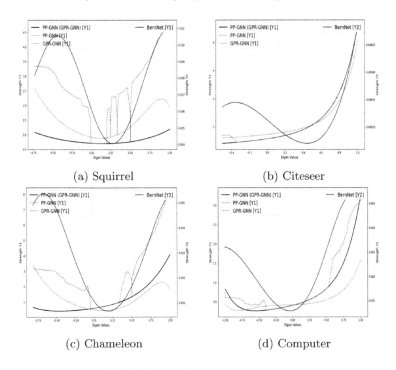

(a) Squirrel

(b) Citeseer

(c) Chameleon

(d) Computer

Fig. 9. Learnt frequency responses

– **Constrained Setting**: In this setting, we impose dropouts as well as L2 regularization on the GFIR model. Both dropouts and L2 regularization were applied on the H_k's (the learnable filter matrices from the above equation).

We also compare PP-GNN (the proposed model) as well as GPR-GNN to the General FIR filter model (GFIR).

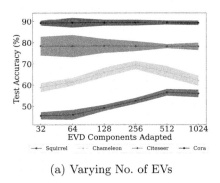

(a) Varying No. of EVs

Fig. 10. Analyzing varying number of eigenvalues on performance

We can make the following observation from the results reported in Table 10:

- Firstly, constrained GFIR performs better than the unconstrained version, with performance lifts of up to ~10%. This suggests that regularization is important for GFIR models.
- GPR-GNN outperforms the constrained GFIR version. It is to be noted that GPR-GNN further restricts the space of graphs explored as compared to GFIR. This suggests that regularization beyond simple L2/dropout kind of regularization (polynomial filter) is beneficial.
- PP-GNN performs better than GPR-GNN. Our model slightly expands the space of graphs explored (as compared to GPR-GNN, but lesser than GFIR), while retaining good performance. This suggests that there is still room for improvement on how regularization is done.

PP-GNN has shown one possible way to constrain the space of graphs while improving performance on several datasets, however, it remains to be seen whether there are alternative methods that can do even better. We hope to study and analyze this aspect in the future.

A.8   More Details on Comparison Against Polynomial Filtering Methods

More details on Sect. 4.4 are given below:

APPNP: The node embeddings are learnt by APPNP as described below:

$$Z = \sum_{k=0}^{K} \gamma_k \tilde{A}_{sym}^k \mathbf{Z}_x(\mathbf{X}, \boldsymbol{\Theta})$$

APPNP uses fixed $\gamma_k = \alpha(1-\alpha)^k$ with $\gamma_K = (1-\alpha)^K$ where α is a hyper-parameter, $\mathbf{Z}_x(\mathbf{X}, \boldsymbol{\Theta})$ are the node features transformed by MLP with parameter $\boldsymbol{\Theta}$.

GPR-GNN: The node embeddings are learnt by GPR-GNN as described below:

$$Z = \sum_{k=0}^{K} \gamma_k \tilde{A}_{sym}^k \mathbf{Z}_x(\mathbf{X}, \boldsymbol{\Theta})$$

where $\gamma_k(\forall k)$ are now learnable parameters, $\mathbf{Z}_x(\mathbf{X}, \boldsymbol{\Theta})$ are the node features transformed by MLP with parameter $\boldsymbol{\Theta}$.

BernNet: The node embeddings for BernNet are learnt as described below:

$$Z = \sum_{k=0}^{K} \frac{\theta_k}{2^K} \binom{K}{k} (2I - L)^{K-k} L^k \mathbf{Z}_x(\mathbf{X}, \mathbf{\Theta})$$

$$= \sum_{k=0}^{K} \frac{\theta_k}{2^K} \binom{K}{k} (A_{sym} + I)^{K-k} (I - A_{sym})^k \mathbf{Z}_x(\mathbf{X}, \mathbf{\Theta})$$

$$= \sum_{q=0}^{K} [\sum_{r=0}^{K} \frac{\theta_k}{2^K} \binom{K}{r} \sum_{p=0}^{q} \binom{K-r}{q-p} \binom{r}{p} (-1)^p] A_{sym}^q \mathbf{Z}_x(\mathbf{X}, \mathbf{\Theta})$$

$$= \sum_{q=0}^{K} [\sum_{r=0}^{K} \theta_r \alpha_{rq}] A_{sym}^q \mathbf{Z}_x(\mathbf{X}, \mathbf{\Theta})$$

$$= \sum_{q=0}^{K} w_q A_{sym}^q \mathbf{Z}_x(\mathbf{X}, \mathbf{\Theta})$$

The following table summarizes recent key polynomial filtering based methods and a short description of the constraints/variant they employ.

A.9 Training Time Analysis

In the following subsections, we provide comprehensive timing analysis.

Computational Complexity: Listed below is the computational complexity for each piece in our model for a single forward pass. Notation n: number of nodes, $|E|$: the number of edges, A: symmetric normalized adjacency matrix, F: features dimensions, d: hidden layer dimension, C: number of classes, e^* denotes the cost of EVD, K: polynomial/hop order, l: number of eigenvalues/vectors in a single partition of spectrum (for implementation, we keep l same for all such intervals), m: number of partitions of a spectrum.

- MLP: $O(nFd + ndC)$
- GPR-term: $O(K|E|C) + O(nKC)$. The first term is the cost for computing $A^K f(X)$ for sparse A. The second term is the cost of summation $\sum_k A^k f(X)$.
- Excess terms for PP-GNN: $O(mnlC)$. This is obtained by the optimal matrix multiplication present in Eq. 5 of the main paper ($\mathbf{U}_i$ is $n \times l$, $H_i(\gamma_i)$ is $l \times l$, $\mathbf{Z}_0()$ is $n \times C$). The additional factor m is because we have m different contiguous intervals/different polynomials. Typically n is much larger than l.
- EVD-term: e^*, the complexity for obtaining the eigenvalues/vectors of the adjacency matrix, which is usually very sparse for the observed graphs. Most publicly available solvers for this task utilize Lanczos' algorithm (which is a specific case of a more general Arnoldi iteration). However, the convergence bound of this iterative procedure depends upon the starting vectors and the underlying spectrum (particularly the ratio of the absolute difference of two largest eigenvalues to the diameter of the spectrum) [33–35].

Lanczos' algorithm is shown to be a practically efficient way for obtaining extreme eigenpairs for a similar and even very large systems. We use ARPACK's built-in implementation to precompute the eigenvalues/vectors for all datasets before training, thus amortizing this cost across training with different hyper-parameters configuration.

Per Component Timing Breakup: In Table 12, we provide a breakdown of cost incurred in seconds for different components of our model. Since the eigenpairs' computation is a one time cost, we amortize this cost over the total hyper-parameters configurations and report the effective training time in the last column on of Table 12.

Average Training Time: In Table 13, we report the training time averaged over 20 hyper-parameter configurations for several models. To understand the relative performance of our model with respect to GCN, we compute the relative time taken and report it in Table 14. We can observe in Table 14 that PP-GNN is $\sim 4x$ slower than GCN, $\sim 2X$ slower than GPR-GNN and BernNET, and $\sim 2X$ faster than AdaGNN. However, it is important to note that in our average training time, the time taken to compute K top and bottom eigenvalues/vectors is amortized across the number of trials (Table 15).

Table 12. PP-GNN's per component timing cost. Training Time refers to the end to end training time (without eigen decomposition) averaged across 20 trials. EVD cost refers to the time taken to obtain **x** top and bottom eigenvalues. This **x** can be found in the 'Number of EV's obtained' column. Since EVD is a one time cost, we average this cost over the total number of trials and add it to the training time. We refer to this cost as the Effective Training Time (ETT).

PP-GNN	Training time	EVD cost	#EV's obtained	ETT
Texas	11.89	0.00747	183 (All EVs)	11.89
Cornell	11.63	0.03271	183 (All EVs)	11.63
Wisconsin	12.08	0.01225	251 (All EVs)	12.08
Chameleon	21.44	3.71883	2048	21.63
Squirrel	31.38	15.8152	2048	32.17
Cora	22.46	54.3684	2048	25.18
Citeseer	20.51	56.9744	2048	23.36
Cora-full	63.98	155.304	2048	71.75
Pubmed	52.54	256.71	2048	65.38
Computers	28.63	76.2738	2048	32.44
Photo	19.3	48.3683	2048	21.72
Flickr	161.16	304.114	2048	176.37
ArXiv	189.94	412.504	1024	210.57
WikiCS	27.92	65.4376	2048	31.19

Table 13. Training time (in seconds) across models

Dataset	GPR-GNN	PP-GNN	MLP	GCN	BernNet	ARMA	AdaGNN
Texas	9.27	11.89	1.08	3.46	5.59	6.00	13.97
Cornell	9.41	11.63	1.06	3.69	5.37	5.51	12.56
Wisconsin	9.67	12.08	1.07	3.42	5.69	5.36	13.57
Chameleon	14.69	21.63	2.60	6.42	12.46	7.84	28.77
Squirrel	18.94	32.17	5.04	7.52	17.82	28.87	90.36
Cora	12.90	25.18	1.95	5.94	12.25	10.67	22.15
Citeseer	10.62	23.36	3.72	4.56	9.52	19.5	35.34
Cora-Full	24.98	71.75	7.77	8.01	31.26	40.21	175.58
Pubmed	14.00	65.38	6.21	11.73	12.64	27.76	162.01
Computers	7.67	32.44	2.24	6.68	7.48	27.76	118.43
Photo	8.58	21.72	1.68	5.1	7.95	14.34	45.46
Flickr	42.64	176.37	21.00	30.4	62.11	119.3	178.7371
ArXiv	118.35	210.57	78.9	102.88	693.92	771.59	307.84
WikiCS	14.37	31.19	3.34	10.8	11.43	30.79	73.63

Table 14. Training time of models relative to the training time of GCN

Dataset	GPR-GNN	PP-GNN	MLP	GCN	BernNet	ARMA	AdaGNN
Texas	2.68	3.44	0.31	1.00	1.62	1.73	4.04
Cornell	2.55	3.15	0.29	1.00	1.46	1.49	3.40
Wisconsin	2.83	3.53	0.31	1.00	1.66	1.57	3.97
Chameleon	2.29	3.37	0.40	1.00	1.94	1.22	4.48
Squirrel	2.52	4.28	0.67	1.00	2.37	3.84	12.02
Cora	2.17	4.24	0.33	1.00	2.06	1.80	3.73
Citeseer	2.33	5.12	0.82	1.00	2.09	4.28	7.75
Cora-Full	3.12	8.96	0.97	1.00	3.90	5.02	21.92
Pubmed	1.19	5.57	0.53	1.00	1.08	2.37	13.81
Computers	1.15	4.86	0.34	1.00	1.12	4.16	17.73
Photo	1.68	4.26	0.33	1.00	1.56	2.81	8.91
Flickr	1.40	5.80	0.69	1.00	2.04	3.92	5.88
ArXiv	1.15	2.05	0.77	1.00	6.74	7.50	2.99
WikiCS	1.33	2.89	0.31	1.00	1.06	2.85	6.82
Average	**2.03**	**4.39**	**0.50**	**1.00**	**2.19**	**3.18**	**8.39**

Table 15. End to end training time (in HH:MM:SS) for optimizing over 20 hyper-parameter configurations

Dataset	Chameleon	Citeseer	Computers	Cora	Cora-Full	Photo
Time	00:03:46	00:10:17	00:34:37	00:05:24	00:59:29	00:10:31

Dataset	Pubmed	Squirrel	Texas	Wisconsin	OGBN-ArXiv
Time	00:57:40	00:10:38	00:02:27	00:02:33	01:03:20

References

1. Kipf, T., Welling, M.: Semi-supervised classification with graph convolutional networks. In: International Conference on Learning Representations (ICLR) (2017)
2. Veličković, P., Cucurull, G., Casanova, A., Romero, A., Liò, P., Bengio, Y.: Graph attention networks. In: International Conference on Learning Representations (ICLR) (2018)
3. Kim, D., Oh, A.: How to find your friendly neighborhood: graph attention design with self-supervision. In: International Conference on Learning Representations (ICLR) (2021)
4. Pei, H., Wei, B., Chang, K., Lei, Y., Yang, B.: Geom-GCN: geometric graph convolutional networks. In: International Conference on Learning Representations (ICLR) (2020)
5. Zhu, J., Yan, Y., Zhao, L., Heimann, M., Akoglu, L., Koutra, D.: Beyond homophily in graph neural networks: current limitations and effective designs. In: Neural Information Processing Systems (NeurIPS) (2020)
6. Zhu, J., et al.: Graph neural networks with heterophily. In: Association for the Advancement of Artificial Intelligence (AAAI) (2021)
7. Bo, D., Wang, X., Shi, C., Shen, H.: Beyond low-frequency information in graph convolutional networks. In: Association for the Advancement of Artificial Intelligence (AAAI) (2021)
8. Chien, E., Peng, J., Li, P., Milenkovic, O.: Adaptive universal generalized pagerank graph neural network. In: International Conference on Learning Representations (ICLR) (2021)
9. Hamilton, W., Ying, R., Leskovec, J.: Inductive representation learning on large graphs. In: Neural Information Processing Systems (NeurIPS) (2017)
10. Klicpera, J., Bojchevski, A., Günnemann, S.: Combining neural networks with personalized PageRank for classification on graphs. In: International Conference on Learning Representations (ICLR) (2019)
11. Bruna, J., Zaremba, W., Szlam, A., LeCun, Y.: Spectral networks and locally connected networks on graphs. In: International Conference on Learning Representations (ICLR) (2014)
12. Defferrard, M., Bresson, X., Vandergheynst, P.: Convolutional neural networks on graphs with fast localized spectral filtering. In: Neural Information Processing Systems (NeurIPS) (2016)
13. Galstyan, S.: MixHop: higher-order graph convolution architectures via sparsified neighborhood mixing. In: International Conference On Machine Learning (ICML) (2019)
14. Lee, S.: N-GCN: multi-scale graph convolution for semi-supervised node classification. In: Conference on Uncertainty in Artificial Intelligence (UAI) (2019)
15. Li, Q., Han, Z., Wu, X.: Deeper insights into graph convolutional networks for semi-supervised learning. In: Association for the Advancement Of Artificial Intelligence (AAAI) (2018)
16. Wu, F., Souza, A., Zhang, T., Fifty, C., Yu, T., Weinberger, K.: Simplifying graph convolutional networks. In: International Conference on Machine Learning (ICML) (2019)
17. Tang, J., Sun, J., Wang, C., Yang, Z.: Social influence analysis in large-scale networks. In: ACM SIGKDD International Conference on Knowledge Discovery and Data Mining (KDD) (2009)

18. Rozemberczki, B., Allen, C., Sarkar, R.: Multi-Scale attributed node embedding. J. Complex Netw. **9**, cnab014 (2021)
19. Kingma, D., Ba, J.: Adam: a method for stochastic optimization. In: International Conference on Learning Representations (ICLR) (2015)
20. Chua, T., Tang, J., Hong, R., Li, H., Luo, Z., Zheng, Y.: NUS-WIDE: a real-world web image database from national university of Singapore. In: Proceedings of ACM Conferen on Image and Video Retrieval (CIVR 2009) (2009)
21. Hu, W., et al.: Open graph benchmark: datasets for machine learning on graphs. ArXiv Preprint ArXiv:2005.00687 (2020)
22. Navarin, N., Erb, W., Pasa, L., Sperduti, A.: Linear graph convolutional networks. In: 28th European Symposium On Artificial Neural Networks, Computational Intelligence And Machine Learning, ESANN 2020, Bruges, Belgium, 2–4 October 2020, pp. 151–156 (2020)
23. Akiba, T., Sano, S., Yanase, T., Ohta, T., Koyama, M.: Optuna: a next-generation hyperparameter optimization framework. ArXiv. abs/1907.10902 (2019)
24. Paszke, A., et al.: PyTorch: an imperative style, high-performance deep learning library. Adv. Neural. Inf. Process. Syst. **32**, 8024–8035 (2019)
25. Tremblay, N., Gonçalves, P., Borgnat, P.: Design of graph filters and filterbanks (2017)
26. Shuman, D., Narang, S., Frossard, P., Ortega, A., Vandergheynst, P.: The emerging field of signal processing on graphs: extending high-dimensional data analysis to networks and other irregular domains. IEEE Signal Process. Mag. **30**, 83–98 (2013)
27. Lim, D., Li, X., Hohne, F., Lim, S.: New benchmarks for learning on non-homophilous graphs. In: The WebConf Workshop on Graph Learning Benchmarks (GLB-WWW) (2021)
28. Lukovnikov, D., Fischer, A.: Improving breadth-wise backpropagation in graph neural networks helps learning long-range dependencies. In: Proceedings of the 38th International Conference on Machine Learning (2021)
29. Chamberlain, B., Rowbottom, J., Gorinova, M., Bronstein, M., Webb, S., Rossi, E.: GRAND: graph neural diffusion. In: Proceedings of the 38th International Conference on Machine Learning, vol. 139, pp. 1407–1418 (2021)
30. Yang, Y., et al.: Graph neural networks inspired by classical iterative algorithms. In: Proceedings of the 38th International Conference on Machine Learning (2021)
31. Wang, Y., Derr, T.: Tree decomposed graph neural network. In: Conference on Information and Knowledge Management (2021)
32. Zheng, X., et al.: How framelets enhance graph neural networks. In: Proceedings of the 38th International Conference on Machine Learning (2021)
33. Saad, Y.: On the rates of convergence of the Lanczos and the block-Lanczos methods. SIAM J. Numer. Anal. **17**, 687–706 (1980)
34. Li, R.: Sharpness in rates of convergence for the symmetric Lanczos method. Math. Comput. **79**, 419–435 (2010)
35. Cullum, J., Willoughby, R.: Lanczos algorithms for large symmetric eigenvalue computations. Society for Industrial (2002)
36. Dong, Y., Ding, K., Jalaian, B., Ji, S., Li, J.: Graph neural networks with adaptive frequency response filter (2021)
37. He, M., Wei, Z., Huang, Z., Xu, H.: BernNet : learning arbitrary graph spectral filters via bernstein approximation (2021)
38. Bianchi, F.M., Grattarola, D., Livi, L., Alippi, C.: Graph neural networks with convolutional ARMA filters. IEEE Trans. Pattern Anal. Mach. Intell. **44**(7), 3496–3507 (2022). https://doi.org/10.1109/TPAMI.2021.3054830

39. Gama, F., Marques, A., Leus, G., Ribeiro, A.: Convolutional neural network architectures for signals supported on graphs. IEEE Trans. Signal Process. **67**, 1034–1049 (2019)
40. Cai, C., Wang, Y.: A note on over-smoothing for graph neural networks (2020)
41. Zhou, K., et al.: Dirichlet energy constrained learning for deep graph neural networks (2021)
42. Davidson, E., Thompson, W.: Monster matrices: their eigenvalues and eigenvectors. Comput. Phys. **7**, 519–522 (1993)
43. Wang, H., Wei, Z., Gan, J., Wang, S., Huang, Z.: Personalized PageRank to a target node, revisited. CoRR. abs/2006.11876 (2020)
44. Stewart, G.: A Krylov-Schur algorithm for large eigenproblems. SIAM J. Matrix Anal. Appl. **23**, 601–614 (2002)
45. Lehoucq, R., Sorensen, D., Yang, C.: ARPACK users guide: solution of large scale eigenvalue problems by implicitly restarted Arnoldi methods (1997)
46. Shchur, O., Mumme, M., Bojchevski, A., Günnemann, S.: Pitfalls of graph neural network evaluation. ArXiv Preprint ArXiv:1811.05868 (2018)

NE-WNA: A Novel Network Embedding Framework Without Neighborhood Aggregation

Jijie Zhang[1], Yan Yang[1,2(✉)], Yong Liu[1,2(✉)], and Meng Han[3,4,5]

[1] School of Computer Science and Technology, Heilongjiang University,
Harbin 150080, China
`{yangyan,liuyong123456}@hlju.edu.cn`
[2] Key Laboratory of Database and Parallel Computing of Heilongjiang Province,
Harbin 150080, China
[3] Zhejiang University, Hangzhou 310027, China
[4] Binjiang Insititute of Zhejiang University, Hangzhou 310053, China
[5] Zhejiang Juntong Intelligence Co. Ltd., Hangzhou 310053, China

Abstract. Graph Neural Networks (GNNs) are powerful tools in representation learning for graphs. Most GNNs use the message passing mechanism to obtain a distinguished feature representation. However, due to this message passing mechanism, most existing GNNs are inherently restricted by over-smoothing and poor robustness. Therefore, we propose a simple yet effective Network Embedding framework Without Neighborhood Aggregation (NE-WNA). Specifically, NE-WNA removes the neighborhood aggregation operation from the message passing mechanism. It only takes node features as input and then obtains node representations by a simple autoencoder. We also design an enhanced neighboring contrastive (ENContrast) loss to incorporate the graph structure into the node representations. In the representation space, the ENContrast encourages low-order neighbors to be closer to the target node than high-order neighbors. Experimental results show that NE-WNA enjoys high accuracy on the node classification task and high robustness against adversarial attacks.

Keywords: Graph Neural Networks · Autoencoder ·
Over-smoothing · Robustness

1 Introduction

In recent years, Graph Neural Networks (GNNs) [1] have received great attention in the data mining community. They have achieved great success in many tasks related to graph representation learning, such as node classification [2,3], graph classification [4], link prediction [5] and so on.

Although GNNs have made significant progress in graph representation learning, most of them suffer from poor robustness and over-smoothing [6,7]. The main idea of GNNs lies in the message passing mechanism to learn expressive node representations. There are two important operations in the message passing mechanism: 1) Feature transformation, which is inherited from traditional

M.-R. Amini et al. (Eds.): ECML PKDD 2022, LNAI 13714, pp. 453–468, 2023.
https://doi.org/10.1007/978-3-031-26390-3_26

neural networks. 2) Neighborhood aggregation, which updates the representations of nodes by aggregating their neighborhood representations. This mechanism will lead to poor robustness and over-smoothing. In terms of robustness, the message passing mechanism forces each node to be highly dependent on its neighborhoods, which makes the node easily misled by potential data noise and thus makes GNNs vulnerable to adversarial disturbances. As a result, GNNs are usually not robust against graph attacks [8,9]. After suffering graph attacks, the neighborhood aggregation operation incorporates the representation of the noisy nodes into the representation of the target node, making the learned node representations underperform in downstream tasks. In terms of over-smoothing, when performing the message passing, the representations of neighboring nodes are aggregated and combined with the representation of the current node to form an updated representation. After this process is iterated multiple times, different nodes will have similar representations, making it difficult to distinguish between nodes with different classes.

To address the above challenge, we propose a novel Network Embedding framework Without Neighborhood Aggregation called NE-WNA to alleviate over-smoothing and enhance the model's robustness. In NE-WNA, we elaborately remove the neighborhood aggregation operation in the messaging mechanism and only preserve the feature transformation operation to reduce the dependence of node representations on the features of their neighbors. To exploit graph structure information in learning the node representations, we design an enhanced neighboring contrastive (ENContrast) loss using the graph structure as a supervision signal. In the representation space, the ENContrast loss considers the importance of different order neighbors, and it encourages low-order neighbors to be closer to the target node than high-order neighbors. The main contributions of this paper are summarized as follows.

- We propose a simple yet effective autoencoder-based graph learning framework. Good node representation can be obtained by using only the basic autoencoder without neighborhood aggregation.
- We design an enhanced neighboring contrastive loss that aims to incorporate graph structure information into node representations.
- Extensive experiments show that the proposed framework outperforms the state-of-the-art baselines. Removing the neighborhood aggregation can alleviate over-smoothing and enhance the robustness against adversarial attacks.

2   Related Work

2.1   Graph Neural Networks

GNNs have boosted research on graph data mining. The key to the success of most GNNs lies in the message passing mechanism, which propagates the neighbor information to the target node in an iterative manner. In the growing number of GNN architectures, the most representative method is Graph Convolutional Network (GCN) [10] and Graph Attention Networks (GAT) [11]. GCN learns the representation of the target node by iteratively aggregating the neighbors of the target node. In the process of learning the target node representation, GAT generates importance scores for all its neighbors, and then uses these importance scores to aggregate the neighboring nodes. GCN and GAT follow coupling feature transformation and neighborhood aggregation together for representation learning. Nevertheless, some recent work [2,7] finds that the coupling of feature transformation and neighborhood aggregation is unnecessary and causes over-smoothing. APPNP [12], SGC [13], SIGN [14] and S^{2}GC [15] achieve good node classification by separating the two operations. DropEdge [16] is introduced to alleviate the over-smoothing by randomly dropping some edges in graph during each training epoch. GCNII [17] obtains better results by applying two simple techniques, initial residuals and identity mapping, to graph convolutional networks. In addition, GNNs can be easily fooled by graph adversarial attacks. Many novel defense approaches, like GCN-Jaccard [18], GCN-SVD [19] and Pro-GNN [20], have been proposed to defend against different types of graph adversarial attacks.

2.2   Graph Contrastive Learning

Contrastive learning is self-supervised learning [21,22] method whose main idea is to train the feature encoder by making the positive samples as close as possible and the negative samples as far away as possible in the representation space. Recently, researchers have been focusing on applying contrastive learning techniques to graph representation learning tasks. This series of techniques have achieved good results and is known as Graph Contrastive Learning (GCL) [23,24]. For a given large amount of unlabeled graph data, the GCL aims to train a graph encoder, which currently generally refers to a GNN. In contrast, instead of using GNN as a feature encoder and data augmentation techniques in GCL, we use the autoencoder as the feature encoder and the adjacency matrix as the supervisory signal for the contrast. GCL's objective aims to pull relevant node representations together while pushing irrelevant node representations away. It fits well with our idea of using the adjacency matrix to guide the contrast of neighboring nodes.

2.3   Auto-Encoder

Currently, more deep models are beginning to be designed for graph-structured data. For example, autoencoders have been extended for graph representation learning on graph-structured data. The autoencoder architecture, which extracts complex features using only unlabeled data, allows deep learning techniques to be applied to a broader range of domains. SDNE [25] uses the deep autoencoder with multiple non-linear layers to capture the first and second-order proximities. SDCN [26] converts raw data into low-dimensional representations and then decodes low-dimensional representations to reconstruct the input node representations. In this work, we only use the basic autoencoder [27] to learn representations for raw data.

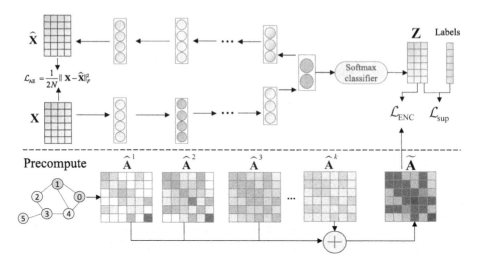

Fig. 1. The framework of our proposed NE-WNA. The input data $\mathbf{X}$ is passed through an autoencoder and a softmax classifier to obtain the prediction probabilities $\mathbf{Z}$ of nodes in different classes. The cross-entropy loss $\mathcal{L}_{\text{sup}}$, the reconstruction loss $\mathcal{L}_{\text{AE}}$ and the enhanced neighboring contrastive loss $\mathcal{L}_{\text{ENC}}$ are used to guide the learning of the model. $\mathcal{L}_{\text{sup}}$ is constructed by $\mathbf{Z}$ and node labels. $\mathcal{L}_{\text{AE}}$ is constructed by the reconstructed data $\hat{\mathbf{X}}$ and the input data $\mathbf{X}$. $\mathcal{L}_{\text{ENC}}$ is constructed by $\mathbf{Z}$ and the enhanced adjacency matrix $\widetilde{\mathbf{A}}$ which is summed by the different order powers of the normalized adjacency matrix, such as $\hat{\mathbf{A}}^1$, $\hat{\mathbf{A}}^2$, $\hat{\mathbf{A}}^3$, $\cdots$, $\hat{\mathbf{A}}^k$.

3 Methodology

In this section, we introduce our proposed NE-WNA, where the overall framework is shown in Fig. 1. NE-WNA is composed of two key components: autoencoder and enhanced neighboring contrastive loss. We will describe our proposed model in detail in the following.

3.1 Preliminaries

Let $\mathcal{G} = (\mathcal{V}, \mathcal{E})$ denote a graph, where $V = \{v_1, v_2, \ldots, v_N\}$ is a set of N nodes, $\mathcal{E} \subseteq \mathcal{V} \times \mathcal{V}$ is a set of $|\mathcal{E}| = M$ edges between nodes. $\mathcal{G}$ is associated with a feature matrix $\mathbf{X} = \{\mathbf{x}_1, \mathbf{x}_2 \ldots, \mathbf{x}_N\}$ in which $\mathbf{x}_i \in \mathbb{R}^d$ represents the feature vector of node v_i, and an adjacency matrix $\mathbf{A} \in \mathbb{R}^{N \times N}$ where $\mathbf{A}_{ij} = 1$ iff $(v_i, v_j) \in \mathcal{E}$ and $\mathbf{A}_{ij} = 0$ otherwise.

3.2 Auto-Encoder Module

In our model, the basic autoencoder is used as a feature encoder in order to reduce the complexity of the model. Assume that the encoder and decoder parts have L layers each and ℓ represents the ℓ-th layer. The processing of the encoder is defined as:

$$\mathbf{H}_e^{(\ell)} = \sigma\left(\mathbf{W}_e^{(\ell)}\mathbf{H}_e^{(\ell-1)} + \mathbf{b}_e^{(\ell)}\right),\tag{1}$$

where $\sigma(\cdot)$ is the activation function and we apply *Relu*. Here $\mathbf{W}_e^{(\ell)}$ and $\mathbf{b}_e^{(\ell)}$ are the trainable weight matrix and bias of the ℓ-th layer in the encoder, respectively. $\mathbf{H}_e^{(\ell-1)}$ and $\mathbf{H}_e^{(\ell)}$ are node embeddings of layer $\ell-1$ and layer ℓ in the encoder respectively while $\mathbf{H}_e^{(0)}$ is set to raw feature matrix $\mathbf{X}$. Similarly, the decoder part is defined as:

$$\mathbf{H}_d^{(\ell)} = \sigma\left(\mathbf{W}_d^{(\ell)}\mathbf{H}_d^{(\ell-1)} + \mathbf{b}_d^{(\ell)}\right),\tag{2}$$

where the input of the decoder part is $\mathbf{H}_d^{(0)} = \mathbf{H}_e^{(L)}$, $\mathbf{W}_d^{(\ell)}$ and $\mathbf{b}_d^{(\ell)}$ are the trainable weight matrix and bias of the ℓ-th layer in the decoder, respectively. $\mathbf{H}_d^{(\ell-1)}$ and $\mathbf{H}_d^{(\ell)}$ are node embeddings of layer $\ell-1$ and layer ℓ in the decoder respectively. The output of the decoder part is the reconstruction of the raw feature matrix $\hat{\mathbf{X}} = \mathbf{H}_d^{(L)}$. The corresponding reconstruction loss is defined as:

$$\mathcal{L}_{\mathrm{AE}} = \frac{1}{2N}\sum_{i=1}^{N}\|\boldsymbol{x}_i - \hat{\boldsymbol{x}}_i\|_2^2 = \frac{1}{2N}\|\mathbf{X} - \hat{\mathbf{X}}\|_F^2.\tag{3}$$

3.3 Enhanced Neighboring Contrastive Loss

Autoencoder is able to learn the useful representations from the data itself while ignoring the graph structure information. To be able to incorporate graph structure information into the autoencoder-specific representation, it is intuitive that connected nodes should be similar to each other and unconnected nodes should be far apart in the representation space. This fits well with the idea of contrastive learning. With this motivation, we propose an Enhanced Neighboring Contrastive (ENContrast) loss, which enables autoencoder-based models to learn graph structure without neighborhood aggregation.

Before describing the ENContrast loss in detail, we first introduce the enhanced adjacency matrix. By summing different order powers of the normalized adjacency matrix, we obtain the enhanced adjacency matrix as:

$$\tilde{\mathbf{A}} = \sum_{i=1}^{k}\hat{\mathbf{A}}^{(i)},\tag{4}$$

where $\hat{\mathbf{A}} = \tilde{\mathbf{D}}^{(-1/2)}(\mathbf{A}+\mathbf{I})\tilde{\mathbf{D}}^{(-1/2)}$, $\tilde{\mathbf{D}} = \mathbf{D}+\mathbf{I}$, $\mathbf{D}_{ii} = \sum_j \mathbf{A}_{ij}$. Many recent studies leverage generalized PageRank matrix [12], which is formulated with the summation of different order powers of the normalized adjacency matrix with coefficients. However, we found through subsequent experiments that a simplified version of PageRank (e.g., all coefficients are 1) achieves better results. Therefore, we use Eq. (4) to obtain the enhanced normalized adjacency matrix. When k tends to ∞, $\hat{\mathbf{A}}^{\infty}$ is

$$\hat{\mathbf{A}}_{i,j}^{\infty} = \frac{(d_i+1)^{1/2}(d_j+1)^{1/2}}{2M+N},\tag{5}$$

where M represents the number of edges and N represents the number of nodes, d_i represents the degree of the node v_i. This shows that after an infinite number of multiplications, the influence of node v_i on v_j is only determined by their degree. As shown in Fig. 2, the connection weights of the target node and its each order neighbors will be very close as k gradually increases. The target node's low-order neighbors usually have more influence on the target node; in other words, the information of the node's low-order neighbors is more important. The enhanced adjacency matrix $\tilde{\mathbf{A}}$ increases the connection weight between the target node and its low-order neighbors as k gradually increases, it can enhance the influence of the target node's low-order neighbors on it during the contrast process.

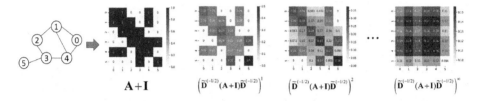

Fig. 2. GNNs smooth the representation of each node via node propagation. As the propagation layers deepen, the connection weights of the target node and its each order neighbors gradually approach.

In the ENContrast loss, for each node, its k-hop neighbors are regarded as the positive samples, while the other nodes are sampled as the negative ones. Since the enhanced adjacency matrix increases the connection weight of the target node with its low-order neighbors, the model will pay more attention to the contrast of the target node with its low-order neighbors in the process of guiding the model learning. In representation space, the loss encourages low-order neighbors in positive samples to be closer to the target node than high-order neighbors in positive samples. At the same time, it pushes negative samples away from the target node. In detail, the ENContrast loss for the node v_i can be formulated as:

$$\ell_i = -\log \frac{\sum_{j=1}^{N} \mathbf{1}_{[j \neq i]} \tilde{\mathbf{A}}_{ij} \exp\left(\text{sim}\left(\mathbf{Z}_i, \mathbf{Z}_j\right)/\tau\right)}{\sum_{q=1}^{N} \mathbf{1}_{[q \neq i]} \exp\left(\text{sim}\left(\mathbf{Z}_i, \mathbf{Z}_q\right)/\tau\right)}, \tag{6}$$

where sim denotes the cosine similarity and τ denotes the temperature parameter. $\mathbf{1}_{[j \neq i]}$ represents the indicator function, which is 1 when i and j are not equal and 0 otherwise. $\tilde{\mathbf{A}}_{ij}$ denotes the strength of the connection between node v_i and v_j and is a non-zero value only if node v_j is the k-hop neighbor of node v_i. $\mathbf{Z}_i$ denotes the prediction probabilities of node v_i on different classes, which is obtained by taking the node representations of the autoencoder $\mathbf{H}_e^{(L)}$ as input to a softmax classifier. The detailed definition is as follows.

$$\mathbf{Z} = \text{softmax}\left(\mathbf{W}\mathbf{H}_e^{(L)}\right), \tag{7}$$

where $\mathbf{Z}$ denotes the prediction probabilities of nodes on different classes, and $\mathbf{W}$ are the trainable weight matrix of softmax classifier.

The corresponding ENContrast loss is defined as:

$$\mathcal{L}_{\text{ENC}} = \frac{1}{N} \sum_{i=1}^{N} \ell_i. \tag{8}$$

Besides the ENContrast loss, we also have a traditional cross-entropy loss for node classification. The cross-entropy loss for labeled noes can be calculated as:

$$\mathcal{L}_{\text{sup}} = - \sum_{i \in \mathcal{V}_l} \sum_{p=1}^{C} \mathbf{Y}_{[i,p]} \ln \mathbf{Z}_{[i,p]}, \tag{9}$$

where $\mathcal{V}_l$ is the set of labeled nodes and $\mathbf{Y} \in \mathbb{R}^{N \times C}$ is the label indicator matrix, C is the number of classes. $\mathbf{Y}_{[i,p]}$ is 1 when node v_i belonging to class p and 0 otherwise. $\mathbf{Z}_{[i,p]}$ denotes the probability of node v_i belonging to class p.

In total, we define the final loss of NE-WNA as a combination of three losses:

$$\mathcal{L} = \mathcal{L}_{\text{sup}} + \alpha \mathcal{L}_{\text{ENC}} + \beta \mathcal{L}_{\text{AE}}, \tag{10}$$

where α and β is the weighting coefficient to balance these losses.

Algorithm 1. NE-WNA

Input: Adjacency matrix $\mathbf{A}$, raw data $\mathbf{X}$, number of layers of the autoencoder L, the order of the adjacency matrix k, learning rate η, balance coefficient α and β, an autoencoder-based model $f(\mathbf{X}, \Theta)$.

Output: Prediction $\mathbf{Z}$.

1: **for** $i = 1$ to k **do**
2:    Precompute $\hat{\mathbf{A}}^{(i)}$.
3: **end for**
4: Compute the enhanced adjacency matrix $\tilde{\mathbf{A}}$ via Eq. (4).
5: **while** not convergence **do**
6:    **for** $\ell = 1$ to L **do**
7:       Generate embeddings $\mathbf{H}_e^{(\ell)}$ via Eq. (1) and $\mathbf{H}_d^{(\ell)}$ via Eq. (2).
8:    **end for**
9:    Compute autoencoder reconstruction loss $\mathcal{L}_{\text{AE}}$ via Eq. (3).
10:    Generate the prediction probabilities $\mathbf{Z}$ via Eq. (7).
11:    Compute the enhanced neighboring contrastive loss $\mathcal{L}_{\text{ENC}}$ via Eq. (6) and Eq.(8).
12:    Compute the cross-entropy loss $\mathcal{L}_{\text{sup}}$ via Eq. (9).
13:    Update the parameters Θ by gradients descending: $\Theta = \Theta - \eta \nabla_\Theta (\mathcal{L}_{\text{sup}} + \alpha \mathcal{L}_{\text{ENC}} + \beta \mathcal{L}_{\text{AE}})$.
14: **end while**

3.4 Algorithm and Complexity Analysis

Algorithm 1 outlines NE-WNA's training process. Line 1–4 represents the pre-computation procedure of the enhanced adjacency matrix. Line 5–14 represents the training process of the model.

Table 1. Complexity Analysis for existing GNNs. N, M, and d are the number of nodes, edges, and feature dimensions (assumed fixed for all layers), respectively. k represents the power of the normalized adjacency matrix. L represents the number of layers of feature transformation. For the coupled GNNs, we always have $k = L$.

Type	Model	Preprocessing	Training	Inference
Coupled GNNs	GCN	–	$\mathcal{O}(LMd^2)$	$\mathcal{O}(LMd^2)$
	GAT	–	$\mathcal{O}(LMd + LNd^2)$	$\mathcal{O}(LMd + LNd^2)$
Decoupled GNNs	SGC	$\mathcal{O}(kMd)$	$\mathcal{O}(Nd^2)$	$\mathcal{O}(Nd^2)$
	S^{2}GC	$\mathcal{O}(kMd)$	$\mathcal{O}(Nd^2)$	$\mathcal{O}(Nd^2)$
	SIGN	$\mathcal{O}(kMd)$	$\mathcal{O}(LNd^2)$	$\mathcal{O}(LNd^2)$
	NE-WNA	$\mathcal{O}(k^2M)$	$\mathcal{O}(LNd^2)$	$\mathcal{O}(LNd^2)$

Table 1 compares the asymptotic complexity of NE-WNA with several representative GNNs. Because the operation of GCN can be efficiently implemented using sparse matrix, the time complexity is linear with the number of edges M. In the stage of the preprocessing, the time complexity of most decoupled models is $\mathcal{O}(kMd)$ and the time complexity of NE-WNA is $\mathcal{O}(k^2M)$. Since $k \ll d$, NE-WNA have smaller preprocessing complexity than decoupled GNNs. Compared with the coupled GNNs, NE-WNA have smaller training and inference complexity, i.e., higher efficiency.

4 Experiments

To evaluate the effectiveness of our proposed NE-WNA, we conduct extensive experiments on node classification tasks. First, we introduced the datasets, experimental environment and parameter settings. Then, we compare NE-WNA with the previous state-of-the-art baselines on node classification to prove the superiority of NE-WNA. Finally, we validate the proposed model further in terms of ablation study, over-smoothing, robustness, and visualization.

4.1 Datasets

We conduct experiments on six datasets: three citation networks (Cora [10], CiteSeer [10], PubMed [10]), two co-purchase networks (Amazon Photo [28], Amazon Computers [28]) and one co-author network (Coauthor CS [28]). We use the same train/validation/test splits as [10] for citation networks. For the other datasets, we randomly select 20 labeled nodes per class as the training set, 30 labeled nodes per class as the validation set, and the remaining nodes as the test set. The statistics of these datasets are summarized in Table 2.

Table 2. Statistics of datasets.

Dataset	Nodes	Edges	Features	Classes	Train/Val/Test
Cora	2708	5278	1433	7	140/500/1000
CiteSeer	3327	4552	3703	6	120/500/1000
PubMed	19717	44324	500	3	60/500/1000
Amazon Computers	13381	245778	767	10	200/300/12881
Amazon Photo	7487	119043	745	8	160/240/7087
Coauthor CS	18333	81894	6805	15	300/450/17583

Table 3. Hyper-parameter specifications.

DataSet	α	β	τ	k	L	Learning rate	Weight decay
Cora	2	3	0.5	4	3	$5e-3$	$5e-4$
CiteSeer	1	2	0.5	4	3	$1e-2$	$5e-4$
PubMed	10	1	0.5	4	3	$1e-2$	$5e-4$
Amazon Computers	30	3	4	6	3	$5e-3$	$5e-4$
Amazon Photo	25	3	4	5	3	$5e-3$	$5e-4$
Coauthor CS	10	1	1	2	3	$1e-2$	$5e-3$

4.2 Implementation and Parameter Settings

The experiments are conducted on a machine with Intel(R) Core(TM) i9-10980XE CPU @ 3.00GHz, and a single NVIDIA GeForce RTX 3090 with 24GB GPU memory. The operating system of the machine is Ubuntu 18.04. As for software versions, we use Python 3.8, Pytorch 1.9.1, Pytorch Geometric 2.0.1 [29], and CUDA 11.1. The hyper-parameters in each baseline are set according to the original paper if available. We perform a grid search to tune the hyper-parameters for NE-WNA based on the accuracy of the validation set. α is obtained from a search of range 1 to 30 with step 1, β is obtained from a search of range 1 to 6 with step 1, τ is obtained from a search of range 1 to 4 with step 0.5, k is obtained from a search of range 1 to 7 with step 1, L is set to 3. Adam optimizer is used on all datasets, the learning rate is chosen from $\{5e-3, 1e-2\}$, the weight decay is chosen from $\{5e-4, 5e-3\}$. The detailed hyper-parameter settings for NE-WNA is in Table 3. Our data and code are publicly available[1].

4.3 Node Classification Results

We choose the following baseline methods: GCN [10], GAT [11], JK-Net [30], APPNP [12], SGC [13], SIGN [14], S^{2}GC [15], DropEdge [16] and GCNII [17]. To alleviate the influence of randomness, we repeat each method 100 times and report the mean performance and the standard deviations. The experimental results are summarized in Table 4. On Cora, NE-WNA is comparable with other

---

[1] https://github.com/YJ199804/NE-WNA.

Table 4. Results on all datasets in terms of classification accuracy.

Model	Cora	CiteSeer	PubMed	Amazon Computers	Amazon Photo	Coauthor CS
GCN	81.3 ± 0.6	71.1 ± 0.8	78.9 ± 0.5	82.6 ± 2.0	91.2 ± 1.2	91.0 ± 0.5
GAT	83.1 ± 0.4	72.5 ± 0.7	79.0 ± 0.3	80.1 ± 0.6	90.8 ± 0.9	90.5 ± 0.6
JK-Net	81.8 ± 0.5	70.8 ± 0.7	78.8 ± 0.5	81.9 ± 0.8	91.9 ± 0.7	89.8 ± 0.7
APPNP	83.3 ± 0.5	71.8 ± 0.4	80.1 ± 0.2	81.7 ± 0.3	91.4 ± 0.3	91.8 ± 0.4
SGC	81.0 ± 0.1	71.3 ± 0.3	78.9 ± 0.2	82.1 ± 0.7	91.5 ± 0.8	90.3 ± 0.5
SIGN	82.2 ± 0.4	72.4 ± 0.5	79.3 ± 0.6	82.8 ± 0.7	91.7 ± 0.8	91.8 ± 0.9
S^2GC	82.6 ± 0.5	72.9 ± 0.2	79.8 ± 0.3	82.9 ± 0.9	91.6 ± 0.6	91.4 ± 0.5
DropEdge	82.8 ± 0.2	72.3 ± 0.4	79.6 ± 0.3	82.4 ± 0.7	91.3 ± 0.5	91.6 ± 0.8
GCNII	**85.5 ± 0.5**	73.4 ± 0.6	80.3 ± 0.4	81.9 ± 0.3	92.1 ± 0.5	92.0 ± 0.5
NE-WNA	82.8 ± 0.8	**74.2 ± 0.6**	**82.5 ± 0.7**	**84.7 ± 1.4**	**93.2 ± 0.7**	**92.5 ± 0.6**

Table 5. Ablation study results in terms of accuracy of node classification.

Ablation	Cora	CiteSeer	PubMed	Amazon Photo	Amazon Computers	Coauthor CS
NE-WNA	**82.8 ± 0.8**	**74.2 ± 0.6**	**82.5 ± 0.7**	**93.2 ± 0.7**	**84.7 ± 1.4**	**92.5 ± 0.6**
w/o ENContrast	81.9 ± 0.8	73.7 ± 0.9	81.2 ± 1.0	92.5 ± 0.6	83.6 ± 1.5	91.7 ± 0.6
w/o AE	80.5 ± 0.3	73.3 ± 0.8	80.7 ± 0.6	92.7 ± 0.6	79.4 ± 0.7	85.3 ± 2.1
w/o AE & ENContrast	79.5 ± 0.8	72.8 ± 0.8	80.2 ± 1.0	90.7 ± 0.7	77.8 ± 0.9	85.6 ± 1.9

methods. On the other datasets, NE-WNA performs better than the representative baselines by significant margins and outperforms the best baseline of each dataset by a margin of 0.5% to 2.9%. While most baselines use a multi-layer perceptron (MLP) for feature transformation, our proposed framework uses an autoencoder for feature transformation, which allows us to better extract information from the data itself. By removing neighborhood aggregation, the target node does not rely excessively on its multi-hop neighbors, which allows it to expand the receptive fields while keeping the node representation distinguishable and able to obtain more structural information.

4.4   Ablation Study

We conduct an ablation study to examine the contributions of different components in NE-WNA. Specifically, we build the following variants:

- **Without Enhanced Neighborhood Contrast (ENContrast):** We only use the k-power of the normalized adjacency matrix to construct the neighboring contrastive loss, i.e., $\tilde{\mathbf{A}} = \hat{\mathbf{A}}^{(k)}$.
- **Without Auto-Encoder (AE):** We use the MLP-based model to replace the AutoEncoder-based model.
- **Without Auto-Encoder and Enhanced Neighborhood Contrast (AE & ENContrast):** We use the MLP-based model to replace the autoencoder-based model and the k-power of the normalized adjacency matrix to construct the neighboring contrastive loss.

In Table 5, we have two observations. First, all NE-WNA variants show a significant performance degradation compared to the full model, indicating that each component contributes to the performance of NE-WNA. Second, the autoencoder does surpass the MLP in feature extraction ability.

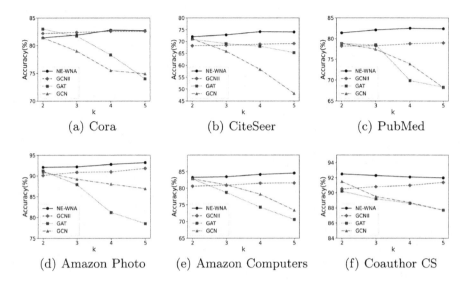

Fig. 3. Over-smoothing comparison

4.5 Over-Smoothing Analysis

We study the NE-WNA's ability to alleviate over-smoothing by using the classification results in the case of stacking different layers. Figure 3 shows the classification accuracy of different models. For the baselines, k denotes the number of layers of the model. Each method makes use of the k-hop neighbors of the target node. As k increases, more and more neighboring nodes are utilized, which inevitably suffers the over-smoothing. Figure 3 suggests that NE-WNA can better alleviate over-smoothing while GCN and GAT show significant performance degradation as the layer deepen. On most datasets, both NE-WNA and the current state-of-the-art method GCNII show an increase in performance with deeper layers. In particular, for a total of 6 datasets, NE-WNA outperforms GCNII for 5 datasets. It is worth mentioning that the optimal effect reported in the original paper on GCNII is achieved at dozens of layers, and our framework can exceed its optimal effect at shallow layers.

4.6 Parameter Analysis

NE-WNA involves a number of parameters and we examine how the different choices of parameters affect the performance of NE-WNA on all datasets. The

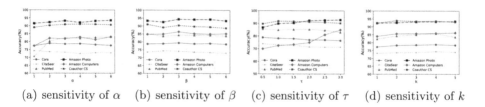

(a) sensitivity of α (b) sensitivity of β (c) sensitivity of τ (d) sensitivity of k

Fig. 4. The performance of NE-WNA with varying different hyperparameters on all datasets.

results of the validation set are shown in Fig. 4. We find that NE-WNA is robust to the k and β. k, β takes smaller values to achieve good results, indicating that the model does not need to expand the contrastive fields of nodes excessively and does not need to excessively extract information from the data itself, which can greatly reduce the training difficulty of the model and speed up the convergence of the model. As the α increases, the performance of the model improves on most of the datasets, indicating that increasing the contrast strength between nodes helps to learn a better node representation. With increasing τ, the model performance improves on the Amazon Photo and Amazon Computers datasets. Still, it decreases on other datasets, probably because other datasets are more sparse than these two datasets. The number of multi-hop neighbors of the target nodes in the Amazon Photo and Amazon Computers datasets is higher. By increasing τ, the data distribution will become flat and more neighbor nodes will be considered in the backpropagation process, so that the model can learn a more comprehensive node representation.

4.7   Robustness Comparison

Recent research has demonstrated that GNNs are vulnerable to adversarial attacks [18,31]. We evaluated the robustness of NE-WNA against two types of attacks based on node classification performance. The two types of attacks are *nettack* [9] and *metattack* [32].

We analyze the robustness of different models on Cora, CiteSeer and PubMed. We randomly divide all nodes into 10%, 10% and 80% for training, validation and testing. We used GCN, GAT, GCN-Jaccard, GCN-SVD and Pro-GNN for comparison. For the current state-of-the-art defense models GCN-Jaccard, GCN-SVD and Pro-GNN, we use DeepRobust [33] to replicate them. We evaluate the node classification accuracy of different methods against *nettack* and *metattack*. For *nettack*, we perturb each target node from 1 to 5 times with a step size of 1. The target nodes are those with degree greater than 10 in the test set. The node classification accuracy on target nodes is shown in Fig. 5. For *metattack*, we perturb the edges from 0 to 25% with a step of 5%. The node classification accuracy is shown in Fig. 6. As shown in Fig. 5 and Fig. 6, NE-WNA shows comparable performance to Pro-GNN on Cora. Our method consistently outperforms other methods under different attacks on CiteSeer and PubMed. For example, on the CiteSeer dataset at 5 perturbations per targeted node, NE-WNA achieves over 31% improvement over the state-of-the-art defense method Pro-GNN. The neighborhood aggregation operation incorporates information from neighboring nodes into the target node, and information from noisy nodes is incorporated into the target node when subjected to adversarial attacks. By removing the neighborhood aggregation operation, the noise generated by adversarial attack is not excessively incorporated into the target node. Therefore, NE-WNA can be robust to adversarial attacks.

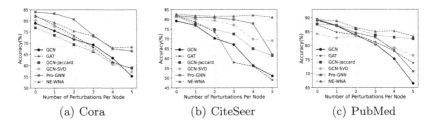

| (a) Cora | (b) CiteSeer | (c) PubMed |

Fig. 5. Results of different models under nettack.

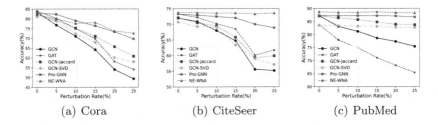

| (a) Cora | (b) CiteSeer | (c) PubMed |

Fig. 6. Results of different models under metattack.

4.8   Loss Validation

To further verify the effectiveness of the ENContrast loss, we use the neighboring contrastive (NContrast) loss for comparison. The NContrast loss is defined as:

$$\mathcal{L}_{\text{NConcrast}} = \frac{1}{N} \sum_{i=1}^{N} \left(-\log \frac{\sum_{j=1}^{N} \mathbf{1}_{[j \neq i]} \hat{\mathbf{A}}_{ij}^{(k)} \exp\left(\text{sim}\left(\mathbf{Z}_i, \mathbf{Z}_j\right)/\tau\right)}{\sum_{q=1}^{N} \mathbf{1}_{[q \neq i]} \exp\left(\text{sim}\left(\mathbf{Z}_i, \mathbf{Z}_q\right)/\tau\right)} \right) \quad (11)$$

The NContrast loss only uses the k-power of the normalized adjacency matrix. It is worth noting that the model with NContrast loss in this section is the same as the first variant in the ablation study. This section is done to further

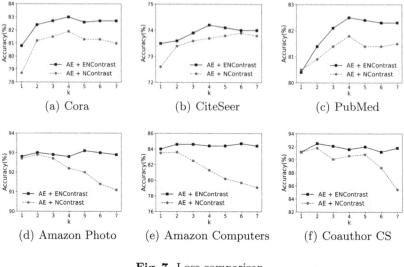

Fig. 7. Loss comparison

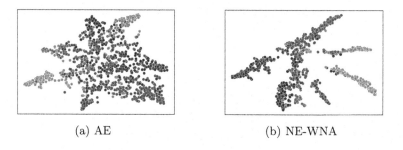

(a) AE   (b) NE-WNA

Fig. 8. t-SNE plots for Cora dataset.

experimental verification of the effectiveness of ENContrast loss under the influence of the different number of layers. As shown in Fig. 7, the ENContast loss consistently improves the performance of the autoencoder than the NContast loss on all datasets. Specifically, the ENContrast loss improves the performance of the autoencoder by a more significant margin than the NContast loss when the order is higher. The results also demonstrate the validity of focusing more on comparing the node's low-order neighbors when the node's different order neighbors are positive samples.

4.9 Visualization of Embeddings

To provide a more intuitive understanding of the learned node embeddings, we visualize node embeddings of AE and NE-WNA by using t-SNE [34]. Each point represents a test node on Cora, and the color represents the node label. The results are shown in Fig. 8. We clearly observe that the nodes are better classified in NE-WNA than AE, which means that NE-WNA captures more detailed class information.

5   Conclusion

In this paper, we propose a simple yet effective autoencoder-based graph learning framework. We remove the neighborhood aggregation commonly used by GNNs. The enhanced neighboring contrastive loss is designed to guide the autoencoder to learn the node representation. By removing the neighborhood aggregation, our framework is significantly more effective in alleviating over-smoothing and is robust to adversarial attacks. Extensive experiments on six benchmark graph datasets demonstrate the high accuracy and robustness of NE-WNA against the state-of-the-art GNNs.

Acknowledgements. This work was supported by the National Natural Science Foundation of China (No. 61972135), and the Natural Science Foundation of Heilongjiang Province in China (No. LH2020F043).

References

1. Waikhom, L., Patgiri, R.: Graph neural networks: methods, applications, and opportunities. arXiv preprint arXiv:2108.10733 (2021)
2. Feng, W., et al.: Graph random neural networks for semi-supervised learning on graphs. In: NIPS (2020)
3. Feng, W., et al.: GRAND+: scalable graph random neural networks. In: WWW, pp. 3248–3258 (2022)
4. Xu, K., Hu, W., Leskovec, J., Jegelka, S.: How powerful are graph neural networks? In: ICLR (2019)
5. Wang, X., He, X., Wang, M., Feng, F., Chua, T.: Neural graph collaborative filtering. In: SIGIR, pp. 165–174 (2019)
6. Li, Q., Han, Z., Wu, X.: Deeper insights into graph convolutional networks for semi-supervised learning. In: AAAI, pp. 3538–3545 (2018)
7. Oono, K., Suzuki, T.: Graph neural networks exponentially lose expressive power for node classification. In: ICLR (2020)
8. Zhu, D., Zhang, Z., Cui, P., Zhu, W.: Robust graph convolutional networks against adversarial attacks. In: SIGKDD, pp. 1399–1407 (2019)
9. Zügner, D., Akbarnejad, A., Günnemann, S.: Adversarial attacks on neural networks for graph data. In: SIGKDD, pp. 2847–2856 (2018)
10. Kipf, T.N., Welling, M.: Semi-supervised classification with graph convolutional networks. In: ICLR (2017)
11. Velickovic, P., Cucurull, G., Casanova, A., Romero, A., Liò, P., Bengio, Y.: Graph attention networks. In: ICLR (2018)
12. Klicpera, J., Bojchevski, A., Günnemann, S.: Predict then Propagate: graph neural networks meet personalized PageRank. In: ICLR (2019)
13. Wu, F., de Souza Jr., A.H., Zhang, T., Fifty, C., Yu, T., Weinberger, K.Q.: Simplifying graph convolutional networks. In: ICML, pp. 6861–6871 (2019)

14. Rossi, E., Frasca, F., Chamberlain, B., Eynard, D., Bronstein, M.M., Monti, F.: SIGN: scalable inception graph neural networks. arXiv preprint arXiv:2004.11198 (2020)

15. Zhu, H., Koniusz, P.: Simple spectral graph convolution. In: ICLR (2021)

16. Rong, Y., Huang, W., Xu, T., Huang, J.: DropEdge: towards deep graph convolutional networks on node classification. In: ICLR (2020)

17. Chen, M., Wei, Z., Huang, Z., Ding, B., Li, Y.: Simple and deep graph convolutional networks. In: ICML, pp. 1725–1735 (2020)

18. Wu, H., Wang, C., Tyshetskiy, Y., Docherty, A., Lu, K., Zhu, L.: Adversarial examples for graph data: deep insights into attack and defense. In: IJCAI, pp. 4816–4823 (2019)

19. Entezari, N., Al-Sayouri, S.A., Darvishzadeh, A., Papalexakis, E.E.: All you need is low (Rank): defending against adversarial attacks on graphs. In: WSDM, pp. 169–177 (2020)

20. Jin, W., Ma, Y., Liu, X., Tang, X., Wang, S., Tang, J.: Graph structure learning for robust graph neural networks. In: SIGKDD, pp. 66–74 (2020)

21. Wu, L., Lin, H., Gao, Z., Tan, C., Li, S.Z.: Self-supervised on graphs: contrastive, generative, or predictive. arXiv preprint arXiv:2105.07342 (2021)

22. Liu, Y., Pan, S., Jin, M., Zhou, C., Xia, F., Yu, P.S.: Graph self-supervised learning: a survey. arXiv preprint arXiv:2103.00111 (2021)

23. Zhu, Y., Xu, Y., Yu, F., Liu, Q., Wu, S., Wang, L.: Deep graph contrastive representation learning. arXiv preprint arXiv:2006.04131 (2020)

24. Zhu, Y., Xu, Y., Yu, F., Liu, Q., Wu, S., Wang, L.: Graph contrastive learning with adaptive augmentation. In: WWW, pp. 2069–2080 (2021)

25. Wang, D., Cui, P., Zhu, W.: Structural deep network embedding. In: SIGKDD, pp. 1225–1234 (2016)

26. Bo, D., Wang, X., Shi, C., Zhu, M., Lu, E., Cui, P.: Structural deep clustering network. In: WWW, pp. 1400–1410 (2020)

27. Hinton, G.E., Salakhutdinov, R.: Reducing the dimensionality of data with neural networks. Science **313**, 504–507 (2006)

28. Shchur, O., Mumme, M., Bojchevski, A., Günnemann, S.: Pitfalls of graph neural network evaluation. arXiv preprint arXiv:1811.05868 (2018)

29. Fey, M., Lenssen, J.E.: Fast graph representation learning with PyTorch geometric. arXiv preprint arXiv:1903.02428 (2019)

30. Xu, K., Li, C., Tian, Y., Sonobe, T., Kawarabayashi, K., Jegelka, S.: Representation learning on graphs with jumping knowledge networks. In: ICML, pp. 5449–5458 (2018)

31. Zhu, Y., Xu, W., Zhang, J., Liu, Q., Wu, S., Wang, L.: Deep graph structure learning for robust representations: a survey. arXiv preprint arXiv:2103.03036 (2021)

32. Zügner, D., Günnemann, S.: Adversarial attacks on graph neural networks via meta learning. In: ICLR (2019)

33. Li, Y., Jin, W., Xu, H., Tang, J.: DeepRobust: a PyTorch library for adversarial attacks and defenses. arXiv preprint arXiv:2005.06149 (2020)

34. Van Der Maaten, L., Hinton, G.: Visualizing data using t-SNE. J. Mach. Learn. Res. **9**(Nov), 2579–2605 (2008)

Transforming PageRank into an Infinite-Depth Graph Neural Network

Andreas Roth$^{(\boxtimes)}$ ⓘ and Thomas Liebig ⓘ

Artificial Intelligence Group, TU Dortmund, Dortmund, Germany
{andreas.roth,thomas.liebig}@tu-dortmund.de

Abstract. Popular graph neural networks are shallow models, despite the success of very deep architectures in other application domains of deep learning. This reduces the modeling capacity and leaves models unable to capture long-range relationships. The primary reason for the shallow design results from over-smoothing, which leads node states to become more similar with increased depth. We build on the close connection between GNNs and PageRank, for which personalized PageRank introduces the consideration of a personalization vector. Adopting this idea, we propose the Personalized PageRank Graph Neural Network (PPRGNN), which extends the graph convolutional network to an infinite-depth model that has a chance to reset the neighbor aggregation back to the initial state in each iteration. We introduce a nicely interpretable tweak to the chance of resetting and prove the convergence of our approach to a unique solution without placing any constraints, even when taking infinitely many neighbor aggregations. As in personalized PageRank, our result does not suffer from over-smoothing. While doing so, time complexity remains linear while we keep memory complexity constant, independently of the depth of the network, making it scale well to large graphs. We empirically show the effectiveness of our approach for various node and graph classification tasks. PPRGNN outperforms comparable methods in almost all cases. (Our code is available at: https://github.com/roth-andreas/pprgnn.)

Keywords: Machine learning · Graph neural networks · PageRank

1 Introduction

Graph-structured data is found in many real-world applications ranging from social networks [26] to biological structures [28]. Steadily growing amounts of data lead to emerging solutions that can extract relevant information from these data types. Tasks like providing recommendations [41], predicting the state of traffic [8] or the classification of entire graphs into distinct categories [39] are some of the tasks of research interest. Approaches based on deep learning have

Supplementary Information The online version contains supplementary material available at https://doi.org/10.1007/978-3-031-26390-3_27.

found great success for grid-structured data, e.g., in image processing [20] and natural language processing [34]. Graph Neural Networks (GNNs) [21] adopt the ideas from convolutions in euclidean space for irregular non-euclidean domains. These methods directly consider the graph structure when performing convolution operations.

One of the challenges of GNNs is to capture long-range dependencies. Recently popular methods use an aggregation scheme, in which k layers of graph convolutions combine the information from k-hops around each node [21,35]. Several issues, most dominantly over-smoothing [24,38] and memory consumption [6,17,42] were found to prevent deep models, as in image processing [20]. Several recent efforts explore options to enable more layers and even formulate infinite-depth equations. However, previous work still only allows a limited depth [30,38] or places hard constraints on the parameters [16] or the architecture [2].

As identified by [22], GNNs are closely related to PageRank [27], which in its basic version only depends on the graph structure, not on the initial distribution. Personalized PageRank [27] introduces a chance to reset PageRank to a teleportation vector, allowing the result to depend not only on the graph structure but also on an initial distribution. We show how the idea of personalization can be adopted to GNNs and propose the Personalized PageRank Graph Neural Network (PPRGNN), an infinitely deep GNN that adds a chance to reset the neighbor aggregation back to the initial state. In order to prove the convergence of PPRGNN to a unique solution when iterating infinitely many times, we modify the chance of resetting to be dynamic based on the distance to the root node. As in personalized PageRank, our approach does not suffer from over-smoothing and the locality of node features around their root nodes is preserved. Due to the large depths, far distant information still influences resulting node representations.

In addition, we provide rich theoretical intuition for the success of our formulation and our design choices. While the depth is theoretically always infinite, the practically effective depth is adaptive and varies depending on the learned parameters, the graph structure, and the observed features. We also provide a way to control the convergence rate since different levels of localization are effective for different types of graphs [1,15]. Furthermore, contrary to previous infinite-depth approaches, we do not impose any constraints on parameters or the model's architecture. To allow scaling to large graphs despite the infinite depth, we design an efficient gradient computation that remains constant in memory and execution time. We validate our proposed approach against comparable methods on various inductive and transductive node and graph classification tasks. Our approach outperforms related methods in almost all cases by considerable margins, while most other approaches are within a competitive range only for individual tasks. The experimental execution time is also improved compared to previous infinite-depth approaches.

The rest of our work is structured as follows. Section 2 introduces our notation and relevant basics in personalized PageRank and GNNs. We describe recent related approaches in Sect. 3. Our method is detailed in Sect. 4, and a comprehensive evaluation is presented in Sect. 5. We discuss our results and potential future directions in Sect. 6.

2   Preliminaries

We represent a graph $G = (V, E)$ as the tuple of n nodes $V = \{v_1, v_2, \ldots, v_n\}$ and a set of edges E between pairs of nodes. We construct an adjacency matrix $\mathbf{A} \in \mathbb{R}^{n \times n}$ describing the connectivity between pairs of nodes from the E. Entries $a_{ij} \in \mathbf{A}$ indicate the strength of an edge between nodes v_i and v_j, a zero-entry indicates the absence of an edge. Our method assumes undirected edges, e.g., $a_{ij} = a_{ji}$, but it is straightforward to apply it to directed graphs. We use a normalized version $\tilde{\mathbf{A}} = \mathbf{D}^{-1/2} \mathbf{A} \mathbf{D}^{-1/2}$ of the adjacency matrix, potentially with self-loops. Each node v_k has a set of F features $\mathbf{u}_k \in \mathbb{R}^F$ associated with them. The feature matrix $\mathbf{U} \in \mathbb{R}^{n \times F}$ contains all nodes' stacked feature vectors $\mathbf{u}_k$. We define the node neighborhood $N_i = \{v_j | \tilde{\mathbf{A}}_{ij} > 0\}$ as the set of all nodes connected to v_i.

2.1   Personalized PageRank

Our approach inherits basic concepts and intuition from personalized PageRank [27], which we briefly describe here. PageRank [27] was originally introduced to score the importance of webpages for web searches. In their work, webpages represent individual nodes in a graph and links on these webpages are modeled as directed edges between these nodes. The solution to PageRank is the fixed point of the equation

$$\mathbf{r} = \mathbf{A}\mathbf{r},\tag{1}$$

with $\mathbf{r} \in \mathbb{R}^n$ being the dominant eigenvector of $\mathbf{A}$. The vector $\mathbf{r}$ can be obtained by power iteration with an arbitrary initial $\mathbf{r}_0$ [27]. For an intuitive interpretation of Eq. (1), we can interpret $\mathbf{A}$ as the stochastic transition matrix over the graph, providing a connection to a random walk. Therefore the stationary probability distribution induced by a random walk is the same as $\mathbf{r}$ in the limit [27]. This also results in r only depending on the graph structure and not on prior information available for nodes. Therefore, the authors also introduce personalized PageRank [27]

$$\mathbf{r} = (1 - \alpha)\mathbf{A}\mathbf{r} + \alpha\mathbf{u}\tag{2}$$

that adds a chance α as a way to teleport back to a personalization vector $\mathbf{u} \in \mathbb{R}^n$ representing an initial distribution over all pages. The corresponding interpretation for a random walk is to introduce a chance to reset the random walk to the personalization vector [27].

2.2   Graph Neural Networks

Another concept we build upon are Graph Neural Networks (GNNs), specifically their subtype of Message-Passing Neural Networks (MPNNs) [13]. GNNs apply permutation equivariant operations to graph structured data in order to identify task-specific features. Originating from spectral graph convolutions [18] as a localized first-order approximation, each message-passing operation updates the node states $\mathbf{h}_i$ by combining the information of the direct neighborhood N_i for each node v_i [36]. The general framework can be described as a node-wise update function

$$\mathbf{h}_i^{(l+1)} = \psi \left(\mathbf{h}_i^{(l)}, \bigoplus_{j \in N_i} \omega(\mathbf{h}_i^{(l)}, \mathbf{h}_j^{(l)}) \right) , \tag{3}$$

for each state $\mathbf{h}_i$, using some functions ω and ψ and a permutation invariant aggregation function $\bigoplus$. In this work, we will demonstrate our approach using the very basic instantiation of this framework, namely the Graph Convolutional Network (GCN) [21]. Making use of the normalized adjacency matrix $\tilde{\mathbf{A}} \in \mathbb{R}^{n \times n}$, the GCN can be expressed in matrix notation

$$\mathbf{H}^{(l+1)} = \phi \left(\tilde{\mathbf{A}} \mathbf{H}^{(l)} \mathbf{W}^{(l)} \right) \tag{4}$$

using a linear transformation $\mathbf{W} \in \mathbb{R}^{d \times d}$ and ϕ as an element-wise activation function. $\mathbf{H}^{(l)} \in \mathbb{R}^{n \times d}$ contains the node states $\mathbf{h}_i^{(l)}$ of all nodes after layer l. Each layer aggregates information only from direct neighborhood N_i for each node v_i. Thus, after k such layers, each node only has access to information a maximum of k hops away. Given this property, choosing any number k of these layers leads to information at $k+1$ hops away being impossible to be considered for making predictions. Moreover, even when the number of layers k can be selected to be sufficient for all potentially considered graphs, a large number k leads to various additional issues that we will describe next.

Over-Smoothing. Recent work found that stacking many layers of Eq. (4) leads to a degradation of experimental performance that is caused by an effect called over-smoothing [21,24,38]. Li et al. [24] show that Eq. (4) is a special form of Laplacian smoothing leading to node representation becoming more similar the more layers are added. They prove that Laplacian smoothing converges to a linear combination of dominant eigenvectors. While some smoothing is needed to share information between nodes, representations eventually become indistinguishable with too much smoothing, thus making accurate data-dependant predictions harder [24].

On a similar note, [38] find a close connection between k layers of Eq. (4) and a k-step random walk. They find that both to converge the limit distribution of the random walk. In the limit, a random walk becomes independent of the root nodes and therefore loses the locality property of individual nodes. Therefore, representations become independent of the starting node and initial node features, thus becoming indistinguishable [38]. In practice, the performance of Eq. (4) already degrades with more than two layers in many cases [21].

Memory Complexity. Another reason that prevents GNNs from being deep models is the memory complexity. Graphs can quickly surpass a million nodes, which leads to out-of-memory issues due to the linear memory requirements $\mathcal{O}(kn)$ in the number of layers k and the number of nodes n. Several approaches try to lower the memory complexity by only considering samples of nodes from a local neighborhood [17]. Due to an effect known as the neighborhood explosion, the number of nodes in the k-hop neighborhood $\mathcal{O}(d^k)$ explodes, with d being the average node degree. Thus, for a large number of layers k, the benefit vanishes. Other approaches cluster the graph into subgraphs and use these for training [6, 42], but cannot leverage the full potential of the entire graphs relationships. Therefore, this issue needs to be considered when designing deep graph neural networks.

3 Related Work

Several approaches aim to increase the depth of MPNNs and simultaneously deal with over-smoothing and memory consumption. Rong et al. [30] found over-smoothing to occur faster for nodes with many incoming edges and propose DropEdge as the equivalent to dropout in regular neural networks. They randomly sample edges to remove during each training epoch and show that the effect of over-smoothing gets slowed down. Klicpera et al. [23] propose a diffusion process that they find to be beneficial for semi-supervised node classification tasks for graphs with high homophily but encounters problems with complex graphs. Zhu et al. [45] further discuss the issue of settings with low homophily. Li et al. [24] co-train a random walk model that explores the global graph topology. Inspired by the findings from ResNet [20], Chen et al. [4] propose GCNII that makes use of residual connections in two ways. In each layer, they add an initial residual connection to the input state $\mathbf{H}^{(0)}$ and an identity mapping to the weights, which were shown to have beneficial properties [19]. Xu et al. [38] combine the results of all intermediate iterations in JKNet. Other works find a rescaling of the weights to alleviate the over-smoothing problem [25,44]. While these approaches help reduce the effect of over-smoothing, they are limited in practical depth and the issue still arises.

3.1 Infinite-Depth Graph Neural Networks

Evaluating the option of repeating iterations infinitely many times have been analyzed in various approaches [11,14,16,22]. These methods iterate some graph convolution until convergence by employing weight-sharing and ensuring the convergence of their formulations. When using an equation for an infinite-depth GNN, the result needs to converge to a unique solution. We summarize this under the following definition of well-posedness.

Definition 1 *(Well-posedness). Given an input matrix* $\mathbf{X} \in \mathbb{R}^{N \times D}$, *an equation* $\mathbf{Y} = g(\mathbf{X})$, *with g being an infinitely recursive function is well-posed, if*

1. *The solution* $\mathbf{Y}$ *is unique*
2. $g(\mathbf{X})$ *converges to the unique solution* $\mathbf{Y}$.

While the GCN (Eq. (4)) is not generally well-posed, our work proposes a similar equation that we prove to be well-posed. We start by reviewing two recent approaches to infinitely deep graph neural networks that serve as the starting point for our contribution. The first [22] is derivated from the PageRank [27] algorithm, the other is the fixed-point solution of an equilibrium equation [16].

APPNP. Klicpera et al. [22] propose a propagation scheme derived from personalized PageRank [27]. They identify the connection between the limit distribution of MPNNs and PageRank, with both losing focus on the local neighborhood of the initial state. As personalized PageRank was introduced as a solution to this issue for PageRank [27], they adopt this idea for MPNNs. They set the personalization vector $\mathbf{r}$ from Eq. (2) to the hidden state of all nodes $\mathbf{H}^{(0)}$. A chance α to teleport back to the root node preserves the local neighborhood with the tunable parameter. Klicpera et al. [22] transfer this idea to MPNNs with Approximate Personalized Propagation of Neural Predictions (APPNP) [22]

$$\mathbf{H}^{(l+1)} = (1 - \alpha)\tilde{\mathbf{A}}\mathbf{H}^{(l)} + \alpha\mathbf{H}^{(0)} \tag{5}$$

that repeatedly, potentially infinitely many times, aggregates the neighborhood. They also add a chance of going back to the initial state $\mathbf{H}^{(0)} = f_\theta(\mathbf{U})$, that is be the output of previous layers f_θ. They show that Eq. (5) is well-posed for any $\alpha \in (0, 1], \mathbf{H}^{(0)} \in \mathbb{R}^{N \times D}, \tilde{\mathbf{A}} \in \mathbb{R}^{N \times N}$ with $\det(\tilde{\mathbf{A}}) \leq 1$. Typical normalizations $\tilde{\mathbf{A}}$ of the adjacency matrix satisfy this property. Notably, Eq. (5) does not utilise any learnable parameters. They rather propose to separate the propagation scheme in Eq. (5) from the learnable part, by making $\mathbf{H}^{(0)} = f_\theta(\mathbf{U})$ as node-wise application of a MLP. This method is proposed only for semi-supervised node classification tasks, with a softmax activation employed to transform the output of the last iteration $\mathbf{H}^{(K)}$ of Eq. (5) to class predictions.

Implicit Graph Neural Networks. Independently, Gu et al. [16] propose the Implicit Graph Neural Network (IGNN) by adapting the general implicit framework [10] for graph convolutions. They obtain the fixed-point solution of a non-linear equilibrium equation

$$\mathbf{X} = \phi(\mathbf{W}\mathbf{X}\tilde{\mathbf{A}} + f_\theta(\mathbf{U})) \tag{6}$$

by iterating it until convergence. While not being well-posed in general, they prove the well-posedness of Eq. (6) for the specific case that $\lambda_{pf}(|\mathbf{A}^T \otimes \mathbf{W}|) < 1$ with λ_{pf} being the Perron-Frobenius (PF) eigenvalue. They make use of the Kronecker product $\otimes$ and the Perron-Frobenius theory [3]. Since $\tilde{\mathbf{A}}$ is fixed, the matrix of parameters $\mathbf{W}$ needs to be strictly constrained to fulfill $\lambda_{pf}(|\mathbf{A}^T \otimes \mathbf{W}|) < 1$. The set $\mathcal{M}$ of allowed matrices $\mathbf{W}$ forms an $\mathcal{L}_1$-ball, with any weight matrix outside the ball not leading to convergence. Remaining inside this ball cannot be guaranteed by regular gradient descent. Instead, after each step of regular gradient descent, they project the result to the closest point on the ball using projected gradient descent, for which efficient algorithms exist [9]. While their inspiring work shows great experimental results, we identify a couple of shortcomings with. Many weight matrices cannot be used given the strict constraint on $\mathbf{W}$, hindering the model capacity. Further, the projection onto the $\mathcal{L}_1$-ball changes the direction of the gradient update away from the steepest descent. Therefore optimization steps are less effective in reducing the models' loss. The strict constraint and the resulting projection step also add complexity to the method's theoretical derivation and practical implementation. Considering different neighborhood sizes was found to be important when applying graph algorithms to varying graph types [1, 15], not having a way to control the effective depth of the model is also unsatisfying.

4   PageRank Graph Neural Network

The solution of PageRank is the stationary probability distribution that is independent of the input. Given the close relation between PageRank (Eq. (1)) and MPNNs (Eq. (4)), the locality of the data and the influence of the input features also diminish with a MPNN, as identified by [22]. As personalized PageRank was introduced to prevent the loss of focus for PageRank [27], we introduce the Personalized PageRank Graph Neural Network (PPRGNN) based on personalized PageRank, that similarly assures the locality of the node states in the limit. Using the initial state $f_\theta(\mathbf{U})$ as personalization matrix for teleportation [22], PPRGNN can be understood as repeatedly applying graph convolutions with a chance to teleport back to this initial state. We assure the convergence of PPRGNN to a unique solution, so our method allows an arbitrary amount of layers - potentially infinitely many - without suffering from over-smoothing. Practically, we iterate graph convolutions until further iterations have negligible impact and our solution is close to the limit distribution. In this work, we adopt GCNs [21], which are the basic version of MPNNs, but these are directly replaceable by more advanced types.

We denote the chance of traversing the graph further by α_l. Rewriting the formulation of the GCN in a similar way to personalized PageRank, we come up with our formulation

$$\mathbf{H}^{(l+1)} = \phi\left(\alpha_l \tilde{\mathbf{A}} \mathbf{H}^{(l)} \mathbf{W} + f_\theta(\mathbf{U})\right) \tag{7}$$

with $\mathbf{H}^{(0)} = \mathbf{0}$ that utilizes shared and unconstrained parameters $\mathbf{W}$. Due to the recursive nature and no constraints, exponential growth in $\mathbf{W}$ prevents well-posedness for any fixed α_l. The issue with having no guarantees for convergence is that the furthest distant nodes are multiplied with the highest exponential of

W, which potentially dominates the result. As in PageRank, these only depend on the graph structure and not on the node features, leading to the loss of locality of the resulting node features.

Our core idea to guarantee convergence of Eq. (7) without constraining the parameters as in [16] is to reduce the chance of expanding further α_l with the distance to the root node. As the message-passing formulation is connected to a random walk on the graph, another interpretation is to increase the chance of resetting the random walk with the number of steps taken. When n is the number of steps taken in that walk, we find using a decay of $\alpha_n = \frac{1}{n}$ to be sufficient for converging to a unique solution. The recursive nature of our formulation leads to a multiplication of all α_n, resulting in the influence to decay by $\frac{1}{n!}$. Because the recursive application of **W** only leads to an exponential $\mathbf{W}^n$ growth, the equation converges. For control over the effective depth, i.e., the speed of convergence and numerical stability, we use $\frac{1}{1+n\epsilon}$ and formally prove its convergence for any $\epsilon > 0$ later. We set the value for teleporting back to $f_\theta(\mathbf{U})$ fixed to 1 because in the limit the chance $(1 - \alpha_l)$ would become very small for close neighbors, leading to the same issues of over-smoothing that we described in Sect. 2.2.

Setting α_l in Eq. (7) accordingly to our findings, the following issue arises: The most distant nodes are processed first in Eq. (7), and the direct neighbors are processed in the last iteration. This results from recursively applying the adjacency matrix $\tilde{\mathbf{A}}$ on the input, leading to the initial state being transformed k times by $\tilde{\mathbf{A}}$. Thus, for calculating $\mathbf{H}^{(1)}$, the expansion factor α_0 needs to be minimal, which is the opposite of using the iteration l as n.

In case we are interested in a fixed number k of total iterations, we can directly set $\alpha_l = \frac{1}{1+(k-l-1)\epsilon}$ for each layer l. When using a fixed number of iterations, this approach is ready for usage directly. Since we are interested in the case when $k \rightarrow \infty$, starting with α_0 poses a challenge.

For a theoretical analysis of the convergence of Eq. (7), an equation that can be iterated infinitely-deep independently of k is desired. We achieve this by setting the index variable to $n = k - l$ resulting in the flipped equation

$$\mathbf{G}^{(n)} = \phi \left(\beta_n \tilde{\mathbf{A}} \mathbf{G}^{(n+1)} \mathbf{W} + f_\theta \left(\mathbf{U} \right) \right) \tag{8}$$

with $\beta_n = \alpha_{k-l-1}$ that is semantically unchanged, i.e., $\mathbf{G}^{(0)} = \mathbf{H}^{(k)}$ for any k used for both $\mathbf{G}$ and $\mathbf{H}$. Calculating $\mathbf{G}^{(n)}$ from a given $\mathbf{G}^{(n+1)}$ can be performed without knowing k beforehand, helping us in the theoretical analysis by expanding the recursive equation infinitely deep without the need to set a fixed value for k. It also leads to a cleaner proof of convergence, which we will provide next. We further simplify our notation by denoting $\mathbf{G}^{(l;k)}$ as the result of k iterations performed by setting $\mathbf{G}^{k+l+1} = \mathbf{0}$, resulting in $\mathbf{G}^{(l)}$.

Theorem 1. *The result of $\mathbf{G}^{(l;k)}$ using the equation $\mathbf{G}^{(n)} = \phi \left(\beta_n \tilde{\mathbf{A}} \mathbf{G}^{(n+1)} \right.$ $\mathbf{W} + \mathbf{B})$ with $\beta_n = \frac{1}{1+n\epsilon}$ converges to a unique solution when $k \rightarrow \infty$ for any $l \in \mathbb{R}$ $\mathbf{W} \in \mathbb{R}^{d \times d}, \tilde{\mathbf{A}} \in \mathbb{R}^{n \times n}, \mathbf{B} \in \mathbb{R}^{d \times n}, n \in \mathbb{N}, \epsilon > 0$, any Lipschitz continuous activation function ϕ.*

We refer to the supplementary material for all proofs.

Practically, for either Eq. (7) or Eq. (8) processing starts at the furthest distant nodes, for which k needs to be known. This a challenge, because we do not know beforehand when our convergence criterion is satisfied. As our interest is in the limit state $\mathbf{G}^{(0;k)}$ when $k \to \infty$, we make use of a convergence threshold ϵ to identify the number of required iterations

$$k = \min\{M \mid \mathbf{G}^{(0;k)} - \mathbf{G}^{(0;k+1)} < \epsilon\} \tag{9}$$

until our solution is close to the limit and iterating further has negligible impact. Because even the initial iteration $\mathbf{G}^{(k-1;1)} \neq \mathbf{G}^{(k;1)}$ is different, intermediate results from $\mathbf{G}^{(0;k-1)}$ cannot be reused for computing $\mathbf{G}^{(0;k)}$. A full recalculation is needed, which requires $k!$ iterations.

Instead, we take a different route to determine k. Determining at which iteration the difference of expanding further on the graph becomes negligible is approximately the same as determining how far the influence of nodes in the graph reach using our message passing scheme. To determining this, we ignore the teleportation term and estimate the influence of the initial state $f_\theta(\mathbf{U})$ on the result of l iterations $\mathbf{G}^{(0;l)}$ with the equation

$$\mathbf{E}^{(l+1)} = \phi(\alpha_{l+1}\tilde{\mathbf{A}}\mathbf{E}^{(l)}\mathbf{W}) \tag{10}$$

by setting $\mathbf{E}^{(0)} = f_\theta(\mathbf{U})$. Unlike in Eq. (7) where we reversed the equation, the result $\mathbf{E}^{(m)}$ is equal for $\alpha_l = \frac{1}{1+(m-l-1)\epsilon}$ and $\alpha_l = \frac{1}{1+l\epsilon}$ when we use ReLU as ϕ. Note, that we start with α_{l+1} because this is the first α that is applied to the teleportation matrix $f_\theta(\mathbf{U})$. Equation (10) converges for similar reasons as Eq. (8), only towards $\mathbf{0} \in 0^{d \times n}$, which we proof with the following theorem.

Theorem 2. *The equation* $\mathbf{E}^{(l+1)} = \phi\left(\alpha_l\tilde{\mathbf{A}}\mathbf{E}^{(l)}\mathbf{W}\right)$ *with* $\alpha_l = \frac{1}{1+l\epsilon}$ *converges to* $\mathbf{0} \in 0^{d \times n}$ *for any* $\mathbf{W} \in \mathbb{R}^{d \times d}, \tilde{\mathbf{A}} \in \mathbb{R}^{n \times n}, l \in \mathbb{N}, \epsilon > 0$, *any initial* $\mathbf{E}^{(0)}$ *and the ReLU activation function* ϕ. *The solution can be obtained by iterating the equation. For any fixed number of iterations* m, *the solution* $\mathbf{E}^{(m)}$ *is the same as using* $\alpha_l = \frac{1}{1+(m-l-1)\epsilon}$.

Since we can evaluate Eq. (10) directly by iterating until our convergence criterion is met, we find the required number of steps with

$$k' = \min\{l \mid \mathbf{E}^{(l)} < \epsilon\}. \tag{11}$$

At this point the effect of the initial state on nodes of distance l is negligible. We use k' as k in Eq. (8) and execute the forward pass. The result $\mathbf{H}^{(k')}$ gets passed onto the next operation in our model, as with other graph convolutions.

4.1   Efficient Optimization

While we do not use Eq. (10) for gradient computation, even tracking only Eq. (7) with autograd software would still lead to memory consumption that is linear in the number of layers. Similarly as in the forward pass, the gradients converge to $\mathbf{0}$ for distant nodes. We iterate the computation of gradients until the same convergence criterion is met. Because of faster converge in the backward pass, this allows the optimization of the model with reduced memory consumption. We will further limit the iterations to guarantee constant complexity, independently of the number of iterations performed.

For calculating derivatives we use the reformulation from Eq. (8). We introduce additional notation and set $\hat{\mathbf{Y}} = f_\theta(\mathbf{G}^{(0)})$ as the output of our model, $\mathbf{Y}$ as the target, and $\mathcal{L} = l(\hat{\mathbf{Y}}, \mathbf{Y})$ to be our loss calculated with any differentiable loss function l. We are interested in the partial derivatives of our loss $\mathcal{L}$ with respect to the parameters $\mathbf{W}$ and the input state $\mathbf{B}$. We let autograd solve the derivation $\frac{\partial L}{\partial \mathbf{G}^{(0)}}$ and apply the chain rule for other partial derivatives. To simplify our notation for the application of the chain rule, we further define $\mathbf{Z}^{(n)} = \alpha_n \tilde{\mathbf{A}} \mathbf{G}^{(n+1)} \mathbf{W} + \mathbf{B}$ and $\mathbf{G}^{(n)} = \phi(\mathbf{Z}^{(n)})$. All further partial derivatives can be calculated by using the following equations:

$$\frac{\partial L}{\partial \mathbf{G}^{(n)}} = \alpha_n \tilde{\mathbf{A}}^T \frac{\partial L}{\partial \mathbf{Z}^{(n-1)}} \mathbf{W}^T \tag{12}$$

$$\frac{\partial L}{\partial \mathbf{Z}^{(n)}} = \phi'\left(\alpha_n \tilde{\mathbf{A}} \mathbf{G}^{(n+1)} \mathbf{W} + \mathbf{B}\right) \odot \frac{\partial L}{\partial \mathbf{G}^{(n)}} \tag{13}$$

$$\frac{\partial L}{\partial \mathbf{W}} = \sum_{n=0}^{\infty} \alpha_n \left(\tilde{\mathbf{A}} \mathbf{G}^{(n+1)}\right)^T \frac{\partial L}{\partial \mathbf{Z}^{(n)}} \tag{14}$$

$$\frac{\partial L}{\partial \mathbf{B}} = \sum_{n=0}^{\infty} \frac{\partial L}{\partial \mathbf{Z}^{(n)}} \tag{15}$$

The partial derivatives $\frac{\partial L}{\partial \mathbf{W}}$ and $\frac{\partial L}{\partial \mathbf{B}}$ converge for similar reasons as before, so we iterate Eq. (14) and Eq. (15) until our convergence criterion is met. The convergence rate turns out to be much faster than the convergence rate of the forward pass, which results in reduced practical memory consumption. To theoretically guarantee constant memory consumption, we only consider a fixed amount n of elements in the sum, similarly to the effectiveness of Truncated Backpropagation Through Time (TBPTT) [33] for sequential data. This also reduces the time complexity of the backward pass to be constant. We found this restriction to have negligible impact even for small values of N. Depending on available memory, we either store the intermediate results for gradient computation or use gradient checkpointing [5] with a few additional forward iterations. Note, that for the backward step the solutions of $\mathbf{G}^{(0)}, \ldots, \mathbf{G}^{(n)}$ are needed explicitly. We assure the convergence of all used $\mathbf{G}^{(i)}$ by using the fact that $\mathbf{G}^{(n;k)} < \mathbf{G}^{(0;k)}$ and therefore compute $\mathbf{G}^{(0;k+n)}$ instead of $\mathbf{G}^{(0;k)}$ in the initial forward pass.

5   Experiments

We evaluate the effectiveness of PPRGNN on various public benchmark datasets and compare these results to popular methods and other infinite-depth

Table 1. Properties of datasets used for evaluation.

Dataset	# of Graphs	Avg. # of nodes	# of classes
Amazon	1	334 863	58
PPI	22	2373	121
MUTAG	188	17.9	2
PTC	344	25.5	2
COX2	467	41.2	2
PROTEINS	1113	39.1	2
NCI1	4110	29.8	2

Table 2. Micro-F_1-Scores for PPI.

Method	Micro-F_1-Score
MLP	46.2
GCN	59.2
SSE	83.6
GAT	97.3
IGNN	97.6
APPNP	44.8
PPRGNN	**98.9**

approaches. We evaluate our approach on an inductive node classification task, a transductive node classification task, and five graph classification tasks. Table 1 shows detailed properties of all used datasets. We closely follow the experimental settings of IGNN [16] and inherit their architectures, only replacing their formulation directly with ours. Thus the number of parameters is the same, so the comparison with their approach is the most meaningful for us. We apply APPNP to all tasks using their setup with 10 iterations. We further compare PPRGNN with a series of other popular methods for the tasks of node classification and graph classification and reuse the results reported in [16]. Due to the increased modeling capacity, we use gradient clipping and weight decay. Additionally, we tune ϵ, the learning rate and whether self-loops are taken into account for $\hat{A}$ for the three different tasks. We set $n = 5$ for the backward pass. We reduce the learning rate when the training loss plateaus. All experiments are executed on a single Nvidia Tesla P100.

PPI. We consider the task of role prediction of proteins in graphs of protein-protein interactions (PPI) [17]. In this inductive node classification task, we use 18 graphs for training our model, 2 for validation, and 2 for testing. Our data split matches that in previous work [17]. As taken over from IGNN, our model consists of 5 stacked layers, each iterating until convergence. We set $\epsilon = 0.25$ and find self-loops detrimental to our approach. In addition to IGNN and APPNP, reference methods are a MLP, GCN [21], SSE [7], GAT [35]. The Micro-F_1-Scores for all considered approaches are presented in Table 2. PPRGNN outperforms all of these approaches and reduces the error by more than 50% compared to IGNN. Our trained PPRGNN uses a total of 82 message passing iterations in testing, while GCN and GAT use a maximum of 3 iterations. We also compare the time needed for PPRGNN to surpass the Micro-F_1-Score of IGNN in Table 3. PPRGNN needs fewer iterations and also takes less time per Epoch. This comes from accurate gradient descent steps without projection and being able to adjust the speed of convergence with ϵ. We find APPNP to underfit the data due to the limited modeling capacity, even when the number of parameters uses all available memory.

Table 3. Time and epochs needed until PPRGNN surpasses the best epoch of IGNN on the validation set.

Dataset	Method	Epochs	Avg. time per epoch	Total time
Amazon (0.05)	IGNN	872	14 s	3 h 21 m
	PPRGNN	**175**	**11 s**	**32 m**
PPI	IGNN	58	26 s	25 m
	PPRGNN	**47**	**18 s**	**14 m**

Amazon. To test the scalability of our approach, we apply it to the Amazon product co-purchasing network data set [40]. Following the settings from [7], product types with at least 5000 different products are selected. This results in 334 863 nodes representing products and 925 872 edges representing products that have been purchased together. The task is to predict the correct product type for each node. Nodes do not have any features, so predictions are made solely based on the graph structure. We use the same data split as [7], leading to a fraction of nodes used for training varying between 5% and 9%. A fixed set consisting of 10% of the nodes is used for training, the rest for validation. The main challenge of this task is not the prediction complexity but rather dealing with the sparsity of the available labels. Our architecture consists of our PPRGNN layer combined with a linear operation before and after. We compare our results with APPNP and reuse the result found in [7,16] for IGNN [16], SSE [7], struct2vec [29] and GCN [21]. Micro-F_1-Scores and Macro-F_1-Scores are shown in Fig. 1 for varying fractions of labels used. While we outperform IGNN, SSE, struct2vec and GCN across all settings by at least 1%, APPNP performs the best. We find the low modeling capacity of APPNP to be better suited for generalizing in this scenario. Again, we compare the execution time needed for PPRGNN to outperform IGNN (Table 3) and find PPRGNN to converge in fewer epochs, with each epoch executing faster. This further adds to our point of benefiting from applying gradient descent without projection and controlling convergence speed.

Graph Classification. We now evaluate our approach for the task of graph classification on five open graph datasets, namely MUTAG, PTC, COX2, PRO-TEINS, and NCI1. Following the same setup from previous work, we conduct a 10-fold cross-validation for each dataset and report the mean and standard deviation of the folds validation sets. We integrate our formulation with $\epsilon = 1$ into the architecture from IGNN, consisting of 3 stacked iterations until convergence. For regularization, we add a weight decay of 1e−6 and gradient clipping of 25 to all datasets. For NCI1, we find removing self-loops to be helpful for generalization. For comparison, we use several graph kernel approaches (GK [32], RW [12], WL [31]) and GNN approaches (DGCNN [43], GCN [21], GIN [37]) in addition to IGNN and APPNP. We reuse reported results from [16]. PPRGNN outperforms

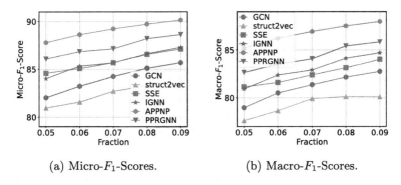

(a) Micro-F_1-Scores.   (b) Macro-F_1-Scores.

Fig. 1. Comparison of results on the Amazon dataset. The fraction of labels used for optimization varies between 0.05 and 0.09.

Table 4. Comparison of accuracies on various graph classification tasks.

Dataset	MUTAG	PTC	COX2	PROTEINS	NCI1
GK	81.4 ±1.7	55.7 ±0.5	–	71.4 ±0.3	62.5 ±0.3
RW	79.2 ±2.1	55.9 ±0.3	–	59.6 ±0.1	–
WL	84.1 ±1.9	58.0 ±2.5	83.2 ±0.2	74.7 ±0.5	**84.5** ±0.5
DGCNN	85.8	58.6	–	75.5	74.4
GCN	85.6 ±5.8	64.2 ±4.3	–	76.0 ±3.2	80.2 ±2.0
GIN	89.0 ±6.0	63.7 ±8.2	–	75.9 ±3.8	82.7 ±1.6
IGNN	89.3 ±6.7	70.1 ±5.6	86.9 ±4.0	77.7 ±3.4	80.5 ±1.9
APPNP	87.7 ±8.6	64.5 ±5.1	82.2 ±5.5	78.7 ±4.8	65.9 ±2.7
PPRGNN	**90.4** ±7.2	**75.0** ±5.7	**89.1** ±3.9	**80.2** ±3.2	83.5 ±1.5

all other approaches across 4 out of 5 datasets by at least 1% and is the second-best performing model with competitive accuracy on the fifth dataset (Table 4). Despite using the same $\epsilon = 1$ across all experiments, the effective depth ranges from 22 to 41 for different datasets. Depth is adaptive even within individual datasets, depending on learned parameters, the examined graph and present node features. These results further demonstrate the effectiveness of our approach and the potential to create deeper models on a wide variety of datasets.

6   Conclusion

We introduced PPRGNN, a reformulation of MPNNs based on personalized PageRank that assures localization and prevents over-smoothing of node features even when using infinitely many layers. Theoretically based on the personalized version of PageRank which allows teleporting back to the initial state, we adopt this idea for MPNNs, specifically for the basic type GCNs [21]. Starting from the classic algorithm, we follow intuitive steps to introduce learnable parameters and still converge to a limit distribution. Compared to previous infinite-depth GNNs, our approach has a higher modeling capacity as we do not place any constraints. Our empirical evaluation on tasks for graph classification, and inductive and transductive node classification confirm our theoretical base. Against regular GCNs that have no way to teleport back to the initial state, we find large improvements across all datasets. We even outperform other comparable approaches, including previous infinite-depth models, across almost all datasets by decent margins. Despite the theoretical infinite-depth, we introduced a path for efficient optimization, running linearly in the number of layers and only using constant memory. Our formulation allows controlling the convergence rate, leading to considerable improvements in experimental execution time compared to IGNN, a previous infinite-depth model. While we show that our formulation allows infinitely many layers, even fixed sized models should benefit from adopting our idea. Our approach is directly applicable to other types of MPNNs, for which our proofs of convergence should hold.

Acknowledgements. Part of the work on this paper has been supported by Deutsche Forschungsgemeinschaft (DFG) - project number 124020371 - within the Collaborative Research Center SFB 876 "Providing Information by Resource-Constrained Analysis", DFG project number 124020371, SFB project B4. The authors are funded by the German Federal Ministry of Education and Research (BMBF) in the course of the 6GEM research hub under grant number 16KISK038.

References

1. Abu-El-Haija, S., Perozzi, B., Al-Rfou, R., Alemi, A.A.: Watch your step: learning node embeddings via graph attention. In: Advances in Neural Information Processing Systems, vol. 31 (2018)
2. Bai, S., Kolter, J.Z., Koltun, V.: Deep equilibrium models. In: Advances in Neural Information Processing Systems, vol. 32 (2019)
3. Berman, A., Plemmons, R.J.: Nonnegative matrices in the mathematical sciences. SIAM (1994)
4. Chen, M., Wei, Z., Huang, Z., Ding, B., Li, Y.: Simple and deep graph convolutional networks. In: International Conference on Machine Learning, pp. 1725–1735. PMLR (2020)
5. Chen, T., Xu, B., Zhang, C., Guestrin, C.: Training deep nets with sublinear memory cost. arXiv preprint arXiv:1604.06174 (2016)
6. Chiang, W.L., Liu, X., Si, S., Li, Y., Bengio, S., Hsieh, C.J.: Cluster-GCN: an efficient algorithm for training deep and large graph convolutional networks. In: Proceedings of the 25th ACM SIGKDD International Conference on Knowledge Discovery & Data Mining, pp. 257–266 (2019)
7. Dai, H., Kozareva, Z., Dai, B., Smola, A., Song, L.: Learning steady-states of iterative algorithms over graphs. In: International Conference On Machine Learning, pp. 1106–1114. PMLR (2018)
8. Derrow-Pinion, A., et al.: Eta prediction with graph neural networks in google maps. In: Proceedings of the 30th ACM International Conference on Information & Knowledge Management, pp. 3767–3776 (2021)

9. Duchi, J., Shalev-Shwartz, S., Singer, Y., Chandra, T.: Efficient projections onto the l 1-ball for learning in high dimensions. In: Proceedings of the 25th International Conference on Machine Learning, pp. 272–279 (2008)

10. El Ghaoui, L., Gu, F., Travacca, B., Askari, A., Tsai, A.: Implicit deep learning. SIAM J. Math. Data Sci. **3**(3), 930–958 (2021)

11. Gallicchio, C., Micheli, A.: Fast and deep graph neural networks. In: Proceedings of the AAAI Conference on Artificial Intelligence, vol. 34, pp. 3898–3905 (2020)

12. Gärtner, T., Flach, P., Wrobel, S.: On graph kernels: hardness results and efficient alternatives. In: Schölkopf, B., Warmuth, M.K. (eds.) COLT-Kernel 2003. LNCS (LNAI), vol. 2777, pp. 129–143. Springer, Heidelberg (2003). https://doi.org/10. 1007/978-3-540-45167-9_11

13. Gilmer, J., Schoenholz, S.S., Riley, P.F., Vinyals, O., Dahl, G.E.: Neural message passing for quantum chemistry. In: International Conference on Machine Learning, pp. 1263–1272. PMLR (2017)

14. Gori, M., Monfardini, G., Scarselli, F.: A new model for learning in graph domains. In: Proceedings. In: 2005 IEEE International Joint Conference on Neural Networks, vol. 2, pp. 729–734 (2005)

15. Grover, A., Leskovec, J.: node2vec: scalable feature learning for networks. In: Proceedings of the 22nd ACM SIGKDD International Conference on Knowledge Discovery and Data Mining, pp. 855–864 (2016)

16. Gu, F., Chang, H., Zhu, W., Sojoudi, S., El Ghaoui, L.: Implicit graph neural networks. In: Advances in Neural Information Processing Systems, vol. 33, pp. 11984–11995 (2020)

17. Hamilton, W., Ying, Z., Leskovec, J.: Inductive representation learning on large graphs. In: Advances in Neural Information Processing Systems, vol. 30 (2017)

18. Hammond, D.K., Vandergheynst, P., Gribonval, R.: Wavelets on graphs via spectral graph theory. Appl. Comput. Harmon. Anal. **30**(2), 129–150 (2011)

19. Hardt, M., Ma, T.: Identity matters in deep learning. In: 5th International Conference on Learning Representations, ICLR 2017, Toulon, France, 24–26 April 2017. Conference Track Proceedings (2017)

20. He, K., Zhang, X., Ren, S., Sun, J.: Deep residual learning for image recognition. In: Proceedings of the IEEE Conference on Computer Vision and Pattern Recognition, pp. 770–778 (2016)

21. Kipf, T.N., Welling, M.: Semi-supervised classification with graph convolutional networks. arXiv preprint arXiv:1609.02907 (2016)

22. Klicpera, J., Bojchevski, A., Günnemann, S.: Predict then propagate: graph neural networks meet personalized pagerank. arXiv preprint arXiv:1810.05997 (2018)

23. Klicpera, J., Weißenberger, S., Günnemann, S.: Diffusion improves graph learning. In: Conference on Neural Information Processing Systems (NeurIPS) (2019)

24. Li, Q., Han, Z., Wu, X.M.: Deeper insights into graph convolutional networks for semi-supervised learning. In: Thirty-Second AAAI Conference on Artificial Intelligence (2018)

25. Oono, K., Suzuki, T.: Graph neural networks exponentially lose expressive power for node classification. In: International Conference on Learning Representations (2020)

26. Otte, E., Rousseau, R.: Social network analysis: a powerful strategy, also for the information sciences. J. Inf. Sci. **28**(6), 441–453 (2002)

27. Page, L., Brin, S., Motwani, R., Winograd, T.: The pagerank citation ranking: Bringing order to the web. Technical report, Stanford InfoLab (1999)

28. Pavlopoulos, G.A., et al.: Using graph theory to analyze biological networks. BioData Mining **4**(1), 1–27 (2011). https://doi.org/10.1186/1756-0381-4-10

29. Ribeiro, L.F., Saverese, P.H., Figueiredo, D.R.: struc2vec: learning node representations from structural identity. In: Proceedings of the 23rd ACM SIGKDD International Conference on Knowledge Discovery and Data Mining, pp. 385–394 (2017)

30. Rong, Y., Huang, W., Xu, T., Huang, J.: Dropedge: towards deep graph convolutional networks on node classification. In: International Conference on Learning Representations (2020)

31. Shervashidze, N., Schweitzer, P., Van Leeuwen, E.J., Mehlhorn, K., Borgwardt, K.M.: Weisfeiler-lehman graph kernels. J. Mach. Learn. Res. **12**(9) (2011)

32. Shervashidze, N., Vishwanathan, S., Petri, T., Mehlhorn, K., Borgwardt, K.: Efficient graphlet kernels for large graph comparison. In: Artificial Intelligence and Statistics, pp. 488–495. PMLR (2009)

33. Sutskever, I.: Training Recurrent Neural Networks. University of Toronto Toronto, ON, Canada (2013)

34. Vaswani, A., et al.: Attention is all you need. In: Advances in Neural Information Processing Systems, vol. 30 (2017)

35. Veličković, P., Cucurull, G., Casanova, A., Romero, A., Liò, P., Bengio, Y.: Graph attention networks. In: International Conference on Learning Representations (2018)

36. Wu, Z., Pan, S., Chen, F., Long, G., Zhang, C., Philip, S.Y.: A comprehensive survey on graph neural networks. IEEE Trans. Neural Netw. Learn. Syst. **32**(1), 4–24 (2020)

37. Xu, K., Hu, W., Leskovec, J., Jegelka, S.: How powerful are graph neural networks? arXiv preprint arXiv:1810.00826 (2018)

38. Xu, K., Li, C., Tian, Y., Sonobe, T., Kawarabayashi, K.i., Jegelka, S.: Representation learning on graphs with jumping knowledge networks. In: International Conference on Machine Learning, pp. 5453–5462. PMLR (2018)

39. Yanardag, P., Vishwanathan, S.: Deep graph kernels. In: Proceedings of the 21th ACM SIGKDD International Conference on Knowledge Discovery and Data Mining, pp. 1365–1374 (2015)

40. Yang, J., Leskovec, J.: Defining and evaluating network communities based on ground-truth. Knowl. Inf. Syst. **42**(1), 181–213 (2015)

41. Ying, R., He, R., Chen, K., Eksombatchai, P., Hamilton, W.L., Leskovec, J.: Graph convolutional neural networks for web-scale recommender systems. In: Proceedings of the 24th ACM SIGKDD International Conference on Knowledge Discovery & Data Mining, pp. 974–983 (2018)

42. Zeng, H., Zhou, H., Srivastava, A., Kannan, R., Prasanna, V.: Graphsaint: Graph sampling based inductive learning method. In: International Conference on Learning Representations (2020)

43. Zhang, M., Cui, Z., Neumann, M., Chen, Y.: An end-to-end deep learning architecture for graph classification. In: Thirty-Second AAAI Conference on Artificial Intelligence (2018)

44. Zhao, L., Akoglu, L.: PairNorm: tackling oversmoothing in GNNs. In: International Conference on Learning Representations (2020)

45. Zhu, J., Yan, Y., Zhao, L., Heimann, M., Akoglu, L., Koutra, D.: Beyond homophily in graph neural networks: current limitations and effective designs. In: Advances in Neural Information Processing Systems, vol. 33, pp. 7793–7804 (2020)

Learning to Solve Minimum Cost Multicuts Efficiently Using Edge-Weighted Graph Convolutional Neural Networks

Steffen Jung[1]([✉]) [iD] and Margret Keuper[1,2] [iD]

[1] Max Planck Institute for Informatics, Saarland Informatics Campus, Saarbrücken, Germany
{steffen.jung,keuper}@mpi-inf.mpg.de
[2] University of Siegen, Siegen, Germany

Abstract. The minimum cost multicut problem is the NP-hard/APX-hard combinatorial optimization problem of partitioning a real-valued edge-weighted graph such as to minimize the total cost of the partition. While graph convolutional neural networks (GNN) have proven to be promising in the context of combinatorial optimization, most of them are only tailored to or tested on positive-valued edge weights, i.e. they do not comply with the nature of the multicut problem. We therefore adapt various GNN architectures including Graph Convolutional Networks, Signed Graph Convolutional Networks and Graph Isomorphic Networks to facilitate the efficient encoding of real-valued edge costs. Moreover, we employ a reformulation of the multicut ILP constraints to a polynomial program as loss function that allows us to learn feasible multicut solutions in a scalable way. Thus, we provide the first approach towards end-to-end trainable multicuts. Our findings support that GNN approaches can produce good solutions in practice while providing lower computation times and largely improved scalability compared to LP solvers and optimized heuristics, especially when considering large instances. Our code is available at https://github.com/steffen-jung/GCN-Multicut.

Keywords: Graph neural network · Graph partitioning

1 Introduction

Recent years have shown great advances of neural network-based approaches in various application domains from image classification [39] and natural language processing [56] up to very recent advances in decision logics [3]. While these successes indicate the importance and potential benefit of learning from data distributions, other domains such as symbolic reasoning or combinatorial optimization are still dominated by classical approaches. Recently, first attempts have been made to address specific NP-hard combinatorial problems in a learning-based setup [12, 42, 48, 50]. Specifically, such papers employ (variants of) message

Supplementary Information The online version contains supplementary material available at https://doi.org/10.1007/978-3-031-26390-3_28.

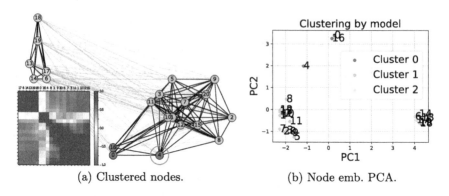

(a) Clustered nodes.   (b) Node emb. PCA.

Fig. 1. (a) Node clustering of the proposed GCN_W_BN on a complete graph ($w = -220.6$) from IrisMP and the ordered cosine similarity between all learned node embeddings. (b) The first two principal components for each node embedding of (a). Node 4 is part of the green cluster in the optimal solution ($w = -222.9$). The closeness of both solutions is reflected in the embedding. (Color figure online)

passing neural networks (MPNN) [20], defined on graphs [37,45,49] in order to model, for example, the boolean satisfiability of conjunctive normal form formulas (SAT) [50] or address the travelling salesman problem [48] - both highly important NP-complete combinatorial problems. These first advances employ the ability of graph convolutional networks to efficiently learn representations of entities in graphs and prove the potential to solve hard combinatorial problems.

In this paper, we address the *minimum cost multicut* problem [5,13], also known as the weighted correlation clustering problem (see Fig. 1). This grouping problem is substantially different from the aforementioned examples as it aims to assign binary edge labels based on a signed edge cost function. Such graph partitioning problems are ubiquitous in practical applications such as image segmentation [1,2,4,31,51], motion segmentation [30,33], stereo matching [27], inpainting [27], object tracking [22,32], pose tracking [24], or entity matching [47]. The minimum cost multicut problem is NP-hard, as well as APX-hard [5], which makes it a particularly challenging subject to explore. Its main difficulty lies in the exponentially growing number of constraints that define feasible solutions, especially whenever non-complete graphs are considered. Established methods solve its binary linear program formulation or linear program relaxations [27]. However, deriving optimal solutions is oftentimes intractable for large problem instances. In such cases, heuristic, iterative solvers are used as a remedy [31]. A significant disadvantage of such methods is that they can not provide gradients that would allow to train downstream tasks in an end-to-end way.

To address this issue, we propose a formulation of the minimum cost multicut problem as an MPNN. While the formulation of the multicut problem as a graph neural network seems natural, most existing GNN approaches are designed to aggregate *node* features potentially under edge constraints [53]. In contrast, instances of the multicut problem are purely defined through their *edge weights*. Graph Convolutional Networks (GCN) [37] rely on diverse node embeddings normalized by the graph Laplacian and an isotropic aggregation function.

Yet, edge weights in general and signed edge weights in particular are not modelled in standard GCNs. In this paper, we propose a simple extension of GCNs and show that the *signed* graph Laplacian can provide sufficiently strong initial node embeddings from signed edge information. This, in conjunction with an anisotropic update function which takes into account signed edge weights, facilitates GCNs to outperform more recent models such as Signed Graph Convolutional Networks (SGCN) [14], Graph Isomorphic Networks (GIN) [58] as well as models that inherently handle real-valued edge weights such as Residual Gated Graph Convolutional Networks (RGGCN) [25] and Graph Transformer Networks (GTN) [52] on the multicut problem.

To facilitate effective training, we consider a polynomial programming formulation of the minimum cost multicut problem to derive a loss function that encourages the network to issue valid solutions. Since currently available benchmarks for the minimum cost multicut problem are notoriously small, we propose two synthetic datasets with different statistics, for example w.r.t. the graph connectivity, which we use for training and analysis. We further evaluate our models on the public benchmarks BSDS300 [44], CREMI [8], and Knott3D [2].

In the following, we first briefly review the minimum cost multicut problem and commonly employed solvers. Then, we provide an overview on GNN approaches and their application in combinatorial optimization. In Subsect. 3.1, we present the proposed approach for solving the minimum cost multicut problem with GNNs including model adaptations and the derivation of the proposed loss function. Section 4 provides an empirical evaluation of the proposed approach.

2   The Minimum Cost Multicut Problem

The minimum cost multicut problem [11,15] is a binary edge labelling problem defined on a graph $G = (V, E)$, where the connectivity is defined by edges $e \in E \subseteq \binom{V}{2}$, i.e. G is not necessarily complete. It allows for the definition of real-valued edge costs $w_e \forall e \in E$. Its solutions decompose G such as to minimize the overall cost. Specifically, the MP can be defined by the following ILP [11]:

Definition 1. *For a simple, connected graph $G = (V, E)$ and an associated cost function $w\colon E \to \mathbb{R}$, written below is an instance of the multicut problem*

$$\min_{\mathbf{y} \in \{0,1\}^{|E|}} \quad c(\mathbf{y}) = \mathbf{y}^T \mathbf{w} = \sum_{e \in E} w_e y_e \tag{1}$$

with y subject to the linear constraints

$$\forall C \in cycles(G), \forall e \in C\colon \ y_e \leq \sum_{e' \in C \setminus \{e\}} y_{e'}, \tag{2}$$

where cycles($\cdot$) enumerates all cycles in graph G. The resulting $\mathbf{y}$ is a vector of binary decision variables for each edge. Equation (2) defines the cycle inequality constraints and ensures that if an edge is cut between two nodes, there can not be another path in the graph connecting them. Chopra and Rao [11] further showed that the facets of the MP can be sufficiently described by cycle inequalities on all chordless cycles of G. The problem in Eq. (1)–(2) can be reformulated in a more compact way as a polynomial program (PP):

$$\min_{y \in \{0,1\}^{|E|}} \sum_{e \in E} w_e y_e + K \sum_{C \in cc(G)} \sum_{e \in C} y_e \prod_{e' \in C \setminus \{e\}} (1 - y_{e'}), \tag{3}$$

for a sufficiently large penalty K. The above problem is well behaved for complete graphs where it suffices to consider all cycles of length three and Eq. (3) becomes a quadratic program. For sparse graphs, sufficient constraints may have arbitrary length $\leq |V|$ and their enumeration might be practically infeasible. Finding an optimal solution is NP-hard and APX-hard [5]. Therefore, exact solvers are intractable for large problem instances. Linear program relaxations as well as primal feasible heuristics have been proposed to overcome this issue, which we will briefly review in the following.

Related Work on Multicut Solvers. To solve the ILP from Definition 1, one can use general purpose LP solvers, like Gurobi [21] or CPLEX, such that optimal solutions might be in reach for small instances if the enumeration of constraints is tractable. However, no guarantees on the runtime can be provided. To mitigate the exponentially growing number of constraints, various cutting-plane [28,34,35] or branch-and-bound [2,27] algorithms exist. For example, [28] employ a relaxed version of the ILP in Eq. (1) without cycle constraints. In each iteration, violated constraints are searched and added to the ILP. This approach provides optimal solutions to formerly intractable instances - yet without any runtime guarantees. Linear program relaxations [27,34,55] increase the tightness of the relaxation, for example using additional constraints, and provide optimality bounds. While such approaches can yield solutions within optimality bounds, their computation time can be very high and the proposed solution can be arbitrarily poor in practice. In contrast, heuristic solvers can provide runtime guarantees and have shown good results in many practical applications. The primal feasible heuristic KLj [31] iterates over pairs of partitions and computes local moves which allow to escape local optima. Competing approaches have been proposed, for example by [7,29] or [6]. The highly efficient Greedy Additive Edge Contraction (GAEC) [31] approach aggregates nodes in a greedy procedure with an $O(|E|\log|E|)$ worst case complexity. While such primal feasible heuristics are highly efficient in practice, they share one important draw-back with ILP solvers that becomes relevant in the learning era: they can not provide gradients that would allow for backpropagation for example to learn edge weights.

In contrast, a third order conditional random field based on the PP in Eq. (3) has been proposed in [54] and adapted in [26], which can be optimized in an end-to-end fashion using mean field iterations. This approach is strictly limited to the optimization on complete graphs. Our approach employs graph neural networks to overcome this limitation and provides a general purpose end-to-end trainable multicut approach.

3   Message Passing Neural Networks for Multicuts

[20] provide a general framework to describe convolutions for graph data spatially as a message-passing scheme. In each convolutional layer, each node is propagating its current node features via edges to all of its neighboring nodes and updates its own features based on the messages it receives. The update is commonly described by an update function

$$\mathbf{h}_{\mathbf{u}}^{(t)} = g^{(t)} \left(\mathbf{h}_{\mathbf{u}}^{(t-1)}, \hat{\sum_{v \in \mathcal{N}(u)}} f^{(t)} \left(\mathbf{h}_{\mathbf{u}}^{(t-1)}, \mathbf{h}_{\mathbf{v}}^{(t-1)}, \mathbf{x}_{\mathbf{v},\mathbf{u}} \right) \right), \quad (4)$$

where $\mathbf{h}_{\mathbf{u}}^{(t)} \in \mathbb{R}^F$ is the feature representation of node u in layer t with dimensionality F, and $\mathbf{x}_{\mathbf{v},\mathbf{u}}$ are edge features. Here, f and g are differentiable functions, and $\hat{\Sigma}$ is a differentiable, permutation invariant aggregation function, mostly sum, max, or mean. Commonly, the message function f and the update function g are parameterized, and apply the same parameters at each location in the graph.

Various formulations have been proposed to define g. Graph Convolutional Network (GCN) [37] normalizes messages with the graph Laplacian and linearly transforms their sum to update node representations. Signed Graph Convolutional Network (SGCN) [14] aggregates messages depending on the sign of the connectivity and keeps two representations per node, one for balanced paths and one for unbalanced paths. Graph Isomorphic Network (GIN) [58] learns an injective function by defining message aggregation as a sum and learning the update function as an MLP. Residual Gated Graph Convolutional Network (RGGCN) [25] computes edge gates to aggregate messages in an anisotropic manner and learns to compute the residuals to the previous representations. Edge conditioned GCNs [53] aggregate node features using dynamic weights computed from high-dimensional edge features. Graph Transformer Network (GTN) [52] also aggregate messages anisotropically by learning a self-attention model based on the transformer models in NLP [56]. While the latter three can directly handle real-valued edge weights, all are tailored towards aggregating meaningful node features. In the following, we review recent approaches to employ such models in the context of combinatorial optimization.

MPNNs and Combinatorial Optimization. Recently, MPNNs have been applied to several hard combinatorial optimization problems, such as the minimum vertex cover [42], maximal clique [42], maximal independent set [42], the satisfiability problem [42], and the travelling salesman problem [25]. Their objective is either to learn heuristics such as branch-and-bound variable selection policies for exact or approximate inference [16,19] or to use attention [57], reinforcement learning [9,12], or both [38,43,46] in an iterative, autoregressive procedure. [25] address the 2D Euclidean travelling salesman problem using the RGGCN model to learn edge representations. Other recent approaches address combinatorial problems by *decoding*, using supervised training such as [10].

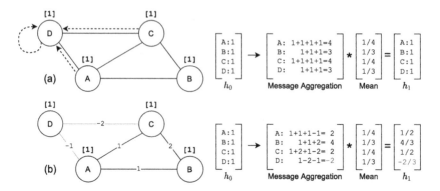

Fig. 2. Message aggregation in an undirected, weighted graph where node features (h_0) are initialized with 1. (a) Standard message aggregation in an isotropic fashion leads to no meaningful node embeddings ($h_1 = h_0$). (b) Our proposed method takes edge weights into account leading to anisotropic message aggregation and meaningful node embeddings. A simple decision boundary at $h = 0$ can now partition the graph.

The proposed approach is related to the work of [25], since we cast the minimum cost multicut problem as a binary edge classification problem that we address using MPNN approaches, including RGGCN. We train our model in a supervised way, yet employing a dedicated loss function which encourages feasible solutions w.r.t. Eq. (2).

3.1   Multicut Neural Network

We cast the multicut problem into a binary edge classification task, where label $y_{u,v} = 1$ is assigned to an edge (u, v) if it is cut, and $y_{u,v} = 0$ otherwise. The task of the model is to learn a probability distribution $\hat{y}_{u,v} = p(y_{u,v} = 1 \mid G)$ over the edges of a given graph, inferring how likely it is that an edge is cut. Based on these probabilities, we derive a configuration of edge labels, $\mathbf{y} = \{0, 1\}^{|E|}$. In contrast to existing autoregressive MPNN-based models in combinatorial optimization, we derive a solution after a *single forward pass* over the graph to achieve an efficient bound on the runtime of the model. In this scenario, our model can be defined by three functions, i.e., $\mathbf{y}_{u,v} = f_r(f_c(f_e(G, w)))$. First, f_e is the edge representation mapping, assigning meaningful embeddings to each edge in the graph given a multicut problem instance. This function is learned by an MPNN. Second, f_c assigns to every edge its probability to be cut. This function is learned by an MLP. Last, function f_r translates the resulting configuration of edge probabilities to a feasible configuration of edge labels, hence, computes a feasible solution.

Edge Representation Mapping. Given a multicut problem instance (G, w), the edge representation mapping f_e learns to assign meaningful edge embeddings via MPNNs. One specific case of MPNN is GCN [37], where the node representation update function is defined as follows:

$$\mathbf{h_u}^{(t)} = g_\theta^{(t)} \left(\mathbf{h_u}^{(t-1)} + \sum_{v \in \mathcal{N}(u)} \mathbf{L}_{[v,u]} \mathbf{h_v}^{(t-1)} \right), \tag{5}$$

where $\mathbf{h_u}^{(t)} \in \mathbb{R}^F$ denotes the feature representation of node u in layer t with channel size F. In each layer, node representations of all neighbors of u are aggregated and normalized by $\mathbf{L}_{[v,u]} = 1/\sqrt{\deg(u)\deg(v)}$, where $\mathbf{L} = \tilde{\mathbf{D}}^{1/2}\tilde{\mathbf{A}}\tilde{\mathbf{D}}^{1/2}$ is the normalized graph Laplacian with additional self-loops in the adjacency matrix $\tilde{\mathbf{A}} = \mathbf{A} + \mathbf{I}$ and degree matrix $\tilde{\mathbf{D}}$. Conventionally, $\mathbf{h_u}^{(0)}$ is initialized with node features $\mathbf{x_u}$. Intuitively, we expect normalization with the graph Laplacian to be beneficial in the MP setting, since i) its eigenvectors encode similarities of nodes within a graph [51] and ii) even sparsely connected nodes can be assigned meaningful representations [4]. However, MP instances consist of real-valued edge-weighted graphs and the normalized graph Laplacian is not defined for negative node degrees. To the best of our knowledge, there is no work enabling GCN to incorporate *real-valued* edge weights so far.

Real-valued Edge Weights. Hence, our first task is to enable negative-valued edge weights in GCN. We can achieve this via the *signed* normalized graph Laplacian [23,40]:

$$\overline{\mathbf{L}}_{[v,u]} = \left(\overline{\mathbf{D}}^{1/2}\tilde{\mathbf{W}}\overline{\mathbf{D}}^{1/2} \right)_{[v,u]} = w_{v,u}/\sqrt{\overline{\deg}(u)\overline{\deg}(v)}, \tag{6}$$

where $\tilde{\mathbf{W}}$ is the weighted adjacency matrix and $\overline{\mathbf{D}}$ is the signed node degree matrix with $\overline{\deg}(u) = \sum_{v \in \mathcal{N}(u)} |w_{u,v}|$. [18] shows that this formulation preserves the desired properties from the graph Laplacian w.r.t. encoding pairwise similarities as well as representation learning on sparsely connected nodes (see i) and ii) above).

Incorporating Eq. (6) into Eq. (5), we get

$$\mathbf{h_u}^{(t)} = g_\theta^{(t)} \left(\mathbf{h_u}^{(t-1)} + \sum_{v \in \mathcal{N}(u)} w_{v,u} \cdot \left(\overline{\deg}(u)\overline{\deg}(v) \right)^{-1/2} \mathbf{h_v}^{(t-1)} \right). \tag{7}$$

Here, we can observe two new terms. First, each message is weighted by the edge weight $w_{v,u}$ between two nodes enabling an anisotropic message-passing scheme. Figure 2 motivates why this is necessary. While [58] show that GNNs with mean aggregation have theoretical limitations, they also note that these limitations vanish in scenarios where node features are diverse. Additionally, [58] only consider the case where neighboring nodes are aggregated in an isotropic fashion. As we show here, diverse node features are not necessary when messages are aggregated in the anisotropic fashion we propose. The resulting node representations enable distinguishing nodes in the graph despite the lack of meaningful node features. This is important in our case, since the multicut problem does not provide node features. Second, we are now able to normalize messages via the Laplacian in real-valued graphs. The normalization acts stronger on messages that are sent to or from nodes whose adjacent edges have weights with large magnitudes. Large magnitudes on the edges usually indicate a confident decision towards joining (for positive weights) or cutting (negative weights). Thus, the normalization will allow nodes with less confident edge cues to converge to a meaningful embedding while, without such normalization, the network would notoriously focus on embedding nodes with strong edge cues, i.e. on easy decisions.

Node Features. Conventionally, node representations at timestep 0, $\mathbf{h}_\mathbf{u}^{(0)}$, are initialized with node features $\mathbf{x_u}$. However, multicut instances describe the magnitude of similarity or dissimilarity between two items via edge weights and provide no node features. Therefore, we initialize node representations with a two-dimensional vector of node degrees as:

$$\mathbf{x_u} = \left(\sum_{v \in \mathcal{N}^+(u)} w_{u,v}, \sum_{v \in \mathcal{N}^-(u)} w_{u,v} \right), \tag{8}$$

where $\mathcal{N}^+(u)$ is the set of neighboring nodes of u connected via positive edges, and $\mathcal{N}^-(u)$ is the set of neighboring nodes of u connected via negative edges.

Node-to-Edge Representation Mapping. To map two node representations to an edge representation, we use concatenation $\mathbf{h_{u,v}} = f_e(\mathbf{h_u}, \mathbf{h_v}) = \binom{\mathbf{h_u}}{\mathbf{h_v}} \in \mathbb{R}^{2 \cdot F}$, where $\mathbf{h_{u,v}}$ is the representation of edge (u, v) and F the dimension of node embedding $\mathbf{h_u}$. Since we consider undirected graphs, the order of the concatenation is ambiguous. Therefore, we generate two representations for each edge, one for each direction. This doubles the number of edges to be classified in the next step. The final classification result is the average computed from both representations.

Edge Classification. We learn edge classification function f_c via an MLP that computes likelihoods $\hat{\mathbf{y}} \in [0, 1]^{|E|}$ for each edge in graph G, expressing the confidence whether an edge should be cut. A binary solution $\mathbf{y} \in \{0, 1\}^{|E|}$ is retrieved by thresholding the likelihoods at 0.5. Since there is no strict guarantee that the edge label configuration $\mathbf{y}$ is feasible w.r.t. Eq. (2), we postprocess $\mathbf{y}$ to *round* it to feasible solutions. Therefore, we compute a connected component labelling on G after removing cut edges from E and reinstate removed edges for which both corresponding nodes remain within the same component. For efficiency, we implement the connected component labelling as a message-passing layer and can therefore assign cluster identifications to each node efficiently on the GPU.

Training. Since we cast the multicut problem to a binary edge labelling problem, we can formulate a supervised training process that minimizes the Binary Cross Entropy (BCE) loss w.r.t. the optimal solution $\tilde{\mathbf{y}}$, which we denote $\mathcal{L}_{BCE}$.

Cycle Consistency Loss. The BCE loss encodes feasibility only implicitly by comparison to the optimal solution. To explicitly learn feasible solutions, we take recourse to the PP formulation of the multicut problem in Eq. (3) and formulate a *feasibility loss*, that we denote *Cycle Consistency Loss* (CCL):

$$\mathcal{L}_{CCL} = \alpha \cdot \sum_{C \in cc(G,l)} \sum_{e \in C} \hat{y}_e \prod_{e' \in C \setminus \{e\}} (1 - \hat{y}_{e'}), \qquad (9)$$

where α is a hyperparameter, balancing BCE and CCL, and $cc(G, l)$ is a function that returns all chordless cycles in G of length at most l. The CCL term effectively penalizes infeasible edge label configurations during training; it adds a penalty of at most α for each chordless cycle that is only cut once. In practice, we only consider chordless cycles of maximum length l, and we only consider a cycle if e is cut, hence $\hat{y}_e \geq 0.5$. This is necessary to ensure practicable training runtimes. The total training loss is given by $\mathcal{L} = \mathcal{L}_{BCE} + \mathcal{L}_{CCL}$. For best results, we train all models using batch normalization.

Training Datasets. While the multicut problem is ubiquitous in many real world applications, the amount of available annotated problem instances is scarce and domain specific. Therefore, in order to train and test a general purpose model, we generated two synthetic datasets, *IrisMP* and *RandomMP*, with complementary connectivity statistics, of 22000 multicut instances each.

The first dataset, IrisMP, consists of multicut instances on complete graphs based on the Iris flower dataset [17]. The generation procedure is described in the Appendix. Each problem instance consists of 120 to 276 edges. Three graphs with their respective optimal solutions are depicted in the Appendix. To complement the IrisMP dataset, we generated a second dataset that contains sparse but larger problem instances with 180 nodes on average, called RandomMP. The generation procedure is described in the Appendix. Examples are depicted in the Appendix.

4   Experiments

We evaluate all models trained on IrisMP and RandomMP and provide runtime as well as objective value evaluations, where we compare the proposed GCN to GIN and SGCN-based, edge-weight enabled models (see appendix for details) as well as to RGGCN [25] and GTN [52]. Then, we provide an ablation study on the proposed GCN-based edge representation mapping and the multicut loss.

4.1   Evaluation on Test Data

We evaluate our models on three segmentation benchmarks: a graph-based image segmentation dataset [1] based on the Berkeley Segmentation Dataset (BSDS300) [44] consisting of 100 test instances, a graph-based volume segmentation dataset [2] (Knott3D) containing 24 volumes, and 3 additional test instances

Table 1. Results on the test datasets. We compare different GNN variants, heuristics (GAEC) [31], LP-solver [27], and ILP-solver [27]. The performance is evaluated as optimal objective ratio ↑ and is averaged over all datasets via harmonic mean to account for generalizability. The last column shows the total runtime ↓ over all datasets in milliseconds. OOM indicates insufficient memory. OOT indicates no termination within 24 h. Neither OOM nor OOT are considered in the runtime (marked with *).

Solver	Variant	Depth	α	l	IrisMP	RandomMP	BSDS300	CREMI	Knott	H.mean	Forward	Total
Proposed learned solvers IrisMP	GCN_W_BN	12	0.001	3	0.9834	0.7188	**0.8912**	**0.7255**	**0.6902**	**0.7865**	0.5	4.4
	GINO_W_BN	12	0.01	3	**0.9905**	**0.7387**	0.8474	0.5464	0.0000	0.0000	0.0	4.0
	Signed_W_BN	12	0.01	3	0.9878	0.2526	0.6451	0.5154	0.3808	0.4510	1.3	5.3
	RGGCN_HE	12	0.01	3	0.7976	0.1449	0.4655	0.1544	0.1735	0.2218	0.1	4.1
	GT	12	0.001	3	0.7940	0.2964	0.6360	0.4037	0.6038	0.4836	0.1	4.0
	LR				0.6769	0.1118	0.6524	0.2689	0.0366	0.1164		N/A
	MLP				0.6626	0.3127	0.7139	0.2789	0.1493	0.3051		N/A
RandomMP	GCN_W_BN	20	0.01	8	**0.9762**	**0.9041**	**0.9204**	0.8440	**0.7870**	**0.8815**	0.9	4.8
	GINO_W_BN	20	0.01	8	0.9528	0.8693	0.9109	0.4812	0.0000	0.0000	0.0	4.0
	Signed_W_BN	20	0.01	8	0.9709	0.8695	0.8825	0.4653	0.6408	0.7120	2.3	6.3
	RGGCN_HE	20	0.01	8	0.9703	0.8787	0.8352	0.5593	OOM	–	0.1	2.7*
	LR				0.8035	0.3938	0.7958	**0.9260**	0.7335	0.6681		N/A
	MLP				0.8985	0.3099	0.6804	0.4845	0.1517	0.3457		N/A
	GAEC				0.9836	0.9780	0.9997	0.9958	0.9968	0.9907		23.2
	Time-bounded GAEC				0.3642	0.0034	0.0000	0.1516	0.0000	0.0000		6.3
	LP solver				0.9882	0.9525	0.9979	0.9998	OOT	–		31 918.8*
	ILP solver				1.0000	1.0000	1.0000	1.0000	1.0000	1.0000		24 361.2

based on the challenge on Circuit Reconstruction from Electron Microscopy Images (CREMI) [8] that contains volumes of electron microscopy images of fly brains. BSDS300 and Knott3D instances are available as part of a benchmark containing discrete energy minimization problems, called OpenGM [27].

Implementation Details. We train the proposed MPNN-based solvers with (adapted) GCN, GIN, SGCN, RGGCN and GT backbones in different settings, where we uniformly set the node representation dimensionality to 128. We set the depth of the MPNN to 12 for IrisMP and 20 for RandomMP. CCL is applied with $\alpha \in \{0, 0.01, 0.001\}$ and chordless cycle length up to 8. All of our experiments are conducted on MEGWARE Gigabyte G291-Z20 servers with NVIDIA Quadro RTX 8000 GPUs. If not stated otherwise, we consider as performance metric m the *optimal objective ratio* achieved, hence $m = \max(0, w(\mathbf{y})/w(\tilde{\mathbf{y}})) \in [0, 1]$.

Results. In Table 1, we show the results on all test datasets of the best models based on the evaluation objective value after rounding, and thereby compare models trained on IrisMP and models trained on RandomMP. In general, sparser problems (RandomMP and established test datasets) are harder for the solvers to generalize to. This is likely due to the longer chordless cycles that the model needs to consider to ensure feasibility. Overall, our GCN-based model provides the best generalizability over all test datasets both when trained on IrisMP and RandomMP. We compare the GNN-based solvers to different baselines. First, we train logistic regression (LR) and MLPs as edge classifiers directly on the training data (concatenation of node features and edge weights). All our learned models outperform these baselines significantly. This indicates that MPNNs provide meaningful topological information to the edge classifier that facilitates solving MP instances. Second, we compare against Branch & Cut LP and ILP solvers as well as GAEC. In terms of objective value, GCN-based solvers are on par with heuristics and LP solvers on complete graphs, even when trained on sparse graphs. On general graphs, ILP solvers and GAEC issue lower energies, and, as expected, training on complete graphs does not generalize well to sparse graphs. However, the wall-clock runtime comparison shows that GCN-based solvers are faster by an order of 10^3 than ILP and LP solvers. They are also significantly faster than the fast and greedy GAEC heuristic. We further compared to a time-constrained version of GAEC, where we set the available time budget to the runtime of the GCN-based solver. The result shows that the trade-off between smaller energies and smaller runtime is in favor of the GCN-based solver. In the Appendix, we report additional experiments for our proposed GCN-based model on domain specific training and show that task specific priors can be learned efficiently from only a few training samples.

Next, we conduct a scalability study on random graphs with an increasing number of nodes, generated according to the RandomMP dataset. Results are shown in Table 2. While the GAEC is fast for small graphs, the GCN-based solver scales better and returns solutions significantly faster for larger graphs. LP and ILP solvers are not able to provide solutions within 24 h for larger instances. It is noteworthy that GNN-based solvers spend 75-99% of their runtime rounding the

Table 2. Wall-clock runtime ↓ and objective values ↓ of MPNN-based solver vs. GAEC, LP and ILP on a growing, randomly-generated graph. OOT indicates no termination within 24 h.

Nodes	GAEC		LP		ILP		GCN_W_BN	
	[ms]	Objective	[ms]	Objective	[ms]	Objective	[ms]	Objective
10^1	0	−29	6	−24	11	−30	29	−29
10^2	4	−327	191	−246	273	−330	26	−276
10^3	24	−3051	6585	−2970	1299	−3093	29	−2643
10^4	228	−32 264	688 851	−31 531	18 604	−32 682	**78**	−27 552
10^5	2534	−323 189	OOT		2 171 134	−328 224	**557**	−269 122
10^6	35 181	−3 401 783	OOT		OOT		**8713**	−2 182 589

solutions. Hence, GNN-based solvers are already more scalable and still have a large potential for improvement in this regard, while GAEC and LP/ILP solvers are already highly optimized for runtime.

Next, we ablate on the GCN aggregation functions, loss and network depths.

Edge-weighted GCNs. First, we determine the impact of each adjustment to the GCN update function. In Table 3 we show the results of this ablation study. While vanilla GCN is not applicable in the MP setting, simply removing the Laplacian from Eq. 5 provides a first baseline. We observe that adding edge weights ($w_{u,v}$) to Eq. 5 improves the performance on the test split of the training data substantially. However, the model is not able to generalize to different graph statistics. By adding the signed normalization term ($(\overline{\deg}(u)\overline{\deg}(v))^{-1/2}$) we arrive at Eq. 7, achieving improved generalizability. Removing edge weights from Eq. 7 deteriorates performance and generalizability. Thus both changes are necessary to enable GCN in the MP setting.

Additionally, we compare GCNs with edge weights and *signed* graph Laplacian normalization, trained with batch normalization, to the plain GCN model [37]. To this end, we train on the IrisMP dataset and set the model width to 128 and its depth to 12. Here, we set $\alpha = 0$, hence, we do not apply CCL. Figure 3 shows the results of this experiment. The corresponding plots for SGCN and GIN are given in the Appendix. The variants with edge features achieve lower losses than without edge features, and batch normalization improves the loss further. In fact, the original GCN is not able to provide any meaningful features for the edge classification network. The proposed extensions enable these networks to find meaningful node representations for the multicut problem.

4.2   Ablation Study

Number of Convolutional Layers. Next, we evaluate the effect of depth of the GCN model when trained on the IrisMP dataset and evaluated on IrisMP, RandomMP as well as BSDS300. Figure 3(c) shows the results after varying the depth in increasing step sizes up to a depth of 40. The results suggest that increasing

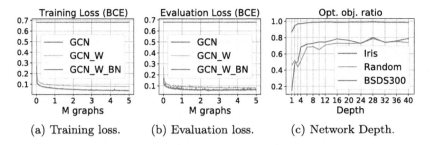

(a) Training loss.   (b) Evaluation loss.   (c) Network Depth.

Fig. 3. (a) Training and (b) evaluation loss while training variants of GCN on IrisMP. Each plot compares the variants with (GCN_W) and without (GCN) edge weights in the aggregation, and GCN_W with batch normalization (GCN_W_BN). (c) Results in terms of optimal objective ratio on the evaluation data when training GCN_W_BN with varying depths.

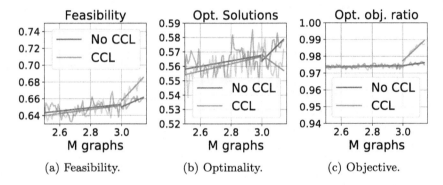

(a) Feasibility.   (b) Optimality.   (c) Objective.

Fig. 4. (a) Ratio of feasible solutions before repairing, (b) Ratio of optimal solutions, and (c) Optimal objective ratio, for GCN_W_BN on RandomMP, applying CCL after $3M$ instances.

the depth improves the objective value up to a certain point. In the case of IrisMP graphs with diameter 1 and lengths of chordless cycles of at most 3, increasing the depth beyond 10 has no obvious effect. This is an important observation, because [41] raise concerns that GCN models can suffer from over-smoothing such that learned representations might become indistinguishable.

Cycle Consistency Loss. Here, we evaluate the effect of applying the cycle consistency loss from Eq. (9) by comparing models where CCL is applied after $3M$ instances to models solely trained without CCL. Figures 4(a) and (b) show the progress of the ratio of feasible solutions and ratio of optimal solutions found during training. As soon as CCL is applied, the ratio of feasible solutions increases while the ratio of optimal solutions decreases. Hence, CCL induces a trade-off between finding feasible and optimal solutions, where the model is forced to find feasible solutions to avoid the penalty, and as a consequence, settles for suboptimal relaxated solutions. However, the objective value after rounding improves, which is most relevant because these values correspond to feasible solutions. This

Table 3. Ablation study with GCN [37] trained on IrisMP without CCL. Additional comparison to vanilla versions of GIN0 [58], and MPNN [20]. We report the performance on the test data in terms of optimal objective ratio ↑.

Variant	IrisMP	RandMP	BSDS300	CREMI	Knott
GCN	Not applicable: Laplacian may not exist.				
− Laplacian	0.41	0.18	0.00	0.49	0.00
+ edge weights	0.95	0.18	0.40	0.57	0.19
+ signed norm.	**0.96**	**0.67**	**0.75**	**0.74**	**0.68**
= GCN_W	**0.96**	**0.67**	**0.75**	**0.74**	**0.68**
− edge weights	0.64	0.05	0.00	0.48	0.00
GIN0	0.41	0.04	0.07	0.48	0.00
MPNN	0.93	0.45	0.48	0.49	0.06

indicates that the model's upper bound on the optimal energy is higher while the relaxation is tighter when CCL is employed. See the Appendix for an ablation on α in Eq. 9.

Meaningful Embeddings. In Fig. 1 we visualize the node embedding space given by our best performing model on an IrisMP instance. Plotting the cosine similarity between all nodes reflects the resulting clusters. This shows that the model is able to distinguish nodes based on their connectivity. We show further examples in the Appendix.

5   Conclusion

In this paper, we address the minimum cost multicut problem using feed forward MPNNs. To this end, we provide appropriate model and training loss modifications. Our experiments on two synthetic and two real datasets with various GCN architectures show that the proposed approach provides highly efficient solutions even to large instances and scales better than highly optimized primal feasible heuristics (GAEC), while providing competitive energies. Another significant advantage of our learning-based approach is the ability to provide gradients for downstream tasks, which we assume will inherently improve inferred solutions.

References

1. Andres, B., Kappes, J.H., Beier, T., Köthe, U., Hamprecht, F.A.: Probabilistic image segmentation with closedness constraints. In: ICCV (2011)
2. Andres, B., et al.: Globally optimal closed-surface segmentation for connectomics. In: Fitzgibbon, A., Lazebnik, S., Perona, P., Sato, Y., Schmid, C. (eds.) ECCV 2012. LNCS, vol. 7574, pp. 778–791. Springer, Heidelberg (2012). https://doi.org/10.1007/978-3-642-33712-3_56

3. Arakelyan, E., Daza, D., Minervini, P., Cochez, M.: Complex query answering with neural link predictors (2021)
4. Arbeláez, P., Maire, M., Fowlkes, C., Malik, J.: Contour detection and hierarchical image segmentation. TPAMI **33**(5), 898–916 (2011)
5. Bansal, N., Blum, A., Chawla, S.: Correlation clustering. Mach. Learn. **56**(1–3), 89–113 (2004)
6. Beier, T., Andres, B., Köthe, U., Hamprecht, F.A.: An efficient fusion move algorithm for the minimum cost lifted multicut problem. In: Leibe, B., Matas, J., Sebe, N., Welling, M. (eds.) ECCV 2016. LNCS, vol. 9906, pp. 715–730. Springer, Cham (2016). https://doi.org/10.1007/978-3-319-46475-6_44
7. Beier, T., Kroeger, T., Kappes, J., Köthe, U., Hamprecht, F.: Cut, glue, & cut: a fast, approximate solver for multicut partitioning. In: CVPR (2014)
8. Beier, T., et al.: Multicut brings automated neurite segmentation closer to human performance. Nat. Methods **14**(2), 101–102 (2017)
9. Bello, I., Pham, H., Le, Q.V., Norouzi, M., Bengio, S.: Neural combinatorial optimization with reinforcement learning. In: ICLR Workshop (2017)
10. Chen, Y., Zhang, B.: Learning to solve network flow problems via neural decoding (2020)
11. Chopra, S., Rao, M.R.: The partition problem. Math. Program. **59**(1), 87–115 (1993)
12. Dai, H., Khalil, E.B., Zhang, Y., Dilkina, B., Song, L.: Learning combinatorial optimization algorithms over graphs. In: NIPS (2017)
13. Demaine, E.D., Emanuel, D., Fiat, A., Immorlica, N.: Correlation clustering in general weighted graphs. Theor. Comput. Sci. **361**(2–3), 172–187 (2006)
14. Derr, T., Ma, Y., Tang, J.: Signed graph convolutional networks. In: ICDM (2018)
15. Deza, M.M., Laurent, M.: Geometry of Cuts and Metrics. Springer, Heidelberg (1997)
16. Ding, J.Y., et al.: Accelerating primal solution findings for mixed integer programs based on solution prediction. In: AAAI, vol. 34, no. 02, pp. 1452–1459 (2020)
17. Fisher, R.A.: The use of multiple measurements in taxonomic problems. Ann. Eugen. **7**(2), 179–188 (1936)
18. Gallier, J.: Spectral theory of unsigned and signed graphs. Applications to graph clustering: a survey (2016)
19. Gasse, M., Chételat, D., Ferroni, N., Charlin, L., Lodi, A.: Exact combinatorial optimization with graph convolutional neural networks. In: NeurIPS (2019)
20. Gilmer, J., Schoenholz, S.S., Riley, P.F., Vinyals, O., Dahl, G.E.: Neural message passing for quantum chemistry. In: ICML (2017)
21. Gurobi Optimization L: Gurobi optimizer reference manual (2020). http://www.gurobi.com
22. Ho, K., Kardoost, A., Pfreundt, F.-J., Keuper, J., Keuper, M.: A two-stage minimum cost multicut approach to self-supervised multiple person tracking. In: Ishikawa, H., Liu, C.-L., Pajdla, T., Shi, J. (eds.) ACCV 2020. LNCS, vol. 12623, pp. 539–557. Springer, Cham (2021). https://doi.org/10.1007/978-3-030-69532-3_33
23. Hou, Y.P.: Bounds for the least Laplacian eigenvalue of a signed graph. Acta Math. Sinica **21**(4), 955–960 (2005)
24. Insafutdinov, E., Pishchulin, L., Andres, B., Andriluka, M., Schiele, B.: DeeperCut: a deeper, stronger, and faster multi-person pose estimation model. In: Leibe, B., Matas, J., Sebe, N., Welling, M. (eds.) ECCV 2016. LNCS, vol. 9910, pp. 34–50. Springer, Cham (2016). https://doi.org/10.1007/978-3-319-46466-4_3

25. Joshi, C.K., Laurent, T., Bresson, X.: An efficient graph convolutional network technique for the travelling salesman problem (2019)
26. Jung, S., Ziegler, S., Kardoost, A., Keuper, M.: Optimizing edge detection for image segmentation with multicut penalties. CoRR abs/2112.05416 (2021)
27. Kappes, J.H., et al.: A comparative study of modern inference techniques for structured discrete energy minimization problems. Int. J. Comput. Vis. **115**(2), 155–184 (2015). https://doi.org/10.1007/s11263-015-0809-x
28. Kappes, J.H., Speth, M., Andres, B., Reinelt, G., Schn, C.: Globally optimal image partitioning by multicuts. In: Boykov, Y., Kahl, F., Lempitsky, V., Schmidt, F.R. (eds.) EMMCVPR 2011. LNCS, vol. 6819, pp. 31–44. Springer, Heidelberg (2011). https://doi.org/10.1007/978-3-642-23094-3_3
29. Kardoost, A., Keuper, M.: Solving minimum cost lifted multicut problems by node agglomeration. In: Jawahar, C.V., Li, H., Mori, G., Schindler, K. (eds.) ACCV 2018. LNCS, vol. 11364, pp. 74–89. Springer, Cham (2019). https://doi.org/10.1007/978-3-030-20870-7_5
30. Keuper, M., Andres, B., Brox, T.: Motion trajectory segmentation via minimum cost multicuts. In: ICCV (2015)
31. Keuper, M., Levinkov, E., Bonneel, N., Lavoué, G., Brox, T., Andres, B.: Efficient decomposition of image and mesh graphs by lifted multicuts. In: ICCV (2015)
32. Keuper, M., Tang, S., Andres, B., Brox, T., Schiele, B.: Motion segmentation multiple object tracking by correlation co-clustering. TPAMI **42**(1), 140–153 (2020)
33. Keuper, M.: Higher-order minimum cost lifted multicuts for motion segmentation. In: ICCV (2017)
34. Kim, S., Nowozin, S., Kohli, P., Yoo, C.D.: Higher-order correlation clustering for image segmentation. In: NIPS (2011)
35. Kim, S., Yoo, C., Nowozin, S., Kohli, P.: Image segmentation using higher-order correlation clustering. TPAMI **36**, 1761–1774 (2014)
36. Kingma, D.P., Ba, J.: Adam: a method for stochastic optimization. In: ICLR (2015)
37. Kipf, T.N., Welling, M.: Semi-supervised classification with graph convolutional networks. In: ICLR (2017)
38. Kool, W., Hoof, H.V., Welling, M.: Attention, learn to solve routing problems! In: ICLR (2019)
39. Krizhevsky, A., Sutskever, I., Hinton, G.E.: ImageNet classification with deep convolutional neural networks. In: NIPS (2012)
40. Kunegis, J., Schmidt, S., Lommatzsch, A., Lerner, J., Luca, E.W.D., Albayrak, S.: Spectral analysis of signed graphs for clustering, prediction and visualization. In: SDM (2010)
41. Li, Q., Han, Z., Wu, X.M.: Deeper insights into graph convolutional networks for semi-supervised learning. In: AAAI (2018)
42. Li, Z., Chen, Q., Koltun, V.: Combinatorial optimization with graph convolutional networks and guided tree search. In: NIPS (2018)
43. Ma, Q., Ge, S., He, D., Thaker, D., Drori, I.: Combinatorial optimization by graph pointer networks and hierarchical reinforcement learning. In: AAAI Workshop on Deep Learning on Graphs: Methodologies and Applications (2020)
44. Martin, D., Fowlkes, C., Tal, D., Malik, J.: A database of human segmented natural images and its application to evaluating segmentation algorithms and measuring ecological statistics. In: ICCV (2001)
45. Micheli, A.: Neural network for graphs: a contextual constructive approach. IEEE Trans. Neural Netw. **20**(3), 498–511 (2009)
46. Nazari, M., Oroojlooy, A., Snyder, L., Takac, M.: Reinforcement learning for solving the vehicle routing problem. In: NIPS (2018)

47. Oulabi, Y., Bizer, C.: Extending cross-domain knowledge bases with long tail entities using web table data. In: Advances in Database Technology, pp. 385–396 (2019)

48. Prates, M.O.R., Avelar, P.H.C., Lemos, H., Lamb, L., Vardi, M.: Learning to solve NP-complete problems - a graph neural network for the decision TSP. In: AAAI (2019)

49. Scarselli, F., Gori, M., Tsoi, A.C., Hagenbuchner, M., Monfardini, G.: The graph neural network model. IEEE Trans. Neural Netw. **20**(1), 61–80 (2009)

50. Selsam, D., Lamm, M., Bünz, B., Liang, P., de Moura, L., Dill, D.L.: Learning a sat solver from single-bit supervision (2019)

51. Shi, J., Malik, J.: Normalized cuts and image segmentation. TPAMI **22**(8), 888–905 (2000)

52. Shi, Y., Huang, Z., Wang, W., Zhong, H., Feng, S., Sun, Y.: Masked label prediction: unified message passing model for semi-supervised classification (2020)

53. Simonovsky, M., Komodakis, N.: Dynamic edge-conditioned filters in convolutional neural networks on graphs. In: CVPR (2017)

54. Song, J., Andres, B., Black, M., Hilliges, O., Tang, S.: End-to-end learning for graph decomposition. In: ICCV (2019)

55. Swoboda, P., Andres, B.: A message passing algorithm for the minimum cost multicut problem. In: CVPR (2017)

56. Vaswani, A., et al.: Attention is all you need. In: NIPS (2017)

57. Vinyals, O., Fortunato, M., Jaitly, N.: Pointer networks. In: NIPS (2015)

58. Xu, K., Hu, W., Leskovec, J., Jegelka, S.: How powerful are graph neural networks? In: ICLR (2019)

Natural Language Processing and Text Mining

AutoMap: Automatic Medical Code Mapping for Clinical Prediction Model Deployment

Zhenbang Wu[1], Cao Xiao[2], Lucas M. Glass[3], David M. Liebovitz[4], and Jimeng Sun[1(✉)]

[1] University of Illinois at Urbana-Champaign, Champaign, USA
{zw12,jimeng}@illinois.edu
[2] Amplitude, San Francisco, USA
danica.xiao@amplitude.com
[3] IQVIA, Durham, USA
lucas.glass@iqvia.com
[4] Northwestern University, Evanston, USA
david.liebovitz@nm.org

Abstract. Given a deep learning model trained on data from a source hospital, how to deploy the model to a target hospital automatically? How to accommodate heterogeneous medical coding systems across different hospitals? Standard approaches rely on existing medical code mapping tools, which have several practical limitations.

To tackle this problem, we propose AutoMap to automatically map the medical codes across different EHR systems in a coarse-to-fine manner: **(1) Ontology-level Alignment:** We leverage the ontology structure to learn a coarse alignment between the source and target medical coding systems; **(2) Code-level Refinement:** We refine the alignment at a fine-grained code level for the downstream tasks using a teacher-student framework.

We evaluate AutoMap using several deep learning models with two real-world EHR datasets: eICU and MIMIC-III. Results show that AutoMap achieves relative improvements up to 3.9% (AUC-ROC) and 8.7% (AUC-PR) for mortality prediction, and up to 4.7% (AUC-ROC) and 3.7% (F1) for length-of-stay estimation. Further, we show that AutoMap can provide accurate mapping across coding systems. Lastly, we demonstrate that AutoMap can adapt to two challenging scenarios: (1) mapping between completely different coding systems and (2) between completely different hospitals.

Keywords: Medical code mapping · Clinical prediction model deployment · Electronic health records

1 Introduction

Deep learning models have been widely used in clinical predictive modeling with electronic health record (EHR) data [33]. These models often leverage medical

---

Supplementary Information The online version contains supplementary material available at https://doi.org/10.1007/978-3-031-26390-3_29.

M.-R. Amini et al. (Eds.): ECML PKDD 2022, LNAI 13714, pp. 505–520, 2023.
https://doi.org/10.1007/978-3-031-26390-3_29

codes as an important data source summarizing patients' health status [5,7,14]. However, in real-world clinical practice, a variety of different coding systems are used across hospital EHR systems [3]. As a result, models trained on data from a source hospital are often hard to adapt to a target hospital where other coding systems are used. A method that can **accommodate different medical coding systems across hospitals for easy model deployment** is highly desirable. Standard approaches rely on existing medical code mapping tools (e.g., Unified Medical Language System (UMLS) [4]), which have significant practical limitations due to the following challenges:

- **Rare coding systems.** Existing commercial and free code mapping tools are only available for a few widely used coding systems (e.g., ICD-9, ICD-10 and SNOMED CT) [32]. Hospitals using rare or private coding systems cannot benefit from these tools.
- **Limited labeled data.** While large hospitals may fine-tune the pre-trained models to adapt to their coding systems, small hospitals with limited labeled data often fail to do so.
- **No access to source data.** Even worse, the source data usually cannot be shared with the target hospital due to privacy and legal concern.

In this paper, we propose `AutoMap` for automatic medical code mapping across different hospitals EHR systems. `AutoMap` constructs appropriate target embeddings unsupervisedly based on the target EHR data and maps the target embeddings to the source embeddings, so that the deep learning model trained on the source data can be seamlessly deployed to the target data without any manual code mapping. More specifically, `AutoMap` learns the mapping across different coding systems in a coarse-to-fine manner:

- **Embedding.** The medical code embeddings will be constructed from the target EHR data unsupervisedly.
- **Ontology-level Alignment.** We leverage the ontology structure to map medical coding groups via iterative self-supervised learning. In this step, we obtain a coarse mapping from groups of target embeddings to the groups of source embeddings.
- **Code-level Refinement.** We refine the coarse mapping at a fine-grained code level via a teacher-student framework. It utilizes a discriminator (teacher A) to align two coding systems at the code level, and the backbone model (teacher B) to optimize the mapping based on the final prediction.

We evaluate `AutoMap` using multiple backbone deep learning models on two real-world EHR datasets: eICU [25] and MIMIC-III [13]. Results show that with a limited set of labeled data, `AutoMap` achieves relative improvements up to 3.9% on AUC-ROC score and 8.7% on AUC-PR score for mortality prediction; and up to 4.7% on AUC-ROC score and 3.7% on F1 score for length-of-stay estimation. Further, we evaluate the mapping accuracy of `AutoMap` and show that `AutoMap` improves the best baseline method by 8.2% in similarity score and 11.3% on hit@10 score. Lastly, we demonstrate that `AutoMap` can still achieve acceptable results under two challenging scenarios: (1) mapping between completely different coding systems: the model is trained on diagnosis codes and deployed on medication codes; (2) mapping between completely different hospitals: the model is trained and deployed in hospitals from different regions.

It is important to note that we do not argue to completely replace existing code mapping tools. Instead, the main contribution of AutoMap is to **evaluate the potential of a novel approach to automatically learn the code mapping from clinical data**, which provides a new direction to support model deployment across different medical coding systems, and complements existing code mapping tools.

2   Related Work

Medical Code Mapping Tools. There exists a variety of commercial and free tools for mapping across different EHR ecosystems. UMLS [4] provides the mapping among ICD-9, ICD-10 and SNOMED CT. Observational Medical Outcomes Partnership (OMOP) [12] and Fast Healthcare Interoperability Resources (FHIR) [20] define the standards for representing clinical data in a consistent format. Relying on these tools, some recent works try to support model deployment across hospitals by transforming the EHR data into a standard format [26,31]. However, creating such tools requires a lot of domain knowledge and human labor [32]. These mapping tools are only available for widely-used coding systems and can be easily outdated due to code updates. To address this, AutoMap proposes to learn the code mapping from clinical data, which complements existing code mapping tools.

Cross-lingual Word Mapping. Our medical code mapping problem has some similarity to the cross-lingual research. Cross-lingual word mapping methods work by mapping the word embeddings in two languages to a shared space using translation pairs [1], shared tokens [29], adversarial learning [9], or the nearest neighbors of similarity distributions [2]. Inspired by [2], AutoMap also leverages the similarity distributions [22] to align medical codes. However, there are significant differences between EHR and natural languages: (1) medical codes often reside in a concept hierarchy; (2) medical codes are often noisier. To address this, instead of directly mapping medical codes, AutoMap adopts a coarse-to-fine method by first performing ontology-level alignment and then code-level refinement.

3   Preliminaries

We first define a few key concepts, and then present our setting in Definition 6. Detailed notations can be found in the appendix.

Definition 1 (EHR Dataset). *In EHR data, a patient has a sequence of visits:* $V_p = [v_p^{(1)}, v_p^{(2)}, \ldots, v_p^{(n_p)}]$, *where n_p is the number of visits of patient p. For model training, each patient has a label $\mathbf{y}_p$ (e.g., mortality or length-of-stay). We will drop the subscript p whenever it is unambiguous. Each visit of a patient is represented by its corresponding medical codes, specified by* $v^{(i)} = \{\mathbf{c}_1, \mathbf{c}_2, \ldots, \mathbf{c}_{m^{(i)}}\}$, *where $m^{(i)}$ is the total number of codes of the i-th visit. Each medical code* $\mathbf{c} \in \{0,1\}^{|C|}$ *is a one-hot vector (i.e., $\|\mathbf{c}\|_1 = 1$), where C denotes the set of all medical codes in the dataset.*

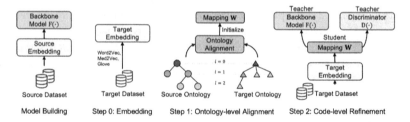

Fig. 1. AutoMap supports model deployment by automatically mapping the medical code embeddings across different coding systems in a coarse-to-fine manner: (0) Embedding that initializes the target code embedding matrix; (1) Ontology-level Alignment that leverages the ontology structure to learn the coarse ontology mapping; (2) Code-level Refinement that refines the mapping at the fine-grained code level for the downstream task with a teacher-student framework.

Our setting involves two datasets: a source dataset $*S$ for pre-training the backbone model but unavailable during deployment, and a mostly unlabeled target dataset $*T$ for deploying the model. The two datasets can have completely different medical codes. We also utilize separate medical ontology structures for source and target medical codes.

Definition 2 (Medical Ontology). *A medical ontology $\mathcal{O}$ specifies the hierarchy of medical codes in the form of a parent-child relationship. Formally, an ontology $\mathcal{O}$ is a directed acyclic graph whose nodes are $\mathcal{C} \cup \overline{\mathcal{C}}$. Here, $\mathcal{C}$ is the set of medical codes (often leaf nodes in the ontology), and $\overline{\mathcal{C}}$ is the set of other intermediate codes (i.e., non-leaf nodes) representing more general concepts.*

For simplicity, we define a function ancestor$(\mathbf{c}, l) : \{0,1\}^{|\mathcal{C}|} \times \mathbb{Z} \to \{0,1\}^{|\overline{\mathcal{C}}|}$, which maps a given medical code $\mathbf{c} \in \{0,1\}^{|\mathcal{C}|}$ to its l-th level ancestor code (i.e., category). For example, in Fig. 1, the root node is the 0-th level ancestor code of all leaf codes.

Definition 3 (Medical Code Embedding). *To fully utilize the code semantic information, it is a common practice to convert the medical code from one-hot vector $\mathbf{c} \in \{0,1\}^{|\mathcal{C}|}$ to a dense embedding vector $\mathbf{e} \in \mathbb{R}^d$ [5, 14], where d is the embedding dimensionality. This can be done via an embedding matrix $\mathbf{E} \in \mathbb{R}^{|\mathcal{C}| \times d}$, where each row corresponds to the embedding for a medical code. The embedding can be computed as $\mathbf{e} = \mathbf{E}^\top \mathbf{c}$.*

We denote the embedding matrices for source and target datasets as $\mathbf{E}_S$ and $\mathbf{E}_T$, respectively. The source embedding $\mathbf{E}_S$ is provided with the trained backbone model as the input. And the target embedding $\mathbf{E}_T$ will be learned using the target dataset. In this work, to deploy the backbone model, we will learn to map the target medical codes to the source.

Definition 4 (Code Embedding Mapping). *We define the mapping from the embedding space of one medical coding system to another as $\phi(\mathbf{E}) : \mathbb{R}^d \to \mathbb{R}^d$.*

We will learn the embedding mapping $\phi(\cdot)$ that maps the target embedding to the source via $\phi(\mathbf{E}_T)$.

Definition 5 (Backbone Deep Learning Model). *The backbone deep learning model $F(\cdot)$ takes EHR sequences and the corresponding medical code embeddings as the input and then outputs the prediction: $\hat{\mathbf{y}} = F([v^{(i)}]_{i=1}^n, \phi(\mathbf{E}))$, where $\hat{\mathbf{y}}$ is the corresponding predictions for label $\mathbf{y}$. The backbone model $F(\cdot)$ is pretrained on source dataset $*S$ and deployed on target dataset $*T$ with a different coding system. Note that the embedding mapping $\phi(\cdot)$ degenerates to the identity function if the backbone model $F(\cdot)$ is trained and deployed on the same coding system.*

Definition 6 (Predictive Model Deployment). *Given a backbone model $F(\cdot)$ and source code embedding matrix $\mathbf{E}_S$, a mostly unlabeled target dataset $*T$ in a different coding system, and the medical ontologies $\mathcal{O}_S, \mathcal{O}_T$ for both coding systems, the goal is to optimize the mapping $\phi(\cdot)$ on the target dataset $*T$, as given by Eq. (1),*

$$\arg\min_{\phi(\cdot)} \mathcal{L}(F(\cdot), \mathbf{E}_S, *T, \mathcal{O}_S, \mathcal{O}_T, \phi(\cdot)), \tag{1}$$

where $\mathcal{L}(\cdot)$ denotes the designated loss function. The prediction on the target dataset can be obtained via $F(V_T, \phi(\mathbf{E}_T))$, where V_T is a sequence of visits from the target dataset $*T$, and $\phi(\mathbf{E}_T)$ is the transformed target embeddings.

In our setting, we can only access the source code embedding $\mathbf{E}_S$ and ontology $\mathcal{O}_S$ instead of the source data $*S$. This is more realistic in deployment setting since the source data often cannot be shared due to legal and privacy concern. In contrast, the source embedding matrix $\mathbf{E}_S$ can be more easily provided along with the backbone model $F(\cdot)$, and the code ontologies are usually publicly accessible. We also assume that the target dataset $*T$ is mostly unlabeled, since the target site may often be some small hospital.

4   AutoMap Method

We propose AutoMap for automatic code mapping across different hospitals EHR systems. The mapping will be done in a coarse-to-fine manner, enabled by the adaptation process shown in Fig. 1. Embedding (step 0) first initializes the target code embedding matrix $\mathbf{E}_T$. Ontology-level alignment (step 1) then derives the initial coarse mapping $\phi(\cdot)$ via iterative self-supervised learning. Code-level refinement (step 2) further fine-tunes the mapping $\phi(\cdot)$ at the code level using a teacher-student framework.

4.1   Step 0: Embedding

As mentioned in Definition 3, we first convert the target medical codes from one-hot vector $\mathbf{c}_T \in \{0,1\}^{|\mathcal{C}|}$ to a corresponding dense embedding vector $\mathbf{e}_T \in \mathbb{R}^d$.

We use GloVe [24] to learn the target code embedding matrix $\mathbf{E}_T$ via a global co-occurrence matrix of medical codes. Other unsupervised learning algorithms such as Med2Vec [7] and Word2Vec [21] can also be used. We employ GloVe because of its computational efficiency. After this, we parameterize $\phi(\cdot)$ by a mapping matrix $\mathbf{W} \in \mathbb{R}^{d \times d}$. The mapping matrix $\mathbf{W}$ can be used to transform the target code embedding via $\mathbf{E}_T \mathbf{W}$.

4.2   Step 1: Ontology-Level Alignment

In this step, we will first learn a coarse mapping $\mathbf{W}$ at the ontology level. This first step is essential because direct code level mapping is difficult and unnecessary: (1) It is difficult due to the large number of medical codes; (2) It is also unnecessary since many codes have similar clinical meanings. Therefore, we follow a common practice to first group similar codes using code ontology [5,7,28] and learn the mapping on groups instead of leaf-level codes. For example, ICD-9 code 438.11 "late effects of cerebrovascular disease, aphasia" corresponds to five ICD-10 codes (I69.020, I69.120, I69.220, I69.320, I69.920). While it is hard to directly align the ICD-9 code to each of these five ICD-10 codes, we can first coarsely map the ICD-9 code to I00-I99 "diseases of the circulatory system", and then gradually refine the mapping to I60-I99 "cerebrovascular diseases", I69 "cerebrovascular diseases", and eventually the five-leaf codes. By leveraging the medical ontology, we can use more general medical concepts as "anchor points" to better align two coding systems.

Next, we introduce the building blocks of the iterative self-supervised learning (i.e., ontology grouping, unsupervised seed induction, Procrustes refinement), and then present the ontology-level alignment algorithm.

Ontology Grouping. At a given hierarchy level l, we group the codes according to their l-th level ontology categories. Specifically, the i-th group $\mathcal{G}_i^{(l)}$ consists of all the leaf medical codes whose l-th level category is $\mathbf{c}_i$, as in Eq. (2),

$$\mathcal{G}_i^{(l)} = \{\mathbf{c}_j \mid \mathrm{ancestor}(\mathbf{c}_j, l) = \mathbf{c}_i, \mathbf{c}_j \in \mathcal{C}\}, \tag{2}$$

where $\mathbf{c}_i \in \overline{\mathcal{C}}$ is the corresponding l-th level category code. We will drop the superscript $^{(l)}$ whenever it is unambiguous. To obtain the group embedding $\mathbf{g}_i$, we first calculate the mean group embedding $\overline{\mathbf{g}}_i$ by averaging all the code embeddings in that group, as in Eq. (3a); then, we represent the group embedding as the closest code embedding, as in Eq. (3b),

$$\overline{\mathbf{g}}_i = \mathrm{mean}\{\mathbf{e}_j \mid \mathbf{c}_j \in \mathcal{G}_i\}, \tag{3a}$$

$$\mathbf{g}_i = \underset{\mathbf{e}_j}{\mathrm{argmin}}\{\mathbf{e}_j \overline{\mathbf{g}}_i^\top \mid \mathbf{c}_j \in \mathcal{C}\}, \tag{3b}$$

where $\mathbf{e}_j$ is the embedding vector for the code $\mathbf{c}_j$, and $\mathbf{e}_j \overline{\mathbf{g}}_i^\top \in \mathbb{R}$ calculates the distance between the code $\mathbf{c}_j$ and the mean group embedding $\overline{\mathbf{g}}_i$. Intuitively, $\mathbf{g}_i$ can be viewed as the "median" group embedding. We select the top-k groups based on the group size, since we want to first learn a coarse mapping while including too many groups may introduce too much granular information. As a result, we have $\mathbf{G}_T, \mathbf{G}_S \in \mathbb{R}^{k \times d}$ for target and source groups, where each row corresponds to an embedding vector for a particular group. We present with the same k to reduce clutter, though it can be different for source and target groups.

We note that when the ontology is not available, `AutoMap` can still apply by using a clustering algorithm (e.g., k-Means) to group the medical codes. Specifically, we provide additional experiments on this setting in the appendix.

Unsupervised Seed Induction. Given the l-th level source and target coding groups $\mathbf{G}_S$ and $\mathbf{G}_T$, we can initialize a coarse alignment in a fully unsupervised way. More specifically, we first calculate the similarity matrices, as in Eq. (4),

$$\mathbf{M}_T = \mathbf{G}_T \mathbf{G}_T^\top; \ \mathbf{M}_S = \mathbf{G}_S \mathbf{G}_S^\top, \tag{4}$$

where $\mathbf{M}_T, \mathbf{M}_S \in \mathbb{R}^{k \times k}$. Each row in the similarity matrices $\mathbf{M}_T, \mathbf{M}_S$ represents the similarities of the corresponding group to all the other groups. Under the ideal case where the embedding spaces between different coding systems are isometric[1], one can permute the rows and columns of $\mathbf{M}_T$ to obtain $\mathbf{M}_S$. We introduce the following heuristics to find the optimal permutation (i.e., a mapping dictionary) of this NP-hard problem. We perform row-wise sort on $\mathbf{M}_T$ and $\mathbf{M}_S$ (i.e., elements in each row are sorted based only on the order in that particular row), as in Eq. (5a). Under the isometric assumption, codes with the same meaning will have exactly the same row vector in $\tilde{\mathbf{M}}_T$ and $\tilde{\mathbf{M}}_S$, suggesting that we can find the mapping dictionary $\mathbf{D} \in \mathbb{R}^{k \times k}$ via nearest neighbor search over row vectors in $\tilde{\mathbf{M}}_T$, as shown in Eq. (5b),

$$\tilde{\mathbf{M}}_T = \mathrm{sorted}(\mathbf{M}_T); \tilde{\mathbf{M}}_S = \mathrm{sorted}(\mathbf{M}_S), \tag{5a}$$

$$\mathbf{D}[i,j] = \begin{cases} 1, & \text{if } j = \mathrm{argmax}((\tilde{\mathbf{M}}_T \cdot \tilde{\mathbf{M}}_S^\top)[i,:]) \\ 0, & \text{otherwise}, \end{cases} \tag{5b}$$

where $\cdot$ denotes matrix multiplication.

Procrustes Optimization. At a given hierarchy level l, we optimize the inducted mapping dictionary $\mathbf{D}$ by iterating the following two steps.

1. The mapping $\mathbf{W} \in \mathbb{R}^{d \times d}$ is obtained by maximizing the similarities for the current dictionary $\mathbf{D}$, as given by Eq. (6a). This optimization problem is known as the Procrustes problem [27] and has a closed form solution, as in Eq. (6b),

$$\underset{\mathbf{W}}{\mathrm{argmin}} \|\mathbf{D} \odot \underbrace{(\mathbf{G}_T \mathbf{W} \mathbf{G}_S^\top)}_{\text{transformed target embedding}}\|_1, \tag{6a}$$

$$\mathbf{W} = \mathbf{U}\mathbf{V}^\top, \text{ where } \mathbf{U}\boldsymbol{\Sigma}\mathbf{V}^T = \mathrm{SVD}(\mathbf{G}_T^\top \mathbf{D} \mathbf{G}_S), \tag{6b}$$

---

[1] In practice, the isometry requirement will not hold exactly, but it can be assumed to hold approximately, or the problem of mapping two code embedding spaces without supervision would be impossible.

where $\odot$ denotes Hadamard product, and SVD denotes Singular Value Decomposition.

2. A new dictionary $\mathbf{D}$ is induced, as in Eq. (7),

$$\mathbf{D}[i,j] = \begin{cases} 1, & \text{if } j = \text{argmax}((\mathbf{G}_T \mathbf{W} \mathbf{G}_S^\top)[i,:]) \\ 0, & \text{otherwise.} \end{cases} \tag{7}$$

Iterative Self-supervised Learning. We now introduce the self-supervised learning strategy, which maps the two coding systems at multiple resolutions iteratively. Starting from a coarse hierarchy level l, we obtain the l-th level medical coding groups $\mathbf{G}_S$ and $\mathbf{G}_T$ with Eq. (2, 3). Then we induct the l-th level seed mapping dictionary $\mathbf{D}^{(l)}$ with Eq. (4, 5). Next, we merge the current and previous level mapping dictionaries, as $\mathbf{D}^{(l)} = \mathbf{D}^{(l)} + \mathbf{D}^{(l-1)}$. Lastly, we optimize the merged mapping dictionary $\mathbf{D}^{(l)}$ using Eq. (6, 7). We gradually increase l (going down in the ontology) during iterative self-supervised learning until we reach the leaf level to learn the mapping at multiple resolutions. We note that source and target codes can use different grouping level l. We present with the same l to reduce clutter.

In this way, we learn a coarse mapping matrix $\mathbf{W}$ between two medical coding systems at the ontology level. This step is inspired by [2]. However, instead of directly mapping medical codes, AutoMap leverages the ontology structure and iteratively maps medical coding groups in a coarse-to-fine manner, allowing AutoMap to better align coding systems with different granularities.

4.3   Step 2: Code-Level Refinement

While we have performed step 1 (ontology-level alignment) to initialize the mapping, the mapping is still too coarse and need further refining. Moreover, there is no guarantee of the performance for the downstream tasks (i.e., mortality prediction and length-of-stay estimation). Thus, it is preferred to further fine-tune the mapping at the code level for downstream tasks.

To do this, we propose a teacher-student framework, where the discriminator $D(\cdot)$ (teacher A) refines the mapping matrix $\mathbf{W}$ (student) via adversarial learning; and the backbone model $F(\cdot)$ (teacher B) optimizes the mapping matrix $\mathbf{W}$ (student) based on the final prediction task. Below we describe the framework in detail.

Teacher A: Discriminator. We leverage the adversarial learning framework by introducing a discriminator $D(\cdot)$, parameterized by a multi-layer neural network. Specifically, the discriminator $D(\cdot)$ tries to classify whether the embeddings are from the target (label 0) or source (label 1) embedding distributions. Formally, discriminator $D(\cdot)$ aims at minimizing the discriminator adversarial loss, as in Eq. (8),

$$\mathcal{L}_D = -\log(D(\mathbf{e}_S)) - \log(D(1 - \mathbf{e}_T \mathbf{W})), \tag{8}$$

where $\mathbf{e}_S$ ($\mathbf{e}_T$) represents the source (target) code embedding sampled randomly from the code embedding matrix $\mathbf{E}_S$ ($\mathbf{E}_T$), and $\mathbf{W}$ maps the target embedding to the source embedding space via $\mathbf{e}_T\mathbf{W}$.

The mapping matrix $\mathbf{W}$ acts as the generator and tries to deceive the discriminator $D(\cdot)$. Formally, we try to minimize the generator adversarial loss, as in Eq. 9,

$$\mathcal{L}_G = -\log(D(\mathbf{e}_T\mathbf{W})). \tag{9}$$

Theoretically, the discriminator $D(\cdot)$ and mapping matrix $\mathbf{W}$ learn to align two coding systems as an adversarial game. Since the minimization happens at the distribution level, we do not require code mapping pairs to supervise training.

Teacher B: Backbone. Here, the backbone model $F(\cdot)$ is leveraged to optimize the ultimate prediction performance based on the transformed target code embeddings. Formally, we aim at minimizing the following classification loss

$$\mathcal{L}_{\mathrm{cls}}(F([v^{(i)}]_{i=1}^n, \mathbf{E}_T\mathbf{W}), \mathbf{y}_T), \tag{10}$$

where the transformed target code embeddings $\mathbf{E}_T\mathbf{W}$ are used to encode patient visits $[v^{(i)}]_{i=1}^n$.

In summary, the mapping matrix $\mathbf{W}$ is fine-tuned by minimizing the combined loss

$$\mathcal{L}_W = \mathcal{L}_{\mathrm{cls}} + \alpha\mathcal{L}_G, \tag{11}$$

where α is a hyper-parameter. The pseudo-code can be found in the appendix.

5 Experiment

5.1 Experimental Setting

We will briefly introduce the experimental settings. Detailed information can be found in the appendix. The code of AutoMap is publicly available[2].

Data. We evaluate the performance of AutoMap extensively with two publicly accessible datasets: eICU [25] and MIMIC-III [13]. eICU [25] is a multi-center database with intensive care unit (ICU) records for over 200K admissions to over 200 hospitals across the United States. MIMIC-III [13] is a single-center database containing 53K ICU records from Beth Israel Deaconess Medical Center.

Baselines. We compare AutoMap with multiple baseline methods ranging from simple methods such as **Direct Training** and **Transfer Learning**, standard method leveraging **Mapping Tools**, to cross-lingual translation methods like **MUSE** [9] and **VecMap** [2]. We also conduct an ablation study of our AutoMap with **Step 1 Only**, **Step 1 Only + Random Ontology**, and **Step 2 Only**.

---

[2] https://github.com/zzachw/AutoMap.

Backbone Models. As `AutoMap` is a general framework that can apply to different backbone models, we incorporate `AutoMap` with the following backbone deep learning models: **MLP, RNN, RETAIN** [5], **GCT** [8], **BEHRT** [14].

Table 1. Results with limited labeled data (100 patients) in the target site. Dataset is eICU [25]. The average scores of two mapping directions between ICD-9 and ICD-10 codes are reported. * indicates that `AutoMap` achieves significant improvement over the best baseline method (i.e., p-value is smaller than 0.05). Experiment results show that `AutoMap` can adapt different backbone models to the target site with limited labeled data.

Backbone	Method	Mortality		Length-of-Stay	
		AUC-PR	AUC-ROC	F1	AUC-ROC
MLP	*Full-Label*	*0.2819*	*0.6531*	*0.5033*	*0.2819*
	Direct Training	0.2524	0.6191	0.2835	0.5345
	Transfer Learning	0.2551	0.6240	0.4584	0.6095
	MUSE	0.2506	0.6276	0.4905	0.6240
	VecMap	0.2820	0.6502	0.4947	0.6341
	`AutoMap`	**0.2934***	**0.6631***	**0.4952**	**0.6350**
RNN	*Full-Label*	*0.2818*	*0.6539*	*0.5030*	*0.2818*
	Direct Training	0.2074	0.5547	0.1222	0.4427
	Transfer Learning	0.2536	0.6234	0.4662	0.6166
	MUSE	0.2455	0.6260	0.4933	0.6367
	VecMap	0.2780	0.6488	**0.5019**	0.6416
	`AutoMap`	**0.2875***	**0.6627***	0.4996	**0.6487***
RETAIN	*Full-Label*	*0.2648*	*0.6190*	*0.4447*	*0.2648*
	Direct Training	0.2031	0.5466	0.1222	0.4427
	Transfer Learning	0.2269	0.5732	0.4455	0.5395
	MUSE	0.2374	0.5838	0.4217	0.5831
	VecMap	0.2744	0.6315	0.4264	0.5963
	`AutoMap`	**0.2835***	**0.6528***	**0.4779***	**0.6007***
GCT	*Full-Label*	*0.2814*	*0.6533*	*0.4986*	*0.2814*
	Direct Training	0.1836	0.5402	0.2680	0.4865
	Transfer Learning	0.2103	0.5967	0.4748	0.5718
	MUSE	0.2242	0.6016	0.4866	0.6129
	VecMap	0.2491	0.6291	0.4863	0.6085
	`AutoMap`	**0.2707***	**0.6539***	**0.4940***	**0.6363***
BEHRT	*Full-Label*	*0.2652*	*0.6673*	*0.3657*	*0.2652*
	Direct Training	0.1740	0.5438	0.3063	0.4730
	Transfer Learning	0.2320	0.6190	0.3291	0.5609
	MUSE	0.2155	0.6040	0.3493	0.5869
	VecMap	**0.2786**	**0.6740**	0.3612	0.6044
	`AutoMap`	0.2712	0.6737	**0.3744***	**0.6328***

Table 2. Results for the scenario where the backbone model is trained on diagnosis code and deployed on medication codes. Dataset is MIMIC-III [13]. Experiment results show that AutoMap can adapt to target data coded in a completely different system.

Method	Mortality AUC-PR	Length-of-Stay F1
Full-Label	*0.7149*	*0.3057*
Direct Training	0.4701	**0.3158**
Transfer Learning	0.5642	0.2999
MUSE	0.4905	0.3022
VecMap	0.3553	0.3014
AutoMap	**0.5902***	0.3022

5.2 Q1: Target Data with Limited Labels

We first evaluate AutoMap in a common setting where the target site has limited labeled data (100 patients). For reference, we also report the performance of the model trained with the fully-labeled target data, as *"Full-Label"* in the table. This can be viewed as an "upper bound" of the model performance. Descriptions of the metrics can be found in the appendix. Results can be found in Table 1.

First, we find that the two simple baselines: direct training and transfer learning methods do not work very well. In most cases, they are much worse compared to the full-label performance. This is expected as the amount of labeled data is insufficient to train or fine-tune the backbone models. Next, code-level mapping methods MUSE [9] and VecMap [2] achieve some improvements, but they are not stable. In some cases, they perform even worse than the two simple baselines. This may because ICD-9 and ICD-10 have different degrees of specificity (e.g., 10K codes in ICD-9 v.s. 68K codes in ICD-10), and directly mapping them at code level does not work very well. Finally, we observe that AutoMap achieves significant improvement over the baseline and can match the full-label performance in most cases. Specifically, AutoMap achieves up to 8.7% relative improvement on AUC-PR score for mortality prediction; for length-of-stay estimation, AutoMap achieves up to and 3.7% relative improvement on F1 score. This demonstrates the effectiveness of coarse-to-fine mapping and the versatility of AutoMap.

5.3 Q2: Completely Different Codes

We then evaluate AutoMap in the challenging case where we train the backbone model on diagnosis code (ICD-9) and deploy it on medication codes (NDC). Due to the limited space, for the rest of the experiments, we only report AUC-PR for mortality and F1 for length-of-stay with backbone model BEHRT [14] using 100 labeled patients in the target data. Results can be found in Table 2.

First, we note that since these two coding systems are so different, no existing mapping tools is available. For mortality prediction, as shown in Table 2, the

Table 3. Results for the scenario where the backbone model is trained and deployed in hospitals from different regions. Dataset is eICU [25]. Experiment results show that AutoMap can adapt to target hospitals from a completely region.

Method	Mortality AUC-PR	Length-of-Stay F1
Full-Label	*0.2578*	*0.4560*
Direct Training	0.1434	**0.4334**
Transfer Learning	0.1860	0.3924
MUSE	0.1314	0.3988
VecMap	0.1305	0.3801
AutoMap	**0.1990***	0.4290

code-level mapping methods perform even worse than direct training and transfer learning. This may due to the large gap between these two coding systems. On the contrary, AutoMap can still give acceptable results, outperforming all baseline methods with 4.6%–66.1% statistically significant improvements. This shows the superiority of AutoMap's coarse-to-fine mapping strategy. For the length-of-stay estimation task, all five methods perform pretty similar to full-label performance. This may indicate that medication codes are not so informational for length-of-stay estimation.

5.4   Q3: Completely Different Hospitals

We next challenge AutoMap under the scenario where we train the backbone model in hospitals from Midwest region (with ICD-9 code) and deploy it in hospitals from South region (with ICD-10 code). Results can be found in Table 3.

For mortality prediction, mapping based methods (MUSE [9] and VecMap [2]) achieve the worst results. This is expected as methods from cross-lingual word mapping do not consider the domain gap between different regions. This also explains why transfer learning perform slightly better (as its training scheme can accommodate some domain gap). Benefit from the refinement step, AutoMap achieves the best result with 7.0%–52.5% statistically significant relative improvements. This shows that AutoMap can adapt to hospitals from different regions. For length-of-stay estimation, all pre-training based methods perform worse than direct training. This may indicate that different hospitals have different decision rules on ICU length-of-stay. As a result, transferring knowledge from other hospitals may not help. Despite this, AutoMap still achieves the best results among all pre-training based methods.

5.5   Q4: Mapping Accuracy

We further evaluate the accuracy of the learnt mapping. The ICD code mapping in the eICU [25] dataset is used as the ground truth. As shown in Table 4,

Table 4. Accuracy of mapping for diagnosis codes (ICD-9 and ICD-10). Dataset is eICU [25]. The average scores of two mapping directions are reported. Experiment results show that AutoMap can learn accurate mapping across medical coding systems.

Method	Similarity	Hit@10
MUSE	0.1633	0.0600
VecMap	0.4612	0.5974
AutoMap	**0.4992***	**0.6657***

Table 5. Ablation study. Dataset is eICU [25]. The average scores of two mapping directions between ICD-9 and ICD-10 codes are reported. R.O. denotes random ontology. Experiment results demonstrate the importance of AutoMap's 2-step coarse-to-fine mapping.

Method	Mortality AUC-PR	Length-of-Stay F1
Step 1 Only	0.2680	0.3623
Step 1 Only + R.O.	0.2054	0.3631
Step 2 Only	0.2038	0.3306
AutoMap	**0.2712***	**0.3744***

VecMap [2] and AutoMap achieve much better performance than MUSE [9]. This supports the isometric assumption used in both methods. Further, AutoMap achieves the best results with statistical significance. This demonstrates that the proposed coarse-to-fine mapping can better map coding systems with different granularities.

5.6 Ablation Study

Finally, we compare AutoMap with three ablated versions. As shown in Table 5, only performing step 2 (code-level refinement) gives the worst results. This is reasonable as the model will easily over-fit the target data with limited labels. Also, since the mapping matrix $\mathbf{W}$ is randomly generated, the adversarial learning module will even harm the downstream tasks. Next, we can see that performing step 1 (ontology-level alignment) only gives better results. This indicates that step 1 contributes most to AutoMap's improvements. This may because the isometric assumption and medical ontology can act as a strong prior to guide the model learning process. This point can also be supported by the performance with randomly-generated ontology. Lastly, AutoMap achieves the best results. This shows the importance of refining the mapping at code-level after the coarse ontology alignment.

6 Conclusion

We propose AutoMap for automatic medical code mapping across different hospitals EHR systems. Benefit from the coarse-to-fine mapping, AutoMap can better align coding systems at different granularities. We evaluate AutoMap extensively using different backbone models with two real-world EHR datasets. Experimental results show that AutoMap outperforms existing solutions on multiple prediction tasks when mapping solutions exist and provides a mapping strategy when conventional solutions do not exist.

References

1. Artetxe, M., Labaka, G., Agirre, E.: Learning bilingual word embeddings with (almost) no bilingual data. In: Proceedings of the 55th Annual Meeting of the Association for Computational Linguistics, Vancouver, Canada, (Volume 1: Long Papers), pp. 451–462. Association for Computational Linguistics (2017). https://doi.org/10.18653/v1/P17-1042. https://www.aclweb.org/anthology/P17-1042
2. Artetxe, M., Labaka, G., Agirre, E.: A robust self-learning method for fully unsupervised cross-lingual mappings of word embeddings. In: Proceedings of the 56th Annual Meeting of the Association for Computational Linguistics, Melbourne, Australia, (Volume 1: Long Papers), pp. 789–798. Association for Computational Linguistics (2018). https://doi.org/10.18653/v1/P18-1073. https://www.aclweb.org/anthology/P18-1073
3. Birkhead, G.S., Klompas, M., Shah, N.R.: Uses of electronic health records for public health surveillance to advance public health. Ann. Rev. Public Health **36**(1), 345–359 (2015). https://doi.org/10.1146/annurev-publhealth-031914-122747, pMID: 25581157
4. Bodenreider, O.: The Unified Medical Language System (UMLS): integrating biomedical terminology. Nucleic Acids Res. **32**(Database issue), D267–270 (2004)
5. Choi, E., Bahadori, M.T., Kulas, J.A., Schuetz, A., Stewart, W.F., Sun, J.: Retain: an interpretable predictive model for healthcare using reverse time attention mechanism. In: Proceedings of the 30th International Conference on Neural Information Processing Systems, NIPS 2016, pp. 3512–3520. Curran Associates Inc., Red Hook (2016)
6. Choi, E., Bahadori, M.T., Schuetz, A., Stewart, W.F., Sun, J.: Doctor AI: predicting clinical events via recurrent neural networks. In: Doshi-Velez, F., Fackler, J., Kale, D., Wallace, B., Wiens, J. (eds.) Proceedings of the 1st Machine Learning for Healthcare Conference. Proceedings of Machine Learning Research, vol. 56, pp. 301–318. PMLR, Northeastern University, Boston, MA, USA (2016). https://proceedings.mlr.press/v56/Choi16.html
7. Choi, E., et al.: Multi-layer representation learning for medical concepts. In: Proceedings of the 22nd ACM SIGKDD International Conference on Knowledge Discovery and Data Mining, KDD 2016, pp. 1495–1504. Association for Computing Machinery, New York (2016). https://doi.org/10.1145/2939672.2939823
8. Choi, E., et al.: Learning the graphical structure of electronic health records with graph convolutional transformer. In: Proceedings of the AAAI Conference on Artificial Intelligence, vol. 34, pp. 606–613 (2020). https://doi.org/10.1609/aaai.v34i01.5400
9. Conneau, A., Lample, G., Ranzato, M., Denoyer, L., Jégou, H.: Word translation without parallel data (2018)
10. Gupta, P., Malhotra, P., Narwariya, J., Vig, L., Shroff, G.: Transfer learning for clinical time series analysis using deep neural networks (2019)
11. Harutyunyan, H., Khachatrian, H., Kale, D.C., Ver Steeg, G., Galstyan, A.: Multitask learning and benchmarking with clinical time series data. Sci. Data **6**(1) (2019). https://doi.org/10.1038/s41597-019-0103-9
12. Hripcsak, G., et al.: Observational health data sciences and informatics (OHDSI): opportunities for observational researchers. Stud. Health Technol. Inform. **216**, 574–578 (2015)
13. Johnson, A.E., et al.: Mimic-iii, a freely accessible critical care database. Sci. Data **3**, 160035 (2016)

14. Li, Y., et al.: BEHRT: transformer for electronic health records. Sci. Rep. **10**(1), 7155 (2020). https://doi.org/10.1038/s41598-020-62922-y
15. Luo, J., Ye, M., Xiao, C., Ma, F.: HiTANet: hierarchical time-aware attention networks for risk prediction on electronic health records, pp. 647–656. Association for Computing Machinery, New York (2020). https://doi.org/10.1145/3394486.3403107
16. Ma, F., Chitta, R., Zhou, J., You, Q., Sun, T., Gao, J.: Dipole: diagnosis prediction in healthcare via attention-based bidirectional recurrent neural networks. In: Proceedings of the 23rd ACM SIGKDD International Conference on Knowledge Discovery and Data Mining, KDD 2017, pp. 1903–1911. Association for Computing Machinery, New York (2017). https://doi.org/10.1145/3097983.3098088
17. Ma, L., et al.: AdaCare: explainable clinical health status representation learning via scale-adaptive feature extraction and recalibration. In: The Thirty-Fourth AAAI Conference on Artificial Intelligence, AAAI 2020, The Thirty-Second Innovative Applications of Artificial Intelligence Conference, IAAI 2020, The Tenth AAAI Symposium on Educational Advances in Artificial Intelligence, EAAI 2020, New York, NY, USA, 7–12 February 2020, pp. 825–832. AAAI Press (2020). https://aaai.org/ojs/index.php/AAAI/article/view/5427
18. Ma, L., et al.: CovidCare: transferring knowledge from existing EMR to emerging epidemic for interpretable prognosis (2020)
19. Maas, A.L., Hannun, A.Y., Ng, A.Y.: Rectifier nonlinearities improve neural network acoustic models. In: ICML Workshop on Deep Learning for Audio, Speech and Language Processing (2013)
20. Mandel, J.C., Kreda, D.A., Mandl, K.D., Kohane, I.S., Ramoni, R.B.: SMART on FHIR: a standards-based, interoperable apps platform for electronic health records. J. Am. Med. Inform. Assoc. **23**(5), 899–908 (2016)
21. Mikolov, T., Chen, K., Corrado, G., Dean, J.: Efficient estimation of word representations in vector space (2013)
22. Mikolov, T., Le, Q.V., Sutskever, I.: Exploiting similarities among languages for machine translation (2013)
23. Nguyen, P., Tran, T., Wickramasinghe, N., Venkatesh, S.: D*eepr*: a convolutional net for medical records. IEEE J. Biomed. Health Inform. **21**(1), 22–30 (2017). https://doi.org/10.1109/JBHI.2016.2633963
24. Pennington, J., Socher, R., Manning, C.D.: Glove: global vectors for word representation. In: Empirical Methods in Natural Language Processing (EMNLP), pp. 1532–1543 (2014). http://www.aclweb.org/anthology/D14-1162
25. Pollard, T.J., Johnson, A.E.W., Raffa, J.D., Celi, L.A., Mark, R.G., Badawi, O.: The eICU collaborative research database, a freely available multi-center database for critical care research. Sci. Data **5**(1), 180178 (2018). https://doi.org/10.1038/sdata.2018.178
26. Rajkomar, A., et al.: Scalable and accurate deep learning with electronic health records. NPJ Digit. Med. **1**(1), 18 (2018). https://doi.org/10.1038/s41746-018-0029-1. http://www.nature.com/articles/s41746-018-0029-1
27. Schönemann, P.H.: A generalized solution of the orthogonal procrustes problem. Psychometrika **31**(1), 1–10 (1966). https://doi.org/10.1007/BF02289451
28. Shang, J., Xiao, C., Ma, T., Li, H., Sun, J.: GameNet: graph augmented memory networks for recommending medication combination. In: The Thirty-Third AAAI Conference on Artificial Intelligence, AAAI 2019, The Thirty-First Innovative Applications of Artificial Intelligence Conference, IAAI 2019, The Ninth AAAI Symposium on Educational Advances in Artificial Intelligence, EAAI 2019,

Honolulu, Hawaii, USA, 27 January–1 February 2019, pp. 1126–1133. AAAI Press (2019). https://doi.org/10.1609/aaai.v33i01.33011126

29. Søgaard, A., Ruder, S., Vulić, I.: On the limitations of unsupervised bilingual dictionary induction. In: Proceedings of the 56th Annual Meeting of the Association for Computational Linguistics, Melbourne, Australia, (Volume 1: Long Papers), pp. 778–788. Association for Computational Linguistics (2018). https://doi.org/10.18653/v1/P18-1072. https://www.aclweb.org/anthology/P18-1072

30. Srivastava, N., Hinton, G., Krizhevsky, A., Sutskever, I., Salakhutdinov, R.: Dropout: a simple way to prevent neural networks from overfitting. J. Mach. Learn. Res. **15**(56), 1929–1958 (2014). http://jmlr.org/papers/v15/srivastava14a.html

31. Tang, S., Davarmanesh, P., Song, Y., Koutra, D., Sjoding, M.W., Wiens, J.: Democratizing EHR analyses with FIDDLE: a flexible data- driven preprocessing pipeline for structured clinical data. J. Am. Med. Inform. Assoc. 14 (2020)

32. Wojcik, B.E., Stein, C.R., Devore, R.B., Hassell, L.H.: The challenge of mapping between two medical coding systems. Mil. Med. **171**(11), 1128–1136 (2006). https://doi.org/10.7205/MILMED.171.11.1128. https://academic.oup.com/milmed/article/171/11/1128-1136/4578127

33. Xiao, C., Choi, E., Sun, J.: Opportunities and challenges in developing deep learning models using electronic health records data: a systematic review. J. Am. Med. Inform. Assoc. **25**(10), 1419–1428 (2018). https://doi.org/10.1093/jamia/ocy068

34. Zhang, C., Gao, X., Ma, L., Wang, Y., Wang, J., Tang, W.: GRASP: generic framework for health status representation learning based on incorporating knowledge from similar patients. In: Proceedings of the AAAI Conference on Artificial Intelligence, vol. 35, no. 1, pp. 715–723 (2021). https://ojs.aaai.org/index.php/AAAI/article/view/16152

35. Zhang, H., Dullerud, N., Seyyed-Kalantari, L., Morris, Q., Joshi, S., Ghassemi, M.: An empirical framework for domain generalization in clinical settings. In: Proceedings of the Conference on Health, Inference, and Learning, CHIL 2021, pp. 279–290. Association for Computing Machinery, New York (2021). https://doi.org/10.1145/3450439.3451878

Hyperbolic Deep Keyphrase Generation

Yuxiang Zhang[1]([✉]), Tianyu Yang[1], Tao Jiang[1], Xiaoli Li[2], and Suge Wang[3]

[1] Civil Aviation University of China, Tianjin, China
{yxzhang,2019051011,2020052049}@cauc.edu.cn
[2] Institute for Infocomm Research/Centre for Frontier AI Research, Singapore,
Singapore
xlli@i2r.a-star.edu.sg
[3] Shanxi University, Taiyuan, China
wsg@sxu.edu.cn

Abstract. Keyphrases can concisely describe the *high-level topics* discussed in a document, and thus keyphrase prediction compresses document's hierarchical semantic information into a few important representative phrases. Numerous methods have been proposed to use the encoder-decoder framework in Euclidean space to generate keyphrases. However, their ability to capture the hierarchical structures is limited by the nature of Euclidean space. To this end, we propose a new research direction that aims to encode the hierarchical semantic information of a document into the low-dimensional representation and then decompress it to generate keyphrases in a *hyperbolic space*, which can effectively capture the underlying semantic hierarchical structures. In addition, we propose a novel *hyperbolic attention mechanism* to selectively focus on the high-level phrases in hierarchical semantics. To the best of our knowledge, this is the first study to explore a hyperbolic network for keyphrase generation. The experimental results illustrate that our method outperforms fifteen state-of-the-art methods across five datasets.

Keywords: Keyphrase generation · Hyperbolic neural network · Hyperbolic attention mechanism

1 Introduction

Keyphrase prediction is to automatically produce a set of representative phrases that are related to the main topics discussed in a given document. Since keyphrases (also referred to as keywords) can provide a *high-level topic description* of a document, they are beneficial for a wide range of natural language processing (NLP) tasks, such as information extraction [32], text summarization [33] and question generation [30]. However, the performance of existing approaches is still far from being satisfactory [16,21]. The main reason is that it is very challenging to determine if a phrase or a set of phrases accurately capture the high-level topics that are presented in a document.

Automatic keyphrase prediction models can be broadly divided into *extraction* and *generation* methods. In particular, traditional *extraction methods* can

© The Author(s), under exclusive license to Springer Nature Switzerland AG 2023
M.-R. Amini et al. (Eds.): ECML PKDD 2022, LNAI 13714, pp. 521–536, 2023.
https://doi.org/10.1007/978-3-031-26390-3_30

only extract *present keyphrases* that appear in a given document, while *genera-tion methods* can generate both present keyphrases as well as *absent keyphrases* that do not appear in a document.

Recently, the sequence-to-sequence (seq2seq) framework has been widely applied in the natural language generation tasks. CopyRNN [23] is the first to employ the attentional seq2seq framework [29] with the copying mechanism [14] to generate both present and absent keyphrases for a document. Following Copy-RNN, several seq2seq-based keyphrase generation methods have been proposed to improve the generation performance [1,6,8,34,36–38,41]. However, all these existing keyphrase generation methods have been proposed to compress the semantic information in a given document into a dense vector in Euclidean space, assuming a *flat* geometry. Although these Euclidean representation models have proved successful for the keyphrase generation task, they still suffer from an inherent limitation: their ability to capture *hierarchical structures* is bounded by the nature of flat geometry of Euclidean space, as mentioned in recent work [27].

As a given document covers different topics and consists of many phrases which could be keyphrases, it is critical to represent it into a *hierarchical seman-tic representation*, facilitating the selection of the most representative keyphrases related to the main topics at the highest level. Figure 1 shows the hierarchical relations among different semantic levels of candidate keyphrases, which can be regarded as the ideal keyphrase generation if viewing it from low-level (bound-ary) to high-level (center) candidates. In Fig. 1, the set of ideal keyphrases should be KP = {cp_1, cp_2, cp_3} at the highest level, covering three topics comprehen-sively. If the set of predicted keyphrases is KP'={cp_{21}, cp_{221}}, it just provides a *local and low-level* topic description of the second topic $Topic_2$ only, ignoring the other two topics and a part of the second topic. This example illustrates that without an effective hierarchical semantic representation, the predicted kephrases will not cover major topics and provide the high-level topic description. As men-tioned in several existing studies [6,16,21,38,41], predicted keyphrases may fall into a single topic and fail to cover all the main topics discussed in a document. In summary, semantic hierarchical relations widely exist among keyphrases, but existing keyphrase generation methods available in *Euclidean space* can not effec-tively capture *semantic hierarchical relations* to improve the topic coverage of predicted keyphrases.

Recently, hyperbolic representation methods [26,27] have been developed to model the latent hierarchical nature of data and demonstrated encourag-ing results. To efficiently utilize hyperbolic embeddings in downstream tasks, researchers have proposed some advanced hyperbolic deep networks, such as hyperbolic neural networks [12] and hyperbolic attention network [15].

Motivated by the above observations, we propose a *hyperbolic seq2seq net-work* for keyphrase generation, which is a novel keyphrase generation frame-work for modeling hierarchical relations. Specifically, we design a *hyperbolic encoder* to compress the hierarchical semantic information discussed in a tar-get document into a hyperbolic embedding, and devise a *hyperbolic decoder* to generate corresponding keyphrases. In the hyperbolic network, we propose an

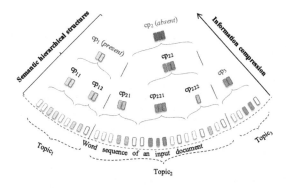

Fig. 1. Ideal semantic hierarchical relations among candidate keyphrases (cp) within a document, in which the dotted line semantically represents a topic segmentation and facilitates understanding of hierarchical structures of topics.

innovative *hyperbolic hierarchy-aware attention mechanism* to enhance the ability to learn semantic hierarchical relations, which can selectively focus on the words with high-level semantics. Different from Euclidean deep generation methods, our proposed hyperbolic hierarchy-aware attention mechanism make our model more effective to capture the semantic hierarchical relations within a target document and thus generate keyphrases based on its semantic understanding with good topic coverage and accuracy. In addition, we propose a new *metric* to measure the degree to which the predicted keyphrases cover the main topics of a target document. To the best of our knowledge, this is the first work to design a new *hyperbolic network* for keyphrase generation.

2   Related Work

2.1   Keyphrase Generation

Following CopyRNN [23], several extensions have been proposed to boost its generation ability. For instance, Ye et al. [36] propose a semi-supervised keyphrase generation model that utilizes both abundant unlabeled data and limited labeled data. Chen et al. [9] propose a title-guided network to sufficiently utilize the already summarized information in given title. In addition, some researches attempted to leverage external knowledge to help reducing duplication and improving coverage, such as syntactic constraints [41] and latent topics [34].

The above-mentioned methods, which utilize the standard seq2seq network, can not generate multiple keyphrases and determine the appropriate number of keyphrases at a time for a target document. To overcome this shortcoming, Yuan et al. [38] introduce the new training and inference setup in the seq2seq network to generate multiple keyphrases and decide on the suitable number of keyphrases for a given document. Ye et al., [37] propose a One2Set paradigm to predict the keyphrases as a set, which eliminates the bias caused by the predefined order in One2Seq paradigm [38]. In addition, some recent works focus on improving the decoding process of seq2seq networks. For example, Chen et al., [8] propose an exclusive hierarchical decoding framework and use either a soft or a hard exclusion mechanism to reduce duplicated keyphrases. More recently, Ahmad et al. [1] design an extractor-generator to jointly extract and generate keyphrases from a document. We observe that almost all existing keyphrase generation methods used the Euclidean seq2seq framework, which cannot provide the most powerful representations for hierarchical structures on keyphrase generation task.

2.2   Hyperbolic Representation

An increasing number of research has shown that many types of complex data exhibit non-Euclidean structures [3]. Recently, hyperbolic embedding methods have been proposed to learn the latent representation of hierarchical data and demonstrated encouraging results. In the field of NLP, hyperbolic representation learning has been successfully applied to generating word embeddings [31] and sentence representations [10], and inferring concept hierarchies from large text corpora [20]. In addition, hyperbolic geometry has been integrated into recent advanced hyperbolic deep learning frameworks, such as hyperbolic neural networks [12], and hyperbolic attention network [15].

3   Preliminaries

Hyperbolic Space. Hyperbolic space, specifically referring to a simply connected manifolds with constant negative curvature [2], can be thought of as a continuous analogue of tree and is more suitable for learning data with hierarchical structures. The hyperbolic space can be constructed using various isomorphic models (*i.e.*, these models can be converted into each other). In this paper, we follow the majority of NLP works and employ the Poincaré ball model with the curvature set as -1, whose distance function is differentiable.

Poincaré Ball Model. The n-dimensional *Poincaré ball model* $\mathcal{P}^n = (\mathcal{B}^n, g^{\mathcal{P}})$ is defined by a Riemannian manifold $\mathcal{B}^n = \{\mathbf{x} \in R^n \mid \|\mathbf{x}\| < 1\}$ with the metric tensor $g^{\mathcal{P}}(\mathbf{x}) = (\frac{2}{1-\|\mathbf{x}\|^2})^2 g^{\mathcal{E}}$, where $\|\cdot\|$ denotes the Euclidean norm, and $g^{\mathcal{E}} = \mathbf{I}_n$ is the Euclidean metric tensor. The *induced distance* between two points $\mathbf{x}, \mathbf{y} \in \mathcal{P}^n$ is defined as

$$d_{\mathcal{P}}(\mathbf{x}, \mathbf{y}) = \cosh^{-1}\left(1 + \frac{2\|\mathbf{x} - \mathbf{y}\|^2}{(1 - \|\mathbf{x}\|^2)(1 - \|\mathbf{y}\|^2)}\right), \tag{1}$$

where $\cosh^{-1}(x) = \ln(x + \sqrt{x^2 - 1})$ is an inverse hyperbolic cosine function.

The induced distance can place root node near the center of the ball and leaf nodes near the boundary of the ball to ensure that the distance from the root node to each of leaf nodes is relatively small while the distance between leaf nodes is relatively large. This explains why hyperbolic space can be seen as a tree-like hierarchical structure.

Klein Model. To define the *hyperbolic average*, we employ the Klein model of hyperbolic space. The n-dimensional *Klein model* $\mathcal{K}^n = (\mathcal{B}^n, g^{\mathcal{K}})$ is also defined in a manifold $\mathcal{B}^n$ with the different metric tensor $g^{\mathcal{K}}$. The Poincaré model and Klein model describe the same hyperbolic space using different coordinates. Thus, these two models can be converted into each other. Given a point $\mathbf{x}^{\mathcal{P}} \in \mathcal{P}^n$ in the Poincaré ball model, we convert it to the Klein model by $\mathbf{x}^{\mathcal{K}} = \frac{2\mathbf{x}^{\mathcal{P}}}{1+\|\mathbf{x}^{\mathcal{P}}\|^2}$. Similarly, a point $\mathbf{x}^{\mathcal{K}} \in \mathcal{K}^n$ in Klein model can be converted into Poincaré ball model as $\mathbf{x}^{\mathcal{P}} = \frac{\mathbf{x}^{\mathcal{K}}}{1+\sqrt{1-\|\mathbf{x}^{\mathcal{K}}\|^2}}$.

Hyperbolic Operations. To make neural networks work in hyperbolic space, Möbius operations including *Möbius addition* and *Möbius matrix-vector multiplication* in the Poincaré ball are used. In addition, the *exponential map* (which maps a Euclidean vector to the hyperbolic space) and the inverse *logarithm map* are also used. The details of these operations can be seen in the work [12].

4 Methodology

4.1 Problem Definition

Let $x = (x_1, \ldots, x_T)$ be a document that is treated as a sequence of words, where T is the length of x. The goal of a keyphrase generation method is to find a model to generate a set of keyphrases $K = \{p_k\}_{k=1}^{|K|}$ for document x, where each keyphrase $p_k = (y_1, ..., y_{|p_k|})$ is also a sequence of words.

To generate multiple keyphrases for an input document, existing approaches provide two different data formats as the predicted keyphrase output (*i.e.*, two training paradigms): *One2One* [23] and *One2Seq* [38]. One2One only predicts a fixed number of keyphrases for all documents, where each training data sample is a pair of source text and one of its keyphrases (x, p). To overcome this drawback, One2Seq can generate a single sequence, which consists of multiple predicted keyphrases and separators, as represented by $K' = p_1 <\text{sep}> p_2 ... <\text{sep}> p_{|K|}$. Each training data sample is a pair of source text and concatenated sequence of its keyphrases and separators (x, K').

4.2 Hyperbolic Encoder-Decoder Model

The basic idea of our keyphrase generation model is to leverage a *hyperbolic deep network* to compress the semantic information of the input document into the low-dimensional representations using the hyperbolic encoder and to generate corresponding keyphrases using the hyperbolic decoder, based on the representations. In this hyperbolic network, we propose a new *hyperbolic attention* mechanism to capture the semantic subordination and select the words with high-level semantics. In addition, a hyperbolic pointer mechanism is used to copy certain out-of-vocabulary words from the input document and paste them into the generated keyphrases. An overview of this method is shown in Fig. 2.

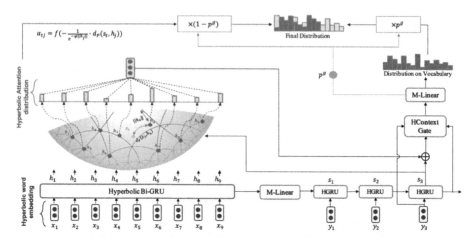

Fig. 2. The overview of the proposed hyperbolic deep model for keyphrase generation.

The encoder and decoder are implemented with a hyperbolic gated recurrent unit (HGRU) [12]. Let $x = (x_1, ..., x_T)$ be a sequence of words within an input document, and $\mathbf{x} = [\mathbf{x}_1, ..., \mathbf{x}_T]$ be its corresponding sequence of hyperbolic word embeddings. The encoder maps the input word sequence x into a set of contextualized hidden representations $\mathbf{h} = [\mathbf{h}_1, ..., \mathbf{h}_T]$, using a bidirectional HGRU $\mathbf{h}_t = [\mathrm{HGRU_f}(\mathbf{x_t}), \mathrm{HGRU_b}(\mathbf{x_t})]$ where $\mathrm{HGRU_f}(\cdot)$ and $\mathrm{HGRU_b}(\cdot)$ are used to learn the forward and backward hidden states around the input text, respectively. HGRU based on Möbius operations in Poincaré model [12] is defined as

$$
\begin{aligned}
\mathbf{r}_t &= \sigma(\log_0(\mathbf{W}^r \otimes \mathbf{h}_{t-1} \oplus \mathbf{U}^r \otimes \mathbf{x}_t \oplus \mathbf{b}^r)) \\
\mathbf{z}_t &= \sigma(\log_0(\mathbf{W}^z \otimes \mathbf{h}_{t-1} \oplus \mathbf{U}^z \otimes \mathbf{x}_t \oplus \mathbf{b}^z)) \\
\tilde{\mathbf{h}}_t &= \varphi((\mathbf{W}^h \mathrm{diag}(\mathbf{r}_t)) \otimes \mathbf{h}_{t-1} \oplus \mathbf{U}^h \otimes \mathbf{x}_t \oplus \mathbf{b}^h) \\
\mathbf{h}_t &= \mathbf{h}_{t-1} \oplus \mathrm{diag}(\mathbf{z}_t) \otimes (-\mathbf{h}_{t-1} \oplus \tilde{\mathbf{h}}_t)
\end{aligned}
\tag{2}
$$

where $\mathbf{r}_t$ is a reset gate, and $\mathbf{z}_t$ is a update gate. $\log_0(\cdot)$, $\otimes$ and $\oplus$ are defined in Subsect. 3. $\sigma(\cdot)$ is a sigmoid function, and $\varphi(\cdot)$ is a pointwise non-linearity. Since the hyperbolic space naturally has non-linearity, $\varphi(\cdot)$ is identity. $\mathrm{diag}(\cdot)$ is a square diagonal matrix. The six weights $\mathbf{W} \in R^{n \times n}$, $\mathbf{U} \in R^{n \times m}$ are trainable parameters in Euclidean space and three biases $\mathbf{b} \in \mathcal{B}^n$ are trainable parameters in hyperbolic space.

The decoder is another forward HGRU which is used to generate the sequence of keyphrases by predicting the next word y_t based on the hidden state $\mathbf{s}_t$. Both y_t and $\mathbf{s}_t$ are conditioned on $\mathbf{y}_{t-1}$ and $\mathbf{c}_t$ of the input sequence. Formally, the hidden state $\mathbf{s}_t$ and decoding function can be written as

$$
\mathbf{s}_t = \mathrm{HGRU_f}(\mathbf{y}_{t-1}, \mathbf{s}_{t-1}, \mathbf{c}_t),
\tag{3}
$$

and

$$
p(y_t \mid y_1, y_2, ..., y_{t-1}, \mathbf{c}) = g(\mathbf{y}_{t-1}, \mathbf{s}_t, \mathbf{c}_t),
\tag{4}
$$

where $g(\cdot)$ is a nonlinear multi-layered function that outputs the probability of y_t. The more details of the decoder are given in the next subsections.

4.3   Hyperbolic Attention Mechanism

The attention mechanism is used to make the network model dynamically focus on the important parts in input data, and consists of two core parts: matching and aggregation. Particularly, the matching part computes attention weight $\alpha_{tj} = \alpha(\mathbf{s}_t, \mathbf{h}_j)$, which reflects the relevance of the hidden states $\mathbf{h}_j$ of input sequence in the presence of the current hidden state $\mathbf{s}_t$ for deciding the next word y_t. The aggregation part, on the other hand, takes a weighted sum of hidden states using these weights, also known as context vector $\mathbf{c}_t$.

A general hyperbolic attention mechanism was first introduced by Gulcehre et al. [15] to build an attentive read operation in the Hyperboloid model. Inspired by this work, we propose a new hyperbolic attention mechanism in the Poincaré ball model *specifically* for the keyphrase generation task. In particular, in the matching part, the most natural way to compute attention weight is to use the hyperbolic distance between points of matching pairs, given as $\alpha_{tj} = \frac{\exp(a(\mathbf{s}_t, \mathbf{h}_j))}{\sum_{k=1}^{T} \exp(a(\mathbf{s}_t, \mathbf{h}_k))}$, where $a(\mathbf{s}_t, \mathbf{h}_j)$ is a soft alignment function that is used to score how well the inputs around position j and the output at position t match (*i.e.*, to measure the relevance between $\mathbf{s}_t$ and $\mathbf{h}_j$), computed as

$$a(\mathbf{s}_t, \mathbf{h}_j) = -\beta d_{\mathcal{P}}(\mathbf{s}_t, \mathbf{h}_j) - d_b, \tag{5}$$

where $d_{\mathcal{P}}(\cdot, \cdot)$ is the distance function in hyperbolic space, and d_b is a parameter learned along with the rest of the network. Note that in the work [15], β is also a learnable coefficient. This causes the attention mechanism to only utilize the hyperbolic distance to measure the relevance between $\mathbf{s}_t$ and $\mathbf{h}_j$, and *ignore* the distance between the center of the Poincaré ball to $\mathbf{h}_j$ (*i.e.*, the norm of $\mathbf{h}_j$), which can reflect the semantic level of an input word at position j (*i.e.*, hidden state $\mathbf{h}_j$) in a tree-like hierarchical structure internalized by the hyperbolic space.

To overcome this drawback and further enhance the ability to capture the high-level semantics, we redefine β as

$$\beta = \frac{1}{\exp(-\varphi \|\mathbf{h}_j\|)}, \tag{6}$$

where φ is a hyper parameter. Thus, the new hyperbolic attention mechanism takes into account not only the semantic relevance between two words but also the *semantic hierarchy* of each word in the semantic tree (as Fig. 1 shows). We name it as the hierarchy-aware attention mechanism.

In the aggregation part, the weighted sum of hidden states is computed by the Einstein midpoint that is defined in Klein model as $\mathbf{c}_t = \sum_{j=1}^{T} \left[\frac{\alpha_{tj} \gamma(\mathbf{h}_j)}{\sum_{l=1}^{T} \alpha_{tl} \gamma(\mathbf{h}_l)} \right] \mathbf{h}_j$, where $\gamma(\mathbf{h}_j) = \frac{1}{\sqrt{1 - \|\mathbf{h}_j\|^2}}$ is a Lorentz factor.

Note that before aggregation process, we first transform the hidden states from Poincaré Ball to Klein model, and transform it back to Poincaré ball model after aggregation. The used formulas are given in subsection Klein model.

4.4   Hyperbolic Pointing Mechanism

To recall some keyphrases which contain out-of-vocabulary words, CopyRNN utilized the copying mechanism [14] to generate out-of-vocabulary words. Here, we use the pointing mechanism (that is a modified copy mechanism) into the Poincaré ball model for the same purpose.

Let $\mathcal{V}$ be a global vocabulary, $\mathcal{V}_s$ be a vocabulary of the source sentences, and unk be any out-of-vocabulary word. It builds an extended vocabulary $\mathcal{V}_e = \mathcal{V} \cup \mathcal{V}_s \cup \{unk\}$. The distribution over $\mathcal{V}_e$ at current time step t is

$$p(y_t) = p_t^g \cdot p_g(y_t) + (1 - p_t^g) \cdot p_c(y_t), \tag{7}$$

where p_t^g is the probability of choosing generate-mode, calculated by

$$p_t^g = \sigma(\log_0(\mathbf{W}^{cg} \otimes \mathbf{d}_t \oplus \mathbf{b}^{cg})). \tag{8}$$

The probability of generate-mode $p_g(\cdot)$ and copy-mode $p_c(\cdot)$ are given by

$$p_g(y_t) = \begin{cases} \mathbf{v}_i^\top \log_0(\mathbf{W}^g \otimes \mathbf{d}_t \oplus \mathbf{b}^g), & y_t \in \mathcal{V} \cup \{unk\}, \\ 0, & \text{otherwise.} \end{cases} \tag{9}$$

$$p_c(y_t) = \begin{cases} \sum_{j:x_j=y_t} \alpha_{tj}, & y_t \in \mathcal{V}_s, \\ 0, & \text{otherwise.} \end{cases} \tag{10}$$

where $\mathbf{v}_i$ is a one-hot indicator vector, $\mathbf{W}$ and $\mathbf{b}$ in Eq. (8) and Eq. (9) are trainable parameters. Finally, we adopt the widely used cross entropy loss function to train the models, both in One2One and One2Seq paradigms.

5   Experiments

5.1   Dataset

We employ KP20k dataset [23], where each example contains a title and an abstract of a scientific paper as source text, and author-assigned keywords as target keyphrases. Following previous works, we use the training dataset of KP20k to train all the models, and use the validation dataset to validate the choice of hyper parameters. In order to evaluate the proposed model comprehensively, we also test on other four widely used public datasets from the scientific domain, namely, Inspec [17], Krapivin [19], SemEval-2010 [18] and NUS [25]. The detailed statistic information of these five datasets are summarized in the work [40].

5.2   Baselines

For the *present* keyphrase prediction, we compare our models with two types of methods, including eight extraction and seven deep generation methods.

Representative *extraction* methods consist of three different types: 1) *statistic-based* unsupervised methods, including (1) **TF-IDF** and (2) **YAKE!** [4], 2) *graph-based* unsupervised methods, including (3) **TextRank** [24], (4) **SingleRank** [32], (5) **PositionRank** [11], (6) **KPRank** [28], and 3) traditional supervised methods, including (7) **KEA** [35] and (8) **Maui** [22]. Due to the limited space, we select the *best-performing method* (BL*) from *each of the three types of baselines* with the best-performing metrics to compare with our method.

The supervised *generation* baselines can be classified into *One2One* and *One2Seq* according to the training paradigm. The One2One baselines include: (1) **CopyRNN** [23], which is the first to use seq2seq network to generate keyphrases, (2) **CorrRNN** [6], which is an extension of CopyRNN integrating the sequential decoding with coverage and review mechanisms, and (3) **KG-KE-KR-M** (abbreviated as **KG-KE**) [7], which is a multi-task learning using extraction and generation models to generate keyphrases.

The One2Seq baselines include: (1) **CatSeq** [38], which has the same framework as CopyRNN, with the key difference between them on the training paradigm, (2) **CatSeqTG-2RF1** (abbreviated as **Cat-2RF1**) [5], which is an extension of CatSeq using reinforcement learning to generate both sufficient and accurate keyphrases, (3) **ExHiRD-h** [8], which uses an exclusive hierarchical decoder to avoid generating duplicated keyphrases, and (4) **SEG-Net** [1], which jointly extracts and generates keyphrases.

The proposed **Hy**perbolic **A**ttentional **N**etwork (HyAN[1]) and its extensions are: (1) HyAN, which is a basic hyperbolic attentional model trained by One2One paradigm, corresponding to CopyRNN, (2) $HyAN_h$, which is an extension of HyAN, in which only the semantic hierarchy is integrated into the hyperbolic attention mechanism, (3) HyANS, which is also an extension of HyAN trained by One2Seq paradigm, corresponding to CatSeq, and (4) $HyANS_h$, which is a composite of HyANS and $HyAN_h$, trained by One2Seq paradigm and incorporated with the semantic hierarchy.

5.3   Evaluation Metrics

We adopt top-N macro-averaged *F-measure* (F_1) and $R@k$ as the evaluation metrics, in which F_1 includes $F_1@k$, $F_1@\mathcal{O}$ and $F_1@\mathcal{M}$. $F_1@k$ is used in almost all existing works, while $F_1@\mathcal{O}$ and $F_1@\mathcal{M}$ proposed in [38] are designed specifically for the One2Seq generation, where $\mathcal{O}$ is the number of author-provided keyphrases and $\mathcal{M}$ is the number of all predicted keyphrases. They are capable of reflecting the nature of variable number of keyphrases for each document. The recall of the top 50 predictions ($R@50$) evaluates prediction of absent keyphrases.

5.4   Implementation Details

We follow the previous works [23,38] to pre-process the experimental data. The top 50,000 most frequently-occurring words in the training data are used as the vocabulary shared in the hyperbolic encoder and decoder.

---

[1] The code of our model is available at https://github.com/SkyFishMoon/HyAN.

Table 1. Results of predicting *present* keyphrases of different methods on five datasets. Best/second-best performing score in each column is highlighted with bold/underline in each of two trained paradigms, and best performing score in both trained paradigms is highlighted with bold and asterisk. CopyRNN$^+$ is re-implemented CopyRNN with best results [38]. Sta-, Gra- and Tra- represent statistic-unsupervised, graph-unsupervised and traditional supervised, respectively.

	Method	Inspec		Krapivin		NUS		SemEval		KP20k	
		F_1@5	F_1@10	F_1@5	F_1@10	F_1@5	F_1@10	F_1@5	F_1@10	F_1@5	F_1@10
Ext.	Sta-BL*	20.4	24.4	21.5	19.6	15.9	19.6	15.1	21.2	14.1	14.6
	Gra-BL*	27.7	32.3	17.7	18.5	21.0	22.3	22.5	25.7	18.1	15.0
	Tra-BL*	10.9	12.9	24.3	20.8	24.9	26.1	4.50	3.90	26.5	22.7
One2One	CopyRNN$^+$	24.4	28.9	30.5	26.6	37.6	35.2	31.8	31.8	31.7	27.3
	CorrRNN	–	–	31.8	27.8	35.8	33.0	32.0	32.0	–	–
	KG-KE	25.7	28.4	27.2	25.0	28.9	28.6	20.2	22.3	31.7	28.2
	HyAN	27.9	29.8	32.2	27.9	38.1	34.7	32.8	32.3	32.9	28.5
	HyAN$_h$	28.8	30.2	33.0	28.9	38.8	36.2	33.3*	32.5	34.0	29.3
		F_1@5	F_1@$\mathcal{M}$	F_1@5	F_1@$\mathcal{M}$	F_1@5	F_1@$\mathcal{M}$	F_1@5	F_1@$\mathcal{M}$	F_1@5	F_1@$\mathcal{M}$
One2Seq	CatSeq	22.5	26.2	26.2	35.4	32.3	39.7	24.2	28.3	29.1	36.7
	Cat-2RF1	25.3	30.1	30.0	36.9	37.5	43.3	28.7	32.9	32.1	38.6
	ExHiRD-h	25.3	29.1	28.6	34.7	–	–	28.4	33.5	31.1	37.4
	SEG-Net	21.6	26.5	27.6	36.6	39.6	46.1	28.3	33.2	31.1	37.9
	HyANS	30.0	33.0	33.9	36.1	40.2	46.5	33.0	34.7	33.9	38.9
	HyANS$_h$	30.8*	34.3	34.6*	36.9	40.7*	47.2	33.2	35.5	34.5*	39.5

The size of hyperbolic word embedding is set as $m = 100$ and the size of hyperbolic hidden state is set as $n = 150$. The word embeddings are initialized first using normal distribution by the method [13], where the gain weight is set as $\sqrt{2}$. Then the embedding is projected into the Poincaré ball by $\exp_0(\cdot)$. In addition, d_b and φ are set as 1.0 and 230 in formula (5) and (6), respectively.

In the training process, we set the batch size as 32. The initial learning rate is set as 0.0008. Early stopping is used when training. In the testing process, our models trained by One2One paradigm use the beam search with a width of 120 and a max depth of 6. Finally our models trained by One2Seq paradigm employ a beam width of 40 and a max depth of 40.

5.5   Results and Analysis

Present Keyphrase Prediction. The results of predicting *present* keyphrases are shown in Table 1. The results show that the generation methods substantially outperform the traditional extraction methods across all the datasets. Among the generation methods, the One2Seq methods can generally achieve better performance than other One2One methods. This improvement may be driven by the inter-relation among keyphrases of each document, which can be effectively captured by the deep models trained by the One2Seq paradigm. In all methods, HyANS$_h$ achieves the best results in term of all metrics on all datasets.

Table 2. Results of predicting *absent* keyphrases of different methods on five datasets.

	Method	Inspec		Krapivin		NUS		SemEval		KP20k	
		F_1@5	R@50	F_1@5	R@50	F_1@5	R@50	F_1@5	R@50	F_1@5	R@50
O2O	CopyRNN$^+$	0.1	8.3	0.9	8.1	1.1	8.1	1.0	2.6	0.8	8.7
	HyAN	**0.3**	8.5	1.1	8.5	1.3	8.5	1.2	2.8	1.2	8.9
	HyAN$_h$	**0.3**	**8.6**	**1.4**	**9.0**	**1.5**	**8.7**	**1.6**	**3.1**	**1.3**	**9.1**
		F_1@5	F_1@$\mathcal{M}$	F_1@5	F_1@$\mathcal{M}$	F_1@5	F_1@$\mathcal{M}$	F_1@5	F_1@$\mathcal{M}$	F_1@5	F_1@$\mathcal{M}$
One2Seq	CatSeq	0.4	0.8	1.8	3.6	1.6	2.8	1.6	2.8	1.5	3.2
	Cat-2RF1	1.2	2.1	3.0	5.3	1.9	3.1	2.1	3.0	2.7	5.0
	ExHiRD-h	1.1	2.2	2.2	4.3	-	-	1.7	2.5	1.6	3.2
	SEG-Net	1.5	0.9	3.6	1.8	3.6	2.1	**3.0***	2.1	3.6	1.8
	HyANS	1.8	2.7	3.9	6.1	4.0	4.8	2.5	2.9	3.9	4.2
	HyANS$_h$	**2.3***	**3.1**	**4.3***	**7.1**	**5.2***	**5.1**	**3.0***	**3.3**	**4.6***	**5.3**

In all the deep models whether trained by the One2One or One2Seq paradigm, the proposed hyperbolic models outperform the corresponding Euclidean baselines across all the datasets. It should be noted that HyAN can be regarded as CopyRNN and HyANS as CatSeq in hyperbolic space, and they do not use any side information or multi-task learning to achieve better performance like almost all extensions of CoypRNN in Euclidean space, so it is only fair to compare HyAN with CopyRNN, and HyANS with CatSeq. The results show that HyAN and HyANS outperform CopyRNN and CatSeq on all datasets, respectively. This demonstrates the superiority of the hyperbolic methods in modeling hierarchical structures for keyphrase prediction.

Absent Keyphrase Prediction. Unlike present keyphrases, absent keyphrases do not appear in the target document, and thus predicting them is very challenging and requires understanding the latent document semantic. The results are presented in Table 2 (where O2O represents One2One paradigm), where recall R@50 is more suitable for evaluating the performance of One2One methods in absent keyphrase prediction (more detailed descriptions are shown in the work [23]). The results indicate no matter which type of model is trained by One2One or One2Seq paradigm, the proposed hyperbolic models can predict absent keyphrases more accurately than the corresponding Euclidean baselines.

Variable-Number Keyphrase Generation. The One2Seq methods can predict a varying number of keyphrases conditioned on the given document, which is one key advantage of this type of method. We conduct experiments on the KP20k dataset to compare the performance of models for generating a varying number of keyphrases in term of both F_1@$\mathcal{O}$ and F_1@$\mathcal{M}$. The results are presented in Table 3. As the results show, HyANS and HyANS$_h$ substantially outperform CatSeq, and HyANS$_h$ achieves the best results in terms of two performance metrics. This indicates that our proposed hierarchy-aware attention mechanism

Table 3. Results of the variable-number keyphrase generation on kp20k dataset.

KP20k	$F_1@\mathcal{O}$	$F_1@\mathcal{M}$
CatSeq	24.3	25.1
HyANS	31.0	32.5
HyANS$_h$	**31.5**	**32.8**

used in HyANS$_h$ is more effective than the primitive hyperbolic attention mechanism [15] used in HyANS.

5.6   Coverage Evaluation of Predicted Keyphrases

As mentioned in the works [21,39], the predicted keyphrases should cover all the main topics discussed in the target document. However, it is challenging to evaluate the degree to which the predicted keyphrases cover the main topics of a target document. To this end, we try to find the ground-truth (*i.e.*, author-provided) keyphrases that are not covered semantically by the predicted keyphrases and use the number of them to measure the *semantic coverage* of predicted keyphrases for a target document.

Specifically, let $G = \{g_i\}_{i=1}^n$ be a set of ground-truth keyphrases of a target document, and $K = \{p_j\}_{j=1}^m$ be its corresponding set of predicted keyphrases. The number of un-covered ground-truth keyphrases (*uck*) is defined as

$$uck = \sum_{i=1}^{n} \mathbf{1}\Big(\sum_{j=1}^{m} \mathbf{1}\big(s_{ij} = \max_{k=1:n}\{s_{kj}\}\big) = 0 \Big), \tag{11}$$

where s_{ij} is the cosine similarity between embeddings of ground-truth keyphrase g_i and predicted keyphrase p_j, produced by the pre-trained BERT[2]. The indicator function $\mathbf{1}(\cdot)$ outputs 1 if the expression evaluates to true and outputs 0 otherwise. This formula is used to count the number of ground-truth keyphrases, each of which has lower similarities with all the predicted keyphrases. A smaller the value of *uck* suggests a better predictor.

As the results shown in Table 4, our hyperbolic deep models indeed outperform the other two Euclidean models, and HyANS$_h$ gets the best result. The results indicate the predicted keyphrases generated by our hyperbolic models can better cover the topics discussed in a target document and reduce duplicated keyphrases generation.

Table 4. Average number of uncovered author-provided keyphrases (*i.e.*, average *uck*) of different methods on KP20k dataset.

CopyRNN$^+$	HyAN	HyAN$_h$	CatSeq	Cat-2RF1	ExHiRD-h	HyANS	HyANS$_h$
1.7784	1.7602	**1.7385**	1.7729	1.7653	1.7542	1.7334	**1.7328**

---

[2] https://github.com/duanzhihua/pytorch-pretrained-BERT.

Table 5. Two examples of generated keyphrases by different methods with the One2Seq training paradigm. Author-assigned (*i.e.*, Gold) keyphrases are shown in bold, and absent keyphrases are labeled by *.

Example 1

Title: Active Learning for Software Defect Prediction (#4445 in KP20k)

Abstract: An **active learning** method, called Two-stage Active learning algorithm (TAL), is developed for software **defect prediction**. Combining the clustering and **support vector machine** techniques, this method improves the performance of the predictor with less labeling effort. Experiments validate its effectiveness

Gold: machine learning*; defect prediction; active learning; support vector machine

HyANS$_h$: **machine learning***; **active learning**; **support vector machine**; support vector machines

Catseq: **active learning**; software defect prediction; clustering; **support vector machine**; software defect prediction

Cat-2RF1: **active learning**; software defect prediction; clustering; **support vector machine**; software metrics

ExHiRD-h: **active learning**; software defect prediction; clustering; **support vector machine**

Example 2

Title: Experience with performance testing of software systems issues, an approach, and case study (#4086 in KP20k)

Gold: performance testing; software performance testing; program testing*; software testing*

HyANS$_h$: **performance testing**; **software performance testing**; **software testing***

Catseq: **performance testing**; **software performance testing**; **software testing***

Cat-2RF1: **performance testing**; software systems; case study; **software testing***

ExHiRD-h: **performance testing**; software systems; case study; **software testing***; performance evaluation

---

5.7 Case Study and Visualization

Here, we select two anecdotal examples of research papers shown in Table 5. The predictions generated by different methods along with human-picked "gold" keyphrases are listed in this table. The first paper (*i.e.*, Example 1) presented an active learning method for software defect prediction and assigned "machine learning" as a *absent* keyphrase, which appears in the first position of the author-assigned keyphrase sequence. Obviously, the keyphrase "machine learning" can be regarded as the root topic description of various machine learning methods, such as active learning discussed in this example. As can be seen from Table 5,

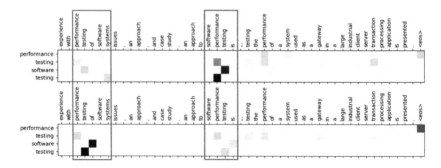

Fig. 3. Attention visualization of hyperbolic $HyANS_h$ (top) and Euclidean Catseq (bottom) on the second example. Deeper shading denotes higher value.

our hyperbolic $HyANS_h$ is capable of understanding the underlying semantic hierarchical structures in this document, and thus can accurately generate this absent root keyphrase while all the baselines in Euclidean space fail to generate it. This example further indicates that hyperbolic space may help to gain better performance in keyphrase generation.

The second paper (*i.e.*, Example 2) proposed an approach to software performance testing. Comparing with the baseline methods, $HyANS_h$ and Catseq achieve best performance and generate the same keyphrases on this example. Figure 3 visualizes the proposed hyperbolic hierarchy-aware attention in $HyANS_h$ and the Euclidean attention in Catseq to further clarify how our model works. Due to space limitation, we only visualize the first present keyphrase "performance testing" and the absent keyphrase "software testing" in the author-assigned keyphrase sequence, and they are already enough to support our analysis. Although these two keyphreses are correctly generated by both $HyANS_h$ and Catseq, from the results shown in Fig. 3, we can clearly see that $HyANS_h$ pays more attention to relevant content words such as "performance" and "testing" while Catseq, to a certain extent, focuses on some irrelevant or functional words such as "is" and "of". This example indicates that compared with the Euclidean space, the hyperbolic space is very helpful for generating keyphrases.

6 Conclusion

In this study, we presented a new solution that aims to predict keyphrases using *hyperbolic* encoder-decoder framework, which can effectively capture the underlying semantic hierarchical structures discussed in a target document. To the best of our knowledge, this is the first study to explore a hyperbolic deep network for keyphrase generation. In addition, we propose a novel hierarchy-aware attention mechanism to further enhance the ability to capture the semantic hierarchical information, and a new metric to measure the degree to which the predicted keyphrases cover the main topics of a target document. Comprehensive experimental results show the proposed hyperbolic models outperform the state-of-the-art Euclidean models across all five datasets. In future, we plan to evaluate the proposed hyperbolic seq2seq model on a large corpus with comprehensive coverage of diverse topics.

Acknowledgements. This work was partially supported by grants from the Scientific Research Project of Tianjin Educational Committee (Grant No. 2021ZD002).

References

1. Ahmad, W.U., Bai, X., Lee, S., Chang, K.W.: Select, extract and generate: neural keyphrase generation with layer-wise coverage attention. In: Proceedings of ACL (2021)
2. Birman, G.S., Ungar, A.A.: The hyperbolic derivative in the Poincaré ball model of hyperbolic geometry. J. Math. Anal. Appl. **254**(1), 321–333 (2001)
3. Bronstein, M.M., Bruna, J., LeCun, Y., Szlam, A., Vandergheynst, P.: Geometric deep learning: going beyond Euclidean data. IEEE Signal Process. Mag. **34**(4), 18–42 (2017)
4. Campos, R., Mangaravite, V., Pasquali, A., Jorge, A., Nunes, C., Jatowt, A.: Yake! keyword extraction from single documents using multiple local features. Inf. Sci. **509**, 257–289 (2020)
5. Chan, H.P., Chen, W., Wang, L., King, I.: Neural keyphrase generation via reinforcement learning with adaptive rewards. In: Proceedings of ACL (2019)
6. Chen, J., Zhang, X., Wu, Y., Yan, Z., Li, Z.: Keyphrase generation with correlation constraints. In: Proceedings of EMNLP (2018)
7. Chen, W., Chan, H.P., Li, P., Bing, L., King, I.: An integrated approach for keyphrase generation via exploring the power of retrieval and extraction. In: Proceedings of NAACL (2019)
8. Chen, W., Chan, H.P., Li, P., King, I.: Exclusive hierarchical decoding for deep keyphrase generation. In: Proceedings of ACL (2020)
9. Chen, W., Gao, Y., Zhang, J., King, I., Lyu, M.R.: Title-guided encoding for keyphrase generation. In: Proceedings of AAAI (2019)
10. Dhingra, B., Shallue, C.J., Norouzi, M., Dai, A.M., Dahl, G.E.: Embedding text in hyperbolic spaces. In: Proceedings of Twelfth Workshop on TextGraphs (2018)
11. Florescu, C., Caragea, C.: Positionrank: an unsupervised approach to keyphrase extraction from scholarly documents. In: Proceedings of ACL (2017)
12. Ganea, O., Bécigneul, G., Hofmann, T.: Hyperbolic neural networks. In: Proceedings of NIPS (2018)
13. Glorot, X., Bengio, Y.: Understanding the difficulty of training deep feedforward neural networks. In: Proceedings of AISTATS (2010)
14. Gu, J., Lu, Z., Li, H., Li, V.O.: Incorporating copying mechanism in sequence-to-sequence learning. In: Proceedings of ACL (2016)
15. Gulcehre, C., et al.: Hyperbolic attention networks. In: Proceedings of ICLR (2019)
16. Hasan, K.S., Ng, V.: Automatic keyphrase extraction: a survey of the state of the art. In: Proceedings of ACL (2014)
17. Hulth, A., Megyesi, B.B.: A study on automatically extracted keywords in text categorization. In: Proceedings of ACL (2006)
18. Kim, S.N., Medelyan, O., Kan, M.Y., Baldwin, T.: Semeval-2010 task: automatic keyphrase extraction from scientific articles. In: Proceedings of Workshop on SemEval (2010)

19. Krapivin, M., Autaeu, A., Marchese, M.: Large dataset for keyphrases extraction. Technical report, University of Trento (2009)
20. Le, M., Roller, S., Papaxanthos, L., Kiela, D., Nickel, M.: Inferring concept hierarchies from text corpora via hyperbolic embeddings. In: Proceedings of ACL (2019)
21. Liu, Z., Huang, W., Zheng, Y., Sun, M.: Automatic keyphrase extraction via topic decomposition. In: Proceedings of EMNLP (2010)
22. Medelyan, O., Frank, E., Witten, I.H.: Human-competitive tagging using automatic keyphrase extraction. In: Proceedings of EMNLP (2009)
23. Meng, R., Zhao, S., Han, S., He, D., Brusilovsky, P., Chi, Y.: Deep keyphrase generation. In: Proceedings of ACL (2017)
24. Mihalcea, R., Tarau, P.: TextRank: bringing order into text. In: Proceedings of EMNLP (2004)
25. Nguyen, T.D., Kan, M.Y.: Keyphrase extraction in scientific publications. In: Proceedings of ICADL (2007)
26. Nickel, M., Kiela, D.: Learning continuous hierarchies in the Lorentz model of hyperbolic geometry. In: Proceedings of ICML (2018)
27. Nickel, M., Kiela, D.: Poincaré embeddings for learning hierarchical representations. In: Proceedings of NIPS (2017)
28. Patel, K., Caragea, C.: Exploiting position and contextual word embeddings for keyphrase extraction from scientific papers. In: Proceedings of ECACL (2021)
29. Sutskever, I., Vinyals, O., Le, Q.V.: Sequence to sequence learning with neural networks. In: Proceedings of NIPS (2014)
30. Tang, Y., Huang, W., Liu, Q., Zhang, B.: Qalink: enriching text documents with relevant Q&A site contents. In: Proceedings of CIKM (2017)
31. Tifrea, A., Bécigneul, G., Ganea, O.E.: Poincaré glove: hyperbolic word embeddings. In: Proceedings of ICLR (2019)
32. Wan, X., Xiao, J.: Single document keyphrase extraction using neighborhood knowledge. In: Proceedings of AAAI (2008)
33. Wang, L., Cardie, C.: Domain-independent abstract generation for focused meeting summarization. In: Proceedings of ACL (2013)
34. Wang, Y., Li, J., Chan, H.P., King, I., Lyu, M.R., Shi, S.: Topic-aware neural keyphrase generation for social media language. In: Proceedings of ACL (2019)
35. Witten, I.H., Paynter, G.W., Frank, E., Gutwin, C., Nevillmanning, C.G.: Kea: practical automatic keyphrase extraction. In: Proceedings of JCDL (1999)
36. Ye, H., Wang, L.: Semi-supervised learning for neural keyphrase generation. In: Proceedings of EMNLP. Proceedings of ACL (2018)
37. Ye, J., Gui, T., Luo, Y., Xu, Y., Zhang, Q.: One2set: generating diverse keyphrases as a set. In: Proceedings of ACL (2021)
38. Yuan, X., Wang, T., Meng, R., Thaker, K., Brusilovsky, P., He, D.: One size does not fit all: Generating and evaluating variable number of keyphrases. In: Proceedings of ACL (2020)
39. Zhang, Y., Chang, Y., Liu, X., Gollapalli, S.D., Li, X., Xiao, C.: Mike: keyphrase extraction by integrating multidimensional information. In: Proceedings of CIKM (2017)
40. Zhang, Y., Jiang, T., Yang, T., Li, X., Wang, S.: HTKG: deep keyphrase generation with neural hierarchical topic guidance. In: Proceedings of SIGIR (2022)
41. Zhao, J., Zhang, Y.: Incorporating linguistic constraints into keyphrase generation. In: Proceedings of ACL (2019)

On the Current State of Reproducibility and Reporting of Uncertainty for Aspect-Based Sentiment Analysis

Elisabeth Lebmeier, Matthias Aßenmacher$^{(\boxtimes)}$ ⓘ, and Christian Heumann ⓘ

Department of Statistics (LMU), Ludwigstr. 33, 80539 Munich, Germany
e.lebmeier@gmx.de, {matthias,chris}@stat.uni-muenchen.de

Abstract. For the latter part of the past decade, Aspect-Based Sentiment Analysis has been a field of great interest within Natural Language Processing. Supported by the Semantic Evaluation Conferences in 2014–2016, a variety of methods has been developed competing in improving performances on benchmark data sets. Exploiting the transformer architecture behind BERT, results improved rapidly and efforts in this direction still continue today. Our contribution to this body of research is a holistic comparison of six different architectures which achieved (near) state-of-the-art results at some point in time. We utilize a broad spectrum of five publicly available benchmark data sets and introduce a fixed setting with respect to the pre-processing, the train/validation splits, the performance measures and the quantification of uncertainty. Overall, our findings are two-fold: First, we find that the results reported in the scientific articles are hardly reproducible, since in our experiments the observed performance most of the time fell short of the reported one. Second, the results are burdened with notable uncertainty, depending on the data splits, which is why a reporting of uncertainty measures is crucial.

Keywords: Natural Language Processing · Sentiment analysis · Pre-trained language models · Reproducibility

1 Introduction

The field of Natural Language Processing (NLP) has profited a lot from technical and algorithmic improvements within the last years. Before the successful times of machine learning and deep learning, NLP was mainly based on what linguists knew about how languages work, i.e. grammar and syntax. Thus, primarily rule-based approaches were employed in the past. Nowadays, far more generalized models based on neural networks are able to learn the desired language features.

The original version of this chapter was previously published non-open access. A Correction to this chapter is available at https://doi.org/10.1007/978-3-031-26390-3_44

Supplementary Information The online version contains supplementary material available at https://doi.org/10.1007/978-3-031-26390-3_31.

M.-R. Amini et al. (Eds.): ECML PKDD 2022, LNAI 13714, pp. 537–552, 2023.
https://doi.org/10.1007/978-3-031-26390-3_31

On the other hand, data in written form is available in huge amounts and thus might be an important source for valuable information. For instance, the internet is full of comparison portals, forums, blogs and social media posts where people state their opinions on a broad range of products, companies and other people. Product developers, politicians or other persons in charge could profit from this information and improve their products, decisions and behavior.

We specifically focus on *Aspect-Based Sentiment Analysis (ABSA)* in our work. ABSA is often used as a generic umbrella term for several unique tasks, which is caused by the inconsistency of terms in literature where many different names are widely used. To be as precise as possible, we explicitly use different terms than ABSA to refer to the exact tasks. The first one (subtask 2 [14]) assumes that in each text, aspect terms are already marked and thus given exactly as written in the text (this differs from so-called aspect categories which do not necessarily appear in the text). Here, the task is to classify the sentiments for those aspect terms. This is why the term *Aspect Term Sentiment Classification (ATSC)* is most accurate.

When referring to ATSC methods, we usually think of *single-task* approaches. These methods are designed to carry out only aspect term sentiment classification as the aspect terms are already given. Whether these were identified manually or by an algorithm is not relevant in this setting. In practice, however, the aspect terms oftentimes are not already known. Thus, approaches dealing with the step of *Aspect Term Extraction (ATE)* have been developed. They can either work on their own or be combined with an ATSC method. For these combined methods, which we refer to as *ATE+ATSC*, one can further distinguish between *pipeline*, *joint* and *collapsed* models. In pipeline models, ATE and ATSC are simply stacked one after another, i.e. the output of the first model is used as input to the second model. The latter two are often also referred to as *multi-task* models, since both tasks are carried out simultaneously or in an alternating way. These models only differ in their labeling mechanisms: There are two label sets for joint models, one to indicate whether a word is part of an aspect term and the other one to state its polarity. For collapsed models, a unified labeling scheme indicates whether a word is part of a positive, negative or neutral aspect term or not.

We re-evaluate four different models for ATSC, covering a variety of different architectures. This encompasses Recurrent neural networks (RNNs), Capsule networks [6,16], networks using a Local Context Focus (LCF [22]), BERT-based approaches [2]), as well as two different ATE+ATSC models, one of which is a pipeline approach while the other one works in a collapsed fashion. All models are re-trained five times using five different (identical) train/validation splits and tested on the respective test sets in order to (i) compare them on a common ground and (ii) quantify the epistemic uncertainty associated with the architectures and the data.

2   Related Work

Related experiments were conducted by Mukherjee et al. [11], yet with a different focus. On the one hand, the authors also try to reproduce results on the benchmark data sets from SemEval-14 about restaurants and laptops. However, they selected six other models than we did for which the implementations are provided in one repository.[1] For these, the authors observed a consistent drop of 1–2 % with respect to both accuracy and macro-averaged F1-Score (F_1^{macro}). Mukherjee et al. [11] report a doubling of this drop when using 15% of the training data as validation data. On the other hand, they executed additional tasks which included the creation of two new data sets about men's t-shirts and television as well as model evaluation on these data sets. Furthermore, they also experimented with cross-domain training and testing. Yet, several important points are not addressed by their work which is why we investigate them in our work. First, while they mostly care about comparing different types of architectures (memory networks vs. BERT), we instead focus on comparing the best performing models for different *tasks* (ATSC vs. ATE+ATSC). Further, we cover a larger variety of types of architectures by selecting the best performing representatives of several different types. Second, we stick closer to the original implementations (by using them, if available) whereas they exclusively rely on community designed implementations, which adds a further potential source of errors. Third, and most important, we provide estimates for the epistemic uncertainty of performance values and are thus able to (at least tentatively) explain performance differences due to different reporting standards.

3   Materials and Methods

This section will introduce the data sets we utilized for training and evaluation as well as the selected model architectures. We start by briefly explaining the data, before the models are described, since (reported) performance values on these data sets partly motivate our choices regarding the models. We selected these data sets as they are either widely known benchmark data sets or interesting adaptations of them. We acknowledge that their sizes are not be that large, yet, the pool of available data sets for this kind of tasks is rather small. Descriptive statistics for all used data sets can be found in Table 1. Note that the data sets we eventually use for training and testing the models are all based on the *original* train/test splits. Further we apply *small* modifications (as described below) which were (a) also applied by some of the authors whose models we re-evaluate and (b) we perceive as reasonable. This allows us to evaluate all of the architectures on a common ground, which is not possible by comparing the reported values from the original publications alone. Nevertheless, we are aware of the fact that this might limit comparability of our results to the original ones to some extent.

3.1   Data Sets

SemEval-14 Restaurants. This data set contains reviews about restaurants in New York. Pontiki et al. [14] chose a subset of the restaurant data from

---

[1] https://github.com/songyouwei/ABSA-PyTorch.

Ganu et al. [4] as training data[2], while collecting test data[3] themselves. Both were labeled for several subtasks in the same way. These data sets were designed for ATSC as well as its equivalent on category level, *Aspect Category Sentiment Classification* (ACSC), but we stick to ATSC samples only. For each identified aspect term within a sentence, the polarity is given as *positive, negative, neutral* or *conflict*. We deleted the labels of the latter category (*conflict*) from the data sets due to their rare appearance. This is similar to previous work [1,3,8,21], yet, they do not all mention or explain the removing process explicitly. Rarely appearing duplicate sentences which occurred in the training set were also removed in our work. Due to their small amount, this procedure should not cause severe problems concerning the over- or underestimation of the applied metrics. We speculate that this rare appearance of duplicates also might be the reason for why a similar preprocessing step was, to the best of our knowledge, only taken in one other work [20].

MAMS. A *Multi-Aspect Multi-Sentiment (MAMS)* data set for the restaurant domain was introduced by Jiang et al. [7] who criticized existing data sets for not being adequate for ATSC. Since the data sets described above mainly consist of sentences which exhibit (i) only one single aspect or (ii) several aspects with the same sentiment, they argued that the task would not be much more difficult than a sentiment prediction on the sentence-level. To circumvent this issue, they extracted sentences of Ganu et al. [4] which comprise at least two aspects with differing sentiments.[4] The data sets have the same structure as the SemEval-14 data sets, with the difference that Jiang et al. [7] provide a fixed validation set for MAMS. The size of the validation split comprises about ten percent of the whole training set, which also inspired our choice when it comes to creating train/validation splits from the two SemEval-14 training data sets.

ARTS. Xing et al. [19] questioned the suitability of existing data sets for testing the aspect robustness of a model, i.e. whether the model is able to correctly identify the words corresponding to the chosen aspect term and predict its sentiment only based on them. Thus, the authors created an automatic generation framework that takes SemEval-14 test data (restaurants and laptops) as input and creates an *Aspect Robustness Test Set (ARTS)*. They used three different strategies to enrich the existing test set: The first one, REVTGT ("*reverse target*"), aims at reversing the sentiment of the chosen aspect term (called "*target aspect*"). This is reached by flipping the opinion using antonyms or adding negation terms like "not". Additionally, conjunctions may be changed in order to make sentences sound more fluent. Another strategy to augment the test set is REVNON ("*reverse non-target*") for which the sentiment of non-target aspects

---

[2] http://metashare.ilsp.gr:8080/repository/browse/semeval-2014-absa-restaurant-reviews-train-data/479d18c0625011e38685842b2b6a04d72cb57ba6c07743b9879d1a 04e72185b8/.

[3] http://metashare.ilsp.gr:8080/repository/browse/semeval-2014-absa-test-data-gold-annotations/b98d11cec18211e38229842b2b6a04d77591d40acd7542b7af823a54 fb03a155/.

[4] https://github.com/siat-nlp/MAMS-for-ABSA.

are (i) changed if they have the same sentiment as the target aspect or (ii) exaggerated if the non-target aspect is of a differing polarity. The third strategy called ADDDIFF ("*add different sentiment*") adds non-target aspects with an opposite sentiment which is intended to confuse the model. These non-target aspects are selected from a set of aspects collected from the whole data set and appended to the end of the sentence. ARTS are only designed to be used as test sets after training an architecture on the respective SemEval-14 training sets. The test sets for both restaurants and laptops are publicly available.[5] During the preparation of the ARTS data for CapsNet-BERT, we noticed that the start and end positions of some aspect terms were not correct. We changed them in order to make the code work properly and we also deleted duplicates (cf. [20]). For these specific test sets, the *Aspect Robustness Score (ARS)* was introduced by Xing et al. [19] in order to measure how well models can deal with variations of sentences. Therefore, each sentence and all its variations are regarded as one unit of observation for which the prediction is only considered to be correct if the predictions for *all* variations are correct. These units alongside with their corresponding predictions are then used to compute the regular accuracy (*ARS accuracy*) on the level of the observational unit.

SemEval-14 Laptops. The second domain-specific subset of the SemEval-14 data is on laptops. The data were collected and annotated by Pontiki et al. [14] for the task of ATE and/or ATSC. The training data set is publicly available,[6] just like the test data (see Footnote 3). Again, there were duplicate sentences in the training data which we deleted (cf. [20]). Unlike other benchmark data sets, both SemEval-14 data sets come without an official train/validation split.

More Data Sets. Recently more data sets have been published in addition to the ones mentioned beforehand. Mukherjee et al. [11] proposed two new data sets about *men's t-Shirts* and *television*. The YASO data set [12] has a different structure as it is a multi-domain collection. This is an interesting approach, yet also the reason for not considering it for our experiments: This data set is far better suited for cross-domain analyses, which is out of the scope of this work.

3.2   Models

MGATN. A *multi-grained attention network (MGATN)* was proposed by Fan et al. [3]. Its *multi-grained attention* is able to take into account the interaction between aspects. We chose MGATN since it is reported to be the best performing representative of RNN-based models on SemEval-14 data sets.

CapsNet-BERT. Capsules networks were initially proposed for the field of computer vision, with the so-called *capsules* being responsible for recognizing certain implicit entities in images. Each capsule performs internal calculations and returns

---

[5] https://github.com/zhijing-jin/ARTS_TestSet.

[6] http://metashare.ilsp.gr:8080/repository/browse/semeval-2014-absa-laptop-reviews-train-data/94748ff4624e11e38d18842b2b6a04d7ca9201ec33f34d74a8551626 be122856.

Table 1. Descriptive Statistics for the five utilized data sets. "*Multi-Sentiment sentences*" are those with at least two different polarities after removing "conflict" polarity. "*Aspect Terms in total*" also exclude "conflict".

Data Set	Subset	Original Sentences in total	Sentences without Duplicates	Sentences for 3-class ATSC	Multi-Sentiment Sentences	Aspect Terms in total	Positive Aspect Terms	Negative Aspect Terms	Neutral Aspect Terms	Removed Conflict Aspect Terms
SemEval-14 Restaurants	Training	3,044	3,038	1,978	320	3,605	2,161	807	637	91
	Test	800	800	600	80	1,120	728	196	196	14
SemEval-14 Laptops	Training	3,048	3,036	1,460	166	2,317	988	866	463	45
	Test	800	800	411	38	638	341	128	169	16
ARTS Restaurants	Test	2,784	2,784	2,784	206	3,528	1,952	1,103	473	0
ARTS Laptops	Test	1,576	1,576	1,576	74	1,877	883	587	407	0
MAMS Restaurant	Training	4,297	4,297	4,297	4,297	11,186	3,380	2,764	5,042	0
	Validation	500	500	500	500	1,332	403	325	604	0
	Test	500	500	500	500	1,336	400	329	607	0

a probability that the corresponding entity appears in the image. A variation of capsule networks for ATSC and its combination with BERT was introduced by Jiang et al. [7]. It was reported to outperform all other capsule networks with respect to their accuracy on the SemEval-14 restaurants data. Additionally, it performed second-best on MAMS, which is why we selected it for this study. Furthermore, we assumed their results on SemEval-14 restaurants data to be for three-class classification, as all the other results they refer to are also three-class. Yet, it is not fully clear to us which makes this experiment even more interesting.

RGAT-BERT. The *Relational Graph Attention Network (RGAT)* was introduced by Bai et al. [1]. It utilizes a dependency graph representing the syntactic relationships between words of a sentence as an additional input. The RGAT encoder creates syntax-aware aspect term embeddings following the representation update procedures from *Graph Attentional Networks (GATs)* [18]. It exhibits the best performance among graph-based models and also performs best on the MAMS data in terms of both accuracy and F_1^{macro}.

LCF-ATEPC. Yang et al. [21] built upon the idea of the LCF mechanism. The local context of an aspect term is defined as a fixed-size window around it, words outside this window are taken into account with lower weights or not at all. For each input token two labels, for aspect and sentiment, are assigned according to the joint labeling scheme described in Sect. 1. We chose LCF-ATEPC to be part of this meta-study since it reached the highest F_1^{macro} and accuracy on SemEval-14 data of all approaches. Yet, this only holds for the variant that is trained using additional domain adaptation.

BERT+TFM. The approach described by Li et al. [9] consists of a BERT model followed by a transformer (TFM [17]) layer for classification. BERT+TFM was the best model on SemEval-14 Laptops among all collapsed models at the time point of its introduction. There were also models using other layers on top instead of the transformer layer, but our variant of choice was TFM as it produced slightly better results than the concurring models.

GRACE. GRACE, a *Gradient Harmonized and Cascaded Labeling* model introduced by Luo et al. [10], belongs to the category of pipeline approaches. It includes a post-training step of the pre-trained BERT model using Yelp[7] and Amazon data [5]. The post-trained model then shares its first l layers between the ATE and the ATSC task. The remaining layers are only used for the former. They are followed by a classification layer for the detected aspect terms. These classification outputs are then used again as inputs for a Transformer decoder which performs sentiment classification. The principle of using the first set of labels as input for the second is called *Cascaded Labeling* here and is assumed to deal with interactions between different aspect terms. *Gradient Harmonization* is applied in order to cope with imbalanced labels during training. GRACE appears to be the best performing one of the pipeline models according to the literature. Furthermore, it is reported to be the best ATE+ATSC model on both SemEval-14 data sets. However, these successes have to be taken into account with care, as their results are based on four-class classification. This means that

---

[7] https://www.yelp.com/dataset.

in comparison to the other authors' settings they did not exclude conflicting reviews of SemEval-14 data. Thus, our analysis contributes to comparability even more since it has not been established yet for the model/data combinations we examine.

4   Experiments

We re-evaluate six models (cf. Sect. 3.2) on the five data sets for the English language presented in Sect. 3.1. Our overall goals are to establish comparability between the models, to examine whether reported performance can be reproduced and to quantify epistemic model uncertainty that might exist due to the lacking knowledge about the train/validation splits. The entire code from our experiments is publicly available on GitHub.[8]

Our proceeding is as follows: First, we re-use the implementations provided by the authors by simply cloning their git repositories and adjusting them to our setup. Subsequently we try to reproduce their results on the data sets they used. Second, we adapt their code to the remaining data sets and conduct the necessary modifications, again sticking as closely as possible to the original hyperparameter settings (cf. Table 2 in the supplementary material). The biggest change we made was increasing the number of training epochs drastically and adding an early stopping mechanism. Apart from that, we did not engage in hyperparameter tuning in order not to modify/falsify the results. For all ATSC models, we selected the optimal model during the training process based on the validation accuracy and/or F_1^{macro}. For performing the experiments, we had one *Tesla V100 PCIe 16GB* GPU at our disposal.

Data Preparation. Unlike other data sets, both SemEval-14 data sets come *without* an official validation split. Thus, we created five different train/validation splits (90/10) for each of the two SemEval-14 training sets. For each split, five training runs with different random initializations were conducted per model. The resulting 25 different versions per model per data set were subsequently evaluated on the two official SemEval-14 test sets (restaurants and laptops) as well as on the ARTS test sets. In Sect. 5 we report overall means per model per test set as well as means and standard deviations per model and test set for each of the different splits. Since there is an official validation set for MAMS, we did not apply the splitting procedure from above when training on this data set. Consequently, the given means and standard deviations are based on five training runs with different random initializations only.

MGATN. As there exists no publicly available implementation provided by its authors, we used the one from a collection of re-implemented ABSA methods from GitHub.[9] We slightly modified the early stopping mechanism from that repository and then implemented it also for the other re-evaluated models.

---

[8] https://github.com/el-ma-le/atsc-experiments-official.
[9] https://github.com/songyouwei/ABSA-PyTorch.

CapsNet-BERT. We used the implementation of CapsNet-BERT provided by its authors.[10]

RGAT-BERT. We relied on the implementation of RGAT-BERT provided by its authors.[11] Since the authors manually created an accuracy score different to the one implemented in `scikit-learn`[12] [13], we substituted their metric by the scikit-learn variant to ensure comparability. For data transformation, we selected the stanza tokenizer [15] over the Deep Biaffine Parser,[13] which was used by Bai et al. [1], since the former provides the necessary syntactic information, whereas the latter failed to produce the syntactic dependency relation tags and head IDs the model requires.

LCF-ATEPC. We were not able to run the best-performing LCF-ATEPC variant based on domain adaptation due to missing pre-trained models. Thus, we decided to go for the second best, LCF-ATEPC-Fusion, using the official implementation of LCF-ATEPC.[14] During our experiments, the authors of LCF-ATEPC started building a new repository[15] based on the existing code which we did not use as it was still subject to changes.

BERT+TFM. We used the implementation of BERT+TFM provided by its authors.[16] Our model selection was based on F_1^{micro} and F_1^{macro}, which were calculated based on *(start position, end position, polarity)*-triples for each identified aspect. Due to the collapsed labeling scheme, these scores account for both ATE and ATSC.

GRACE. We used the post-trained BERT model provided by Luo et al. [10].[17] Our model selection was based on ATSC-F_1^{micro} and -F_1^{macro} as well as on ATE-F_1^{micro}, with their calculations being slightly adjusted in order to match the calculation of those from BERT+TFM.

5   Results

In general, reported values were not reproducible. Figure 1 shows a comparison of our results (averaged over all 25 runs, including 95% confidence intervals) to the reported results from the original publications on the two SemEval-14 data sets. For all architectures there exists a notable gap between the blue (reproduced) and the orange (reported) values. In general, the gap tends to be larger for the ATSC models compared to the two ATE+ATSC models, where we were even able to reach a better performance for BERT+TFM within our replication study.[18]

---

[10] https://github.com/siat-nlp/MAMS-for-ABSA.

[11] https://github.com/muyeby/RGAT-ABSA.

[12] https://scikit-learn.org/stable/modules/generated/sklearn.metrics.accuracy_score. html.

[13] https://github.com/yzhangcs/parser.

[14] https://github.com/yangheng95/LCF-ATEPC.

[15] https://github.com/yangheng95/pyabsa.

[16] https://github.com/lixin4ever/BERT-E2E-ABSA.

[17] https://github.com/ArrowLuo/GRACE.

[18] We do not give a similar figure for MAMS or ARTS as there are not enough reported values to display the results in a meaningful way.

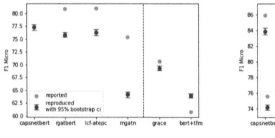
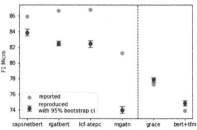

(a) SemEval-14 Laptops   (b) SemEval-14 Restaurants

Fig. 1. Comparison of reported and reproduced performance. The reproduced value is the mean of all 25 runs per model in total. Further, 95% bootstrap ($n = 2000$) confidence intervals are displayed. Note that absolute performance of GRACE (four classes) and BERT+TFM cannot be compared to the other models due to different tasks. No F_1^{micro} was reported for CapsNet-BERT on SemEval-14 Laptops.

It is also interesting to see how different runs can lead to rather broad ranges of results, although having done only five training runs per model and data split. An example for this phenomenon is the Accuracy of MGATN on SemEval-14 Laptops (cf. Figure 2). For the first, the fourth and fifth split, all of the values lie very close together (within mean $\pm$ std), whereas the results of the other two splits show a rather high variance.

MGATN. For MGATN, our reproduced results fell short of the reported values for accuracy, around five to ten percentage points for SemEval-2014 laptops and restaurants, respectively (cf. Table 5 and 6 in the supplementary material). Figure 2 depicts the results on the laptops test set, the difference between reported and reproduced performance on the restaurant data (not shown) looks similar. A reason for this behavior might be that we could not use the official implementation of the authors, but had to rely on a re-implementation from the community. In terms of ARS accuracy on ARTS Restaurants, MGATN was the only model that reached only a single-digit value which means that it is not good at dealing with perturbed sentences.

CapsNet-BERT. Comparing all the selected models on the ATSC task, CapsNet-BERT performed best on all data sets regarding all the metrics except for ARS accuracy on the ARTS restaurant test set (cf. Table 5 and 6 in the supplementary material). For ARTS, it seems as if the reported ARS accuracy for laptops matched our result for restaurants, and vice versa, as Fig. 3 illustrates. As far as we can tell, we did not mix up the data sets during our calculations which makes this look quite peculiar. The difference between the reported and reproduced values on SemEval-14 Restaurants data (as shown in Fig. 1b) may be explained by the fact that we did three-class classification and we only assumed so for the reported value.

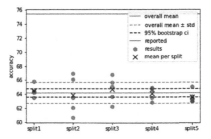

Fig. 2. Example for high differences between data splits: Accuracy of MGATN on SemEval-14 Laptops.

RGAT-BERT. For both SemEval-14 and MAMS we missed the reported values by around five percentage points (cf. Table 5 and 6 in the supplementary material). ARTS restaurants is the only data set on which the best ARS accuracy was not reached by CapsNet-BERT, but RGAT-BERT. Regarding MAMS, Bai et al. [1] provided accuracy as well as F_1^{macro}, which is why we also compare these results here. Figure 4 shows the all five values of the four different measures as well as the average. For accuracy and F_1^{macro}, reported values from Bai et al. [1] were added.

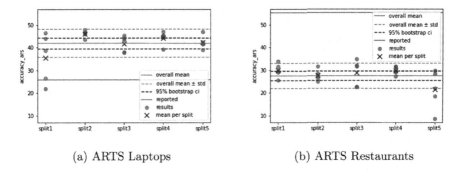

(a) ARTS Laptops   (b) ARTS Restaurants

Fig. 3. Aspect Robustness Score (ARS) Accuracy of CapsNet-BERT.

LCF-ATEPC. Our experiments on average resulted in about five percentage points lower accuracies for LCF-ATEPC than were reported. Yet, LCF-ATEPC reached the best ARS accuracy value on ARTS restaurant data in our analysis.

BERT+TFM. In contrast to the majority of the other models, for BERT+TFM the (average) performance of our runs surpassed the reported performance values on the SemEval-14 data. As Fig. 5 indicates, this holds for all runs (laptop domain) and on average (restaurant domain). The reasons for our improved values may lie in the chosen hyperparameters, yet we cannot tell for sure.

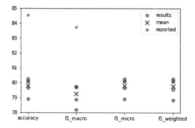

Fig. 4. Performance of RGAT-BERT on MAMS.

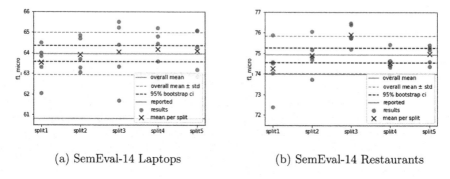

(a) SemEval-14 Laptops   (b) SemEval-14 Restaurants

Fig. 5. F_1^{micro} of BERT+TFM.

GRACE. During our experiments with GRACE, we were able to produce results approximately in the same range as the reported values. Regarding SemEval-14 restaurants our results on average were better than the reported ones (cf. Figure 6b), while for laptops we could not quite reach the reported performance (cf. Figure 6a). For the latter case, our results of single runs were better than (or at least equal to) the reported one, which is kind of a symptom of the problem. If we only reported the best of all runs, our conclusion would have been that we were able to outperform the original model. However, as we have already mentioned, reported results were based on four-class classification, whereas our results were made for three-class. This might be the reason for different results. In the ATE+ATSC task, GRACE outperformed BERT+TFM on all data sets except for MAMS (cf. Table 3 and 4 in the supplementary material).

6   Discussion

6.1   General Takeaways

Results differing from the reported values can be explained by various reasons. First, we often do not know how the reported values were created, i.e. whether the authors took the best or an average value of their runs. In Fig. 6a, it becomes clearly visible that taking the best value compared to the mean over multiple runs yields a difference of about almost three percentage points. Unfortunately

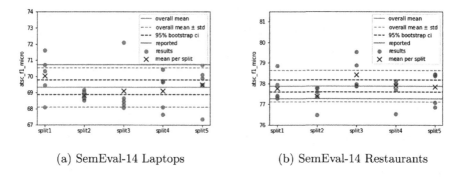

(a) SemEval-14 Laptops   (b) SemEval-14 Restaurants

Fig. 6. ATSC F_1^{micro} of GRACE.

there are also, to the best of our knowledge, no clear guidelines for how to properly report the uncertainty resulting from different data splits. Second, our data are usually not exactly identical to the data sets used for the original papers due to the preprocessing steps we explained beforehand. Also, training and validation splits are probably different from ours. Some models required additional syntactical information which we (potentially) inferred from other packages than indicated, because either none were given or because the ones that were given did not work as stated. Third, hyperparameter configurations are often not totally clear due to a lack of concise descriptions in the original work. In these cases we took those that were chosen by default in the implementations we used. Since those were not necessarily always provided by the authors of the models, we have no information about how close they are to the original configurations. What we could find out regarding hyperparameters can be found in Table 2 in the supplementary material. Consequently, it is not surprising that we were not able to exactly reproduce given results, since hyperparameter tuning often has a large impact on the model performance. This insight is also shared by Mukherjee et al. [11], although they tested other models in a different setup.

6.2   Possible Guidelines

Taking all considerations into account, we want to tentatively propose some guidelines that might be beneficial for making NLP research reproducible and for quantifying different types of uncertainty. First, it is not enough to purely open-source your code but it also requires a thorough documentation and explanation. This should also include all the information about hyperparameters, additional training data, custom data splits (if applicable), and non-standard pre-processing, since all of this can have a (potentially) large impact on the results. Second, every information about potential randomness/variation in the results has to be acknowledged, ideally even researched further and reported/displayed properly. One potential starting point could be to *always* perform multiple runs on multiple different splits and use the results to report standard deviations between and within splits. While the former gives an impression for the uncertainty induced by data heterogeneity, the latter rather reflects the model's share of the overall uncertainty. This would of course to some extent mean, to move away from (overly confidently) reporting single performance values. A reporting convention indicating a common procedure combined with already prepared data sets with all possible labels could improve the comparability between models a lot.

7 Conclusion and Future Work

Our experiments revealed that reproducing reported results is hardly possible, given the current practice of performance reporting (at least for this subset of selected models). A tendency towards lower results is visible in our experiments, sometimes even five to ten percentage points lower than the original values. The only exception was BERT+TFM for which given values were surpassed. The reasons for these observations may lay in the data preprocessing steps, in the hyperparameters or in the absence of a convention on which values to report (best or mean of several runs). This discovery of models hardly being comparable based on their performance measures is a very important one from our point of view. When new models are proposed, one of the main aspects during their evaluation is the improvement with respect to the state-of-the-art. But when the performance of a single model can vary between single runs, the question is which results to take into account for model rankings. Also a huge practical meta-analysis of all models on several data sets would clarify the situation.

Acknowledgements. This work has been partially funded by the Deutsche Forschungsgemeinschaft (DFG, German Research Foundation) as part of BERD@NFDI - grant number 460037581.

References

1. Bai, X., Liu, P., Zhang, Y.: Investigating typed syntactic dependencies for targeted sentiment classification using graph attention neural network. In: IEEE/ACM Transactions on Audio, Speech, and Language Processing **29**, 503–514 (2020). http://dx.doi.org/10.1109/TASLP.2020.3042009
2. Devlin, J., Chang, M.W., Lee, K., Toutanova, K.: BERT: pre-training of deep bidirectional transformers for language understanding. In: Proceedings of the 2019 Conference of the North American Chapter of the Association for Computational Linguistics: Human Language Technologies, Volume 1 (Long and Short Papers), pp. 4171–4186. Association for Computational Linguistics, Minneapolis, Minnesota (Jun 2019). https://doi.org/10.18653/v1/N19-1423, https://www.aclweb.org/anthology/N19-1423
3. Fan, F., Feng, Y., Zhao, D.: Multi-grained attention network for aspect-level sentiment classification. In: Proceedings of the 2018 Conference on Empirical Methods in Natural Language Processing. pp. 3433–3442. Association for Computational Linguistics, Brussels, Belgium (Oct-Nov 2018). https://doi.org/10.18653/v1/D18-1380, https://aclanthology.org/D18-1380

4. Ganu, G., Elhadad, N., Marian, A.: Beyond the stars: improving rating predictions using review text content. In: Twelfth International Workshop on the Web and Databases (WebDB 2009), vol. 9, pp. 1–6. Citeseer (2009)

5. He, R., McAuley, J.: Ups and downs: modeling the visual evolution of fashion trends with one-class collaborative filtering. In: Proceedings of the 25th International Conference on World Wide Web, pp. 507–517. WWW 2016, International World Wide Web Conferences Steering Committee, Republic and Canton of Geneva, CHE (2016). https://doi.org/10.1145/2872427.2883037,https://doi.org/10.1145/2872427.2883037

6. Hinton, G.E., Krizhevsky, A., Wang, S.D.: Transforming auto-encoders. In: Honkela, T., Duch, W., Girolami, M., Kaski, S. (eds.) ICANN 2011. LNCS, vol. 6791, pp. 44–51. Springer, Heidelberg (2011). https://doi.org/10.1007/978-3-642-21735-7_6

7. Jiang, Q., Chen, L., Xu, R., Ao, X., Yang, M.: A challenge dataset and effective models for aspect-based sentiment analysis. In: Proceedings of the 2019 Conference on Empirical Methods in Natural Language Processing and the 9th International Joint Conference on Natural Language Processing (EMNLP-IJCNLP), pp. 6280–6285. Association for Computational Linguistics, Hong Kong, China (Nov 2019). https://doi.org/10.18653/v1/D19-1654, https://aclanthology.org/D19-1654

8. Li, X., Bing, L., Li, P., Lam, W.: A unified model for opinion target extraction and target sentiment prediction. In: Proceedings of the AAAI Conference on Artificial Intelligence, vol. 33, pp. 6714–6721 (2019)

9. Li, X., Bing, L., Zhang, W., Lam, W.: Exploiting BERT for end-to-end aspect-based sentiment analysis. In: Proceedings of the 5th Workshop on Noisy User-generated Text (W-NUT 2019), pp. 34–41. Association for Computational Linguistics, Hong Kong, China (Nov 2019). https://doi.org/10.18653/v1/D19-5505, https://aclanthology.org/D19-5505

10. Luo, H., Ji, L., Li, T., Jiang, D., Duan, N.: GRACE: Gradient harmonized and cascaded labeling for aspect-based sentiment analysis. In: Findings of the Association for Computational Linguistics: EMNLP 2020, pp. 54–64. Association for Computational Linguistics, Online (Nov 2020). https://doi.org/10.18653/v1/2020.findings-emnlp.6, https://aclanthology.org/2020.findings-emnlp.6

11. Mukherjee, R., Shetty, S., Chattopadhyay, S., Maji, S., Datta, S., Goyal, P.: Reproducibility, replicability and beyond: assessing production readiness of aspect based sentiment analysis in the wild. arXiv preprint arXiv:2101.09449 (2021)

12. Orbach, M., Toledo-Ronen, O., Spector, A., Aharonov, R., Katz, Y., Slonim, N.: YASO: a new benchmark for targeted sentiment analysis. arXiv preprint arXiv:2012.14541 (2020)

13. Pedregosa, F., et al.: Scikit-Learn: machine learning in Python. J. Mach. Learn. Res. **12**, 2825–2830 (2011)

14. Pontiki, M., Galanis, D., Pavlopoulos, J., Papageorgiou, H., Androutsopoulos, I., Manandhar, S.: SemEval-2014 task 4: aspect based sentiment analysis. In: Proceedings of the 8th International Workshop on Semantic Evaluation (SemEval 2014), pp. 27–35. Association for Computational Linguistics, Dublin, Ireland (Aug 2014). https://doi.org/10.3115/v1/S14-2004, https://aclanthology.org/S14-2004

15. Qi, P., Zhang, Y., Zhang, Y., Bolton, J., Manning, C.D.: Stanza: a Python natural language processing toolkit for many human languages. In: Proceedings of the 58th Annual Meeting of the Association for Computational Linguistics: System Demonstrations (2020), https://nlp.stanford.edu/pubs/qi2020stanza.pdf

16. Sabour, S., Frosst, N., Hinton, G.E.: Dynamic routing between capsules. In: Guyon, I., et al. (eds.) Advances in Neural Information Processing Systems. vol. 30. Curran Associates, Inc. (2017). https://proceedings.neurips.cc/paper/2017/file/2cad8fa47bbef282badbb8de5374b894-Paper.pdf

17. Vaswani, A., et al.: Attention is all you need. CoRR abs/1706.03762 (2017). http://arxiv.org/abs/1706.03762

18. Veličković, P., Cucurull, G., Casanova, A., Romero, A., Liò, P., Bengio, Y.: Graph attention networks. In: International Conference on Learning Representations (2018). https://openreview.net/forum?id=rJXMpikCZ

19. Xing, X., Jin, Z., Jin, D., Wang, B., Zhang, Q., Huang, X.: Tasty burgers, soggy fries: probing aspect robustness in aspect-based sentiment analysis. In: Proceedings of the 2020 Conference on Empirical Methods in Natural Language Processing (EMNLP), pp. 3594–3605 (2020). https://doi.org/10.18653/v1/2020.emnlp-main.292, https://www.aclweb.org/anthology/2020.emnlp-main.292

20. Xue, W., Li, T.: Aspect based sentiment analysis with gated convolutional networks. In: Proceedings of the 56th Annual Meeting of the Association for Computational Linguistics (Volume 1: Long Papers), pp. 2514–2523. Association for Computational Linguistics, Melbourne, Australia (Jul 2018). https://doi.org/10.18653/v1/P18-1234, https://aclanthology.org/P18-1234

21. Yang, H., Zeng, B., Yang, J., Song, Y., Xu, R.: A multi-task learning model for chinese-oriented aspect polarity classification and aspect term extraction. Neurocomputing **419**, 344–356 (2021). https://doi.org/10.1016/j.neucom.2020.08.001, https://www.sciencedirect.com/science/article/pii/S0925231220312534

22. Zeng, B., Yang, H., Xu, R., Zhou, W., Han, X.: LCF: a local context focus mechanism for aspect-based sentiment classification. Appl. Sci. **9**, 3389 (2019)

An Ion Exchange Mechanism Inspired Story Ending Generator for Different Characters

Xinyu Jiang[1], Qi Zhang[2,3], Chongyang Shi[1(✉)], Kaiying Jiang[4], Liang Hu[3,5], and Shoujin Wang[6]

[1] Beijing Institute of Technology, Beijing, China
cy_shi@bit.edu.cn
[2] University of Technology Sydney, Ultimo, Australia
[3] Deepblue Academy of Sciences, Changzhou, China
zhangqi_cs@bit.edu.cn
[4] University of Science and Technology Beijing, Beijing, China
[5] Tongji University, Shanghai, China
[6] Macquarie University, Sydney, Australia

Abstract. Story ending generation aims at generating reasonable endings for a given story context. Most existing studies in this area focus on generating coherent or diversified story endings, while they ignore that different characters may lead to different endings for a given story. In this paper, we propose a Character-oriented Story Ending Generator (CoSEG) to customize an ending for each character in a story. Specifically, we first propose a character modeling module to learn the personalities of characters from their descriptive experiences extracted from the story context. Then, inspired by the ion exchange mechanism in chemical reactions, we design a novel vector breaking/forming module to learn the intrinsic interactions between each character and the corresponding context through an analogical information exchange procedure. Finally, we leverage the attention mechanism to learn effective character-specific interactions and feed each interaction into a decoder to generate character-orient endings. Extensive experimental results and case studies demonstrate that CoSEG achieves significant improvements in the quality of generated endings compared with state-of-the-art methods, and it effectively customizes the endings for different characters.

Keywords: Story ending generation · Character-oriented · Neural network

1 Introduction

Story ending generation aims to deliver a comprehensive understanding of the context and predict the next plot for a given story [10,16,29,31]. Some studies in this field generate coherent stories by modeling the sequence of events or verbs

---

X. Jiang and Q. Zhang—The first two authors contribute equally to this work.

M.-R. Amini et al. (Eds.): ECML PKDD 2022, LNAI 13714, pp. 553–570, 2023.
https://doi.org/10.1007/978-3-031-26390-3_32

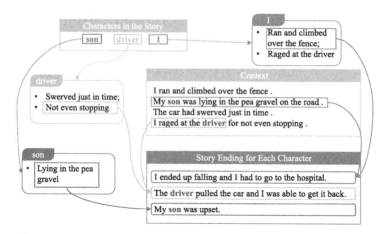

Fig. 1. An example of the story context in the ROCStories corpus, and the endings generated by our model for different characters.

[5,19], or diversify story generation by introducing common senses or vocabulary information [8,18]. Others focus on controlling the sentiment of story endings [16, 27] or generating the missing plot for an incomplete story [1,29]. These methods generally ignore the relation and interaction between story plots and characters and simplify the influence of character personality on story generation, leading to desirable but character-irrelevant story endings.

Intuitively, stories are derived from characters, and character personality directly determines the plot and direction of the story. Figure 1 shows an example of a typical story in the ROCStories corpus [21] and the endings generated for different characters. From the figure, we can observe that: **1)** each character has its unique personality depicted by its character token and character experience, i.e., the character-related descriptions in a story. For example, the character **son** is depicted by the token "son" and the description "lying in the pea gravel"; **2)** naturally, different characters with different personalities interact with the story context and thus affect the story plot, leading to different story endings (see the different endings for 'son', 'driver' and 'I' in the example).

Customizing the endings for different characters in a story is a novel but challenging task since there is no one-to-many dataset (i.e., one story corresponds to many ground-truth endings). To the best of our knowledge, most previous methods for the story ending generation aim to generate a single ending or missing plot rather than diverse coherent endings of different characters, for a given story context [10,29,31]. The main challenges in customizing character-oriented story endings are 1) *to model the personality of each character*, and 2) *to learn the diverse interactions between different characters and the story context*. Intuitively, a story context contains a character's experiences, i.e., the multiple descriptions of the character, which depict the personalities of the character. It would be helpful to extract the related descriptions of each character from the

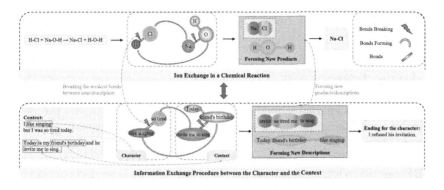

Fig. 2. A character-context information exchange mechanism to learn the interaction between character and context, which is inspired by the ion exchange mechanism in chemical reactions.

story content and build its experience sequence via organizing the descriptions in chronological order for modeling each character's personalities.

Inspired by recent studies using deep learning to plan and predict chemical reactions [25,26], we model the personalities of characters by analogizing the interactions between characters and context to chemical reactions. Specifically, we believe that the information exchange between different characters and context in generating new situational (character-specific) semantics during the interaction is similar to the ion exchange [2] to form new products in chemical reactions (cf. Fig. 2). Derived from this observation, it is promising to learn the interaction between a character and the corresponding context following an information exchange procedure. As depicted in Fig. 2, by exchanging related descriptions of the character and context, a new informative and character-related description "invite so tired me to sing" is formed by putting "so tired" (from character) and "invite me to sing" (from context) together. Consequently, the newly formed character-related description leads to an ending "refused his invitation" customized for the character "I" with a high probability.

Accordingly, we propose a Character-oriented Story Ending Generator (CoSEG) to customize an ending for each character in a story. Specifically, the proposed model first learns a representation of each character's personality by modeling its experiences with a Character Modeling module (CMM) and a context representation by modeling the story content. Then, a novel Vector Breaking/Forming module (VBF) is introduced to effectively learn the interaction between each character and the context through multiple information exchanges. Finally, a character-specific interaction representation is generated by adaptively picking out the most effective interaction via a Character-Context Attention module (C-CA), and each interaction representation is further utilized to customize the ending for the corresponding character. Note that CoSEG adopts LSTM-based encoder-decoder architecture, and the proposed key modules are network-agnostics and are also suitable for other prevailing networks, e.g., CNN [14] and Transformer [32]. The main contributions of our paper are summarized below:

- We propose a Character-oriented Story Ending Generator (CoSEG) model to tackle the challenging task of customizing story endings for different characters.
- We introduce a character modeling module to effectively model the personality of each character and learn a personalized and informative character representation.
- Inspired by the ion exchange process in chemical reactions, we propose a novel VBF module to learn the interaction between the character and context based on the information exchange mechanism.

Extensive experimental results on the ROCStories dataset show that our proposed CoSEG not only generates more coherent and diversified story endings compared with state-of-the-art and/or representative baseline methods, but also customizes effective endings for each character in a story. The superiority of CoSEG also demonstrates the effectiveness of the proposed CMM and VBF module in customizing story endings.

2 Related Work

Neural network-based models are the current mainstream in story generation methods owing to their impressive generation performance [6,18,22,30,31]. In recent years, there have been many innovations that utilize the encoder and decoder framework to generate coherent and diversified story endings. [33] applies a hierarchical attention architecture to encode text information to generate the context representation. [19] predicts the next event by extracting the event represented from the sentence, thereby ensuring the coherence of the story. [5] uses one head of the decoder's self-attention to attend only to previously generated verbs in order to generate a coherent story. [4] learns a second seq2seq model, which is led by the first model to focus on what the first model failed to learn. [10] introduces external knowledge and utilizes an incremental encoding scheme to ensure the diversity of stories. In addition, recent work proposes a character-centric neural storytelling model to generate stories for a given character [15]. Excited with the excellent performance of attention-based models [7,34] like Transformer [28] and BERT [3] in recent years, many story-ending generation and completion models leverage self-attention mechanism and Transformer architecture to enhance the quality of generated story endings [9,32]. In this work, we adopt LSTM as the backbone in the model design and experiments. Other architectures, e.g., attention networks and Transformer, can easily be incorporated into the proposed CoSEG.

However, most of the aforementioned generation methods cannot generate multiple coherent and diverse endings for a single story context. Moreover, only a few works focus on generating multiple endings or responses given a single context. [8] uses several unobserved latent variables z to generate different responses.

This method, however, relies on a one-to-many dataset. [16] applies an additional sentiment analyzer to first predict the sentiment intensity s of the ground truth ending y, then constructs paired data $(x, s; y)$ for training, where x is the story context. In the generation stage, the model receives the sentiment variable s from users to generate a sentiment-specific ending. Recent work [20,29] introduce prevalent Transformers to learn story representation for generating a missing plot or a story ending. All these methods assume that the plot has little relation or interaction with the personality of the characters in the story. Unlike these models, our proposed model can customize an ending for each character in the story context without relying on a one-to-many dataset.

3   Character-oriented Story Ending Generator (CoSEG)

3.1   Problem Definition and Architecture

In this section, we formulate the task of customizing the ending for each character in a story. Given a story content $x = (x_1, ..., x_l)$, which contains l sentences, and m characters $(c_1, ..., c_m)$. The task is to predict customized endings $y = (y_1, ..., y_m)$ for all the m characters.

A story generally corresponds to only one ground truth ending, which may consist of the actions or opinions of a particular character; the endings of other characters are unavailable. To tackle the issue, in the training stage, we extract the experience sequence of the character in the ground-truth ending, then train our proposed model to generate an ending related to the extracted character (ground-truth ending). In the generating stage, we extract the experience sequences of all characters who appear in the story and then apply the proposed model to generate an ending for the characters.

The architecture of our proposed CoSEG model is depicted in Fig. 3. Our model consists of three modules-a CMM module, a VBF module and a C-CA module-as well as an encoder to encode the story context and a decoder to generate the story ending. As shown in Fig. 3, the CMM module generates the character representation cc for each character c_i by modeling the character's experiences; the VBF module learns the interaction between the character representation and the story context through multiple information exchanges and generates multiple interaction results, namely product candidates $(\mathbf{p}_0^c, ..., \mathbf{p}_n^c)$; the C-CA module generates a character-related story context representation by picking out the most effective product candidates. The context representation is further used as the initial state of the decoder to predict an ending for the character. The following sections present the details of each module.

3.2   Character Experience Sequences

We construct the experience sequence for each character in the story; the experience sequence is further fed into CMM to generate the character representation.

We take the sentence *She knew a discount store near her sold socks* as an example to illustrate how to extract a character experience with the following four steps:

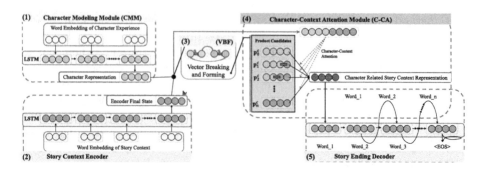

Fig. 3. Character-oriented Story Ending Generator (CoSEG).

1) Construct a **dependency tree** for the given sentence, and get the headword *knew* of the character *She*.
2) Extract the **context words**, namely *knew, discount, store, sold, socks*, as the first part of the character experience. This part is the background of the story.
3) Extract the **entity words**, namely the character *She*, the headword *knew* and the corresponding object *store*, as the second part of the character experience.
4) Connect the above two parts of character experience with a token *OBJ* to obtain the final character experience [*knew, discount, store, sold, socks, OBJ, knew, store, She*]. The token *OBJ* is utilized to separate the two parts, explicitly telling the model which words are context information and which are directly related to the character.

In this way, we can extract each character's experience from each sentence. Subsequently, we build the experience sequence $(e_1^i, ..., e_s^i)$ of the character c_i via organizing the character's experience in chronological order, where s is the number of the sentences that contain c_i.

3.3 Character Modeling Module (CMM)

We generate the character representation by modeling the character experience sequence. Formally, given a character c_i and a corresponding character experience sequence $(e_1^i, ..., e_S^i)$, the module encodes the token sequence $(w_1^s, ..., w_{T_s}^s)$ inside each experience e_s^i to obtain the hidden states $(\mathbf{h}_1^s, ..., \mathbf{h}_{T_s}^s)$, where T_s is the length of e_s^i, and the superscript s of $\mathbf{h}$ or w represents it is for the sth character experience.

We choose the final hidden state $\mathbf{h}_{T_s}^s$ as the representation of the character c_i who went through the character experience e_s^i. The character representation $\mathbf{h}_{T_s}^s$ is then used as the initial state of the next encoder to generate further enriched character representation $\mathbf{h}_{T_{s+1}}^{s+1}$ as follows:

$$\mathbf{h}_t^{s+1} = \mathbf{LSTM}(\mathbf{h}_{t-1}^{s+1}, \mathbf{w}_t^{s+1}, \mathbf{h}_{T_s}^s), \tag{1}$$

where T_{s+1} is the length of experience e^i_{s+1}, w^{s+1}_t is the tth token in the sequence $(w^{s+1}_1, ..., w^{s+1}_{T_{s+1}})$ inside the experience e^i_{s+1}, $\mathbf{w}^{s+1}_t$ denotes the embedding of w^{s+1}_t and the superscript $s + 1$ of $\mathbf{h}$ or w represents it is for the $(s + 1)^{th}$ character experience.

Finally, the CMM Module will generate the Sth character representation $\mathbf{h}_{T_S}$, which has gone through all the character experiences $(e^i_1, ..., e^i_S)$.

3.4 Vector Breaking and Forming (VBF)

Since both character and context are represented with high-dimensional vectors, we propose a novel VBF module to learn the interaction between the character and the story context representation based on the information exchange procedure (cf. Fig. 2) and then generate multiple product candidates.

Vector Breaking: Assume that there are invisible bonds between the adjacent elements of a vector, and VBF breaks the bonds between adjacent elements. For a vector of size n, there are $n+1$ potential bond-breaking positions. For example, given a vector $v_1 = [0.1, 0.2]$, the size of v_1 is 2 and we have 3 bond-breaking positions, as follows:

$$
\begin{aligned}
v^l_1, v^r_1 &= [], [0.1, 0.2] = VecB_0(v_1), \\
v^{l'}_1, v^{r'}_1 &= [0.1], [0.2] = VecB_1(v_1), \\
v^{l''}_1, v^{r''}_1 &= [0.1, 0.2], [] = VecB_2(v_1),
\end{aligned}
\tag{2}
$$

where $VecB_k$ represents breaking the bond in the position k, and the superscripts l and r of the breaking results represent the *left* and *right* parts of v_1, respectively.

Vector Forming: The two interaction vectors break at each position respectively. To keep the size of interaction results constant, for each bond-breaking position, VBF integrates the left part of the first vector and the right part of the second vector to generate a product candidate[1]. In this way, two vectors of size n can interact to obtain a total of $n + 1$ product candidates. For example, let v_1 interact with $v_2 = [0.3, 0.4]$, and we can obtain such three product candidates $\mathbf{p}^c_0 = [0.3, 0.4]$, $\mathbf{p}^c_1 = [0.1, 0.4]$, $\mathbf{p}^c_2 = [0.1, 0.2]$. As shown in Fig. 3 part (3), the character representation and the encoder final state $\mathbf{h}_t$ (i.e., story context representation) interact in the VBF module to generate the product candidates $(\mathbf{p}^c_0, ..., \mathbf{p}^c_n)$.

3.5 Character-Context Attention (C-CA)

As shown in Fig. 3 part (4), the C-CA Module aims to pick out the most effective product candidates. Specifically, we utilize the sth character representation $\mathbf{h}_{T_s}$

---

[1] There is no order between the two interaction vectors, which vector as the first one has little influence on the experimental results.

and the encoder final state $\mathbf{h}_t$ to obtain the attention weight of each product candidate $\mathbf{p}_k^c$:

$$\mathbf{a}^s = \sigma(\mathbf{W}_a[\mathbf{h}_{T_s}; \mathbf{h}_t] + \mathbf{b}_a),$$

$$\mathbf{r}^s = \sum_{k=0}^{n} \mathbf{a}_k^s \mathbf{p}_k^c, \tag{3}$$

where σ is the softmax function, $\mathbf{W}_a$ is the weight matrix, $\mathbf{b}_a$ is the bias, att is the attention weight, and $\mathbf{a}^s$ stands for the character-related story context representation of the sth character. As shown in Fig. 3 part (5), the $\mathbf{r}^s$ is further used as the initial state of the decoder (note that we omit the superscript s in the following for simplicity):

$$\mathbf{h}_t^y = \text{LSTM}(\mathbf{h}_{t-1}^y, \mathbf{w}_{t-1}^y, \mathbf{r}), \tag{4}$$

where $\mathbf{h}_t^y$ is the tth hidden state of the decoder, which is further utilized to generate the tth word w_t^y, $\mathbf{w}_{t-1}^y$ is the embedding of word w_{t-1}^y.

4 Experiment

In this section, we conduct extensive experiments to investigate the quality of CoSEG and the comparative baselines.

4.1 Dataset

We evaluated our model on the ROCStories corpus [21]. This corpus contains 98,162 five-sentence stories. Our task is to generate an ending for each character that appears in a given four-sentence story[2]. For each story, we extract the experience sequence for the character who appears in the ground truth ending. We select 66,881 stories in which the length of the ground-truth character's experience sequence is no less than 2 and treat these stories as the training set. We elaborately design two test sets, each with 3073 stories. Specifically, the two sets are called sufficient test set and inadequate test set. In the sufficient test set, the length of the ground-truth character's experience sequence is no less than 2 for all stories, while the length is less than 2 for all stories in the inadequate test set. The two test sets are applied to evaluate the performance of our proposed model when the character information is sufficient and inadequate, respectively.

---

[2] We identify characters in a macro way. We extract the subject of each sentence in the story. We regard *name entity* or *noun* as the character of the sentence. In principle, in this way, we can generate an ending for any *noun*. Since there are few endings regarding nouns as the characters in the training data (such as the ending with "car" as the character), the proposed model is difficult to generate high-quality endings for those characters.

4.2   Experimental Settings

We use the GloVe.6B [24] pre-trained word embedding, and the number of dimensions is 200. The hidden size of the LSTM cell is 512. Since the size of the character representation and the encoder's final state both are 512, the number of product candidates will be $512 + 1 = 513$. A larger dimension size brings large computation costs for the model and the device. In summary, the detailed experimental settings are provided as follows:

- We use 66,881 stories for training.
- We use 3073 stories for validation, which is shared by the two test sets.
- We have two test sets with 3073 stories each. We refer to the two test sets as the sufficient test set and the inadequate test set respectively. In the sufficient test set, the character in the ground truth ending also appears multiple times in the story context, and in the inadequate test set is the opposite. These two test sets evaluate the performance of our proposed model when the character information is sufficient and inadequate, respectively.
- We use Momentum Optimizer to update parameters when training and empirically set the momentum to be 0.9.
- The size of the character representation and the encoder hidden state are 512.
- We select the product candidates generated using $VecB_0$, $VecB_{128}$, $VecB_{256}$, $VecB_{384}$ and $VecB_{512}$ five breaking operation.
- The number of product candidates will be $512 + 1 = 513$.

Note that the reason for the selection of the product candidates is analyzed in *Combination Analysis on Product Candidates* in Sect. 4.5.

4.3   Baselines

We compared our model with the following state-of-the-art baseline methods:

Seq2Seq [17]: A vanilla encoder-decoder model with an attention mechanism. The model treats the story context as a single sentence.

HAN [33]: A hierarchical attention architecture is applied to encode text information so as to generate the context representation.

IE [10]: It adopts an incremental encoding scheme to represent context clues and applies commonsense knowledge by multi-source attention.

T-CVAE [29]: It proposes a conditional variational autoencoder based on Transformer for missing plot generation.

MGCN-DP [11]: It leverages multi-level graph convolutional networks over dependency parse trees to capture dependency relations and context clues.
    In addition, we introduce two variations of the proposed CoSEG model:

CoADD: We replace the VBF Module in CoSEG with an element-wise summation.

CoCAT: We concatenate the character representation and the encoder final state, and pass the concatenated vector through a linear layer to obtain the character-related story context representation.

4.4   Evaluation Metrics

We evaluate our model from two perspectives: the quality of the generated endings and the ability to customize endings.

Quality Evaluation. We adopt two kinds of evaluations to investigate the ability of the proposed method and the baselines in generating high-quality story endings.

Automatic Evaluation: We use perplexity (PPL) and BLEU (BLEU-1, BLEU-2 and BLEU-3) [23] to evaluate the quality of the generated endings. A smaller PPL and a higher BLEU indicate a better ending.

Manual Evaluation: We hire three evaluators, who are experts in English, to evaluate the generated story endings. We randomly sampled 200 stories from the two test sets and obtained 1400 endings from the seven models for each test set. Evaluators need to score the generated endings in terms of two criteria: coherency and grammar. The coherency score measures whether the endings are coherent with the story context; specifically, the score of 3 denotes coherency, the score of 1 denotes coherency to some extent, and the score of 0 denotes no coherency at all. In addition, the grammar score measures whether there are grammatical errors in generated endings; a grammar score is 0 if endings have errors, and 1 otherwise.

Ability to Customize Endings. We randomly sample 200 stories from the two test sets and generate ending for one random character in each story. To evaluate the ability to customize endings for different characters, we propose three evaluation metrics:

Success Rate (SucR): SucR measures whether the subject of the generated ending is the selected character.

Rationality: We adopt three levels to evaluate whether the generated ending matches the selected character given the story context: level 3 denotes perfect matching, level 1 denotes partial matching, and level 0 for mismatching.

Discrimination Degree (DiscD): Given an ending generated by our proposed model, we further hire three evaluators to choose which character is the ending generated for. If the character chosen by the evaluator is consistent with the selected character, it scores 1; and 0 otherwise.

4.5   Evaluation Results

Automatic Evaluation. The automatic evaluation results for the sufficient and inadequate test sets are shown in Table 1. From the table, we can observe the following:

1) In both the sufficient and inadequate test sets, our model has lower perplexity and higher BLEU scores than the baselines. Specifically, in terms of perplexity,

Table 1. Automatic evaluation results of the sufficient and the inadequate test set.

Sufficient				
Model	PPL	BLEU-1 (%)	BLEU-2 (%)	BLEU-3 (%)
Seq2Seq	13.26	22.46	6.88	4.21
HAN	13.43	22.43	6.96	4.47
IE	12.08	23.08	7.43	4.67
T-CVAE	11.21	23.72	8.05	5.11
MGCN-DP	11.01	23.90	8.11	5.34
CoADD	12.14	23.92	7.74	4.68
CoCAT	11.45	24.26	8.53	5.41
CoSEG	**9.99**	**25.28**	**9.10**	**5.93**
Inadequate				
Model	PPL	BLEU-1 (%)	BLEU-2 (%)	BLEU-3 (%)
Seq2Seq	21.81	17.13	3.76	1.76
HAN	24.26	17.15	4.08	2.32
IE	16.90	18.40	4.89	2.78
T-CVAE	17.08	22.10	7.05	4.22
MGCN-DP	18.16	20.89	5.90	3.68
CoADD	14.53	21.83	6.99	4.03
CoCAT	15.08	24.50	9.09	5.26
CoSEG	**11.45**	**26.06**	**9.80**	**5.70**

CoSEG outperforms **MGCN-DP, T-CVAE, IE, HAN** and **Seq2Seq** by 1.02/ 1.22/ 2.09/ 3.44/ 3.27 respectively in the sufficient test set, and by 6.01/ 5.63/ 5.45/ 12.81/ 10.36 respectively in the inadequate test set. In addition, in terms of BLEU-1, **CoSEG** outperforms **MGCN-DP, T-CVAE, IE, HAN** and **Seq2Seq** by 1.38%/ 1.56%/ 2.2%/ 2.85%/ 2.82% respectively in the sufficient test set, and by 5.17%/ 3.96%/ 7.66%/ 8.91%/ 8.93% respectively in the inadequate test set.

2) Our **CoSEG** model has the smallest performance gap between the two test sets, which illustrates the performance of our model is not easily affected by the amount of information. Specifically, the perplexity increased by 1.55 in the inadequate test set based on the sufficient test set, and the BLEU-1 increased by 0.78%.

3) In both the sufficient and inadequate test set, the **CoSEG** model outperforms the **CoADD** and the **CoCAT** a lot, which illustrates the interaction ability of the VBF Module is much stronger than the addition and concatenation.

Manual Evaluation. The manual evaluation results for the sufficient and inadequate test sets are shown in Table 2, where we can observe:

Table 2. Manual evaluation results of the sufficient and the inadequate test set.

Model	Sufficient		Inadequate	
	Coherency	Grammar	Coherency	Grammar
Seq2Seq	1.395	0.655	0.905	0.780
HAN	1.160	0.685	0.600	0.785
IE	1.360	0.760	1.210	0.820
T-CVAE	1.750	0.785	1.440	0.815
MGCN-DP	1.760	0.780	1.315	0.795
CoADD	1.690	0.775	1.220	0.760
CoCAT	1.855	0.605	0.965	0.705
CoSEG	**1.880**	**0.805**	**1.620**	**0.835**

In both the sufficient and inadequate test set, the **CoSEG** model obtains the best coherency score and the best grammar score. Specifically, in terms of Coherency, **CoSEG** outperforms **CoCAT, CoADD, MGCN-DP,T-CVAE, IE, HAN** and **Seq2Seq** by 0.025/ 0.19/ 0.12/ 0.13/ 0.52/ 0.72/ 0.485 respectively in sufficient test set, and by 0.655/ 0.4/ 0.305/ 0.18/ 0.41/ 1.02/ 0.715 respectively in inadequate test set. Moreover, in terms of Grammar, **CoSEG** outperforms **CoCAT, CoADD, MGCN-DP, T-CVAE, IE, HAN** and **Seq2Seq** by 0.2/ 0.03/ 0.025/ 0.02/ 0.045/ 0.12/ 0.15 respectively in sufficient test set, and by 0.13/ 0.075/ 0.04/ 0.02/ 0.015/ 0.05/ 0.055 respectively in inadequate test set.

Table 3. Ability to Customize Endings.

Testset	SucR	Rationality	DiscD
Sufficient	0.855	1.965	0.755
Inadequate	0.605	1.980	0.555

Ability to Customize Endings. The ability to customize endings for different characters of our model is shown in Table 3, where we can observe:

1) In the sufficient test set, the success rate (SucR) and the discrimination degree (DiscD) are 85.5% and 75.5% respectively, which indicates that our model is able to identify the differences between characters. The SucR and DiscD in the inadequate test set are lower than there in the sufficient test set, which illustrates that the amount of character information has a certain influence on distinguishing different characters.
2) In both the sufficient and inadequate test sets, the rationality of the customized endings is close to 2.0 on the premise that the maximum score is 3.0. It indicates that our model has a high probability of 66% to predict a reasonable ending for each character.

Table 4. Experimental results of several different combinations.

Model	PPL
CoSEG (0)	12.08
CoSEG (128)	14.74
CoSEG (256)	13.06
CoSEG (0-256-512)	10.93
CoSEG (0-128-256-384-512)	**9.99**

Combination Analysis on Product Candidates. We conduct experiments on several different combinations of product candidates. Specifically, the **CoSEG (n)** selects the product candidate generated using the $VecB_n$ breaking operation; the **CoSEG (0-256-512)** selects the product candidates generated using $VecB_0$, $VecB_{256}$ and $VecB_{512}$ three breaking operations; the **CoSEG (0-128-256-384-512)** selects the product candidates generated using $VecB_0$, $VecB_{128}$, $VecB_{256}$, $VecB_{384}$ and $VecB_{512}$ five breaking operations.

The experimental results are shown in Table 4. We can observe that **CoSEG (0-128-256-384-512)** achieves the best performance. The result explains that we finally selected the product candidates generated using $VecB_0$, $VecB_{128}$, $VecB_{256}$, $VecB_{384}$ and $VecB_{512}$ five breaking operations and utilize the selected five product candidates as inputs to C-CA module. In addition, the result is attributed to the fact that **CoSEG (0-128-256-384-512)** involves more and smaller candidate which facilitate generating fine-grained semantic elements and providing more semantic combinations.

5   Case Study

5.1   Ground-Truth Endings

We present several examples of ground-truth story endings generated by baselines and our model in Table 5 to demonstrate that our model is able to generate more natural and more character-related endings than the baselines. Specifically, in the first story in Table 5, the ending generated by baseline **T-CVAE** makes the reader feel that the character is not smart enough, it has begun to *rain*, and it should not be very useful to *look at the sky* at this time; the baseline **IE** only describes *He had to go home*, without other details. And other baselines predicted wrong endings. By contrast, our model **CoSEG** allows the character to take an appropriate action, *get an umbrella*, according to the given story context, *started raining*.

In addition, the second story in Table 5 is derived from the inadequate test set. In this example, the context is a story about a wrong lamp. The baselines **HAN**, **IE** and **MGCN-DP** describe *She/We had to clean it up*, and **IE** further describes *get a new one*. Our model **CoSEG** not only allows the character to have an appropriate emotion, *I was so upset*, but also let the character take a reasonable action, *decided to get her a new one*. Obviously, the ending generated

Table 5. Case study: endings generated by different models.

Context:	Ned was walking in the park one day. He noticed the sky started to turn gray. Ned turned back toward his house. He didn't quite get home before it started raining
Gold:	Ned ran inside, a little wet, but happy to be home
Seq2Seq:	He went back to his car and bought with his warm
HAN:	He had a clean mess
IE:	He had to go home and go home
T-CVAE:	Ned looked around and saw the sky
MGCN-DP:	he was very happy with his new car!
CoADD:	He decided to go to the store to buy more umbrella
CoCAT:	He went outside to find his car
CoSEG:	He decided to go to the store to get an umbrella
Context:	We removed a lamp last week during carpet cleaning After the cleaners left **my wife** put the lamp back The lighting seemed much dimmer Then **she** realized she had brought out the wrong lamp
Ground truth:	I said I had not even noticed
Seq2Seq:	She was so relieved
HAN:	She had to clean it up
IE:	She had to clean it up and get a new one
T-CVAE:	I bought a new lamp
MGCN-DP:	We had to clean it up
CoADD:	I was so mad that she was able to get it fixed
CoCAT:	She was so happy to have a new lamp
CoSEG:	I was so **upset, I decided** to **get her** a new one
Context:	Larry **needed a job** He was let go from a job a month ago and hadn't found another one His **bills were starting to mount up** Larry was determined and applied to many jobs every day
Ground truth:	He was finally hired
Seq2Seq:	He was able to get a job in time and help for his job
HAN:	He was able to get a job in a new city
IE:	He was hired for a new job
T-CVAE:	He have a job
MGCN-DP:	He said he was going to be more careful
CoADD:	He was able to get to work and get a job
CoCAT:	He was hired at the store and bought a brand new car
CoSEG:	He finally **got a job** offer and **paid his bills**

Table 6. Customizing endings for each character using our proposed CoSEG model.

Context:	**I** ran and climbed over the fence
	My son was lying in the pea gravel on the road
	The car had swerved just in time
	I raged at the **driver** for not even stopping
Ground truth:	I called 911 to come get my child
Endings for each character:	
For **I**:	I ended up **falling** and I had to **go to the hospital**
For **driver**:	The driver **pulled the car** and I was able to get it back
For **son**:	My son was **upset**
Context:	**I** had a dental appointment I had to go to today
	While getting my teeth checked, my **dentist** told **I** had a cavity
	He said it's probably because **I**'ve been using subpar toothpaste
	I've been using the same toothpaste **he** recommended six months ago
Ground truth:	Thanks a lot for the recommendation, doc
Endings for each character:	
For **I**:	I am **glad I have a new toothpaste**
For **dentist**:	The dentist **told me** that he **had to get a new toothpaste**
Context:	**John** was awakened by a phone call
	Answering, **John** realized it was his buddy, **Rich**
	Rich said he was stranded on a highway just outside of town
	John drove out to pick up **Rich**
Ground truth:	John drove Rich home, where they both fell asleep on the couch
Endings for each character:	
For **Rich**:	He **drove to the mall** and **bought a new car**
For **John**:	John and his friends **went to the park** and had a great time

by **CoSEG** takes the character's emotions (***upset***), behaviors (***decided...get...***), and relationships (*get **her** a...*) into account, which illustrates the ability of our model to obtain character's personality.

The third example in Table 5 is sampled from the sufficient test set. In this example, the context is a story about a man's bills mount up and he needs a job. Most of the baselines describe *He get a job*, as well as the ground truth and our proposed model. In addition, different from all baselines and the ground truth endings, our model further describes the man's purpose of looking for a job, ***paid bills***, which also demonstrates that our model is able to generate a more character-related ending.

5.2 Character-Orient Endings

In this section, we present three examples with character-orient endings (including the ground-truth endings) generated by our method in Table 6, to illustrate the ability of our model to customize endings for different characters.

In the first example in Table 6, our model customizes endings for the characters *I*, *driver* and *son*. The context is a story about a car accident that happened to a father and son. The endings generated by our model are that the *father* had to go to the hospital, the *driver* ran away, and the *son* was upset. These three endings generated by our model describe the behavior of the *father* and the *driver* and the mood of the *son*, which demonstrate the effectiveness of our proposed model to customize endings for different characters. The second example in Table 6 is a story in that a man has a cavity because he uses the toothpaste recommended by the dentist, and our model customizes endings for the characters *I* and *dentist*. The third example in Table 6 is a story in that Rich is stranded on a highway and he calls John to pick him up. Our model identifies the differences between Rich and John, generating endings, *He drove to the mall and bought a new car*, for Rich, and *John and his friends went to the park and had a great time*, for John. Rich **need a new car**, because his car is **stranded on the highway**.

6   Conclusion

To tackle the challenging task of customizing story endings for different characters, we propose a Character-oriented Story Ending Generator (CoSEG). Experimental results demonstrate that our proposed model can not only generate more coherent and diversified story endings compared with state-of-the-art methods but also effectively customize the ending for each character in a story.

Acknowledgements. This work is supported by the National Key Research and Development Program of China(No. 2019YFB1406300), National Natural Science Foundation of China (No. 61502033) and the Fundamental Research Funds for the Central Universities.

References

1. Chen, J., Chen, J., Yu, Z.: Incorporating structured commonsense knowledge in story completion. In: AAAI, pp. 6244–6251. AAAI Press (2019)
2. Crittenden, J.C., Trussell, R.R., Hand, D.W., Howe, K.J., Tchobanoglous, G.: Ion Exchange. John Wiley, New York (2012)
3. Devlin, J., Chang, M., Lee, K., Toutanova, K.: BERT: pre-training of deep bidirectional transformers for language understanding. In: NAACL-HLT (1), pp. 4171–4186 (2019)
4. Fan, A., Lewis, M., Dauphin, Y.N.: Hierarchical neural story generation. In: ACL, pp. 889–898 (2018)
5. Fan, A., Lewis, M., Dauphin, Y.N.: Strategies for structuring story generation. In: ACL, pp. 2650–2660 (2019)
6. Fedus, W., Goodfellow, I.J., Dai, A.M.: Maskgan: Better text generation via filling in the _ _ _ _ _ _ _. In: ICLR (2018)
7. Feng, C., Shi, C., Hao, S., Zhang, Q., Jiang, X., Yu, D.: Hierarchical social similarity-guided model with dual-mode attention for session-based recommendation. Knowl. Based Syst. **230**, 107380 (2021)

8. Gao, J., Bi, W., Liu, X., Li, J., Shi, S.: Generating multiple diverse responses for short-text conversation. In: AAAI, pp. 6383–6390 (2019)
9. Guan, J., Huang, F., Huang, M., Zhao, Z., Zhu, X.: A knowledge-enhanced pre-training model for commonsense story generation. Trans. Assoc. Comput. Linguistics **8**, 93–108 (2020)
10. Guan, J., Wang, Y., Huang, M.: Story ending generation with incremental encoding and commonsense knowledge. In: AAAI, pp. 6473–6480 (2019)
11. Huang, Q., et al.: Story ending generation with multi-level graph convolutional networks over dependency trees. In: AAAI, pp. 13073–13081 (2021)
12. Kennedy, R.H.: Elution of uranium values from ion exchange resins (1959)
13. Kim, J., Benjamin, M.M.: Modeling a novel ion exchange process for arsenic and nitrate removal. Water Res. **38**(8), 2053–2062 (2004)
14. Kim, Y.: Convolutional neural networks for sentence classification. In: EMNLP, pp. 1746–1751 (2014)
15. Liu, D., et al.: A character-centric neural model for automated story generation. In: AAAI, pp. 1725–1732 (2020)
16. Luo, F., et al.: Learning to control the fine-grained sentiment for story ending generation. In: ACL, pp. 6020–6026 (2019)
17. Luong, T., Pham, H., Manning, C.D.: Effective approaches to attention-based neural machine translation. In: EMNLP, pp. 1412–1421 (2015)
18. Mao, H.H., Majumder, B.P., McAuley, J.J., Cottrell, G.W.: Improving neural story generation by targeted common sense grounding. In: EMNLP, pp. 5987–5992 (2019)
19. Martin, L.J., et al.: Event representations for automated story generation with deep neural nets. In: AAAI, pp. 868–875 (2018)
20. Mo, L.: Incorporating sentimental trend into gated mechanism based transformer network for story ending generation. Neurocomputing **453**, 453–464 (2021)
21. Mostafazadeh, N., Vanderwende, L., Yih, W., Kohli, P., Allen, J.F.: Story cloze evaluator: Vector space representation evaluation by predicting what happens next. In: Proceedings of the 1st Workshop on Evaluating Vector-Space Representations for NLP, pp. 24–29 (2016)
22. Mou, L., Song, Y., Yan, R., Li, G., Zhang, L., Jin, Z.: Sequence to backward and forward sequences: a content-introducing approach to generative short-text conversation. In: COLING, pp. 3349–3358 (2016)
23. Papineni, K., Roukos, S., Ward, T., Zhu, W.: Bleu: a method for automatic evaluation of machine translation. In: ACL, pp. 311–318 (2002)
24. Pennington, J., Socher, R., Manning, C.D.: Glove: global vectors for word representation. In: EMNLP, pp. 1532–1543 (2014)
25. Schütt, K.T., Gastegger, M., Tkatchenko, A., Müller, K.R., Maurer, R.J.: Unifying machine learning and quantum chemistry - a deep neural network for molecular wavefunctions. Nat. Commun. (2019)
26. Segler, M.H.S., Preuss, M., Waller, M.P.: Planning chemical syntheses with deep neural networks and symbolic AI. Nature **555**(7698), 604 (2018)
27. Tu, L., Ding, X., Yu, D., Gimpel, K.: Generating diverse story continuations with controllable semantics. In: NGT@EMNLP-IJCNLP, pp. 44–58 (2019)
28. Vaswani, A., et al.: Attention is all you need. In: NIPS, pp. 5998–6008 (2017)
29. Wang, T., Wan, X.: T-CVAE: transformer-based conditioned variational autoencoder for story completion. In: IJCAI, pp. 5233–5239 (2019)
30. Welleck, S., Brantley, K., III, H.D., Cho, K.: Non-monotonic sequential text generation. In: ICML Proceedings of Machine Learning Research, vol. 97, pp. 6716–6726 (2019)

31. Xu, J., Ren, X., Zhang, Y., Zeng, Q., Cai, X., Sun, X.: A skeleton-based model for promoting coherence among sentences in narrative story generation. In: EMNLP, pp. 4306–4315 (2018)
32. Xu, P., et al.: MEGATRON-CNTRL: controllable story generation with external knowledge using large-scale language models. In: EMNLP (1), pp. 2831–2845 (2020)
33. Yang, Z., Yang, D., Dyer, C., He, X., Smola, A.J., Hovy, E.H.: Hierarchical attention networks for document classification. In: NAACL, pp. 1480–1489 (2016)
34. Zhang, Q., Cao, L., Shi, C., Niu, Z.: Neural time-aware sequential recommendation by jointly modeling preference dynamics and explicit feature couplings. IEEE Trans. Neural Netw. Learn. Syst. 1–13 (2021)

Vec2Node: Self-Training with Tensor Augmentation for Text Classification with Few Labels

Sara Abdali[1]([✉]), Subhabrata Mukherjee[2], and Evangelos E. Papalexakis[3]

[1] Georgia Institute of Technology, Atlanta, USA
sabdali3@gatech.edu, sabda005@ucr.edu
[2] Microsoft Research, Redmond, USA
Subhabrata.Mukherjee@microsoft.com
[3] University of California, Riverside, USA
epapalex@cs.ucr.edu

Abstract. Recent advances in state-of-the-art machine learning models like deep neural networks heavily rely on large amounts of labeled training data which is difficult to obtain for many applications. To address label scarcity, recent work has focused on data augmentation techniques to create synthetic training data. In this work, we propose a novel approach of data augmentation leveraging tensor decomposition to generate synthetic samples by exploiting local and global information in text and reducing concept drift. We develop `Vec2Node` that leverages self-training from in-domain unlabeled data augmented with tensorized word embeddings that significantly improves over state-of-the-art models, particularly in low-resource settings. For instance, with only 1% of labeled training data, `Vec2Node` improves the accuracy of a base model by 16.7%. Furthermore, `Vec2Node` generates explicable augmented data leveraging tensor embeddings.

Keywords: Text augmentation · Tensor decomposition · Self-training

1 Introduction

In recent years, neural network models have obtained state-of-the-art performance in several language understanding tasks employing non-contextualized `FastText` [4] as well as contextualized `BERT` [5] word embeddings. Even though these models have been greatly successful, they rely on large amounts of labeled training data for their state-of-the-art performance. However, labeled data is not only difficult to obtain for many applications, especially for tasks dealing with sensitive information, but also requires time consuming and costly human annotation efforts. To mitigate label scarcity, recent techniques such as self-training [6,11] and few

---

S. Abdali—This research work was conducted while the first author was a Ph.D. student at the University of California, Riverside.

Supplementary Information The online version contains supplementary material available at https://doi.org/10.1007/978-3-031-26390-3_33.

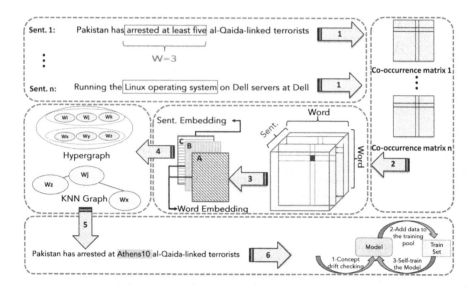

Fig. 1. Overview of the proposed approach.

shot learning [24, 28] methods have been developed to learn from large amounts of in-domain unlabeled or augmented data. The core idea of self-training is to augment the original labeled dataset with pseudo-labeled data [11] in an iterative teacher-student learning paradigm. Traditional self-training techniques are subject to gradual concept drift and error propagation [24, 29]. In general, data augmentation techniques aim to generate synthetic data with similar characteristics as the original ones. While data augmentation has been widely used for image classification tasks [18] leveraging techniques like image perturbation (e.g., cropping, flipping) and adding stochastic noise, there has been limited exploration of such techniques for text classifiers. Recent work on data augmentation for text classification like [28] rely on auxiliary resources like an externally trained Neural Machine Translation (NMT) system to generate back-translations[1] for consistency learning.

In contrast to the above works, we solely rely on the available in-domain unlabeled data for augmentation without relying on external resources like an NMT system. To this end, we develop `Vec2Node` that employs tensor embeddings to consider both the global context and local word-level information. In order to do so, we leverage the association of words and their tensor embeddings with a graph-based representation to capture local and global interactions. Additionally, we learn this augmentation and the underlying classification task jointly to bridge the gap between self-training and augmentation techniques that are learned in separate stages in prior works.

Our contributions can be summarized as follows:

- A novel tensor embedding based data augmentation technique for text classification with few labels.

---

[1] Process of translating a text to another language and translating it back to the original language.

- A dynamic augmentation technique for detecting concept drift learned jointly with the downstream task in a self-training framework.
- Extensive evaluation on benchmark text classification datasets demonstrate the effectiveness of our approach, particular in low-resource settings with limited training labels along with interpretable explanations.

2   Background

In this section, first, we present mathematical background; then we discuss the problem formulation followed by the details of the proposed method.

2.1   Tensor

A data tensor $\mathcal{D} \in \mathbb{R}^{I_1 \times I_2 \times \cdots \times I_M}$ is a multi-way array i.e., an array with three or more than three dimensions where the dimensions are usually referred to as modes [13].

2.2   Singular Value Decomposition (SVD)

Singular Value Decomposition (SVD) is a decomposition technique which factorizes a matrix X into the following three matrices [13]:

$$X = \mathbf{U}\mathbf{\Sigma}\mathbf{V}^T \tag{1}$$

where the columns of $\mathbf{U}$ and $\mathbf{V}$ are orthonormal and $\mathbf{\Sigma}$ is a diagonal matrix with positive real entries. A matrix can be estimated by a rank-R SVD as a sum of R rank-1 matrices:

$$X \approx \Sigma_{r=1}^{R}\sigma_{\mathbf{r}}\mathbf{u}_r \circ \mathbf{v}_r \tag{2}$$

2.3   Canonical Polyadic (CP) Decomposition

Canonical Polyadic (CP) or PARAFAC is an extension of SVD for higher mode arrays i.e., tensors [10]. CP/PARAFAC factorizes a tensor into a sum of rank-1 tensors. For instance, a 3-mode tensor is decomposed into a sum of outer products of three vectors:

$$\mathcal{X} \approx \Sigma_{r=1}^{R}\mathbf{a}_r \circ \mathbf{b}_r \circ \mathbf{c}_r \tag{3}$$

where $\mathbf{a}_r \in \mathbb{R}^I$, $\mathbf{b}_r \in \mathbb{R}^J$, $\mathbf{c}_r \in \mathbb{R}^K$ and the outer product is given by [19,20]:

$$(\mathbf{a}_r, \mathbf{b}_r, \mathbf{c}_r)(i, j, k) = \mathbf{a}_r(i)\, \mathbf{b}_r(j)\, \mathbf{c}_r(k)\, \forall i, j, k \tag{4}$$

Factor matrices are defined as $\mathbf{A} = [\mathbf{a}_1\ \mathbf{a}_2 \ldots \mathbf{a}_R]$, $\mathbf{B} = [\mathbf{b}_1\ \mathbf{b}_2 \ldots \mathbf{b}_R]$, and $\mathbf{C} = [\mathbf{c}_1\ \mathbf{c}_2 \ldots \mathbf{c}_R]$ where R is the rank of the decomposition, which is also the number of columns in the factor matrices. PARAFAC optimization problem is formulated as [13]:

$$\min_{A,B,C} = \|\mathcal{X} - \Sigma_{r=1}^{R} \mathbf{a}_r \circ \mathbf{b}_r \circ \mathbf{c}_r\|_F^2 \tag{5}$$

One effective way to optimize the above problem is to use Alternating Least Squares (ALS) which solves for each one of the factor matrices by fixing the others and cycles over all matrices iteratively until convergence [13].

2.4   KNN Tensor Graph

A k-nearest-neighbor (KNN) graph is a model for representing the nodes in a given feature space such that the k most similar nodes are connected with edges, weighted by a similarity measure [9]. In this work, we use a co-occurrence tensor to map words into an embedding space such that each word (represented by a vector) is a node in the embedding space and then we measure the similarity of the nodes using the Euclidean distance between the corresponding vectors.

2.5   Hypergraph

Hypergraphs [7,31] are an extension of graphs where an edge may connect more than two nodes to indicate higher-order relationships between the nodes. In contrast to a single weighted connection in traditional graphs, an edge in a hypergraph is a subset of nodes that are similar in terms of features or distance.

3   Vec2Node Framework

3.1   Problem Formulation

Given a corpus D of labeled data, we aim to **generate** D' that augments D and improves the performance of a classification model M on the downstream task i.e. $f(M(D)) > f(M(D + D'))$, where f is an evaluation measure (e.g., accuracy).

To address the above problem, we propose a novel tensor-based approach for generating synthetic texts from the corpus D. The details of the proposed method, henceforth referred to as Vec2Node, are described in the following section.

3.2   Data Augmentation

Vec2Node leverages tensor decomposition to find word and text embeddings. These are further used for graph-based representations of the word vectors in order to find similar ones as replacement candidates to generate synthetic samples while minimizing the concept drift. Vec2Node consists of the following steps:

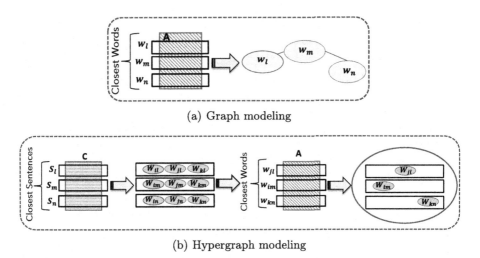

(a) Graph modeling

(b) Hypergraph modeling

Fig. 2. Graph and hypergraph modeling for representing words' homophily.

Tensor-Based Corpus Representation. Textual content of documents can be represented by a co-occurrence tensor [1,8] which embeds the patterns shared between different topics or classes. These patterns are formed by words that are more likely to co-occur in documents of the same class. We leverage similar principles to capture existing similarities within a given text. To this end, given a set of samples, we first slide a window of size w across the text of each sample and capture the co-occurring words to represent them in a co-occurrence matrix. Furthermore, we stack the co-occurrence matrices of all samples to form a 3-mode tensor of dimension $T \times T \times S$ where T is the number of terms or words in the entire corpus and S is the number of samples. This process is demonstrated in Fig. 1. The rationale behind this approach is to capture the context (words) for a given target word. In the experimental section, we demonstrate how this approach captures contextually related words.

Decomposing Tensors into Word and Text Embeddings. The objective of this step is to embed the words and the texts of the corpus into rank-R representations which are later used for calculating word similarities. As explained in Sect. 2, we use CP/PARAFAC to decompose our 3-mode tensor as:

$$\mathcal{X} \approx \Sigma_{r=1}^{R} \mathbf{a}_r \circ \mathbf{b}_r \circ \mathbf{c}_r \tag{6}$$

where $\mathbf{A} = [\mathbf{a}_1 \ \mathbf{a}_2 \dots \mathbf{a}_R]$, $\mathbf{B} = [\mathbf{b}_1 \ \mathbf{b}_2 \dots \mathbf{b}_R]$, and $\mathbf{C} = [\mathbf{c}_1 \ \mathbf{c}_2 \dots \mathbf{c}_R]$ are embedding representations of word, word and text respectively. The word co-occurrence $\mathbf{A}$ and $\mathbf{B}$ are symmetric. Thus, they capture the same information.

Tensor Embeddings for KNN and Hypergraph Homophily Representation. In this step, we exploit tensor embedded representations $\mathbf{A}$ and $\mathbf{C}$ to estimate words and texts homophilies (similarities) to find the best candidates for replacement in a given text and generate new synthetic samples. We leverage the following two graph based modelings:

K Nearest Neighbor Graph Modeling. Consider the factor matrix $\mathbf{A}$ (or $\mathbf{B}$, as they are symmetric and capture the same information) of dimension $N \times R$ where each row is a tensor word embedding in R-dimensional space $\mathbb{R}^R$. We represent the ith row of this matrix which corresponds to word i as a node in R dimensional space. This allows for calculating the Euclidean distance between the nodes and represent the similarity between the nodes (words) as a weighted undirected edge. The Euclidean distance between rows i and j measures the similarity of these two vectors in R-dimensional space.

Hypergraph Modeling. Spitz et al. [23] propose a hypergraph modeling of the documents where hyperedges are defined by consecutive sentences and words within the text. In that work, the similarity is considered based on spatial closeness. However, in this work, we first leverage the factor matrix $\mathbf{C}$ corresponding to text embedding to find K closest texts and then we use factor matrix $\mathbf{A}$ to find K' closest words within these K samples. Thus, a hyperedge in this hypergraph is the set of K' closest words. The details of this process are shown in Fig. 2. It is worth mentioning that *our proposed model uses KNN tensor graph for modeling word similarities*. However, for comparison purposes we implement `Vec2Node` framework with hypergraph modeling as well.

3.3   Learning with Data Augmentation and Limited Labels

Contextualized Word Replacement. Modeling the corpus using graph or hypergraph representations allows for finding similar words by sorting the edge weights i.e., the Euclidean distances between the nodes, and picking the ones with the smallest distance (i.e., closest words) as the best candidates for replacement and generation of synthetic samples. This process is fully unsupervised given that the tensor decomposition method does not require any labels. Also, it considers local and global contextual information given the graph and tensorial representation of words and texts.

Self-training with Consistency Learning. In order to eliminate noisy samples, we check for *concept drift* between the original samples and the synthetic ones using consistency learning in a self-training framework. Given a few labeled samples $\{x_l, y_l\} \in D_l$ for the downstream task, we first fine-tune a base model with parameter θ.

Consider x_u to be the target augmented pair for a source instance x_l generated using the augmentation technique described before. We can use the current parameters θ of the model to predict the pseudo-label for the target x_u as:

$$y_u = argmax_y \ p(y|x_u; \theta) \tag{7}$$

Since the objective of data augmentation is to generate semantically similar instances for the model, we expect the output labels for the source-target

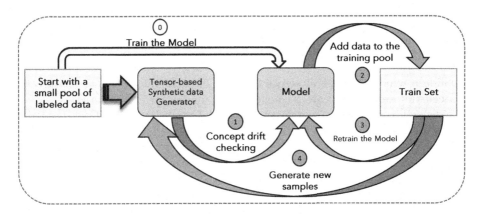

Fig. 3. Few-shot self-training with data augmentation and consistency learning to prevent concept drift.

augmented pair $\{x_l, x_u\}$ to be similar as well; otherwise, we designate this as a concept drift and discard augmented pairs where $y_l \neq y_u$.

We add the remaining target pseudo-labeled data with consistent model predictions with the source data as our augmented training set $\{x_u, y_u\} \in D_u$ and re-train the base model to update θ. The above steps are repeated with iterative training of the base model with pseudo-labeled augmented data until convergence. The optimization objective for the above self-training process can be formulated as:

$$
min_\theta \; \mathbb{E}_{x_l, y_l \in D_l}[-log \; p(y_l|x_l; \theta)]
$$
$$
+ \lambda \; \mathbb{E}_{x_u \in D_u} \mathbb{E}_{y_u \sim p(y|x_u; \theta^*)}[-log \; p(y_u|x_u; \theta)] \quad (8)
$$

where $p(y|x; \theta)$ is the conditional distribution under model parameters θ. θ^* is given by the model parameters from the last iteration and fixed in the current iteration. Similar optimization functions have been used recently in variants of self-training for neural sequence generation [11], data augmentation [28] and knowledge distillation. The details of this process are shown in Fig. 3 with the pseudo-code in Algorithm 1.

3.4   Complexity Analysis

In the proposed `Vec2Node` pipeline, the main computation bottleneck is CP decomposition (CPD). In general, CPD is shown to be in the order of the number of non-zero elements [2] of a tensor. In fact, CPD is very fast and efficient for sparse tensors which is the case in this work due to sparsity of the word co-occurrences. Meanwhile, some methods have been proposed for CPD which are amenable to hundreds of concurrent threads while maintaining load balance and low synchronization costs [21]. Moreover, CPD is an offline step in the `Vec2Node` framework i.e., we only execute it once to obtain the embeddings and there is no need to repeat it while training the model.

Algorithm 1. Self-train `Vec2Node`

Input : Base model M, small labeled set D_l.
Return : Self-trained M.

1. Slide a window of size w across the text of each sample in D_l, capture co-occurring words to create a co-occurrence matrix for each sample.
2. Stack all co-occurrence matrices to create a 3-mode tensor $\mathcal{X}$ of size $T \times T \times S$.
3. Decompose $\mathcal{X}$ into **A,B,C**
4. Use **A,C** to model the corpus using graph/hypergraph representations.
5. Calculate Euclidean distances between the nodes to find the closest words.
6. Train M using $D_l = \{x_l, y_l\}$. Set $D = D_l$.
7. While not converged
   - For $\{x_l, y_l\} \in D$, generate augmented samples D'_u by replacing closest words.
   - Assign pseudo-label y_u to each sample $x_u \in D'_u$ using Eq. 7.
   - If $y_l = y_u$ then $D = D \bigcup \{x_u, y_u\}$.
   - Retrain M using augmented data D using Eq. 8.
8. Return model M

---

4   Experimental Evaluation

In this section, we assess performance of `Vec2Node` against baselines we further introduce and then we conduct an ablation study to evaluate components of `Vec2Node`.

4.1   Baselines

- **Base classifiers to asses the effectiveness of augmentation** We compare against the following base classifiers:
  - `FastText-Softmax` `FastText` is an efficient word embedding which is an extension of Word2Vec. It represents each word as an n-gram of characters. Thus, in contrast to other non-contextualized embeddings such as GloVe and Word2Vec, provides representations for unseen words [4,12]. Considering this advantage of `FastText` over mentioned embeddings, we choose `FastText` with a softmax layer (`FastText-Softmax`), as one of our base classifiers.
  - `BERT` leverages contextualized representations using deep bidirectional transformers. We experiment with the pre-trained checkpoints of HuggingFace[2] [26].
- **Neural Machine Translation (NMT) to assess the effectiveness of `Vec2Node` augmentation** An Encoder-Decoder architecture with recurrent neural networks (RNN) has become an effective and standard approach for Neural Machine Translation (NMT), sequence-to-sequence prediction and data

---

[2] https://github.com/huggingface/transformers.

augmentation. NMT is the core of the Google translation service [27]. We use NMT to translate original sentences into French and then translate them back into English. This process results in synthetic sentences which will be added to the original dataset.

- **GPT-3 to assess the effectiveness of** Vec2Node **augmentation** Generative Pre-trained Transformer 3 (GPT-3) is an autoregressive language model that generates human-like text. In this work, for each training sample, we generate multiple sentences using GPT-3 and train a base classifier on the training set, leveraging classic self-training to assign pseudo labels to the generated samples.
- **NLP Word embeddings to assess the effectiveness of tensor embedding** We experiment with the following word embeddings to investigate the efficacy of the tensor embedding in our proposed Vec2Node framework. For a fair comparison, for all of the following baselines, we retain KNN graph, self-training and concept drift check components of the proposed Vec2Node and only substitute tensor embedding with the following:
  - FastText **embedding.** Not only do we use FastText for classification, but also we replace the tensor embedding by FastText embedding to find the most similar words. we retain other components as mentioned above.
  - Word2Vec **embedding.** A shallow 2-layers neural network proposed in [14]. We use Word2Vec instead of tensor embedding to find the most similar words using cosine similarity. Similarly, we retain other components in Vec2Node.
  - **Random replacement.** We replace tensor-based similarity strategy by random word replacement while retaining self-training and consistency learning.
- **Matrix modeling (**tf-idf**) to compare the effectiveness of tensor modeling against matrix modeling**: First, we create a tf-idf matrix and decompose it into word embeddings using SVD decomposition. Similar to the previous setup, we retain other components in Vec2Node and only replace tensor embedding by tf-idf embedding. Both random replacement and tf-idf, with strong data augmentation and self-training techniques have been shown to obtain very competitive results for text classification [25,28].
- **Hypergraph similarity representation to assess the effectiveness of KNN graph modeling** We investigate the efficacy of KNN graph modeling against hypergraph modeling proposed in [23]. Similar to the above setup, we only replace KNN tensor graph by hypergraph while keeping other components of Vec2Node.
- Vec2Node **with and without self-training and consistency learning** We remove the self-training and consistency learning from the Vec2Node pipeline to assess the effectiveness of aforementioned mechanisms.

Table 1. Dataset statistics.

Dataset	Class	Train	Test	Avg. Words/Doc
SST2	2	67340	872	17
IMDB	2	25000	25000	235
AG News	4	12000	7600	40
DBpedia	14	560000	70000	51

4.2 Evaluation

We experiment on SST2 [22], IMDB [16], AG News [30] and DBpedia [3] with statistics in Table 1, to assess the efficacy of Vec2Node on short, long and multi-label datasets respectively. We report results on the corresponding test splits as available from the mentioned works. To facilitate easy comparison, we report relative accuracy improvement (↑) for all the methods over the base model without augmentation.

Base Classifiers. From Table 3, we observe that Vec2Node with tensor data augmentation obtains on average 16.75% and 10.5% improvement over FastText-Softmax with no augmentation, using only 1% and 5% of labeled training data respectively. In this experiment, Vec2Node is built on top of FastText-Softmax to demonstrate the strength of augmentation. We also observe the relative improvement with augmentation to significantly increase with longer text. For example, the improvement in accuracy for IMDB is 16% more than that on SST2 dataset using 5% of labels. This could be attributed to the shorter context samples not being able to generate diverse variety of synthetic samples that are significantly different from the original ones. However, we still demonstrate significant accuracy improvement with augmentation on SST2 as well. In case of DBpedia classification, which is relatively a hard task, Vec2Node improves the accuracy of base FastText-Softmax by 3–4% using only 1–5% of training labels. As illustrated, when we use 100% of the training data, we still observe improvement in classification accuracy which demonstrates the effectiveness of tensor augmentation in both low and high-resource settings.

In contrast to FastText-Softmax which is trained from the scratch, the BERT model we use here is pre-trained over massive amounts of unlabeled data thereby, works well even in the low-data regime. Thus, to demonstrate the strength of our tensor augmentation i.e., Vec2Node, we choose the few-shot setting with only 0.5% of labeled training data. From Table 2, we observe that Vec2Node using BERT as an encoder along with tensor augmentation to obtain 3% improvement on average over the base BERT model using very few training labels. Meanwhile, augmenting SST2, using BERT as a classifier, improves the overall performance of Vec2Node, where we observe 7.2% improvement of accuracy after augmentation. In case of DBpedia, since it is a very large dataset, even with 0.5% of the labels a pretrained BERT achieves its maximum accuracy. Thus we skip it for this experiment. It is worth emphasizing that the pre-trained BERT outperforms

Table 2. Performance of `FastText-Softmax` classifier with and without `Vec2Node` augmentation.

Dataset	%Train	#Train	w/o `Vec2Node`	w/ `Vec2Node`	Average ↑
SST2	1	673	0.509 ± 0.000	$\mathbf{0.638 \pm 0.0007}$	
	5	3367	0.710 ± 0.100	$\mathbf{0.740 \pm 0.004}$	**(5.46↑)**
	100	67340	0.818 ± 0.0018	$\mathbf{0.823 \pm 0.0006}$	
IMDB	1	250	0.499 ± 0.000	$\mathbf{0.605 \pm 0.004}$	
	5	1250	0.522 ± 0.012	$\mathbf{0.718 \pm 0.001}$	**(10.26↑)**
	100	25000	0.857 ± 0.0007	$\mathbf{0.863 \pm 0.002}$	
AG News	1	1200	0.295 ± 0.003	$\mathbf{0.687 \pm 0.023}$	
	5	6000	0.663 ± 0.001	$\mathbf{0.825 \pm 0.002}$	**(18.56 ↑)**
	100	12000	0.900 ± 0.0003	$\mathbf{0.903 \pm 0.0008}$	
DBpedia	1	5600	0.566 ± 0.000	$\mathbf{0.603 \pm 0.000}$	
	5	28000	0.589 ± 0.015	$\mathbf{0.619 \pm 0.000}$	**(3.06 ↑)**
	100	56000	0.602 ± 0.013	$\mathbf{0.627 \pm 0.000}$	

Table 3. Performance of `FastText-Softmax` classifier with and without `Vec2Node` augmentation.

Dataset	%Train	#Train	w/o `Vec2Node`	w/ `Vec2Node`
SST2	0.5	336	0.754	**0.826(7.2↑)**
IMDB	0.5	125	0.776	**0.783(0.7↑)**
AG News	0.5	600	0.869	**0.880(1.1↑)**

`FastText-Softmax` which is trained from the scratch. However, in both base model settings, `Vec2Node` improves the performance.

Neural Machine Translation (NMT): As reported in Table 4, `Vec2Node` outperforms NMT augmentation strategy as well. We observed that in contrast to synthetic samples of `Vec2Node`, the majority of the synthetic samples created by NMT are quite identical with the original ones and as a result, they do not add diversity to the datasets. `GPT-3` ***Text Generation***: Table 4 also illustrates performance of `Vec2Node` against `GPT-3` on `FastText-Softmax` classifier. while `GPT-3` outperforms `Vec2Node` by only 1.53% on average (all four datasets), it is also significantly larger with 175 billion parameters compared to `Vec2Node` with only few hyper-parameters (i.e., R, w and K) as well as pre-trained over massive amount of web corpora.

Ablation Study. In this part, we conduct an ablation study to evaluate different components of `Vec2Node` namely, tensor embedding, KNN tensor graph, and self-training mechanism for few label classification.

Table 4. Performance of `FastText-Softmax` classifier with augmentations from NMT, GPT-3 and Vec2Node.

Dataset	%Train	#Train	w/o Aug.	w/ NMT	w/ GPT3	w/ Vec2Node
SST2	5	3367	0.710 ± 0.100	$0.715 \pm 0.008(0.50\uparrow)$	$0.700 \pm 0.005(0.01\downarrow)$	$\mathbf{0.740 \pm 0.004(3.00\uparrow)}$
IMDB	5	1250	0.522 ± 0.012	$0.692 \pm 0.016(17.00\uparrow)$	$\mathbf{0.795 \pm 0.001(27.3\uparrow)}$	$0.718 \pm 0.001(19.06\uparrow)$
AG News	5	6000	0.663 ± 0.001	$0.786 \pm 0.021(12.30\uparrow)$	$0.801 \pm 0.001(13.8\uparrow)$	$\mathbf{0.825 \pm 0.002(16.20\uparrow)}$
DBpedia	5	28000	0.589 ± 0.015	$0.610 \pm 0.005(2.10\uparrow)$	$\mathbf{0.667 \pm 0.060(7.8\uparrow)}$	$0.619 \pm 0.000(3.00\uparrow)$

NLP Word Embeddings vs. Tensor Embeddings. Table 5 demonstrates performance of `Vec2Node` with different replacement strategies including `FastText` and `Word2Vec`. As illustrated, with longer texts as in IMDB and AG News, `Vec2Node` with tensor embedding, outperforms other word embeddings due to more tangible word co-occurrences in the texts. In case of SST2, where samples are short phrases with fewer co-occurring non-stop words, we observe less diverse synthetic samples. However, we may conclude that tensor embedding outperform other embeddings in general.

Tensor Modeling vs. Matrix Modeling and `tf-idf` Embedding. In addition, Table 5 illustrates the performance of `Vec2Node` against Random and `tf-idf` word replacement strategies. Random and `tf-idf` do not consider the local and global contextual information of the target word during replacement, and, consequently, generate noisy samples. `Vec2Node` captures both local and global context to outperform these strategies. In case of large datasets such as DBpedia, we observe that matrix modeling results in a very large and memory inefficient representation and suffers from compute bottleneck for SVD decomposition, whereas tensor modeling is memory efficient due to the fact that it breaks down a large co-occurrence matrix into multiple, yet smaller ones.

KNN Graph vs. Hypergraph for Word Similarities. From Table 6, we observe that `Vec2Node` with KNN graph representation to capture word similarities, outperform hypergraph representation on all four datasets. The KNN graph captures globally similar words, whether or not they co-occur in similar sentences, whereas the hypergraph representation confines the similarity search to words that co-occur in similar texts. This may lead to situations in which all words in a given sentence are replaced by the same word due to lack of candidates in the pool. Moreover, similar words may occur in different contexts and in such cases hypergraph does not capture them.

`Vec2Node` *with and without Self-Training and Consistency Learning.* In this experiment, we ablate the self-training and consistency learning components in `Vec2Node` to analyze their contribution to the results in Table 7. We observe the self-training component where the model leverages augmented data and pseudo-labels for consistency learning to further improve the performance of `Vec2Node` by 8.2% on all datasets. Also, this component along with augmentation jointly contributes to 10.45% improvement of `Vec2Node` over that of `FastText-Softmax`.

Table 5. Vec2Node with different word strategies on FastText-Softmax classifier

Dataset	%Train	Matrix (tf-idf)	Random	Word2Vec	FastText	Tensor (Our)
SST2	5	0.733 ± 0.004 (2.3↑)	0.737 ± 0.001(2.7↑)	**0.759 ± 0.03(4.9↑)**	0.730 ± 0.025(2↑)	0.740 ± 0.004(3.0↑)
IMDB	5	0.602 ± 0.021(7.9↑)	0.659 ± 0.013(13.7↑)	0.663 ± 0.01(14.1↑)	0.680 ± 0.045(15.8↑)	**0.718 ± 0.001(19.6↑)**
AG News	5	0.807 ± 0.002(14.3↑)	0.799 ± 0.002(13.6↑)	0.806 ± 0.042(14.3↑)	0.810 ± 0.054(14.7↑)	**0.825 ± 0.001(16.2↑)**
DBpedia	5	Out of Memory	0.619 ± 0.000(3.0↑)	0.619 ± 0.000(3.0↑)	0.619 ± 0.000(3.0↑)	0.619 ± 0.000(3.0↑)
Average↑		6.125↑	8.25↑	9.07↑	8.87↑	10.45↑

Table 6. Vec2Node with KNN vs. hypergraph on `FastText-Softmax` classifier.

Dataset	%Train	FastText	Hypergraph	KNN
SST2	5	0.710 ± 0.100	$0.722 \pm 0.003(1.2\uparrow)$	$\mathbf{0.740 \pm 0.004(3.0\uparrow)}$
IMDB	5	0.522 ± 0.012	$0.664 \pm 0.004(14.2\uparrow)$	$\mathbf{0.718 \pm 0.001(19.6\uparrow)}$
AG News	5	0.663 ± 0.001	$0.811 \pm 0.002(14.8\uparrow)$	$\mathbf{0.825 \pm 0.001(16.2\uparrow)}$
DBpedia	5	0.589 ± 0.015	$0.615 \pm 0.000(2.6\uparrow)$	$\mathbf{0.619 \pm 0.000(3.0\uparrow)}$

Table 7. Vec2Node with and without self-training & consistency learning (ST & CL) on `FastText-Softmax` classifier.

Dataset	%Train	FastText	w/o ST & CL	w/ ST & CL
SST2	5	0.710 ± 0.100	$0.720 \pm 0.006(1.0\uparrow)$	$\mathbf{0.740 \pm 0.006(3.0\uparrow)}$
IMDB	5	0.522 ± 0.012	$0.686 \pm 0.005(16.4\uparrow)$	$\mathbf{0.718 \pm 0.001(19.6\uparrow)}$
AG News	5	0.663 ± 0.001	$0.791 \pm 0.001(12.8\uparrow)$	$\mathbf{0.825 \pm 0.001(16.2\uparrow)}$
DBpedia	5	0.589 ± 0.015	$0.614 \pm 0.000(2.5\uparrow)$	$\mathbf{0.619 \pm 0.000(3.0\uparrow)}$

4.3   Interpretability and Examples

Table 8 in Appendix 7, demonstrates synthetic examples from the AG news and SST2 datasets, generated by Vec2Node using different word replacement strategies i.e., random, `tf-idf` and tensor embedding. We observe Vec2Node to generate better samples with the following features.

Preserving Context for Word Replacement. In contrast to random selection which blindly substitutes words, the co-occurrence based structure of the tensor embedding preserves the context, and selects candidate words that are contextually similar to the original ones. For instance, in example #1 the entity "Jermain Defoe" is replaced by "Owen Michael" as they are more likely to co-occur in a Sport text related to "Real Madrid". As illustrated, the other approaches replace words quite randomly. This feature helps to minimize the concept drift that might happen in the replacement process.

Paraphrasing Context. Vec2Node leverages a sliding window to capture co-occurring concepts in a sentence, such that non-adjacent words that occur within the same context can be substituted with each other. This contributes to paraphrased sentences generated during augmentation as illustrated in example #2 with re-ordered proper nouns "Samsung" and "SCH-S250".

Tensor Embedding Preserves Word-Level Similarities. Tensor embedding not only preserves the context-level similarity, but also retains the semantics of the replaced concept. More precisely, it is more likely that a number gets replaced by another number (# 3) or an adverb by another adverb (# 5), and so on and so forth. We observe that not only numbers and verbs, but also prepositions like "a", "an", and "the" are replaced by similar concepts in the synthetic samples while preserving the context.

4.4   Related Work

Self-Training and Consistency Learning. Self-training is one of the well-known semi-supervised approaches which has been widely used to minimize the need for annotation leveraging large-scale unlabeled data [11,15,17,24]. For instance, Wang et al. leverage self-training and meta-learning for few-shot training of neural sequence taggers [24]. Moreover, a recent work, a.k.a UDA [28] exploits consistency learning with paraphrasing and back-translation from Neural Machine Translation systems for few-shot learning. In this work, we do not use any external resources such as an NMT system. In fact, we aim to bridge the gap between self-training and augmentation techniques, while solely relying on in-domain unlabeled data for tensor-based augmentation.

5   Conclusion

In this work, we propose a novel tensor-based technique i.e., `Vec2Node`, to augment textual datasets leveraging local and global information in corpus. `Vec2Node` leverages tensor data augmentation with self-training and consistency learning for text classification with few labels. Our experiments demonstrate that synthetic data generated by `Vec2Node` are interpretable and improve the classification accuracy over different datasets significantly in low-resource settings. For instance, `Vec2Node` improves the accuracy of `FastText` by 16.75% while using only 1% of labeled data. Overall, we demonstrate `Vec2Node` to work well both in low and high-data regime with improved performance when built on top of different encoders (e.g., `FastText`, `BERT`).

Acknowledgments. The GPUs used for this research were donated by the NVIDIA Corp. Research was partly supported by a UCR Regents Faculty Fellowship. Research was also supported by the National Science Foundation grant no. 1901379, CAREER grant no. IIS 2046086 and grant no. 2127309 to the Computing Research Associate for the CIFellows project.

References

1. Abdali, S., Shah, N., Papalexakis, E.E.: Hijod: semi-supervised multi-aspect detection of misinformation using hierarchical joint decomposition. In: ECML/PKDD (2020)
2. Bader, B., Kolda, T.: Algorithm 862: matlab tensor classes for fast algorithm prototyping. ACM Trans. Math. Softw. **32**, 635–653 (2006)
3. Bizer, C., et al.: Dbpedia - a crystallization point for the web of data. J. Web Semant. **7**(3), 154–165 (2009). https://doi.org/10.1016/j.websem.2009.07.002
4. Bojanowski, P., Grave, E., Joulin, A., Mikolov, T.: Enriching word vectors with subword information. arXiv preprint arXiv:1607.04606 (2016)
5. Devlin, J., Chang, M.W., Lee, K., Toutanova, K.: BERT: pre-training of deep bidirectional transformers for language understanding. In: Proceedings of the 2019 NAACL, pp. 4171–4186. ACL, Minneapolis, Minnesota (2019). https://doi.org/10.18653/v1/N19-1423

6. Du, J., et al.: Self-training improves pre-training for natural language understanding (2020)
7. Gallo, G., Longo, G., Pallottino, S., Nguyen, S.: Directed hypergraphs and applications. Discrete Appl. Math. **42**, 177–201 (1993). https://doi.org/10.1016/0166-218X(93)90045-P
8. Guacho, G.B., Abdali, S., Shah, N., Papalexakis, E.E.: Semi-supervised content-based detection of misinformation via tensor embeddings, pp. 322–325 (2018). https://doi.org/10.1109/ASONAM.2018.8508241
9. Han, J., Kamber, M., Pei, J.: Data Mining: Concepts and Techniques, 3rd edn. Morgan Kaufmann Publishers Inc., San Francisco (2011)
10. Harshman, R.A.: Foundations of the PARAFAC procedure: models and conditions for an explanatory multi-modal factor analysis. UCLA Working Pap. Phonetics **16**(1), 84 (1970)
11. He, J., Gu, J., Shen, J., Ranzato, M.: Revisiting self-training for neural sequence generation (2020)
12. Joulin, A., Grave, E., Bojanowski, P., Mikolov, T.: Bag of tricks for efficient text classification (2016)
13. Kolda, T.G., Bader, B.W.: Tensor decompositions and applications. SIAM Rev. **51**(3), 455–500 (2009). https://doi.org/10.1137/07070111X
14. Le, Q., Mikolov, T.: Distributed representations of sentences and documents. In: ICML 2014, vol. 4 (2014)
15. Li, X., et al.: Learning to self-train for semi-supervised few-shot classification (2019)
16. Maas, A.L., Daly, R.E., Pham, P.T., Huang, D., Ng, A.Y., Potts, C.: Learning word vectors for sentiment analysis. In: Proceedings of the 49th Annual Meeting of the Association for Computational Linguistics: Human Language Technologies, pp. 142–150. ACL, Portland, Oregon, USA (2011)
17. Meng, Y., Shen, J., Zhang, C., Han, J.: Weakly-supervised neural text classification. In: Proceedings of the 27th ACM International Conference on Information and Knowledge Management. ACM (2018). https://doi.org/10.1145/3269206.3271737
18. P. Liu, X. Wang, C.X., Meng, W.: A survey of text data augmentation (2020)
19. Papalexakis, E.E., Faloutsos, C., Sidiropoulos, N.D.: Tensors for data mining and data fusion: models, applications, and scalable algorithms. ACM Trans. Intell. Syst. Technol. **8**(2), 16:1–16:44 (2016). https://doi.org/10.1145/2915921
20. Sidiropoulos, N., De Lathauwer, L., Fu, X., Huang, K., Papalexakis, E., Faloutsos, C.: Tensor decomposition for signal processing and machine learning. IEEE Trans. Signal Process. **65**(13), 3551–3582 (2016). https://doi.org/10.1109/TSP.2017.2690524
21. Smith, S., Ravindran, N., Sidiropoulos, N.D., Karypis, G.: Splatt: efficient and parallel sparse tensor-matrix multiplication. In: IPDPS, pp. 61–70 (2015)
22. Socher, R., et al.: Recursive deep models for semantic compositionality over a sentiment treebank. In: Proceedings of the 2013 Conference on Empirical Methods in Natural Language Processing, pp. 1631–1642. ACL, Seattle, Washington, USA (2013)
23. Spitz, A., Aumiller, D., Soproni, B., Gertz, M.: A versatile hypergraph model for document collections. In: SSDBM 2020 (2020)
24. Wang, Y., et al.: Adaptive self-training for few-shot neural sequence labeling. ArXiv: abs/2010.03680 (2020)
25. Wei, J., Zou, K.: EDA: easy data augmentation techniques for boosting performance on text classification tasks. In: EMNLP-IJCNLP, pp. 6383–6389. Association for Computational Linguistics, Hong Kong (2019)

26. Wolf, T., et al.: Transformers: state-of-the-art natural language processing. In: Proceedings of the 2020 Conference on Empirical Methods in Natural Language Processing, pp. 38–45. ACL (2020)

27. Wu, Y., et al.: Google's neural machine translation system: Bridging the gap between human and machine translation. ArXiv: abs/1609.08144 (2016)

28. Xie, Q., Dai, Z., Hovy, E.H., Luong, M., Le, Q.V.: Unsupervised data augmentation. CoRR abs/1904.12848 (2019)

29. Zhang, C., Bengio, S., Hardt, M., Recht, B., Vinyals, O.: Understanding deep learning requires rethinking generalization. In: ICLR (2017)

30. Zhang, X., Zhao, J., LeCun, Y.: Character-level convolutional networks for text classification. In: Cortes, C., Lawrence, N., Lee, D., Sugiyama, M., Garnett, R. (eds.) Advances in Neural Information Processing Systems, vol. 28, pp. 649–657. Curran Associates, Inc. (2015)

31. Zhou, D., Huang, J., Schölkopf, B.: Learning with hypergraphs: Clustering, classification, and embedding, vol. 19, pp. 1601–1608 (2006)

"Let's Eat Grandma": Does Punctuation Matter in Sentence Representation?

Mansooreh Karami[✉][iD], Ahmadreza Mosallanezhad[iD],
Michelle V. Mancenido[iD], and Huan Liu[iD]

Arizona State University, Tempe, AZ, USA
{mkarami,amosalla,mmanceni,huanliu}@asu.edu

Abstract. Neural network-based embeddings have been the mainstream approach for creating a vector representation of the text to capture lexical and semantic similarities and dissimilarities. In general, existing encoding methods dismiss the punctuation as insignificant information; consequently, they are routinely treated as a predefined token/word or eliminated in the pre-processing phase. However, punctuation could play a significant role in the semantics of the sentences, as in "Let's eat, grandma" and "Let's eat grandma". We hypothesize that a punctuation-aware representation model would affect the performance of the downstream tasks. Thereby, we propose a model-agnostic method that incorporates both syntactic and contextual information to improve the performance of the sentiment classification task. We corroborate our findings by conducting experiments on publicly available datasets and provide case studies that our model generates representations with respect to the punctuation in the sentence.

Keywords: Sentiment analysis · Representation learning · Structural embedding · Punctuation

1 Introduction

According to a famous legend, Julius Caesar had decided to grant amnesty to one of his unscrupulous generals, who had been fated to be executed. "Execute not, liberate," Caesar had ordered his guards. However, the message had been delivered with a small but calamitous error: "Execute, not liberate."

The recent paradigm shift to pre-training the NLP models with language modeling has gained tremendous success across a wide variety of downstream tasks. Word and sentence embeddings from these pre-trained language models have revolutionized the modern NLP research and reduced the non-trivial computational time of training NLP-related tasks. BERT [4], an example of a

---

M. Karami and A. Mosallanezhad—Authors contributed equally to this work.

M.-R. Amini et al. (Eds.): ECML PKDD 2022, LNAI 13714, pp. 588–604, 2023.
https://doi.org/10.1007/978-3-031-26390-3_34

pre-trained language model, addresses limitations of other methods by incorporating context from both directions to capture the semantic concepts more accurately [33].

In pre-trained language models, punctuation is often treated as an ordinary word or as a predefined token in the data or, in some cases, filtered out during the pre-processing phase [7,11,20]. The lack of considerable attention to punctuation in NLP models stems from the fact that punctuation has long been considered as cues that only aid text's readability, thus not providing additional semantic value to the sentence's coherence [5]. However, studies show that the misplacement or elimination of these symbols can change the original meaning or obscure a text's implicit sentiment [2,31] as it conveys rich information about structural relations among the elements of a text. For example, "No investments will be made over three years" and "No, investments will be made over three years" have drastically different meanings and implications. But BERT, as a representation tool, will assign a fixed predefined token to the punctuation treating it as an ordinary word in the data; under BERT, the vector representations of these two sentences are nearly the same. On the other hand, methods that account for punctuation are typically model-specific and cannot be integrated into SOTA representation models.

In this work, we hypothesize that trivializing the role of punctuation in *sentiment analysis tasks* results in the degraded quality of representations which consequently, affects traditional measures of classifier performance. To provide evidence, we propose a model-agnostic module for representing the syntactic and contextual information that could be derived from punctuation. Our approach is based on an encoder that integrates structural and textual embedding to capture sentence-level semantics accurately through the use of parsing trees. Previous works on parsing trees have shown that there is an association between a text's punctuation and syntactic structure [11].

The following summarizes the major contributions of this work:

- We conduct preliminary experiments to show that the state-of-the-art representation learning models do not distinguish between sentences with and without punctuation (Sect. 5);
- We develop (Sect. 4) and evaluate (Sect. 5) a model-agnostic methodology for sentiment analysis that augments the structure of the sentences to the original sentence embedding which can be integrated into SOTA representation models;
- We provide case studies to demonstrate that the proposed model yields proper representation for cases when punctuation change and do not change the meaning of the sentences (Sect. 7).

2 Related Work

The proposed methodology spans the subject domains of word and sentence embeddings, punctuation in NLP tasks such as sentiment analysis, and tree-structured encoding. Current state-of-the-art in these areas are discussed in this section.

2.1 Embeddings

Word and sentence embeddings are techniques used to map text data to vector representations so that the distance between the vectors corresponds to their semantic proximity. Word2vec had been applied on many tasks since it was introduced in 2013 [18]. Although this neural network-based model could effectively encode the semantic and syntactic meaning of the text into vectors, word2vec is sub-optimal for syntax-based problems such as Part-of-Speech (POS) tagging or dependency parsing [13]. In recent years, embeddings such as BERT [4] improved on term-based embeddings by not only encoding the semantic information of words but also their contextualized meanings (i.e. terms and related contexts). Despite proving its usefulness across a wide range of tasks in NLP, BERT has been shown lacking in some aspects, such as common sense, pragmatic inferences, and the meaning of negations [6].

One prevailing issue in sentiment analysis is that these representations typically fail to distinguish between words with similar contexts but opposite sentiment polarities (e.g., wonderful vs. terrible) because they were mapped to vectors that were closely contiguous in the latent space [34]. Thus, researchers proposed various word embedding methods to encode sentiments [3,10,12,17]. In this work, we propose a novel sentence embedding as an improvement over current methods for sentiment analysis tasks.

2.2 Punctuation in NLP

Punctuation has long been considered the visual equivalent of spoken-language prosody, thus only providing cues that aid a text's readability. However, Nunberg [21] argued that punctuation has a more important role. He defines it as a linguistic subsystem related to grammar that conveys rich information about the structural relations among the elements of a text [21].

In NLP, the inclusion of punctuation marks has been shown to be useful in syntactic processing [15] and could be used to enhance grammar induction in unsupervised dependency parsing. As an example, Spitkovsky et al. [28] showed improved performance by splitting sentences at their punctuation to impose parsing restrictions over their fragments. Additionally, in the context of sentiment analysis, punctuation marks have been shown to add extra value to the sentiment [2,22,31] and could be used to create more meaningful syntax trees [1,11].

Despite evidence that incorporating punctuation improves aspects of an NLP's performance, very few NLP models make significant use of these symbols, which we concurrently address in the proposed methodology. Moreover, we investigate sentiments at the sentence-level.

2.3   Tree-Structured Encoders

Tree-structured encoders, which have been shown to perform as well as their sequential counterparts, are representations constructed from the syntactic structure of groups of words or sentences. An example of a tree-structured encoder is the Tree-LSTM, a generalization of the long short term memory (LSTM) architecture that accounts for the topological structure of sentences [29]. Each unit in the Tree-LSTM consists of values provided by the input vector and the hidden states of its children (as derived from the syntactic tree); in contrast, the standard LSTM only considers hidden states from the previous time step. Tree-LSTM was inspired by an RNN-based compositional model that captured the parent representation in syntactic trees [26, 27].

In addition to changing the LSTM architecture, another method to capture the syntactic structure of sentences is by directly using the LSTM architecture to code the syntactic structures. Liu et al. [14] encoded the variable-length syntactic information, i.e. the path from leaf node to the root node in the constituency or dependency tree, into a fixed-length vector representation to embed the structural characteristics of the sentences on neural attention models for machine comprehension tasks. To jointly learn syntax and lexicon, Shen et al. [25] proposed a Parsing-Reading-Predict neural language model (PRPN) that learns the syntactic structure from an unannotated corpus and uses the learned structure to form a premier language model. There has also been some work that extended the Transformer [30] architecture for syntactic coding.

The work in this paper augments constituency trees to the original word embedding to record the position of the punctuation by capturing structural information.

3   Problem Statement

Let $\mathcal{X} = \{(\mathbf{x_1}, y_1), (\mathbf{x_2}, y_2), ..., (\mathbf{x_N}, y_N)\}$ denote a set of N textual data with text $\mathbf{x_i}$ and the sentiment label y_i for sample i. Each text $\mathbf{x_i}$ consists of sequence of words/punctuation $\mathbf{x_i} = [w_1\ w_2\ ...\ w_M]$, where M represents the number of words and punctuation in the text. Since the punctuation and their position affects the structure of the sentence and its meaning, we focus on generating a robust sentence embedding for sentiment analysis with respect to the structure of the sentence. Formally, this problem can be stated as follows:

Problem 1. Given a set of textual data $\mathcal{X}$ comprising of words and punctuation, learn an embedding $\mathbf{E}$ which accounts for the constituency tree structure of the sentences and finds a function $\mathbf{F}$ for sentiment classification.

4   Proposed Model

We hypothesize that due to the effect of punctuation on the constituency structures of the sentences, adding the structural embedding of the sentences could improve the vector representation of sentences. The general framework of the proposed model is shown in Fig. 1. The proposed model has three major components: (1) a sentence encoder, (2) a structural encoder, and (3) a text classifier. In the following discussion, we describe in detail the sentence and structural encoders and discuss how these two methods are integrated into a robust framework that improves embedding and classification performance.

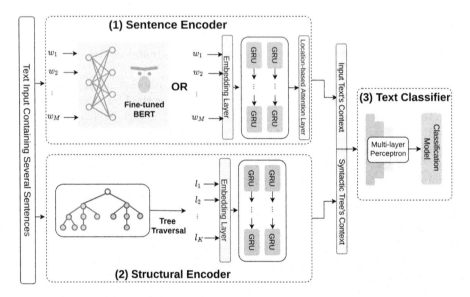

Fig. 1. The three components of the model: (1) the sentence encoder that captures the input context, (2) the syntactic tree encoder which accounts for the structural content, and (3) the sentiment analysis classifier.

4.1   Sentence Embedding for Sentiment Analysis

In sentiment analysis, textual data is first converted into vectors or matrices. The ability of recurrent neural networks (RNNs) to model order-sensitive data makes it an effective choice for modeling textual data, where the order of words alter the contextual meaning. Our framework uses a bi-directional gated recurrent unit (BiGRU), an RNN that models contextual meanings more effectively than uni-directional networks [9]. However, later as demonstrated in the experiments, we also considered a fine-tuned BERT instead of the BiGRU module in creating the text embeddings. This will ensure the generalization of our method for other representation models.

To create the text embeddings, a sample, $\mathbf{x}_i = [\mathbf{w}_1 \ \mathbf{w}_2 \ ... \ \mathbf{w}_M]$, is passed through an embedding layer which converts each word $\mathbf{w}_j$ to its representation. This layer has a tensor of dimension $|V| \times d_w$, where V is the vocabulary and

d_w is the dimension of the word embeddings. The representations will be fed to a BiGRU that yields the following M outputs:

$$(\overrightarrow{\mathbf{h}}_m, \overleftarrow{\mathbf{h}}_m) = \text{BiGRU}(\mathbf{w}_m, (\overrightarrow{\mathbf{h}}_{m-1}, \overleftarrow{\mathbf{h}}_{m-1})) \tag{1}$$

where $\overrightarrow{\mathbf{h}}_m$ and $\overleftarrow{\mathbf{h}}_m$ are, respectively, the forward and backward outputs of the BiGRU at time step $m \in M$. These BiGRU's outputs are then concatenated to form a fixed-length context vector:

$$\mathbf{H}_m = \text{Concat}(\overrightarrow{\mathbf{h}}_m, \overleftarrow{\mathbf{h}}_m) \tag{2}$$

Further, to establish a comprehensive context vector, an attention mechanism was included by augmenting a location-based attention layer [16]. The weighted average of the importance values $a_m \in \mathbf{H_m}$ provided by the attention layer creates the final context vector:

$$\mathbf{H}' = \sum_i a_i \mathbf{H}_i \tag{3}$$

Using the context vector $\mathbf{H}'$ with a Multi-Layer Perceptron (MLP) classifier yields good performance on sentiment analysis tasks [24,32].

Information learned from BiGRU/BERT, as described in this subsection, will be combined with the encoded syntactic structure of the sentence. This will enhance the context vector to include salient information provided by punctuation.

4.2   Enhanced Embedding

We use a constituency tree to analyze sentence structure and organize words into nested constituents. In the constituency tree, words are represented by the leaves while the internal nodes show the phrasal (e.g. S, NP and VP) or pre-terminal Part-Of-Speech (POS) categories. Edges in the tree indicate the set of grammar rules. Figure 2 shows an example of a constituency tree that demonstrates the parsing of a sample sentence. Subsequent to the generation of the syntactic tree, we adopt the word-level approach in Liu et al. [14] to capture syntactic information but in a sentence-level manner. We use the traversal of the syntactic tree T to pass it through a bi-directional GRU and create a representation of T. Because the order of the nodes in a tree impact the traversal result, we use BiGRU to create a correct representation:

$$(\overrightarrow{\mathbf{h}}_t, \overleftarrow{\mathbf{h}}_t) = \text{BiGRU}(\mathbf{l}_t, (\overrightarrow{\mathbf{h}}_{t-1}, \overleftarrow{\mathbf{h}}_{t-1})) \tag{4}$$

where $\mathbf{l}$ is the value of the tree node and $\mathbf{h}_t$ shows the BiGRU output. We consider the last output of the BiGRU, $\mathbf{H}_T = \text{Concat}(\overrightarrow{\mathbf{h}}_t, \overleftarrow{\mathbf{h}}_t)$, as the context of the syntactic tree.

Finally, to balance the effect of the extracted contexts, the context of the text $\mathbf{H}'$ and the context of its syntactic tree $\mathbf{H}_T$ are passed through a feed-forward neural network to create the enhanced text representation:

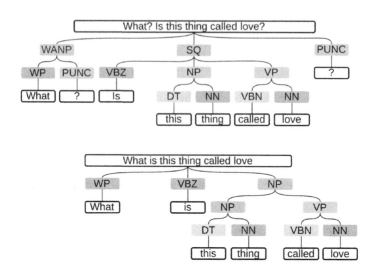

Fig. 2. The constituency tree of a text with and without punctuation, "what is this thing called love" versus "what? is this thing called love?"

$$\mathbf{H_F} = \mathrm{MLP}(\mathbf{H'}, \mathbf{H}_T) \qquad (5)$$

where $\mathbf{H_F}$ is the enhanced text representation containing the text's semantic information and information about its syntactic tree. This enhanced representation could now be used for sentiment analysis tasks.

5   Experimental Settings

In this section, we present details about the datasets, the implementation details, as well as the baseline methods used for the experiments.

5.1   Datasets

Four publicly available datasets – IMDB, Rotten Tomatoes (RT), Stanford Sentiment Treebank (SST), and Yelp Polarity (Yelp P.) – were used to evaluate and compare the proposed method with other baselines. The IMDB movie reviews dataset contains $50,000$ movie reviews, with each review labeled as 'positive' or 'negative'. In a similar fashion, the Rotten Tomatoes dataset contains $480,000$ movie reviews from the Rotten Tomatoes website, labeled as 'fresh' (positive) or 'rotten' (negative). As a more challenging task, we consider the SST-2 dataset, which consists of $10,754$ samples having a binary label of positive and negative sentiment. Finally, we utilized a subsample of $100,000$ reviews from Yelp Polarity dataset which uses 'negative' and 'positive' labels instead of the five point star

Table 1. The statistics of the datasets.

Dataset	# of samples	Avg. text length (# of words)	# of sentences
IMDB	50,000	231.1 ± 171.3	536,641
RT	480,000	21.8 ± 9.3	601,787
SST-2	10,754	19.4 ± 9.3	11,855
Yelp P	100,000	133 ± 122.5	814,596

scale [35]. Table 1 summarizes some key statistics of each dataset. We used 10-fold cross-validation with 45/5/50 for Train/Validation/Test split configuration to compare the proposed model with other baselines.

5.2 Implementation Details

In this subsection, we discuss the parameters and implementation details of the proposed model for conducting the experiments[1]. Based on the average number of the words in the datasets (Table 1), we truncate every textual data to 128 words. Next, we extract the syntactic tree for each sentence, in the spirit of Liu et al. [14] but in a sentence-level manner using Spacy toolkit[2] Finally, to combine all trees related to a text, an empty root was added as the parent of all the other roots of the syntactic trees. Children are arranged based on the order of the sentences in the text (Fig. 1).

We use GloVe 100d [23] to replace each word with its corresponding word vector to convert sentences into matrices. For words and POS tags that are not included in GloVe, a trainable random vector was used as a proxy. We use a 1-layer BiGRU with 256 hidden neurons to generate the text's context vector and a 128-hidden neuron BiGRU for the syntactic tree's context vector. To combine both context vectors, we use a simple neural network with 512 output neurons. The output of this neural network is the final context vector $\mathbf{H_F}$ containing both semantic and syntactic information of the input text:

$$\mathbf{o} = \tanh(\mathbf{W}_F^{(1)}(\mathbf{H'}||\mathbf{H_T}) + \mathbf{b}_F^{(1)}) \tag{6}$$

$$\mathbf{H_F} = \tanh(\mathbf{W}_F^{(2)}\mathbf{o} + \mathbf{b}_F^{(2)}), \tag{7}$$

where $||$ is the concatenation operator, $(\mathbf{W}, \mathbf{b})$ are the learnable weights, and $\mathbf{H'}$, $\mathbf{H_T}$ are the input's context and the syntactic tree's context vectors, respectively.

The integrated context vector $\mathbf{H_F}$ is used for text classification. The neural network classifier includes three layers with 512, 128, and C number of neurons, respectively, where C is the number of classes. Model parameters θ and the data labels y are updated using a cross-entropy loss function in the training phase:

$$L(\theta) = -\frac{1}{N}\sum_{i=1}^{N}\sum_{j=1}^{C} y_{ij}\log(p_{ij}), \tag{8}$$

---

[1] The code for this work is available at: https://github.com/mansourehk/Grandma.
[2] Available at https://spacy.io/.

where N is the number of samples. We use the Adam optimizer [8] to update the parameters of the network.

5.3   Baseline Methods

Several embedding methods are implemented to generate sentence representations for comparison with the proposed model. The vectors created by these sentence encoders are used as inputs to the three-layered neural network classifier. Each sentence representation method is described below.

- **BERT** [4]: Bidirectional Encoder Representations from Transformers is a model used for various NLP tasks, including sentiment analysis. In this paper, a pre-trained base BERT is used to extract the sentence embeddings.
- **BiGRU**: similar to the approach in Mosallanezhad et al. [19], we design a baseline that uses a bidirectional GRU to create a context vector based on the input text. Each word is replaced by its corresponding GloVe vector and passed through a bidirectional Gated Recurrent Unit. The final output of the BiGRU is then considered as the context vector.
- **BiGRU+Attn**: similar to the BiGRU method, but uses a location-based attention layer [16] to create the context vector.
- **SEDT–LSTM** [14]: creates a word-level embedding by including the dependency tree of the sentences. For each word w in the text, this method merges the GloVe vector of w with the fixed-length context vector extracted from the dependency tree. To create this context vector, all the words in the path from w to the root node in the dependency tree are fed to an LSTM.

We integrated our model-agnostic module (i.e., the syntactic tree encoder) to the BiGRU, BiGRU+Attn, and BERT.

6   Discussion and Experimental Results

In this section, we conduct experiments to evaluate the effectiveness of our method in sentiment analysis tasks. We propose two major research questions:

(Q1) How do other methods behave in terms of the embeddings and perform in terms of the sentiment classification task when punctuation is included in the input text?

(Q2) How well does the proposed method incorporate the effect of the punctuation in the sentence embeddings?

Figure 3 shows the similarity between sentence embeddings with and without punctuation in the text. To calculate the similarity between embeddings, we use the cosine similarity measure:

$$\text{CosineSim}(\mathbf{E}_w, \mathbf{E}_{wo}) = \frac{\mathbf{E}_w \cdot \mathbf{E}_{wo}}{||\mathbf{E}_w||||\mathbf{E}_{wo}||}, \tag{9}$$

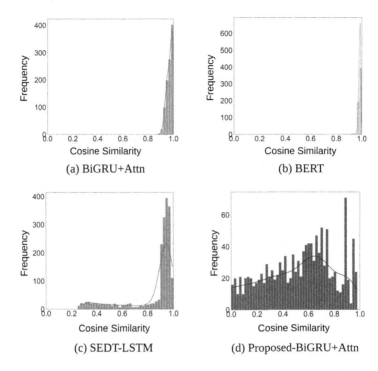

Fig. 3. The histogram of cosine similarities between sentence embeddings with and without punctuation. Higher similarity means that the embeddings are close to each other.

where $\mathbf{E}_w$ and $\mathbf{E}_{wo}$ are the sentence embeddings with and without punctuation, respectively. The cosine similarity measure is close to 1.0 when context vectors are close to each other.

Q1. In Fig. 3 (a–b), it is observable that BERT and Recurrent Neural Networks (BiGRU+Attn) have higher cosine similarity measures, implying that they do not produce different embeddings for sentences with and without punctuation. The minimum similarity between embeddings for these models is approximately 0.9. This finding corroborates our hypothesis that these models *consider punctuation as just another fixed word/token in the data*, strongly justifying the development of an enhanced representation method.

Additionally, Table 2 shows the accuracy of the baseline models on the aforementioned datasets when punctuation is excluded. By comparing it with the first section of Table 3, it is evident that the performance of the baselines are *agnostic to the use of punctuation* due to their similar representation vectors in both cases. For the baselines, inclusion of the punctuation is almost irrelevant and even lowers performance in some cases providing evidence why most researchers exclude punctuation in the preprocessing phase.

Q2. Fig. 3 (c–d) shows the trend of cosine similarity when the syntactic information is augmented with the word embedding. The lower similarity values,

Table 2. *Without Punctuation:* Performance (accuracy) of the baseline models on the datasets.

Model	Datasets			
	IMDB	RT	SST-2	Yelp P.
BiGRU	88.0	69.1	86.9	84.8
BiGRU+Attn	88.8	70.0	87.4	84.6
BERT	92.3	71.6	91.7	90.6

Table 3. *With Punctuation:* Performance (accuracy) of SEDT-LSTM and our added module to different representation baselines when punctuation is included.

Model	Datasets			
	IMDB	RT	SST-2	Yelp P.
BiGRU	88.1	70.3	87.3	84.8
BiGRU+Attn	88.2	70.5	88.1	84.8
BERT	92.1	71.5	91.7	90.6
SEDT-LSTM	91.1	72.0	90.5	85.1
Proposed-BiGRU	92.7	74.2	90.1	87.1
Proposed-BiGRU+Attn	93.0	74.3	91.3	88.3
Proposed-BERT	**94.6**	**74.8**	**92.4**	**91.7**

ranging from as low as 0.10 to only as high as 0.90, indicate that the representation vectors of sentences with and without punctuation are distinct. While the SEDT-LSTM model shows promising results, the proposed model still outperforms SEDT-LSTM in the sentiment analysis task (Table 3). This difference is due to the fact that our model operates in *a sentence-level manner and provides a richer structural embedding*, while SEDT-LSTM works as a word-level approach and does not account for the whole structure of the syntactic tree.

7   Case Studies

The cosine similarity of several sentences were also calculated to investigate how the methods compare when punctuation is removed. We combined the IMDB and Rotten Tomatoes datasets into a larger dataset, which is justifiable due to the similarity in the purpose and structure of the two datasets. The combined dataset was used to train the proposed model and the baseline methods.

Tables 4, 5 and 6 shows the cosine similarity measures of sample sentences with and without punctuation for all models. What is interesting in the results is that *the proposed model clearly distinguishes the syntax between sentences where punctuation is necessary* (similarity measures are lower). Specifically, this is apparent in sentences provided in Table 4.

Table 4. Examples of sentences in which punctuation change the meaning of the sentence. The proposed method distinguishes between the two versions, with and without punctuation. In this experiment, we use both inputs on a single model.

Examples in which punctuation change the meaning of the sentence		Cosine similarity			
With punctuation	Without punctuation	Proposed	SEDT-LSTM	BERT	BiGRU+Attn
1 Now, my friends, listen to me	Now my friends listen to me	0.56	0.67	0.97	0.95
2 Help. wanted	Help wanted	0.51	0.45	0.99	0.99
3 What? Is this thing called 'love'?	What is this thing called love	0.75	0.78	0.98	0.99
4 No, investments will be made in United States	No investments will be made in United States	0.57	0.55	0.96	0.96
5 If you go, pack your knitting needles	If you go pack your knitting needles	0.43	0.67	0.97	0.98
6 When the plot kicks in, the film loses credibility	When the plot kicks in the film loses credibility	0.48	0.78	0.96	0.94

Further, *if the context of the sentence is agnostic with respect to the punctuation, our proposed model still performs relatively well* (high cosine similarity measure). This is evident in sentences 7–11 in Table 5. In a specific example,

Table 5. The cosine similarity of sentences with and without punctuation in which the punctuation do not change the meaning of the sentence using different embedding methods. The proposed method can incorporate the syntactic tree's information better than the baselines. In this experiment, we use both inputs on a single model.

Examples in which punctuation do not change the meaning of the sentence		Cosine similarity			
With punctuation	Without punctuation	Proposed	SEDT-LSTM	BERT	BiGRU+Attn
7 A gorgeously strange movie, heaven is deeply concerned with morality, but it refuses to spell things out for viewers	A gorgeously strange movie heaven is deeply concerned with morality but it refuses to spell things out for viewers	0.89	0.91	0.98	0.99
8 But, like silence, it's a movie that gets under your skin	But like silence its a movie that gets under your skin	0.96	0.98	0.98	0.99
9 You will be required to work twenty-four hour shifts	You will be required to work twenty four hour shifts	0.99	0.99	0.99	0.99
10 The talents of the actors helps "Moonlight Mile" rise above its heart-on-its-sleeve writing	The talents of the actors helps Moonlight Mile rise above its heart on its sleeve writing	0.97	0.95	0.97	0.98
11 It's a fine, old - fashioned - movie. movie, which is to say it's unburdened by pretensions to great artistic significance	It s a fine old fashioned movie which is to say it s unburdened by pretensions to great artistic significance	0.95	0.94	0.98	0.99
12 Her favorite pies were lemon meringue, apple, and pecan	Her favorite pies were lemon meringue apple and pecan	0.83	0.93	0.98	0.97

sentence 12 shows a case where the Oxford/serial comma helps in preventing ambiguity. Without the serial comma, 'apple and pecan' could be interpreted as a pie containing both apples and pecans. By looking into the cosine similarities, the proposed method seems to distinguish this nuance.

Additionally, to confirm our hypothesis that baselines such as BERT do not differentiate among different kinds of punctuation, we randomly replaced the punctuation in sentences with other types. It is evident from Table 6 results that *the proposed method creates different representations when punctuation changes* while BERT and BiGRU provided nearly similar representations.

Table 6. Examples of sentences with random punctuation alongside their cosine similarity using different embedding methods. The proposed method can incorporate the syntactic tree's information better than the baselines. In this experiment, we use both inputs on a single model.

Examples in which random punctuation may change the meaning of the sentence		Cosine similarity			
With punctuation	Without punctuation	Proposed	SEDT-LSTM	BERT	BiGRU+Attn
13 Now, my friends, listen to me.	Now. my friends! listen to me,	0.52	0.59	0.96	0.98
14 Help. wanted.	Help, wanted?	0.67	0.61	0.96	0.98
15 What? Is this thing called 'love'?	What. Is this thing called 'love'!	0.82	0.78	0.99	0.99
16 A gorgeously strange movie, heaven is deeply concerned with morality, but it refuses to spell things out for viewers.	A gorgeously strange movie? heaven is deeply concerned with morality. but it refuses to spell things out for viewers,	0.91	0.94	0.98	0.99
17 But, like silence, it's a movie that gets under your skin.	But! like silence. it?s a movie that gets under your skin?	0.74	0.77	0.97	0.98
18 You will be required to work twenty-four hour shifts.	You will be required to work twenty!four hour shifts,	0.94	0.95	0.99	0.99

8 Conclusion and Future Work

In this paper, we proposed a model-agnostic methodology for sentence embeddings that consider punctuation as a salient feature of textual data. By leveraging on the association between punctuation and syntactic trees, our model yielded embeddings that were consistently able to convey the contextual meaning of sentences more accurately. We integrate our proposed module into state-of-the-art representation models, including BERT, the gold standard for NLP tasks. The proposed model in this paper outperformed the baselines in distinguishing between sentences with and without punctuation, especially those that require punctuation to be sensical. Moreover, as task performance, it performed accurately on classifying opinions for the IMDB, Rotten Tomatoes, SST-2, and Yelp P. datasets. A possible direction for future research is to use syntactic trees for other NLP-related tasks, such as automated chatbots and machine comprehension.

Acknowledgment. The authors would like to thank Sarath Sreedharan (ASU) and Sachin Grover (ASU) for their comments on the manuscript. This material is, in part, based upon works supported by ONR (N00014-21-1-4002) and the U.S. Department of Homeland Security (17STQAC00001-05-00) (Disclaimer: "The views and conclusions contained in this document are those of the authors and should not be interpreted as necessarily representing the official policies, either expressed or implied, of the U.S. Department of Homeland Security.").

References

1. Agarwal, A., Xie, B., Vovsha, I., Rambow, O., Passonneau, R.J.: Sentiment analysis of Twitter data. In: Proceedings of the workshop on language in social media (LSM 2011), pp. 30–38 (2011)
2. Altrabsheh, N., Cocea, M., Fallahkhair, S.: Sentiment analysis: towards a tool for analysing real-time students feedback. In: 2014 IEEE 26th International Conference on Tools with Artificial Intelligence, pp. 419–423. IEEE (2014)
3. Bespalov, D., Bai, B., Qi, Y., Shokoufandeh, A.: Sentiment classification based on supervised latent n-gram analysis. In: Proceedings of the 20th ACM International Conference on Information and Knowledge Management, pp. 375–382 (2011)
4. Devlin, J., Chang, M.W., Lee, K., Toutanova, K.: BERT: pre-training of deep bidirectional transformers for language understanding. In: NAACL-HLT (1) (2019)
5. Ek, A., Bernardy, J.P., Chatzikyriakidis, S.: How does punctuation affect neural models in natural language inference. In: Proceedings of the Probability and Meaning Conference (PaM 2020), pp. 109–116 (2020)
6. Ettinger, A.: What BERT is not: lessons from a new suite of psycholinguistic diagnostics for language models. Trans. Assoc. Comput. Linguist. **8**, 34–48 (2020)
7. Karami, M., Nazer, T.H., Liu, H.: Profiling fake news spreaders on social media through psychological and motivational factors. In: Proceedings of the 32nd ACM Conference on Hypertext and Social Media, pp. 225–230 (2021)
8. Kingma, D.P., Ba, J.: Adam: a method for stochastic optimization. In: ICLR (Poster) (2015)
9. Kiperwasser, E., Goldberg, Y.: Simple and accurate dependency parsing using bidirectional LSTM feature representations. Trans. Assoc. Comput. Linguist. **4**, 313–327 (2016)
10. Labutov, I., Lipson, H.: Re-embedding words. In: Proceedings of the 51st Annual Meeting of the Association for Computational Linguistics (Volume 2: Short Papers), pp. 489–493 (2013)
11. Li, X.L., Wang, D., Eisner, J.: A generative model for punctuation in dependency trees. Trans. Assoc. Comput. Linguist. **7**, 357–373 (2019)
12. Lin, Z., Feng, M., Santos, C.N., Yu, M., Xiang, B., Zhou, B., Bengio, Y.: A structured self-attentive sentence embedding. In: International Conference on Learning Representations (ICLR) (2017)
13. Ling, W., Dyer, C., Black, A.W., Trancoso, I.: Two/too simple adaptations of word2vec for syntax problems. In: Proceedings of the 2015 Conference of the North American Chapter of the Association for Computational Linguistics: Human Language Technologies, pp. 1299–1304 (2015)
14. Liu, R., Hu, J., Wei, W., Yang, Z., Nyberg, E.: Structural embedding of syntactic trees for machine comprehension. In: Proceedings of the 2017 Conference on Empirical Methods in Natural Language Processing, pp. 815–824 (2017)
15. Lou, P.J., Wang, Y., Johnson, M.: Neural constituency parsing of speech transcripts. In: NAACL-HLT (1) (2019)
16. Luong, T., Pham, H., Manning, C.D.: Effective approaches to attention-based neural machine translation. In: EMNLP (2015)

17. Maas, A., Daly, R.E., Pham, P.T., Huang, D., Ng, A.Y., Potts, C.: Learning word vectors for sentiment analysis. In: Proceedings of the 49th Annual Meeting of the Association for Computational Linguistics: Human Language Technologies, pp. 142–150 (2011)

18. Mikolov, T., Chen, K., Corrado, G., Dean, J.: Efficient estimation of word representations in vector space. In: Bengio, Y., LeCun, Y. (eds.) 1st International Conference on Learning Representations, ICLR 2013, Scottsdale, Arizona, USA, 2–4 May 2013, Workshop Track Proceedings (2013)

19. Mosallanezhad, A., Beigi, G., Liu, H.: Deep reinforcement learning-based text anonymization against private-attribute inference. In: Proceedings of the 2019 Conference on Empirical Methods in Natural Language Processing and the 9th International Joint Conference on Natural Language Processing (EMNLP-IJCNLP), pp. 2360–2369 (2019)

20. Mosallanezhad, A., Karami, M., Shu, K., Mancenido, M.V., Liu, H.: Domain adaptive fake news detection via reinforcement learning. In: Proceedings of the ACM Web Conference 2022, pp. 3632–3640 (2022)

21. Nunberg, G.: The Linguistics of Punctuation. Center for the Study of Language (CSLI) (1990)

22. Pang, B., Lee, L., Vaithyanathan, S.: Thumbs up? Sentiment classification using machine learning techniques. In: In proceedings of EMNLP (2002)

23. Pennington, J., Socher, R., Manning, C.D.: GloVe: global vectors for word representation. In: Proceedings of the 2014 Conference on Empirical Methods in Natural Language Processing (EMNLP), pp. 1532–1543 (2014)

24. Sachin, S., Tripathi, A., Mahajan, N., Aggarwal, S., Nagrath, P.: Sentiment analysis using gated recurrent neural networks. SN Comput. Sci. $1(2)$, 1–13 (2020)

25. Shen, Y., Lin, Z., Huang, C., Courville, A.: Neural language modeling by jointly learning syntax and lexicon. In: International Conference on Learning Representations (2018)

26. Socher, R., Lin, C.C., Manning, C., Ng, A.Y.: Parsing natural scenes and natural language with recursive neural networks. In: Proceedings of the 28th International Conference on Machine Learning (ICML-2011), pp. 129–136 (2011)

27. Socher, R., et al.: Recursive deep models for semantic compositionality over a sentiment treebank. In: Proceedings of the 2013 Conference on Empirical Methods in Natural Language Processing, pp. 1631–1642 (2013)

28. Spitkovsky, V.I., Alshawi, H., Jurafsky, D.: Punctuation: making a point in unsupervised dependency parsing. In: Proceedings of the Fifteenth Conference on Computational Natural Language Learning, pp. 19–28 (2011)

29. Tai, K.S., Socher, R., Manning, C.D.: Improved semantic representations from tree-structured long short-term memory networks. In: Proceedings of the 53rd Annual Meeting of the Association for Computational Linguistics and the 7th International Joint Conference on Natural Language Processing (Volume 1: Long Papers), pp. 1556–1566 (2015)

30. Vaswani, A., et al.: Attention is all you need. In: Advances in Neural Information Processing Systems, pp. 5998–6008 (2017)

31. Wang, H., Liu, L., Song, W., Lu, J.: Feature-based sentiment analysis approach for product reviews. J. Softw. $9(2)$, 274–279 (2014)

32. Yang, Z., Yang, D., Dyer, C., He, X., Smola, A., Hovy, E.: Hierarchical attention networks for document classification. In: Proceedings of the 2016 Conference of the North American Chapter of the Association For Computational Linguistics: Human Language Technologies, pp. 1480–1489 (2016)

33. Yenicelik, D., Schmidt, F., Kilcher, Y.: How does BERT capture semantics? A closer look at polysemous words. In: Proceedings of the Third Blackbox NLP Workshop on Analyzing and Interpreting Neural Networks for NLP, pp. 156–162 (2020)
34. Zhang, L., Wang, S., Liu, B.: Deep learning for sentiment analysis: a survey. Wiley Interdisc. Rev. Data Min. Knowl. Disc. **8**(4), e1253 (2018)
35. Zhang, X., Zhao, J., LeCun, Y.: Character-level convolutional networks for text classification. Adv. Neural. Inf. Process. Syst. **28**, 649–657 (2015)

Contextualized Graph Embeddings
for Adverse Drug Event Detection

Ya Gao[1], Shaoxiong Ji[1]([✉]), Tongxuan Zhang[2], Prayag Tiwari[1],
and Pekka Marttinen[1]

[1] Aalto University, 02150 Espoo, Finland
{ya.gao,shaoxiong.ji,prayag.tiwari,pekka.marttinen}@aalto.fi
[2] Tianjin Normal University, Tianjin 300387, China
txzhang@tjnu.edu.cn

Abstract. An adverse drug event (ADE) is defined as an adverse reaction resulting from improper drug use, reported in various documents such as biomedical literature, drug reviews, and user posts on social media. The recent advances in natural language processing techniques have facilitated automated ADE detection from documents. However, the contextualized information and relations among text pieces are less explored. This paper investigates contextualized language models and heterogeneous graph representations. It builds a contextualized graph embedding model for adverse drug event detection. We employ different convolutional graph neural networks and pre-trained contextualized embeddings as the building blocks. Experimental results show that our methods can improve the performance by comparing recent ADE detection models, suggesting that a text graph can capture causal relationships and dependency between different entities in a document.

Keywords: Adverse drug events · Graph neural networks · Contextualized embeddings

1 Introduction

Adverse Drug Events (ADEs) are injuries resulting from medical intervention related to a drug [7]. A typical way to detect ADEs is to conduct a clinical trial. However, there are many settings where a drug would be used, and we cannot check all of them during the clinical trial. Besides, some ADEs have long latency, making them hard to be discovered by an ordinary clinical trial [29]. Post-marketing drug safety surveillance, also called pharmacovigilance, is conducted to solve these problems. Pharmacovigilance activities mostly depend on Spontaneous Reporting Systems, which collect users' voluntary ADE reports [18]. However, the number of people willing to report their experiences through the official systems is negligible. Furthermore, these systems are limited due to biased and incomplete reports.

Compared with reports using Spontaneous Reporting Systems, more people often talk about their adverse reactions on social media platforms. Recent

M.-R. Amini et al. (Eds.): ECML PKDD 2022, LNAI 13714, pp. 605–620, 2023.
https://doi.org/10.1007/978-3-031-26390-3_35

publications collect documents from social media such as Twitter and Reddit to obtain more reliable data and detect ADEs automatically using Nature Language Processing (NLP) techniques. The detection of ADEs can be seen as a text classification task or a sequence-labeling problem, where we need to identify documents including ADEs [8]. The early studies include lexicon-based and rule-based methods [28]. These methods focus on string-matching, which is less effective for social media text and consumes many resources to build rules. Machine learning algorithms are also used to solve this task, such as Support Vector Machine (SVM) [4], Recurrent Neural Network (RNNs) [5] and Convolutional Neural Networks (CNNs) [10]. These approaches can process text with manual feature engineering or enable automated feature learning with deep learning methods, facilitating automated ADE detection from biomedical text or social content. However, the existing approaches and models have two limitations: (1) some works are limited in capturing the rich context information in the text. (2) some do not fully consider the causal relationship and dependency between different entities in a document. Effective text encoding should be considered for the ADE detection task to capture rich semantic and contextualized information. Note that detecting causal relationships does not here refer to causal inference as in the field of machine learning focusing on causality [24], but rather expressing or indicating the relationship between the cause, e.g. a drug taken, and the respective individual's adverse health outcome as reported in the text sample.

Graphs are commonly used for different data representations because of their strong expressivity. Text data can be represented by heterogeneous graphs, where different words, phrases, and documents are seen as nodes, and their relations are shown using edges. Text graphs and graph neural networks are widely used in many NLP applications for healthcare tasks such as sentiment classification and review rating [20,35]. Graph Neural Networks (GNNs) [33] can be applied to graph representation learning and capture the causal relationships and dependency of objects, making them more suitable for representing text with adverse drug events. However, no existing studies on ADE detection employ graph representation and graph neural networks. Besides, contextualized representations of text facilitate various NLP applications and boost the performance of NLP systems with minimal architecture engineering. In the medical domain, contextualized embeddings with domain knowledge are also in need. Pretrained contextualized language embeddings have been applied to various medical applications such as medical code assignment [11] and biomedical knowledge graph construction [12].

This paper presents a contextualized graph embedding model for ADE detection. We build contextualized language embeddings to capture contextualized information. With a heterogeneous graph built to embody word and document relations from the ADE corpus, we use graph neural networks to learn causal relations between word and document nodes to improve adverse drug reaction detection. This paper deploys different GNN-based models and pre-trained contextualized embeddings. The performance of these models is evaluated and compared with state-of-the-art models on three public benchmarks for ADE detection. Our model outperforms several strong ADE detection models in most cases. We also analyze the experiment results to discuss some potential challenges and explore the potential for improving the ADE detection tasks. The code will be made publicly available on acceptance.

Our contributions include the following folds.

- We develop a contextualized graph embedding model (CGEM) that introduces text graphs to capture the cause-effect relation for drug adverse event detection.
- The CGEM model utilizes contextualized embeddings pre-trained in large-scale domain-specific corpora for capturing context information, convolutional GNNs for text graph encoding, and an attention classifier for ADE classification.
- Experimental results show our approach outperforms recent advanced ADE detection models in three public datasets from the biomedical domain and social media.

2   Related Work

The rapid development of deep learning makes neural network-based approaches predominant in ADE detection. RNN can process sequence information and capture the sequential dependency, making it is suitable for ADE detection from text. Many studies on the ADE detection task employ RNN-based models. Cocos et al. [5] developed a Bidirectional Long Short-Term Memory (BiLSTM) network to label different parts of a sequence for ADE detection. Information from recognition of concepts and relations can benefit each other, enabling this joint modeling technique to obtain more useful information during learning. However, inaccurate recognition in the first step will affect the following steps, known as the error propagation issue. To address this issue, Wei et al. [31] proposed a joint learning model which can recognize entities of ADE, the reason, and their relations simultaneously. In the recognition phase, the joint model employs CRF and BiLSTM. To achieve relation classification, it uses CNN-RNN and SVM.

Some studies also developed models with other neural network architectures, such as capsule networks and the self-attention mechanism. Zhang et al. [38] presented a model called Gated iterative capsule network (GICN), which applies CNN to obtain the complete phrase information and extracts deep semantic information using a capsule network with a gated iteration unit. This unit can remember contextual information by clustering features. However, they did not consider the wights of different parts of a document. With attention mechanisms, more critical parts of a document get higher weights. Ge et al. [9] employed Multi-Head Self-Attention in their model to distinguish the importance of different words. Wunnava et al. [34] developed a dual-attention mechanism with BiLSTM to capture both task-specific and semantic information in the sentence. However, they did not fully consider the causal relationship between entities in a document.

3   Methods

3.1   Overall Architecture

This paper defines ADE detection as a classification task. We develop the contextualized graph embedding model as illustrated in Fig. 1. There are three components of the model. **(1) Graph Construction with Contextualized Embeddings.** We construct a heterogeneous graph to represent words and documents in the whole dataset, following TextGCN [35], and use pre-trained language models, specifically BERT [6] and its domain-specific variants, to obtain the contextualized text representation. **(2) Graph-based Text Encoding.** To capture neighborhood information in the heterogeneous graph, the feature matrix obtained from the embedding layer and the adjacency matrix from the constructed graph are fed into graph encoders. The feature embeddings are iteratively updated in the heterogeneous relational networks of words and documents. **(3) ADE Classification.** We follow the BertGCN model [20] to fuse contextualized embedding and graph networks with a weight coefficient to balance these two branches. Furthermore, we build an attentive classification layer to allow more critical content to contribute more to predictions. Figure 1 shows the overall model architecture. The details of these components are introduced in the following sections.

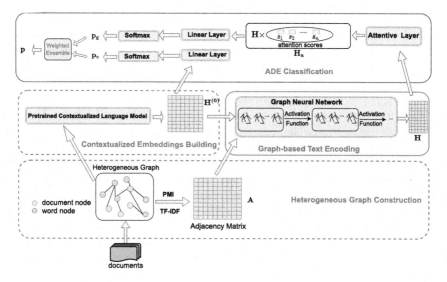

Fig. 1. The illustration of the model architecture with contextualized graph embeddings for ADE detection

3.2   Graph Construction

Heterogeneous Graph. We first represent text as a graph before feeding it to neural networks. Representing text in a heterogeneous graph can provide different perspectives for text encoding and improve ADE detection. The process of graph construction follows TextGCN [35]. Nodes in the graph represent documents and different words. The number of nodes n equal to the number of documents n_d plus the number of unique words n_w in the whole dataset, i.e., $n = n_d + n_w$. There are two types of edges, i.e., word-word and document-word edges. We use the term frequency-inverse document frequency (TF-IDF) of one word in the document to represent the weight of a document-word edge, while the weight of a word-word edge is based on positive point-wise mutual information (PMI) of two words. We can represent the weight between the node i and the node j as:

$$\mathbf{A}_{ij} = \begin{cases} \text{PMI}(i,j), \ \text{PMI} > 0; \text{i, j: words} \\ \text{TF-IDF}_{ij}, \ \text{i: document, j: word} \\ 1, \qquad\qquad\qquad\qquad i = j \\ 0, \qquad\qquad\qquad\qquad \text{otherwise} \end{cases} \tag{1}$$

Contextualized Embeddings. We used three pre-trained contextualized language models to obtain embeddings for documents. They are all BERT-based models but pre-trained with different strategies or corpora collected from different domains. The pre-trained language embeddings include:

- RoBERTa [21]: a pre-trained model with masked language modeling (MLM) objective on English language. In this paper, we used the base version.
- BioBERT [17]: a BERT-based model trained with biomedical corpora including PubMed abstracts and PubMed Central full-text articles.
- ClinicalBERT [2]: another domain-specific BERT-based model which is trained on clinical notes from the MIMIC-III database [13].

Given the dimension of embeddings denoted as d, the final output of contextualized text encoding are denoted as $\mathbf{H}_{doc} \in \mathbb{R}^{n_d \times d}$. We then apply a zero matrix as the initialization of word nodes to get the feature matrix input to GNN:

$$\mathbf{H}^{(0)} = \begin{pmatrix} \mathbf{H}_{doc} \\ \mathbf{0} \end{pmatrix} \tag{2}$$

where $\mathbf{H}^{(0)} \in \mathbb{R}^{(n_d + n_w) \times d}$.

3.3   Graph-Based Text Encoding

This section employs a graph-based model for text encoding and capturing complex heterogeneous relationships. Graph neural networks are powerful models to mine and capture the relations and dependencies of graph data. Specifically, we apply two graph neural networks, i.e., Graph Convolutional Network (GCN) [16] and Graph Attention Network (GAT) [30], which are commonly used in different tasks. Graph convolution encodes the topological structure of the heterogeneous graph, enables label influence propagation, and achieves effective modeling of ADE corpora. In this section, we introduce their principles.

GCN is a category of Convolutional Graph Neural Networks (ConvGNNs) models. It is a spectral-based model which incorporates nodes' feature information from their neighbors. It can be seen as a multilayer neural network limited to undirected graphs where the number of layers is fixed. Each layer has different weights to better process cyclic mutual dependencies. GCN is the approximations and simplifications of Spectral CNN. It approximates spectral graph convolutions using convolutional architecture to get a localized first-order representation.

A graph G consists of nodes set V, and edge sets E. $\mathbf{A}$ is the adjacency matrix obtained from the step of graph construction, and $\hat{\mathbf{A}}$ is its normalized form. $\mathbf{D}$ is the degree matrix, where $\mathbf{D}_{ij} = \sum_j \mathbf{A}_{ij}$. In the GCN model, multiple layers are stacked to integrate information about higher-order neighborhoods. In the m-th layer, the feature matrix is updated as:

$$\mathbf{H}^{(m)} = f(\hat{\mathbf{A}}\mathbf{H}^{(m-1)}\mathbf{W}^{(m-1)}), \qquad \hat{\mathbf{A}} = \mathbf{D}^{-\frac{1}{2}}\mathbf{A}\mathbf{D}^{\frac{1}{2}}, \qquad (3)$$

where $\mathbf{H}^{(m)} \in \mathbb{R}^{n \times d_m}$, and $\mathbf{W}^{(m-1)} \in \mathbb{R}^{d_{m-1} \times d_m}$ is the weight matrix, $\mathbf{H}^{(0)}$ is the output from contextualized language models, and $f(\cdot)$ is an activation function.

Being similar to GCN, GAT is also a ConvGNNs model. However, it is spatial-based neural networks, where node information is propagated within edges and graph convolutions are finally decided by the spatial relation. It employs the message passing process and attention mechanism to learn relations between nodes. Graph attention layers in GAT assign different attention scores to one node's distant neighbors and prioritize the importance of different types of nodes.

3.4   Classification Layers

The GNN-based text encoding produces hidden feature representations $\mathbf{H} \in \mathbb{R}^{n \times d_c}$. We propose to use an attention mechanism (Eq. 4) to put more attention on nodes with more important information related to positive or negative ADE classes, denoted as

$$\mathbf{s} = \mathrm{softmax}(\mathbf{w_a}\mathbf{H}^{\mathrm{T}}), \qquad (4)$$

where $\mathbf{w_a} \in \mathbb{R}^{d_c}$ and $\mathbf{s} = (s_1, s_2, \cdots, s_n) \in \mathbb{R}^n$ is the attention weight vector containing attention score of each node. Attention scores from the attentive classification layer are different from the attention layer of GAT. Here, attention scores measure which nodes are more important to the graph, while in the attention layer of GAT, attention scores decide the importance of one node to the other node in the neighborhood. The weight is assigned to feature matrix to obtained attentive hidden representation weighted by attention scores, i.e., $\mathbf{H_a} = [s_1 \times \mathbf{h'_1}, s_2 \times \mathbf{h'_2}, ..., s_n \times \mathbf{h'_n}]$.

Then, we apply the softmax classifier over the graph-based encoding and obtain the probability of each class as:

$$\mathbf{p}_g = \mathrm{softmax}(\mathbf{W_f}\mathbf{H_a}^{\mathrm{T}}\mathbf{v_f}), \qquad (5)$$

where $\mathbf{W_f} \in \mathbb{R}^{n \times d_h}$ and $\mathbf{v_f} \in \mathbb{R}^{n \times 2}$ are the weight matrices. We apply the same calculation as Eq. 5 but with different weight matrices to pretrained contextualized embeddings $\mathbf{H}^{(0)}$. Finally, we get $\mathbf{p}_c$ as the prediction probability from the contextualized embeddings. A weight coefficient $\lambda \in [0, 1)$ is introduced to balance the result from graph-based encoding models and the result from BERT-based contextualized models:

$$\mathbf{p} = \lambda \mathbf{p}_g + (1 - \lambda)\mathbf{p}_c. \tag{6}$$

This weighted strategy can also be viewed as an ensemble of two classifiers or the interpolation of the prediction probability of two classifiers.

3.5 Model Training

We apply the negative log-likelihood loss function as the training objective. Because data in one of the datasets used in our study is imbalanced and the number of instances of this dataset is not large where the downsampling method is not suitable, we use the weighted negative log-likelihood loss function to solve the data imbalance problem [27]. Assuming that the number of documents containing ADE is N_1 and the number of documents not containing ADE is N_2, the weight w_+ for documents predicted as positive samples is $\frac{N_2}{N_1+N_2}$ and the weight w_- for documents predicted as negative samples is $\frac{N_1}{N_1+N_2}$. The weighted loss function is:

$$L = -\frac{1}{N} \sum_{i=1}^{N} (w_+ y_i \log(p_i) + w_-(1 - y_i) \log(1 - p_i)), \tag{7}$$

where N is the number of documents in one batch and y_i is the true label of a document. When a document contains ADE, y_i equals to 1; otherwise, y_i equals to 0. The Adam optimizer [15] is used for model optimization. To control the learning rate, we use the multiple-step learning rate scheduler. The learning rate scheduler decays the learning rate by the parameter γ when the number of epochs reaches a specific number.

4 Experiment

4.1 Data and Pre-processing

We used three datasets from the biomedical domain and social media to evaluate the performance of baselines and our model. The details of these datasets are shown in Table 1. We perform data pre-processing before building graph representation. Specifically, stop words, punctuation, and numbers are removed. For the data collected from Twitter, we use the tweet-preprocessor Python package[1] to remove URLs, emojis, and some reserved words for tweets.

---

[1] https://pypi.org/project/tweet-preprocessor/.

Table 1. A statistical summary of datasets

Dataset	Documents	ADR	non-ADR
SMM4H	2418	1209	1209
TwiMed-Pub	1000	191	809
TwiMed-Twitter	625	232	393

TwiMed-Twitter and TwiMed-Pub[2]. The TwiMed dataset [3] includes two sets collected from different domains, i.e., TwiMed-Twitter and TwiMed-Pub. They consist of documents from Twitter and PubMed, respectively. People with different backgrounds annotate diseases, symptoms, drugs, and their relations in each document. There are three types of relations: Outcome-negative, Outcome-positive, and Reason-to-use. When a document is annotated as outcome-negative, it is marked as ADE (positive). Otherwise, we mark it as non-ADE (negative). The TwiMed-Pub has a small number of documents containing ADEs. The weighted loss function is used to solve the issue of imbalanced classification. Models are evaluated by 10-fold cross-validation.

SMM4H Dataset[3] [22,26]. The dataset is from Social Media Mining for Health Applications (#SMM4H) shared tasks. Documents collected from Twitter contain a description of drugs and diseases. The dataset contains 17,385 tweets for training and 915 tweets for testing. In our experiment, since this dataset is large enough, we conduct downsampling to mitigate the problem of imbalance, where we only use 2418 tweets, half of which are negative (non-ADE) and the other half are positive (ADE). The training tweets are split into train and validation sets, with a ratio of 9:1. We use the official validation set to evaluate the model performance for a fair comparison with baseline models developed in the SMM4H shared task, such as [14,25,36].

4.2   Baselines, Evaluation and Setup

Precision (P), Recall (R), and F1-score are commonly used to measure different models in a classification task. We report these three metrics in our results and mainly use the F1-score to compare models' performance in our experiments. We consider two sets of baseline models for performance comparison: 1) models explicitly designed for ADE detection and 2) pre-trained contextualized models.

Customized models for ADE detection include:

- CNN-Transfer [19] (CNN-T for short): a CNN-based model with transfer learning module. It has two sentence classifiers and a shared feature extractor based on CNN.

---

[2] https://github.com/nestoralvaro/TwiMed.
[3] https://healthlanguageprocessing.org/smm4h-2021/task-1/.

- HTR-MSA [32]: a model with hierarchical tweet representation and multi-head self-attention. This model learns word representations and tweet representations with CNN and Bi-LSTM. The multi-head self-attention mechanism is also applied.
- ATL [19]: a model based on adversarial transfer learning for the ADE detection, where corpus-shared features are exploited.
- MSAM [37]: a model with the multihop self-attention mechanism. It captures contextual information using Bi-LSTM and applies an attention mechanism in multiple steps to generate semantic representations of sentences.
- IAN [1]: interactive attention networks, a model to interactively learn attentions in the context and model targets and context separately.

We compare our model with pre-trained language models on the SMM4H dataset as it is a recent dataset not studied by the aforementioned ADE detection baselines. We use the base version of pretrained models in our experiments for a fair comparison, which is the same setting as in the compared baselines.

- BERT [6]: a language representation models pre-training with unlabeled text. Yaseen et al. [36] proposed a model that combined LSTM with a BERT encoder for ADE detection, denoted as BERT-LSTM in this paper.
- RoBERTa [21]: a BERT-based model on the English language with slightly different pre-training strategies. Pimpalkhute et al. [25] developed a data augmentation method with RoBERTa text encoder for ADE detection, denoted as RoBERTa-aug in this paper.
- BERTweet [23]: a domain-specific model for English Tweets with the same architecture as BERT-base. Kayastha et al. [14] built a model with BERTweet and single-layer BiLSTM for ADE detection, denoted as BERTweet-LSTM in this paper.

We use Python 3.7 and PyTorch 1.7.1 to implement the model. The hyper-parameters we tuned in our experiments are presented in Table 2. In our experiment, we set the hyper-parameter of the learning rate scheduler γ and the milestone of epoch number to 0.1 and 30, respectively.

Table 2. Choices of hyper-parameters

Hyper-parameters	Choices
Learning rate for text encoder	$2e^{-5}$, $3e^{-5}$, $1e^{-4}$
Learning rate for classifier	$1e^{-4}$, $5e^{-4}$, $1e^{-3}$
Learning rate for graph-based models	$1e^{-3}$, $3e^{-3}$, $5e^{-3}$
Hidden dimension for GNN	200, 300, 400
Weight coefficient λ	0, 0.1 0.3, 0.5, 0.7, 0.9

4.3   Main Results

We compared our model with baseline models for the ADE detection task to validate the performance of our model. Tables 3 and Table 4 show the results of TwiMed and SMM4H dataset, respectively. Our model achieves the best performance for all datasets compared with other methods in terms of F1-score. The best result of TwiMed-Pub is obtained with ClinicalBERT embeddings and a GAT encoder. As for SMM4H and TwiMed-Twitter, the best combination of building blocks is RoBERTa embeddings and GCN encoder.

Table 3. Results of TwiMed datasets

Datasets	Metrics	HTR-MSA [32]	CNN-T [19]	MSAM [37]	IAN [1]	ATL [19]	Ours
TwiMed-Pub	P (%)	75.0	81.3	85.8	87.8	81.5	**88.4**
	R (%)	66.0	63.9	**85.2**	73.8	67.0	85.0
	F1 (%)	70.2	71.6	85.3	79.2	73.4	**86.7**
TwiMed-Twitter	P (%)	60.7	61.8	74.8	83.6	63.7	**84.2**
	R (%)	61.7	60.0	**85.6**	81.3	63.4	83.7
	F1 (%)	61.2	60.9	79.9	82.4	63.5	**83.9**

Table 4. Results of SMM4H dataset

Methods	P (%)	R (%)	F1 (%)
BERT-LSTM [36]	77.0	72.0	74.0
BERTweet-LSTM [14]	81.2	86.2	83.6
RoBERTa-aug [25]	82.1	85.7	84.3
Ours	**86.7**	**93.4**	**89.9**

As shown in Table 3, performances of HTR-MSA, ATL, and CNN-Transfer are lower than others. The network structures of these three models are complex, resulting in a large amount of data being required. Thus, it performs worse than other models on small corpora. MSAM achieves the best performance on recall, while our model performs the best on precision and F1-score. Our model can balance precision and recall better. The competitive performance on the three datasets also shows the high generalization ability of our model. In Table 3, the performances of most models on the two datasets are significantly different. It is challenging to detect ADEs from tweets since tweets are informal text and contain much colloquial language. However, our model performs well on the TwiMed-Twitter dataset, showing that it can effectively encode information from the informal text and better capture relationships of entities in a document. From Table 4, we can find that other models are all BERT-based models. In contrast, our model employs GNN architectures, which suggests GNN can significantly improve models' performance on this task.

4.4   Analyses and Discussion

We further analyze the contextualized graph embedding model in this section, discuss the choice of different building blocks, and conduct a case study.

Choice of Graph Encoders. Our experiment examines GCN and GAT to study which one is more suitable for the ADE detection task. We record the best result under different graph encoders. For both GCN and GAT, we obtain the best result from RoBERTa for the SMM4H dataset and TwiMed-Twitter. For TwiMed-Pub, the best result is obtained using ClinicalBERT. From Table 5, we can find the results from the two GNNs are similar, showing that they both performed well on this task.

Table 5. Comparison on the choices of graph encoders, i.e., GCN and GAT

Graph encoder	SMM4H			TwiMed-Pub			TwiMed-Twitter		
	P (%)	R (%)	F1 (%)	P (%)	R (%)	F1 (%)	P (%)	R (%)	F1 (%)
GCN	86.7	93.4	**89.9**	88.6	84.3	86.4	84.2	83.7	**83.9**
GAT	84.8	92.3	88.4	88.4	85.0	**86.7**	83.1	81.9	82.5

Choice of Pretrained Embeddings. We examine three contextualized language models in our experiment. We record the best results with different language models. When using RoBERTa, the best results for the SMM4H dataset, TwiMed-Pub, and TwiMed-Twitter are from GCN, GAT, and GCN, respectively. When using ClinicalBERT, the best results for the SMM4H dataset and TwiMed-Pub are from GAT, and for TwiMed-Twitter, the best result is from GCN. When using BioBERT, the choice of GNNs for best results is the same as using ClinicalBERT.

From Table 6, we can find that, for TwiMed-Pub, there is little difference among the three pre-trained language models. However, for the SMM4H dataset and TwiMed-Twitter, RoBERTa performs better than others. The SMM4H dataset and TwiMed-Twitter dataset contain documents with many non-medical terms, while ClinicalBERT and BioBERT are trained with many medical terms. Therefore, when there are insufficient medical terms in the text, ClinicalBERT and BioBERT are unsuitable. RoBERTa is a better choice for informal text for this task.

Ablation Study on the Attention Classifier. To examine the effect of the attention classifier, we conduct an ablation study in our experiment. We remove the attentive classification layer and check the performance change in F1 scores.

From Table 7, we can find that after removing the attentive classification layer, values of F1-scores get decreased for all three datasets. It suggests that

Table 6. The effect of contextualized text embeddings obtained pretrained from different domains

Pretrained embeddings	SMM4H			TwiMed-Pub			TwiMed-Twitter		
	P (%)	R (%)	F1 (%)	P (%)	R (%)	F1 (%)	P (%)	R (%)	F1 (%)
RoBERTa	86.7	93.4	**89.9**	88.2	84.3	86.2	84.2	83.7	**83.9**
ClinicalBERT	81.5	92.3	86.6	88.4	85.0	**86.7**	80.1	80.7	80.4
BioBERT	80.9	93.4	86.7	88.2	84.0	86.0	81.2	80.6	80.9

the attentive classification layer can improve the model to prioritize information in the heterogeneous graph. More meaningful content, such as the description of symptoms and drugs, medical terms, and other relevant information related to ADEs, can contribute more to final predictions by employing attention mechanisms in the classification layer.

We also notice that F1 scores increase with the attentive classification layer, while precision scores for the SMM4H and TwiMed-Twitter datasets decrease. The documents of these two datasets are both from Twitter. Tweets are informal texts that do not follow the logical order, and their structures are unclear. They lack medical terms, and some content that seems not to be related to ADEs may also help determine whether a document contains ADEs or not. After applying the attentive classification layer, the model puts more attention to parts directly related to the description of symptoms, resulting in a tendency where a tweet is more easily to be predicted as a positive sample. Therefore, the precision value decreases after employing the attention classification layer. Besides, we can find that the F1 score on the SMM4H dataset decreases to a greater extent without an attentive classification layer. This dataset contains more documents compared to others. It suggests that the attentive classification layer works better for larger datasets. For small corpora, models with simpler architectures also perform well.

Table 7. Comparison between our model and the model without attentive classification layer

	SMM4H			TwiMed-Pub			TwiMed-Twitter		
	P (%)	R (%)	F1 (%)	P (%)	R (%)	F1 (%)	P (%)	R (%)	F1 (%)
Our model	86.7	93.4	89.9	88.4	85.0	86.7	84.2	83.7	83.9
- Attentive layer	87.0	90.1	88.5	87.8	83.9	85.8	84.6	82.2	83.3

Effect of Weight Coefficient λ. The weight coefficient λ's value controls the trade-off between the contextualized language models and graph neural networks. When λ equals zero, only BERT-based pre-trained contextualized embeddings are considered. In 2, dashed lines show the values of the F1-score when λ equals to zero. After employing GNNs ($\lambda = 0.1, 0.3, 0.5, 0.7, 0.9$), we can find that the value of the F1-score increases on all three datasets. It demonstrates that convolutional GNNs can improve the performance of our model significantly. Determining whether a symptom description is about the disease itself or adverse reactions resulting from the disease is a challenge in ADE detection. Utilizing GNNs helps solve this issue since GNNs can better capture the cause-effect relation and dependency between different entities of documents.

We can find the trend of the three lines are similar in respective plots of Fig. 2. In terms of F1-score, the best choices of the value of λ for three datasets are 0.5 (SMM4H), 0.9 (TwiMed-Pub), and 0.7 (TwiMed-Twitter). It suggests how to choose the value of λ depending on which datasets we use and other model hyper-parameters. Also, when values of λ are greater than 0.5, the F1 scores are relatively high. Therefore, we can first choose a high value of λ to allow graph embeddings to contribute more.

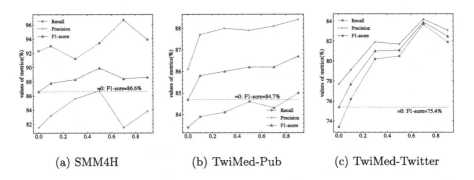

(a) SMM4H   (b) TwiMed-Pub   (c) TwiMed-Twitter

Fig. 2. The effect of weight coefficient λ on values of metrics

Case Study. We conduct a case study to explore the effect of the attention mechanism in Eq. 4. We choose two documents classified as positive samples in the SMM4H test dataset, where one is classified correctly while the other one does not contain ADE. We record the attention scores of words of these two tweets and utilize a heap map to show the value of different words' attention scores in a document, illustrated in Fig. 3. Figure 3a of a correctly classified tweet shows nouns (such as medication, sideaffects and seroquel), verbs (such as jolting), and sentiment words (such as hard and bad) related to drugs and symptoms get high attention scores. It helps the model put more attention on these important words. However, assigning high attention scores to such words does not ensure correct predictions. Figure 3b shows the attention scores of a tweet incorrectly classified as a positive sample. We can find that words related to symptoms, negative sentiment, and drugs are still getting high scores, while the tweet does not talk about ADE directly.

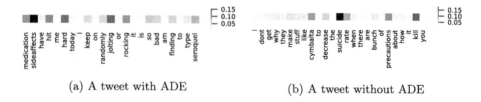

(a) A tweet with ADE   (b) A tweet without ADE

Fig. 3. Case study of the attention scores in two tweets: (a) with ADE; and (b) without ADE

5   Conclusion

The automated detection of adverse drug events from social content or biomedical literature requires the model to encode text information and capture the causal relation efficiently. This paper utilizes contextualized graph embeddings to learn contextual information and causal relations for ADE detection. We equip different convolutional graph neural networks with pre-trained language representation, develop an attention classifier to detect ADEs in documents and study the effects of different building components in our model. By comparing our model with other baseline methods, experiment results show that graph-based embeddings can better capture causal relationships and dependency between different entities in documents, leading to better detection performance.

Acknowledgment. We thank Professor Hongfei Lin for his kind support of this work. We acknowledge the computational resources provided by the Aalto Science-IT project and CSC - IT Center for Science, Finland. This work was supported by the Academy of Finland (grants 315896) and EU H2020 (grant 101016775).

References

1. Alimova, I., Solovyev, V.: Interactive attention network for adverse drug reaction classification. In: Ustalov, D., Filchenkov, A., Pivovarova, L., Žižka, J. (eds.) AINL 2018. CCIS, vol. 930, pp. 185–196. Springer, Cham (2018). https://doi.org/10.1007/978-3-030-01204-5_18
2. Alsentzer, E., et al.: Publicly available clinical BERT embeddings. In: Proceedings of the 2nd Clinical Natural Language Processing Workshop, pp. 72–78 (2019)
3. Alvaro, N., Miyao, Y., Collier, N.: Twimed: Twitter and PubMed comparable corpus of drugs, diseases, symptoms, and their relations. JMIR Public Health Surveill. **3**(2), e6396 (2017)
4. Bollegala, D., Sloane, R., Maskell, S., Hajne, J., Pirmohamed, M.: Learning causality patterns for detecting adverse drug reactions from social media. J. Med. Internet Res. (2018)
5. Cocos, A., Fiks, A.G., Masino, A.J.: Deep learning for pharmacovigilance: recurrent neural network architectures for labeling adverse drug reactions in twitter posts. JAMIA **24**(4), 813–821 (2017)
6. Devlin, J., Chang, M.W., Lee, K., Toutanova, K.: Bert: pre-training of deep bidirectional transformers for language understanding. In: NAACL-HLT (2019)

7. Donaldson, M.S., Corrigan, J.M., Kohn, L.T., et al.: To Err is Human: Building a Safer Health System (2000)
8. Duan, L., Khoshneshin, M., Street, W.N., Liu, M.: Adverse drug effect detection. IEEE J. Biomed. Health Inform. **17**(2), 305–311 (2012)
9. Ge, S., Qi, T., Wu, C., Huang, Y.: Detecting and extracting of adverse drug reaction mentioning tweets with multi-head self attention. In: Proceedings of SMM4H Workshop, pp. 96–98 (2019)
10. Huynh, T., He, Y., Willis, A., Rüger, S.: Adverse drug reaction classification with deep neural networks. In: COLING (2016)
11. Ji, S., Hölttä, M., Marttinen, P.: Does the magic of BERT apply to medical code assignment? A quantitative study. Comput. Biol. Med. **139**, 104998 (2021)
12. Jiang, T., et al.: Biomedical knowledge graphs construction from conditional statements. IEEE/ACM Trans. Comput. Biol. Bioinf. **18**(3), 823–835 (2020)
13. Johnson, A.E., et al.: Mimic-iii, a freely accessible critical care database. Sci. Data **3**(1), 1–9 (2016)
14. Kayastha, T., Gupta, P., Bhattacharyya, P.: BERT based adverse drug effect tweet classification. In: Proceedings of SMM4H Workshop, pp. 88–90 (2021)
15. Kingma, D.P., Ba, J.: Adam: a method for stochastic optimization. In: ICLR (2015)
16. Kipf, T.N., Welling, M.: Semi-supervised classification with graph convolutional networks. In: International Conference on Learning Representations (2017)
17. Lee, J., et al.: BioBERT: a pre-trained biomedical language representation model for biomedical text mining. Bioinformatics **36**(4), 1234–1240 (2020)
18. Li, H., et al.: Adverse drug reactions of spontaneous reports in shanghai pediatric population. PLoS ONE **9**(2), e89829 (2014)
19. Li, Z., Yang, Z., Luo, L., Xiang, Y., Lin, H.: Exploiting adversarial transfer learning for adverse drug reaction detection from texts. J. Biomed. Inform. **106**, 103431 (2020)
20. Lin, Y., et al.: BertGCN: Transductive Text Classification by Combining GCN and BERT. arXiv preprint arXiv:2105.05727 (2021)
21. Liu, Y., et al.: Roberta: a robustly optimized BERT pretraining approach. arXiv preprint arXiv:1907.11692 (2019)
22. Magge, A., et al.: Overview of the sixth social media mining for health applications (# smm4h) shared tasks at NAACL 2021. In: Proceedings of SMM4H Workshop, pp. 21–32 (2021)
23. Nguyen, D.Q., Vu, T., Nguyen, A.T.: BERTweet: a pre-trained language model for english tweets. In: Proceedings of the 2020 Conference on Empirical Methods in Natural Language Processing: System Demonstrations, pp. 9–14 (2020)
24. Pearl, J.: Causality. Cambridge University Press, Cambridge (2009)
25. Pimpalkhute, V., Nakhate, P., Diwan, T.: IIITN NLP at SMM4H 2021 tasks: transformer models for classification on health-related imbalanced twitter datasets. In: Proceedings of SMM4H Workshop, pp. 118–122 (2021)
26. Sarker, A., Gonzalez-Hernandez, G.: Overview of the second social media mining for health (SMM4H) shared tasks at AMIA 2017. Training. **1**(10,822), 1239 (2017)
27. Shimodaira, H.: Improving predictive inference under covariate shift by weighting the log-likelihood function. J. Statist. Plann. Inference **90**(2), 227–244 (2000)
28. Sohn, S., Clark, C., Halgrim, S.R., Murphy, S.P., Chute, C.G., Liu, H.: MedXN: an open source medication extraction and normalization tool for clinical text. JAMIA **21**(5), 858–865 (2014)
29. Sultana, J., Cutroneo, P., Trifirò, G.: Clinical and economic burden of adverse drug reactions. J. Pharmacol. Pharmacotherap. **4**(Suppl1), S73 (2013)

30. Veličković, P., Cucurull, G., Casanova, A., Romero, A., Liò, P., Bengio, Y.: Graph attention networks. In: International Conference on Learning Representations (2018)
31. Wei, Q., et al.: A study of deep learning approaches for medication and adverse drug event extraction from clinical text. JAMIA **27**(1), 13–21 (2020)
32. Wu, C., Wu, F., Liu, J., Wu, S., Huang, Y., Xie, X.: Detecting tweets mentioning drug name and adverse drug reaction with hierarchical tweet representation and multi-head self-attention. In: Proceedings of SMM4H Workshop, pp. 34–37 (2018)
33. Wu, Z., Pan, S., Chen, F., Long, G., Zhang, C., Philip, S.Y.: A comprehensive survey on graph neural networks. IEEE Trans. Neural Netw. Learn. Syst. **32**(1), 4–24 (2020)
34. Wunnava, S., Qin, X., Kakar, T., Kong, X., Rundensteiner, E.: A dual-attention network for joint named entity recognition and sentence classification of adverse drug events. In: Proceedings of the 2020 Conference on Empirical Methods in Natural Language Processing: Findings, pp. 3414–3423 (2020)
35. Yao, L., Mao, C., Luo, Y.: Graph convolutional networks for text classification. In: Proceedings of AAAI, vol. 33, pp. 7370–7377 (2019)
36. Yaseen, U., Langer, S.: Neural text classification and stacked heterogeneous embeddings for named entity recognition in SMM4H 2021. In: Proceedings of SMM4H Workshop, pp. 83–87 (2021)
37. Zhang, T., et al.: Adverse drug reaction detection via a multihop self-attention mechanism. BMC Bioinform. **20**(1), 1–11 (2019)
38. Zhang, T., et al.: Gated iterative capsule network for adverse drug reaction detection from social media. In: 2020 IEEE International Conference on Bioinformatics and Biomedicine (BIBM), pp. 387–390. IEEE (2020)

Bi-matching Mechanism to Combat Long-tail Senses of Word Sense Disambiguation

Junwei Zhang[1,2], Ruifang He[1,2(✉)], and Fengyu Guo[3(✉)]

[1] Tianjin Key Laboratory of Cognitive Computing and Application,
College of Intelligence and Computing, Tianjin University, Tianjin, China
{junwei,rfhe}@tju.edu.cn
[2] State Key Laboratory of Communication Content Cognition,
People's Daily Online, Beijing, China
[3] College of Computer and Information Engineering, Tianjin Normal University,
Tianjin, China
fyguo@tjnu.edu.cn

Abstract. The long-tail phenomenon of word sense distribution in linguistics causes Word Sense Disambiguation (WSD) to face both head senses with a large number of samples and tail senses with only a few samples. Traditional recognition methods are suitable for head senses with sufficient training samples, but they cannot effectively deal with tail senses. Inspired by the diverse memory and recognition abilities of children's linguistic behavior, we propose a bi-matching mechanism approach for WSD. Considering that tail senses are often presented in the form of fixed collocations, a collocation feature matching method suitable for tail senses is designed; the traditional definition matching method is used for head senses; finally, the two matching methods are combined to construct a WSD model with the bi-matching mechanism (called Bi-MWSD). Bi-MWSD can effectively combat the difficulty of identifying the tail senses due to insufficient training samples. The experiments are implemented in the standard English all-words WSD evaluation framework and the training data augmented evaluation framework. The experimental results outperform the baseline models and achieve state-of-the-art performance under the data augmentation evaluation framework.

Keywords: Word sense disambiguation · Long tail senses · Bi-matching mechanism

1 Introduction

Word Sense Disambiguation (WSD) is to assign the correct sense to the target word according to the given context [1,2]. WSD occupies an important position in the field of Natural Language Processing (NLP) [3], and the correct identification of word senses has a direct and profound impact on subsequent semantic

M.-R. Amini et al. (Eds.): ECML PKDD 2022, LNAI 13714, pp. 621–637, 2023.
https://doi.org/10.1007/978-3-031-26390-3_36

understanding tasks, such as machine translation [4,5] and natural language understanding [6,7].

However, due to the long-tail phenomenon of word sense distribution in linguistics, the WSD model needs to face both head senses with a large number of samples and tail senses with only a few samples [8,9]. For example, the verb form of the word *Play*[1] has 35 senses in WordNet 3.1, of which the most commonly used is "*Participate in games or sports*", and the vast majority are rarely used tail senses, such as "*Contend against an opponent in a sport, game, or battle*". In addition, due to the long-tail phenomenon of vocabulary usage frequency in linguistics, the occurrence frequency of tail senses is severely reduced, which makes it more difficult for the WSD model to identify long-tail senses. Note that the long-tail senses here refer to the tail senses under the long-tailed distribution.

Traditional recognition methods can effectively deal with head senses with sufficient training samples, but it is difficult to take into account tail senses with insufficient training samples. BEM, proposed by Blevins et al. [10], attempts to employ BERT [11] to obtain a context-based embedding of the target word, and then determines possible sense by calculating the similarity between this embedding and the textual embedding of each gloss. For head senses, this method can obtain effective sense representations, but for tail senses, it is difficult to obtain highly recognizable representations. The reason is that embeddings of all senses can be easily obtained based on glosses, but it is difficult to effectively improve the accuracy of embeddings when training samples are lacking or not. GlossBERT, proposed by Huang et al. [12], combines the sentence containing the target word with each gloss separately to obtain shared embeddings, and then treats the WSD task as a sentence-level classification task to achieve word sense recognition. This method has similar drawbacks to BEM, that is, it is difficult to obtain reliable representations when training samples are lacking or not. In addition, some researchers attempt to treat the WSD task as a few-shot learning problem to deal with insufficient training samples for tail senses. For example, Holla et al. [13] propose a meta-learning framework to deal with few-shot WSD, which aims to learn features from labeled instances to disambiguate unseen words. See also Refs. [8,9,14].

Inspired by the diverse memory and recognition abilities of children's linguistic behavior [15] (see Sect. 3.2 for a detailed analysis), we propose a bi-matching mechanism approach for WSD. Analysis of a large number of tail senses finds that tail senses are mostly presented in the form of fixed collocations, that is, they often appear together with fixed words or often appear in fixed contexts. This is also the main reason for insufficient samples of tail senses. Considering that the collocation words of tail senses are fixed, and the collocation words are clear, this paper proposes a collocation feature matching method to combat the challenge of insufficient training samples of tail senses. This paper extracts collocation words from the example sentences provided by the corresponding word senses in the dictionary, and collectively calls them the collocation feature. When there are multiple example sentences, the collocation feature integrates

---

[1] http://wordnetweb.princeton.edu/perl/webwn?s=play.

all the collocation words in the example sentences; when there is no example sentence, the collocation feature directly uses the gloss instead. Considering the outstanding performance of definition matching in traditional recognition methods, this paper adopts traditional definition matching to deal with head senses. Finally, the two matching methods together constitute a WSD model with the bi-matching mechanism.

The contributions of this paper are summarized as follows:

- By mining the characteristics of long-tail senses, a collocation feature matching method against insufficient training samples of tail senses is proposed.
- Inspired by the diverse memory and recognition abilities of children's linguistic behavior, a WSD model with the bi-matching mechanism is constructed, which fills the gap of using different matching methods for head and tail senses.
- The experiments are carried out under the evaluation framework of English all-words WSD, and the experimental results are better than the baseline models. Moreover, state-of-the-art performance is achieved under data-augmented evaluation framework.

Codes and pre-trained models are available at https://github.com/yboys0504/wsd.

2   Related Work

In the early development of WSD, researchers did not focus on long-tail senses, but more on dealing with all senses by adopting a unified approach. During this period, WSD models used a single recognition method to complete the recognition process at the end of the model [1,3]. These recognition methods are also often used in other tasks in NLP, so we call them **traditional recognition methods**. Subsequently, with the continuous improvement of the overall level of WSD models, long-tail senses became the bottleneck of development, and researchers began to focus on **few-shot learning methods** to combat long-tail senses [14,16].

2.1   Traditional Recognition Methods for WSD

According to the classical classification method, WSD models can be roughly divided into two categories, namely supervised models and knowledge-based models.

Supervised models usually employ a deep network structure to process the target word with context, and connect a classifier at the end of the model to calculate the probability of each sense [17,18]. For example, Recurrent Neural Network (RNN) suitable for sequence features is often used to build the core network structure of the WSD models, and a fully connected layer with normalization constraints is added as a classifier in the output layer [19,20]. Subsequent WSD models based on pre-trained language models only replace the core network structure with pre-trained models, but the classifiers are still implemented using a traditional fully connected layer [21–23]. The reason why supervised models are accustomed to this design is that the model can be trained end-to-end as a whole.

Knowledge-based models attempt to employ external knowledge to improve the recognition rate of WSD models, such as dictionary knowledge [10,24], semantic network knowledge [25,27], and multilingual knowledge [21,28]. Among them, glosses in the dictionary are often trained as text embeddings to replace word sense labels [9,10,26]. Such definition matching methods are good for identifying head senses, but they are not good for identifying tail senses. The fundamental reason is that tail senses often appear in the form of fixed collocations and they are difficult to give a clear definition.

2.2   Few-shot Learning Methods for WSD

Subsequently, the researchers realized the importance of long-tail senses in WSD, and adopted some targeted solutions for tail senses, such as meta-learning, zero-shot learning, reinforcement learning, etc. Holla et al. [13] proposed a meta-learning framework for few-shot WSD, where the goal is to learn features from labeled instances to disambiguate unseen words. See also Refs. [14,16]. Blevins et al. [10] noticed the long-tail phenomenon of word sense distribution, and proposed a dual encoder model, that is, one BERT is used to extract the word embedding of the target word with contextual information, and another BERT is used to obtain the text embeddings of the glosses. The innovation of this work is that the model adopts a joint training mechanism of dual encoders, but the disadvantage is that the model still adopts a single matching method to deal with both head and tail senses.

3   Methodology

In this section, we first formalize the WSD task, then clarify the cognitive basis of the bi-matching mechanism derived from children's literacy behavior, and finally describe the structure of our model in the formal language.

3.1   Word Sense Disambiguation

WSD is to predict the senses of the target word in a given context [1,2]. The formal definition can be expressed as: the possible sense $s \in S_{\hat{w}}$ of the target word $\hat{w}$ in the given context $C_{\hat{w}}$ is formally described as

$$f(\hat{w}, C_{\hat{w}}) = s \in S_{\hat{w}} \tag{1}$$

where $f(\cdot)$ refers to the WSD model, and $S_{\hat{w}}$ is the candidate list of the senses of the target word.

All-words WSD is to predict all ambiguous words in a given context [1,2]. This means that the WSD model may predict the noun, verb, adjective, and adverb forms of ambiguous words. In this case, the input and output of the WSD model are defined as $C = (..., w_i, ...)$ and $S = (..., s^x_{w_i}, ...)$, respectively, where $s^x_{w_i}$ represents the xth sense of the target word w_i.

3.2 Cognitive Basis of Bi-matching Mechanism

Masaru Ibuka [15], a Japanese educator, pointed out that children's literacy behavior is mainly based on mechanical memory and recognition ability in the early stage, and then gradually develops concept-oriented memory and recognition ability in the later stage. The mechanical method rigidly remembers the structure of the word itself and its application scenarios, such as collocation features of words. The concept-oriented method establishes the relationship between the structure, meaning, and usage of words through analysis and comparison, such as the definitions given in the dictionary.

For the WSD task, we should not only pay attention to head senses with a large number of samples, but also tail senses with only a few samples, because long-tail senses are an important bottleneck for the development. For head senses, it is reasonable to distinguish senses through the definition system, because theoretically, the definition system of word senses can clearly distinguish different head senses. But for tail senses, it is difficult to define a clear and non-confusing definition system for each sense. For example, "*Go to plant fish*", where the word *plant* means "*Place into a river*". This sense of the word *plant* mostly appears in such a collocation form. Therefore, considering the characteristics of tail senses, the collocation feature matching method is more suitable for identifying tail senses.

In this paper, we propose a bi-matching mechanism approach to construct a WSD model (called **Bi-MWSD**), namely the **collocation feature matching method** for tail senses and the **definition matching method** for head senses. We describe the construction details and operation process of Bi-MWSD in Sect. 3.3.

3.3 Bi-matching Mechanism for WSD

The architecture of Bi-MWSD is shown in Fig. 1. Bi-MWSD uses two pre-trained language models as text feature encoders, and the pre-trained model adopts the widely used BERT [11]. One encoder is used to extract the collocation features of the target word in the training samples and the example sentences, which is called the **collocation feature encoder**. The other is used to learn the definition system in the glosses of the target word, which is called the **definition encoder**. The example sentences and glosses come from the examples and definitions corresponding to each sense in WordNet. The last step is the matching process of head senses and tail senses, which is called **word sense matching**.

Collocation Feature Encoder: The function of the collocation feature encoder is to memorize the collocation features of the target word, such as the

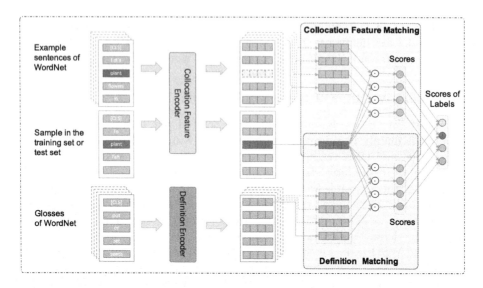

Fig. 1. Schematic diagram of the Bi-MWSD architecture, which illustrates the disambiguation process of the target word *Plant*. The collocation feature encoder is used to encode target words and example sentences; the definition encoder is only used to encode glosses. The symbol ⊙ represents the dot product of matrices.

structure and relationship between the target word and the collocation words, and the entire application scenario. The encoder process two kinds of texts:

- One is the example sentences corresponding to each sense of the target word in WordNet, $E^x = (..., e^x_k, ...)$ where e^x_k represents the kth word of the example sentence E^x of the xth sense of the target word.
- And the other is the training samples containing the target word, $C = (..., w_i, ...)$ where w_i represents the ith word.

The texts are encoded using BERT standard processing rules, that is, adding $[CLS]$ and $[SEP]$ marks at the beginning and end of the text respectively, such as

$$E^x = ([CLS], ..., e^x_k, ..., [SEP]) \qquad (2)$$
$$= (e^x_{cls}, ..., e^x_k, ..., e^x_{sep}). \qquad (3)$$

The processing method of the training samples is also the same. The encoder encodes each word, including the added $[CLS]$ and $[SEP]$, to obtain a corresponding 768-dimensional vector.

The reason why we use one encoder to process two kinds of texts here is that both the example sentences and the training samples contain the target word, which can all be considered that there are collocation features of the target word. Moreover, the advantage of this processing is that the training sample will truly reflect the frequency of each sense of the target word, and the example sentences can provide the collocation features of tail senses. Processing them together can make up for the lack of scene information of tail senses, but it will not (seriously) change their frequency. In WordNet 3.0, sometimes multiple example sentences are given for one sense, and we integrate all the example sentences by default; when no example sentences are given, we use the embedded representation of the gloss instead.

After processing by the collocation feature encoder, we can get the vector representation of the target word in the training sample, which is defined as $v_{\hat{w}}$, and the vector representation of the collocation features of each sense x provided by the example sentences, which is defined as V_{E^x}. $v_{\hat{w}}$ is the vector representation corresponding to the target word in the output of the pretrained model BERT. For V_{E^x}, we here provide two calculation methods, namely the overall text vector minus the target word vector,

$$V_{E^x} = v_{e^x_{cls}} - v_{e^x_{\hat{w}}}, \tag{4}$$

and the vectors except the target word vector are added,

$$V_{E^x} = \sum_k v_{e^x_k} - v_{e^x_{\hat{w}}}. \tag{5}$$

Through experimental analysis of these two methods, we find that the first one is relatively better. The possible reason is that it can not only characterize the collocation features of the target word, but also remember the entire text, namely the application scenario.

Definition Encoder: The definition encoder constructs the definition system of the target word by learning the glosses G^x for each sense x in WordNet, $G^x = (..., g^x_j, ...)$ where g^x_j represents the jth word of the gloss text of the xth sense of the target word. The glosses are simple and accurate generalizations of word senses and are therefore suitable for refining the definition system of the target word. What needs to be emphasized here is that the target word itself is not included in the glosses, so glosses cannot be used to extract the collocation features of the target word. Following standard processing rules of BERT, $[CLS]$ and $[SEP]$ marks are also added for the glosses,

$$G^x = ([CLS], ..., g^x_j, ..., [SEP]) \tag{6}$$

$$= (g^x_{cls}, ..., g^x_j, ..., g^x_{sep}). \tag{7}$$

The encoder encodes each word, including the added $[CLS]$ and $[SEP]$, to obtain a corresponding 768-dimensional vector. Here we choose the output vector corresponding to $[CLS]$, i.e., $v_{g^x_{cls}}$, to represent the entire gloss text, i.e., $V_{G^x} = v_{g^x_{cls}}$. This method is a common practice in the industry.

Word Sense Matching: At this point, we can calculate the score of each sense of the target word $\hat{w}$ in a given context C,

$$Score(\hat{w}|C) = F(\{v_{\hat{w}} \odot (\alpha V_{G^x} + \beta V_{E^x})\}^x) \tag{8}$$

where α and β respectively represent the proportion of the definition matching method and the collocation feature matching method. $F(\cdot)$ can be a standard *Softmax* or other distribution function. When $F(\cdot)$ is selected as *Softmax*, $Score(\hat{w}|C)$ is a probability distribution of all senses of the target word in a given context. Finally, we can conclude that the one with the highest probability is the most likely sense.

Here α and β can be the weights learned by the model itself, or they can be the proportions of each sense provided by WordNet. Through experimental analysis, we find that they work best when they are set to the same value. It needs to be explained that it is difficult to know in advance which sense of the target word is, so it is appropriate to use the equal probability method, that is, the possibility of the head sense or the tail sense is the same.

Parameter Optimization: We use a cross-entropy loss on the scores of the candidate senses of the target word to train Bi-MWSD. The loss function is

$$Loss(Score, index) \tag{9}$$

$$= -\log\left(\frac{\exp(Score^{[index]})}{\sum_{i=1}\exp(Score^{[i]})}\right) \tag{10}$$

$$= -Score^{[index]} + \log\sum_{i=1}\exp(Score^{[i]}) \tag{11}$$

where *index* is the index of the list of the candidate senses of the target word.

Bi-MWSD employs an Adam optimizer [29] to update the parameters of the model, and the specific settings of the optimizer are given in the experimental section.

4   Experiments

4.1   Datasets and Evaluation Metrics

Bi-MWSD adopts the unified evaluation framework of English all-words WSD proposed by Raganato et al. [1] to implement training and evaluation. In the **standard evaluation experiment**, the training set is SemCor[2]; in the **evaluation experiment under data augmentation**, the training set is SemCor and WNGT[3] (WordNet Gloss Tagged). Following common practice, SemEval-2007 (SE07; [30]) is designated as the development set, and Senseval-2 (SE2; [31]), Senseval-3 (SE3; [32]), SemEval-2013 (SE13; [33]), and SemEval-2015 (SE15; [34]) are used as the test sets. The statistical information of each dataset is shown in Table 1. Also, we concatenate the development set and all the test sets to reconstruct the test sets of verbs (V), nouns (N), adjectives (A), and adverbs (R), and treat them as a whole as a test set (**ALL**).

---

[2] http://lcl.uniroma1.it/wsdeval/training-data.
[3] https://wordnetcode.princeton.edu/glosstag.shtml.

Table 1. Statistics of the datasets: the number of documents (Docs), sentences (Sents), tokens (Tokens), sense annotations, sense types covered, annotated lemma types covered and ambiguity level in each dataset, where the ambiguity level implies the difficulty of the dataset.

Dataset	Docs	Sents	Tokens	Annotations	Sense types	Lemma types	Ambiguity
SE2	3	242	5,766	2,282	1,335	1,093	5.4
SE3	3	352	5,541	1,850	1,167	977	6.8
SE07	3	135	3,201	455	375	330	8.5
SE13	13	306	8,391	1,644	827	751	4.9
SE15	4	138	2,604	1,022	659	512	5.5

In this paper, we select all word senses in WordNet 3.0 [35] as candidate senses of the target word. All experimental results in the figures and tables are reported as a percentage of the F1-score.

4.2 Baseline Models

To evaluate the comprehensive performance of Bi-MWSD in the community, we select state-of-the-art models in the past three years, including LMMS [36], EWISE [9], and GlossBERT [12] in 2019, SREF [37], ARES [26], EWISER [38], BEM [10], and SparseLMMS [39] in 2020, and COF [40], ESR [41], Multi-Label [42], and SACE [43] in 2021. All experimental results of the above models are taken from the data published in the original paper.

From these, we select three most comparable models as baseline models, which are GlossBERT [12] with similar external resources, BEM [10] with similar framework structure, and Multi-Label [42] with multi-label classification method. GlossBERT and BEM employ typical and traditional word sense recognition methods. GlossBERT employs a fully connected layer with normalization constraints as the output layer of the model. BEM implements word sense matching by calculating the similarity between the target word vector and the definition vectors. Multi-Label designs the WSD model as a multi-label classification task. Although this method has the ability to match multiple times, it is not the same as the bi-matching mechanism proposed in this paper.

In addition, we select three models as baselines for the evaluation experiment under data augmentation, which are SparseLMMS [39], EWISER [38], and ESR [41].

4.3 Experimental Setting

The hardware platform of Bi-MWSD is Ubuntu 18.04.3, which installs two GPUs whose version is NVIDIA Tesla P40. The development platform is Python 3.8.3[4], and the learning framework is Pytorch 1.8.1[5]. The pre-trained language model

---

[4] https://www.python.org/.
[5] https://pytorch.org/.

is provided by Transformers 4.5.1[6]. Under the **standard evaluation experiment**, the encoders of Bi-MWSD use *BERT-base-uncased*; under the **evaluation experiment of data augmentation**, the encoders of Bi-MWSD use *BERT-large-uncased*. The hyperparameter *Learning Rate, Context Batch Size, Gloss Batch Size, Epochs, Context Maximum Length* and *Gloss Maximum Length* of the model are set to $[1E\text{-}5, 5E\text{-}6, 1E\text{-}6]$, 4, 256, 20, 128 and 32, respectively. Super-parameters not listed are given in the published code.

Table 2. F1-score (%) on the English all-words WSD task. Dev refers to the development set, and N, V, A, R, and **ALL** refer to the nouns, verbs, adjectives, adverbs, and overall datasets constructed by concatenating the development set and the test sets, respectively. The experimental results are organized according to the standard evaluation experiment (that is, Training data: SemCor) and the evaluation experiment under data augmentation (that is, Training data: SemCor + WNGT). The underlined and the bolded results refer to the overall and the regional best results, respectively.

Model	Dev	Test sets				Concatenation				
	SE07	SE2	SE3	SE13	SE15	N	V	A	R	ALL
Training data: SemCor										
Prior work										
LMMS (ACL, 2019, [36])	68.1	76.3	75.6	75.1	77.0	–	–	–	–	75.4
EWISE (ACL, 2019, [9])	67.3	73.8	71.1	69.4	74.5	74.0	60.2	78.0	82.1	71.8
SREF (EMNLP, 2020, [37])	72.1	78.6	76.6	78.0	80.5	80.6	66.5	82.6	84.4	77.8
ARES (EMNLP, 2020, [26])	71.0	78.0	77.1	77.3	83.2	80.6	68.3	80.5	83.5	77.9
EWISER (ACL, 2020, [38])	71.0	78.9	78.4	78.9	79.3	81.7	66.3	81.2	85.8	78.3
COF (EMNLP, 2021, [40])	69.2	76.0	74.2	78.2	80.9	80.6	61.4	80.5	81.8	76.3
ESR (EMNLP, 2021, [41])	75.4	80.6	78.2	79.8	82.8	82.5	69.5	82.5	87.3	79.8
SACE (ACL, 2021, [43])	74.7	80.9	79.1	82.4	<u>84.6</u>	83.2	71.1	<u>85.4</u>	87.9	80.9
Baseline models										
GlossBERT (EMNLP, 2019, [12])	72.5	77.7	75.2	76.1	80.4	79.8	67.1	79.6	87.4	77.0
BEM (ACL, 2020, [10])	74.5	79.4	77.4	79.7	**81.7**	81.4	68.5	**83.0**	<u>**87.9**</u>	79.0
Multi-Label (EACL, 2021, [42])	72.2	78.4	77.8	76.7	78.2	80.1	67.0	80.5	86.2	77.6
Bi-MWSD	**75.2**	**80.2**	**78.0**	**79.8**	81.4	**82.8**	**69.5**	82.5	87.5	**79.4**
Training data: SemCor + WNGT										
SparseLMMS (EMNLP, 2020, [39])	73.0	79.6	77.3	79.4	81.3	–	–	–	–	78.8
EWISER (ACL, 2020, [38])	75.2	80.8	79.0	80.7	81.8	81.7	66.3	81.2	85.8	80.1
ESR (EMNLP, 2021, [41])	<u>**77.4**</u>	<u>**81.4**</u>	78.0	81.5	<u>**83.9**</u>	83.1	71.1	<u>**83.6**</u>	<u>**87.5**</u>	80.7
Bi-MWSD$_{large}$	77.3	80.8	<u>**79.9**</u>	<u>**83.8**</u>	83.7	<u>**84.0**</u>	<u>**71.7**</u>	81.5	86.5	<u>**81.5**</u>

4.4 Experimental Results

The experimental results are shown in Table 2, where according to common practice, all results are presented as a percentage of the F1-score. The experimental

---

[6] https://huggingface.co/transformers/v4.5.1/.

results are organized according to the **standard evaluation experiment** and the **evaluation experiment under data augmentation**.

- **In the standard evaluation experiment**, compared with previous work, Bi-MWSD is in an upper-middle position; compared with baseline models, Bi-MWSD achieves state-of-the-art in multiple metrics. The experimental results confirm that the bi-matching mechanism is indeed beneficial to improve the recognition ability of the model. Compared with GlossBERT [12], it shows that the matching mechanism of Bi-MWSD is superior to the recognition method constructed by a fully connected layer with normalization constraints. The possible reason is that the recognizer constructed by a fully connected layer has a large number of parameters that need to be learned, and the lack of training samples of long-tail senses makes it difficult to learn the parameters effectively. Compared with BEM [10], it shows that the bi-matching mechanism of Bi-MWSD will improve the recognition ability compared with the single-matching mechanism model with a similar structure. For the contribution of the collocation feature matching method, we will give an analysis in the ablation study.
- **In the evaluation experiment under data augmentation**, Bi-MWSD also achieves state-of-the-art performance in multiple metrics, indicating that Bi-MWSD has great potential. Moreover, it also shows that when the training sample size of tail senses is expanded, it is beneficial to improve the performance of Bi-MWSD.

Analysis of poor performance on indicators A (adjectives) and R (adverbs) of Table 2: In linguistics, nouns and verbs are words with a serious long-tail, and adjectives and adverbs are relatively weaker. In other words, there are fewer tail senses in adjectives and adverbs. For datasets where the proportion of tail senses is not high, the method of not distinguishing or ignoring tail senses has advantages.

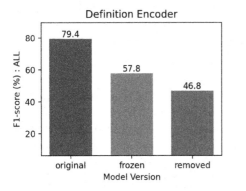 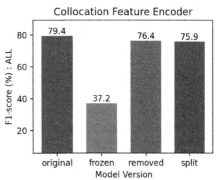

Fig. 2. Experimental results of ablation studies on the definition encoder and the collocation feature encoder. All values are experimental results under the test set **ALL** and are presented as a percentage of the F1-score.

4.5   Ablation Study

Bi-MWSD employs a bi-matching mechanism to replace the traditional single-matching mechanism of the WSD model, namely **definition matching** and **collocation feature matching**. To clarify the contribution of various matching mechanisms to the overall representation, and to determine their value for the target task, we perform ablation experiments.

Ablation Study for Definition Matching: For the analysis of the definition matching mechanism, we use the method of **ablation function** (i.e., freeze the encoder) and **ablation module** (i.e., directly remove the encoder). The method of freezing the encoder will prevent the encoder from fine-tuning the parameters on the training set, that is, preventing the encoder from learning more semantic information on the training set. We know that tail senses are marked in the training set. Preventing the encoder from fine-tuning the parameters on the training set will hinder the encoder's ability to recognize tail senses. Compared with the original model, this method will directly reflect the contribution of the definition encoder to solving tail senses. The method of removing the encoder is more direct, which directly reflects the contribution of the definition matching method to the overall representation.

We separately freeze and remove the definition encoder on the original model, and adjust the hyperparameters to get the best results. The experimental results are shown in Fig. 2.

1. Comparing the original version and the frozen version, it can be seen that the definition encoder can indeed learn new semantic knowledge by fine-tuning the parameters on the training set, and it can greatly improve the overall representation.
2. Comparing the original version and the removed version, it can be seen that the contribution of the definition encoder to the overall representation is huge. This result is in line with reality, because head senses are indeed far greater than the usage rate of tail senses in life, and the function of the definition encoder is reflected in the recognition of head senses. Again, comparing the frozen version with the deleted version confirms this conclusion.

Ablation Study for Collocation Feature Matching: For the analysis of the collocation feature matching mechanism, in addition to the **ablation function** and **ablation module**, we also need to **disassemble the two functions** of the collocation feature encoder, that is, target word vectorization and example sentence vectorization. It should be emphasized that the removed version here only removes the example sentence learning function of the encoder.

We fine-tune the hyperparameters of the modified versions to obtain the best results. The experimental results are shown in Fig. 2.

1. Comparing the original version and the frozen version, it can be seen that the model shows the worst case without fine-tuning the parameters under the

training set. The main reason is that the encoder is responsible for the learning of the target word vector. If there is no good target word representation, it will directly affect the overall representation.

2. Comparing the original version with the removed version, that is, removing the collocation feature matching method, it can be seen that introducing this matching mechanism can indeed improve the effectiveness of the model. Although there is only two percentage point improvement, considering the difficulty of tail sense recognition, it also shows that the bi-matching mechanism does contribute to the recognition of tail senses.

3. Regarding whether the training process of merging the target word and the collocation feature can improve the overall representation of the model, we can compare the results of the original version and the split version. An improvement of close to 3% proves that this design is reasonable. Example sentences of tail senses in the dictionary improve the ability of the pre-trained model to represent low-tail words.

5   Experiments Under Head and Tail Senses

To confirm the effectiveness of the bi-matching mechanism for various word senses, namely, head senses and tail senses, we conduct experiments under the reconstructed head sense and tail sense test sets respectively. The ablation experiments focus more on analyzing the effectiveness of each module, while the experiments here can more clearly present the specific contribution of the bi-matching mechanism to various word senses.

Datasets: The training set and development set still employ the settings of the standard evaluation experiment. The test sets are divided into head sense (HS) and tail sense (TS) datasets obtained by reconstructing **ALL**.

- The construction method of the head sense datasets is to obtain the dataset by **removing** the specified word sense samples in **ALL**. We construct two head sense datasets: a dataset constructed by removing data with only one sample (called Removed 1-shot TS); and a dataset constructed by removing data with less than three samples (called Removed 2-shot TS).
- The construction method of the tail sense datasets is to obtain the dataset by **retaining only** the specified word sense samples in **ALL**. We construct two tail sense datasets: a dataset constructed by retaining only data with only one sample (called Retained 1-shot TS); a dataset constructed by retaining only data with less than three samples (called Retained 2-shot TS).

Experimental Setting and Baseline Models: The experimental setting is still carried out according to the setting method of the standard evaluation experiment. The baseline models select the most comparable GlossBERT [12] and BEM [10] as the control group. Bi-MWSD adopts the setup of the standard evaluation experiment.

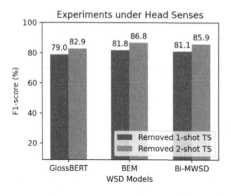 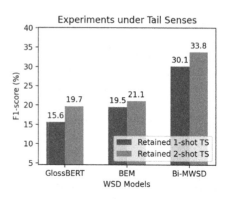

Fig. 3. Experimental results on the head sense and the tail sense datasets reconstructed by **ALL**. All values are presented as a percentage of the F1-score. *Removed *-shot TS* and *Retained *-shot TS* refer to different kinds of head sense (HS) and tail sense (TS) datasets, respectively.

5.1   Bi-MWSD for Head Senses

The experimental results under the head sense datasets are shown in Fig. 3. From the overall data performance, Bi-MWSD outperforms GlossBERT but is inferior to BEM on both head sense datasets, indicating that the bi-matching mechanism is stronger than the single-matching mechanism constructed by the fully connected layer but weaker than the single-matching mechanism constructed by the definition identification method on datasets with all head senses. This conclusion shows that there is a certain interference between the double matching mechanisms, and it is difficult to obtain the best performance when only one class of word senses is processed.

5.2   Bi-MWSD for Tail Senses

The experimental results under the tail sense datasets are shown in Fig. 3. From the overall data performance, Bi-MWSD outperforms the control models on both tail sense datasets, indicating that the bi-matching mechanism has significant advantages in dealing with tail senses. This conclusion fully proves that the collocation feature matching method can effectively deal with the long-tail senses; the multi-matching mechanism (not limited to the bi-matching mechanism proposed in this paper) can be used to achieve the purpose of dealing with various word senses in a targeted manner.

6   Conclusion

Inspired by the diverse memory and recognition abilities of children's linguistic behavior, this paper proposes a method of bi-matching mechanism to deal with the head and tail senses in Word Sense Disambiguation (WSD). We design a collocation feature matching method for tail senses, and leverage traditional definition matching method to deal with head senses, which together constitute a WSD model with the bi-matching mechanism (called Bi-MWSD). Bi-MWSD can effectively combat the difficulty of insufficient tail sense training samples caused by the long tail distribution of word sense. In addition, Bi-MWSD outperforms baseline models and achieves state-of-the-art performance under data-augmented evaluation framework. The contribution of this work is to fill the gap of bi-matching mechanism in WSD, and moreover explore the feasibility of bi-matching mechanism against insufficient training samples.

In future work, we will build a hierarchical multi-matching mechanism to better address the imbalance of training samples caused by the long-tailed phenomenon of word sense distribution. Moreover, we will further subdivide the word senses, and employ this multi-matching method to deal with various word senses in a targeted manner to improve the accuracy of word sense recognition.

Acknowledgements. Our work is supported by the National Natural Science Foundation of China (61976154), the National Key R&D Program of China (2019YFC1521200), the State Key Laboratory of Communication Content Cognition, People's Daily Online (No. A32003), and the National Natural Science Foundation of China (No. 62106176).

References

1. Navigli, R., Camacho-Collados, J., Raganato, A.: Word sense disambiguation: a unified evaluation framework and empirical comparison. In: EACL (2017)
2. Navigli, R.: Word sense disambiguation: a survey. ACM Comput. Surv. **41**, 1–69 (2009)
3. Bevilacqua, M., Pasini, T., Raganato, A., Navigli, R.: Recent trends in word sense disambiguation: a survey. In: IJCAI (2021)
4. Neale, S., Gomes, L.-M., Agirre, E., Lacalle, O.-L., Branco, A.-H.: Word sense-aware machine translation: including senses as contextual features for improved translation models. In: LREC (2016)
5. Rios Gonzales, A., Mascarell, L., Sennrich, R.: Improving word sense disambiguation in neural machine translation with sense embeddings. In: WMT (2017)
6. Dewadkar, D.-A., Haribhakta, Y.-V., Kulkarni, P.-A., Balvir, P.-D.: Unsupervised word sense disambiguation in natural language understanding. In: ICAI (2010)
7. Mills, M.-T., Bourbakis, N.-G.: Graph-based methods for natural language processing and understanding-a survey and analysis. IEEE Trans. Syst. Man Cybern. Syst. **44**, 59–71 (2014)
8. Li, W., Madabushi, H.-T., Lee, M.-G.: UoB_UK at SemEval 2021 Task 2: Zero-shot and few-shot learning for multi-lingual and cross-lingual word sense disambiguation. In: SEMEVAL (2021)
9. Kumar, S., Jat, S., Saxena, K., Talukdar, P.-P.: Zero-shot word sense disambiguation using sense definition embeddings. In: ACL (2019)
10. Blevins, T., Zettlemoyer, L.: Moving down the long tail of word sense disambiguation with gloss informed bi-encoders. In: ACL (2020)
11. Devlin, J., Chang, M., Lee, K., Toutanova, K.: BERT: pre-training of deep bidirectional transformers for language understanding. In: NAACL (2019)

12. Huang, L., Sun, C., Qiu, X., Huang, X.: GlossBERT: BERT for word sense disambiguation with gloss knowledge. In: EMNLP (2019)
13. Holla, N., Mishra, P., Yannakoudakis, H., Shutova, E.: Learning to learn to disambiguate: meta-learning for few-shot word sense disambiguation. In: EMNLP (2020)
14. Du, Y., Holla, N., Zhen, X., Snoek, C.-G., Shutova, E.: Meta-learning with variational semantic memory for word sense disambiguation. In: ACL (2021)
15. Ibuka, M.: Kindergarten is Too Late!. Souvenir Press, London (1977)
16. Chen, H., Xia, M., Chen, D.: Non-parametric few-shot learning for word sense disambiguation. In: NAACL (2021)
17. Yuan, D., Richardson, J., Doherty, R., Evans, C., Altendorf, E.: Semi-supervised word sense disambiguation with neural models. In: COLING (2016)
18. Raganato, A., Bovi, C.-D., Navigli, R.: Neural sequence learning models for word sense disambiguation. In: EMNLP (2017)
19. Le, M.-N., Postma, M., Urbani, J., Vossen, P.: A Deep dive into word sense disambiguation with LSTM. In: COLING (2018)
20. Kågebäck, M., Salomonsson, H.: Word sense disambiguation using a bidirectional LSTM. In: COLING (2016)
21. Scarlini, B., Pasini, T., Navigli, R.: SensEmBERT: context-enhanced sense embeddings for multilingual word sense disambiguation. In: AAAI (2020)
22. Hadiwinoto, C., Ng, H.-T., Gan, W.-C.: Improved word sense disambiguation using pre-trained contextualized word representations. In: EMNLP (2019)
23. Du, J., Qi, F., Sun, M.: Using BERT for word sense disambiguation. arXiv:1909.08358 (2019)
24. Luo, F., Liu, T., Xia, Q., Chang, B., Sui, Z.: Incorporating glosses into neural word sense disambiguation. In: ACL (2018)
25. Fernandez, A.-D., Stevenson, M., Martínez-Romo, J., Araujo, L.: Co-occurrence graphs for word sense disambiguation in the biomedical domain. Artif. Intell. Med. **87**, 9–19 (2018)
26. Scarlini, B., Pasini, T., Navigli, R.: With more contexts comes better performance: contextualized sense embeddings for all-round word sense disambiguation. In: EMNLP (2020)
27. Dongsuk, O., Kwon, S., Kim, K., Ko, Y.: Word sense disambiguation based on word similarity calculation using word vector representation from a knowledge-based graph. In: COLING (2018)
28. Pasini, T.: The knowledge acquisition bottleneck problem in multilingual word sense disambiguation. In: IJCAI (2020)
29. Kingma, D.-P., Ba, J.: Adam: A Method for Stochastic Optimization. CoRR, abs/1412.6980 (2015)
30. Pradhan, S., Loper, E., Dligach, D., Palmer, M.: SemEval-2007 Task 2017: English lexical sample. In: SRL and All Words, Fourth International Workshop on Semantic Evaluations (2007)
31. Edmonds, P., Cotton, S.: SENSEVAL-2: Overview. *SEMEVAL (2001)
32. Snyder, B., Palmer, M.: The English all-words task. In: ACL (2004)
33. Navigli, R., Jurgens, D., Vannella, D.: SemEval-2013 task 12: multilingual word sense disambiguation. In: *SEMEVAL (2013)
34. Moro, A., Navigli, R.: SemEval-2015 Task 13: multilingual all-words sense disambiguation and entity linking. In: *SEMEVAL (2015)
35. Fellbaum, C.-D.: WordNet: An Electronic Lexical Database. Language. MIT Press, Cambridge (2000)

36. Loureiro, D., Jorge, A.-M.: Language modelling makes sense: propagating representations through wordnet for full-coverage word sense disambiguation. In: ACL (2019)

37. Wang, M., Wang, Y.: A synset relation-enhanced framework with a try-again mechanism for word sense disambiguation. In: EMNLP (2020)

38. Bevilacqua, M., Navigli, R.: Breaking Through the 80% Glass Ceiling: Raising the State of the Art in Word Sense Disambiguation by Incorporating Knowledge Graph Information. ACL (2020)

39. Berend, G.: Sparsity Makes Sense: Word Sense Disambiguation Using Sparse Contextualized Word Representations. EMNLP (2020)

40. Wang, M., Zhang, J., Wang, Y.: Enhancing the Context Representation in Similarity-based Word Sense Disambiguation. EMNLP (2021)

41. Song, Y., Ong, X.C., Ng, H.T., Lin, Q.: Improved Word Sense Disambiguation with Enhanced Sense Representations. EMNLP (2021)

42. Conia, S., Navigli, R.: Framing Word Sense Disambiguation as a Multi-Label Problem for Model-Agnostic Knowledge Integration. EACL (2021)

43. Wang, M., Wang, Y.: Word Sense Disambiguation: Towards Interactive Context Exploitation from Both Word and Sense Perspectives. ACL (2021)

FairDistillation: Mitigating Stereotyping in Language Models

Pieter Delobelle[1,2(✉)] and Bettina Berendt[1,2,3,4]

[1] Department of Computer Science, KU Leuven, Leuven, Belgium
pieter.delobelle@kuleuven.be
[2] Leuven.AI Institute, Leuven, Belgium
[3] Faculty of Electrical Engineering and Computer Science,
TU Berlin, Berlin, Germany
[4] Weizenbaum Institute, Berlin, Germany

Abstract. Large pre-trained language models are successfully being used in a variety of tasks, across many languages. With this ever-increasing usage, the risk of harmful side effects also rises, for example by reproducing and reinforcing stereotypes. However, detecting and mitigating these harms is difficult to do in general and becomes computationally expensive when tackling multiple languages or when considering different biases. To address this, we present FAIRDISTILLATION: a cross-lingual method based on knowledge distillation to construct smaller language models while controlling for specific biases. We found that our distillation method does not negatively affect the downstream performance on most tasks and successfully mitigates stereotyping and representational harms. We demonstrate that FairDistillation can create fairer language models at a considerably lower cost than alternative approaches.

Keywords: Knowledge distillation · Fairness · BERT · Language models

1 Introduction

Pre-trained transformer-based Language Models (LMs), like BERT [14], are not only pushing the state-of-the-art across many languages, they are also being deployed in various services, ranging from machine translation to internet search [14,22,34]. However, these deployed language models have been shown to exhibit problematic behaviour. For instance, BERT and other models (i) replicate gender stereotypes [1,12,32], (ii) exhibit dubious racial correlations [32] and (iii) reproduce racial stereotypes [26]. These behaviours are all present in pre-trained models that are used in a wide range of applications, which are referred to as *downstream tasks*.

Without precautions, downstream tasks could use such problematic behaviour to make biased predictions. LMs are generally finetuned for such tasks, where allocation harms (i.e. allocating or withholding a resource) might occur [3]. These can originate from the fine-tuning dataset or the pre-trained model or a

© The Author(s), under exclusive license to Springer Nature Switzerland AG 2023
M.-R. Amini et al. (Eds.): ECML PKDD 2022, LNAI 13714, pp. 638–654, 2023.
https://doi.org/10.1007/978-3-031-26390-3_37

combination of both. We focus on the pre-training, where representation harms (i.e. encoding stereotypes) can occur in the pre-trained LMs [3,36].

Multiple methods have been proposed to reduce representational harms in language models [1,36], These methods are based on *pre-processing* of the data, for example *Counterfactual Data Augmentation* (CDA) [23] or *Counterfactual Data Substitution* (CDS) [16]. In both cases, gendered words in input sequences are replaced by a predefined counterfactual, e.g. *"He is a doctor"* → *"She is a doctor"*. CDA can significantly increase the training dataset, with longer training times as a consequence, so CDS-based methods replace input sequences instead. Nevertheless, both techniques require retraining the model with an augmented dataset, instead of leveraging the efforts done to train the original model.

We propose a framework for mitigating representational harms based on knowledge distillation [17], which we demonstrate on gender stereotypes. Our approach uses existing language models as a teacher, which provides a richer training signal and does not require retraining from scratch. To prevent the transfer of learnt correlations to new LMs, our framework replaces CDA's augmentation strategy with probabilistic rules between tokens. Since our approach can be performed at a fraction of the original training cost and also creates smaller models, it becomes more feasible to create domain-specific bias-controlled LMs.

In this paper, we start in Sect. 2 with an overview of language models and fairness interventions (Sect. 2.1). In Sect. 3, we present our method to create debiased language models, which we call FairDistillation. Section 4 describes the evaluation set-up and Sect. 5 presents the results. Section 6 gives an overview of future work and ethical considerations and we conclude in Sect. 7.

2   Background

BERT [14] is a language model that is trained in two phases: (i) self-supervised *pre-training* with a Masked Language Modeling (MLM) objective and afterwards (ii) supervised *finetuning* for downstream tasks. The intuition behind the first learning task is that learning to reconstruct missing words in a sentence helps with capturing interesting semantics—and because this relies on co-occurrences it also captures stereotypes. Formally, a token x_m in the input sequence $x_1, \ldots, x_N$ is replaced by a masked token (`<mask>`) and the MLM objective is to predict the original token x_m based on the context $\mathbf{x} = x_1, \ldots, x_{m-1}, x_{m+1}, \ldots, x_N$, following

$$\max_{\theta} \sum_{i=1}^{N} \mathbf{1}_{x_i = x_m} \log\left(P\left(x_i \mid \mathbf{x}; \theta\right)\right)$$

with $\mathbf{1}_{x_i = x_m}$ as an indicator function whether the token is correctly predicted. This training setup results in a good estimator of the contextualized probability of a word $P(x_i \mid \mathbf{x}; \theta)$. Aside from the MLM objective, the original BERT model also incorporated a Next Sentence Prediction (NSP) objective. Liu et al. [22] later concluded that the NSP objective did not improve training and removed it when constructing RoBERTa. Because of this, we do not further consider this objective during distillation or evaluation.

After pre-training, the newly obtained model can be reused and finetuned for different classification and regression tasks, like sentiment analysis. Finetuning requires different datasets, that can also introduce biases that are referred to as *extrinsic* biases [11]. Mitigating extrinsic biases in downstream tasks is out of scope for this work. Nevertheless, since LMs are used both for downstream tasks and for generating contextualized embeddings, mitigating intrinsic biases is still crucial.

2.1   Mitigating Intrinsic Biases

Bolukbasi et al. [5] presented two intrinsic debiasing methods based on removing the observed *gender axis* in static word embeddings. Mitigating problematic correlations is more challenging for LMs because of the contextualization that models like BERT incorporate. This means that word representations from LMs cannot be considered in isolation, so mitigation strategies for word embeddings cannot be applied.

Models like BERT can only generate meaningful representations for a given sequence, so for this reason, mitigation strategies have mostly been based on Counterfactual Data Augmentation (CDA) [1,16,23,36]. This strategy augments the pre-training dataset with sequences where certain words, like pronouns or names, are swapped.

Unfortunately, this requires re-training the model from scratch, which can be extremely costly and with many negative side effects [2].

One of few mitigation strategies that does not alter the training data was also presented by Webster et al. [36], namely using dropout as a regularisation method against problematic correlations. Regularisation as a means to mitigate problematic correlations thus seems a feasible option, but albeit effective, the method still requires retraining the model from scratch. It should also be noted that these efforts are mostly focused on English. Results of performing CDA on a German model were less successful, likely due to gender marking [1].

All previous methods require retaining a language model. Lauscher et al. [21] presents a unique approach, ADELE, that addresses this issue by using adapters [18,29,30]. These adapters are inserted after each attention layer and are the only trainable parameters, so the majority of parameters of a language model are shared over different tasks. ADELE trains these adapters on a subset (1/3th) of the original BERT corpus[1] with the MLM objective and CDA to mitigate biases. Although ADELE works very different from our distillation method, both methods aim to reduce the computational requirements and associated costs, by reusing existing models.

2.2   Knowledge Distillation

Knowledge distillation is a method to transfer learnt knowledge from one model—originally proposed as an ensemble of models—to another, usually smaller

---

[1] The original BERT corpus is a concatenation of Wikipedia and the Toronto Book-corpus [14].

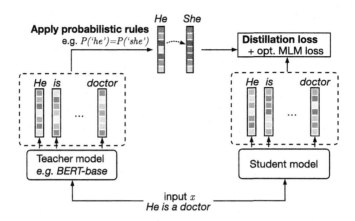

Fig. 1. Overview of the training procedure with FairDistillation for a single input sequence in English.

model [7,17]. Bucilă et al. [7] introduced this technique as model compression with an ensemble of models that are used to label a dataset. This was later adapted for neural networks [17]. The teacher outputs a label probability distribution z_i where some labels have higher probabilities, for example names or pronouns are more likely than verbs in the sentence "`<mask>` is a doctor.". To incorporate this information, a variation of the softmax function (Eq. 1) can be used with a temperature T to tune the importance of these labels.

$$p_i = \frac{\exp\left(\frac{z_i}{T}\right)}{\sum_j \exp\left(\frac{z_j}{T}\right)}. \tag{1}$$

Sanh et al. [31] focus on the distillation of the MLM task from pre-trained LMs. Their models, DistilBERT and DistilRoBERTa, are trained on a linear combination of a distillation loss $\mathcal{L}_{ce}$ with the softening function from Eq. 1, the original MLM loss $\mathcal{L}_{mlm}$, and additionally a cosine loss $\mathcal{L}_{cos}$ for the last hidden states, following

$$\mathcal{L} = \alpha_{ce}\mathcal{L}_{ce} + \alpha_{mlm}\mathcal{L}_{mlm} + \alpha_{cos}\mathcal{L}_{cos}.$$

TinyBERT [19] takes the same approach but also proposes a set of loss functions that perform distillation on (i) the embeddings layer, (ii) each of the transformer layers, and (iii) the prediction layer for specific tasks. These different loss functions make TinyBERT perform slightly better than DistilBERT, but these functions require additional transformations to be learnt. In addition, if the student and teacher have a different number of layers, a mapping function is also required to transfer the knowledge between both.

3   FairDistillation

In this section, we introduce FairDistillation, a method to mitigate problematic correlations in pre-trained language models. We first present the distillation architecture (Sect. 3.1) and afterwards, we will discuss the probabilistic rules that our method relies on (Sect. 3.2).

3.1   Architecture

Our method trains a newly initialized model (the student) from an already trained model (the teacher). Often, the teacher model has already been evaluated for biases, for example stereotypical gender norms for professions, which can lead to representational harms [3]. In this example, models like BERT-base [14] predict that the input sentence "<mask> *is a doctor.*" should be filled with '*He*' instead of '*She*'. The LM encoded that the token '*He*' is more frequent in the training dataset, both in isolation and in combination with words like '*doctor*'.

To prevent representational harms from being encoded in the final model, we apply a set of user-specified rules to the predictions of the original model. By doing so, we can train a new model with these predictions. Predictions of a teacher model provide a richer training signal and thus require less training time compared to CDA and CDS [17,19,31]. Moreover, we can simultaneously reduce the student's model size to improve both training and inference times, which boils down to knowledge distillation as is done for DistilBERT [31].

Figure 1 illustrates our method, which consists of 5 steps. First, an input sequence x is passed to both the teacher and the student model, both with an MLM prediction head. Second, the MLM predictions of the teacher model are passed to the rule engine. Third, the predictions for certain tokens, like '*He*', are modified based on the provided rules. Figure 1 demonstrates how a rule where we assume equal probabilities $P('He' \mid x) = P('She' \mid x)$ alters the MLM prediction, which we discuss more in-depth later in this section. Fourth, both MLM outputs, after applying possible rules to the teacher outputs, are used to calculate the distillation loss $\mathcal{L}_{ce}$ between the teacher and student outputs. Finally, the MLM outputs of the student model can also be used to calculate an additional loss term $\mathcal{L}_{mlm}$ to train the student model in the same manner as the original model.

Student Architecture. The student models use the same base architecture as the teacher models, but with 6 attention layers instead of the typical 12 layers, following Sanh et al. [31]. The weights are initialized at random, which we prefer over smarter initialization strategies [31] to prevent an accidental transfer of problematic correlations. We also reuse the teacher's tokenizer for the student, since these are already specifically constructed for the targeted language and no complex token translation is needed.

Applying Probabilistic Rules. The MLM head outputs a vector for each position in the input sequence, so for BERT-base this means at most 512 vectors. Each value in this vector represents the probability that a token fits in this position. Consequently, there will be 30,522 values for BERT-base-uncased. We assume that some probabilities should be equal, like $P('He' \mid "<mask>$ is a doctor") $= P('She' \mid "<mask>$ is a doctor"), so our method can enforce these kind of equality rules.

During distillation, our method applies these equality rules to all the MLM outputs of the teacher. For efficiency reasons, the tokens of interest are translated into a small lookup table at the start of the distillation loop so that applying each rule only requires a few lookup operations. The corresponding values of the tokens are set to the mean of both values. Consequently, the outcome is also normalized and each prediction still sums up to 1.

Currently, our method only supports equalization between two or more tokens. We did experiment with implementing these and more complex rules in ProbLog, a probabilistic logic programming language [10], but this proved to be unfeasible because of inference times that frequently exceeded 0.5 s per training example. Nevertheless, future work could focus on adding more complex rules that also depend for example on context or on part-of-speech tags to distinguish between adjectives ('His car' → 'Her car') or pronouns ('. . . is his' → '. . . is hers').

Knowledge Distillation. We follow the DistilBERT [31] distillation method, as discussed in Sect. 2. FairDistillation applies a set of rules to affect the distillation loss, but the student not only learns from the distillation task, but also from the MLM task. It is possible to concurrently train on this MLM objective for little additional cost. Although this can be another source of problematic correlations, we opted to use this loss without correcting any associations. We reason that the contextual probability for a single input sequence can also be a useful signal.

3.2   Obtaining Probabilistic Rules

Until now, we used a running example of a probabilistic rule where the contextualized probability, as generated by the teacher LM, has to be equal for two tokens, namely '*He*' and '*She*'. CDA achieves something similar by augmenting the dataset based on word mappings [15,23]. These mappings are very similar to our probabilistic rules; in fact, AugLy, a popular data augmentation framework [28], has the same mapping[2] that we use for our running example in the context of gender bias.

Depending on which biases one wants to mitigate, different sets of rules are required. We focus in this work on gender bias, so we rely on the same kind of rules as CDA. Simple rules to balance predictions highlight the robustness of our method and do not require lists of professions, which come with their own issues and biases [4]. However, creating more fine-grained, domain-specific rules might improve our results. Such rules could aim at balancing, for example, profession titles or proper names.

4   Experimental Setup

We evaluate our method in two Indo-European languages: (i) English and (ii) Dutch, of which the results are discussed further in Sect. 5. Both languages have their own set of models, pre-training corpora and evaluation datasets, which we briefly cover in this section. The evaluation of gender biases is also highly language-dependent and to illustrate generalization of our method beyond

---

[2] https://raw.githubusercontent.com/facebookresearch/AugLy/main/augly/assets/ text/gendered_words_mapping.json.

English, we also used a monolingual model for Dutch [12] with an architecture similar to RoBERTa [22]. We opted for this language since it has some interesting, challenging characteristics, namely it is one of only two languages with cross-serial dependencies that make it non-context free, with the other one being Swiss-German [6]. It also has gendered suffixes for some, yet not all, nouns. This affects such evaluations since these rely on implicit associations between nouns (e.g. for professions). However, grammatical gender can also be an opportunity to evaluate how e.g. gendered professions align with the workforce [1] or with equal opportunity policies. We compare our method based on three popular metrics that we discuss in this section, Delobelle et al. [11] provides a more comprehensive overview of intrinsic fairness measures.

SEAT. The Word Embedding Association Test (WEAT) [8] measures associations between target words ('*He*', '*She*', ...) and attribute words ('doctor', 'nurse', ...). Between the embeddings of each target and attribute word, a similarity measure like cosine similarity can be used to quantify the association between word pairs. To add context, SEAT uses some 'semantically bleached' template sentences [26].

LPBS. Kurita et al. [20] observe that using SEAT for the learned BERT embeddings fails to find many statistically significant biases, which is addressed in the presented *log probability bias score* (LPBS). This score computes a probability p_{tgt} for a target token t (e.g. '*He*' or '*She*') from the distribution of the masked position X_m following

$$p_{tgt} = P\left(X_m = t \mid \mathbf{x}; \theta\right),$$

for a template sentence, e.g. "`<mask>` is a doctor". Since the prior likelihood $P(X_m = t)$ can skew the results, the authors correct for this by calculating a template prior p_{prior} by additionally masking the token(s) with a profession or another attribute x_p, following

$$p_{prior} = P\left(X_m = t \mid \mathbf{x} \backslash \{x_p\}; \theta\right).$$

Both probabilities are combined in a measure of association $\log \frac{p_{tgt}}{p_{prior}}$ and the bias score is the difference between the association measures for two targets, like '*He*' and '*She*'. Kurita et al. [20] applied their method to the original English BERT model [14] and found statistically significant differences for all categories of the WEAT templates.

DisCo. Webster et al. [36] also utilize templates to evaluate possible biases which their approach also mitigates (see Sect. 2.1). As an intrinsic measure, the authors present *discovery of correlations* (DisCo). Compared to previously discussed metrics, this metric measures the difference in predictions for the attribute token x_p when varying gendered tokens (i.e. '`<P>` *is a* `<mask>`' for different pronouns or names instead of '`<mask>` *is a* `<P>`' with different professions).

We experimented with the original DisCo metric, which performs statistical tests between predicted tokens, but we found that it didn't produce any statistically significant tokens. So, we simplified the metric to measure the differences in probabilities for the predicted tokens. The resulting score of our DisCo implementation can therefore also be negative, while the original version has a lower bound of 0 as it counts the number of statistically significant fills.

In the remainder of this section, we discuss our evaluations of English and Dutch in their respective subsections, where we define the used datasets, models and language-specific evaluation aspects.

4.1   English Setup

The first model we use as a teacher is the original uncased BERT model (`BERT-base-uncased`) as released by Devlin et al. [14], which is also the most-studied LM with regard to gender stereotypes. This model was trained on the Toronto Bookcorpus and Wikipedia, but the Toronto Bookcorpus is no longer publicly available anymore and thus hinders reproduction. For this reason, Jiao et al. [19] use only Wikipedia. We used a portion of the English section of the OSCAR corpus [27] to keep the training dataset size similar. More specifically, we used the first two shards of the unshuffled version. We recognise that there is a mismatch between the domains of the Bookcorpus and OSCAR, but we believe this is acceptable to increase reproducibility.

As introduced in Sect. 3.2, we use a set of gendered pronouns and define which ones should have the same probability. Since we use the uncased variant, we only need to define one set of rules, since '*She*' and '*she*' result in the same token.

The tokenization method used by BERT, WordPiece, splits words and adds a merge symbol (e.g. 'word' + '##piece'), so no special care is required. For RoBERTa [22], which uses Byte Pair Encoding (BPE), a word boundary symbol is used. Consequently, a word can have different tokens and representations depending if a space, punctuation mark, mask token or sequence start token are in front of the target token. Since the Dutch model uses BPE, we will revisit this issue in Sect. 4.2.

Evaluation. To evaluate our method's performance trade-off, we finetune the obtained model on the Internet Movie Database (IMDB) sentiment analysis task [24], which was also done by BERT [14] and DistilBERT [31]. The dataset contains 25k training examples, from which we used 5k as a separate validation set, and another 25k test sequences. This is a high-level task where no gendered correlations should be used for predictions. Predicting entailment is a high-level task covered multiple times in GLUE [35], on which we also evaluated our method with the pre-trained model. For a description of this benchmark and all datasets, we refer to Wang et al. [35].

Bias Evaluation. To evaluate possible problematic correlations with regard to gender stereotypes, we compute DisCo and LPBS, which we introduce at the beginning of this section. We use the Employee Salary dataset[3] [20]. Following

---

[3] https://github.com/keitakurita/contextual_embedding_bias_measure/blob/master/notebooks/data/employeesalaries2017.csv.

Kurita et al. [20], we filter on the top 1000 highest-earning instances as a proxy for prestigious jobs and test this for the same two templates ('<mask> is a <P>' and '<mask> can do <P>'). However, we additionally filter digits from the job titles and remove duplicate titles, to not skew the results towards more popular professions.

4.2 Dutch Setup

We use a Dutch RoBERTa-based model called RobBERT [12] as a teacher, more specifically `robbert-v2-dutch-base`. This model was pre-trained on the Dutch section of the shuffled version of OSCAR[4]. Similar to the distilled version of RobBERT [13], we select a 1GB portion of the OSCAR corpus (using `head`, 2.5%) to illustrate the ability to perform successful knowledge distillation with only a small fraction of the data required in comparison to the pre-trained model.

To create our model, we used a defined a set of rules based on the gendered pronouns 'Hij' and 'Zij' ('*He*' and '*She*'). The tokens corresponding to these pronouns were grouped based on capitalization and included spaces, since the BPE tokenizer includes a word boundary character at the beginning of some tokens. Our method then used these rules to equalise the distributions predicted by the teacher during distillation, which we performed for 3 epochs. This took approximately 40 h per epoch on a Nvidia 1080 Ti and a traditionally distilled model required the same time, indicating our method has very limited effect on training time.

Evaluation. We compare the model created with FairDistillation to RobBERT and RobBERTje [12,13] on the same set of benchmark tasks: (i) sentiment analysis on book reviews (DBRD) [33], (ii) NER, (iii) POS tagging, and (iv) natural language inference with SICK-NL [37]. These tasks are fairly high-level sequence-labelling tasks that can exhibit allocational harms, such as the predictive difference for sentiment analysis that was illustrated by Delobelle et al. [12].

Bias Evaluation. We also evaluate numerically using the LPBS and DisCo metrics, but the RobBERT LM has also been evaluated by the authors on gender stereotyping using a different technique. This evaluation technique is based on a set of templates and a translated set of professions[5] from Bolukbasi et al. [5]. These professions have a perceived gender (e.g. 'actress' is a female profession and 'surveyor' is neutral), which can be correlated with the predictions by the model. The authors rank the tokens based on the predicted probability instead of using this probability directly. Interestingly, a correlation was not considered problematic, but male pronouns were predicted higher on average, even for by definition female professions (e.g. 'nun'). To compare these results, we recreate the same plot and report the Mean Ranking Difference (MRD). We focus on the gendered pronouns '*zij*' ('*she*') and '*hij*' ('*he*') for our evaluation.

---

[4] https://oscar-corpus.com.
[5] https://people.cs.kuleuven.be/~pieter.delobelle/data.html.

Table 1. English results on IMDB (sentiment analysis), GLUE [35], and two bias measures. Following Devlin et al. [14], we report F_1 scores for QQP and MRPC, Spearman correlations for STS-B, and accuracy for all other tasks. Results reported by Devlin et al. [14] on the GLUE dashboard are indicated with an obelisk (†), while the results from [31] are also on the GLUE dev set, indicated with an asterisk (*). For LPBS, positive values represent more stereotypical associations, and for DisCo, lower values are more favorable.

Model	IMDB	GLUE								Bias	
		MNLI	QQP	QNLI	SST-2	CoLA	STS-B	MRPC	RTE	LPBS	DisCo
BERT [14]	93.5	84.0†	71.2†	90.5†	93.5†	52.1†	85.8†	88.9†	66.4†	1.16	−0.48
DistilBERT [31]	92.82	82.2*	88.5*	89.2*	91.3*	51.3*	86.9*	87.5*	59.9*	−0.27	−0.55
FairDistillation	85.5 ± 0.4	80.1	82.1	86.6	90.6	38.5	84.0	85.1	59.6	**−0.16**	**0.25**

5 Results

In this section, we present the results of the experiments (Sect. 4). We discuss English (Sect. 5.1) and Dutch (Sect. 5.2) results separately. We also performed experiments on French using the CamemBERT model [25], but we chose to ommit those results due to our limited understanding of the language, which we address further in Sect. 6.

To eliminate any possible effect from hyperparameter assignments on the results, we ran each finetuning training 10 times with random hyperparameter assignments. We varied the (i) learning rate, (ii) weight decay, and (iii) the number of gradient accumulation steps to effectively scale the batch size while still fully utilizing the GPU. The full set of hyperparameters is listed in Table 3 in the supplementary materials. For the Dutch benchmarks and for the English IMDB, we select the best-performing model based on the *validation* set and present the results on the held-out *test* set. The results from the GLUE benchmark are from the dev set, which were also the results reported by [31].

Unless indicated otherwise, all training runs are done on a single Nvidia 1080 Ti with 11 GB VRAM. All models are also comparably sized, with 66M trainable parameters each. This is 50% of the model size of the teachers.

5.1 English Results

We observe that problematic correlations are reduced on all three metrics, as is shown in Table 1. One interesting observation—which also holds for Dutch—is that distillation in itself is already successful in mitigating these correlations. This might be related to regularization as a method to control correlations [36], but we leave this for a future study.

On the IMDB task, our model suffers a 10% accuracy drop, which is significant. However, as noted in Sect. 4, we used a smaller training set for finetuning than BERT and DistilBERT, because we created a separate validation set from the original training set.

For GLUE, the results are in line with distilBERT. We do observe some diminished scores, notably CoLA, but the overall trade-off is limited.

Table 2. Dutch results on several benchmarks, namely Dutch Book Reviews (DBRD, sentiment analysis), named entity recognition (NER), part-of-speech (POS), tagging, and language inference (SICK-NL). We report bias as measured with LPBS and DisCo and additionally the mean ranking difference (MRD), which measures the preference of a language model to fill in male tokens (negative score) or female tokens (positive score). Benchmarks are reported with accuracy with 95% CI, except for the NER task, where we report the F_1 score. Results indicated with † were reported by Delobelle et al. [12]. For MRD, smaller ranking differences are more favorable, for LPBS, positive values represent more stereotypical associations, and for DisCo, lower values are more favorable.

Model	Params	Benchmark scores				Bias		
		DBRD	NER	POS	SICK-NL	MRD	LPBS	DisCo
RobBERT [12]	116 M	$94.4 \pm 1.0^\dagger$	$89.1^\dagger$	$96.4 \pm 0.4^\dagger$	84.2 ± 1.0	−7.47	1.13	−0.29
RobBERTje [13]	74 M	92.5 ± 1.1	82.7	$\mathbf{95.6 \pm 0.4}$	$\mathbf{83.4 \pm 1.0}$	−6.66	**−0.45**	−0.41
FairDistillation	74 M	92.1 ± 1.1	82.7	95.4 ± 0.4	82.4 ± 1.1	**−3.98**	1.14	**−0.08**

Unlike the other models, we performed our FairDistillation method on 4 Nvidia V100's for 3 epochs, which took 70h per epoch. Finetuning was done on an Nvidia 1080 Ti for 4 epochs for IMDB, which took approximately 1h per run and was replicated 10 times. For GLUE, we report the dev results and did not do any hyperparameter search. We used the same hyperparameters as distilBERT [31], who also report the development set results.

5.2 Dutch Results

Both the distilled RobBERT model and the model obtained with FAIRDISTIL-LATION perform only slightly worse (within 97.5% of the original model) for both downstream tasks (see Table 2). Both models have only half the parameters compared to the original RobBERT model and are thus faster to train and deploy, making this a decent trade-off between model size and predictive performance. With no significant differences in performance between the distilled model and our FAIRDISTILLATION model, this highlights the potential of our method.

With regards to the bias evaluation, we observe a reduction between the original model and ours (Table 2): correlations are significantly reduced as measured by DisCo and the mean ranking of female associated tokens improved by 3.5 tokens. The only exception is LPBS [20], which incorporates a correction based on the prior probability of a token. Our method effectively corrects this prior, while still allowing the context to affect individual results with the MLM objective. Since Dutch has gendered nouns for some professions, a correlation is not necessarily undesirable, but the prior is (e.g. assuming all physicians are men). Further graphical analysis of the predicted rankings for the third person singular pronouns confirms this, as shown in Fig. 2. These charts reveal that most professions are now less associated with the masculine pronoun. When considering

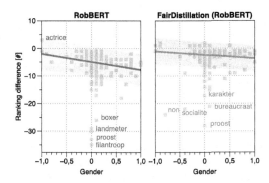

Fig. 2. Differences in predictions for the Dutch template '`<mask>` *is een* `<P>`' for our and the original RobBERT model. The 'gender' axis ranges from words associated with female professions (left) to words associated with male professions (right). A positive ranking difference indicates '*She*' is predicted before '*He*'.

which pronoun is ranked higher (i.e. above or below $y = 0$), this result is even more pronounced. RobBERT only predicted a feminine pronoun for a single profession [12], while with our method this increased to 15 professions.

6   Limitations and Ethical Considerations

Despite the promising results, there are several potential improvements possible to our methods, as well as some ethical considerations. First, we rely on facts that express probabilities for a single token at a time. For gender stereotyping, this is sufficient as the vocabulary usually contains the tokens of interest. However, this is not the case for many other problematic correlations, especially those affecting minority groups. Tokens that are interesting here, like names, are not in the tokenizer's vocabulary because this is created based on occurrence counts. Addressing this limitation would require extending our method to support facts that span multiple tokens.

Second, the effects of our method on 'low-level' grammatical tasks require further study in English, as we focused on GLUE and sentiment analysis. The used Dutch benchmarks do cover more tasks and seem to indicate favourable reductions.

Third, our work focuses on binary gender stereotypes. This leaves out a wide range of people who do not identify as such and although our method supports equalization over more than two tokens, this might be challenging if the intended words span multiple tokens. This also poses a challenge for generalizing our method beyond gender bias, since this frequently involves words that are not a single token in BERT's vocabulary.

Fourth, since none of the authors is a native speaker of a language like French, we only performed a limited, exploratory evaluation of our methods with Camem-BERT [25]. Our method appears successful, with improved scores of 0.04 (DisCo) and -0.85 (LPBS) compared to -1.15 (DisCo) and 1.99 (LPBS) for the original CamemBERT model [25]. The performance of the LM was also still high, with XNLI [9] scores 75.6 compared to 82.5 for CamemBERT. However, constructing correct probabilistic rules and evaluating them is tricky for non-native speakers. For example, the female variant of a profession can refer to a woman practising said profession, but also to the spouse of a man with this profession. When discussing these results with native French speakers from Wallonia, Belgium and from northern France, we realised that we are not well-suited to address this. We thus leave a more comprehensive evaluation across languages as future work.

Finally, by presenting a method to remove correlations with gender stereotypes in pre-trained language models, we risk it being used as a 'rubber stamp' to absolve model creators from their responsibilities. Therefore, we urge creators to critically analyse LMs within the social context that these models will be deployed in, both with respect to both the pre-trained and the finetuned model.

7   Conclusion

We introduced a method called FairDistillation that allows to use probabilistic rules during knowledge distillation. We showed that this can effectively mitigate gender stereotypes in language models. Our method demonstrates that knowledge distillation of language models with probabilistic rules is a possible alternative to re-training in order to reduce representational harms. Even though comes at a slight cost for some downstream tasks, but we find that the overall cost is limited and can mostly be attributed to the distillation process.

Acknowledgements. Pieter Delobelle was supported by the Research Foundation - Flanders (FWO) under EOS No. 30992574 (VeriLearn) and received funding from the Flemish Government under the "Onderzoeksprogramma Artificiële Intelligentie (AI) Vlaanderen" programme. Bettina Berendt received funding from the German Federal Ministry of Education and Research (BMBF) - Nr. 16DII113. Some resources and services used in this work were provided by the VSC (Flemish Supercomputer Center), funded by the Research Foundation - Flanders (FWO) and the Flemish Government.

A   Hyperparameters

Table 3. The hyperparameter space used for finetuning.

Hyperparameter	Value
adam_epsilon	10^{-8}
fp16	False
gradient_accumulation_steps	$i \in \{1, 2, 3, 4\}$
learning_rate	$[10^{-6}, 10^{-4}]$
max_grad_norm	1.0
max_steps	-1
num_train_epochs	3
per_device_eval_batch_size	4 (16 for XNLI)
per_device_train_batch_size	4 (16 for XNLI)
max_sequence_length	512 (128 for XNLI)
seed	1
warmup_steps	0
weight_decay	$[0, 0.1]$

References

1. Bartl, M., Nissim, M., Gatt, A.: Unmasking contextual stereotypes: measuring and mitigating BERT's gender bias. arXiv:2010.14534 [cs] (2020)
2. Bender, E.M., Gebru, T., McMillan-Major, A., Shmitchell, S.: On the dangers of stochastic parrots: can language models be too big? In: Proceedings of the 2021 ACM Conference on Fairness, Accountability, and Transparency, pp. 610–623 (2021)
3. Blodgett, S.L., Barocas, S., Daumé III, H., Wallach, H.: Language (technology) is power: a critical survey of "bias" in NLP. In: Proceedings of the 58th Annual Meeting of the Association for Computational Linguistics, pp. 5454–5476, Association for Computational Linguistics (2020). https://doi.org/10.18653/v1/2020.acl-main.485, https://www.aclweb.org/anthology/2020.acl-main.485
4. Blodgett, S.L., Lopez, G., Olteanu, A., Sim, R., Wallach, H.: Stereotyping Norwegian salmon: an inventory of pitfalls in fairness benchmark datasets. In: Proceedings of the 59th Annual Meeting of the Association for Computational Linguistics and the 11th International Joint Conference on Natural Language Processing (Volume 1: Long Papers), pp. 1004–1015 (2021)
5. Bolukbasi, T., Chang, K.W., Zou, J., Saligrama, V., Kalai, A.: Man is to computer programmer as woman is to homemaker? Debiasing word embeddings. arXiv:1607.06520 [cs, stat] (2016)
6. Bresnan, J., Kaplan, R.M., Peters, S., Zaenen, A.: Cross-serial dependencies in Dutch. In: Savitch, W.J., Bach, E., Marsh, W., Safran-Naveh, G. (eds.) The Formal Complexity of Natural Language. Studies in Linguistics and Philosophy, vol. 33, pp. 286–319. Springer, Dordrecht (1982). https://doi.org/10.1007/978-94-009-3401-6_11

7. Buciluǎ, C., Caruana, R., Niculescu-Mizil, A.: Model compression. In: Proceedings of the 12th ACM SIGKDD International Conference on Knowledge Discovery and Data Mining, pp. 535–541 (2006)

8. Caliskan, A., Bryson, J.J., Narayanan, A.: Semantics derived automatically from language corpora contain human-like biases. Science **356**(6334), 183–186 (2017). https://doi.org/10.1126/science.aal4230, https://science.sciencemag.org/content/356/6334/183. ISSN 0036-8075

9. Conneau, A., et al.: XNLI: evaluating cross-lingual sentence representations. In: Proceedings of the 2018 Conference on Empirical Methods in Natural Language Processing. Association for Computational Linguistics (2018)

10. De Raedt, L., Kimmig, A., Toivonen, H.: ProbLog: a probabilistic prolog and its application in link discovery. In: IJCAI, Hyderabad, vol. 7, pp. 2462–2467 (2007)

11. Delobelle, P., Tokpo, E., Calders, T., Berendt, B.: Measuring fairness with biased rulers: a comparative study on bias metrics for pre-trained language models. In: Proceedings of the 2022 Conference of the North American Chapter of the Association for Computational Linguistics: Human Language Technologies, pp. 1693–1706. Association for Computational Linguistics, Seattle (2022)

12. Delobelle, P., Winters, T., Berendt, B.: RobBERT: a dutch RoBERTa-based language model. In: Findings of ACL: EMNLP 2020 (2020)

13. Delobelle, P., Winters, T., Berendt, B.: RobBERTje: a distilled Dutch BERT model. Comput. Linguist. Netherlands J. **11**, 125–140 (2022). https://www.clinjournal.org/clinj/article/view/131

14. Devlin, J., Chang, M.W., Lee, K., Toutanova, K.: BERT: pre-training of deep bidirectional transformers for language understanding. In: Proceedings of the 2019 Conference of the North American Chapter of the Association for Computational Linguistics: Human Language Technologies, Volume 1 (Long and Short Papers), pp. 4171–4186. Association for Computational Linguistics, Minneapolis (2019). https://doi.org/10.18653/v1/N19-1423

15. Feng, S.Y., et al: A survey of data augmentation approaches for NLP. arXiv preprint arXiv:2105.03075 (2021)

16. Hall Maudslay, R., Gonen, H., Cotterell, R., Teufel, S.: It's all in the name: mitigating gender bias with name-based counterfactual data substitution. In: Proceedings of the 2019 Conference on Empirical Methods in Natural Language Processing and the 9th International Joint Conference on Natural Language Processing (EMNLP-IJCNLP), pp. 5267–5275. Association for Computational Linguistics, Hong Kong (2019). https://doi.org/10.18653/v1/D19-1530, https://www.aclweb.org/anthology/D19-1530

17. Hinton, G., Vinyals, O., Dean, J.: Distilling the knowledge in a neural network. arXiv:1503.02531 [cs, stat] (2015)

18. Houlsby, N., et al.: Parameter-efficient transfer learning for NLP. In: Chaudhuri, K., Salakhutdinov, R. (eds.) Proceedings of the 36th International Conference on Machine Learning, Proceedings of Machine Learning Research, vol. 97, pp. 2790–2799. PMLR (2019). https://proceedings.mlr.press/v97/houlsby19a.html

19. Jiao, X., et al.: TinyBERT: distilling BERT for natural language understanding. In: Findings of ACL: EMNLP 2020 (2020)

20. Kurita, K., Vyas, N., Pareek, A., Black, A.W., Tsvetkov, Y.: Measuring bias in contextualized word representations. In: Proceedings of the First Workshop on Gender Bias in Natural Language Processing, pp. 166–172. Association for Computational Linguistics, Florence (2019). https://doi.org/10.18653/v1/W19-3823, https://www.aclweb.org/anthology/W19-3823

21. Lauscher, A., Lueken, T., Glavaš, G.: Sustainable modular debiasing of language models. In: Findings of the Association for Computational Linguistics: EMNLP 2021, pp. 4782–4797. Association for Computational Linguistics, Punta Cana (2021). https://doi.org/10.18653/v1/2021.findings-emnlp.411, https://aclanthology.org/2021.findings-emnlp.411

22. Liu, Y., et al.: RoBERTa: a robustly optimized BERT pretraining approach. arXiv:1907.11692 [cs] (2019)

23. Lu, K., Mardziel, P., Wu, F., Amancharla, P., Datta, A.: Gender bias in neural natural language processing. In: Nigam, V., et al. (eds.) Logic, Language, and Security. LNCS, vol. 12300, pp. 189–202. Springer, Cham (2020). https://doi.org/10.1007/978-3-030-62077-6_14. ISBN 978-3-030-62077-6

24. Maas, A.L., Daly, R.E., Pham, P.T., Huang, D., Ng, A.Y., Potts, C.: Learning word vectors for sentiment analysis. In: Proceedings of the 49th Annual Meeting of the Association for Computational Linguistics: Human Language Technologies, pp. 142–150. Association for Computational Linguistics, Portland (2011). http://www.aclweb.org/anthology/P11-1015

25. Martin, L., et al.: CamemBERT: a tasty french language model. In: Proceedings of the 58th Annual Meeting of the Association for Computational Linguistics, pp. 7203–7219. Association for Computational Linguistics (2020). https://doi.org/10.18653/v1/2020.acl-main.645

26. May, C., Wang, A., Bordia, S., Bowman, S.R., Rudinger, R.: On measuring social biases in sentence encoders. In: Proceedings of the 2019 Conference of the North American Chapter of the Association for Computational Linguistics: Human Language Technologies, Volume 1 (Long and Short Papers), pp. 622–628. Association for Computational Linguistics, Minneapolis (2019). https://doi.org/10.18653/v1/N19-1063

27. Ortiz Suárez, P.J., Romary, L., Sagot, B.: A monolingual approach to contextualized word embeddings for mid-resource languages. In: Proceedings of the 58th Annual Meeting of the Association for Computational Linguistics, pp. 1703–1714. Association for Computational Linguistics (2020). https://www.aclweb.org/anthology/2020.acl-main.156

28. Papakipos, Z., Bitton, J.: AugLy: data augmentations for robustness. arXiv preprint arXiv:2201.06494 (2022)

29. Pfeiffer, J., Kamath, A., Rücklé, A., Cho, K., Gurevych, I.: AdapterFusion: nondestructive task composition for transfer learning. In: Proceedings of the 16th Conference of the European Chapter of the Association for Computational Linguistics: Main Volume, pp. 487–503. Association for Computational Linguistics (2021). https://doi.org/10.18653/v1/2021.eacl-main.39, https://aclanthology.org/2021.eacl-main.39

30. Pfeiffer, J., et al.: AdapterHub: a framework for adapting transformers. In: Proceedings of the 2020 Conference on Empirical Methods in Natural Language Processing: System Demonstrations, pp. 46–54. Association for Computational Linguistics (2020). https://doi.org/10.18653/v1/2020.emnlp-demos.7, https://aclanthology.org/2020.emnlp-demos.7

31. Sanh, V., Debut, L., Chaumond, J., Wolf, T.: DistilBERT, a distilled version of BERT: smaller, faster, cheaper and lighter. In: NeurIPS EMC^2 Workshop (2019)

32. Tan, Y.C., Celis, L.E.: Assessing social and intersectional biases in contextualized word representations. In: Wallach, H., Larochelle, H., Beygelzimer, A., dAlché-Buc, F., Fox, E., Garnett, R. (eds.) Advances in Neural Information Processing Systems, vol. 32, pp. 13230–13241. Curran Associates, Inc. (2019)

33. van der Burgh, B., Verberne, S.: The merits of universal language model fine-tuning for small datasets - a case with dutch book reviews. arXiv:1910.00896 [cs] (2019)

34. Vaswani, A., et al.: Attention is all you need. In: Guyon, I., et al. (eds.) Advances in Neural Information Processing Systems, vol. 30, pp. 5998–6008. Curran Associates, Inc. (2017)

35. Wang, A., Singh, A., Michael, J., Hill, F., Levy, O., Bowman, S.R.: GLUE: a multitask benchmark and analysis platform for natural language understanding. In: the Proceedings of ICLR (2019)

36. Webster, K., et al.: Measuring and reducing gendered correlations in pre-trained models. arXiv:2010.06032 [cs] (2020)

37. Wijnholds, G., Moortgat, M.: SICKNL: a dataset for dutch natural language inference. arXiv preprint arXiv:2101.05716 (2021)

Self-distilled Pruning of Deep Neural Networks

James O' Neill[(✉)], Sourav Dutta, and Haytham Assem

Huawei Ireland Research Center, Dublin, Ireland
james.o.neil@huawei-partners.com,
{sourav.dutta2,haytham.assem}@huawei.com

Abstract. Pruning aims to reduce the number of parameters while maintaining performance close to the original network. This work proposes a novel *self-distillation* based pruning strategy, whereby the representational similarity between the pruned and unpruned versions of the same network is maximized. Unlike previous approaches that treat distillation and pruning separately, we use distillation to inform the pruning criteria, without requiring a separate student network as in knowledge distillation. We show that the proposed *cross-correlation objective for self-distilled pruning* implicitly encourages sparse solutions, naturally complementing magnitude-based pruning criteria. Experiments on the GLUE and XGLUE benchmarks show that self-distilled pruning increases mono- and cross-lingual language model performance. Self-distilled pruned models also outperform smaller Transformers with an equal number of parameters and are competitive against (6 times) larger distilled networks. We also observe that self-distillation (1) maximizes class separability, (2) increases the signal-to-noise ratio, and (3) converges faster after pruning steps, providing further insights into why self-distilled pruning improves generalization.

Keywords: Iterative pruning · Self-distillation · Language models

1 Introduction

Neural network pruning [16,29,33] zeros out weights of a pretrained model with the aim of reducing parameter count and storage requirements, while maintaining performance close to the original model. The criteria to choose which weights to prune has been an active research area over the past three decades [3,10,16,19,28]. Lately, there has been a focus on pruning models in the transfer learning setting whereby a self-supervised pretrained model trained on a large amount of unlabelled data is fine-tuned to a downstream task while weights are simultaneously pruned, referred to as *fine-pruning*. In this context, recent work proposes to learn important scores over weights with a continuous mask and prune away those that having the smallest scores [25,36]. However, these learned continuous masks double the number of parameters and gradient updates in the network [36]. Ideally, we aim

M.-R. Amini et al. (Eds.): ECML PKDD 2022, LNAI 13714, pp. 655–670, 2023.
https://doi.org/10.1007/978-3-031-26390-3_38

to perform task-dependent fine-pruning *without* adding more parameters to the network, or at least far fewer than twice the count. Additionally, we desire pruning methods that can recover from performance degradation directly after pruning steps, faster than current pruning methods while encoding task-dependent information into the pruning process. To this end, we hypothesize self-distillation may recover performance faster after consecutive pruning steps, which becomes more important with larger performance degradation at a higher compression regime. Additionally, self-distillation has shown to encourage sparsity as the training error tends to 0 [27]. This implicit sparse regularization effect complements magnitude-based pruning.

Hence, this paper proposes to combine self-distillation and magnitude-based pruning to achieve task-dependent pruning efficiently. This is achieved by *maximizing the cross-correlation* between output representations of the fine-tuned pretrained network and a pruned version of the same network – referred to as *self-distilled pruning* (SDP). Cross-correlation maximization reduces redundancy and encourages sparse solutions [49], naturally fitting with magnitude-based pruning. Unlike typical knowledge distillation (KD) where the student is a separate network trained from random initialization, here the student is initially a masked version of the teacher. We find that SDP sets state-of-the-art results when compared to alternative magnitude-based pruning methods and equivalently sized distilled networks. We also provide three insights as to why self-distillation leads to more generalizable pruned networks. We observe that self-distilled pruning (1) *recovers performance faster* after pruning steps (i.e., improves convergence), (2) *maximizes the signal-to-noise ratio* (SNR), where pruned weights are considered as noise, and (3) *improves the fidelity* between pruned and unpruned representations as measured by mutual information of the respective penultimate layers. We focus on pruning fine-tuned monolingual *and* cross-lingual transformer models, namely BERT [6] and XLM-RoBERTa [5]. To our knowledge, this is the first study that introduces the concept of *self-distilled pruning*, analyzes iterative pruning in the mono-lingual *and* cross-lingual settings on the GLUE and XGLUE benchmarks respectively and the only work to include an evaluation of pruned model performance in the cross-lingual transfer setting.

2   Background and Related Work

Regularization-based pruning can be achieved by using a weight regularizer that encourages network sparsity. Three well-established regularizers are L_1, L_2 and L_0 weight regularization [23,24,47] for weight sparsity [10,11]. For structured pruning, Group-wise Brain Damage [18] and SSL [45] propose to use Group LASSO [48] to prune whole structures (e.g., convolution blocks or blocks within standard linear layers). Park et al. [31] avoid pruning small weights if they are connected to larger weights in consecutive layers and vice-versa, by penalizing the Frobenius norm between pruned and unpruned layers to be small.

Importance-based pruning assigns a score for each weight in the network and removes weights with the lowest importance score. The simplest scoring criteria is magnitude-based pruning (MBP), which uses the lowest absolute value

(LAV) as the criteria [10,11,33] or L_1/L_2-norm for structured pruning [23]. MBP can be seen as a zero-th order pruning criteria. However higher order pruning methods approximate the difference in pruned and unpruned model loss using a Taylor series expansion up until 1st order [12,19] or the 2nd order, which requires approximating the Hessian matrix [26,37,44] for scalability. Lastly, the regularization-based pruning is commonly used with importance-based pruning e.g. using L_2 weight regularization alongside MBP.

Knowledge Distillation (KD) transfers the knowledge of an already trained network, such as the logit outputs [13]), and uses them as soft targets to optimize a student network. The student network is typically smaller than the teacher network and benefits from the additional information soft targets provide. There has been various extensions that involve distilling intermediate representations [34], distributions [14], maximizing mutual information between student and teacher representations [1], using pairwise interactions for improved KD [32] and contrastive representation distillation [30,39].

Self-Distillation is a special case of KD whereby the student and teacher networks have the same capacity. Interestingly, self-distilled students often generalize better than the teacher [9,46], however the mechanisms by which self-distillation leads to improved generalization remain somewhat unclear. Recent works have provided insightful observations of this phenomena. For example, Stanton et al. [38] have shown that soft targets make optimization easier for the student when compared to the task-provided one-hot targets. Allen et al. [2] view self-distillation as implicitly combining ensemble learning and KD to explain the improvement in test accuracy when dealing with multi-view data. The core idea is that the self-distillation objective results in the network learning a unique set of features that are distinct from the original model, similar to features learned by combining the outputs of independent models in an ensemble. Given this background on pruning and distillation, we now describe our proposed methodology for *SDP*.

3   Proposed Methodology

We begin by defining a dataset $\mathcal{D} := \{(X_i, y_i)\}_{i=1}^D$ with single samples $s_i = (X_i, \boldsymbol{y}_i)$, where each X_i (in the D training samples) consists of a sequence of vectors $X_i := (\boldsymbol{x}_1, \ldots, \boldsymbol{x}_N)$ and $\boldsymbol{x}_i \in \mathbb{R}^d$. For structured prediction (e.g., NER, POS) $y_i \in \{0,1\}^{N \times C}$, and for single and pairwise sentence classification, $y_i \in \{0,1\}^C$, where C is the number of classes. Let $\boldsymbol{y}^S = f_\theta(X_i)$ be the output prediction ($y^S \in \mathbb{R}^C$) from the student $f_\theta(\cdot)$ with pretrained parameters $\theta := \{\mathbf{W}_l, \boldsymbol{b}_l\}_{l=1}^L$ for L layers. The intermediate input to each subsequent layer is denoted as $\boldsymbol{z}_l \in \mathbb{R}^{n_l}$ where $\boldsymbol{z}_0 := \boldsymbol{x}$ for n_l number of units in layer l and the corresponding output activation as $\boldsymbol{A}_l = g(\boldsymbol{z}_l)$. The loss function for standard classification fine-tuning is defined as the cross-entropy loss $\ell_{CE}(\boldsymbol{y}^S, \boldsymbol{y}) := -\frac{1}{C} \sum_{i=1}^c \boldsymbol{y}_c \log(\boldsymbol{y}_c^s)$.

For self-distilled pruning, we also require an already fine-tuned teacher network f_Θ, that has been tuned from the pretrained state f_θ, to retrieve the soft teacher labels $y^T := f_\Theta(\boldsymbol{x})$, where $y^T \in \mathbb{R}^C$ and $\sum_c^C y_c^T = 1$. The soft label $\boldsymbol{y}^T$

can be more informative than the one-hot targets $\boldsymbol{y}$ used for standard classification as they implicitly approximate pairwise class similarities through logit
probabilities. The Kullbeck-Leibler divergence ℓ_{KLD} is then used with the main
task cross-entropy loss ℓ_{CE} to express $\ell_{\mathrm{SDP-KLD}}$ as shown in Eq. 1,

$$\ell_{\mathrm{SDP\text{-}KLD}} = (1 - \alpha)\ell_{\mathrm{CE}}(\boldsymbol{y}^S, \boldsymbol{y}) + \alpha\tau^2 D_{\mathrm{KLD}}(\boldsymbol{y}^S, \boldsymbol{y}^T) \tag{1}$$

where $D_{\mathrm{KLD}}(\boldsymbol{y}^S, \boldsymbol{y}^T) = \mathbb{H}(\boldsymbol{y}^T) - \boldsymbol{y}^T \log(\boldsymbol{y}^S)$, $\mathbb{H}(\boldsymbol{y}^T) = \boldsymbol{y}^T \log(\boldsymbol{y}^T)$ is the entropy
of the teacher distribution and τ is the softmax temperature. Following [13], the
weighted sum of cross-entropy loss and KLD loss shown in Eq. 1 is used as our
main SDP-based KD loss baseline, where $\alpha \in [0,1]$. After each pruning step during iterative pruning, we aim to recover the immediate performance degradation
by minimizing $\ell_{\mathrm{SDP-KLD}}$. In our experiments, we use weight magnitude-based
pruning as the criteria for SDP given MBP's flexibility, scalability and miniscule
computation overhead (only requires a binary tensor multiplication to be applied
for each linear layer at each pruning step). However, D_{KLD} only distils the knowledge from the soft targets which may not propagate enough information about
the intermediate dynamics of the teacher, nor does it penalize representational
redundancy. This brings us to our proposed SDP objective.

3.1   Cross-Correlation Between Pruned and Unpruned Embeddings

Iterative pruning can be viewed as progressively adding noise $\mathbf{M}_l \in \{0,1\}^{n_{l-1} \times n_l}$
to the weights $\mathbf{W}_l \in \mathbb{R}^{n_{l-1} \times n_l}$. Thus, as the pruning steps increase, the outputs become noisier and the relationship between the inputs and outputs becomes
weaker. Hence, a correlation measure is a natural choice for dealing with such
pruning-induced noise. To this end, we use a cross-correlation loss to maximize
the correlation between the output representations of the last hidden state of the
pruned network and the unpruned network to reduce the effects of this pruning
noise. The proposed *cross-correlation SDP loss function*, ℓ_{CC}, is expressed in Eq. 2,
where λ controls the importance of minimizing the non-adjacent pairwise correlations between z^S and z^T in the correlation matrix $\mathcal{C}$. Here, m denotes the sample
index in a mini-batch of M samples. Unlike ℓ_{KLD}, this loss is applied to the outputs of the last hidden layer as opposed to the classification logit outputs. Thus,
we have,

$$\ell_{\mathrm{CC}}(\boldsymbol{z}^S, \boldsymbol{z}^T) := \sum_i (1 - \mathcal{C}_{ii})^2 + \lambda \sum_i \sum_{j \neq i} \mathcal{C}_{ij}^2 \tag{2}$$

such that $\mathcal{C}_{ij} := \dfrac{\sum_m z_{m,i}^S z_{m,j}^T}{\sqrt{\sum_m (z_{m,i}^S)^2} \sqrt{\sum_m (z_{m,j}^T)^2}}$.

Maximizing correlation along the diagonal of $\mathcal{C}$ makes the representations
invariant to pruning noise, while minimizing the off-diagonal term decorrelates
the components of the representations that are batch normalized. To reiterate,
$\boldsymbol{z}^S$ is obtained from the pruned version of the network (f_{θ_p}) and $\boldsymbol{z}^T$ is obtained
from the unpruned version (f_θ).

Since the learned output representations should be similar if their inputs are
similar, we aim to address the problem where a correlation measure may produce

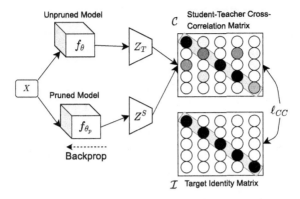

Fig. 1. Self-distilled pruning with a cross-correlation knowledge distillation loss.

representations that are instead *proportional* to their inputs. To address this, batch normalization is used across mini-batches to stabilize the optimization when using the cross-correlation loss, avoiding local optima that correspond to degenerate representations that do not distinguish proportionality. In our experiments, this is used with the classification loss and KLD distillation loss as shown in Eq. 3.

$$\ell_{\text{SDP--CC}} = (1 - \alpha)\ell_{\text{CE}}(\boldsymbol{y}^S, \boldsymbol{y}) + \alpha\tau^2 D_{\text{KLD}}(\boldsymbol{y}^S, \boldsymbol{y}^T) + \beta\ell_{\text{CC}}(\boldsymbol{z}^S, \boldsymbol{z}^T) \quad (3)$$

Figure 1 illustrates the proposed framework of *self-distilled pruning with cross-correlation loss* (SDP-CC), where $\mathcal{I}$ is the identity matrix.

3.2 A Frobenius Distortion Perspective of Self-distilled Pruning

To formalize the objective being minimized when using MBP with self-distillation, we take the view of *Frobenius distortion minimization* (FDM) [7] which says that layer-wise MBP is equivalent to minimizing the Frobenius distortions of a single layer. This can be described as $\min_{\mathbf{M}:||\mathbf{M}||_0=p} ||\mathbf{W} - \mathbf{M} \odot \mathbf{W}||_F$, where $\odot$ is the Hadamard product and p is a constraint of the number of weights to remove as a percentage of the total number of weights for a layer. Therefore, the output distortion is approximately the product of single layer Frobenius distortions. However, this minimization only defines a 1st order approximation of pruning induced Frobenius distortions which is a loose approximation for deep networks. In contrast, the $\boldsymbol{y}^T$ targets provide higher-order information outside of the l-th layer being pruned in this FDM framework because Θ encodes information of all neighboring layers. Hence, we reformulate the FDM problem for SDP as an approximately higher-order MBP method as in Eq. 4 where $\mathbf{W}^T$ are the weights in f_Θ.

$$\min_{\mathbf{M}:||\mathbf{M}||_0=p} \left[||\mathbf{W} - \mathbf{M} \odot \mathbf{W}||_F + \lambda||\mathbf{W}^T - \mathbf{M} \odot \mathbf{W}||_F \right] \quad (4)$$

As described in [7, 12], the difference in error can be approximated with a Taylor Series (TS) expansion as $\delta \mathcal{E}_l \approx \left(\frac{\partial \mathcal{E}_l}{\partial \mathbf{W}^l}\right)^\top \delta \mathbf{W}_l + \frac{1}{2}\delta \mathbf{W}_l^\top \mathbf{H}_l \delta \mathbf{W}_l + O(||\delta \mathbf{W}_l||^3)$ where $\mathbf{H}$ is the Hessian matrix. When using SDP with a 1st TS, we can further express the TS approximation for SDP as shown in Eq. 5, where $\mathcal{E}_l^S$ is the error of the pruned network for task provided targets and $\mathcal{E}_l^T$ are the errors of the pruned network with distilled logits.

$$\left(\mathcal{E}_l - \mathcal{E}_l^S\right)^2 + \lambda\left(\mathcal{E}_l - \mathcal{E}_l^T\right)^2 \approx \delta \mathcal{E}_l^S + \delta \mathcal{E}_l^T \approx \left(\frac{\partial \mathcal{E}_l^S}{\partial \theta_l}\right)^\top \delta \theta_l + \lambda\left(\frac{\partial \mathcal{E}_l^T}{\partial \theta_l}\right)^\top \delta \theta_l \quad (5)$$

3.3 How Does Self-distillation Improve Pruned Model Generalization?

We put forth the following insights as to the advantages provided by self-distillation for better pruned model generalization, and later experimentally demonstrate their validity.

Recovering Faster From Performance Degradation After Pruning Steps. The first explanation for why self-distillation leads to better generalization in iterative pruning is that the soft targets bias the optimization and smoothen the loss surface through implicit similarities between the classes encoded in the logits. We posit this too holds true for performance recovery after pruning steps, as the classification boundaries become distorted due to the removal of weights. Faster convergence is particularly important for high compression rates where the performance drops become larger.

Implicit Maximization of the Signal-to-Noise Ratio. One explanation for faster convergence is that optimizing for soft targets translates to maximizing the margin of class boundaries given the implicit class similarities provided by teacher logits. Intuitively, task provided one-hot targets do not inform SGD of how similar incorrect predictions are to the correct class, whereas the teacher logits do, to the extent they have learned on the same task. To measure this, we use a formulation of the signal-to-noise ratio[1] (SNR) to measure the class separability and compactness differences between pruned model representations trained with and without self-distillation. We formulate SNR as Eq. 6, where for a batch of inputs $\mathbf{X}$, we obtain $\mathbf{Z}$ output representations from the pruned network, which contain samples with C classes where each class has the same N number of samples. The numerator expresses the average ℓ_2 inter-class distance between instances of each class pair and the denominator expresses the intra-class distance between instances within the same class.

$$\frac{1/N(C-1)^2 \sum_n^N \sum_{c=1}^C \sum_{i \neq c}^C ||\sqrt{\mathbf{Z}_{c,n}} - \sqrt{\mathbf{Z}_{i,n}}||_2}{1/C(N-1)^2 \sum_{c=1}^C \sum_n^N \sum_{j \neq n} ||\sqrt{\mathbf{Z}_{c,n}} - \sqrt{\mathbf{Z}_{c,j}}||_2} \quad (6)$$

---

[1] A measure typically used in signal processing to evaluate signal quality.

This estimation is $C-1\binom{C+1}{2}$ in the number of pairwise distances to be computed between the inter-class distances for the classes. For large output spaces (e.g., language modeling) we recommend defining the top k-NN classes for each class and estimate their distances on samples from them.

Quantifying Fidelity Between Pruned Models Trained With and Without Self-distillation. A natural question to ask is *how much generalization power does the distilled soft targets provide when compared to the task provided one-hot targets ?* If best generalization is achieved when $\alpha = 1$ in Eq. 1, this implies that the pruned network should have as high fidelity as possible with the unpruned network. However, as we will see there is a bias-variance trade-off between fidelity and generalization performance, i.e., $\alpha = 1$ is not optimal in most cases. To measure fidelity between SDP representations and standard fine-tuned representations, we compute their *mutual information* (MI) and compare this to the MI between representations of pruned models without self-distillation and standard fine-tuned models. The MI between continuous variables can be expressed as,

$$\hat{I}(\mathbf{Z}^T; \mathbf{Z}^S) = H(\mathbf{Z}^T) - H(\mathbf{Z}^T | \mathbf{Z}^S)$$
$$= -\mathbb{E}_{\mathbf{Z}^T}[\log p(\mathbf{Z}^T)] + \mathbb{E}_{\mathbf{Z}^T, \mathbf{Z}^S}[\log p(\mathbf{Z}^T | \mathbf{Z}^S)] \tag{7}$$

where $H(\mathbf{Z}^T)$ is the entropy of the teacher representation and $H(\mathbf{Z}^T | \mathbf{Z}^S)$ is the conditional entropy that is derived from the joint distribution $p(\mathbf{Z}^T, \mathbf{Z}^S)$. This can also be expressed as the KL divergence between the joint probabilities and product of marginals as $I(Z^T; Z^S) = D_{\mathrm{KLD}}[p(Z^S, Z^T) \| p(Z^S) p(Z^T)]$. However, these theoretical quantities have to be estimated from test sample representations. We use a k-NN based MI estimator [8,17,41,42] which partitions the supports into a finite number of bins of equal size, forming a histogram that can be used to estimate $\hat{I}(Z^S; Z^T)$ based on discrete counts in each bin. This MI estimator is given as,

$$I(z^S; z^T) \approx \epsilon\left(\log \frac{\phi_{[z^S]}(i, k_{[z^S]}) \phi_{[z^T]}(i, k_{[z^T]})}{\phi_z(i, k)} \right) \tag{8}$$

where $\phi_{z^S}(i, k_{[z^S]})$ is the probability measure of the k-th nearest neighbour ball of $\mathbf{z}^S \in \mathbb{R}^{n_L}$ and $\omega_{[z^T]}(i, k_{[z^T]})$ is the probability measure of the k_y-th nearest neighbour ball of $\mathbf{z}^T \in \mathbb{R}^{n_L}$ where n_L is the dimension of the penultimate layer. In our experiments, we use 256 bins for the histogram with Gaussian smoothing and $k = 5$ (see [17] for further details).

4   Experimental Setup

Datasets. We perform experiments on monolingual tasks within the GLUE [43] benchmark[2] with pretrained BERT$_{\mathrm{Base}}$ and multilingual tasks from the XGLUE benchmark [22] with pretrained XLMR$_{\mathrm{Base}}$. In total, this covers 18 different

---

[2] WNLI is excluded for known issues, see the Q. 12 on the GLUE benchmark FAQ.

Table 1. GLUE benchmark results for pruned models @10% (or @20%) remaining weights.

Compression method	Score (avg.)	Single sentence		Similarity and paraphrase			Natural language inference		
		CoLA (mcc)	SST-2 (acc)	MNLI (acc)	MRPC (f1/acc)	STS-B (pears./spear.)	QQP (f1/acc)	RTE (acc)	QNLI (acc)
BERT$_{Base}$ (Ours)	84.06	53.24	90.71	80.27	80.9/77.7	83.5/83.8	83.9/88.0	68.59	86.91
Knowledge distilled baselines (% parameters w.r.t. original BERT)									
DistilBERT (60%)	82.85	51.3	91.3	82.2	87.5/-.-	86.9/-.-	-.-/85.5	59.9	89.2
BERT-Medium (44.4%)	81.54	38.0	89.6	80.0	86.6/81.6	80.4/78.4	69.6/87.9	62.2	87.7
BERT-Small (20%)	79.02	27.8	89.7	77.6	83.4/76.2	78.8/77.0	68.1/87.0	61.8	86.4
BERT-Mini (10%)	76.97	0.0	85.9	75.1	74.8/74.3	75.4/73.3	66.4/86.2	57.9	84.1
BERT-Tiny (3.6%)	73.32	0.0	83.2	70.2	81.1/71.1	74.3/73.6	62.2/83.4	57.2	81.5
Pruning baselines		20%	10%	10%	10%	10%	10%	10%	10%
Random	66.03	6.50	78.44	69.55	77.5/67.1	27.4/26.9	77.07/81.86	52.70	74.66
L_0-MBP	77.25	31.68	83.37	75.61	78.4/68.2	75.9/75.7	81.56/86.49	**64.26**	82.62
L_2-MBP	76.48	29.51	83.37	76.19	78.4/68.2	75.3/75.6	77.50/82.98	62.09	82.61
L_2-Global-MBP	77.16	29.25	82.83	76.40	81.2/69.9	75.1/75.5	82.77/86.70	62.01	82.24
L_2-Gradient-MBP	74.84	15.46	82.91	72.51	81.0/73.7	73.8/73.6	80.41/85.19	56.31	79.33
1st-order Taylor	76.31	28.88	83.26	74.64	**83.0/74.8**	76.7/76.6	80.09/85.29	57.76	81.20
Lookahead	76.40	28.15	82.80	75.31	79.8/70.5	71.9/71.9	81.84/86.53	60.29	81.80
LAMP	74.03	20.31	83.26	74.27	72.3/63.7	73.7/74.1	79.32/85.07	58.84	81.09
Proposed methodology									
L_2-MBP + SDP-COS	77.83	31.80	86.00	75.68	81.6/72.2	76.4/76.3	81.39/86.68	61.73	83.07
L_2-MBP + SDP-KLD	78.34	36.74	**87.96**	77.94	80.5/68.2	77.1/77.3	83.21/85.58	63.18	83.54
L_2-MBP + SDP-CC	**78.90**	**36.77**	87.84	**78.04**	81.1/71.0	**77.3/77.5**	**83.79/86.37**	62.64	**84.20**

BERT- results reported from prior work [15,35,40] and MNLI results are for the matched dataset.

datasets, covering pairwise classification, sentence classification, structured prediction and question answering. To our knowledge, this work is the first to analyse iterative pruning in the context of cross-lingual models and their application on multilingual datasets. Further dataset statistics can be found in supplementary material.

Iterative Pruning Baselines. For XGLUE tasks, we perform 15 pruning steps on XLM-RoBERTA$_{Base}$, one per 15 epochs, while for the GLUE tasks, we perform 32 pruning steps on BERT$_{Base}$. The compression rate and number of pruning steps is higher for GLUE tasks compared to XGLUE, because GLUE tasks involve evaluation in the *supervised classification* setting; whereas in XGLUE we report in the more challenging *zero-shot cross-lingual transfer* setting with only a single language used for training (i.e., English). At each pruning step, we uniformly pruning 10% of the parameters for both the models. Although prior work suggests non-uniform pruning schedules (e.g., cubic schedule [50]), we did not see any major differences to uniform pruning. We compare the performance of the proposed SDP-CC method against the following baselines:

– **Random Pruning** (*MBP-Random*) - prunes weights uniformly at random across all layers. Random pruning can be considered as a lower bound on iterative pruning performance.

- **Layer-wise Magnitude Based Pruning** (*MBP*) - for each layer, prunes weights with the LAV.
- **Global Magnitude Pruning** (*Global-MBP*) - prunes the LAV of all weights in the network.
- **Layer-wise Gradient Magnitude Pruning** (*Gradient-MBP*) - for each layer, prunes the weights with the LAV of the accumulated gradients evaluated on a batch of inputs.
- 1st **Taylor Series Pruning** (*TS*) - prunes weights based on the LAV of |gradient × weight|.
- L_0 **norm MBP** [24] - uses non-negative stochastic gates that choose which weights are set to zero as a smooth approximation to the non-differentiable L_0-norm.
- L_1 **norm MBP** [21] - applies L_1 weight regularization and uses MBP.
- **Lookahead pruning (LAP)** [31] - prunes weight paths that have the smallest magnitude across blocks of layers, unlike MBP that does not consider neighboring layers.
- **Layer-Adaptive MBP (LAMP)** [20] - adaptively computes the pruning ratio for each layer.

For all above pruning methods we exclude weight pruning of the embeddings, layer normalization parameters and the last classification layer, as they play an important role for generalization and account for less than 1% of weights in both BERT and XLM-R$_{\text{Base}}$.

Knowledge Distillation. We also compare against a class of smaller knowledge distilled versions of the BERT model with varying parameter sizes on the GLUE benchmark. We report prior results of *DistilBERT* [35] and also mini-BERT models including *TinyBERT* [15], *BERT-small* [40] and *BERT-medium* [40]. In addition, we consider maximizing the cosine similarity between pruned and unpruned representations in the SDP loss, as $\ell_{\text{SDP-COS}} := \alpha \ell_{\text{CE}}(\boldsymbol{y}^S, \boldsymbol{y}) + \beta \big(1 - \frac{\mathbf{z}^S \cdot \mathbf{z}^T}{\|\boldsymbol{z}^S\| \|\|\boldsymbol{z}^T\|}\big)$. Unlike cross-correlation, there is no decorrelation of non-adjacent features in both representations for SDP-COS. This helps identify whether the redundancy reduction in cross-correlation is beneficial compared to the correlation loss that does not directly optimize this.

5   Empirical Results

Pruning Results on GLUE. Table 1 shows the test performance across all GLUE tasks of the different models with varying pruning ratios, up to *10% remaining weights* of original BERT$_{\text{Base}}$ along with mini-BERT models [35, 40] of varying size. However, for the CoLA dataset, we report at 20% pruning as nearly all compression methods have an MCC score of 0, making the compressed method performance indistinguishable. For this reason, the GLUE score (**Score**) is computed for all tasks and methods @10% apart from CoLA. The best performing compression method per task is marked in **bold**. We find that our proposed

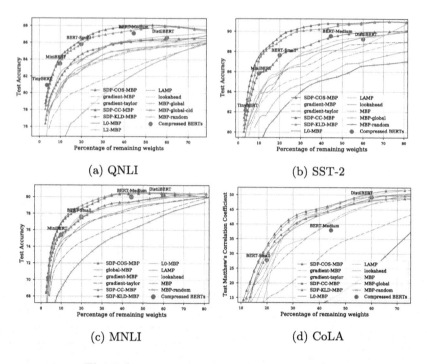

Fig. 2. Iterative pruning results on GLUE tasks.

SDP approaches (all three variants) outperform against baseline pruning methods, with *SDP-CC* performing the best across all tasks. We note that for the tasks with fewer training samples (e.g., CoLA has 8.5k samples, STS-B has 7k samples and RTE has 3k samples), the performance gap is larger compared to $BERT_{Base}$, as the pruning step interval is shorter and less training data allows lesser time for the model to recover from pruning losses and also less data for teacher model to distil in the case of using SDP.

Smaller dense versions of BERT require more labelled data in order to compete with unstructured MBP and higher-order pruning methods such as 1st order Taylor series and Lookahead pruning. For example, we see BERT-Mini (@10%) shows competitive test accuracy with our proposed SDP-CC on QNLI, MNLI and QQP, the three datasets with the most training samples (105k, 393k and 364k respectively). Overall, L_2-MBP + SDP-CC achieves the highest GLUE score for all models at 10% remaining weights when compared to BERT-Base parameter count. Moreover, we find that L_2-MBP + SDP-CC achieves best performance for 5 of the 8 tasks, with 1 of the remaining 3 being from L_2MBP+SDP-KLD. This suggests that redundancy reduction via a cross-correlation objective is useful for SDP and clearly improve over SDP-COS which does not minimize correlations between off-diagonal terms. Figure 2 shows the performance across all pruning steps. Interestingly, for QNLI we observe the performance notably improves between 30–70% for SDP-CC and SDP-KLD. For SST-2, we observe a

Table 2. XGLUE iterative pruning @ 30% remaining weights of XLM-R$_{base}$ - zero shot cross-lingual performance per task and overall average score (avg).

Prune method	XNLI	NC	NER	PAWSX	POS	QAM	QADSM	WPR	Avg.
XLM-R$_{Base}$	73.48	80.10	82.60	89.24	80.34	68.56	68.06	73.32	76.96
Random	51.22	70.19	38.19	57.37	52.57	53.85	52.34	70.69	55.80
Global-Random	50.97	69.88	38.30	56.74	53.02	54.02	53.49	69.11	55.69
L_0-MBP	64.75	78.98	56.22	72.09	71.38	59.31	53.35	71.70	65.97
L_2-MBP	64.30	78.79	54.43	77.99	70.68	59.24	60.33	71.52	67.16
L_2-Global-MBP	65.12	78.64	54.47	79.13	71.37	59.26	60.61	71.80	67.55
L_2-Gradient-MBP	61.11	73.77	53.25	79.56	65.89	57.35	59.33	71.59	65.23
1st-order Taylor	64.26	79.34	63.60	**82.83**	68.94	61.69	62.42	72.28	69.09
Lookahead	60.84	79.18	54.44	71.05	68.76	55.94	53.41	71.26	64.36
LAMP	58.04	63.64	51.92	66.05	67.43	55.36	52.42	71.09	60.74
L_2-MBP + SDP-COS	64.96	79.02	62.77	78.70	72.88	60.21	60.94	72.04	68.94
L_2-MBP + SDP-KLD	65.94	**80.72**	64.50	79.25	73.18	61.66	61.09	71.84	**69.77**
L_2-MBP + SDP-CC	**66.47**	79.73	**66.34**	80.03	**73.45**	**63.73**	**62.78**	**72.59**	**70.76**

significant gap between SDP-KLD and SDP-CC compared to the pruning baselines and smaller versions of BERT, while TinyBERT becomes competitive at extreme compression (<4%). **Pruning Results on _XGLUE_.** We show the per task test performance and the _average task understanding_ score on XGLUE for pruning baselines and our proposed SDP approaches in Table 2. Our proposed cross-correlation objective for SDP again achieves the best average (Avg.) score and achieves the best task performance in 6 out of 8 tasks, while standard SDP-KLD achieves best performance on one (news classification) of the remaining two. Most notably, we outperform methods which use higher order gradient information (1st-order Taylor) at 30% remaining weights, which tends to be a point at which XLM-R$_{Base}$ begins to degrade performance below 10% of the original fine-tuned test performance for SDP methods and competitive baselines. In Fig. 3, we can observe this trend from the various tasks within XGLUE. We note that the number of training samples used for retraining plays an important role in the rate of performance degradation. For example, of the 6 presented XGLUE tasks, NER has the lowest number of training samples (15k) of all XGLUE tasks and also degrades the fastest in performance (from 90% to 50% Test F1 at 30% remaining weights). In comparison, XNLI has the most training samples for retraining (433k) and maintains performance relatively well, keeping within 10% of the original fine-tuned model at 30% remaining weights. **Summary of Results.** From our experiments on GLUE and XGLUE task, we find that SDP consistently outperforms pruning, KD and smaller BERT baselines. SDP-KLD and SDP-CC both outperform larger sized BERT models (BERT-Small), somewhat surprisingly, given that BERT-Small (and the remaining BERT models) have the advantage of large-scale self-supervised pretraining, while pruning only has supervision from the downstream task. For NER in XGLUE, higher order pruning methods such as Taylor-Series pruning have an advantage at high compression rates mainly due to lack of training samples (only 15k). Apart from this low training sample regime, SDP with MBP dominates at high compression rates.

Measuring Fidelity to the Fine-Tuned Model. We now analyse the empirical evidence that soft targets used in SDP may force higher fidelity with the representations of the fine-tuned model when compared to using MBP without self-distillation. As described in Subsect. 3.3 we measure mutual dependencies between both representations of models with the best performing hyperparameter settings of α, β and the softmax temperature τ. We note that increasing the temperature τ translates to "peakier" teacher logit distributions, encouraging SGD to learn a student with high fidelity to the teacher. From the LHS of Fig. 4, we can see that SDP models have higher mutual information (MI) with the teacher compared to MBP, which performs worse for PAWS-X (similar on remaining tasks, not shown for brevity). In fact, the rank order of the best performing pruned models at each pruning step has a direct correlation with MI, e.g., SDP-COS-MBP maintains highest MI and the highest test accuracy for PAWS-X for the same α. However, too high fidelity ($\alpha = 1$.) led to worse generalization compared to a balance between the task provided targets and the teacher logits ($\alpha = 0.5$).

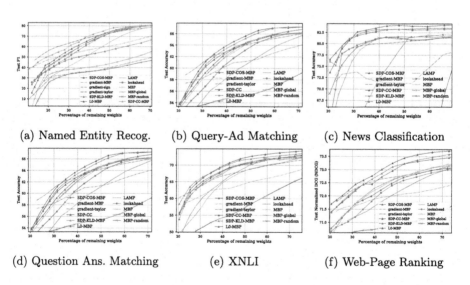

(a) Named Entity Recog.   (b) Query-Ad Matching   (c) News Classification

(d) Question Ans. Matching   (e) XNLI   (f) Web-Page Ranking

Fig. 3. Zero-shot results after iteratively fine-pruning XLM-R$_{\text{Base}}$ on XGLUE tasks.

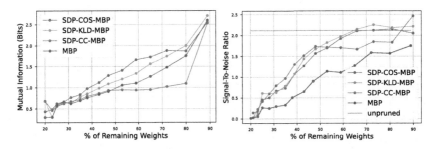

Fig. 4. Mutual information between unpruned and pruned representations (left) and signal-to-noise ratio (right)

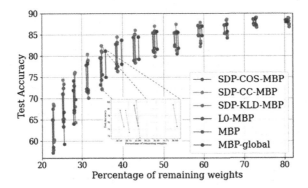

Fig. 5. PAWS-X development set representations and (right) pruning performance recovery with self-distilled pruning.

Self-distilled Pruning Increases Class Separability and the Signal-to-Noise Ratio (SNR). We also find that the SNR is increased at each pruning step as formulated in Sect. 3.3. From this observation, we find that *SDP-CC-MBP* using cross-correlation loss does particularly well in the 30%–50% remaining weights range. More generally, all 3 SDP losses clearly lead to better class separability and class compactness across all pruning steps compared to MBP (i.e., no self-distillation).

Self-distilled Pruning Recovers Faster Performance Degradation Directly After Pruning Steps. In Fig. 5 we show how SDP with Magnitude pruning (SDP-MBP) recovers during training in between pruning steps. The top of each vertical bar is the recovery development accuracy and the bottom is the initial performance degradation prior to retrainng. We see that SDP pruned models degrade in performance more than magnitude pruning without self-distillation. This suggests that SDP-MBP may force weights to be closer, as there is more initial performance degradation if weights are not driven to zero. However, the recovery is faster. This may be explained by recent work that suggests the stability generalization tradeoff [4].

6   Conclusion

In this paper, we proposed a novel *self-distillation* based pruning technique based on a *cross-correlation* objective. We extensively studied the confluence between pruning and self-distillation for masked language models and its enhanced utility on downstream tasks in both monolingual and multi-lingual settings. We find that self-distillation aids in recovering directly after pruning in iterative magnitude-based pruning, increases representational fidelity with the unpruned model and implicitly maximize the signal-to-noise ratio. Additionally, we find our cross-correlation based self-distillation pruning objective minimizes neuronal redundancy and achieves state-of-the-art in magnitude-based pruning baselines, and even outperforms KD based smaller BERT models with more parameters.

References

1. Ahn, S., Hu, S.X., Damianou, A., Lawrence, N.D., Dai, Z.: Variational information distillation for knowledge transfer. In: Proceedings of the IEEE/CVF Conference on Computer Vision and Pattern Recognition, pp. 9163–9171 (2019)
2. Allen-Zhu, Z., Li, Y.: Towards understanding ensemble, knowledge distillation and self-distillation in deep learning. arXiv preprint arXiv:2012.09816 (2020)
3. Anwar, S., Hwang, K., Sung, W.: Structured pruning of deep convolutional neural networks. ACM J. Emerg. Technol. Comput. Syst. (JETC) **13**(3), 1–18 (2017)
4. Bartoldson, B.R., Morcos, A.S., Barbu, A., Erlebacher, G.: The generalization-stability tradeoff in neural network pruning. arXiv preprint arXiv:1906.03728 (2019)
5. Conneau, A., et al.: Unsupervised cross-lingual representation learning at scale. arXiv preprint arXiv:1911.02116 (2019)
6. Devlin, J., Chang, M.W., Lee, K., Toutanova, K.: BERT: pre-training of deep bidirectional transformers for language understanding. arXiv preprint arXiv:1810.04805 (2018)
7. Dong, X., Chen, S., Pan, S.J.: Learning to prune deep neural networks via layer-wise optimal brain surgeon. arXiv preprint arXiv:1705.07565 (2017)
8. Evans, D.: A computationally efficient estimator for mutual information. Proc. R. Soc. A Math. Phys. Eng. Sci. **464**(2093), 1203–1215 (2008)
9. Furlanello, T., Lipton, Z., Tschannen, M., Itti, L., Anandkumar, A.: Born again neural networks. In: International Conference on Machine Learning, pp. 1607–1616. PMLR (2018)
10. Han, S., Mao, H., Dally, W.: Compressing deep neural networks with pruning, trained quantization and huffman coding. arXiv preprint (2015)
11. Han, S., Pool, J., Tran, J., Dally, W.J.: Learning both weights and connections for efficient neural networks. arXiv preprint arXiv:1506.02626 (2015)
12. Hassibi, B., Stork, D.G.: Second Order Derivatives for Network Pruning: Optimal Brain Surgeon. Morgan Kaufmann (1993)
13. Hinton, G., Vinyals, O., Dean, J.: Distilling the knowledge in a neural network. arXiv preprint arXiv:1503.02531 (2015)
14. Huang, Z., Wang, N.: Like what you like: knowledge distill via neuron selectivity transfer. arXiv preprint arXiv:1707.01219 (2017)

15. Jiao, X., et al.: TinyBERT: distilling bert for natural language understanding. arXiv preprint arXiv:1909.10351 (2019)
16. Karnin, E.D.: A simple procedure for pruning back-propagation trained neural networks. IEEE Trans. Neural Netw. **1**(2), 239–242 (1990)
17. Kraskov, A., Stögbauer, H., Grassberger, P.: Estimating mutual information. Phys. Rev. E **69**(6), 066138 (2004)
18. Lebedev, V., Lempitsky, V.: Fast convnets using group-wise brain damage. In: Proceedings of the IEEE Conference on Computer Vision and Pattern Recognition, pp. 2554–2564 (2016)
19. LeCun, Y., Denker, J.S., Solla, S.A.: Optimal brain damage. In: Advances in Neural Information Processing Systems, pp. 598–605 (1990)
20. Lee, J., Park, S., Mo, S., Ahn, S., Shin, J.: Layer-adaptive sparsity for the magnitude-based pruning. In: International Conference on Learning Representations (2020)
21. Li, H., Kadav, A., Durdanovic, I., Samet, H., Graf, H.P.: Pruning filters for efficient convnets. arXiv preprint arXiv:1608.08710 (2016)
22. Liang, Y., et al.: XGLUE: a new benchmark dataset for cross-lingual pre-training, understanding and generation. In: Proceedings of the 2020 Conference on Empirical Methods in Natural Language Processing (EMNLP), pp. 6008–6018 (2020)
23. Liu, Z., Li, J., Shen, Z., Huang, G., Yan, S., Zhang, C.: Learning efficient convolutional networks through network slimming. In: Proceedings of the IEEE International Conference on Computer Vision, pp. 2736–2744 (2017)
24. Louizos, C., Welling, M., Kingma, D.P.: Learning sparse neural networks through l_0 regularization. arXiv preprint arXiv:1712.01312 (2017)
25. Mallya, A., Davis, D., Lazebnik, S.: Piggyback: adapting a single network to multiple tasks by learning to mask weights. In: Proceedings of the European Conference on Computer Vision (ECCV), pp. 67–82 (2018)
26. Martens, J., Grosse, R.: Optimizing neural networks with kronecker-factored approximate curvature. In: International Conference on Machine Learning, pp. 2408–2417. PMLR (2015)
27. Mobahi, H., Farajtabar, M., Bartlett, P.L.: Self-distillation amplifies regularization in hilbert space. arXiv preprint arXiv:2002.05715 (2020)
28. Molchanov, D., Ashukha, A., Vetrov, D.: Variational dropout sparsifies deep neural networks. In: International Conference on Machine Learning, pp. 2498–2507. PMLR (2017)
29. Mozer, M.C., Smolensky, P.: Skeletonization: a technique for trimming the fat from a network via relevance assessment. In: Advances in Neural Information Processing Systems, pp. 107–115 (1989)
30. Neill, J.O., Bollegala, D.: Semantically-conditioned negative samples for efficient contrastive learning. arXiv preprint arXiv:2102.06603 (2021)
31. Park, S., Lee, J., Mo, S., Shin, J.: Lookahead: a far-sighted alternative of magnitude-based pruning. arXiv preprint arXiv:2002.04809 (2020)
32. Park, W., Kim, D., Lu, Y., Cho, M.: Relational knowledge distillation. In: Proceedings of the IEEE/CVF Conference on Computer Vision and Pattern Recognition, pp. 3967–3976 (2019)
33. Reed, R.: Pruning algorithms-a survey. IEEE Trans. Neural Netw. **4**(5), 740–747 (1993)
34. Romero, A., Ballas, N., Kahou, S.E., Chassang, A., Gatta, C., Bengio, Y.: FitNets: hints for thin deep nets. arXiv preprint arXiv:1412.6550 (2014)
35. Sanh, V., Debut, L., Chaumond, J., Wolf, T.: DistilBERT, a distilled version of BERT: smaller, faster, cheaper and lighter. arXiv preprint arXiv:1910.01108 (2019)

36. Sanh, V., Wolf, T., Rush, A.M.: Movement pruning: adaptive sparsity by fine-tuning. arXiv preprint arXiv:2005.07683 (2020)
37. Singh, S.P., Alistarh, D.: WoodFisher: efficient second-order approximations for model compression. arXiv preprint arXiv:2004.14340 (2020)
38. Stanton, S., Izmailov, P., Kirichenko, P., Alemi, A.A., Wilson, A.G.: Does knowledge distillation really work? arXiv preprint arXiv:2106.05945 (2021)
39. Tian, Y., Krishnan, D., Isola, P.: Contrastive representation distillation. arXiv preprint arXiv:1910.10699 (2019)
40. Turc, I., Chang, M.W., Lee, K., Toutanova, K.: Well-read students learn better: on the importance of pre-training compact models. arXiv preprint arXiv:1908.08962 (2019)
41. Ver Steeg, G.: Non-parametric entropy estimation toolbox (NPEET). Technical report (2000). https://www.isi.edu/~gregv/npeet_doc.pdf
42. Ver Steeg, G., Galstyan, A.: Information-theoretic measures of influence based on content dynamics. In: Proceedings of the Sixth ACM International Conference on Web Search and Data Mining, pp. 3–12 (2013)
43. Wang, A., Singh, A., Michael, J., Hill, F., Levy, O., Bowman, S.R.: GLUE: a multi-task benchmark and analysis platform for natural language understanding. arXiv preprint arXiv:1804.07461 (2018)
44. Wang, C., Grosse, R., Fidler, S., Zhang, G.: EigenDamage: structured pruning in the kronecker-factored eigenbasis. In: International Conference on Machine Learning, pp. 6566–6575. PMLR (2019)
45. Wen, W., Wu, C., Wang, Y., Chen, Y., Li, H.: Learning structured sparsity in deep neural networks. arXiv preprint arXiv:1608.03665 (2016)
46. Yang, C., Xie, L., Qiao, S., Yuille, A.L.: Training deep neural networks in generations: a more tolerant teacher educates better students. In: Proceedings of the AAAI Conference on Artificial Intelligence, vol. 33, pp. 5628–5635 (2019)
47. Ye, J., Lu, X., Lin, Z., Wang, J.Z.: Rethinking the smaller-norm-less-informative assumption in channel pruning of convolution layers. arXiv preprint arXiv:1802.00124 (2018)
48. Yuan, M., Lin, Y.: Model selection and estimation in regression with grouped variables. J. R. Stat. Soc. Ser. B (Stat. Methodol.) **68**(1), 49–67 (2006)
49. Zbontar, J., Jing, L., Misra, I., LeCun, Y., Deny, S.: Barlow twins: self-supervised learning via redundancy reduction. arXiv preprint arXiv:2103.03230 (2021)
50. Zhu, M., Gupta, S.: To prune, or not to prune: exploring the efficacy of pruning for model compression. arXiv preprint arXiv:1710.01878 (2017)

MultiLayerET: A Unified Representation of Entities and Topics Using Multilayer Graphs

Jumanah Alshehri(✉)ⓘ, Marija Stanojevicⓘ, Parisa Khan, Benjamin Rapp, Eduard Dragutⓘ, and Zoran Obradovicⓘ

Center for Data Analytics and Biomedical Informatics, Temple University, Philadelphia, PA, USA
{shehri.j,marija.stanojevic,edragut,zoran.obradovic}@temple.edu

Abstract. Many online news outlets, forums, and blogs provide a rich stream of publications and user comments. This rich body of data is a valuable source of information for researchers, journalists, and policy-makers. However, the ever-increasing production and user engagement rate make it difficult to analyze this data without automated tools. This work presents MultiLayerET, a method to unify the representation of entities and topics in articles and comments. In MultiLayerET, articles' content and associated comments are parsed into a multilayer graph consisting of heterogeneous nodes representing named entities and news topics. The nodes within this graph have attributed edges denoting weight, i.e., the strength of the connection between the two nodes, time, i.e., the co-occurrence contemporaneity of two nodes, and sentiment, i.e., the opinion (in aggregate) of an entity toward a topic. Such information helps in analyzing articles and their comments. We infer the edges connecting two nodes using information mined from the textual data. The multilayer representation gives an advantage over a single-layer representation since it integrates articles and comments via shared topics and entities, providing richer signal points about emerging events. MultiLayerET can be applied to different downstream tasks, such as detecting media bias and misinformation. To explore the efficacy of the proposed method, we apply MultiLayerET to a body of data gathered from six representative online news outlets. We show that with MultiLayerET, the classification F1 score of a media bias prediction model improves by 36%, and that of a state-of-the-art fake news detection model improves by 4%.

Keywords: News mining · Multilayer graphs · Text mining · Social network analysis

1 Introduction

The amount of published articles is steadily increasing, and readers are shifting toward online platforms because of the affordable technology costs and the ability to share their opinions. News articles are conveniently accessed either via news outlets' websites or news aggregator platforms, like Google News, that collect

M.-R. Amini et al. (Eds.): ECML PKDD 2022, LNAI 13714, pp. 671–687, 2023.
https://doi.org/10.1007/978-3-031-26390-3_39

articles and recommend a subset of them to readers according to their interests. The current news ecosystem escalates the competition between platforms and motivates them to scale and enhance their systems. Such growth makes it harder for readers and analysts to get a complete picture of a particular event or entity without falling into the news source bias or the contrasting opinions hidden between the lines of articles across sources. Therefore, building an automated system able to represent and model semantic data is essential to help readers, researchers, journalists, and decision-makers understand emerging events and their associated entities. Such a system will benefit downstream applications, such as news popularity, media bias, news recommendations, and fake news detection. We propose a unified representation of news and comments in our MultiLayerET system based on shared entities and topics.

To our knowledge, most news research focuses on extracting topics and entities from the news articles and omits the user-generated content associated with news items in comment sections. In this study, we formulate the problem as follows, *having a large set of documents in the form of articles and their associated comments. We aim to extract a rich graph representation of emerging events or topics.* To achieve this objective, we introduce our system MultiLayerET. The output of our system is a heterogeneous *attributed multilayer graph.* A multilayer graph is a graph with a set of nodes, each assigned a type from a set of types [36]; our graph has two types of edges: intra-layer edges, which connect nodes within the same layer, and inter-layer edges, which connect nodes across layers. In MultiLayerET, nodes are entities and topics, and edges are attributed. We consider three types of attributes: 1) co-occurrence, which is the co-occurrence frequency of a pair of nodes, 2) contemporaneity, which is the published times of the documents where the two nodes co-occur (note that articles and comments have temporal information, published and posted times, respectively), and 3) sign, which denotes the aggregated sentiment of the text where two nodes co-occur. The extracted graph consists of two layers, the G_a layer where nodes and edges are extracted from articles, and the G_c layer where the nodes and edges are obtained from comment sections related to the articles in G_a.

The multilayer approach provides a unique representation of emerging events. To illustrate, Fig. 1 shows both layers with their attributes on a set of documents from Washington Post (WP) during the 2016 US election, we unfold the graph based on time to show the contemporaneity attribute over three months period from May - July 2016. Comparing G_a and G_c, we can see that G_c complements G_a greatly including entities that are not mentioned in G_a. For example, the comment layer mentions other candidates such as *Bernie Sanders* during the 2016 election that are from the same party and their relationship with the topic *Hillary Clinton's email story.* Moreover, *James Comey* the director of the FBI's investigation, appears in the comment layer around July when the investigation started, and *Colin Powell* and his relationship with Clinton email controversy appeared in early stages in G_c. Related to topics, we see that the topic *election debates* and its relationship with the *Clinton's email story* appeared much earlier in G_c compared to G_a. This indicates that information mined from comments

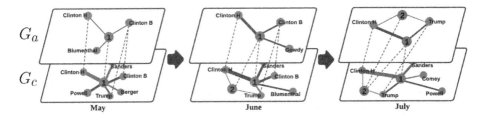

Fig. 1. Monthly representation of a two-layer graph of Washington Post articles and comments during the 2016 US election over three months (May - July 2016). Gray node 1 represents the topic "Hillary Clinton's email story", and gray node 2 represents the topic "2016 elections debates". Orange nodes represent associated entities in articles and comments layers $(G_a, G_c) \in WP$. Blue and red edges represent the intra-layer connections, while green edges represent the inter-layer connections. The width of the link represents the weight, red edges represent a negative sentiment, while blue edges represent positive sentiment. (Color figure online)

complements that mined from articles; we capture even richer (latent) information if we mine concomitantly articles and comments from multiple sources.

MultiLayerET can be applied to other areas of study besides news like blogs and their comments and research papers discussed on social media platforms. In this work, our focus is on online news and its application. We apply our system to six English online news sources, Washington Post (WP), Cable News Network (CNN), Wall Street Journal (WSJ), British Broadcasting Corporation (BBC), Fox News (FN), and Daily Mail (DM). The number of articles and comments varies across sources, resulting in graphs with a different number of nodes per source. The number of nodes varies between $4K$ and $200K$, producing small to extensive graphs. For this study, we focus on entities and topics related to politics, and we consider people entities and associated topics. We make the following contribution in this work:

1. We introduce MultiLayerET, a system that represents entities and topics in articles and comments as a heterogeneous attributed multilayer graph. Such representation assists in highlighting significant events and their associated entities; it enhances the analysis of emerging events reported in news streams.
2. We analyze the topological graph structure to show the unique information encoded within the network.
3. We show that MultiLayerET improves downstream applications, e.g., media bias classification by 36% and fake news detection by 4%.
4. We build a dictionary of over $1M$ people entities; the dictionary contains useful data such as aliases crucial to downstream applications.

2   Related Work

Here we give an overview of the related work to properly position MultiLayerET within the literature on topic modeling and entity extraction from articles and their comments.

Comments carry valuable information concerning public opinion [46–48,52, 53]. Utilizing user comments can enhance the performance of many models for downstream tasks, such as, fake news detection [26,32,49,50], news popularity prediction [21,23,51], media bias [22,25,45], news recommendations [37–39], and news summarization [41].

Latent Dirichlet Allocation (LDA) [1] is a hierarchical Bayesian model that generates probabilities of corpora for a given document. LDA is the Bayesian version of the Probabilistic Latent Semantic Analysis (PLSA) model [13]. It is considered the foundation of many other models such as Topic-link LDA [12], Labeled LDA [11], and Spatial LDA [14]. LDA is utilized in many applications (e.g., information retrieval, topics overlapping, and visualization) where the extraction of topics is needed. For a given set of news sources, a line of work aims to link entities, and topics using statistical [5,10] and other techniques [29–31,33,34]. Other studies, construct an entity-centric graph and the topic associated with it [6–9]. [16] uses an attribute proximity graph to mine events reported in the news.

Most works in news mining either work with news articles and ignore comments or with comments and ignore articles. There are a few exceptions. For example, [40,42] shows that comments combined with articles improve topic discovery and [18] shows that comments improve explainability in fake news detection. Our MultiLayerET approach creates a unified representation of articles and comments, giving a richer graph representation of topics discussed in the news and their associated entities. It is argued that entities are an essential component significantly affecting the comprehension level of a given document [15]. Therefore, entities are first-class citizens in MultiLayerET as the candidate documents to be mined are added to the graph according to an input set of entities. We extract topics from a set of documents, which assists in producing a coherent list of terms for each extracted topic. To our knowledge, our work is the first to propose a heterogeneous attributed multilayer graph to represent information in news streams.

3   MultiLayerET

This section defines the problem and the notations used in this paper. We also introduce our system MultiLayerET to represent entities and topics in articles and comments. The system pipeline is shown in Fig. 2.

3.1   Problem Formulation

Having a source $s \in S$ that consists of several articles and their comments $s_i = \{\langle a_1, (c_{11}, .., c_{1n}) \rangle, .., \langle a_n, (c_{n1}, .., c_{nm}) \rangle\}$, MultiLayerET's objective is to extract

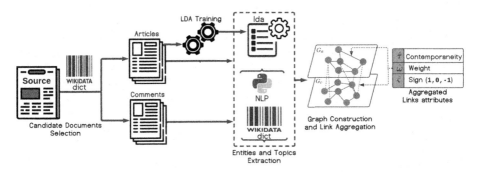

Fig. 2. MultiLayerET pipeline to represent entities and topics in articles and comments

(major) topics and entities mentioned in articles and comments. MultiLayerET produces an undirected heterogeneous attributed multilayer graph $G = \{G_a, G_c\}$ where G_a is the article graph and G_c is the comment graph. The nodes in the graph are sets of entities E and topics T, respectively, along with attributed edges contemporaneity $\hat{\tau}$, co-occurrence count $\hat{\omega}$, and sign $\hat{\varsigma}$. Table 1 describes the notations used in this paper.

Table 1. Notations used in this paper.

Symbol	Description	Symbol	Description
S	Set of sources	T	Set of topics $\in \{D_a, D_c\}$
s_i	The $i^{th} \in S$ where $i = 1..n$	t	Topic node $\in T$
A	Set of articles $\in s$	G_a	Articles graph
a_i	The ith article $\in A$	G_c	Comments graph
C	Set of comments section	G	Multilayer graph $\{G_a, G_c\}$
c_{ij}	The jth comment for an article a_i	τ	Contemporaneity, link attribute
D_a	Set of candidate articles $\in A$	ω	Weight, link attribute
D_c	Set of candidate comments $\in C$	ς	Sign, link attribute
$dict$	Entities dictionary	$\hat{\tau}$	Aggregated contemporaneity
lda	Best LDA model	$\hat{\omega}$	Aggregated weights
E	Set of entities in $D \in \{D_a, D_c\}$	$\hat{\varsigma}$	Aggregated signs
e	Entity node $\in E$		

3.2   Selection of Candidate Documents

Importing a large set of articles and comments is a challenge in itself. In this work, we assume one can access an extensive collection of documents. We focus on the challenge of gleaning knowledge from such data. An essential step in the MultiLayerET pipeline is collecting representative documents; this preliminary curation process helps eliminate noise, enabling gaining meaningful and interpretable information.

In order to fulfil this objective, we build an entity dictionary *dict* from *wikidata*[1]. The dictionary contains more than one million person entities along with their names, alias names, affiliations, descriptions, and URLs to their Wikidata pages. The list of aliases contains the most commonly misspelled names to capture better the varied ways entities are mentioned in articles and comments. In *dict*, entities are divided into six categories according to their current or most recent affiliation, such as politicians and officeholders, military figures, sports figures, musicians and actors, writers, and social media personalities.

For a specific source s, article a and comment c are added to D_a, the set of candidate articles, and D_c, the set of candidate comments, respectively, if any entity token of e is present in the entity dictionary *dict*. For example, we add the article "Bernie Sanders FINALLY unloaded on Hillary Clinton for not being 'qualified.' Here's why."[2] from WP and its subset of comments since the entities *Hillary Clinton* and *Bernie Sander* in our *dict* are present in that article. In this work, we focus on politics-related documents; therefore, we use the politicians and office holders *dict* to extract coherent topic terms.

3.3   Entity Extraction and Topic Mining

Once candidate documents D_a and D_c are obtained, we preprocess the data to construct the graph where $nodes = \{E, T\}$. In this section we describe how to extract entity nodes E and topic nodes T from D_a and D_c.

Entity Nodes: To obtain meaningful information from the selected articles and comments, we focus on entities present in our entity dictionary *dict*. We utilize it along with NLTK [44] and TextBlob[3] to extract entity name phrases. Entities may appear multiple times in the same D_a and D_c; therefore, we keep track of their frequencies.

Topic Nodes: To extract topics, we train an LDA [1] model for each D_a in s. The LDA model looks into the bag of words from D_a and returns a set of terms with their probabilities. Each probability represents how much each word contributes to that cluster. We evaluate the quality of the *lda* model by measuring 1) Coherence score [3], which is the semantic degree of similarity between high-scoring words in a given cluster (topic), and 2) Perplexity score; normalized log-likelihood. MultiLayerET chooses the best *lda* model according to the coherence value; a higher coherence value returns interpretable topics. If two LDA models have the same coherence score, we choose the one with the lower perplexity score. We only train *lda* on D_a, assuming that D_c will align with at least one of the extracted topics. We select the most relevant topic for a given article in D_a and comment D_c based on the topic probability given by the trained *lda* model.

---

[1] https://www.wikidata.org/.

[2] Full article: https://wapo.st/3yOMYdO.

[3] https://textblob.readthedocs.io/en/dev/.

Algorithm 1: Single-layer graph construction

1: **procedure** GC(D, lda, dict)
   /* Nodes Extraction */
2:   $E = \{token \in D \wedge token \in dict\}$
3:   $T = \{lda(d) \; where \; d \in D\}$
   /* Intra-layer Edges */
4:   $Initialize \; G \leftarrow [\;]$
   for $(e,t) \in \langle E,T \rangle$ **do**
      for $d \in D$ **do**
         if (e,t) *co-occur ind* **then**
            $\varsigma_d \leftarrow Majority(\varsigma_i)$
            $G[e,t] \leftarrow \langle \tau_d, \omega_d, \varsigma_d \rangle$

/* Link aggregation */
5: **for** $(e,t) \in G$ **do**
    $\hat{\tau}_d \leftarrow \cup_{\langle \tau_d, \omega_d, \varsigma_d \rangle \in G[e,t]} \tau_d$
    $\hat{\omega}_d \leftarrow \sum_{\langle \tau_d, \omega_d, \varsigma_d \rangle \in G[e,t]} \omega_d$
    $\hat{\varsigma}_d \leftarrow Majority(\varsigma_d)$
    $G[e,t] \leftarrow \langle \hat{\tau}_d, \hat{\omega}_d, \hat{\varsigma}_d \rangle$
6: **return** Single layer $\in G$

Algorithm 2: Multilayer graph construction

1: **procedure** MGC(D_a, D_c, lda, dict)
   /* Graph construction */
2:   $G_a \leftarrow$ GC(D_a, lda, dict)
3:   $G_c \leftarrow$ GC(D_c, lda, dict)
   /* Inter-layer Edges */
4:   $Initialize \; G \leftarrow [\;]$
   /* Entity-entity edges */

5: **for** $(e_a, e_c) \in \{G_a, G_c\}$ **do**
    if $(e_a = e_c) || (e_a, e_c) \in \{a_i, c_{ij}\}$
    then
       $G[e_a, e_c] \leftarrow \langle \hat{\tau}_d, \hat{\omega}_d \rangle$
    /* Topic-topic edges */
6: **for** $(t_a, t_c) \in \{G_a, G_c\}$ **do**
    if $(t_a = t_c) || (t_a, t_c) \in \{a_i, c_{ij}\})$
    then
       $G[t_a, t_c] \langle \hat{\tau}_d, \hat{\omega}_d \rangle$
7:   **return** G

3.4 Graph Construction

In our setting, G is a multilayer, undirected, weighted, attributed graph. The first layer in the graph G is G_a, which represents the article graph, and the second layer G_c is the comment graph. The nodes in our graph are heterogeneous; they consist of 1) a set of entities E and 2) a set of topics T that maps to one of the extracted topics. Some nodes may appear in G_c but not in G_a and vice-versa. This phenomenon depends on the commenters' behavior; they tend to discuss or leave out entities and topics that might be mentioned in the article. Algorithm 1 summarizes the graph construction for a single layer in G, and Algorithm2 shows the process of constructing the multilayer graph G.

Edges Construction: Since we are dealing with a multilayer graph, we have two types of edges: intra-layer edges, which represent edges within the same layer, and inter-layer edges, which represent edges between the layers. **Intra-layer edges** are (e, t), which links $e \in E$ with $t \in T$. This link captures the relationship between (e, t) pairs in news articles and comments. The **inter-layer edges** are *entity-entity links* (e_a, e_c), which link entities together; this link gives an intuition of how entities are connected between articles and their comments. Another type of inter-layer edge is *topic-topic* (t_a, t_c), which links topics together; this helps in projecting topics in G and analyzing the level of relevancy between events in news articles and comments. Edges are formed between nodes v and w if they co-occur in a single document d.

Edge Attributes and Aggregation: Once a link is formed between a pair of nodes, we compute its attributes ω, τ, and ς. The first attribute is ω, representing the link weight. For two nodes v and w, we compute the co-occurrence frequency in a single document D, where D can be an article or comment. This attribute represents the connection strength. Second, τ contemporaneity, is a concatenated list of all publish times where node v and w co-occurred. This attribute assists in understanding the temporal evolution of pair of nodes. Finally, ς is the sign attribute representing the text sentiment between a pair of nodes co-occurring in d. This attribute is different as it is only found in intra-layer edges *(e,t)*. The value of ς is 1 if the sentiment of d is positive, -1 if it is negative, and 0 if it is neutral. We calculate the aggregated set $\langle \hat{\tau}, \hat{\omega}, \hat{\varsigma} \rangle$ of edges as follows, $\hat{\tau}$ is a concatenation of all τ for a given pair of nodes. $\hat{\omega}$ is the sum of all ω for a given pair of nodes, and finally, $\hat{\varsigma}$ is the majority vote of all ς for a given pair of nodes.

3.5   Graph Construction Complexity

The graph construction complexity of Algorithm 2 is $\mathcal{O}(|S| \cdot |T| \cdot (|D_a|^2 \cdot max_{len}(D_a)^2 + |D_c|^2 \cdot max_{len}(D_c)^2)$. Here, $|S|$ is the number of sources, $|T|$ is the number of unique topics, $|D_a|$ is the total number of the selected candidate articles, $|D_c|$ is the total number of selected candidate comments, $max_{len}(D_a)$ is the maximum number of tokens calculated over all selected candidate articles, and $max_{len}(D_c)$ is the maximum number of tokens calculated over all selected candidate comment sections. The graph construction is quadratic, where $|D_a|$ and $|D_c|$ play larger roles in controlling complexity compared to $|S|$ and $|T|$. This indicates that construction runtime grows gracefully with the number of articles and comments.

4   Graph Analysis

Here, we describe the dataset and analyze the topological structure of the multilayer graph G.

4.1   Data

News articles and comments were collected from Google News [2] between 2015 and 2017; the database contains over one million articles and 33 million comments from 22 thousand difference English and Spanish news sources. For this

Table 2. Candidate documents D statistics for each source, showing the total number of articles and comments used in the study.

Dataset	WP	DM	FN	CNN	WSJ	BBC
No. articles	$36K$	$14K$	$9K$	$4K$	$2K$	101
No. comments	$290K$	$90K$	$410K$	$14K$	$69K$	$105K$

study, we draw six English news sources, Washington Post (WP), Cable News Network (CNN), Wall Street Journal (WSJ), British Broadcasting Corporation (BBC), Fox News (FN), and Daily Mail (DM). We selected all articles and comments for this study between January 2016 and July 2016. Table 2 shows the datasets statics.

4.2 Topological Graph Structure Analysis

We selected sources where the number of nodes and edges varies across sources, leading to the construction of small to extensive graphs as shown in Table 3. In addition, the size of the graph varies between layers; for example, CNN and WSJ have a similar number of nodes in G_a. However, WSJ has a 3 times larger number of nodes in G_c compared to CNN. This phenomenon will aid in better understating the structural differences across sources.

Table 3. Topological structure properties for each layer G_a and G_c, and multilayer graph G. N_a = Number of nodes in article layer, N_c = Number of nodes in comments layer, E_a= Number of edges in article layer, E_c= Number of edges in comments layer, avg N_d = Average node degree for each layer G_a and G_c, $Diameter$= Diameter of the layer largest component in each layer G_a and G_c, $Inter$= The number of multilayer graph inter-layer edges, avg C = Multilayer average clustering coefficient, and r = Multilayer assortativity coefficient.

Dataset	N_a	N_c	E_a	E_c	Avg N_d		Diameter		Inter	Avg C	r
					G_a	G_c	G_a	G_c			
WP	1.2K	2K	3.6K	18K	3.50	5.01	7	6	200K	0.82	-0.34
DM	1K	1.3K	1.3K	2.8K	2.53	4.15	6	5	56K	0.81	-0.33
FN	515	1.7K	3K	10K	2.96	4.81	6	4	115K	0.80	-0.35
WSJ	308	2K	475	6K	3.08	4.98	6	4	149K	0.83	-0.34
CNN	297	534	447	10K	3.01	3.76	6	4	16K	0.82	-0.35
BBC	59	176	75	436	2.54	4.95	7	5	4K	0.81	-0.19

We analyze each layer, the multilayer graph structure, and edge sentiment to understand the topological graph structure. In terms of topological structure, we consider the following properties: *1) Average Node Degree*: that helps understand the connectivity differences between layers in G. *2) Diameter*: that shows the graph connectivity, which indicates how many steps we need to take to traverse the graph; we calculated the diameter of the largest component in each layer. *3) avg C*: which measures how well the nodes tend to form clusters [4]; a value close to 1 means that nodes have a high tendency to form clusters, while a value near to 0 means otherwise. *4) Assortativity Coefficient*: this property indicates the tendency of nodes to be connected, whether they have the same degree magnitude, large, or low-degree. Assortativity is calculated as the Pearson correlation coefficient of nodes at either side of an edge. The assortativity value ranges from -1 to $+1$; positive values mean that nodes of similar degrees connect, while negative values mean that large-degree nodes tend to attach to low-degree nodes. The degree sequence of the graph heavily influences the measure. Finally, *5) Edge Sentiment Distribution* which is the edge sign $\hat{\varsigma}$, gives the stance of an entity toward a topic.

Across Sources Analysis: Table 3 shows that the largest graph is *WP* with more than $1K$ nodes and $3K$ intra-layer edges in G_a and $2K$ nodes and $18K$ intra-layer edges in G_c, respectively. The smallest graph is *BBC* with 59 nodes and around 176 intra-layer edges in G_a, and around 75 nodes and 436 intra-layer edges in G_c. The average N_d indicates that the nodes tend to be connected similarly across sources, which suggests that the size of the graph does not affect the connectivity. The diameter for the largest component across sources is similar, and we can see that the diameter of G_a is larger than that of G_c.

Multilayer Graph Analysis: Comparing G_a and G_c together, we can see that G_c is always larger than G_a. This indicates that users tend to mention entities and discuss topics that are not present in articles. In other words, users tend to drift into topics unrelated to that of the article, but still mentioning entities present in the article along with new ones. The number of inter-layer edges is much greater than the number of intra-layer edges resulting of high average clustering coefficient across all outlets. Although *BCC* is the smallest graph, it has the highest density among all sources, which suggests that *BCC*'s topics and entities are highly connected compared to other sources. The analysis of assortativity coefficient shows that nodes with high degree tend to be connected with nodes with smaller degree; this is a sign of the existence of hubs.

Signed Edge Analysis: Signed edges give an intuition of the sentiment difference between articles and comments. In G_a, we observe that the number of positive and neutral edges represent 78%–85% of the total signed edges in all sources. The percentages indicate that most entities have positive or neutral sentiments towards a particular topic. In G_c, neutral edges represent between 6%–10%, while the percentages of both positive and negative edges are between 90%–94%. This phenomenon indicates that even though sources may have some inherent (political) bias, the articles are written in a way that their text conveys positive or neutral sentiment. Users, however, tend to express more polarized opinions, which explains the low percentage of neutral edges.

5   System Evaluation

In this section, we present the added benefit of MultiLayerET on two downstream applications, media bias classification, and fake news detection. We compare methods that only utilize the textual representation against the same methods when combined with our MultiLayerET system.

5.1   Experimental Setup

We pre-process the text by removing stop words, punctuation, and digits. To obtain the base of the words, we utilize NLTK [44] to perform lemmatization, which removes the conjugation ending of the word. Our comparison models are: 1) Doc2Vec [43], which is trained to predict words in the text; it produces a dense vector for a given text. We train Doc2Vec for 100 epochs to produce 300 dimension vectors for articles and comments. 2) BERT [27], which generates an expressive feature embedding for a given text using a self-attention mechanism and bidirectional cross attention. We utilize $BERT_{base}$ to get the text feature representation. Then we feed the feature vectors to a feed-forward network with a softmax function to get the predictions.

In MultilayerET, we use Node2Vec [17] to obtain the graph feature representation. Node2Vec maps the graph nodes to low-dimensional vectors while preserving the graph structure. We train Node2Vec and produce a vector feature representation of 100 dimensions. Then we concatenate the graph features with text features obtained with BERT. Once we obtain the feature vectors, we feed them to a feed-forward network with a softmax function to get the predictions. We investigate the performance of MultiLayerET by analyzing the performance of each layer of MultiLayerET. We run experiments 1) using only the graph representation of articles layer G_a along with the representation of the article obtained by BERT, 2) using the graph representation of comments layer G_c along with the comments representation obtained by BERT, and 3) using the multilayer representation of MultiLayerET along with the articles and comments representation obtained by BERT.

In both applications, we repeat experiments five times and test on a different fold not used for training. The dataset split is 80:20 ratios for training, and testing, respectively. We report the average accuracy, precision, recall, F1 score, and standard deviation.

5.2   Media Bias Classification

To perform this experiment, we selected articles and comments from six news sources mentioned in Sect. 4.1 that discuss well-known politicians from different political parties, such as Donald Trump and Hillary Clinton. We label all examples from each source as left, right, or center based on their media bias rank given by AllSides[4]. To get a balanced dataset, we randomly selected around $1K$

---

[4] https://www.allsides.com/media-bias/media-bias-ratings.

Table 4. Performance on MediaBias. The first row is the input representation; G_a and G_c use only the articles or the comments layer, respectively. MultiLayerET means using the representation of the multilayer graph. We report the average scores for each metric and standard deviation

Representation	Doc2Vec	BERT	G_a	G_c	MultiLayerET
Accuracy	0.547 (0.004)	0.687 (0.006)	0.785 (0.009)	0.806 (0.009)	**0.912 (0.005)**
Precision	0.525 (0.009)	0.656 (0.009)	0.736 (0.008)	0.791 (0.009)	**0.880 (0.007)**
Recall	0.538 (0.008)	0.669 (0.007)	0.739 (0.008)	0.795 (0.006)	**0.894 (0.003)**
F1 score	0.545 (0.008)	0.686 (0.007)	0.774 (0.007)	0.797 (0.006)	**0.901 (0.003)**

article and $5K$ comments from each outlet except for BBC, where we have only 101 articles. This result in having a smaller sample of the center class. The total number of articles and comments is about $6K$ and $30K$ respectively, and the proportion of classes are: left $= 34\%$, center $= 10\%$, and right $= 56\%$.

The worst performance was obtained using only Doc2Vec representation, where the accuracy is 54% (Table 4). Using any part of MultiLayerET enhanced the model performance from 21% to 37%. In sign edge analysis (Sect. 4.2) we show that journalists tend to write articles mostly with positive or neutral sentiments, which makes it harder to understand the hidden bias in online news from the textural representation itself. We also observe that using the G_a or G_c already enhances the prediction results across all metrics. From the last column of Table 4, one notices that the best accuracy and F1 score are obtained when the full MultiLayerET system is used. This supports our hypothesis that the comment section is important in increasing the prediction accuracy in downstream tasks.

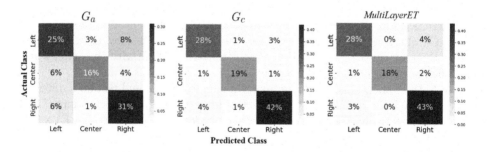

Fig. 3. Confusion matrix on media bias classification for proposed methods

To better understand the advantage of using MultiLayerET compared to separate layers G_a and G_c, we plot the confusion matrix for each of these experiments as shown in Fig. 3. MultiLayerET performs the best in predicting each class. The left class with the highest error rate in all cases is the most challenging class to predict. MultiLayerET and G_c have a lower error rate in predicting the center class than G_a. Looking at the misclassified examples, when MultiLayerET is not sure about the prediction, it consistently predicts the right class and avoids predicting the center class. On the contrary, G_a and G_c randomly assign a class to miss classified examples. This observation indicated that MultiLayerET learns better in imbalanced data cases compared to G_a and G_c.

5.3   Fake News Detection

To evaluate MultiLayerET on Fake News Detection we utilize the benchmark dataset FakeNewsNet [19,20]. We focus on political news retrieved from *Politi-Fact* articles and comments (tweets) and use it to perform a binary classification, where texts are labeled as fake and real. *PolitiFact* consists of 415 news articles with around $89K$ comments, where 35% are real news and 65% are fake. We compare MultiLayerET to dEFEND [18] since it utilizes both articles and comments. It reports the best results in FakeNewsNet compared to alternative models for fake news detection.

Table 5 shows that our MultiLayerET-based method for fake news detection outperforms dEFEND's reported accuracy between 1%–6% in F1 score. The results highlight the importance of utilizing the interaction between entities and topics in this classification task. We note that when we use G_a alone, our prediction performs similarly to that of dEFEND; dEFEND outperforms our prediction model when we use G_c alone.

Table 5. Performance on fake news detection. The first row is the input representation; G_a and G_c mean that only articles and comment layers are utilized, respectively. MultiLayerET means that the entire multilayer graph is used. We report the average scores for each metric and standard deviation. dEFEND results are reported as in [18]; we did not report dEFEND standard deviation since we do not have access to the results.

Model	dEFEND	G_a	G_c	MultiLayerET
Accuracy	0.904	0.939 (0.007)	0.895 (0.008)	**0.972 (0.003)**
Precision	0.902	0.919 (0.009)	0.878 (0.007)	**0.942 (0.003)**
Recall	0.956	0.919 (0.009)	0.880 (0.007)	**0.959 (0.003)**
F1 score	0.928	0.929 (0.009)	0.892 (0.008)	**0.960 (0.004)**

The problem appears to be complex; the articles carry latent information that is useful to distinguish between real and fake news compared to comments. We should mention that dEFEND is an explainable fake news detection model that indicates the article sentences and comments that lead to a specific prediction. However, in this work, we focus on the performance of the models; we believe that appending MultiLayerET to dEFEND will enhances dEFEND's performance while maintaining its explainability power. We leave this for future work.

6 Conclusion

We propose a novel system, MultiLayerET, to create a unified representation of entities and topics in online news using multilayer graphs. The layers of the graph are the articles and comments, respectively. This study is the first to consider the comments representing topics and entities and analyze them from a multilayer perspective. Our proposed system encodes novel interactions between articles and comments, which proves beneficial to downstream tasks. MultiLayerET is not limited to online news articles and their comments; it can be applied to many areas such as blogs and their comments, research papers, and discussion such as in Twitter. To characterize the capabilities of our proposed system on real applications, we provided a detailed analysis of MultiLayerET on six representative online news sources. We showed how MultiLayerET assisted in highlighting significant events and their associated entities to better understand and extract information from large-scale online news. We applied MultiLayerET to two downstream tasks. The results obtained on the media bias classification showed that MultiLayerET enhanced the textual representation and helped in better understanding the bias across sources. We also showed that MultiLayerET outperforms a state-of-the-art fake news detection model that considers both articles and comments. In the future, we will focus on expanding the dictionary of entities to include organizations, locations, subjects, materials, and other entities. We also plan to study the multilayered graph from a temporal aspect.

Acknowledgements. This research was supported in part by the U.S. NSF awards 2026513 and 1838145, and the ARL subaward 555080-78055 under Prime Contract No. W911NF2220001 and Temple University office of the Vice President for Research 2022 Catalytic Collaborative Research Initiative Program. AI & ML Focus Area. In addition, this research includes calculations carried out on HPC resources supported in part by the U.S. NSF through major research instrumentation grant number 1625061 and by the U.S. Army Research Laboratory under contract number W911NF-16-2-0189.

References

1. Blei, D., Ng, A., Jordan, M.: Latent Dirichlet allocation. J. Mach. Learn. Res. **3**, 993–1022 (2003)
2. He, L., Han, C., Mukherjee, A., Obradovic, Z., Dragut, E.: On the dynamics of user engagement in news comment media. Wiley Interdiscip. Rev. Data Min. Knowl. Discov. **10**, e1342 (2020)
3. Röder, M., Both, A., Hinneburg, A.: Exploring the space of topic coherence measures. In:WSDM (2015)
4. Watts, D., Strogatz, S.: Collective dynamics of 'small-world' networks. Nature **393**, 440–442 (1998)
5. Newman, D., Chemudugunta, C., Smyth, P., Steyvers, M.: Analyzing entities and topics in news articles using statistical topic models. In: Mehrotra, S., Zeng, D.D., Chen, H., Thuraisingham, B., Wang, F.-Y. (eds.) ISI 2006. LNCS, vol. 3975, pp. 93–104. Springer, Heidelberg (2006). https://doi.org/10.1007/11760146_9

6. Spitz, A., Gertz, M.: Exploring entity-centric networks in entangled news streams. In: TheWebConf (2018)

7. Spitz, A., Gertz, M.: Entity-centric topic extraction and exploration: a network-based approach. In: Pasi, G., Piwowarski, B., Azzopardi, L., Hanbury, A. (eds.) ECIR 2018. LNCS, vol. 10772, pp. 3–15. Springer, Cham (2018). https://doi.org/10.1007/978-3-319-76941-7_1

8. Spitz, A., Almasian, S., Gertz, M.: Entity-centric network topic exploration in news streams. In: WSDM (2019)

9. Wu, C., Kanoulas, E., Rijke, M.: Learning entity-centric document representations using an entity facet topic model. Inf. Process. Manage. **57**, 102216 (2020)

10. Kim, H., Sun, Y., Hockenmaier, J., Han, J.: ETM: entity topic models for mining documents associated with entities. In: ICDM (2012)

11. Ramage, D., Hall, D., Nallapati, R., Manning, C.: Labeled LDA: a supervised topic model for credit attribution in multi-labeled corpora. EMNLP (2009)

12. Liu, Y., Niculescu-Mizil, A., Gryc, W.: Topic-link LDA: joint models of topic and author community. In: ICML (2009)

13. Hofmann, T.: Probabilistic latent semantic analysis. In: UAI (1999)

14. Wang, X., Grimson, E.: Spatial latent dirichlet allocation. In: NeurIPS, vol. 20 (2008)

15. Wu, C., Kanoulas, E., Rijke, M.: It all starts with entities: a salient entity topic model. Nat. Lang. Eng. **26**, 531–549 (2020)

16. Kim, H., El-Kishky, A., Ren, X., Han, J.: Mining news events from comparable news corpora: a multi-attribute proximity network modeling approach. In: IEEE BigData (2019)

17. Grover, A., Leskovec, J.: Node2vec: scalable feature learning for networks. In: SIGKDD (2016)

18. Shu, K., Cui, L., Wang, S., Lee, D., Liu, H.: DEFEND: explainable fake news detection. In: SIGKDD (2019)

19. Shu, K., Sliva, A., Wang, S., Tang, J., Liu, H.: Fake news detection on social media: a data mining perspective. ACM SIGKDD. **19**, 22–36 (2017)

20. Shu, K., Mahudeswaran, D., Wang, S., Lee, D., Liu, H.: FakeNewsNet: a data repository with news content, social context, and spatiotemporal information for studying fake news on social media. Big Data **8**, 171–188 (2020)

21. Tatar, A., Leguay, J., Antoniadis, P., Limbourg, A., Amorim, M., Fdida, S.: Predicting the popularity of online articles based on user comments. In: WIMS (2011)

22. Yigit-Sert, S., Altingovde, I., Ulusoy, Ö.: Towards detecting media bias by utilizing user comments. In: WebSci (2016)

23. Rizos, G., Papadopoulos, S., Kompatsiaris, Y.: Predicting news popularity by mining online discussions. In: The Web Conference (2016)

24. Tsagkias, M., Weerkamp, W., de Rijke, M.: News comments: exploring, modeling, and online prediction. In: Gurrin, C., et al. (eds.) ECIR 2010. LNCS, vol. 5993, pp. 191–203. Springer, Heidelberg (2010). https://doi.org/10.1007/978-3-642-12275-0_19

25. Lee, E.: That's not the way it is: how user-generated comments on the news affect perceived media bias. J. Comput.-Mediat. Comm. **18**, 32–45 (2012)

26. Yanagi, Y., Orihara, R., Sei, Y., Tahara, Y., Ohsuga, A.: Fake news detection with generated comments for news articles. In: INES (2020)

27. Devlin, J., Chang, M., Lee, K., Toutanova, K.: BERT: pre-training of deep bidirectional transformers for language understanding. In: NAACL (2019)

28. Reimers, N., Gurevych, I.: Sentence-BERT: sentence embeddings using siamese BERT-networks. In: EMNLP (2019)

29. Leban, G., Fortuna, B., Brank, J., Grobelnik, M.: Event registry: learning about world events from news. In: TheWebConference (2014)
30. Watanabe, K., Ochi, M., Okabe, M., Onai, R.: Jasmine: a real-time local-event detection system based on geolocation information propagated to microblogs. In: CIKM (2011)
31. Sankaranarayanan, J., Samet, H., Teitler, B., Lieberman, M., Sperling, J.: TwitterStand: news in tweets. In: GIS (2009)
32. Panagiotou, N., Saravanou, A., Gunopulos, D.: News monitor: a framework for exploring news in real-time. Data 7, 3 (2022)
33. Saravanou, A., Stefanoni, G., Meij, E.: Identifying notable news stories. In: Jose, J.M., et al. (eds.) ECIR 2020. LNCS, vol. 12036, pp. 352–358. Springer, Cham (2020). https://doi.org/10.1007/978-3-030-45442-5_44
34. Mathioudakis, M., Koudas, N.: TwitterMonitor: trend detection over the twitter stream. In: SIGMOD (2010)
35. Syed, M., et al.: Unified representation of twitter and online news using graph and entities. Front. Big Data 4, 699070 (2021)
36. Barabási, A.: Network science. Philos. Trans. R. Soc. A Math. Phys. Eng. Sci. 371, 20120375 (2013)
37. Trevisiol, M., Aiello, L., Schifanella, R., Jaimes, A.: Cold-start news recommendation with domain-dependent browse graph. In: RecSys (2014)
38. Bach, N., Hai, N., Phuong, T.: Personalized recommendation of stories for commenting in forum-based social media. Inf. Sci. 352–353 (2016)
39. Li, Q., Wang, J., Chen, Y., Lin, Z.: User comments for news recommendation in forum-based social media. Inf. Sci. 180, 4929–4939 (2010)
40. Guo, W., Li, H., Ji, H., Diab, M.: Linking tweets to news: a framework to enrich short text data in social media. In: ACL (2013)
41. Wei, Z., Gao, W.: Gibberish, assistant, or master? Using tweets linking to news for extractive single-document summarization. In: SIGIR (2015)
42. Li, M., et al.: EKNOT: event Knowledge from news and opinions in Twitter. In: AAAI (2016)
43. Le, Q., Mikolov, T.: Distributed representations of sentences and documents. In: ICML, vol. 32 (2014)
44. Loper, E., Bird, S.: NLTK: the natural language toolkit. In: ACL (2004)
45. Stanojevic, M., Alshehri, J., Dragut, E., Obradovic, Z.: Biased news data influence on classifying social media posts. In:sIR@ SIGIR (2019)
46. Stanojevic, M., Alshehri, J., Obradovic, Z.: Surveying public opinion using label prediction on social media data. In: ASONAM (2019)
47. Alshehri, J., Stanojevic, M., Dragut, E., Obradovic, Z.: Stay on topic, please: aligning user comments to the content of a news article. In: Hiemstra, D., Moens, M.-F., Mothe, J., Perego, R., Potthast, M., Sebastiani, F. (eds.) ECIR 2021. LNCS, vol. 12656, pp. 3–17. Springer, Cham (2021). https://doi.org/10.1007/978-3-030-72113-8_1
48. Yang, F., Dragut, E., Mukherjee, A.: Predicting personal opinion on future events with fingerprints. In: COLING (2020)
49. Yang, F., Dragut, E., Mukherjee, A.: Claim verification under positive unlabeled learning. In: ASONAM (2020)
50. Yang, F., Dragut, E., Mukherjee, A.: Improving evidence retrieval with claim-evidence entailment. In: RANLP (2021)
51. He, L., Shen, C., Mukherjee, A., Vucetic, S., Dragut, E.: Cannot Predict comment volume of a news article before (a few) users read it. In: ICWSM (2021)

52. Hosseinia, M., Dragut, E., Boumber, D., Mukherjee, A.: On the usefulness of personality traits in opinion-oriented tasks. In: RANLP (2021)
53. Tumarada, K., Zhang, Y., Yang, F., Dragut, E., Gnawali, O., Mukherjee, A.: Opinion prediction with user fingerprinting. arXiv (2021)

Conversational Systems

MFDG: A Multi-Factor Dialogue Graph Model for Dialogue Intent Classification

Jinhui Pang[1(✉)], Huinan Xu[1], Shuangyong Song[2], Bo Zou[2], and Xiaodong He[2]

[1] Beijing Institute of Technology, Beijing 100081, China
{pangjinhui,xuhuinan}@bit.edu.cn
[2] JD AI Research, Beijing 100176, China
{songshuangyong,cdzoubo,hexiaodong}@jd.com

Abstract. Interest in speaker intent classification has been increasing in multi-turn dialogues, as the intention of a speaker is one of the components for dialogue understanding. While most existing methods perform speaker intent classification at utterance-level, the dialogue-level comprehension is ignored. To obtain a full understanding of dialogues, we propose a **M**ulti-**F**actor **D**ialogue **G**raph Model (MFDG) for Dialogue Core Intent (DCI) classification. The model gains an understanding of the entire dialogue by explicitly modeling multi factors that are essential for speaker-specific and contextual information extraction across the dialogue. The main module of MFDG is a heterogeneous graph encoder, where speakers, local discourses, and utterances are modelled in a graph interaction manner. Based on the framework of MFDG, we propose two variants, MFDG-EN and MFDG-EE, to fuse domain knowledge into the dialogue graph. We apply MFDG and its two variants to a real-world online customer service dialogue system on the e-commerce website, JD, in which the MFDG can help achieving an automatic intent-oriented classification of finished service dialogues, and the MFDG-EE can further promote dialogue comprehension with a well-designed knowledge graph. Experiments on this in-house JD dataset and a public DailyDialog dataset demonstrate that MFDG performs reasonably well in multi-turn dialogue classification.

Keywords: Dialogue classification · Core intent classification · Graph neural network

1 Introduction

There are increasing number of internet firms and platforms providing online customer services, thus creating lots of available multi-turn dialogues between customer service staffs and customers, which could be explored further for enhancing the user experience and satisfaction. Especially, the ability to recognize speakers' intentions, which is officially called Dialogue Intent (DI) classification [24,25], is essential to perceive the customers' requests across the dialogue. Most of works

M.-R. Amini et al. (Eds.): ECML PKDD 2022, LNAI 13714, pp. 691–706, 2023.
https://doi.org/10.1007/978-3-031-26390-3_40

focus on the utterance-level DI classification, ignoring the dialogue-level comprehension. To promote the full understanding of dialogues, we bring forward a task, Dialogue Core Intent (DCI) classification, aiming to infer the core intention of the entire dialogue, such as refund promoting, product consultation, service complaint and etc. Early works regarded multi-turn dialogues as ordinary texts [1]. They simply concatenated the utterances in the dialogue, preventing them from learning the dialogue-level contextual dependency among utterances.

Dialogues have their own specific characteristics. As an example shown in Table 1, the speakers of a dialogue talk in a random order, breaking up the continuity of adjacent utterances in the dialogue. Moreover, topic transitions are common in human-human dialogues, which brings a new challenge of modeling the dependency among remote but interrelated utterances. The key point to address dialogue classification is adopting speaker-specific and contextual modeling [2].

Firstly, the speaker-specific dialogue modeling considers speaker information contained in the dialogue. It consists of two aspects: intra and inter speaker dependency. Intra-speaker dependency is used to reflect the affect that speakers have on themselves, which can contribute to the understanding of individual speakers. Inter-dependency implies the dynamic interactions among speakers. Modeling intra and inter speaker dependency in dialogues relies on plenty of factors, such as topic, speakers' personality and viewpoint [2]. Secondly, the contextual information coming from both neighbouring and distant utterances is indispensable for dialogue understanding. While the importance of neighbouring utterances is generally considered, it should be stressed that distant utterances can sometimes offer supplementary information when speakers refer to the same word that appears at former utterances.

In term of the above two points, DialogueGCN was proposed in [3], which built a directed graph to model both speaker dependency and contextual information in the dialogue. Later, other works inherited the graph modeling pattern and introduced discourse relations [4,5], position encoding [6] to the dialogue graph for enhancing the understanding of utterances in the dialogue.

It reminds us to acquire the comprehension of the dialogue based on a dialogue graph. Besides, considering the lack of additional factors' annotation, we focus on the very nature of multi-turn dialogues and build a multi factor dialogue graph. Not like prior methods using edges to inject speaker dependency, we explicitly define speaker nodes to represent the contextual information of speakers in the dialogue.

Moreover, we find the consecutive utterances spoken by the same speaker are generally highly correlated and supplementary to each other. A real example is shown in Table 1. The customer speaks U1, U2 and U3 continuously to explain the problem he (she) faces, thus it's helpful to integrate them to know the background information of the customer. We add local discourse nodes to aggregate such consecutive utterances for generating a dialogue representation later. The multi factor dialogue graph we build has three types of nodes, namely utterance nodes, speaker nodes and local discourse nodes. And the graph has five different

Table 1. An example dialog from JD dataset. The core intent label of this dialogue is 'refund urging'. Bold font denotes the pre-defined entities coming from a well-designed knowledge graph for JD dataset.

	Speaker	Utterance
U1	Customer	Hello? I can not reach the merchant
U2	Customer	I bought some bread with a shelf life of a week, and it has been 4 days after I ordered
U3	Customer	I haven't receive the bread, but it is probably expired at that time
U4	Staff	We are sorry for our neglect. We will connect the merchant right now
U5	Customer	I demand for return
U6	Staff	You can apply for **Refund of unreceived goods** on the app
U7	Customer	I have tried. This needs the permission of the merchant and I can not reach him
U8	Staff	We can apply for **Order dispute** for you
U9	Staff	Then you need to provide some evidence, after that we help you with the refund
U10	Customer	All right

types of edges, i.e., speaker edges, local edges, local-speaker edges, utterance-order edges and local-order edges. By applying a Graph Convolution Network (GCN) to this graph, we propagate contextual information among multi factors and obtain a multi-factor representation of the dialogue.

To sum up, we propose the Multi-Factor Dialogue Graph model (MFDG) by explicitly modeling the relations among speakers, utterances and local discourses in dialogues. We believe that the representation contains richer information relevant to core intent than other graph-based and text classification models. The results are shown in Sect. 5.

Furthermore, we discover there exist entities that contain domain-specific knowledge in online customer service dialogues. As shown in Table 1, pre-defined entities 'Refund of unreceived goods' and 'Order dispute' always appear with the refund demand of the customer. It will be helpful to take advantage of such domain knowledge. We explore two ways to fuse the fine-grained entity information into our original model MFDG, namely MFDG-Entity Node (MFDG-EN) and MFDG-Entity Embedding (MFDG-EE). On the basis of MFDG, MFDG-EN adds entity nodes to the dialogue graph and considers the inclusion relations among utterances and entities, MFDG-EE combine the token-level and entity-level representations and generate knowledge-aware initial representations for utterance nodes.

In summary, our main contributions are as follows:

- We propose a novel model, MFDG, to infer the core intention of a multi-turn dialogue by obtaining a full understanding of the entire dialogue.
- We build a heterogeneous dialogue graph to model the interactions among multi factors in the dialogue. Especially, we create local discourse nodes to aggregate consecutive utterances spoken by the same speaker and add speaker nodes to explicitly capture the speaker information.

- Additionally, we propose two variants of MFDG to explore an appropriate way to fuse domain knowledge into the dialogue graph.

2   Related Work

In this section, we firstly introduce current deep learning models for text classification, as dialogue core intent classification is a specific type of text classification. Considering the lack of dialogue-level classification models, we then introduce related works for utterance-level dialogue classification from the following two perspectives: recurrence-based models and graph-based models.

- **Text Classification.** Deep learning models have achieved state-of-the-art results in many domains, including a wide variety of NLP applications. TextCNN [7] firstly migrated Convolutional Neural Networks (CNN) from computer vision to natural language processing. CNN makes use of convolution kernel to generate latent semantic features across the sentences, and performs much better than traditional feature-based text classification models. However, CNN does not take sequential information among sentences into consideration. As Recurrent Neural Network (RNN) is designed to recognize the sequential characteristics of data, it's a powerful model for text, string and sequential data classification [8]. Furthermore, an attention-based Long Short-Term Memory (LSTM) network [9] was proposed to dynamically integrate text information.
- **Recurrence-based Models.** Utterances of the dialogue are inherently sequential, then [10] proposed RNN and LSTM models for utterance intent classification task. DialogueRNN [11] used two Gate Recurrent Units (GRU) to track individual speaker states and global context across the dialogue. COSMIC [12] shared a similar network with DialogueRNN and incorporated different elements of commonsense to learn interactions between speakers participating in a dialogue.
- **Graph-based Models.** Many recent utterance-level dialogue classification models utilized graph-based neural networks to adopt speaker-specific and contextual modeling. DialogueGCN [3] firstly leveraged self and inter-speaker dependency of the speakers in a graph-based framework to model a conversational context, treating each dialogue as a graph where each utterance is connected to its surrounding utterances. Based on DialogueGCN, RGAT [15] added relational positional encodings that provide RGAT with sequential information implying in the dialogue. Besides, some other methods [4,5] draw support from pre-defined discourse relations between utterances. Lately, DAG [21] attempted to combine the advantages of recurrence-based models and graph-based models, which designs a directed acyclic graph to model the connections between nearby and distant utterances.

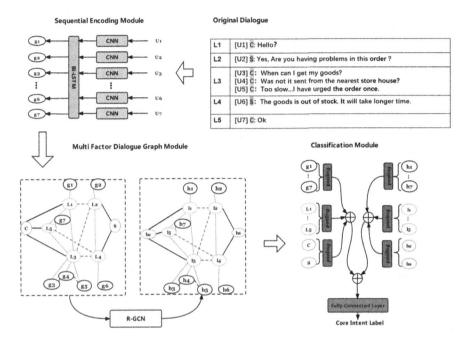

Fig. 1. The overall framework of MFDG. First, a sequential encoding module is used to obtain the initial representations of utterances in the dialogue. Then, in multi factor dialogue graph module we construct a dialogue graph consisting of three types of nodes and five types of edges. We utilize a graph convolutional network to update the nodes' features. Finally, three types of nodes are pooled and then concatenated together to be the dialogue representation which is fed to a fully connected layer for dialogue-level core intent classification.

3   Methodology

3.1   Problem Definition

In the following sections, let $D = \{U_1, U_2, ..., U_{N_u}\}$ be a dialogue with N_u utterances, and let there be N_s speakers $S = \{s_1, s_2, ..., s_{N_s}\}$ in dialogue D, where each utterance U_i is associated with the ID of its corresponding speaker by a mapping function $P(U_i)$. Given D, S and P, DCI attempts to predict the core intention label I of the dialogue.

3.2   Model

Now we present our Multi-Factor Dialogue Graph model (MFDG), which mainly consists of three modules as shown in Fig. 1.

- **Sequential Encoding Module.** This module is used to produce context-dependent representations for utterances without considering speaker-specific information, which will be used as the initial node features for the dialogue graph.

– **Multi Factor Dialogue Graph Module.** In this module, we organize the dialogue context as a heterogeneous graph. The detailed process of dialogue graph construction is below. Then a Relational Graph Convolutional Network (R-GCN) is applied to integrate contextual and speaker-specific information from multi factors in the dialogue graph.
– **Classification Module.** The last module predicts the core intention of a dialogue over the multi-factor involved dialogue representation.

Sequential Encoding Module. Firstly, we follow [7] to make use of a CNN to extract features for each utterance. We use a simple CNN with one layer of convolution followed by max-pooling and a fully connected layer to learn the representations for the utterances.

Then, in order to obtain inherent contextual information among utterances, we feed the output of CNN to a Bidirectional Long Short-Term Memory (Bi-LSTM). Let $H = \{g_1, g_2, ...g_{N_u}\}$ be the output of the former CNN, the output features of utterances through Bi-LSTM can be represented as:

$$u_i = \left[\overleftarrow{LSTM(g_i)}; \overrightarrow{LSTM(g_i)} \right] \tag{1}$$

for $i = 1, 2, ...N_u$, where u_i is the sequential contextual feature for utterance U_i. Then $u_1, u_2, ..., u_{N_u}$ are used to initialize the node features of the dialogue graph.

Multi Factor Dialogue Graph Module. In view of the characteristics of dialogues mentioned before, we explicitly model the interactions between utterances, speakers and local discourses. A heterogeneous graph with these three types of nodes is built to model the dialogue. Figure 2 is an example of dialogue graph for the original dialogue in Fig. 1.

Same as prior works, each utterance in a dialogue is viewed as a node to represent the information of each turn in this dialogue, and the number of utterance nodes in a dialogue graph is same as that of turns in the dialogue. Then, speaker nodes are added for obtaining speaker information. The number of speaker nodes is decided by the speakers involved in the dialogue. In the online customer service scenario, there are usually two speakers, staff and customer. Besides, local discourse nodes denote the aggregated information for the sets of local longest continuous utterances uttered by the same speaker.

The initial representations of utterance nodes are the outputs of sequential encoding module. In addition, each speaker node initializes itself by averaging the representations of utterance nodes uttered by this speaker. Similarly, the mean of the representations of local longest continuous utterance nodes is used as the initial representation of the corresponding local discourse node. The number of speaker and local discourse nodes is denoted as N_s, N_l, respectively.

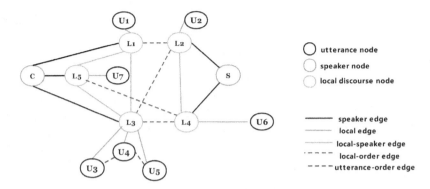

Fig. 2. Dialogue graph of the original dialogue in Fig. 1. For brevity, we omit the self-loop edges. We set both utterance-level and local discourse-level context window to [1, 1].

In this heterogeneous graph, we define several different types of edges to indicate different aspects of knowledge. Here is the introduction of edges in the dialogue graph.

- **speaker edge**: Each of the speaker nodes is connected to all of its spoken local discourse nodes with the speaker edge so that the speaker node can learn speaker information in the dialogue.
- **local edge**: To strengthen the connections among local continuous utterances, we create the local discourse node for each of the sets of local longest continuous utterance nodes and connect the local discourse node with every utterance nodes in the set by the local edge.
- **local-speaker edge**: Despite using speaker edges to explicitly include the speaker information, local discourse nodes spoken by the same speaker are fully connected with the local-speaker edge to inject the intra-speaker dependency into the graph.
- **utterance-order edge** and **local-order edge**: To obtain the contextual information that comes from both neighbouring and distant utterances, two types of edges are created to introduce utterance-level and local discourse-level contextual information, respectively. Each utterance is connected to its contextual utterances by the utterance-order edge, and we set a utterance-level context window $[p, q]$ so that each utterance node has an edge with its past p utterances and latter q utterances. Besides, it should be emphasized that an utterance node only has utterance-order edges with utterance nodes which connect to the same discourse node with it. Likewise, each local discourse node is connected to its contextual local discourse nodes by local-order edges with a local discourse-level window size $[m, n]$, which promotes the message passing among distant utterances.

Apart from above five types of edges, we also add self-loop edges for each node in the dialogue graph to facilitate effective computation. Therefore, there are totally six types of edges in the dialogue graph.

After acquiring the initial representation h_k for each node n_k and the edges among nodes, we feed the node features and the adjacent matrix into a graph neural network to obtain structural and semantic information of the dialogue. We apply R-GCN [14] to acquire the high-level hidden features with multi factors considered. The graph convolutional operation for node n_v at the $l+1$ layer can be defined as:

$$h_v^{(l+1)} = RELU \left(\sum_{r \in \mathcal{R}} \sum_{a \in N_r(v)} W_r^{(l)} h_a^{(l)} + b_r^{(l)} \right) \qquad (2)$$

where $\mathcal{R}$ denotes different types of edges, $N_r(v)$ is the set of one-hop neighbors of node n_v under edge r, $W_r^{(l)}$ and $b_r^{(l)}$ denote the edge-specific learnable parameters at the l-th layer. Furthermore, $h_k^{(0)} = h_k$, for $k = 1, 2, ..., N$, where $N = N_u + N_s + N_l$ denotes the total number of nodes in the dialogue graph.

In addition, it is a natural thought that different types of edges can not be treated equally. We make use of the gating mechanism when aggregating information from different relations [13]. The simple idea is to compute a coefficient between 0 and 1 for each relation:

$$c_v^{(l)} = Sigmod(h_v^{(l)} W_r^{(l)}) \qquad (3)$$

Therefore the message passing process for node n_v at the $l+1$ layer in the R-GCN can be overwrote as:

$$h_v^{(l+1)} = RELU \left(\sum_{r \in \mathcal{R}} \sum_{a \in N_r(v)} c_v^{(l)} W_r^{(l)} h_a^{(l)} + b_r^{(l)} \right) \qquad (4)$$

Classification Module. Finally, we concatenate the pooling results of output features of speaker nodes, utterance nodes and local discourse nodes at each GCN layer as hidden graph features. Here the pooling operation can be either max or mean pooling. Then, we concatenate the hidden graph features of all the GCN layers as the representation of the dialogue and makes the prediction using a fully-connected network.

3.3 Domain Knowledge Integration

Utterances of online customer service dialogues contain lots of domain-specific entities. An example is shown in Fig. 1, where the entities come from a well-designed knowledge graph for JD dataset. Here we design two approaches to take advantage of the fine-grained entity information based on MFDG, MFDG-EN and MFDG-EE. Both of them utilize pre-trained knowledge graph embedding so we firstly give a brief introduction to knowledge graph embedding and then detail the two variant models.

Knowledge Graph Embedding. Knowledge Graph (KG) is composed of triples in the form of *(head entity, relation, tail entity)*. Given all the triples in a KG, knowledge graph embedding aims to learn representation for each entity and relation that preserves structural information of the KG. There exist many translation-based knowledge graph embedding methods, such as TransE [17], TransH [18], TransR [19]. Considering those methods lack the ability of using the graph structures to enforce the local/global smoothness in the embedding spaces for entities and relations [20], we apply a simple R-GCN to acquire entity embedding from a pre-defined KG. Let us denote the pre-trained entity embedding as $[E_1, E_2, ..., E_j, ..., E_k]$, where K is the total number of entities of the KG and E_j is the generated embedding for entity e_j in the KG.

MFDG-EN. The first variant of MFDG is proposed by adding entity nodes to the dialogue graph, named as MFDG-Entity Node (MFDG-EN). That is, every individual entity appearing in a dialogue is treated as a entity node in the dialogue graph, and each utterance node is connected to entity nodes that contained in the corresponding utterances by entity-utterance edges. Besides, the entity embedding generated from above is used to initialize the entity node. In this way, utterances containing the same entities can be indirectly connected by two consecutive entity-utterance edges, which was designed to promote message passing of domain knowledge in the dialogue graph.

MFDG-EE. MFDG-Entity Embedding (MFDG-EE) leaves the dialogue graph unchanged, combining the token-level and entity-level representations and generating knowledge-aware initial representations for utterance nodes.

Here we use $U = t_{1:n} = [t_1, t_2, ..., t_n]$ to denote the raw sequence of an utterance in dialogue D, where n is the number of tokens in U. Then the token-level vectors for U can be obtained from a look-up word embedding table, which is denoted as $W = [w_1 w_2 ... w_n]$.

The entity-level vectors $E = [g_1 g_2 ... g_n]$ for U is generated as below:

$$g_i = \begin{cases} E_j, & \text{if } t_i \text{ is in the span of entity } e_j (j = 1, 2, ..., K) \\ \mathbf{0}, & \text{else} \end{cases}$$

Considering entity vectors are not in the same vector space with token vectors, we introduce a transformation function F for entity vectors:

$$F(E) = [F(g_1) F(g_2) ... F(g_n)], \tag{5}$$

where F can be either linear or non-linear mapping function.

Then we align and stack the token-level and entity-level embedding matrices as $M = [[w_1 F(g_1)][w_2 F(g_2)]...[w_n F(g_n)]]$. M will be fed into the sequential encoding module to compute knowledge-aware utterance representations. The rest of MFDG-EE is same as MFDG.

4 Experiment Setting

4.1 Datasets

We investigate several public dialogue datasets and find little information is available about dialogue-level labels. For this reason, we evaluate our MFDG model and its two variants on JD and DailyDialog datasets. The statistical information of them is shown in Table 2. Both the two datasets are composed of multi-turn dialogues where at least two speakers involve.

- **JD Dialogue dataset.** This dataset is supplied by the customer service department of JD. Dialogues in this dataset are produced when customers consult the online customer service staffs about a series of issues. Each dialogue consists of several utterances with speaker annotations. The dialogues are annotated with one of 50 core intent labels, which are carefully designed by experts to summarize the essential intention of the customer during conversation. The dataset has 20,000 samples of dialogues, with a total of 437,060 utterances. We use 18,000 dialogues for training, 1,000 for validation, and the remaining for test.
- **DailyDialog.** This dataset [22] reflects our daily communication way and covers various topics. Each dialogues in DailyDialog is annotated with one of the 10 certain topics, ranging from ordinary life to financial. It totally has 13,118 dialogues and 102,979 utterances. We use 11,118 dialogues for training, 1,000 for validation, and the remaining for test. Despite it does not have speaker annotations for utterances, we assume the utterances are spoken by two speakers one by one like previous works did.

Table 2. Statistical information of datasets. #Turn refers to the average number of utterances in a dialogue.

Dataset	#Dialogue	#Utterances	#Turn	#Class
JD dataset	20,000	473,060	23.65	50
DailyDialog	13,118	102,979	7.85	10

4.2 Evaluation Metrics

We adopt several widely used evaluation metrics, which are accuracy, H@3, H@5, macro-F1 and weighted-F1, to evaluate the performance of MFDG. Besides, we remove H@5 for DailyDialog, as there are just 10 classes in this dataset.

4.3 Baseline Methods

For the lack of dialogue-level classification model, we compare our model with several baseline methods for text classification, pre-trained models and some modified models of utterance-level classification models.

- **TextCNN** [7]. This is a convolutional neural network based model for sentence classification. To acquire the features for dialogue-level classification, we add a max pooling layer to aggregate the utterances in the dialogue.
- **TextRNN** [8]. In this method, a Bi-LSTM network is used to capture the contextual information from surrounding tokens in a text. We concatenate the utterances in a dialogue as an input of this model.
- **TextRNN-Att** [9]. This model uses a Bi-LSTM with attention mechanism to automatically focus on the most informative words in a text. Likewise, we concatenate the utterances in a dialogue as an input of this model.
- **BERT-base**[1], **Roberta-base**[2], **ERNIE**[3]. We use each of these three pre-trained models as an encoder for dialogues, following with a fully connected layer to acquire the dialogue-level labels.
- **Dialog-BERT** [26]. Dialog-BERT designs three pre-training strategies to sufficiently capture dialogue exclusive features. We use the pre-trained model[4] as an encoder for dialogues, following with a fully connected layer to acquire the dialogue-level labels.
- **DialogueGCN** [3]. DialogueGCN builds a graph for the dialogue where nodes represent individual utterances and the edges represent both the speaker and temporal dependency across the dialogue. DialogueGCN uses R-GCN as its graph encoder and initializes utterance features by using a CNN following a GRU. We modify DialogueGCN to a dialogue-level classification model by adding a max pooling layer to the graph neural network for acquiring representations of dialogues.
- **RGAT** [15]. Based on the dialogue graph DialogueGCN builds, this module introduces position encodings to the graph to retain the sequential information contained in dialogues. RGAT uses the pre-trained BERT-base model to acquire the initial representations of utterance nodes. The modified operation is same as above.
- **DAG** [21]. This model builds a directed acyclic graph for the dialogue with several carefully designed constraints on speaker dependency and positional relations. DAG introduces a directed acyclic graph neural network for utterance-level emotion recognition. Initial utterance embeddings in DAG is acquired form the pre-trained Bert-base model. The modified operation is same as above.

4.4   Other Settings

We choose cross entropy as the loss function for our model on two datasets. We take advantage of a cosine annealing schedule to dynamically modify the learning rate, and the initial learning rate is set to 1e-4. Adam optimizer is used in the training process with a batch size of 32 on both of the two datasets. JD dataset

---

[1] https://huggingface.co/bert-base-cased.
[2] https://github.com/pytorch/fairseq/tree/main/examples/roberta.
[3] https://github.com/nghuyong/ERNIE-Pytorch.
[4] https://github.com/xyease/Dialog-PrLM.

take the 300 dimensional Chinese Word Vectors [16] and DailyDialog use 300 dimensional pretrained 840B Glove vectors [23] as word embeddings. Then we set the CNN filter size to (3, 4, 5) with 50 out channels in each, following is a fully connected layer to get a 100 dimensional feature for each utterance. The hidden size of Bi-LSTM in the sequential encoding module is set to 100. We use 2-layer R-GCN to perform message passing on the dialogue graph. The utterance-level and local discourse-level window sizes are set to [5, 5] and [2,2], respectively. And We choose dropout rate that achieved the best score on each dataset by using validation data. Each training and testing process were conducted on a single Tesla P40 GPU. Every training process contain 60 epochs. The presented results are averages of 5 turns.

Besides, as for the knowledge graph resource that the two variant models MFDG-EN and MFDG-EE demand, we use a well-designed KG built by experts for JD Dialogue dataset. DailyDialog consists of daily communication dialogs and it's hard to design a KG for it, so we just extract general entities by $spaCy^5$ without pre-defined relations between entities and use the word embeddings of entities as the initial features of entity nodes.

5   Results and Analysis

5.1   MFDG Comparing with Baseline Methods

Table 3. Comparison with baseline methods on the JD Dialogue dataset; Bold font denotes the best performances.

Model	Acc	Top-3	Top-5	Macro-F1	Weighted-F1
TextRNN	49.10	74.90	84.00	38.62	46.52
TextRNN-Att	55.30	79.00	86.10	47.77	52.83
TextCNN	63.80	85.80	92.20	57.72	62.58
Bert-base	62.90	83.40	88.60	57.48	61.39
Robert-base	61.60	83.10	88.70	56.52	61.10
ERNIE	64.80	83.30	87.90	60.89	64.04
Dialog-BERT	63.70	87.70	93.40	55.09	61.21
DialogueGCN	61.30	83.90	90.80	52.48	58.70
RGAT	63.50	89.40	93.90	59.02	63.53
DAG	63.20	86.40	93.20	58.52	61.69
MFDG	66.50	89.30	94.40	60.64	65.30
MFDG-EN	65.00	88.60	94.00	60.71	64.00
MFDG-EE	**67.70**	**90.40**	**95.20**	**61.06**	**66.48**

We show the performance of MFDG and its variants with other baseline methods in Table 3 and Table 4. Our model outperforms text classification baseline methods and other graph-based models. On the JD dataset, apart from

---

[5] https://spacy.io/.

MFDG-EE, MFDG achieves best Macro-F1 of 60.64%, Top-3 of 89.3%, Top-5 of 94.4%, and accuracy of 66.5%, which is 4.2% better than RGAT, and 2.9% better than the pre-trained model ERNIE. On the DailyDialog, MFDG achieves best Macro-F1 of 61.41% and Weighted-F1 of 72.22%.

It shows that graph-based models outperform most of the text classification models, as they adopt speaker-specific and contextual modeling for dialogue understanding, whereas text classification models treat the dialogue as an ordinary text without consider the characteristics of the dialogue. Besides, we notice that DialogueGCN perform worse than TextCNN, It demonstrates that DialogueGCN can obtain a good understanding of utterances, however mere modeling interactions between surrounding utterances leads to obvious losses of dialogue-level contextual information.

Besides, we notice MFDG underperforms Dialog-BERT on DailyDialog, otherwise outperforms Dialog-BERT on JD dataset. As the dialogues in JD dataset contain more utterances and speakers of dialogues in it talk in a random order, which is differ from dialogues in DailyDialog as the speakers talk one by one, we consider our MFDG shows its superiority in the real human-to-human multi-turn conversation scenarios.

With regard to the gap in performance between MFDG and other three graph-based models, it is important to understand the nature of these models. All of them build a dialogue graph and apply a GNN to train the model, whereas, other graph-based models only capture contextual information among utterances. MFDG adds other factors, speaker and local discourse, to the dialogue graph, modeling the contextual information of the dialogue form different levels, acquiring a more comprehensive understanding of the dialogue.

In addition, we notice that MFDG performs much better than other graph-based models on the real-world e-commerce dialogue dataset. As the dialogue in JD dataset contains more turns and is more complicated than that of DailyDialog, we believe our model MFDG contributes to enhancing the understanding of complex multi-turn dialogues in a real world scenario.

5.2   Ablation Study

We conduct ablation studies to evaluate the effectiveness of speaker nodes and local discourse nodes we add to the dialogue graph. The results are shown in Table 5.

Firstly, we remove speaker nodes and local discourse nodes from the dialogue graph in MFDG, leaving only the utterance-order edges accordingly. Without the two types of nodes, the performance of MFDG drops by 5.3% accuracy score on JD dataset and 3.44% accuracy score on DailyDialog. Besides, it should be mentioned that we find MFDG without considering speaker and local discourse nodes shares a close accuracy score with DialogueGCN, which can be rationally explained, as both of them model interactions between surrounding utterances.

Secondly, we remove the speaker nodes from the dialogue graph in MFDG, thus removing speaker edges accordingly. Without speaker nodes, the performance of MFDG drops by 0.9% accuracy score on JD dataset and 2.1% accuracy

Table 4. Comparison with baseline methods on DailyDialog.

Model	Acc	Top-3	Macro-F1	Weighted-F1
TextRNN	53.12	84.38	42.62	47.20
TextRNN-Att	68.20	93.40	50.17	66.36
TextCNN	71.60	93.30	55.29	69.39
Bert-base	70.20	93.00	59.00	69.01
Robert-base	72.90	95.20	60.67	71.54
ERNIE	71.90	93.30	53.12	70.55
Dialog-BERT	**74.00**	**94.90**	59.09	72.13
DialogueGCN	70.30	93.70	52.64	68.30
RGAT	72.30	93.40	55.32	70.31
DAG	72.30	92.90	56.18	70.58
MFDG	73.70	94.20	**61.41**	**72.22**
MFDG-EN	70.90	93.50	47.09	69.51
MFDG-EE	71.00	94.10	55.47	68.90

Table 5. Nodes ablation on two datasets. ✗ and ✓denotes nodes removed and added respectively.

Speaker node	Local discourse node	Acc(JD)	Acc(DailyDialog)
✗	✗	61.20(−5.3%)	70.26(−3.44%)
✗	✓	65.60(−0.9%)	71.60(−2.1%)
✓	✗	61.40(−5.1%)	73.60(−0.1%)
✓	✓	**66.50**	**73.70**

score on DailyDialog. This shows that speaker nodes help aggregating speaker-specific information in message passing of dialogues.

Lastly, we remove the local discourse nodes from the dialogue graph in MFDG, thus leaving only the utterance-order edges. In order to keep speaker nodes function in MFDG, we add utterance-speaker edges, which connect each speaker node with its corresponding spoken utterance nodes. Without local discourse nodes, the performance of MFDG drops by 5.1% accuracy score on the JD dataset and 0.1% accuracy score on the DailyDialog. The tiny drop on DialyDialog is because that speakers of the dialogue in DailyDialog talk one by one, forcing each local discourse node connect to only one utterance node, which can not show its superiority. And the drop on JD dataset shows that local discourse nodes are effective at aggregating multiple consecutive utterances spoken by the same speaker.

5.3   Variants of MFDG

As shown in Table 3 and Table 4, MFDG-EN obtains the accuracy score of 65.00% on JD dataset and 70.09% on DailyDialog, underperforming MFDG on the two datasets. It indicates that the addition of entity nodes leads to information loss of the dialogue graph, as the features of entity nodes generated from KG are not in the same semantic space with other nodes in the graph.

For JD dataset, MFDG-EE outperforms MFDG on all the metrics, with a 1.2% promotion of accuracy score and 1.18% improvement of weighted-F1. The results prove the effectiveness of commonsense sense integration on dialogue classification. And it also shows the knowledge-aware representation method we design in MFDG-EE is an appropriate way to integrate entity information. However, MFDG-EN underperforms MFDG on DailyDialog. This is a predictable result as we use general entities for DailyDialog because of the lack of a well-designed KG.

6   Conclusion

In summary, we propose MFDG for dialogue core intent classification. MFDG is designed to obtain a full understanding of the dialogue by building a multi factor graph. Experimental results on two datasets demonstrate that MFDG outperforms other baseline methods. Furthermore, we propose MFDG-EE and MFDG-EN to fuse domain knowledge into the dialogue graph, the experiment results show that MFDG-EE can promote dialogue comprehension with a well-designed knowledge graph.

Acknowledgement. This work was supported by the National Key R&D Program of China under Grant No. 2020AAA0108600 and Guizhou Province Science and Technology Plan Project-Research on Knowledge Management Technology Based on KG.

References

1. Ortega, D., Vu, N.T.: Neural-based context representation learning for dialog act classification. In: Proceedings of SIGDIAL 2017 (2017)
2. Ghosal, D., Majumder, N., Poria, S., et al.: Utterance-level Dialogue Understanding: An Empirical Study. CoRR abs/2009.13902 (2020)
3. Ghosal, D., Majumder, N., Poria, S., et al.: DialogueGCN: a graph convolutional neural network for emotion recognition in conversation. In: Proceedings of EMNLP-IJCNLP, pp. 154–164. ACL (2019)
4. Feng, X., Feng, X., Qin, B., et al.: Dialogue discourse-aware graph model and data augmentation for meeting summarization. In: Proceedings of IJCAI 2021, pp. 3808–3814 (2021)
5. Li, J., Liu, M., Zheng, Z., et al.: DADgraph: a discourse-aware dialogue graph neural network for multiparty dialogue machine reading comprehension. In: Proceedings of IJCNN, pp. 1–8. IEEE (2021)
6. Ishiwatari, T., Yasuda, Y., Miyazaki, T., et al.: Relation-aware graph attention networks with relational position encodings for emotion recognition in conversations. In: Proceedings of EMNLP 2020, pp. 7360–7370. ACL (2020)
7. Kim, Y.: Convolutional neural networks for sentence classification. In: Proceedings of EMNLP 2014, pp. 1746–1751. ACL (2014)

8. Liu, P., Qiu, X., Huang, X.: Recurrent neural network for text classification with multi-task learning. In: Proceedings of IJCAI 2016, pp. 2873–2879. IJCAI/AAAI Press (2016)

9. Zhou, P., Shi, W., Tian, J., et al.: Attention-based bidirectional long short-term memory networks for relation classification. In: Proceedings of ACL 2016, Volume 2: Short Papers (2016)

10. Ravuri, S.V., Stolcke, A.: Recurrent neural network and LSTM models for lexical utterance classification. In: INTERSPEECH 2015, 16th Annual Conference of the International Speech Communication Association, pp. 135–139. ISCA (2015)

11. Majumder, N., Poria, S., Hazarika, D., et al.: DialogueRNN: an attentive RNN for emotion detection in conversations. In: Proceedings of the AAAI Conference on Artificial Intelligence, vol. 33, pp. 6818–6825 (2019)

12. Ghosal, D., Majumder, N., Gelbukh, A., et al.: COSMIC: COmmonSense knowledge for eMotion identification in conversations. In: Proceedings of EMNLP 2020, pp. 2470–2481. ACL (2020)

13. Marcheggiani, D., Titov, I.: Encoding sentences with graph convolutional networks for semantic role labeling. In: Proceedings of EMNLP, pp. 1506–1515 (2017)

14. Schlichtkrull, M., Kipf, T.N., Bloem, P., van den Berg, R., Titov, I., Welling, M.: Modeling relational data with graph convolutional networks. In: Gangemi, A., et al. (eds.) ESWC 2018. LNCS, vol. 10843, pp. 593–607. Springer, Cham (2018). https://doi.org/10.1007/978-3-319-93417-4_38

15. Ishiwatari, T., Yasuda, Y., Miyazaki, T., et al.: Relation-aware graph attention networks with relational position encodings for emotion recognition in conversations. In: Proceedings of EMNLP 2020, pp. 7360–7370 (2020)

16. Li, S., Zhao, Z., Hu, R., et al.: Analogical reasoning on Chinese morphological and semantic relations. In: Proceedings of the 56th Annual Meeting of the Association for Computational Linguistics (Volume 2: Short Papers), pp. 138–143 (2018)

17. Bordes, A., Usunier, N., Garcia-Duran, A., et al.: Translating embeddings for modeling multi-relational data. In: Advances in Neural Information Processing Systems, pp. 2787–2795 (2013)

18. Wang, Z., Zhang, J., Feng, J., et al.: Knowledge graph embedding by translating on hyperplanes. In: AAAI, pp. 1112–1119 (2014)

19. Lin, Y., Liu, Z., Sun, M., et al.: Learning entity and relation embeddings for knowledge graph completion. In: AAAI, pp. 2181–2187 (2015)

20. Yu, D., Yang, Y., Zhang, R., et al.: Knowledge embedding based graph convolutional network. In: Proceedings of the Web Conference 2021, pp. 1619–1628 (2021)

21. Shen, W., Wu, S., Yang, Y., et al.: Quan, directed acyclic graph network for conversational emotion recognition. In: Proceedings of ACL/IJCNLP, pp. 1551–1560 (2021)

22. Li, Y., Su, H., Shen, X., et al.: DailyDialog: a manually labelled multi-turn dialogue dataset. In: Proceedings of IJCNLP, pp. 986–995 (2017)

23. Pennington, J., Socher, R., Manning, C.D.: Glove: global vectors for word representation. In: Proceedings of EMNLP, pp. 1532–1543 (2014)

24. Guo, D., Tur, G., Yih, W., et al.: Joint semantic utterance classification and slot filling with recursive neural networks. In: IEEE Spoken Language Technology Workshop (SLT), pp. 554–559. IEEE (2014)

25. Ravuri, S., Stoicke, A.A.: Comparative study of neural network models for lexical intent classification. In: IEEE Workshop on Automatic Speech Recognition and Understanding (ASRU), pp. 368–374. IEEE (2015)

26. Xu, Y., Zhao, H.: Dialogue-oriented pre-training. Findings of the Association for Computational Linguistics, Online Event, 1–6 August 2021 (2021)

Contextual Information and Commonsense Based Prompt for Emotion Recognition in Conversation

Jingjie Yi[1] ⓘ, Deqing Yang[1(✉)] ⓘ, Siyu Yuan[1] ⓘ, Kaiyan Cao[1] ⓘ,
Zhiyao Zhang[2] ⓘ, and Yanghua Xiao[2] ⓘ

[1] School of Data Science, Fudan University, Shanghai, China
{jjyi20,yangdeqing,yuansy17,kycao20}@fudan.edu.cn
[2] School of Computer Science, Fudan University, Shanghai, China
{zhiyaozhang19,shawyh}@fudan.edu.cn

Abstract. Emotion recognition in conversation (ERC) aims to detect the emotion for each utterance in a given conversation. The newly proposed ERC models have leveraged pre-trained language models (PLMs) with the paradigm of pre-training and fine-tuning to obtain good performance. However, these models seldom exploit PLMs' advantages thoroughly, and perform poorly for the conversations lacking explicit emotional expressions. In order to fully leverage the latent knowledge related to the emotional expressions in utterances, we propose a novel ERC model *CISPER* with the new paradigm of prompt and language model (LM) tuning. Specifically, CISPER is equipped with the prompt blending the contextual information and commonsense related to the interlocutor's utterances, to achieve ERC more effectively. Our extensive experiments demonstrate CISPER's superior performance over the state-of-the-art ERC models, and the effectiveness of leveraging these two kinds of significant prompt information for performance gains. To reproduce our experimental results conveniently, CISPER's source code and the datasets have been shared at https://github.com/DeqingYang/CISPER.

Keywords: Emotion recognition · Prompt · Pre-trained language model

1 Introduction

Emotion recognition in conversation (ERC) aims to judge the emotion category expressed by each interlocutor in a given conversation. In recent years, ERC has been widely studied in natural language processing (NLP), and applied in many fields, including dialogue robots (such as chat and self-help psychological diagnosis), sentiment and opinion mining in the conversations on social media.

This paper was supported by Shanghai Science and Technology Innovation Action Plan No. 21511100401.

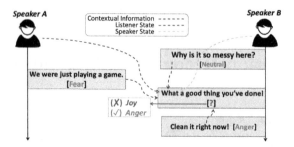

Fig. 1. A toy example of recognizing utterance emotion based on the cues from the contextual information and commonsense related to the states of speaker and listener.

Most previous ERC models are implemented through encoding the dialogue's text into semantic embeddings at first, followed by regarding each round of dialogue as a step or node. Then, they employ recurrent neural networks (RNNs) [21] or graph neural networks (GNNs) [5,39] to obtain utterance representations for the final sentiment prediction. The encoders of dialogue texts in the earlier models include Glove [24] and Word2Vec [23]. Recently, inspired by the power of pre-trained language models (PLMs) [4,19] on encoding text semantics, PLMs are also employed as the encoders to obtain enhanced recognition performance [6,27].

Despite the achievements, the previous PLM-based ERC models seldom fully exploit PLMs' latent knowledge, resulting in limited performance gains. More recently, some researchers have proposed the *prompt-based* learning paradigm to utilize PLMs on various downstream NLP tasks, in which an appropriate prompt is designed to guide the PLM to better take advantage of the knowledge related to the downstream task. As a result, the PLM's performance on the downstream task is improved. Given that PLMs also contain rich semantic and emotional knowledge related to the utterances in a human dialogue at pre-training stage, we are inspired to leverage the prompt about such knowledge to guide the PLM to achieve the ERC task more effectively.

However, it is nontrivial to apply the prompt-based learning paradigm on a PLM to achieve ERC. Although prompt-based PLMs have been employed for generic sentimental analysis successfully [14], ERC is an entirely different task posing new challenges. In ERC, multiple utterances in a conversation are semantically similar to or logically correlated with each other. Thus the *contextual information* is helpful to the emotion recognition of the current utterance in a conversation [21]. Besides, due to the lack of the *commonsense* related to emotional expressions, the colloquial and obscure expressions in a conversation make it difficult for PLMs to understand the real utterance emotions. We use a conversation example in Fig. 1 to explain the significance of these two kinds of significant information. Without any prompt, it is difficult to identify *Anger* is the real emotion of Speaker B's utterance "What a good thing you've done!", because "good thing" is obscure that is actually an irony in this conversation.

While it would be recognized correctly if the cues from contextual utterances were provided, such as "so messy" and "playing a game". Furthermore, the states of Speaker A and B when expressing these utterances are also helpful in identifying the emotion.

Therefore, *it is challenging but crucial for ERC to design a valid and effective prompt to leverage the contextual information and commonsense*. To tackle this challenge, we propose a PLM-based ERC model with prompt, namely **CIS-PER** (**C**ontextual **I**nformation and common**S**ense based **P**rompt for **E**motion **R**ecognition). Specifically, we adopt the trainable embeddings of pseudo-tokens as the *continuous* prompt to cue the PLM, which blends two kinds of significant information. One is the contextual information in the conversation, and the other is the inferential commonsense related to the emotional expression in the utterance, which is extracted from a famous commonsense base ATOMIC [28]. Compared with the explicit discrete prompt [29] in previous models, the trainable continuous prompt in CISPER blends these two kinds of information more flexibly, and makes the model converge more quickly with the learning paradigm of prompt + LM tuning (language model tuning). In fact, these prompt embeddings can be regarded as some informative "sentences" with crucial emotional cues of the conversation, which are then attached with the utterance text and fed into the PLM to achieve ERC.

In summary, the main contributions of our paper include:

1. To the best of our knowledge, this is the first to successfully practice the prompt-based learning paradigm on the ERC task. Unlike previous work focusing on task-specific model design, we focus more on prompt template mining.

2. We propose a novel ERC model built with the trainable continuous prompt from the contextual information and commonsense related to the emotional expressions in utterances. The prompt provides the model with significant cues, and thus enhances the model's ERC performance effectively.

3. Our extensive experimental results on two benchmark ERC datasets prove that, our CISPER outperforms the state-of-the-art (SOTA) baselines, especially in the emotion categories with fewer instances. Meanwhile, the rationality of incorporating contextual information and commonsense for enhanced ERC performance is also verified.

2   Related Work

Emotion Recognition in Conversation. Emotion recognition (including ERC) has been widely applied in many fields, such as man-machine dialogue and psychological and emotional intervention [26]. Previous work in ERC generally adopts fine-tuning paradigm. Specifically, the utterance embeddings are first extracted by PLMs (such as Bert [4] and Roberta [19]), and then fed into the ERC model for emotion identification. Most of previous works based on fine-tuning paradigm design sophisticated deep neural networks to model various hidden states in the conversation, which can be divided into *RNN-based methods* [21], and *GNN-based methods* [5,10]. However, those methods with fine-tuning focus on identifying utterance emotions through downstream model designing, that implicitly model related elements in a conversation but ignore incorporating the latent knowledge in the PLM.

Commonsense Knowledge. Commonsense knowledge benefits many NLP tasks such as dialogue generation [35] and story ending generation [7]. Widely used commonsense knowledge graphs (CKGs) include ATOMIC [28], Concept-Net [33], etc. Commonsense knowledge is essential for ERC, since the colloquial expressions often occur in a conversation, making it difficult for the model to understand the semantics of sentences. Therefore, the CKGs containing abundant commonsense, are leveraged to incorporate such commonsense into the ERC model to improve ERC performance. For example, COSMIC [6] adopts COMET [2] to generate several types of commonsense for each utterance from ATOMIC, and achieves SOTA performance. Inspired by those works, we also incorporate commonsense knowledge into our ERC model.

Language Prompting. In recent years, as a new paradigm, "pre-training, prompting, and predicting" has been proposed to directly exploit the knowledge in pre-trained language models (PLMs), which greatly bridges the gap between the pre-training and fine-tuning of PLMs in downstream tasks. The construction methods of language prompts can be classified into manually constructed prompts and automatically constructed prompts [17]. *Manual constructed prompts* are manually created based on human insights into the task and widely used in machine translation, and text classification [29,30]. Constructing an appropriate prompt template for a certain downstream task is still a challenge even for experienced prompt designers. *Automatically constructed prompts* are automatically generated to address the shortcomings of manual prompts. Some efforts have exploited natural language phrases to discover discrete prompts [11,37]. In addition, given the inherent continuous characteristics of neural networks, others focused on implementing prompts directly in vector spaces rather than designing the human-interpretable template of prompts [16,18]. These continuous prompts are trainable and, therefore, optimal for downstream tasks. The training strategies of the prompt-based models can be divided into four categories: *Tuning-free Prompting* [3], *Fixed-LM Prompt Tuning* [8,16], *Fixed-prompt LM Tuning* [29,30] and *Prompt+LM Tuning* [1,18]. The third category does not need to train the prompts, and the last category takes the prompts as the parameters to fine-tune. In our CISPER, we also adopt *Prompt+LM Tuning* paradigm to train the model given its good flexibility and performance on ERC.

3   Methodology

3.1   Task Formalization

Given a conversation containing L utterances $\{u_1, u_2, ..., u_L\}$, suppose that the t-th utterance $u_t (1 \leq t \leq L)$ is spoken by the speaker q_t and has K_t words,

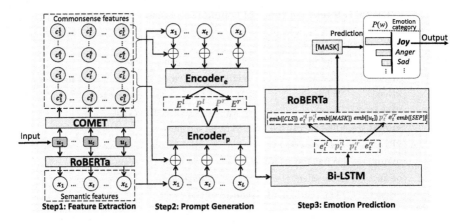

Fig. 2. The overall framework of our proposed CISPER. It has three main steps: feature extraction, prompt generation and emotion prediction.

i.e., $u_t = \{w_1^t, w_2^t, ..., w_{K_t}^t\}$. The task of ERC is to identify each utterance u_t's emotion m_t based on the features of u_t and q_t, as well as any other important cues. In other words, ERC is achieved at the utterance level.

3.2 Framework

Compared with the previous of ERC models with the fine-tuning paradigm, we adopt the prompts+LM-tuning paradigm for our CISPER, and focus more on how to mine an appropriate and effective prompt template to guide the PLM to achieve better ERC. As we claimed before, although the emotional expressions seldom appear in most conversations, the potential information derived from contextual utterances and commonsense reasoning are highly related to the emotional expression of the current utterance. It implies that these two kinds of information are informative for the PLM to infer the current utterance's emotion. Therefore, we pay more attention to the generation of the appropriate prompt based on these two kinds of significant information. To this end, we adopt a trainable continuous prompt that can be updated during training to blend contextual information and commonsense better. Our CISPER's architecture is depicted in Fig. 2, of which the pipeline can be mainly divided into the following three steps (components):

1. *Feature Extraction*: The information features related to a conversation are extracted by the language models at first, including the semantics of the utterances in the conversation and the various inferential relations of commonsense.
2. *Prompt Generation*: The trainable continuous prompt in CISPER is generated based on the features extracted in the first step.
3. *Emotion Prediction*: The continuous prompt embeddings generated in the previous step and the target utterance's embeddings are together fed into the PLM to predict the token indicating the utterance's emotion.

Table 1. 9 relation types of commonsense used in CISPER.

Notation	Type token	Relation meaning
r_1	xIntent	The reason why speaker would cause the event
r_2	xAttr	How the speaker might be described given their part in the event
r_3	xNeed	What speaker might need to do before the event
r_4	xWant	What speaker may want to do after the event
r_5	xEffect	The effect that the event would have on speaker
r_6	xReact	The reaction that speaker would have to the event
r_7	oWant	What **listener** may want to do after the event
r_8	oEffect	The effect the event has on **listener**
r_9	oReact	The reaction of **listener** to the event

3.3   Information Feature Extraction

This step aims to obtain the embeddings encoding the semantics of the utterances and the commonsense related to the utterances. These semantic embeddings will be subsequently used to generate the prompt in our model.

Semantic Features Extraction. For each utterance in a conversation, we directly use a PLM to generate its semantic embeddings. In our experiments, we adopted a RoBERTa-large model [20] as the PLM in this step, consisting of 24 Transformer encoder layers with 16-head self-attentions.

Specifically, we append two special tokens [CLS], [SEP] to the token sequence of a given utterance $u_t = \{w_1^t, w_2^t, ..., w_{K_t}^t\}$, to constitute RoBERTa's input sequence as $[CLS][w_1^t w_2^t ... w_{K_t}^t][SEP]$. As verified in previous work [6], the special token $[CLS]$ in such input format generally encodes the whole sequence's semantics through the PLM's encoding. Thus, among the output embeddings of RoBERTa, we only use $[CLS]$'s embeddings in the last 4 layers, denoted as $\mathbf{v}_1$, $\mathbf{v}_2$, $\mathbf{v}_3$ and $\mathbf{v}_4$. Then, we average these 4 embeddings as u_t's semantic embedding $\mathbf{x}_t \in \mathbb{R}^{d_u}$. All utterances' semantic embeddings are obtained by this method.

Commonsense Features Extraction. Similar to COSMIC [6], the commonsense related to the utterances in the conversation is extracted from COMET [2]. COMET is a Transformer-based model that constructs commonsense through training the language model on a seed set of knowledge triplets from ATOMIC [28]. ATOMIC is one representative commonsense graph with 880K triplets of everyday inferential knowledge, covering 9 relations about entities and events. In CISPER, we select the 9 relation types of commonsense from ATOMIC, as listed in Table 1. In these types, the former six types are related to the inference of different states of the speaker in the conversion, while the latter three types are related to the states of the listener.

The procedure of extracting the features of commonsense is presented as follows. Suppose $r_j (1 \leq j \leq 9)$ is the token of one relation type in the 9 inferential commonsense types, we concatenate it with the token sequence of given utterance

u_t and feed it into the COMET encoder. Then, we extract the hidden state (embedding) of the encoder's last layer, namely $\mathbf{c}_j^t \in \mathbb{R}^{d_c}$, as the embedding of the j-th commonsense type for u_t. So we have

$$\mathbf{c}_j^t = COMET(w_1^t w_2^t ... w_{K_t}^t r_j). \tag{1}$$

All embeddings of the 9 commonsense types are used together with $\mathbf{x}_t$ for generating the prompt in CISPER in the next step.

3.4 Continuous Prompt Generation

In general, there are two types of language prompts, i.e., discrete and continuous. As we mentioned in Sect. 2, a continuous prompt may be more appropriate and effective for deep models, since deep neural networks are inherently continuous. Inspired by P-tuning [18], we also adopt some trainable embeddings as the continuous prompt in CISPER. These trainable embeddings are generated by the encoders fed with the contextual information and commonsense related to the current utterance in the conversation.

Previous research [21] has found that, the utterance emotion is highly related to the states of this utterance's speaker (such as speaker's intent, reaction, etc.) and listener (listeners' effect, reaction, etc.), which has also been illustrated in Fig. 1. Inspired by it, we generate two groups of continuous prompt embeddings from the perspective of speaker and listener, respectively, which are denoted as $\mathbf{E}$ and $\mathbf{P}$. $\mathbf{E}$ corresponds to the speaker-related conversational information while $\mathbf{P}$ corresponds to the listener-related conversational information. Furthermore, the inferential commonsense related to speaker and listener is blended with the contextual information in the conversation and encoded into these embeddings, which are finally leveraged as the emotional prompts for the PLM to predict the utterance's emotion. The details of the generation of these prompt embeddings are described as follows.

Encoding Contextual Information and Commonsense. At first, we build a Transformer encoder to encode the contextual information and commonsense related to a conversation, which is fed with the semantic embeddings and the commonsense type embeddings of the utterances in the conversation obtained in the previous step.

Specifically, given a conversation consisting of L utterances, for each commonsense type $j(1 \leq j \leq 9)$, we concatenate its embeddings related to all utterances that are computed by Eq. 1, as

$$\mathbf{c}_j = \mathbf{c}_j^1 \oplus \mathbf{c}_j^2 \oplus ... \oplus \mathbf{c}_j^L \in \mathbb{R}^{Ld_c}, \tag{2}$$

where $\oplus$ is concatenation operation. Then, suppose $\mathbf{x} = \mathbf{x}_1 \oplus \mathbf{x}_2 \oplus ... \oplus \mathbf{x}_L \in \mathbb{R}^{Ld_u}$ represent this conversation's contextual information, the two hidden embedding matrices about the conversation are obtained as

$$\mathbf{H}_e = \text{Transformer}_e\big(\mathbf{x} \oplus (\mathbf{W}_e[\mathbf{c}_1 \oplus ... \oplus \mathbf{c}_6])\big) \in \mathbb{R}^{L \times d_T},$$
$$\mathbf{H}_p = \text{Transformer}_p\big(\mathbf{x} \oplus (\mathbf{W}_p[\mathbf{c}_7 \oplus \mathbf{c}_8 \oplus \mathbf{c}_9])\big) \in \mathbb{R}^{L \times d_T}, \tag{3}$$

where $\mathbf{W}_e, \mathbf{W}_p$ are two linear projection matrices, and d_T is the dimension of hidden embeddings.

The encoding operations from Eq. 1 to Eq. 3 indicate that all contextual information in the conversation and the commonsense are blended and encoded into $\mathbf{H}_e$ and $\mathbf{H}_p$ with respect to (w.r.t.) speaker and listener, respectively, which are subsequently used as the basis of generating the final prompt embeddings.

Generating Prompt Embeddings of Pseudo Tokens. In the last prediction step of CISPER, a PLM identifies the target utterance's emotion by predicting the special middle token based on its surrounding (contextual) tokens' embeddings. In order to better fit with such a prediction mechanism, we adopt a symmetrical prompt template to simultaneously insert the pseudo (prompt) tokens of the same number on the left side and the right side of utterance tokens.

Accordingly, based on either $\mathbf{H}_e$ or $\mathbf{H}_p$, we respectively generate two sets of prompt embeddings of the pseudo tokens by a multi-layer perceptron (MLP) followed by reshape operation. Specifically, suppose $\mathbf{E} \in \mathbb{R}^{L \times (2N_e d_T)}, \mathbf{P} \in \mathbb{R}^{L \times (2N_p d_T)}$ are the continuous embedding matrices containing the speaker-related and listener-related conversational information, respectively, where N_e and N_p are the number of prompt embeddings. Then, we have

$$\begin{aligned} \mathbf{E} &= [\mathbf{E}^l, \mathbf{E}^r] = \text{Reshape}_e\big(\text{MLP}_e(\mathbf{H}_e)\big), \\ \mathbf{P} &= [\mathbf{P}^l, \mathbf{P}^r] = \text{Reshape}_p\big(\text{MLP}_p(\mathbf{H}_p)\big), \end{aligned} \tag{4}$$

where $\mathbf{E}^l(\mathbf{E}^r) \in \mathbb{R}^{L \times (N_e d_T)}$ is the left (right) half of $\mathbf{E}$ used as the continuous embeddings for the left (right) pseudo tokens. So is $\mathbf{P}^l(\mathbf{P}^r)$.

Finally, for utterance $u_t (1 \le t \le L)$, we take the t-th vectors in the continuous embedding matrices to constitute its hidden prompt embeddings of pseudo tokens, denoted as $\mathbf{e}_t^l, \mathbf{p}_t^l, \mathbf{p}_t^r, \mathbf{e}_t^r$. Note that the current continuous prompt embeddings are not encoded with sequential correlations among the tokens. It is not satisfied with the requirement that the input token embeddings of PLMs should encode sequential features. As a result, we further use Bi-LSTM [9] to obtain the final prompt embeddings of pseudo tokens as:

$$[\mathbf{e'}_t^l, \mathbf{p'}_t^l, \mathbf{p'}_t^r, \mathbf{e'}_t^r] = \text{Bi-LSTM}([\mathbf{e}_t^l, \mathbf{p}_t^l, \mathbf{p}_t^r, \mathbf{e}_t^r]). \tag{5}$$

3.5   Utterance Emotion Prediction

Recall that ERC task is to identify the emotion of a given conversation $\{u_1, ..., u_L\}$ at utterance level. In the last step, we leverage a PLM to predict the emotion of utterance. To guide the PLM to better take advantage of the knowledge related to the utterances which is obtained from its pre-training, we convert the original emotion recognition task into a cloze task that meets the masked PLM's pre-training task. Specifically, in the PLM pre-training, some tokens in the original corpus are masked by a special token [MASK] with a certain probability. Then, the PLM predicts the masked tokens based on their contextual tokens.

According to this task's principle, we feed a [MASK] corresponding to m_t along with u_t's token sequence and the prompt pseudo tokens, into a RoBERTa with the following format as

$$[CLS][E_t^l][P_t^l][MASK][w_1^t w_2^t ... w_{K_t}^t][P_t^r][E_t^r][SEP] \qquad (6)$$

where $[E_t^l], [E_t^r]$ are two sequences of N_e pseudo tokens w.r.t. speaker, and $[P_t^l], [P_t^r]$ are two sequences of N_p pseudo tokens w.r.t. listener. Fed with such token sequence, the RoBERTa can predict the word that would most probably appear at the position of [MASK], based on the embeddings of all input tokens. Formally, the predicted word corresponding to [MASK] is

$$\hat{w} = \arg\max_{w \in \mathbb{V}} P([MASK] = w) \qquad (7)$$

where $P([MASK] = w)$ is the predicted probability of w appearing at the position of [MASK] and w is one word in the tokenizer's vocabulary $\mathbb{V}$. Since the predicted word may be any word in the vocabulary, we maintain a thesaurus to map the predicted word $\hat{w}$ into one emotion category, i.e., m_t. Hence, the prediction of u_t's emotion is achieved.

Please note that, in order to exert the continuous prompt's effect, the embeddings of $[E_t^l], [E_t^r], [P_t^l], [P_t^r]$ used in the RoBERTa are just $\mathbf{e'}_t^l, \mathbf{p'}_t^l, \mathbf{p'}_t^r, \mathbf{e'}_t^r$, which are generated by Eq. 5. The embeddings of the rest input tokens in Eq. 6 are obtained directly for RoBERTa's pre-training results.

3.6 Model Training

We adopt the cross entropy loss to train our ERC model as follows,

$$\mathcal{L} = -\frac{1}{\sum_{q \in \mathcal{Q}} L_q} \sum_{q \in \mathcal{Q}} \sum_{t=1}^{L_q} w_t \log P(w_t) \qquad (8)$$

where q is one conversation from the training set $\mathcal{Q}$, and L_q is the utterance number in q. w_t is the word corresponding to the true emotion category of utterance u_t, while $P(w_t)$ is the estimated probability of w_t appearing at the position of [MASK] for u_t. In addition, we use ADAM [13] as the optimizer to update the model's parameters based on the error inverse propagation strategy.

4 Experiments

4.1 Datasets

MELD [25]: It has 1,432 conversations with more than 13,000 utterances in total, which were extracted from the famous TV show *Friends*. All utterances are labeled with seven emotion categories: anger, disgust, sadness, joy, surprise, fear and neutral, as well as three sentiment classes of positive, negative or neutral. We only evaluated the models' performance of recognizing the emotion categories.

EmoryNLP [38]: It is another dataset also extracted from the TV show *Friends*. The utterances in this dataset are also annotated on seven emotion categories and three sentiment classes. The emotion categories are neutral, joyful, peaceful, powerful, scared, mad and sad. To create three sentiment classes, joyful, peaceful, and powerful are grouped to constitute the positive class; scared, mad and sad are grouped to constitute the negative class; and neutral is the rest class.

We divided the two datasets into training, validation and test set according to the size the same as the previous work [6]. Table 2 lists the sample number statistics of the three sample sets in these two datasets.

Table 2. The statistics of sample division for the two datasets.

Model	Dataset					
	Conversation			Utterance		
	Train	Validation	Test	Train	Validation	Test
MELD	1,039	114	280	9,989	1,109	2,610
EmoryNLP	659	89	79	7,551	954	984

4.2 Baselines

CNN [12]: It is constructed based on convolutional neural networks, where Glove is used to obtain word embeddings. This model has no conversation modeling.

KET [40]: It uses knowledge-enriched Transformer, hierarchical self-attention and context-aware graph attention to maintain the commonsense of emotions.

ConGCN [39]: It first treats speakers and utterances as the nodes in a conversation graph and then uses GCN to achieve emotion recognition.

DialogueRNN [21]: It uses three different GRUs to update the situations of global states, speaker states and emotion states.

DialogueGCN [5]: It also treats the utterances in a conversation as the nodes in the graph and uses different edge types to model dialogue context for emotion detection.

SenticGAT [34]: It proposes a context/sentiment-aware network based on contextual&sentiment-based graph attention to link relevant entities with similar sentiment.

BERT+MTL [15]: It obtains utterance embeddings by BERT which are fed into RNNs to recognize emotions as well as identify speakers. It also adopts a multi-task learning framework.

DialogXL [31]: It modifies the recurrence mechanism in XLNet [36], and uses the dialog-aware self-attention to model conversational data better.

DialogueTRM [22]: It first utilizes a hierarchical transformer to generate features maintaining utterance-level and individual context and then utilizes a multi-modal transformer for Multi-Grained Interactive Fusion in the ERC task.

DAG-ERC [32]: It proposes a directed acyclic graph network to simulate better the internal structure of a conversation, which provides a more intuitive way to model the information flow between the background of the conversation and nearby context.

COSMIC [6]: It utilizes commonsense Transformer COMET [2] to extract commonsense from ATOMIC [28] graph for each utterance, and uses RNNs to blend those knowledge with contextual information.

P-tuning [18]: It is a framework using Bi-LSTM to generate trainable continuous prompt that would be fed along with utterances into the PLM. We apply this baseline in ERC to examine its difference from our model.

We also specially designed several methods with fixed prompt templates to be compared with CISPER. Notice that we have tested some manual templates and finally chose the best effective fixed template "my emotion is [MASK]" as the prompt template for a prompt-based baseline, denoted as **FixedTemplate**. In Sect. 3.4, we have mentioned the reason of adopting a symmetrical prompt template in CISPER. To justify such a symmetrical template's advantage, we further compared CISPER with its two variants which are equipped with the same size prompt only on the left or right side, denoted as **CISPER (left)** and **CISPER (right)**.

4.3   Important Settings

We used the following score as the metric to evaluate all compared models as,

$$\text{weighted-F1} = \frac{1}{\sum_{m \in M} N_m} \sum_{m \in M} N_m \text{F1}(m) \tag{9}$$

where M is the set of all emotion categories, N_m is the number of utterances with emotion category m, and $\text{F1}(m)$ is the F1-score on m.

In CISPER, we adopted the Roberta-large model from https://huggingface.co/, and used the Transformer comes from https://pytorch.org/ as the encoder in Eq. 3. We used ADAM [13] as the optimizer to update our model's parameters and set the learning rate and weight decay to 5×10^{-6} and 10^{-2} respectively. The batch size was set to 64. In addition, we set $N_e = N_p = 3$, which was determined by our tuning studies displayed subsequently. In addition, the dimension of commonsense type embedding d_C is 768, the dimensions of utterance's semantic embedding d_u and prompt embeddings d_T were both set to 1,024. All these settings were decided as the optimum through our tuning studies.

4.4   ERC Performance Comparisons

We compared our CISPER with the baselines regarding macro (overall) ERC performance and micro performance on each emotion category level.

Macro Comparisons. We first display the overall performance (weighted-F1) of all compared models on the two datasets in Table 3, where all models are

Table 3. ERC performance (weighted-F1) comparisons of all compared models.

Paradigm	Language model	ERC model	MELD	EmoryNLP
Fine-tuning	Glove-based	CNN	55.02%	32.59%
		KET	58.18%	34.19%
		ConGCN	57.40%	–
		DialogueRNN	57.03%	31.70%
		DialogueGCN	58.13%	–
		SenticGAT	58.31%	35.45%
Fine-tuning	BERT&RoBERTa -based	BERT+MTL	61.90%	35.92%
		DialogXL	62.41%	34.73%
		DialogueTRM	63.55%	–
		DialogueRNN	63.61%	37.44%
		DAG-ERC	63.65%	39.02%
		COSMIC	65.21%	38.11%
Prompt+LM tuning	RoBERTa-based	FixedTemplate	65.12%	38.67%
		P-tuning	64.90%	37.97%
		CISPER (left)	65.92%	39.46%
		CISPER (right)	65.88%	39.39%
		CISPER	**66.10%**	**39.86%**

divided into three groups according to their learning paradigms and used language models. Except for DialogueGCN and P-tuning for the baselines, their performance scores were directly obtained from their original papers. For the rest models, we ran each one for 4 times and reported its average scores.

As we mentioned, MELD was extracted from the famous TV show *Friends* where the utterances are very colloquial and hardly contain explicit emotional expressions. As shown in Table 3, all types of prompts can help the PLM obtain good ERC performance, since all prompt-based models in the third group almost outperform the rest baselines. CISPER and CISPER(left/right) both outperform FixedTemplate, justifying the advantage of the trainable continuous prompt over the fixed prompt template. Specifically, our CISPER has a performance improvement of 0.89% over COSMIC, which is the current SOTA ERC model except for the prompt-based ones. We attribute this improvement to the employment of prompt in the ERC task and the effective way of incorporating contextual information and commonsense into the prompt.

Compared with MELD, all models' performance on EmoryNLP declines apparently, due to the more "obscure" emotional expressions in the utterances of this dataset. Nonetheless, our CISPER outperforms COSMIC with an improvement of 1.75%, which is more significant than that in MELD. Please note that COSMIC also leverages contextual information and commonsense as CISPER. Thus, CISPER's superior performance shows that leveraging these two important pieces of information through our proposed prompt is more effective than the solution in COSMIC for enhancing ERC performance. Especially for the conversations with more "obscure" emotional expressions as MELD, the prompt in CISPER can guide the PLM to recall its latent knowledge related to the emotional cues, which has been learned in the PLM's pre-training.

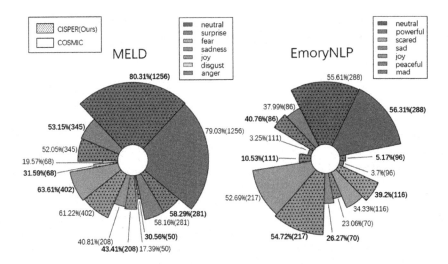

Fig. 3. Micro performance comparisons of CISPER and COSMIC on emotion category level (better viewed in color). It shows that the two models perform better in the categories with more samples, while our CISPER outperforms COSMIC in the categories with fewer samples, justifying its capability of few-shot learning. (Color figure online)

Micro Comparisons. Since COSMIC is the current SOTA model, we further compared the performance of CISPER and COSMIC on the level of seven emotion categories. In Fig. 3, each sector of a certain color corresponds to a certain emotion category, of which the size of the sectorial area quantifies the sample number. Meanwhile, the sample proportion and the corresponding category number are also listed beside the sector. Figure 3 shows that, on MELD, our CISPER has the performance nearly equivalent to COSMIC on neutral, surprise, joy and anger, while has better performance on sadness, fear and disgust. CISPER's advantage is more obvious in the categories of fear (+13.17%) and disgust (+12.02%). Compared with MELD, CISPER's performance on EmoryNLP is better than COSMIC in more emotion categories, i.e., powerful (+1.47%), peaceful (+7.28%), mad (+2.77%) and scared (+4.87%).

In addition, from Fig. 3 we can easily find that both of the two compared models have high weighted-F1 on neutral, joy, anger, mad, surprise and scared. In general, these popular categories of emotions are obviously expressed in the utterances. Both models perform poorly on sad (sadness) fear, disgust, peaceful and powerful. The reason is two-fold: on the one hand, the utterances' emotions are inherently obscure. On the other hand, the utterances of these emotions are relatively rare in the conversations, so the models cannot obtain satisfactory performance only with sparse training data. Notably, with the help of contextual semantics and commonsense-based prompt, our CISPER can take full advantage of its latent knowledge related to the emotional expressions in utterances. Particularly, CISPER outperforms COSMIC, especially in the emotion categories with fewer samples. Such results also prove CISPER's capability of few-shot learning, which is consistent with the findings in previous work [17] about the advantage of prompt-based models in few-shot learning tasks.

4.5 Ablation Studies

The main innovation of our work is to use two Transformer encoders to blend and encode two types of significant information, i.e., contextual information and commonsense, to generate an effective continuous prompt that guides the PLM to achieve ERC better. To verify the effectiveness of either type of prompt information, we added three ablated variants of CISPER into performance comparisons, as shown in Table 4 where all models' weighted-F1 scores and the improvements w.r.t. that of the variant without prompt are both listed. If one type of prompt information is not incorporated, we use randomly initialized embeddings (denoted as "no") to replace our proposed continuous prompt (denoted as "yes"). The results in Table 4 show that, although the random embeddings have the same amount of parameters as the continuous prompt, they can not help the model sufficiently since they contain no meaningful information. We also find that either contextual information or commonsense is helpful for the model to improve ERC performance on these two datasets. Specifically, contextual information brings a more apparent performance improvement than commonsense on both datasets. Furthermore, incorporating these two types of information results in more performance improvement. Even without those two types of information, our model still outperforms P-tuning, justifying the advantage of our model structure. In addition, as shown in Table 3, CISPER's superiority over CISPER(left) and CISPER(right) shows that the symmetrical prompt structure is better than the one side structure.

4.6 Prompt Length Decision

Unlike manually designed prompt templates with fixed length and explicit semantics, the continuous prompt in our model is in fact a group of embeddings and has no explicit semantics. To investigate the influence of prompt length on model performance, we compared CISPER's performance when setting $N_e = N_p = 1 \sim 5$ (the corresponding pseudo token number is 4, 8, 12, 16, 20). According to the results in Fig. 4, we set $N_e = N_p = 3$ when comparing CISPER

Table 4. CISPER's performance comparisons of different prompt information selections.

Commonsense	Contextual info.	MELD	EmoryNLP
No	No	65.12%	38.02%
No	Yes	65.95% (+0.83%)	39.42% (+1.40%)
Yes	No	65.78% (+0.66%)	38.97% (+0.95%)
Yes	**Yes**	**66.10% (+0.98%)**	**39.86% (+1.84%)**

with the baselines. In fact, the small value of N_e/N_p can not ensure the prompt to bring adequate emotional cues for the PLM. While the large value of N_e/N_p may result in redundant information that disturbs the model.

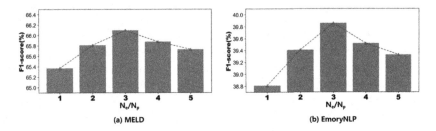

Fig. 4. CISPER's performance with different prompt lengths. The X-axis is the value of $N_e(=N_p)$. It shows that $N_e = N_p = 3$ is the best setting for our model.

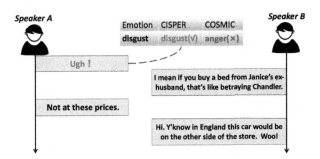

Fig. 5. A conversation case of MELD. The baseline COSMIC can not correctly identify the emotion of "Ugh!", while our CISPER recognizes it correctly with the prompt.

4.7   Case Study

In actual conversation scenarios, many utterances contain very few words, making their real emotions hard to be recognized. We illustrate such a situation by an actual case from our test set, as shown in Fig. 5 where the emotion of Speaker A's utterance: "Ugh!" is disgust in fact. Obviously, it is tough to identify Speaker A's emotion expressed in this utterance only with such a single word. For this case, COSMIC failed to recognize the emotion of "Ugh!", although it has leveraged contextual semantics and commonsense information. Comparatively, CISPER can thoroughly exploit the contextual semantics in the conversation and the speaker/listener's state through the sophisticated prompt generation. Furthermore, with the prompt+LM tuning paradigm, CISPER successfully identifies the emotion of this utterance as disgust.

5   Conclusion

In this paper, we propose an ERC model *CISPER* which blends contextual information and common sense related to the utterances in a conversation into the continuous prompt for enhanced ERC performance. Unlike previous ERC methods adopting fine-tuning paradigm, our CISPER achieves ERC with the paradigm of prompt+LM tuning, which explicitly brings the information related to emotional expressions in the conversation to the PLM. With the help of contextual information and commonsense-based prompts, our model can well handle the challenge of recognizing the implicit emotional expressions in the utterances. Our experiments show that our CISPER significantly outperforms the state-of-the-art ERC models, especially on some critical emotion categories.

References

1. Ben-David, E., Oved, N., Reichart, R.: Pada: a prompt-based autoregressive approach for adaptation to unseen domains. arXiv preprint arXiv:2102.12206 (2021)
2. Bosselut, A., Rashkin, H., et al.: Comet: commonsense transformers for automatic knowledge graph construction. arXiv preprint arXiv:1906.05317 (2019)
3. Brown, T.B., Mann, B., et al.: Language models are few-shot learners. arXiv preprint arXiv:2005.14165 (2020)
4. Devlin, J., Chang, M., et al.: BERT: pre-training of deep bidirectional transformers for language understanding. CoRR (2018)
5. Ghosal, D., Majumder, N., et al.: Dialoguegcn: a graph convolutional neural network for emotion recognition in conversation. arXiv preprint arXiv:1908.11540 (2019)
6. Ghosal, D., Majumder, N., et al.: Cosmic: commonsense knowledge for emotion identification in conversations. arXiv preprint arXiv:2010.02795 (2020)
7. Guan, J., Wang, Y., Huang, M.: Story ending generation with incremental encoding and commonsense knowledge. In: Proceedings of AAAI (2019)
8. Hambardzumyan, K., Khachatrian, H., May, J.: Warp: word-level adversarial reprogramming. arXiv preprint arXiv:2101.00121 (2021)
9. Hochreiter, S., Schmidhuber, J.: Long short-term memory. Neural Comput. (1997)
10. Hu, J., Liu, Y., et al.: MMGCN: multi-modal fusion via deep graph convolution network for emotion recognition in conversation. arXiv preprint arXiv:2107.06779 (2021)
11. Jiang, Z., Xu, F.F., et al.: How can we know what language models know? Trans. Assoc. Comput. Linguist. (2020)
12. Kim, Y.: Convolutional neural networks for sentence classification. Eprint Arxiv (2014)
13. Kingma, J., Ba, J.: Adam: a method for stochastic optimization. In: Proceedings of ICLR (2015)
14. Li, C., Gao, F., et al.: Sentiprompt: sentiment knowledge enhanced prompt-tuning for aspect-based sentiment analysis. CoRR (2021)
15. Li, J., Zhang, M., et al.: Multi-task learning with auxiliary speaker identification for conversational emotion recognition. CoRR (2020)
16. Li, X.L., Liang, P.: Prefix-tuning: optimizing continuous prompts for generation. arXiv preprint arXiv:2101.00190 (2021)

17. Liu, P., Yuan, W., et al.: Pre-train, prompt, and predict: a systematic survey of prompting methods in natural language processing. arXiv preprint 2107.13586 (2021)

18. Liu, X., Zheng, Y., et al.: GPT understands, too. arXiv preprint arXiv:2103.10385 (2021)

19. Liu, Y., Ott, M., et al.: RoBERTa: a robustly optimized BERT pretraining approach. CoRR (2019)

20. Liu, Y., Ott, M., et al.: RoBERTa: a robustly optimized BERT pre-training approach. arXiv preprint arXiv:1907.11692 (2019)

21. Majumder, N., Poria, S., et al.: DialogueRNN: an attentive RNN for emotion detection in conversations. In: Proceedings of AAAI (2019)

22. Mao, Y., Sun, Q., et al.: DialogueTRM: exploring the intra-and inter-modal emotional behaviors in the conversation. arXiv preprint arXiv:2010.07637 (2020)

23. Mikolov, T., Chen, K., Corrado, G., Dean, J.: Efficient estimation of word representations in vector space. arXiv preprint arXiv:1301.3781 (2013)

24. Pennington, J., Socher, R., Manning, C.D.: Glove: global vectors for word representation. In: Proceedings of EMNLP (2014)

25. Poria, S., Hazarika, D., et al.: Meld: a multi-modal multi-party dataset for emotion recognition in conversations. arXiv preprint arXiv:1810.02508 (2018)

26. Poria, S., Majumder, N., et al.: Emotion recognition in conversation: research challenges, datasets, and recent advances. IEEE Access (2019)

27. Qin, L., Che, W., et al.: DCR-net: a deep co-interactive relation network for joint dialog act recognition and sentiment classification. In: Proceedings of AAAI (2020)

28. Sap, M., Le Bras, R., et al.: Atomic: an atlas of machine commonsense for if-then reasoning. In: Proceedings of AAAI (2019)

29. Schick, T., Schütze, H.: Exploiting cloze questions for few shot text classification and natural language inference. arXiv preprint arXiv:2001.07676 (2020)

30. Schick, T., Schütze, H.: Few-shot text generation with pattern-exploiting training. arXiv preprint arXiv:2012.11926 (2020)

31. Shen, W., Chen, J., et al.: DialogXL: all-in-one XLnet for multi-party conversation emotion recognition. arXiv preprint arXiv:2012.08695 (2020)

32. Shen, W., Wu, S., et al.: Directed acyclic graph network for conversational emotion recognition. arXiv preprint arXiv:2105.12907 (2021)

33. Speer, R., Chin, J., Havasi, C.: Conceptnet 5.5: an open multilingual graph of general knowledge. In: Proceedings of AAAI (2017)

34. Tu, G., Wen, J., Liu, C., Jiang, D., Cambria, E.: Context-and sentiment-aware networks for emotion recognition in conversation. IEEE Trans. Artif. Intell. (2022)

35. Wu, S., Li, Y., et al.: Diverse and informative dialogue generation with context-specific commonsense knowledge awareness. In: Proceedings of ACL (2020)

36. Yang, Z., Dai, Z., et al.: XLnet: generalized autoregressive pre-training for language understanding. In: Proceedings of NeurIPS (2019)

37. Yuan, W., Neubig, G., Liu, P.: BartScore: evaluating generated text as text generation. arXiv preprint arXiv:2106.11520 (2021)

38. Zahiri, S.M., Choi, J.D.: Emotion detection on tv show transcripts with sequence-based convolutional neural networks. In: Proceedings of AAAI (2018)

39. Zhang, D., Wu, L., et al.: Modeling both context-and speaker-sensitive dependence for emotion detection in multi-speaker conversations. In: Proceedings of IJCAI (2019)

40. Zhong, P., Wang, D., Miao, C.: Knowledge-enriched transformer for emotion detection in textual conversations. arXiv preprint arXiv:1909.10681 (2019)

Do You Know My Emotion? Emotion-Aware Strategy Recognition Towards a Persuasive Dialogue System

Wei Peng[1,2], Yue Hu[1,2]([✉]), Luxi Xing[1,2], Yuqiang Xie[1,2], and Yajing Sun[1,2]

[1] Institute of Information Engineering, Chinese Academy of Sciences, Beijing, China
huyue@iie.ac.cn
[2] School of Cyber Security, University of Chinese Academy of Sciences, Beijing, China

Abstract. Persuasive strategy recognition task requires the system to recognize the adopted strategy of the persuader according to the conversation. However, previous methods mainly focus on the contextual information, little is known about incorporating the psychological feedback, i.e. emotion of the persuadee, to predict the strategy. In this paper, we propose a Cross-channel Feedback memOry Network (CFO-Net) to leverage the emotional feedback to iteratively measure the potential benefits of strategies and incorporate them into the contextual-aware dialogue information. Specifically, CFO-Net designs a feedback memory module, including strategy pool and feedback pool, to obtain emotion-aware strategy representation. The strategy pool aims to store historical strategies and the feedback pool is to obtain updated strategy weight based on feedback emotional information. Furthermore, a cross-channel fusion predictor is developed to make a mutual interaction between the emotion-aware strategy representation and the contextual-aware dialogue information for strategy recognition. Experimental results on PERSUASIONFORGOOD confirm that the proposed model CFO-Net is effective to improve the performance on M-F1 from 61.74 to 65.41.

Keywords: Persuasive dialogue · Emotional feedback · Strategy recognition

1 Introduction

Persuasive conversation is an essential area in dialogue systems and has embraced a boom in recent NLP research [4,10,32,35]. In a dyadic persuasive dialogue, one party, the persuader, tries to induce another party, the persuadee, to believe something or to do something [14] by a series of persuasion strategies [34]. However, recognizing the persuasion strategy is challenging in the field of natural language understanding since it needs a deeper understanding of conversation, semantic information, and even the psychological feedback of speakers [4,25,27]. Furthermore, dialogue systems can utilize the predicted historical strategy chains to guide the dialogue generation task.

© The Author(s), under exclusive license to Springer Nature Switzerland AG 2023
M.-R. Amini et al. (Eds.): ECML PKDD 2022, LNAI 13714, pp. 724–739, 2023.
https://doi.org/10.1007/978-3-031-26390-3_42

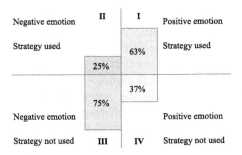

Fig. 1. Statistics in the dataset to show relationships between the emotion and strategy.

To make persuasive strategy prediction, mainstream studies [4,7] focused on the conversational context to recognize strategies. Some work considered resistance strategies to model the strategy conversations, such as [6] and [30]. However, analyzing and understanding speaker's psychological emotion is an essential job [22,23] to fully understand the conversation and help persuader to adopt appropriate strategies. Previous methods do not take the persuadee-aware emotional feedback into account thereby fail to model the emotion-aware human persuasive dialogue systems. To illustrate the importance of emotional feedback, the statistics in the dataset have shown the relationships between the emotion and strategy in Fig. 1. The whole plane is divided into four quadrants. As shown in quadrant-I, if the persuadee shows positive emotion after using the strategy $\mathcal{X}$, the probability of strategy $\mathcal{X}$ continuing to be used is 63% in the following conversation. Similarly, in quadrant-III, when the persuadee shows negative emotion, the probability of the strategy not being used in the subsequent conversation is 75%.

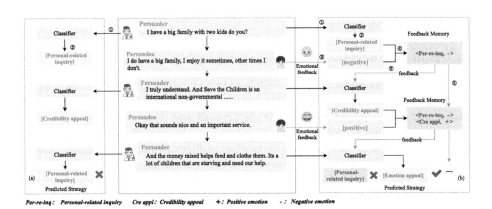

Fig. 2. An example is to compare previous work (a) that utilized the contextual information and our work (b) that considers emotional feedback of the persuadee to recognize the strategy. ⓝ indicates the order of processes.

Statistical results indicate that if the strategy obtains positive feedback, it can be given priority in the future. On the contrary, the strategy should be paid less attention [2,28]. To present the discrepancy between the previous work (a) and ours (b), an example is shown in Fig. 2. Specifically, in the third turn, (a) outputs the wrong prediction *personal-related inquiry* which has received the negative emotional feedback in the previous turn. Therefore, it would be more appropriate to give priority to a different strategy based on the emotional feedback. This leaves us with: *How to model and incorporate emotional feedback into the contextual dialogue information to achieve a better strategy recognition?*

In this paper, the proposed Cross-channel Feedback memOry Network (CFO-Net) leverages persuadee's emotional feedback to iteratively measure the potential benefit of historical strategies, and further the updated representations of strategies are used to guide the strategy recognition. Specifically, the novel feedback memory module designs strategy pool and feedback pool to process and store the historical strategies and update the strategy weight based on the emotional feedback, respectively. Furthermore, the emotion-aware strategy representation and the contextual information are interacted by the designed cross-channel fusion predictor to make the final strategy recognition.

The contributions can be summarized as follows:

- We propose a CFO-Net to leverage persuadee's emotional feedback to measure the potential benefit of historical strategies, and incorporate them into context with cross-channel fusion predictor for persuasive strategy recognition.
- A novel feedback memory module is presented to keep track of the historical strategies and further to obtain the emotion-aware strategy representation in a dynamic and iterative manner.
- Experiments on the dataset show that the CFO-Net has strong competitiveness with baselines and improves the performance of strategy recognition significantly.

2   Related Work

2.1   Non-collaborative Dialogue

In collaborative dialogue, systems collaborate and communicate with each other to achieve a common goal [8]. A large number of researches [3,16,32] have shown remarkable advancement in the collaborative setting. However, they are out of scope when applied to non-collaborative settings like negotiation or persuasion. For the negotiation task, two agents have a conflict of interest but must strategically communicate to reach an agreement like a bargaining scenario [9]. In this paper, the main focus is on the persuasive scenario, where the persuader tries to induce people to donate [34]. The persuasion strategies are identified as ten categories in [34] that can be divided into two types, 1) persuasive appeal and 2) persuasive inquiry. Specifically, persuasive appeal contains seven strategies (Logical appeal, Emotion appeal, Credibility appeal, Foot-in-the-door, Self-modeling, Personal story and Donation information). For example, personal story refers to the strategy of using narrative examples to state someone's donation experiences or the beneficiaries' positive outcomes, which can encourage others to follow the actions. In addition, the three strategies (Task-related inquiry, Personal-related inquiry and Source-related inquiry) belong to persuasive inquiry, which builds better interpersonal relationships by asking questions. For example, source-related inquiry asks whether the persuadee knows about the organization (i.e., the source in our specific donation task).

2.2 Persuasive Dialogue Systems

Persuasive dialogue systems, which have come to increasing attention, aim to change people's behaviors by persuasive strategies [1,12,21,36]. For instance, [11] proposed a two-tiered annotation scheme to distinguish claims in an online persuasive forum. [10] proposed to predict persuasiveness by modeling argument sequence in social media. [35] designed a hierarchical neural network to identify persuasion strategies. Furthermore, some work focused on the contextual information and modeled the utterances to recognize the strategy. [7] explored and quantified the role of context for different aspects of dialogue for strategy prediction. [4] introduced a transformer-based approach coupled with Conditional Random Field for strategy recognition. A few work considered the resistance strategies to model the strategy conversations like [30] and [6]. The Hybrid-RCNN [34] extracted sentiment embedding features (pos, neg, neu) but did not include the emotion in the history modeling, and ignored the corresponding strategy. To overcome these defects, we present the CFO-Net to leverage the emotional feedback to iteratively measure the potential benefits of strategies and incorporate them into the context.

3 Approach

As shown in Fig. 3, the proposed model consists of three components: **(a) a hierarchical encoder**, which encodes the contextual dependency with the multi-head attention to capture the semantic information, **(b) a feedback memory module**, which models the interaction between the strategy pool and the feedback pool to obtain emotion-aware strategy representation, and **(c) a cross-channel fusion predictor**, which makes an interaction between the emotional feedback and the contextual information, and outputs the final result. Each component is described in the following.

3.1 Hierarchical Encoder

The hierarchical encoder uses a **Bi-directional LSTM** (BiLSTM) [13] or **BERT-style encoder** [5,17,18], which capture the temporal features within the words. Then, the **Multi-head Attention** aims to explore the semantic information at different granularity.

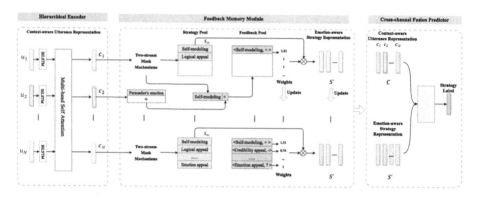

Fig. 3. The overview of CFO-Net, which consists of hierarchical encoder, feedback memory module and cross-channel fusion predictor. Green and blue vertical bars mean the utterances of persuader and persuadee. The emotion-aware strategy representation is updated iteratively based on the strategy pool and feedback pool. (Color figure online)

Utterance Encoder with BiLSTM. The Utterance Encoder vectorizes an input utterance. Given a historical conversation $C = (u_1, u_2, \ldots, u_N)$ a set of N utterances, where $u_i = (x_{i,1}, x_{i,2}, \ldots, x_{i,T})$ that consists of a sequence of T words, u_N indicates the utterance of the persuader, which is used to predict the persuasion strategy. A BiLSTM is utilized to encode each word $x_{i,t}$ in the utterance $u_i \in C$, leading to a series of context-aware hidden states $(\mathbf{h}_{i,1}, \mathbf{h}_{i,2}, \ldots, \mathbf{h}_{i,T})$, $\mathbf{h}_{i,t} = \text{concat}[\ \overrightarrow{\mathbf{h}_{i,t}}\ ;\ \overleftarrow{\mathbf{h}_{i,t}}\ ]$.

The last hidden state $\mathbf{h}_{i,T}$ is considered to get the utterance-level representation. (Note: the representation of the [CLS] is used as the utterance-level representation in BERT-style encoders). Therefore, the set of N utterances in C can be represented as $\mathbf{H} = (\mathbf{h}_{1,T}, \mathbf{h}_{2,T}, \ldots, \mathbf{h}_{N,T})$.

Utterance-Level Multi-head Attention. To explore the semantic information at different granularity, the multi-head attention [31] is adopted as shown in Eq. (1). $\mathbf{c}_i$ indicates the representation of i-th utterance:

$$\mathbf{c}_i = \text{Multi-head Attention}(\mathbf{h}_{i,T}) \tag{1}$$

3.2 Feedback Memory Module

The proposed feedback memory module is composed of three novel factors. **(i) Strategy Embedding** represents the features of strategies which will be continuously updated to capture persuasive strategy features. **(ii) Strategy Pool** temporarily processes and stores all the possible historical strategies for future reference. **(iii) Feedback Pool** considers the emotional feedback of the persuadee to measure the potential benefits of strategies and updates the strategy weight γ. Finally, the strategy pool and feedback pool are interacted to obtain the emotion-aware strategy representation for later strategy recognition.

Strategy Embedding. In the feedback memory module, a randomly initialized strategy embedding is defined to represent the strategy features as $\mathbf{S} \in \mathbb{R}^{L \times d}$, where L is the number of the strategy labels and d indicates the dimension. The strategy embedding will be continuously updated to capture persuasive strategy features. Specifically, CFO-Net selects the appropriate strategies (i.e. top-k) based on the context from strategy embedding and stores them into the strategy pool with the context-aware softmax function that shown in Eq. (2).

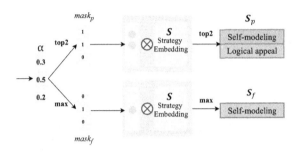

Fig. 4. The two-stream mask mechanisms are defined in the feedback memory module.

Strategy Pool. Strategy pool aims to process and store the possible historical strategies for future reference. As shown in Fig. 4, to achieve the selection of strategies and prevent gradient truncation, **two-stream mask** mechanisms are defined in the following:

- $mask_p$: The selected strategies (i.e. top-k) are stored into the strategy pool to reserve the possible historical strategies (here, size is set to 10).
- $mask_f$: The best strategy of the current moment is stored into the feedback pool.

Specifically, the module first outputs a probability distribution α of the strategy based on the contextual information, as:

$$\alpha = \text{softmax}(\text{MLP}([\mathbf{c}_1; \dots; \mathbf{c}_N])) \qquad (2)$$

Then, the $mask_p$ is obtained based on the α with the top-k operation where k is a hyper-parameter, and $mask_f$ is obtained when $k = 1$. The strategies $\mathbf{S}_p$ which contain multiple possible strategies are stored into the strategy pool, as:

$$\mathbf{S}_p = \mathbf{S} \odot (mask_p \otimes \mathbf{e}_d) \qquad (3)$$

where $\odot$ is element-wise multiplication, $(\cdot \otimes \mathbf{e}_d)$ produces a matrix by repeating the vector on the left for d times [33].

The strategies $\mathbf{S}_m$ in the strategy pool are obtained by making a concatenation with the stored strategies $\mathbf{S}_p$. Similarly, the strategy $\mathbf{S}_f$ that stored into the feedback pool is formulated as:

$$\mathbf{S}_f = \mathbf{S} \odot (mask_f \otimes \mathbf{e}_d) \qquad (4)$$

Feedback Pool. The purpose of the feedback pool is to update the strategy weight γ dynamically to record the emotional feedback of the persuadee towards the strategy. The tuple $\{strategy, emotion\}$ stored in the pool calculates the strategy weight $\gamma \in \mathbb{R}^L$ that is used to obtain the subsequent emotion-aware strategy representation. Firstly, the representation of utterance c_i is considered to predict the emotional label $\mathbf{y}^e \in \{pos, neu, neg\}$ of the persuadee, as:

$$\mathbf{y}^e = \text{softmax}(\text{MLP}([\mathbf{c}_1; \ldots ; \mathbf{c}_{N-1}])) \tag{5}$$

where $\mathbf{c}_{N-1}$ indicates the $(N-1)^{th}$ utterance spoken by the persuadee.

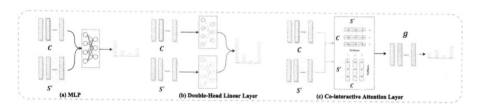

Fig. 5. The three different cross-channel fusion mechanisms that include (a) MLP [20], (b) double-head linear layer and (c) co-interactive attention layer.

Then, the weight γ is assigned based on the score of the predicted emotion and stream $mask_f$. To enhance the generalization of the model, soft weight $\gamma \in \mathbb{R}^L$ (randomly initialized with an all-one vector at the first) can be defined as:

$$\gamma_i = \begin{cases} \gamma_i + mask_f \cdot \mu \exp^{-\zeta} & if \quad pos; \\ \gamma_i & if \quad neu; \\ \gamma_i - mask_f \cdot \mu \exp^{-\zeta} & if \quad neg; \end{cases} \tag{6}$$

where the scalar parameter μ controls the proportion of $\exp^{-\zeta}$ that guarantees to be greater than zero. For the first condition, the weight of γ increases when ζ becomes smaller. To this end, we intuitively set the confidence factor ζ that depends on the score of emotion $\mathbf{y}^e$, as:

$$\zeta = (1 - y_x^e) \tag{7}$$

where y_x^e is a scalar that indicates the score of the $x \in \{pos, neu, neg\}$ emotion. Finally, the emotion-aware strategy representation $\mathbf{S}'$ is modeled as:

$$\mathbf{S}' = \gamma \cdot \mathbf{S}_m \tag{8}$$

3.3   Cross-Channel Fusion Predictor

In this section, the predictor aims to make a recognition of the strategy. Three main types of fusion mechanisms are designed for horizontal comparison in Fig. 5. The mechanisms are introduced to fully interact the psychological feedback with the contextual dialogue information. And the predictor outputs the fusion distribution which captures the profound relationships between two sources.

Multi-layer Perceptron. An MLP can obtain the integrated representation automatically in a simple fashion, as:

$$\mathbf{g} = \text{MLP}([\mathbf{c}_1; \ldots; \mathbf{c}_N; \mathbf{s}_1'; \ldots; \mathbf{s}_L']) \tag{9}$$

The predicted distribution of the strategy $\mathbf{y}^s$ can be defined as follows:

$$\mathbf{y}^s = \text{softmax}(\mathbf{W}^s \mathbf{g} + \mathbf{b}_s) \tag{10}$$

where $\mathbf{W}^s \in \mathbb{R}^{L \times 2d}$ is transformation matrices, $\mathbf{b}_s \in \mathbb{R}^L$ is the bias vector, L is the number of the labels.

Double-Head Linear Layer. To achieve the fusion of two probability distribution, a double-head linear layer is designed for prediction. Specifically, we introduce two MLPs to calculate respective probabilities and then combine them, as:

$$\mathbf{y}_1^s = \text{softmax}(\text{MLP}\,([\mathbf{c}_1; \ldots; \mathbf{c}_N])) \tag{11}$$
$$\mathbf{y}_2^s = \text{softmax}(\text{MLP}\,([\mathbf{s}_1'; \ldots; \mathbf{s}_L'])) \tag{12}$$
$$\mathbf{y}^s = \text{softmax}(\,\mathbf{y}_1^s + \mathbf{y}_2^s\,) \tag{13}$$

where $\mathbf{y}_1^s \in \mathbb{R}^L$ and $\mathbf{y}_2^s \in \mathbb{R}^L$, $\mathbf{y}^s$ is the final predicted distribution of the strategy.

Co-interactive Attention Layer. Motivated by attention mechanism [19, 26, 29], the co-interactive attention layer is proposed to effectively model mutually relational dependency. In this layer, attentions are computed in two directions: from $\mathbf{C} = (\mathbf{c}_1, \ldots, \mathbf{c}_N)$ to $\mathbf{S}' = (\mathbf{s}_1', \ldots, \mathbf{s}_L')$ as well as from $\mathbf{S}'$ to $\mathbf{C}$.

Specifically, the layer first yields a shared similarity matrix $A \in \mathbb{R}^{N \times L}$, between $\mathbf{C}$ and $\mathbf{S}'$. $\mathbf{A}_{ij}$ indicates the similarity between i-th context-aware utterance and j-th emotion-aware strategy, as:

$$\mathbf{A}_{ij} = \mathcal{F}(\mathbf{C}_{:i}, \mathbf{S}_{:j}') \tag{14}$$

where $\mathcal{F}$ is a dot product function, $\mathbf{C}_{:i}$ is i-th row vector of $\mathbf{C}$, and $\mathbf{S}_{:j}'$ is j-th row vector of $\mathbf{S}'$.

The attention weights and the attended vectors can be obtained in both directions. Firstly, considering the direction from $\mathbf{S}'$ to $\mathbf{C}$, the attention weight is computed by $\mathbf{a}_i = \text{softmax}(\mathbf{A}_{i:}) \in \mathbb{R}^L$, and subsequently context-aware utterance vector is $\tilde{\mathbf{C}}_{:i} = \sum_j \mathbf{a}_{ij} \mathbf{S}_{:j}'$. Similarly, the attention weight $\mathbf{b}_j = \text{softmax}(\mathbf{A}_{:j}) \in \mathbb{R}^N$, and updated emotion-aware strategy vector is $\tilde{\mathbf{S}}_{:j}' = \sum_i \mathbf{b}_{ij} \mathbf{C}_{i:}$.

Finally, the context-aware utterance representation and emotion-aware strategy representation are combined to yield $\mathbf{g}$ and $\mathbf{y}^s$ like Eq. (9) and Eq. (10), as:

$$\mathbf{y}^s = \text{softmax}(\mathbf{W}^s \mathbf{g} + \mathbf{b}_s) \tag{15}$$

3.4 Training

The objective of strategy and emotion prediction can be formulated as:

$$\mathcal{L}_s = -\sum_{i=1}^{D} \hat{\mathbf{y}}_i^s \log\left(\mathbf{y}_i^s\right) \tag{16}$$

$$\mathcal{L}_e = -\sum_{i=1}^{D} \hat{\mathbf{y}}_i^e \log\left(\mathbf{y}_i^e\right) \tag{17}$$

where D is the number of the training data, $\hat{\mathbf{y}}_i^s$ and $\hat{\mathbf{y}}_i^e$ are gold strategy label and sentiment label, respectively. The joint objective function $\mathcal{L}_\theta$ is formulated with the hyper-parameters β as, $\mathcal{L}_\theta = \beta_1 \mathcal{L}_s + \beta_2 \mathcal{L}_e$.

4   Experiments

4.1   Experimental Setting

Dataset & Evaluation Metric. Considering there is no emotional score in other dataset, we focus on the PERSUASIONFORGOOD [34][1] whose sentiment label can be obtained based on the manually annotated score. The persuader strategies are identified to ten categories (detail in Sect. 2) and one none category. As for the evaluation metric, Precision, Recall, and Macro F1 (M-F1) are used for the strategy recognition and emotion prediction as the dataset is highly imbalanced [4].

Implementation Details. The BERT-style baselines have the same hyper parameters given on the paper [5,18]. Adam optimizer [15] is used for training, with a start learning rate from {2e-5, 4e-5, 6e-5, 8e-5} and mini-batch size from {32, 64}. The epoch of training is set from {3, 5, 7, 9}. The scalar parameter μ is set from {0.2, 0.5}. k is set to 2 based on the parameter analysis. The historical strategies and emotion will be preprocessed to the two pool. To coordinate the joint training of the two training objectives, we set $\beta_1 = \beta_2 = 0.5$. Tesla V-100 GPU and PyTorch [24] are used to implement our experiments.

4.2   Experimental Results

Baselines. State-of-the-art models are used as baselines to test the performance. Considering the advantages of pre-trained language models (PLMs), we replace the Bi-LSTM with RoBERTa [18] to strengthen the baseline for fair comparison, as with the work [4]. The base and large PLMs are used in the main experiments for a complete comparison. To increase training speed, the base PLMs are utilized in other experiments. Other baselines are shown in Table 1, [34] considered a hybrid RCNN model to extract textual features. [4] combined the PLMs with some state-of-the-art models to recognize the strategy of the persuader.

---

[1] The data are available at: https://gitlab.com/ucdavisnlp/persuasionforgood.

Table 1. Experiments on PERSUASIONFORGOOD for strategy recognition and emotion prediction. – indicates the baselines don't take emotional feedback into account, therefore the results are none. * indicates the experiments are implemented by ourselves.

	Strategy recognition			Emotion prediction		
	P ↑	R ↑	M-F1 ↑	P ↑	R ↑	M-F1 ↑
Hybrid RCNN + All features [34]	62.17*	59.80*	58.76*	–	–	–
RoBERTa$_{large}$ LogReg [4]	64.88*	68.32*	63.15*	–	–	–
RoBERTa$_{large}$ cLSTM [4]	/	/	64.10	–	–	–
RoBERTa$_{large}$ DialogueRNN [4]	/	/	64.30	–	–	–
RoBERTa$_{base}$ [18]	59.58	64.39	58.35	53.21	72.05	60.41
CFO-Net$_{base}$	63.29	67.74	62.41	53.08	75.22	61.94
RoBERTa$_{large}$ [18]	62.69	69.91	61.74	55.49	71.30	62.11
CFO-Net$_{large}$	**66.81**	**72.28**	**65.41**	**58.11**	**75.88**	**63.91**

Main Results. As depicted in Table 1, compared with state-of-the-art models and RoBERTa, the performance of our CFO-Net (with double-head linear layer) has gained a lot. The CFO-Net achieves 4.12% gain on Precision, 2.37% gain on Recall and 3.67% gain on M-F1 score compared with RoBERTa$_{large}$, which demonstrates that the psychological feedback of the persuadee is beneficial for the strategy recognition. The M-F1 reaches the decent performance with the RoBERTa DialogueRNN where four tasks are jointly trained, which shows that the CFO-Net can achieve better performance with fewer tasks. As for the emotion prediction task, the CFO-Net also improves the performance, which shows that jointly training the tasks can provide benefits and boost each other. This phenomenon illustrates that the emotional feedback of the persuadee has the potential to help the process of strategy recognition task. Our code will be released in the link.[2]

4.3   Ablation Study

To get a better insight into the components of the CFO-Net, the ablation study is performed in the Table 2. The experiments demonstrate that either component is beneficial to the final results. Note that by removing the feedback memory module, configuration (1) reduces to the RB-base model.

w/o Feedback Memory Module. In this setting, the feedback memory module is abandoned for exploring the effectiveness of the psychological feedback. From the result, the performance has declined significantly in all metrics, which confirms our hypothesis that introducing the emotion of the persuadee to the strategy recognition is important.

---

[2] The codes are available at: https://github.com/pengwei-iie/CFONETWORK.

Table 2. The results of ablation study on model components.

	Strategy recognition				Emotion prediction			
	P ↑	R ↑	M-F1 ↑	$\Delta_{(M-F1)}$	P ↑	R ↑	M-F1 ↑	$\Delta_{(M-F1)}$
CFO-Net + RoBERTa$_{base}$	63.29	67.74	62.41	–	53.08	75.22	61.94	–
(1) w/o feedback memory module	59.58	64.39	58.35	−4.06	53.21	72.05	60.41	−1.54
(2) w/o multi-task learning	59.04	65.17	58.50	−3.91	53.06	72.62	60.52	−1.42
(3) w/o cross-channel fusion	62.44	66.38	60.53	−1.88	53.12	72.97	60.84	−1.10

w/o Multi-task Learning. Multi-task learning considers the mutual connection between tasks by sharing latent representations. Here, the emotion prediction task is removed to see the performance of strategy recognition. In Table 2, the multi-task learning that is jointly training ($\mathcal{L}_s$ and $\mathcal{L}_e$) can provide benefits, which shows that the training objectives are closely related and boost each other.

w/o Cross-channel Fusion. The cross-channel fusion combines the persuader-aware contextual dependency with persuadee-aware emotional dependency. In this setting, these representations are concatenated directly to make a prediction. The results indicate the fusion mechanisms make a contribution to the overall performance.

4.4   Performances on the Fusion Mechanism

The fusion mechanism is adopted to exploit the two types of the interaction, including persuader-aware contextual dependency and persuadee-aware emotional dependency. To further investigate the effectiveness of these mechanisms, a couple of experiments are conducted from two perspectives, as shown in Fig. 6 and Fig. 7. One is the comparison between three fusion methods and baselines, the other is to consider the horizontal comparison of the fusion mechanisms.

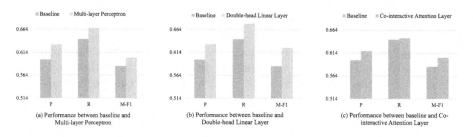

Fig. 6. The performances on the fusion mechanism. (a), (b), (c) represent the results between the baseline and the MLP, Double-head Linear Layer and Co-interactive Attention Layer, respectively.

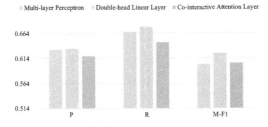

Fig. 7. The performances and comparisons on the three different fusion mechanisms.

As shown in Fig. 6, the results conclude that the fusion mechanisms incorporating persuadee-aware emotional dependency into persuader-aware contextual dependency can bring consistent improvements and surpass baselines on all evaluation metrics. In addition, Fig. 7 presents the performances of different fusion mechanisms, in which the double-head linear layer performs best, with the M-F1 score achieving 62.41%. Surprisingly, the co-interactive attention layer underperforms the double-head linear layer. The phenomenon could be attributed to the fact that the strategy representation and the utterance-level dialogue information belong to different levels of abstract semantic information, leading to the introduction of noise after co-attention operation.

4.5 Parameter Analysis

In the feedback memory module, k is a key hyper-parameter. As shown in Table 3, the model will introduce more noise when k is set too large, and the confidence score will become lower, leading to worse performance. On the contrary, the enriched semantic representations of the strategy will be ignored when k is set to one. It shows that although the confidence score is higher, the performance is not the best. The analysis validates that an appropriate k is crucial to the experimental results.

Table 3. Performance on the hyper-parameter k. Confidence score indicates k^{th} average predicted probability.

Top-k	Top-1	Top-2	Top-3	Top-4
M-F1	60.68	**62.41**	60.42	58.68
Confidence score	**0.877**	0.473	0.326	0.242

4.6 Case Study

A case study is conducted with the example in Fig. 8 to demonstrate how CFO-Net works when recognizing a strategy. We list the possible strategies, the state

Turn	Utterance	Possible Strategy	Strategy Pool	Feedback Pool	Weight	Output
1	ER: I have a big family with two kids do you?	P,S	P / S	<>	<1, 1, ..., 1>	P
	EE: I do have a big family, I enjoy it sometimes, other times I don't.		P / S	<P, ->	<1, 0.736, ..., 1>	-
2	ER: I truly understand. And Save the Children is an international non-governmental organization that promotes children's rights, provides relief and helps support children	L,C	L / C / P / S	<P, ->	<1, 0.736, ..., 1>	L
	EE: Okay that sounds nice and an important service.		C / L / P / S	<P, -> <L, +>	<1.215, 0.736, ..., 1>	+
3	ER: And the money raised helps feed and clothe them. Its a lot of children that are starving and need our help. Would like to help?	P,E	P (highest score) / E / C / L / P / S	<P, -> <L, +>	<1.215, 0.736, ..., 1>	E

ER: Persuader EE: Persuadee S: source-related inquiry P: Personal-related inquiry C: Credibility appeal L: Logical appeal F: foot-in-door E: Emotion appeal

Fig. 8. An example to illustrate the process of novel feedback memory module. The red marker indicates the changes. (Color figure online)

of the strategy pool and feedback pool, and the updated weights. In this case, two possible strategies are selected to the strategy pool at a time. Then, the predicted emotion and the strategy with the highest score are stored into the feedback pool in a tuple fashion, such as <*Personal-related inquiry, A*> where *A* represents *positive* or *neutral* or *negative*. Finally, weights γ will be calculated with Eq. (6). During the conversation, the strategy recognition not only depends on the contextual dialogue information, but also the emotional feedback of the persuadee. The weights are utilized to compute the emotion-aware strategy representation for the final prediction. In the third turn, the CFO-Net outputs a correct prediction *emotion appeal* rather than *personal-related inquiry* with the highest score calculated by the contextual dialogue information, which indicates that incorporating the emotion-aware strategy representation into the contextual dialogue information is of great importance.

5   Conclusion

This paper concentrates on incorporating the psychological feedback (emotion of the persuadee) into the recognition of strategies in the persuasive dialogue. In this paper, we propose a novel Cross-channel Feedback memOry Network (CFO-Net), with a feedback memory module and three different cross-channel fusion mechanisms, to model and explore the historical emotional feedback of persuadee. Experimental results and analysis demonstrate that the CFO-Net has strong competitiveness with baselines and significantly improves the performance of strategy recognition. For the future work, some other categories of psychological feedback will be considered with BiLSTM-CRF, such as personal character, educational background and so on. These cognitive factors are still worth researching for persuasion dialogue systems. Furthermore, dialogue systems can utilize the predicted historical strategy chains to guide the dialogue generation task.

Acknowledgment. We thank all anonymous reviewers for their constructive comments and we have made some modifications. This work is supported by the National Natural Science Foundation of China (No. U21B2009).

References

1. André, E., Rist, T., Van Mulken, S., Klesen, M., Baldes, S.: The automated design of believable dialogues for animated presentation teams. In: Embodied Conversational Agents, pp. 220–255 (2000)
2. Baron-Cohen, S., Wheelwright, S.: The empathy quotient: an investigation of adults with asperger syndrome or high functioning autism, and normal sex differences. J. Autism Dev. Disord. **34**, 163–175 (2004)
3. Bowden, K.K., Oraby, S., Wu, J., Misra, A., Walker, M.A.: Combining search with structured data to create a more engaging user experience in open domain dialogue. CoRR abs/1709.05411 (2017). http://arxiv.org/abs/1709.05411
4. Chen, H., Ghosal, D., Majumder, N., Hussain, A., Poria, S.: Persuasive dialogue understanding: the baselines and negative results. Neurocomputing **431**, 47–56 (2021). https://doi.org/10.1016/j.neucom.2020.11.040
5. Devlin, J., Chang, M., Lee, K., Toutanova, K.: BERT: pre-training of deep bidirectional transformers for language understanding. In: NAACL-HLT, pp. 4171–4186 (2019)
6. Dutt, R., et al.: RESPER: computationally modelling resisting strategies in persuasive conversations. CoRR abs/2101.10545 (2021). https://arxiv.org/abs/2101.10545
7. Ghosal, D., Majumder, N., Mihalcea, R., Poria, S.: Utterance-level dialogue understanding: an empirical study. CoRR abs/2009.13902 (2020). https://arxiv.org/abs/2009.13902
8. He, H., Balakrishnan, A., Eric, M., Liang, P.: Learning symmetric collaborative dialogue agents with dynamic knowledge graph embeddings. In: Barzilay, R., Kan, M. (eds.) ACL, pp. 1766–1776. Association for Computational Linguistics (2017). https://doi.org/10.18653/v1/P17-1162
9. He, H., Chen, D., Balakrishnan, A., Liang, P.: Decoupling strategy and generation in negotiation dialogues. In: Riloff, E., Chiang, D., Hockenmaier, J., Tsujii, J. (eds.) EMNLP, pp. 2333–2343. Association for Computational Linguistics (2018). https://doi.org/10.18653/v1/d18-1256
10. Hidey, C., McKeown, K.R.: Persuasive influence detection: the role of argument sequencing. In: McIlraith, S.A., Weinberger, K.Q. (eds.) AAAI, pp. 5173–5180. AAAI Press (2018). https://www.aaai.org/ocs/index.php/AAAI/AAAI18/paper/view/17077
11. Hidey, C., Musi, E., Hwang, A., Muresan, S., McKeown, K.: Analyzing the semantic types of claims and premises in an online persuasive forum. In: Habernal, I., et al. (eds.) Proceedings of the 4th Workshop on Argument Mining, ArgMining@EMNLP 2017, Copenhagen, Denmark, 8 September 2017, pp. 11–21. Association for Computational Linguistics (2017). https://doi.org/10.18653/v1/w17-5102
12. Fogg, B.J.: Persuasive Technology: Using Computers to Change What We Think and Do (2003)
13. Hochreiter, S., Schmidhuber, J.: Long short-term memory. Neural Comput. **9**, 1735–1780 (1997)
14. Iyer, R.R., Sycara, K.P.: An unsupervised domain-independent framework for automated detection of persuasion tactics in text. CoRR abs/1912.06745 (2019). http://arxiv.org/abs/1912.06745
15. Kingma, D.P., Ba, J.: Adam: a method for stochastic optimization. In: ICLR (2015)
16. Larionov, G., et al.: Tartan: a retrieval-based socialbot powered by a dynamic finite-state machine architecture. arXiv preprint arXiv:1812.01260 (2018)

17. Li, Y., et al.: Enhancing Chinese pre-trained language model via heterogeneous linguistics graph. In: Muresan, S., Nakov, P., Villavicencio, A. (eds.) Proceedings of the 60th Annual Meeting of the Association for Computational Linguistics (Volume 1: Long Papers), ACL 2022, Dublin, Ireland, 22–27 May 2022, pp. 1986–1996. Association for Computational Linguistics (2022). https://aclanthology.org/2022.acl-long.140

18. Liu, Y., et al.: Roberta: a robustly optimized BERT pretraining approach. CoRR abs/1907.11692 (2019)

19. Luong, T., Pham, H., Manning, C.D.: Effective approaches to attention-based neural machine translation. In: Màrquez, L., Callison-Burch, C., Su, J., Pighin, D., Marton, Y. (eds.) EMNLP, pp. 1412–1421. The Association for Computational Linguistics (2015). https://doi.org/10.18653/v1/d15-1166

20. Nguyen, D., Okatani, T.: Improved fusion of visual and language representations by dense symmetric co-attention for visual question answering. In: CVPR, pp. 6087–6096. IEEE Computer Society (2018). https://doi.org/10.1109/CVPR.2018.00637

21. Oinas-Kukkonen, H., Harjumaa, M.: Towards deeper understanding of persuasion in software and information systems. In: First International Conference on Advances in Computer-Human Interaction, pp. 200–205 (2008)

22. Pamungkas, E.W.: Emotionally-aware chatbots: a survey. CoRR abs/1906.09774 (2019). http://arxiv.org/abs/1906.09774

23. Partala, T., Surakka, V.: The effects of affective interventions in human-computer interaction. Interact. Comput. **16**(2), 295–309 (2004). https://doi.org/10.1016/j.intcom.2003.12.001

24. Paszke, A., et al.: Automatic differentiation in pytorch (2017)

25. Peng, W., Hu, Y., Xing, L., Xie, Y., Sun, Y., Li, Y.: Control globally, understand locally: a global-to-local hierarchical graph network for emotional support conversation. CoRR abs/2204.12749 (2022). https://doi.org/10.48550/arXiv.2204.12749

26. Peng, W., Hu, Y., Yu, J., Xing, L., Xie, Y.: APER: adaptive evidence-driven reasoning network for machine reading comprehension with unanswerable questions. Knowl.-Based Syst. **229**, 107364 (2021)

27. Prendinger, H., Ishizuka, M.: The empathic companion: a character-based interface that addresses users' affective states. Appl. Artif. Intell. **19**(3–4), 267–285 (2005). https://doi.org/10.1080/08839510590910174

28. Scott, J.: Understanding contemporary society: theories of the present - rational choice theory- complexity theory (2000)

29. Seo, M.J., Kembhavi, A., Farhadi, A., Hajishirzi, H.: Bidirectional attention flow for machine comprehension. In: ICLR (2017). https://openreview.net/forum?id=HJ0UKP9ge

30. Tian, Y., Shi, W., Li, C., Yu, Z.: Understanding user resistance strategies in persuasive conversations. In: Cohn, T., He, Y., Liu, Y. (eds.) EMNLP, pp. 4794–4798. Association for Computational Linguistics (2020). https://doi.org/10.18653/v1/2020.findings-emnlp.431

31. Vaswani, A., et al.: Attention is all you need. In: Guyon, I., et al. (eds.) Advances in Neural Information Processing Systems 30: Annual Conference on Neural Information Processing Systems, 4–9 2017 December, Long Beach, CA, USA, pp. 5998–6008 (2017). https://proceedings.neurips.cc/paper/2017/hash/3f5ee243547dee91fbd053c1c4a845aa-Abstract.html

32. Wang, Q., Cao, Y., Jiang, J., Wang, Y., Tong, L., Guo, L.: Incorporating specific knowledge into end-to-end task-oriented dialogue systems. In: 2021 International Joint Conference on Neural Networks (IJCNN), pp. 1–8 (2021)

33. Wang, S., Jiang, J.: Machine comprehension using match-LSTM and answer pointer. In: ICLR (2017). https://openreview.net/forum?id=B1-q5Pqxl

34. Wang, X., et al.: Persuasion for good: towards a personalized persuasive dialogue system for social good. In: Korhonen, A., Traum, D.R., Màrquez, L. (eds.) ACL, pp. 5635–5649. Association for Computational Linguistics (2019). https://doi.org/10.18653/v1/p19-1566

35. Yang, D., Chen, J., Yang, Z., Jurafsky, D., Hovy, E.H.: Let's make your request more persuasive: modeling persuasive strategies via semi-supervised neural nets on crowdfunding platforms. In: Burstein, J., Doran, C., Solorio, T. (eds.) NAACL-HLT, pp. 3620–3630. Association for Computational Linguistics (2019). https://doi.org/10.18653/v1/n19-1364

36. Yuan, T., Moore, D.J., Grierson, A.: A human-computer dialogue system for educational debate: a computational dialectics approach. Int. J. Artif. Intell. Educ. **18**(1), 3–26 (2008). http://content.iospress.com/articles/international-journal-of-artificial-intelligence-in-education/jai18-1-02

Customized Conversational Recommender Systems

Shuokai Li[1,2], Yongchun Zhu[1,2], Ruobing Xie[3], Zhenwei Tang[4], Zhao Zhang[1], Fuzhen Zhuang[5,6(✉)], Qing He[1,2(✉)], and Hui Xiong[7]

[1] Institute of Computing Technology, Chinese Academy of Sciences, Beijing 100190, China
{lishuokai18z,zhuyongchun18s,zhangzhao2021,heqing}@ict.ac.cn
[2] University of Chinese Academy of Sciences, Beijing 100049, China
[3] WeChat Search Application Department, Tencent, Shenzhen, China
ruobingxie@tencent.com
[4] King Abdullah University of Science and Technology, Thuwal, Saudi Arabia
zhenwei.tang@kaust.edu.sa
[5] Institute of Artificial Intelligence, Beihang University, Beijing 100191, China
zhuangfuzhen@buaa.edu.cn
[6] SKLSDE, School of Computer Science, Beihang University, Beijing 100191, China
[7] Artificial Intelligence Thrust, The Hong Kong University of Science and Technology, Guangzhou, China
xionghui@ust.hk

Abstract. Conversational recommender systems (CRS) aim to capture user's current intentions and provide recommendations through real-time multi-turn conversational interactions. As a human-machine interactive system, it is essential for CRS to improve the user experience. However, most CRS methods neglect the importance of user experience. In this paper, we propose two key points for CRS to improve the user experience: (1) *Speaking like a human*, human can speak with different styles according to the current dialogue context. (2) *Identifying fine-grained intentions*, even for the same utterance, different users have diverse fine-grained intentions, which are related to users' inherent preference. Based on the observations, we propose a novel CRS model, coined **C**ustomized **C**onversational **R**ecommender **S**ystem (CCRS), which customizes CRS model for users from three perspectives. For human-like dialogue services, we propose multi-style dialogue response generator which selects context-aware speaking style for utterance generation. To provide personalized recommendations, we extract user's current fine-grained intentions from dialogue context with the guidance of user's inherent preferences. Finally, to customize the model parameters for each user, we train the model from the meta-learning perspective. Extensive experiments and a series of analyses have shown the superiority of our CCRS on both the recommendation and dialogue services.

Keywords: Conversational recommendation · Knowledge graph · Customization · Meta learning

S. Li, Y. Zhu, Q. He—The authors are at the Key Lab of Intelligent Information Processing of Chinese Academy of Sciences.

M.-R. Amini et al. (Eds.): ECML PKDD 2022, LNAI 13714, pp. 740–756, 2023.
https://doi.org/10.1007/978-3-031-26390-3_43

1   Introduction

Recently, conversational recommender systems (CRS) [2,16,17,19,24,31,33], which capture the user current preference and recommend high-quality items through real-time dialog interactions, have become an emerging research topic. They [2,16,33] mainly require a dialogue system and a recommender system. The dialogue system elicits the user intentions through conversational interaction and responses to the user with reasonable utterances. On the other hand, the recommender system provides high-quality recommendations by user's intentions and inherent preferences. CRS has not only a high research value but also a broad application prospect [7], such as "Siri", "Cortana" etc.

Fig. 1. An motivating example of our CCRS. The left part shows the various human speaking styles on different topics. The right part presents different users have the same coarse-grained intentions but with various fine-grained intentions.

As a kind of human-machine interactive system, improving user experience is of vital importance. However, existing CRS methods neglect the importance of user experience. Some methods [24,31] not only require lots of labor to construct rules or templates but also make results rely on the pre-processing, which hurt user experience as the constrained interaction [7]. Some other approaches [2,16,20,33,34] generate inflexible and fixed-style responses which could make users uncomfortable. Besides, some methods [2,33] only identify coarse-grained intentions and cannot provide customized recommendations. In this paper, to improve the user experience in CRS, we propose two key points:

- *Speaking like a human:* Facing various dialogue scenes, people's responses may be diverse largely in terms of speaking styles. Figure 1 (left) shows an example when the topics are about horror films and romantic movies. Obviously, a human-like dialogue system is expected to: (1) generate utterances that fit the current content semantics and topics, rather than using a fixed template; (2) generate vivid and attractive conversations, rather than short, dull and boring expressions. In this way, the user experience would be improved and user engagement would also increase, which helps identify the user intentions more accurately.
- *Identifying fine-grained intentions:* The same utterance from different users could reflect diverse fine-grained intentions. For example, in Fig. 1 (right),

if users mentioned the movie "*Titanic*" during the conversation, they would have the same coarse-grained intentions to find movies related to "*Titanic*". Nevertheless, they may have different fine-grained intentions. This is because of the diversity of the inherent preferences: some users prefer movies of the actor "*Leonardo DiCaprio*", while others prefer movies of the director "*James Cameron*". Thus, modeling the fine-grained intentions which are related to the users' inherent preferences helps provide high-quality recommendations.

Along this line, we design a novel model Customized Conversational Recommender System (CCRS), which customizes CRS model from three perspectives. Firstly, for the recommender service, given user mentioned entities, the key idea is to highlight user fine-grained intentions with the guidance of user's inherent preferences (i.e. the preferences on different relations of entity). Then, for the dialogue service, we generate customized utterances with the guidance of content semantics. In detail, multiple styles are modeled as style embeddings, and we aggregate multiple style embeddings into the customized speaking style embedding according to the dialogue context. Finally, we further customize the (recommendation and generation) models for each user, with the advantage of meta-learning. As user fine-grained intentions and speaking styles are sparse in CRS, we choose Model-Agnostic Meta-Learning (MAML) algorithm [6], which can rapidly learn customized model parameters.

To summarize, the contributions of this paper are as follows:

- We propose a novel customized conversational recommender system CCRS, which consists of customized recommendation and dialog services. We further customize the model parameters for each user with the advantage of meta-learning.
- We model user fine-grained intentions on entity relations, and propose multi-style generation to provide human-like dialogue service, which improves user experience.
- Extensive quantitative and qualitative experiments demonstrate that the proposed approach can significantly outperform baseline methods.

2   Method

In this section, we present our novel method to provide customized services for users, coined Customized Conversational Recommender System (CCRS). First, we extract the fine-grained intentions of users with the guidance of user inherent preferences. Then, we design the multi-style embeddings and generate personalized responses based on the extracted fine-grained user intentions. Finally, we adopt meta training to learn customized model parameters for each user rapidly. The overview illustration of the proposed model is presented in Fig. 2.

2.1   Preliminary and Formulations

Recommender Module in CRS. Given a user $u \in \mathcal{U}$ with his identifier u_{id} and his mentioned entities $e \in \mathcal{E}$, a recommender system aims to retrieve a subset of items that meet the customized user needs. To be noticed, the entities consist of item entities and non-item entities. For example, in a movie recommender system, the item entities denote the movies and the non-item entities can be the actor/actress, director, and genre. To better model user mentioned entities, external knowledge graphs $\mathcal{G}$ are often incorporated.

Specifically, the knowledge graph $\mathcal{G}$ consists of triples (e, r, e') where the entities $e, e' \in \mathcal{E}$ and the entity relation $r \in \mathcal{R}$. $\mathcal{E}$ and $\mathcal{R}$ denote the entity set and entity relation set, respectively. Following [2,33], we utilize the knowledge graph from DBpedia [15] to learn the entity representations. As the original graph consists of redundant information, we collect all the entities appearing in the dialogue corpus and extract the subgraph following [2,33].

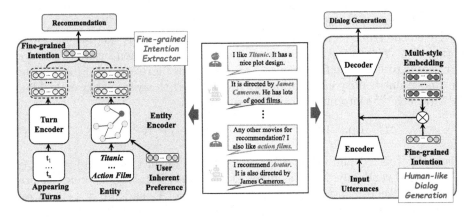

Fig. 2. The overview of our CCRS, which consists of a customized recommender part and a personalized dialogue generation part. Moreover, we use MAML to learn customized model parameters for both recommendation and dialogue generation modules.

Dialogue Module in CRS. The dialogue system is designed to generate proper utterance responses in natural languages. At the T-th turn of the conversation, the dialogue system receives the dialogue history $\mathcal{H} = \{s_t\}_{t=1}^{T-1}$ and user mentioned entities $\{e_1, e_2, ..., e_n\}$. Here the entities are extracted from the dialogue utterances using entity linking. For simplify, the items are replaced by a special token "<unk>". When generating "<unk>", it means to provide recommendations for users. Then the items are provided by the recommendation part.

2.2 Fine-Grained User Intentions Extraction

In this section, we extract the fine-grained user intentions by considering the entity relations and the appearing turns of the entities. These two factors are leveraged to learn the importance of user mentioned entities.

Entity Encoder. Given the user mentioned entities and corresponding knowledge graph, previous methods [2,33] mainly leverage Relational Graph Convolutional Networks (R-GCNs) [23] to incorporate the structural information and learn entity representations. Though R-GCNs keep a distinct linear projection for each type of entity relation, the projections are fixed for different users, regardless of the fine-grained user intentions. Indeed, the fixed projections could only model the coarse-grained preferences (i.e. *knowing what entities do users like*), and ignore the customized fine-grained user intentions on entity relations (i.e. *not knowing what types of entity relations do users like*).

Motivated by [11], we proposed to capture the fine-grained user intentions by modeling personalized preference on entity relations. For each user, with the guidance of the user inherent embedding, we learn the customized attention weights on various relations and aggregated the neighboring entities according to the customized attention weights.

Given the triple (e, r, e') in knowledge graph, where e is the user mentioned entity, $e' \in N(e)$ is the neighbour of e and r is the relation between e and e', our goal is to learn the fine-grained user attentions on relation r. First, to learn the fine-grained entity information, we project the l-th layer's entity embeddings $(h^{l-1}(e) \in \mathbb{R}^{d \times 1})$ into multi-head representations:

$$T_i^{l-1}(e) = W_i^T h^{l-1}(e), \qquad S_i^{l-1}(e') = W_i^S h^{l-1}(e'), \qquad (1)$$

where $W_i^T \in \mathbb{R}^{\frac{d}{k} \times d}$ and $W_i^S \in \mathbb{R}^{\frac{d}{k} \times d}$ are trainable weights, and $h^0(e) = \mathrm{Emb}_e(e)$ is initialized randomly. Next, as different users may have various inherent preferences, we keep a distinct relation-aware matrix A_r^i and use $\gamma^i_{<r,u>}$ to represent the user intentions on distinct entity relation r. Concretely, $\gamma^i_{<r,u>} = \mathrm{Vec}(A_r^i) W_u U_{id}$ is the similarity between A_r^i and the inherent user embedding U_{id}, here $\mathrm{Vec}(A_r^i)$ represents flattening the matrix into a vector. Then the fine-grained user intentions on the relation r for i-th head is calculated as:

$$g_i(r, u) = S_i^{l-1}(e')^{\mathbf{T}} A_r^i T_i^{l-1}(e) \cdot \frac{\gamma^i_{<r,u>}}{\sqrt{d}}. \qquad (2)$$

Finally, the overall user fine-grained intention on entity relation r is calculated by concatenating k heads together:

$$G(r, u) = \underset{N(e)}{\mathrm{Softmax}}(\mathrm{Concat}(g_1, ..., g_k))[e']. \qquad (3)$$

Here the "Softmax" is the normalization of the source nodes, and "$[e']$" denotes the e'-th element. By this, user fine-grained intentions on relations $G(r, u)$ are learned with the guidance of user inherent preferences U_{id}.

Similar to the attention calculation procedure, we model the information of source nodes message using k heads projections and concatenated the multi-head information like Eq. 3:

$$f_i(e') = M_r^i W_i^M h^{l-1}(e'), \qquad F(e') = \mathrm{Concat}(f_1, ..., f_k), \qquad (4)$$

where $M_r^i \in \mathbb{R}^{\frac{d}{k} \times \frac{d}{k}}$ and $W_i^M \in \mathbb{R}^{\frac{d}{k} \times d}$ denote the i-th head message matrices.

Then we aggregate the message from the source nodes to the target node with the guidance of user inherent attention $G(r, u)$, which finally leads to personalized entity representations and captures the customized and fine-grained user intentions:

$$h_*^l(e) = \sum_{\forall e' \in N(e)} (G(r, u) \cdot F(e')). \tag{5}$$

Finally, the entity embedding is finally updated by residual connection [9]:

$$h^l(e) = \sigma(W^A h_*^l(e)) + h^{l-1}(e), \tag{6}$$

where $W^A \in \mathbb{R}^{d \times d}$ is the aggregation matrix and $\sigma(\cdot)$ denotes the activation functions (in practice, we use GELU [10]). In the following, we utilize the last layer's (layer L) representation $h^L(e)$ as the entity representation and denote $h^L(e)$ as $h(e)$ for simplification.

Now given the user mentioned entities $\{e_1, e_2, ..., e_n\}$ through the conversations, we encode them into fine-grained entity representations $H_u = (h(e_1), h(e_2), ..., h(e_n))$, where $h(e_j) \in \mathbb{R}^d$ denotes the entity embedding of e_j.

Turn Encoder. By the entity encoder, we learn the fine-grained user intentions H_u on mentioned entities. However, these entities are not equally important. In CRS, the importance of entities is also influenced by the temporal factor. That is, the entities that appeared in the later turns are prone to be more important than early entities. Motivated by the position embedding technique [4,25] in NLP, we take the appearing turns of entities into consideration:

$$\boldsymbol{\mu}_u^o = \text{Attn}(O_u) = \text{Softmax}(\boldsymbol{w}_2^O \text{Tanh}(W_1^O O_u)), \tag{7}$$

where $o_i = \text{Emb}_t(t_i)$ is the turn embedding and $O_u = (o_1, ..., o_n)$ is the combination of o_i. Here t_i is a scalar, which denotes the appearing turn of the entity e_i, and Emb_t is the turn embedding layer. We then use the turn importance (i.e. $\boldsymbol{\mu}_u^o$) of entities to better learn fine-grained user intentions.

Fine-Grained Intention Encoder. Actually, the importance of entities is also influenced by the entities themselves. Thus we calculate the self-importance of entities like Eq. 7: $\boldsymbol{\mu}_u^r = \text{Attn}(H_u)$. Finally, we calculate the user representations $\boldsymbol{p}_u$ in terms of entity and turn importance, and recommend items according to user intentions:

$$\boldsymbol{p}_u = \frac{1}{2}(\boldsymbol{\mu}_u^r + \boldsymbol{\mu}_u^o)H_u, \qquad \boldsymbol{p}_{rec}(i) = \text{Softmax}(\boldsymbol{p}_u \tilde{H})[i], \tag{8}$$

where $\tilde{H}$ is the embedding matrix of the whole items and i is the index of items.

2.3 Customized Dialogue Generation

Sequence-to-Sequence Model. The seq2seq framework has been verified in NLP and recommendation [1,27,28]. It consists of an encoder that encodes the input utterances into high-level representations and a decoder that generates the responses. Following [2,33], we leverage the Transformer [25] as base architecture.

Given the input utterance $x = (x_1, ..., x_{n_c})$ with dialogue history, the encoder extracts information from x. Then the decoder receives the encoder outputs and generates a representation q at each decoding time step. According to the representation q, the generator calculates a probability distribution over the whole vocabulary to determine the generated tokens.

Multi-style Generation. Actually, the speaking style depends on the current user intention. That is to say, when talking about horror films and romance movies, the speaking style varies definitely. Thus we would like to model the speaking styles of users and perform customized generation.

First, we pre-define n_s *latent speaking style embeddings* $\mathrm{L} = \{l_1, ..., l_{n_s}\}$ $\in \mathbb{R}^{d \times n_s}$. Then for each user, the corresponding styles vary according to current dialogue contexts and user fine-grained intention. Thus the multi-style vocabulary bias g_u is learned by proper speaking styles:

$$\boldsymbol{\mu}_u^m = \boldsymbol{p}_u W^C \mathrm{L}, \qquad \boldsymbol{g}_u = \boldsymbol{\mu}_u^m \mathrm{L}^\mathbf{T}, \qquad (9)$$

where $\boldsymbol{p}_u$ is the user fine-grained intentions learned by Eq. 8 and W^C is the similarity matrix. The selected style embeddings fit the context semantics and user inherent preferences. Finally, we add the vocabulary bias to the original generator to perform personalized utterances generation:

$$\boldsymbol{p}_{dial} = \mathrm{softmax}(W^G \boldsymbol{q} + \mathcal{F}(\boldsymbol{g}_u) + \boldsymbol{b}), \qquad (10)$$

where $\mathcal{F} : \mathbb{R}^d \rightarrow \mathbb{R}^{|V|}$ maps the vocabulary bias vector into a $|V|$-dimension vector, which consists of two fully connected layers.

2.4 Customized Model Training

In previous sections, we describe the details of our model and capture the customized user preferences from the *design of network* perspective. In this section, we will model the customized user preferences from the *training* perspective.

Training Loss. A common way to train the parameters is the back-propagation algorithm. For the recommendation part, we could leverage a common cross-entropy loss to train the recommendation part:

$$\mathcal{L}_{rec} = \frac{1}{|\mathcal{U}|} \sum_{u \in \mathcal{U}} \frac{1}{N_u} \sum_{n=1}^{N_u} \log \boldsymbol{p}_{rec}(y_{u,n}), \qquad (11)$$

where y_{un} denotes the actual preference of user u and N_u is the number of the movies in which the user u is interested.

For the dialogue generation module, the common training loss is also the cross-entropy loss:

$$\mathcal{L}_{dial} = \frac{1}{|\mathcal{U}|} \sum_{u \in \mathcal{U}} \frac{1}{N_u^g} \sum_{n=1}^{N_u^g} \log \boldsymbol{p}(s_{n,u} | s_{(n-1),u}, ..., s_{1,u}, x_{n,u}), \tag{12}$$

where N_u^g denotes the whole utterance of user u, and $s_{n,u}$ is for the gold response tokens.

Nevertheless, it lacks the personality for various users and could not tune the network according to the customized and fine-grained user preferences. Motivated by [14], we adopt the meta-learning framework [6] to learn the communal user preferences and customized user preferences for a specific user.

Algorithm 1. The training algorithm of recommendation and dialogue generation parts.

Require:
    The training model: $m = $ rec or dial,
    β_m and ν_m: inner and outer learning rates.
    Randomly initialized the inner θ_m^{inner} and outer θ_m^{outer} params
    while not converge **do**
        Sample batch of users $\mathcal{U} \sim \mathcal{D}_{tr}$
        for user u in $\mathcal{U}$ **do**
            $(\mathcal{D}_u^{sup}, \mathcal{D}_u^{qu}) \sim \mathcal{D}_{tr,u}$
            $\mathcal{L}_1(u) = \mathcal{L}_{m, \mathcal{D}_u^{sup}}(f_{\theta_m^{inner}, \theta_m^{outer}})$
            Inner update: $\phi(u)_m^{inner} \leftarrow \theta_m^{inner} - \beta_m \nabla_{\theta_m^{inner}} \mathcal{L}_1(u)$
            $\mathcal{L}_2(u) = \mathcal{L}_{m, \mathcal{D}_u^{qu}}(f_{\phi(u)_m^{inner}, \theta_m^{outer}})$
        end for
        Global update
        $\theta_m^{outer} \leftarrow \theta_m^{outer} - \nu_m \sum_{u \in \mathcal{U}} (\nabla_{\theta_m^{outer}} \mathcal{L}_2(u) + \nabla_{\theta_m^{outer}} \mathcal{L}_1(u))$
        $\theta_m^{inner} \leftarrow \theta_m^{inner} - \nu_m \sum_{u \in \mathcal{U}} (\nabla_{\theta_m^{inner}} \mathcal{L}_2(u) + \nabla_{\theta_m^{inner}} \mathcal{L}_1(u))$
    end while

Meta Training. The meta training includes the inner update and global update. We first define predicting each user's preference as an individual task and sample a set of records as support set $\mathcal{D}_u^{sup}$, while others as the query set $\mathcal{D}_u^{qu}$ for each user. In the inner update, the model updates the inner parameters θ^{inner} (see Section Parameters Setting for details) to learn customized user tastes, according to the user's unique item-consumptions (i.e., the support set). It takes a single gradient step with inner learning rate β:

$$\phi(u)^{inner} \leftarrow \theta^{inner} - \beta \nabla_\theta \mathcal{L}_{\mathcal{D}_u^{sup}}(\mathcal{M}), \tag{13}$$

where $\mathcal{M}$ denotes the training parameters of the whole model. As the inner parameters are updated in different directions for different users, it captures the user customized preferences. The following procedure is global update and its goal is to learn a communal parameters initialization, such that each of the users' customized preferences would be met from the common initialization with a few update steps, i.e., capturing the customized preferences rapidly:

$$\theta \leftarrow \theta - \nu \sum_{u \in \tilde{U}} \nabla_\theta \mathcal{L}_{\mathcal{D}_u^{qu}}(\mathcal{M}(\phi(u)^{\text{inner}})), \tag{14}$$

where θ includes the whole parameters of $\mathcal{M}$, and ν is the outer learning rate.

In practice, we first train the recommender part to learn the entity embeddings and take the customized recommendation. When the recommender part converges, the dialogue generation module is optimized with the guidance of user intentions on entities, and personalized utterances are generated. The detailed training algorithm of recommendation and dialogue generation is shown in Algorithm 1.

Parameters Setting. For the recommendation part, the inner parameters are set to $\theta_{\text{rec}}^{\text{outer}} = \{\text{Emb}_u, \text{Emb}_e\}$, and the outer parameters are $\theta_{\text{rec}}^{\text{inner}} = \theta_{\text{rec}} \setminus \theta_{\text{rec}}^{\text{outer}}$, where $\theta_{\text{rec}} = \{\theta | \theta \in \mathcal{M}_{\text{rec}}\}$. For the dialogue generation part, The parameters are also divided into two categories: the inner parameters $\theta_{\text{dial}}^{\text{inner}} = \{\text{Encoder}', W_j^L|_{j=1}^{n_s}\}$ and the outer parameters $\theta_{\text{dial}}^{\text{outer}} = \theta_{\text{dial}} \setminus \theta_{\text{dial}}^{\text{inner}}$, where $\theta_{\text{dial}} = \{\theta | \theta \in \mathcal{M}_{\text{dial}}\}$.

3 Experiment

In this section, we first introduce the details of our experiments and then answer the following questions: **RQ1**: How does our CCRS perform on recommendation and dialogue generation compared with the SOTA baselines? **RQ2**: Does our CCRS capture the fine-grained user intentions? **RQ3**: Does our CCRS generate human-like responses? **RQ4**: How do different components of CCRS benefit the performances?

3.1 Experimental Setup

Dataset. To evaluate the effectiveness of CCRS, we conduct experiments on real-world dataset ReDial. It contains 10,006 conversations and 182,150 utterances. The total number of users and movies are 956 and 51699, respectively.

For meta-learning, we define each task as a user's interactive conversations. Then we group the train, validation, and test sets by the user id, and the ratio of samples is about 8:1:1. For each user, half of the user conversations are used as the query set, and the remaining conversations are used as the support set. During the meta test phase, CCRS is fine-tuned on the support set $\mathcal{D}_{test}^{sup}$ and tested on the query set $\mathcal{D}_{test}^{qu}$. For a fair comparison with baseline methods, when training these models, we add the meta test support set $\mathcal{D}_{test}^{sup}$ into the training data.

Baselines. We consider the following baselines: (1) *Popularity* ranks the items according to the historical recommendation frequencies. (2) *TextCNN* [12]: encodes utterances to extract user intent by CNN-based model. (3) *ReDial* [16] is a CRS method which adopts an auto-encoder for recommendation. (4) *KBRD* [2] adopts the external knowledge graph DBpedia to enhance the user representations. (5) *KGSF* [33] incorporates both the semantic KG ConceptNet and the entity KG DBpedia for modeling user preferences. (6) *KECRS* [30] constructs a high-quality KG and it proposes the Bag-of-Entity loss and the infusion loss to better integrate KG with CRS for generation. (7) *RevCore* [20] collects user reviews on movies to enhance the recommendation and dialogue generation modules.

Evaluation Metrics. For the recommendation task, we evaluate whether it recommends high-quality items. So we adopt Recall@k (following [33]), MRR@k and NDCG@k (following [34]) for evaluation ($k = 10, 50$). For the dialogue generation task, we evaluate the performance by the automatic and human evaluations. In the automatic evaluations, we adopt BLEU [22] and F1 to estimate the generation quality, and Distinct n-gram ($n = 3, 4$) to measure the diversity at sentence level. In the human evaluations, we invite three annotators to score whether the generations are fluent (*Fluency*) and plenty of useful information (*Informativeness*). The range of scores is 0 to 2, and the final result is the average scores of the three annotators.

Implementation Details. We implement our approach based on PyTorch framework. For the recommendation part, the entity embedding size is set to 128. We choose the number of entity relation extractor layers $L = 1$ and the number of heads $k = 4$. The network parameters are initialized by glorot uniform [8]. For training, the inner and outer learning rates are set to 0.006 and 0.003, respectively. For the dialogue part, the dimension of word embeddings is set to 300, and the number of styles n_s equals 4. For training, the inner and outer learning rates are set to 0.0003 and 0.001, respectively. For both parts, we use Adam [13] optimizer with default parameter setting, and gradient clipping restricts the gradients within $[0, 0.1]$.

3.2 Overall Performance (RQ1)

Recommendation. The recommendation results on ReDial are shown in Table 1. From Table 1, we have the following observations. (1) First, our CCRS outperforms CRS baselines (ReDial, KBRD, KGSF, KECRS, and RevCore) by a large margin, which shows the superiority of the fine-grained user intentions extraction and meta training framework. With the guidance of inherent user preferences, the fine-grained user intentions are modeled, and MAML further customizes user preferences by the local update. Moreover, though CCRS does not consider ConceptNet and external user reviews, it outperforms KGSF and RevCore significantly, which also shows the effectiveness of our customization recommendation module. (2) Second, CCRS beats the non-CRS method

TextCNN. The reason is that TextCNN directly learns user representation by the whole history, which suffers from the sparsity and noise of the utterances. (3) Finally, KBRD, KGSF, and KECRS perform better than ReDial, which proves the effectiveness of incorporating external knowledge graphs. Besides, KGSF outperforms KBRD, as KGSF leverages an extra word-oriented KG ConceptNet.

Table 1. The recommendation results. The marker * indicates that the improvement is statistically significant compared with the best baseline (t-test with p-value < 0.05).

Dataset	ReDial					
Method	HR@10	HR@50	MRR@10	MRR@50	NDCG@10	NDCG@50
Population	6.47	17.68	0.0158	0.0204	0.0346	0.0617
Text CNN	6.57	16.51	0.0235	0.0275	0.0425	0.0661
ReDial	11.79	30.11	0.0551	0.0628	0.0895	0.1273
KBRD	15.20	33.26	0.0593	0.0677	0.0979	0.1382
KGSF	17.00	35.72	0.0637	0.0717	0.1072	0.1497
KECRS	11.57	31.95	0.0573	0.0639	0.0911	0.1291
RevCore	17.36	36.66	0.0659	0.0741	0.1103	0.1528
CCRS	**19.09***	**38.41***	**0.0717***	**0.0808***	**0.1193***	**0.1618***
w/o TE	17.26	36.58	0.0661	0.0751	0.1092	0.1520
w/o RE&TE	16.49	35.23	0.0633	0.0720	0.1063	0.1463

Table 2. Evaluation results on dialogue generation. Flu. and Inf. stand for Fluency and Informativeness, respectively. The marker * indicates that the improvement is statistically significant compared with the best baseline (t-test with p-value < 0.05).

Dataset	ReDial						
Method	BLEU	F1	Dist-2	Dist-3	Dist-4	Flu	Inf.
ReDial	1.213	0.183	0.089	0.393	0.798	0.83	0.96
KBRD	1.287	0.192	0.118	0.571	1.212	0.95	1.03
KGSF	1.629	0.227	0.123	0.647	1.583	1.23	1.32
KECRS	1.088	0.125	0.078	0.351	0.761	0.85	0.99
RevCore	1.236	0.186	0.105	0.553	1.321	1.21	1.33
CCRS	**2.386***	**0.267***	**0.146***	**0.776***	**1.924***	**1.36***	**1.43***
Only-Meta	2.129	0.258	0.131	0.692	1.772	1.30	1.40

Dialogue Generation. We also evaluate our CCRS on the dialogue generation task and the main results are listed in Table 2. From Table 2, we have the following observations. (1) First, our CCRS outperforms the baselines significantly. The improvement on BLEU shows that the generation of CCRS is more consistent with the ground truth, and the high value of Distinct n-gram reflects the diversity of CCRS's results. (2) Then, from the human evaluation perspective, CCRS also generates the most fluent and informative responses, which are more human-like responses. The main reason is that CCRS considers multi-style generations according to the current content semantics and inherent user preferences. The human-like generations then improve the user experience, and users are more willing to chat with CCRS. (3) Besides, compared with ReDial, we can see that the external knowledge graph also contributes to the generation.

3.3   Qualitative Results on Recommendation (RQ2)

Fine-Grained User Intentions on Entity Relations. In this part, we present qualitative examples to show that we capture fine-grained customized user intentions.

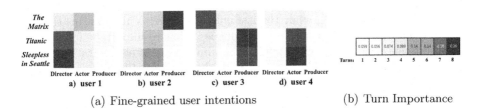

(a) Fine-grained user intentions   (b) Turn Importance

Fig. 3. The visualization of fine-grained user intentions on different relations (left), and the importance of the dialogue turns (right).

We first choose three movies "The Matrix", "Titanic" and "Sleepless in Seattle" and three different relations "Director", "Actor" and "Producer". Then we randomly pick up four users to show their fine-grained intentions on relations. The results are shown in Fig. 3 left. (1) Firstly, we can see that the user's preferences on relations vary. For example, user 1 pays more attention to the "Director" relation, while user 4 cares about "Actor". So it is necessary to learn the user attention on the entity relation, and we aggregate more information from the "Director" relation for user 1, while the "Actor" relation for user 4. (2) Secondly, even for a fixed user, the intentions on different movies also vary. For user 1, he likes the *director* of "Sleepless in Seattle" and the *actor* of "The Matrix". (3) Last but not least, in the original corpus, user 4 mentioned that he loves Tom Hanks, which is the star of "Sleepless in Seattle", and he also likes the movie "Castaway" which also includes Tom Hanks. These observations show that our CCRS correctly captures the fine-grained user intentions on entity relations and provides high-quality recommendations.

Turn Encoder. For the turn encoder, we first conduct an experiment to prove that the appearing turn of the entities contributes to the recommendation performances. Given a user's mentioned entities, we try to rank them purely according

Table 3. The human-like responses for different types of movies.

Topic 1: Horror film	User: I recently watched *The Shining* and it was great
	KGSF: Maybe you will also like *Scream*
	CCRS: I'm not sure if you have seen *The Conjuring*. I would say it's pretty horror. It scared me for whatever reason
Topic 2: Romantic movie	User: Hello! Can you suggest some movies like *Roman Holiday*?
	KGSF: *50 First Date* is a good choice
	CCRS: What about *The Notebook*? The characters suffer lots of hardships and get an happy-ending. It tells a moving story!

to the appearing turn. Then we use the first and last entity's first-order neighbors as the candidate pool. If the candidate pool includes the target movie, this recommendation is viewed as correct. After the calculation, we find the accuracy of the first entity is 2.32%, while the last entity is 3.50%. This confirms that the entities near the current turn of the dialogue are more important for learning user preference. Secondly, we visualize the importance of each turn learned by CCRS in Fig. 3 right. We observe that with the increase of turns, the attention weights become larger, which helps to recommend high-quality items according to the customized user preferences.

3.4   Qualitative Results on Dialogue (RQ3)

In this part, we present some qualitative examples to illustrate the personalized generation. Firstly, as shown in Table 3, when talking about different topics of movies, our CCRS could generate multi-style responses according to the current content semantics. While the speaking style of KGSF is monotonous. Thus our CCRS improves the user experience, and the user would feel like chatting with a real person.

Fig. 4. The probability μ_u^m of choosing the different styles. We can see that for different type of movies, the style probability varies.

Then, we visualize the style distribution vector μ_u^m (see Eq. 9) in Fig. 4, which controls the choice of the four speaking style embeddings. On the one hand, for different types of movies, the style distribution vector varies significantly. For example, for horror film, $\boldsymbol{\mu}_u^m$ is prone to choose Style 3, while the romantic movie is Style 1. On the other hand, for the same type of movie, the style distribution is similar among different users, but not identical. In a word, CCRS chooses customized style vectors for various kinds of movie topics, which leads to human-like generations.

3.5   Ablation Study (RQ4)

Recommendation. In our recommendation part, we propose to learn the fine-grained user intentions, temporal factors of entities, and the meta-learning framework to achieve the customized recommendations. Here we would like to examine the effectiveness of each part and show the results in Table 1. For meta training, CCRS w/o RE&TE denotes training KBRD from the meta-learning perspective (KBRD leverages RGCN to learn the entity representations). CCRS w/o RE&TE beats KBRD, which shows the effectiveness of meta training. For the remaining two parts, CCRS w/o TE denotes training CCRS with only the entity relation encoder, and it outperforms without entity relation encoder (i.e. CCRS w/o RE&TE). CCRS incorporates both the entity relation encoder and the turn encoder, and it beats without turn encoder (i.e. CCRS w/o TE) significantly. These results show the usefulness of each part of CCRS. With the help of the fine-grained user intention, the appearing turns of entities, and meta-learning, CCRS achieves good performances.

Dialogue Generation. In our generation part, we adopt the multi-style generator and the meta training to generate personalized utterances. We also conduct ablation studies to examine the effectiveness of each part. First, the Only-Meta denotes training KBRD with MAML. We can see that MAML significantly improves the quality (BLEU) and diversity (Distinct n-gram) simultaneously. This is not surprising, as we update the network parameters in the user specific direction in the inner update, which leads to the customized generations. Then, CCRS beats Only-Meta, which shows the effectiveness of the multi-style generator. Compared with the Only-Meta, though the improvements on BLEU and F1 scores are marginal, CCRS generates more personalized responses.

4   Related Work

Conversational Recommender Systems. Conversational Recommender Systems (CRS) [2,16,17,19,24,31,33] aim to provide high-quality items through the interactive conversations. It mainly consists of a recommender system that identifies the user customized preferences given the item-consumption history and a dialogue system that converses with the users and collects their preferences. Early conversational recommender systems [3,24] mainly collects the user preference via the pre-defined questions. They pay more attention to recommendation accuracy, and the dialogue system is an auxiliary part, which is implemented by simple or pre-defined patterns. Recently, several works focus on building end-to-end CRS models. [16] proposed a standard CRS dataset named ReDial and an end-to-end model which consists of a hierarchical recurrent encoder, a switching decoder, an RNN-based sentiment analysis module, and an autoencoder-based recommendation module. Moreover, [2,33] proposed to incorporate the external knowledge to improve the CRS performances. [2] mainly leverages the knowledge graph (KG) of user mentioned entities, which includes movies, actors, etc., while [33] incorporates both word-oriented and entity-oriented KGs, and adopts Mutual Information Maximization to align word-level and entity-level representations. Other works focus on selecting proper inter-active action (policy) during the conversations. [19] constructs a goal sequence which includes question answering, chit-chat, and recommendation phrase. Then they characterized the goal planning process and focus type switch during conversations. [34] focused on topic-guided CRS and proposed a topic prediction model. Furthermore, [32] proposed an open-source CRS toolkit CRSLab, which provides a unified and extensible framework for previous CRS works. Moreover, [20] proposes a review-enhanced framework, in which user reviews are incorporated to enhance the CRS performances. [18] further learns templates automatically for utterances generation.

Meta-Learning. Meta-learning is also named learning to learn, which aims to improve new tasks' performance with several similar tasks. The work can be grouped into three clusters, e.g. metric-based [26], model-based [37], and optimization-based [6] approaches. Recently, meta-learning has also been applied to recommender systems [14,21,29,35,36,38] and natural language processing [5]. In this paper, we leverage meta-learning to learn customized conversational recommendation model, and provide personalized service, which is different from previous works. To the best of our knowledge, this is the first attempt in adapting meta-learning to conversational recommendation for model customization and improving user experience.

5   Conclusion

In this paper, we proposed a novel approach CCRS, which aims to improve the user experience and explore the customization in CRS from three perspectives. For the customized recommendation results, we capture fine-grained intentions by exploring user inherent preference on entity relations and the appearing turn of the entities. For the customized dialogue interactions, we proposed a multi-style generator, which generates responses in customized speaking styles. Finally, for the model training, we adopted the meta-learning framework to enhance the recommendation and dialogue generation via customized model parameters. The extensive experiments show that our CCRS outperforms competitive baselines.

Acknowledgments. This work is also supported by the National Natural Science Foundation of China under Grant (No. 61976204, U1811461, U1836206). Zhao Zhang is supported by the China Postdoctoral Science Foundation under Grant No. 2021M703273.

References

1. Bahdanau, D., Cho, K.H., Bengio, Y.: Neural machine translation by jointly learning to align and translate. In: ICLR (2015)
2. Chen, Q., et al.: Towards knowledge-based recommender dialog system. In: EMNLP, pp. 1803–1813 (2019)
3. Christakopoulou, K., Radlinski, F., Hofmann, K.: Towards conversational recommender systems. In: SIGKDD, pp. 815–824 (2016)
4. Devlin, J., Chang, M.W., Lee, K., et al.: Bert: pre-training of deep bidirectional transformers for language understanding. In: NAACL, pp. 4171–4186 (2019)
5. Dong, B., et al.: Meta-information guided meta-learning for few-shot relation classification. In: COLING, pp. 1594–1605 (2020)
6. Finn, C., Abbeel, P., Levine, S.: Model-agnostic meta-learning for fast adaptation of deep networks. In: ICML (2017)
7. Gao, C., Lei, W., He, X., de Rijke, M., Chua, T.S.: Advances and challenges in conversational recommender systems: a survey. AI Open **2**, 100–126 (2021)
8. Glorot, X., Bengio, Y.: Understanding the difficulty of training deep feedforward neural networks. In: Proceedings of the Thirteenth International Conference on Artificial Intelligence and Statistics, pp. 249–256. JMLR Proceedings (2010)
9. He, K., Zhang, X., Ren, S., Sun, J.: Deep residual learning for image recognition. In: CVPR, pp. 770–778 (2016)
10. Hendrycks, D., Gimpel, K.: Gaussian error linear units (GELUs). arXiv preprint arXiv:1606.08415 (2016)
11. Hu, Z., Dong, Y., Wang, K., Sun, Y.: Heterogeneous graph transformer. In: WWW, pp. 2704–2710 (2020)
12. Kim, Y.: Convolutional neural networks for sentence classification. In: EMNLP 2014, pp. 1746–1751 (2014)
13. Kingma, D.P., Ba, J.: Adam: a method for stochastic optimization. arXiv (2014)
14. Lee, H., Im, J., Jang, S., Cho, H., Chung, S.: MeLU: meta-learned user preference estimator for cold-start recommendation. In: SIGKDD, pp. 1073–1082 (2019)
15. Lehmann, J., Isele, R., Jakob, M., et al.: Dbpedia-a large-scale, multilingual knowledge base extracted from Wikipedia. Semant. Web **6**(2), 167–195 (2015)
16. Li, R., Ebrahimi Kahou, S., Schulz, H., Michalski, V., Charlin, L., Pal, C.: Towards deep conversational recommendations. In: NeurIPS, vol. 31, pp. 9725–9735 (2018)
17. Li, S., Xie, R., Zhu, Y., Ao, X., Zhuang, F., He, Q.: User-centric conversational recommendation with multi-aspect user modeling. In: SIGIR (2022)
18. Liang, Z., Hu, H., Xu, C., et al.: Learning neural templates for recommender dialogue system. In: EMNLP, pp. 7821–7833 (2021)
19. Liu, Z., Wang, H., Niu, Z.Y., Wu, H., Che, W., Liu, T.: Towards conversational recommendation over multi-type dialogs. In: ACL, pp. 1036–1049 (2020)
20. Lu, Y., et al.: Revcore: review-augmented conversational recommendation. arXiv preprint arXiv:2106.00957 (2021)
21. Pan, F., Li, S., Ao, X., Tang, P., He, Q.: Warm up cold-start advertisements: improving CTR predictions via learning to learn id embeddings. In: SIGIR, pp. 695–704 (2019)
22. Papineni, K., Roukos, S., Ward, T., Zhu, W.J.: Bleu: a method for automatic evaluation of machine translation. In: ACL, pp. 311–318 (2002)
23. Schlichtkrull, M., Kipf, T.N., Bloem, P., van den Berg, R., Titov, I., Welling, M.: Modeling relational data with graph convolutional networks. In: Gangemi, A., et al. (eds.) ESWC 2018. LNCS, vol. 10843, pp. 593–607. Springer, Cham (2018). https://doi.org/10.1007/978-3-319-93417-4_38

24. Sun, Y., Zhang, Y.: Conversational recommender system. In: SIGIR, pp. 235–244 (2018)
25. Vaswani, A., et al.: Attention is all you need. In: NeurIPS, pp. 5998–6008 (2017)
26. Vinyals, O., Blundell, C., Lillicrap, T., Wierstra, D., et al.: Matching networks for one shot learning. In: NeurIPS, vol. 29, pp. 3630–3638 (2016)
27. Wu, Y., et al.: Selective fairness in recommendation via prompts. In: SIGIR (2022)
28. Wu, Y., et al.: Personalized prompts for sequential recommendation. arXiv (2022)
29. Xie, R., et al.: Long short-term temporal meta-learning in online recommendation. In: WSDM (2022)
30. Zhang, T., Liu, Y., Zhong, P., et al.: KECRS: towards knowledge-enriched conversational recommendation system. arXiv preprint arXiv:2105.08261 (2021)
31. Zhang, Y., Chen, X., Ai, Q., Yang, L., Croft, W.B.: Towards conversational search and recommendation: system ask, user respond. In: CIKM, pp. 177–186 (2018)
32. Zhou, K., et al.: CRSLab: an open-source toolkit for building conversational recommender system. In: ACL: System Demonstrations, pp. 185–193 (2021)
33. Zhou, K., Zhao, W.X., Bian, S., Zhou, Y., Wen, J.R., Yu, J.: Improving conversational recommender systems via knowledge graph based semantic fusion. In: SIGKDD, pp. 1006–1014 (2020)
34. Zhou, K., Zhou, Y., Zhao, W.X., Wang, X., Wen, J.R.: Towards topic-guided conversational recommender system. In: COLING, pp. 4128–4139 (2020)
35. Zhu, Y., et al.: Transfer-meta framework for cross-domain recommendation to cold-start users. In: SIGIR, pp. 1813–1817 (2021)
36. Zhu, Y., et al.: Learning to expand audience via meta hybrid experts and critics for recommendation and advertising. In: KDD, pp. 4005–4013 (2021)
37. Zhu, Y., et al.: Personalized transfer of user preferences for cross-domain recommendation. In: WSDM, pp. 1507–1515 (2022)
38. Zhu, Y., et al.: Learning to warm up cold item embeddings for cold-start recommendation with meta scaling and shifting networks. In: SIGIR, pp. 1167–1176 (2021)

Correction to: On the Current State of Reproducibility and Reporting of Uncertainty for Aspect-Based Sentiment Analysis

Elisabeth Lebmeier, Matthias Aßenmacher,
and Christian Heumann

Correction to:
**Chapter "On the Current State of Reproducibility
and Reporting of Uncertainty for Aspect-Based Sentiment
Analysis" in: M.-R. Amini et al. (Eds.): *Machine Learning
and Knowledge Discovery in Databases*, LNAI 13714,
https://doi.org/10.1007/978-3-031-26390-3_31**

Chapter "On the Current State of Reproducibility and Reporting of Uncertainty for Aspect-Based Sentiment Analysis" was previously published non-open access. It has now been changed to open access under a CC BY 4.0 license and the copyright holder updated to 'The Author(s)'.

The updated original version of this chapter can be found at
https://doi.org/10.1007/978-3-031-26390-3_31

© The Author(s) 2023
M.-R. Amini et al. (Eds.): ECML PKDD 2022, LNAI 13714, p. C1, 2023.
https://doi.org/10.1007/978-3-031-26390-3_44

Author Index